ACS SYMPOSIUM SERIES **429**

Geochemistry of Sulfur in Fossil Fuels

Wilson L. Orr, EDITOR
Mobil Research and Development Corporation

Curt M. White, EDITOR
Pittsburgh Energy Technology Center

Developed from a symposium sponsored
by the Division of Geochemistry
at the 197th National Meeting
of the American Chemical Society,
Dallas, Texas,
April 9–14, 1989

American Chemical Society, Washington, DC 1990

Library of Congress Cataloging-in-Publication Data

Geochemistry of sulfur in fossil fuels
  Wilson L. Orr, editor, Curt M. White, editor

  p.   cm.—(ACS Symposium Series, 0097–6156; 429).

  Developed from a symposium sponsored by the division of
Geochemistry at the 197th National Meeting of the American
Chemical Society, Dallas, Texas, April 9–14.

  Includes bibliographical references and indexes.

  ISBN 978-0-8412-1804-8
  1. Petroleum—Sulphur content—Congresses. 2. Coal—
Sulphur content—Congresses. 3. Geochemistry—Congresses.

  I. Orr, Wilson L., 1923–   . II. White, C. M. (Curt M.)
III. American Chemical Society. Division of Geochemistry.
IV. American Chemical Society. Meeting (197th: 1989: Dallas,
Tex.). V. Series

TN870.5.G3864   1990
665.5—dc20
                                              90–839
                                                CIP

Foreword

The ACS SYMPOSIUM SERIES was founded in 1974 to provide a medium for publishing symposia quickly in book form. The format of the Series parallels that of the continuing ADVANCES IN CHEMISTRY SERIES except that, in order to save time, the papers are not typeset but are reproduced as they are submitted by the authors in camera-ready form. Papers are reviewed under the supervision of the Editors with the assistance of the Series Advisory Board and are selected to maintain the integrity of the symposia; however, verbatim reproductions of previously published papers are not accepted. Both reviews and reports of research are acceptable, because symposia may embrace both types of presentation.

Contents

Preface ix

OVERVIEWS AND GENERAL TOPICS

1. **Geochemistry of Sulfur in Petroleum Systems** 2
   Wilson L. Orr and Jaap S. Sinninghe Damsté

2. **Geochemistry of Sulfur in Coal** 30
   C.-L. Chou

3. **Environmental Aspects of the Combustion of Sulfur-Bearing Fuels** 53
   B. Manowitz and F. W. Lipfert

4. **Polysulfide Reactions in the Formation of Organosulfur and Other Organic Compounds in the Geosphere** 68
   Robert T. LaLonde

5. **Isolation of Sulfur Compounds from Petroleum** 83
   O. P. Strausz, E. M. Lown, and J. D. Payzant

6. **Microbial Metabolism of Organosulfur Compounds in Petroleum** 93
   Phillip M. Fedorak

STUDIES OF DEPOSITIONAL ENVIRONMENTS

7. **Geochemistry of Organic and Inorganic Sulfur in Ancient and Modern Lacustrine Environments: Case Studies of Freshwater and Saline Lakes** 114
   Michele L. Tuttle, Cynthia A. Rice,
   and Martin B. Goldhaber

8. Characterization of Organic Matter in Sulfur-Rich Lacustrine Sediments of Miocene Age (Nördlinger Ries, Southern Germany) 149
   Jürgen Rullkötter, Ralf Littke, and Rainer G. Schaefer

9. Incorporation of Sulfur into Recent Organic Matter in a Carbonate Environment (Abu Dhabi, United Arab Emirates) 170
   F. Kenig and A. Y. Huc

10. Sulfur and Pyrite in Precursors for Coal and Associated Rocks: A Reconnaissance Study of Three Modern Sites 186
    A. M. Bailey, J. F. Sherrill, J. H. Blackson, and E. C. Kosters

11. Formation of Iron Sulfides in Modern Salt Marsh Sediments (Wallops Island, Virginia) 204
    T. S. White, J. L. Morrison, and L. R. Kump

CHARACTERIZATION OF SULFUR
IN FOSSIL FUEL MATERIALS

12. Sulfur K-Edge X-ray Absorption Spectroscopy of Petroleum Asphaltenes and Model Compounds..220
    Graham N. George, Martin L. Gorbaty, and Simon R. Kelemen

13. Direct Determination of Total Organic Sulfur in Coal 231
    J. T. Riley, G. M. Ruba, and C. C. Lee

14. Elemental Sulfur in Bituminous Coals 241
    Leon M. Stock and Ryszard Wolny

15. Multiple-Heteroatom-Containing Sulfur Compounds in a High Sulfur Coal..249
    Randall E. Winans and Paul H. Neill

16. Organosulfur Constituents in Rasa Coal..261
    Curt M. White, L. J. Douglas, R. R. Anderson, C. E. Schmidt, and R. J. Gray

17. Distribution of Organic-Sulfur-Containing Structures in High Organic Sulfur Coals 287
    R. J. Torres-Ordoñez, W. H. Calkins, and M. T. Klein

18. Characterization of Organic Sulfur Compounds in Coals and Coal Macerals...296
Stephen R. Palmer, Edwin J. Hippo, Michael A. Kruge, and John C. Crelling

19. Spatial Variation of Organic Sulfur in Coal.......................................316
Elizabeth Ge and Charles Wert

20. Characterization of Organosulfur Compounds in Oklahoma Coals by Pyrolysis–Gas Chromatography...326
Allen J. Bakel, R. Paul Philp, and A. Galvez-Sinibaldi

21. Coal Desulfurization by Programmed-Temperature Pyrolysis and Oxidation...345
Jean-Paul Boudou

MOLECULAR STRUCTURE OF SULFUR
COMPOUNDS AND THEIR GEOCHEMICAL SIGNIFICANCE

22. Nature and Geochemistry of Sulfur-Containing Compounds in Alberta Petroleums...366
O. P. Strausz, E. M. Lown, and J. D. Payzant

23. Identification of Alkylthiophenes Occurring in the Geosphere by Synthesis of Authentic Standards..397
T. M. Peakman and A. C. Kock-van Dalen

24. Organic Sulfur Compounds and Other Biomarkers as Indicators of Palaeosalinity...417
Jan W. de Leeuw and Jaap S. Sinninghe Damsté

25. Alkylthiophenes as Sensitive Indicators of Palaeoenvironmental Changes: A Study of a Cretaceous Oil Shale from Jordan..............444
Mathieu E. L. Kohnen, Jaap S. Sinninghe Damsté, W. Irene C. Rijpstra, and Jan W. de Leeuw

26. Characterization of Organically Bound Sulfur in High-Molecular-Weight, Sedimentary Organic Matter Using Flash Pyrolysis and Raney Ni Desulfurization..486
Jaap S. Sinninghe Damsté, Timothy I. Eglinton, W. Irene C. Rijpstra, and Jan W. de Leeuw

vii

27. Analysis of Maturity-Related Changes in the Organic Sulfur Composition of Kerogens by Flash Pyrolysis—Gas Chromatography 529
    Timothy I. Eglinton, Jaap S. Sinninghe Damsté, Mathieu E. L. Kohnen, Jan W. de Leeuw, Steve R. Larter, and Richard L. Patience

SPECIAL STUDIES

28. Isotopic Study of Coal-Associated Hydrogen Sulfide 568
    J. W. Smith and R. Phillips

29. Pyrolysis of High-Sulfur Monterey Kerogens: Stable Isotopes of Sulfur, Carbon, and Hydrogen 575
    Erdem F. Idiz, Eli Tannenbaum, and Isaac R. Kaplan

30. Sulfur Isotope Data Analysis of Crude Oils from the Bolivar Coastal Fields (Venezuela) 592
    B. Manowitz, H. R. Krouse, C. Barker, and E. T. Premuzic

31. Distribution of Organic Sulfur Compounds in Mesozoic and Cenozoic Sediments from the Atlantic and Pacific Oceans and the Gulf of California 613
    H. Lo ten Haven, Jürgen Rullkötter, Jaap S. Sinninghe Damsté, and Jan W. de Leeuw

32. Carbon Isotope Fractionation during Oxidation of Light Hydrocarbon Gases: Relevance to Thermochemical Sulfate Reduction in Gas Reservoirs 633
    Y. Kiyosu, H. R. Krouse, and C. A. Viau

BIBLIOGRAPHY AND INDEXES

Bibliography 644

Author Index 694

Affiliation Index 694

Subject Index 695

Preface

From exploration for new resources to the final utilization and resulting effects on our environment from combustion of sulfur-containing fuels, sulfur is a major concern in the science and technology of fossil fuels. The word sulfur provokes diverse opinions and concerns depending on individual interest and background. All fossil fuels contain sulfur, but the sulfur content varies from traces to greater than 10%, and the chemical forms are themselves diverse. The presence of sulfur exacts economic penalties at all intermediate stages of exploitation, from recovery (oil and gas production, mining, storage, transportation) to processing (refining, cleaning, upgrading), the extent depending on sulfur's abundance and chemical form. In short, removing sulfur from fuels means greater costs to industry, consumers, and government. These costs can be minimized by increasing our basic understanding of the sulfur system in fossil fuels.

This volume was developed from the ACS symposium Geochemistry of Sulfur in Fossil Fuels, which encompassed a wide range of disciplines and subject areas and offered papers from leading research groups worldwide. The broad scope of the symposium makes evident the similarities and differences in the various fuel systems and reflects the current specific interest and the recent advances in understanding the multidisciplinary aspects of the sulfur cycle in the geosphere. In many respects, a geochemical focus helps make clear the reasons for variations in the sulfur system in different fossil fuels.

Major advances have been made in the past decade, largely because of improved analytical techniques that have made easier the identification of organic sulfur compounds at the detailed molecular-structure level. These advances have established sulfur-compound biomarkers that have greatly clarified our understanding of how and when most of the sulfur is introduced into the surviving biogenic materials that eventually become fossil fuels. The latest advances included in this volume have many implications for practical utility. Quantitative evaluations of compound abundances and distributions in specific fossil fuels (in relation to their geo logic history and geologic settings) and case studies of sedimentary depositional environments will help to integrate and consolidate our understanding of the geochemistry of sulfur in fossil fuels.

The broad scope and diverse nature of the chapters in this volume made the process of dividing them into sections or subject areas somewhat difficult. Many chapters relate to more than one section. The common methods of subject classification according to fuel type (petroleum, natural gas, coal, etc.) was intentionally avoided so as to emphasize geochemical principles and controls that have common chemistry for all fuel types.

The first section contains overview and introductory material, some historical background, summaries of the present state of knowledge, and other general topics applicable to all fossil fuel. These include environmental consequences of fossil fuel combustion, polysulfide chemistry, microbial metabolism of sulfur compounds, and a review of methods for isolating sulfur compounds from petroleum.

The second section groups chapters that are case studies of specific sedimentary environments and focus on environmental factors that influence the behavior of sulfur in the formation of sediments and fossil fuels.

Chapters in section three characterize sulfur in selected petroleums and coals using various methods and techniques. These studies either address specific questions or illustrate the scope or application of a particular technique.

The chapters in the fourth section report some of the major advances that have established a new era of understanding at the molecular level in sulfur geochemistry. Organic sulfur compounds have now acquired a level of geochemical significance rivaling hydrocarbons as fossil biomarkers. Expansion of these concepts using pyrolysis and other techniques shows great promise for improved characterization of sulfur in macromolecules (kerogen and asphaltenes).

The final section contains chapters with diverse subject areas that did not fit neatly into the other sections. Several of these studies as well as a few in other sections deal with stable isotopes.

Finally, an overall bibliography was compiled, an addition to the usual ACS Symposium Series book format. Arranged alphabetically by author, the bibliography supplements the shorter literature citations in each chapter by providing the titles of works cited.

Acknowledgments

The major credit for the scientific merit of this volume goes to the authors and their supporting institutions for the quality of their research, their enthusiasm at the symposium, and their cooperation in the preparation of final manuscripts. Of course, for various reasons, a number of interesting papers given in the symposium could not be included. Hopefully, some of these will be published elsewhere.

We thank Simon C. Brassell of Stanford University, Ed L. Obermiller of Consolidation Coal Company, and Otto P. Strausz of the University of Alberta for serving on the organizing committee for the symposium and as advisers during the preparation of this volume.

A large number of technical reviewers are acknowledged and thanked by both the editors and the authors for their expert reviews and many helpful suggestions and criticisms that greatly improved the quality of the book.

Finally, financial support that made it possible for many foreign speakers to participate in the symposium is most gratefully acknowledged. Substantial contributions were made by the following: Amoco Production Company, Arco Oil and Gas Company, Conoco Inc., Consolidation Coal Company, Mobil Research and Development Corporation, Unocal Corporation, and the American Chemical Society Petroleum Research Fund.

We also express gratitude to our managements for encouragement during the preparation of the volume.

WILSON L. ORR
Mobil Research and Development Corporation
Dallas Research Laboratory
137877 Midway Road
Dallas, TX 75244–4312

CURT M. WHITE
Pittsburgh Energy Technology Center
U.S. Department of Energy
P.O. Box 10940
Pittsburgh, PA 15236

March 5, 1990

OVERVIEWS AND GENERAL TOPICS

Chapter 1

Geochemistry of Sulfur in Petroleum Systems

Wilson L. Orr[1] and Jaap S. Sinninghe Damsté[2]

[1]Dallas Research Laboratory, Mobil Research & Development Corporation, 13777 Midway Road, Dallas, TX 75244–4312
[2]Faculty of Chemical Engineering and Materials Science, Organic Geochemistry Unit, Delft University of Technology, De Vries van Heystplantsoen 2, 2628 RZ Delft, Netherlands

A renaissance in the 1980s concerning geochemistry of sulfur in fossil fuels makes an update of the subject timely. Papers developed from the 1989 ACS Symposium in Dallas provide a cross-section of recent research and progress in our understanding of the abundance and nature of organically bound sulfur in fossil fuels. This chapter furnishes an overview of our understanding of sulfur introduction into the geologic systems and biogenic materials that produce petroleum and natural gas. Many of the geochemical considerations also apply to coal and lignite deposits. A brief historical account leads to a summary of the most recent advances, an appraisal of current understanding, and some remaining challenges and opportunities for further research.

This chapter provides a historical perspective and overview of the status of knowledge regarding the geochemistry of sulfur in fossil fuels with emphasis on petroleum related systems. The major reason for considering the diverse fossil fuel systems in a single symposium is that many aspects of the geochemistry are common to all classes of fossil fuels (petroleum, natural gas, coal, lignite and oil shales). However, the technologies, terminologies, methods of study, and technical problems are sufficiently different that some topics are specific to a given fuel class while others are relevant to all fuel classes. The broad scope of this volume will make it easy to appreciate the similarities and differences in the various fuel systems.

The chemistry of sulfur in fossil fuels is of both practical and academic interest. The behavior of sulfur during processing high-sulfur crude oils, coals and natural gas to make environmentally acceptable fuels poses special

0097–6156/90/0429–0002$08.00/0
© 1990 American Chemical Society

engineering problems and economic penalties. Sulfur must be reduced to appropriate levels for each intended application. The *acid rain problem*, resulting largely from combustion of sulfur in fossil fuels, is a pressing national and international concern. The magnitude of sulfur dioxide emissions in the United States is discussed by Manowitz and Lipfert (1). Pending U.S. legislation will mandate major reductions in emissions of sulfur dioxide and nitrogen oxides into our air in the near future. The fuel and utility industries will make major expenditures to meet these requirements. Although naturally formed biogenic sulfur compounds are recognized as harmful environmental pollutants, they appear to make up only about 0.13% of total sulfur emissions into the atmosphere (2,3).

The chemistry of desulfurization involved in processing fuels is, of course, dependent on molecular structures or form of sulfur in the fuel material. This is true for lignite, coal, and oil shales as well as for petroleum, petroleum fractions, and natural gas. Desulfurization of natural gas presents few problems because the technology for removing hydrogen sulfide and traces of thiols is relatively simple and well-developed. In contrast, crude oils and other fossil fuels vary considerably both in total sulfur content and in modes of combination that make sulfur removal more difficult. The more we know about chemical structure, thermal stability, and reactivity of sulfur in a given fuel, the more guidance we have for improving sulfur removal processes.

The emphasis of this symposium is on the geochemistry of sulfur rather than on processing, utilization or environmental consequences. Nevertheless the factual information, generalizations and insights provided by these current research topics should be informative and should suggest practical applications for many concerns of the fossil fuel industry.

Although petroleum processing is not treated in this symposium, it is evident that industrial treatment of crude oils at relatively high temperatures and short times involves many of the same chemical reactions that occur in the geologic environment at lower temperature on a much longer time-scale. Therefore, experience from refining and geochemistry support each other in a symbiotic relationship. A number of studies reported here utilize pyrolysis or thermal treatments for either the simulation of geochemical processes or the analytical characterization of materials. These studies supply information relating to the chemistry of high temperature reactions.

Refining engineers and chemists are most interested in the ease of desulfurizing petroleum using thermal and thermocatalytic treatments. The sulfur is removed primarily as hydrogen sulfide. Thermal and thermocatalytic studies have established that non-thiophenic sulfur (aliphatic as in thiols, acyclic and cyclic sulfides) evolve H_2S much more readily than thiophenic sulfur (aromatic heterocyclic compounds). Thus, the relative abundances of nonthiophenic (aliphatic) and thiophenic (aromatic) sulfur is a critical characteristic for all fuels with respect to ease of desulfurization. Analytical methods were developed in the 1960s for classifying the total sulfur in crude oils into types or fractions that correlate with ease of desulfurization (4-7). Although these analytical methods left a significant fraction of the total sulfur (commonly 30-60%) as "sulfur not recovered" or "an unknown fraction", this fraction was considered to be part of the more resistant category (complex thiophenic components).

Application of these methods showed that the distribution of types of organic sulfur compound(s) (OSC) in crude oils is variable. In other words, the ratio of thermally labile to more resistant sulfur has a large range in crude oils. Geochemically, the large range is expected and understandable because of variations in source, maturity and other alteration processes.

The industry prefers to refine low-sulfur crude oils, and usually handles high-sulfur crudes by blending them with much lower sulfur oils. Refinery design and operation determine the sulfur level that can be processed. In the future, more high-sulfur oil containing 3-6% sulfur will be exploited, and refining processes must be modified accordingly. The shift to higher sulfur crudes will be gradual and the time frame for the transition is not evident. However, the world's potential reserves of "heavy oils" and "tar sands" (generally high in sulfur content) greatly exceed known reserves of conventionally produced low-to-moderate-sulfur crude oils (8, 9). Increasing use of these potential resources will be required unless alternative energy sources replace fossil fuels.

The bottom line for the fuel industry is that a more complete understanding of the origin and evolution of sulfur in fossil fuels can lead to improved practical applications in a number of areas such as:

1. Guiding exploration for fossil fuels, particularly for oil and gas with a concern for fuel quality.
2. Providing clues that help rationalize the wide range of oil and gas compositions and properties found in natural occurrences (fuel quality).
3. Suggesting reasons why different fuels and fuel fractions behave differently during processing.
4. Developing improved processes and uses for fossil fuels.

The Geochemical Focus

Geochemists attempts to understand the chemistry that takes place in the geologic environments where fossil fuels originate and evolve over periods of millions of years. With respect to sulfur this science is concerned with: (1) how and when sulfur is introduced into the natural biogenic materials that lead to fossil fuels; (2) why and how the abundance and forms of sulfur differ in various geologic environments and; (3) how abundance and forms of sulfur evolve (change) with subsequent geologic history.

Limitations in methods for analysis of sulfur in complex materials and complex mixtures determine the extent of our knowledge and understanding of sulfur in fossil fuels. Historically, as new and improved methods of analysis became available, increased knowledge and insights followed quickly. Therefore, improvements in analytical methods continue to be critical research topics. A major current limitation is our inability to characterize adequately the forms of sulfur (functional groups) in solids (kerogen, coals, etc.) and in the higher-molecular-weight fractions of petroleum (resins, asphaltenes and refinery residues).

The scope of geological and geochemical considerations related to sedimentary organic matter and fossil fuels embraces many different disciplines including oceanography, sedimentology, biology (especially microbiology), mineralogy, petrology, tectonics of basin development, etc., and concerns inorganic/organic interactions involving water, minerals, biogenic organic materials and gases. The literature on these subjects is extensive and many specialized reference works and reviews are available.

Although A. Treibs is regarded as the founder of modern organic geochemistry for his work on porphyrins in the 1930s ([10], [11]), organic geochemistry did not become established as an active science until the 1950s and 1960s and interest has increased significantly since 1970. The first reference book on organic geochemistry appeared in 1963 ([12]) (edited by the late I. Breger) and the second in 1969 ([13]) (edited by Eglinton and Murphy). Erdman made significant contributions to petroleum geochemistry in the 1950s and 1960s; his 1965 paper ([14]) on the origin of petroleum summarized various historical views and the accepted view at that time. Silverman ([15]) pioneered studies of carbon isotopes in petroleum geochemistry and made many important contributions. A later review of the use of carbon isotopes in exploration for petroleum was given by Fuex in 1977 ([16]). The pioneering work of Thode on sulfur isotopes in petroleum, which started in the 1950s, had a major impact on sulfur geochemistry ([17,18]). The use of sulfur isotopes in petroleum exploration was reviewed by Krouse in 1977 ([19]), and broader aspects of sulfur isotopes were covered by Nielsen ([20]). Sulfur isotopes continue to make important contributions in studies of the geochemistry of sulfur.

An early comprehensive review of the organic geochemistry of sulfur was provided by the late Claude E. Zobell in 1963 ([21]) with some emphasis on microbiological involvement. The sulfur cycle in oceans and sediments was treated by Goldhaber and Kaplan in 1974 ([22]). Berner and others have published extensively on sulfur in sediments with emphasis on pyrite ([23,24]). Interestingly, biogenic sedimentary organic matter has often been considered mainly as the substrate necessary for bacteria to reduce sulfate to hydrogen sulfide which is the major source of sulfur forming pyrite in early sedimentary environments. The reaction of H_2S with the organic matter to form sulfur-rich precursors of fossil fuels has often been ignored or has been treated as insignificant in many discussions of sediments and of the global sulfur cycle. This is unfortunate from our viewpoint because the sulfurization of organic matter is the major concern regarding sulfur in fossil fuels. A number of papers in this symposium relate to the early introduction of sulfur into sediments in both organic and inorganic forms and the environmental factors controlling the operative processes. A recent book dealing with biogenic sulfur in the environment also reports major progress in understanding sulfur interactions with organic matter in sediments and in biological systems (2).

Comprehensive monographs by Tissot and Welte (25) and Hunt (26) are available on geology and geochemistry of oil and gas occurrences. These works have incorporated concepts developed in the 1960s and 1970s with later works to derive general schemes for the origin of petroleum, coal, and natural gas. These schemes are comprehensive and generally accepted. Sedimentary environmental conditions necessary for source-rock deposition are well known. The oil generation process is understood to be largely a thermal conversion of kerogen and associated immature bitumens into petroleum and gas (see later discussion). This evolutionary stage (catagenesis) occurs with increasing temperature and time as source-rocks are progressively buried. The nature of the migration process, which moves oil generated in a source-rock to oil trapped in a reservoir, remains the least understood process. Subsequent petroleum alteration processes are reasonably well understood but remain current topics of investigation. These include: (1) in-reservoir maturation that changes oil composition and properties, and in the extreme includes cracking of oil to gas, and (2) biodegradation and water-washing of petroleum in shallow reservoirs, which not only decrease oil quality but also destroy large amounts of hydrocarbons. The general scheme allows for the formation of kerogens with different compositions that reflect different biogenic source materials and different conditions in the initial sedimentary environments. The composition and nature of the kerogen then determine the nature of the oil and/or gas that it generates; i.e., differences in initial oil properties and gas/oil ratio. Subsequent alteration processes account for additional changes in composition and properties after generation. Current research continues to examine and refine these concepts.

The importance of the heteroatoms nitrogen, sulfur and oxygen (NSO) is recognized in all of the processes involved in the origin and alteration of petroleum (25,26). The nitrogen containing tetrapyrrole pigments (porphyrins and the vanadyl and nickel metalloporphyrins) have received more attention than most NSO compounds (10,11). The abundance of vanadium and/or vanadyl porphyrins is highly correlated with sulfur content of bitumens and crude oils, but it is not clear whether this correlation reflects mainly chemical associations or sensitivity to preservation processes with different oxic-anoxic sedimentary conditions (25-27). Aside from porphyrins, major emphasis over the years has been on studies of hydrocarbons (i.e., compounds that contain only carbon and hydrogen). Hydrocarbons are the major useful fuel components and are the simplest compounds to characterize. Naturally, hydrocarbon geochemistry has advanced more rapidly than that of heteroatomic (NSO) compounds.

Sulfur in Petroleum and Related Bitumens

Sulfur content in crude oils and natural bitumens varies from less than 0.05 to more than 14 weight percent, but few commercially produced crude oils exceed 4% sulfur. Tissot and Welte (25) show a frequency distribution of crude oils based on over 9,000 samples and report the average sulfur content as 0.65%. The distribution is clearly bimodal with a minimum at about 1% sulfur. Oils with less than 1% sulfur are classified as low-sulfur, and those above 1% as high-sulfur. In general, high-sulfur oils are derived from carbonate or carbonate-evaporite rock sequences whereas low-sulfur oils are derived largely from clay-rich clastic sequences (25-26, 28-29).

Most of the sulfur present in crude oils and bitumens is organically bound, (i.e., bound to carbon) because dissolved hydrogen sulfide and elemental sulfur usually represent only a minor part of the total sulfur. Organic sulfur is present in low- to medium-molecular-weight molecules, but the largest fraction is in the high-molecular-weight components (25-26, 28-29).

OSC vary in polarity and chromatographic behavior such that some elute during liquid chromatography with the aromatic hydrocarbon fraction while others elute with the more polar NSO fractions that include the resins and asphaltenes. Until recently, precise organic molecular structures have been established only for relatively low molecular weight sulfur compounds, generally with fewer than 15 carbon atoms and from fractions with boiling points below 250-300 °C. This unsatisfactory state of knowledge is aggravated by the fact that generally 60-80% of the sulfur is in fractions boiling above 300 °C. We now know a great deal about the types of OSC and functional groups in perhaps 20-40% of the total sulfur. The same types of functional groups (and perhaps others) are believed to extend into uncharacterized fractions in increasingly complex combinations with many of the more polar and higher molecular-weight compounds containing more than one heteroatom. The most thoroughly characterized OSC are those which separate chromatographically with lower-molecular-weight aromatic hydrocarbons.

A number of reviews related to the identification of sulfur compounds in petroleum should be mentioned. Dean and Whitehead (30) summarized work on separation and identification of sulfur compounds in petroleum and shale oil in 1967. Drushel (7) reviewed sulfur compound types with an emphasis on the available analytical methods as of 1970. Mehmet (31) and Gal'pern (32) also reviewed sulfur compounds in petroleum in 1971 with some speculation on their origin. The major contributions of API Project 48 reported by Coleman et al. in 1971 , Rall et al. in 1972, and Thompson in 1981 will be discussed below (33-35). More recent summaries of sulfur compounds by Aksenov and Kamyanov in 1981 (36) and by Gal'pern in 1985 (37), discuss OSC in petroleum, processed petroleum fractions, shale oil, coal derived liquids and related products. These reviews include much of the Soviet literature. In 1975 and 1978 Orr (28-29) discussed sulfur in the petroleum system with somewhat more emphasis on geochemistry. For the most part, all of these papers preceded advances made in the identification of OSC in sediments and bitumens during the last decade by GC-MS analysis.

Rall et al. (34) gave a detailed account of the U. S. Bureau of Mines work under The American Petroleum Institute (API) Project 48, which devoted about 50 man-years of research to sulfur in petroleum during the time from 1948 to 1966. Their report is a comprehensive summary of the project accomplishments and also reviewed most prior work with 469 literature citations. Only 25 sulfur compounds had been reported in petroleum prior to 1948. Structures of these 25 compounds were confirmed and 176 additional compounds were isolated and identified during the life of this project (Table I). Figures 1 and 2 illustrate major structural types for lower-molecular-weight OSC in petroleum. Structures of higher-molecular-weight OSC compounds in more complex aromatic ring systems and in higher boiling fractions of petroleum and bitumen are largely unknown. Most characterizations of higher molecular weight fractions are by group type analysis (mass spectrometry supported by other techniques) rather than by isolation and rigorous structural determination (_38-40_).

The API project settled a number of questions that had been debated for years. API 48 took special precautions to isolate and identify sulfur compounds from virgin crude oils using mild thermal conditions. They investigated the thermal stability of crude oils and reactions of oils and hydrocarbons with elemental sulfur to settle long standing debates about whether the OSC identified were in fact present in virgin crude oils or were produced by processing practices. They made quantitative or semiquantitative estimates of the abundance of many of the OSC identified. New techniques for separation, analysis and characterization of the compounds were investigated and developed. Limitations of the project were mainly that the work was slow and tedious, and the state-of-the-art limited separations and identifications to low-molecular-weight compounds.

The question of the presence of elemental sulfur in crude oils was also a matter of controversy for many years. Many virgin crude oils contain small but significant amounts of H_2S that is easily oxidized to elemental sulfur by exposure to air. API 48 confirmed that elemental sulfur indeed does exist in some crude oils but it is not common; in most cases dissolved S_8 is a minor part of the total sulfur.

API 48 not only established that the types of sulfur compounds listed in Table I were common in virgin crude oils, but also found that the relative abundance of these compounds was highly variable in the different crude oils studied. Geochemical reasons for these differences are more easily rationalized now than they were in 1972.

Ho et al. (41) applied "sulfur-compound-type" analytical methods (_4-7_) to 79 crude oils having a range of sulfur contents from 0.05 to 7.82 % with results that were very informative from a geochemical viewpoint. Sulfur type

Table I. Number and Types of Sulfur Compounds Identified by API Project 48[a]

Compound Type	Wasson W. Texas	Wilmington California	Agha Jari Iran	Deep River Michigan	Total
THIOLS					
Alkyl-	40	0	6		46
Cyclic-	6				6
Aromatic-	1				1
SULFIDES					
Dialkyl-	38		5		43
Alkylcycloalkyl-	5				5
Alkylaryl-	4				4
Cyclic-	21	19			40
Thiaindans	18				18
DISULFIDES	1			3	4
THIOPHENES					
Alkyl-	0	11			11
Benzo-	22				22
Thieno-	2				2
Dibenzo-	2				2
TOTAL IDENTIFICATIONS	160	30	11	3	204
UNDUPLICATED IDENTIFICATIONS	160	14	0	2	176

(a) From Rall et al. (34)

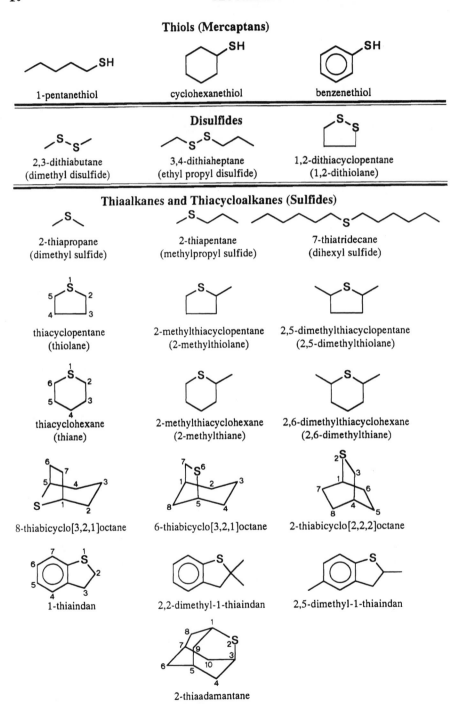

Figure 1. Examples of non-thiophenic sulfur compounds in petroleum and related materials.

Thiophenes

thiophene   2-methylthiophene   2,5-dimethylthiophene   3,4,5-trimethyl-2(1-thiaethyl)-
thiophene

Condensed Thiophenes

benzo [*b*] thiophene

2-methylbenzo [*b*] thiophene   2,4-methylbenzo [*b*] thiophene

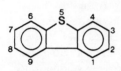   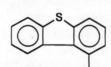   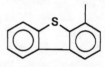

dibenzo [*bd*] thiophene   1-methyldibenzo [*bd*] thiophene   4-methyldibenzo [*bd*] thiophene

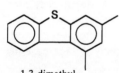   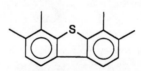   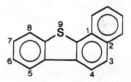

1,3-dimethyl-
dibenzo [*bd*] thiophene

3,4,6,7-tetramethyl-
dibenzo [*bd*] thiophene

9-thia-1,2-benzofluorene

3-ethyl-6,8-dimethylnaphtho
[1,2-*b*] thiophene

thieno [3,2-*b*] thiophene

thieno [2,3-*b*] thiophene

Figure 2. Examples of thiophenic sulfur compounds in petroleum and related
materials.

distributions were determined for the following compound types: (1) nonthiophenic sulfur, (2) thiophenes, (3) benzothiophenes, (4) dibenzo-thiophenes, (5) benzonaphthothiophenes, and (6) sulfur not recovered.

They classified the oils into three groups based on the distribution of types of sulfur compounds. These groups were immature (I), altered (II) and mature (III). This classification by sulfur distributions was consistent with geological data such as depth and age and with conventional bulk maturity parameter such as API gravity, %S, molecular-weight-distribution of n-alkanes, etc. The least mature group (I) had the highest sulfur content and contained the largest amount of nonthiophenic sulfur (the thermally labile type). The most mature group (III) was from deeper high-temperature reservoirs, contained less total sulfur but relatively more of the thermally stable OSC such as dibenzothiophenes. They showed that the ratio of benzothiophenes to dibenzothiophenes decreased with increasing maturity and suggested that this ratio is a useful maturity indicator.

One important conclusion was that the mature group oils (III) could be divided into two types. One was relatively low in thiols and the other quite high. They pointed out that the high-thiol type is characteristic of very mature oils or condensates from deep carbonate reservoirs containing abundant hydrogen sulfide. Such reservoirs are associated with evaporitic sequences that contain sulfate. In these relatively high temperature reservoirs, Orr (42) postulated that thermochemical sulfate reduction leads to the abundant hydrogen sulfide (and polysulfides) and to conditions that result in a dynamic sulfur system where desulfurization and sulfurization reactions are competing. Reactions of H_2S and/or polysulfides with the organic compounds (oil, condensate or gas) can result in a steady-state concentration of reactive thiols that may be an appreciable fraction of the low total-sulfur content. Thermal maturation in the absence of sulfate follows a different course resulting in a continual decrease in total sulfur and a low thiol content in mature oils and condensates (42).

More recently, increasing interest has been shown in the distributions of dibenzothiophenes (DBT) and methyl dibenzothiophenes (MDBT) (43-49). Radke et al. (43) observed that the relative abundance and distribution of DBT and MDBT varied with increasing depth of burial in a well in a western Canada Basin containing Type III kerogen. The most suggestive maturity indication was shown as a decrease in the 4-MDBT/DBT ratio as a function of depth. A related maturity trend was suggested by Leythaeuser et al. (49) from observing a regular decrease in the 1-MDBT/DBT ratios with depth for extracts from two wells in the North Sea basin. Maturity related changes are also observed within the MDBT isomers. The decrease in 1-MDBT isomer relative to the 2-, 3- and 4-MDBT isomers with increasing maturity is attributed to the relatively lower thermal stability of the 1-isomer (44-48). A decrease in the 1-MDBT/4-MDBT ratio with maturity appears to be common for all types of organic matter, but the value of this ratio at a certain vitrinite reflectance level varies with the type of organic matter (44). Hughes (45) reported that a characteristic distribution pattern of MDBT isomers was useful to identify oils derived from carbonate source rocks. However, Radke et al. (44) found that this MDBT pattern was also observed in bitumens from siliciclastic rocks containing immature to marginally mature Type II kerogen.

These examples illustrate recent interest in investigating the potential of sulfur compound distributions for improving our understanding of geologic influences in determining variations in composition of crude oils with relation to source and evolution history.

Sulfur in Kerogens and Asphaltenes

The sulfur content of oil is determined primarily by the sulfur content of the source kerogen. The classic paper by Gransch and Posthuma (50) clearly demonstrated this relationship which has been confirmed by all subsequent work. The oil formed at the beginning of oil generation has the maximum sulfur content. Oil sulfur content decreases with increasing maturity as a result of dilution by further generation of non-sulfur compounds as well as by removal of sulfur. The sulfur loss is partly as H_2S and partly by retention of sulfur in insoluble reservoir bitumens or pyrobitumens formed by disproportionation of NSO compounds including resins and asphaltenes into lower-molecular-weight liquid and gaseous hydrocarbons and an insoluble carbonaceous residue.

Kerogen is defined as the fraction of sedimentary organic matter that is insoluble in common organic solvents. This operational definition provides a label or category for the material that is regarded as the major source material for oil and gas but does not imply a unique substance (25-26,51). Kerogens, in fact, have a wide range of composition that depends on the original nature of the organic matter, conditions in the depositional environment, and the stage of evolution or maturity. Kerogens are considered as mixtures of complex macromolecules derived in part from modified biopolymers and in part from new macromolecules formed by condensation, polymerization and crosslinking reactions of initial biogenic precursor materials (51-52). Sulfur may react with the biopolymers and precursor molecules and also may enter into the condensation cross-linking reactions forming kerogen with variations in sulfur content, depending on the depositional environment. As sediment burial increases, kerogens undergo evolution (diagenesis and catagenesis) involving disproportionation into lower molecular weight molecules (oil and gas) on the one hand, and a more condensed carbonaceous residue (residual kerogen) on the other.

Commonly, immature sedimentary organic matter consists of 85-97% kerogen accompanied by 3-15% of extractable organic matter (EOM, or bitumen). Both kerogen and bitumen contribute to petroleum by thermal degradation reactions (catagenesis) as sediments are heated during burial, but quantitatively, the bulk of the oil must come from kerogen.

Petroleum geochemists need the best possible insights into kerogen structure because the amount and composition of oils formed during kerogen evolution depend on its chemical composition. The sulfur content of a crude oil is determined primarily by the sulfur content of the source kerogen and secondarily by the maturity level. High-sulfur kerogens produce high-sulfur oils (42,50,53). In spite of this important relationship, little is known about the amount and forms of sulfur in most kerogens. Even total organic sulfur in kerogens is not routinely determined, mainly because of common contamination by pyrite that makes the analysis difficult. Nevertheless, determination of the total organic sulfur content is possible and the analyses supply very useful information (51,53,54). Examples of complete elemental analysis of kerogen (corrected for pyrite impurities) are given in Table II and show a wide range in organic sulfur content.

Various investigators have proposed that high-sulfur kerogens start to generate oil at lower thermal exposures than classical Type II kerogens, probably because of weaker carbon-sulfur and sulfur-sulfur bonds (25, 53, 55-59). Furthermore, it has been proposed that oil-prone kerogens with an atomic S_{org}/C ratio greater than 0.04 have sufficiently different properties from the classical Type II kerogens that they should be given a distinctive designation "Type II-S" kerogen (53).

Not only the total organic sulfur content but also exact structures of sulfur-containing functional groups in kerogens and their relative abundances may be key factors in determining kinetics of petroleum generation from high-sulfur kerogens. However, satisfactory functional group analysis of sulfur in kerogens is not presently possible. Therefore, this property is not used in establishing kinetic models except for the recognition that different kinetic parameters are required for different types of kerogen (56-59).

Organic sulfur in kerogens is thought to be present as sulfide, disulfide, and heterocyclic moieties; i.e., structures similar to those found in smaller extractable bitumen and oil components. However, the high atomic S_{org}/C ratios in the range of 0.04 to 0.09 for Type II-S kerogens (Table II) suggests either that disulfide and/or polysulfide moieties must be significant or that a large number of sulfur cross-linkages (e.g., >C-S-C< bonds) must be present. At the atomic S_{org}/C ratio of 0.04 (ca. 7-8 wt %S), there is one sulfur atom for every 25 carbon atoms; at an atomic S_{org}/C ratio of 0.08 to 0.09 (ca. 13-14 wt %S), this is one sulfur atom for every 11 to 13 carbon atoms. Trying to accommodate these high sulfur abundances into a kerogen molecular stuctural model, such as proposed by Behar and Vandenbrocke (52), is difficult and informative.

Relationship between kerogens and asphaltenes have taken on new meaning in the last few years especially for relatively low maturity oils. Asphaltene composition can give inferences about the source kerogen composition when kerogen data are unavailable (53,58-59). Asphaltenes are defined as materials soluble or peptized in oil or bitumen that precipitate when

Table II. Examples of Organic Sulfur Content of Oil–Prone Kerogens

| | Elemental Analysis of Kerogens (Normalized Ash–Free) | | | | | | | | | Sample Information | |
| | Weight Percent | | | | | Atomic Ratio | | | | | |
	C	H	O	N	S	H/C	O/C	N/C	S/C	Formation or Area	Geologic Age
1	56.40	6.80	20.10	2.50	14.2	1.44	0.268	0.038	0.094	Monterey	Miocene
2	59.28	6.62	18.53	2.26	13.2	1.33	0.235	0.033	0.084	Monterey	Miocene
3	63.95	7.15	12.19	2.44	14.3	1.33	0.143	0.033	0.084	Monterey	Miocene
4	69.44	5.20	8.53	2.18	14.6	0.91	0.092	0.027	0.079	Monterey	Miocene
5	73.58	7.37	4.40	1.85	12.8	1.19	0.045	0.022	0.065	Monterey	Miocene
6	74.89	8.01	3.75	1.77	11.6	1.27	0.038	0.020	0.058	Monterey	Miocene
7	75.57	6.65	4.83	1.59	11.4	1.05	0.048	0.018	0.056	Kimmeridge	Jurassic
8	74.26	8.40	4.64	2.00	10.7	1.35	0.047	0.023	0.054	Monterey	Miocene
9	76.19	7.89	3.59	2.06	10.3	1.23	0.035	0.023	0.051	Monterey	Miocene
10	74.27	8.40	5.50	2.02	9.8	1.35	0.056	0.023	0.050	Monterey	Miocene
11	74.50	7.50	6.20	2.80	9.0	1.20	0.063	0.032	0.045	Phosphoria	Permian
12	68.70	7.81	13.48	3.65	6.5	1.32	0.147	0.046	0.036	Sisquoc	Miocene
13	70.47	6.97	13.76	2.58	6.4	1.15	0.147	0.031	0.034	Phosphoria	Permian
14	67.77	7.33	15.63	3.75	5.5	1.29	0.173	0.047	0.031	Sisquoc	Miocene
15	72.60	7.90	12.40	2.10	4.9	1.30	0.128	0.025	0.025	Toarcian	Jurassic
16	78.20	7.70	7.00	2.10	5.0	1.17	0.067	0.023	0.024	Woodford	Devonian
17	70.85	7.47	13.58	4.03	4.1	1.26	0.144	0.049	0.022	Sisquoc	Miocene
18	77.25	7.51	8.03	3.09	4.1	1.16	0.078	0.034	0.020	Nodular Sh.	Miocene
19	75.95	8.15	8.46	3.62	3.8	1.28	0.084	0.041	0.019	Sisquoc	Miocene
20	77.52	8.05	7.76	2.97	3.7	1.24	0.075	0.033	0.018	McDonald Sh.	Miocene
21	78.08	7.28	9.13	2.09	3.4	1.11	0.088	0.023	0.016	Kimmeridge	Jurassic
22	85.40	3.50	5.60	2.10	3.3	0.49	0.492	0.021	0.015	N. Sahara	Silurian
23	75.90	9.10	8.40	3.90	2.6	1.43	0.083	0.044	0.013	Green River	Eocene
24	73.50	8.30	15.60	0.40	2.2	1.35	0.139	0.005	0.011	Kukersite	Ordovician
25	69.30	8.30	18.00	2.60	1.8	1.43	0.195	0.032	0.010	Messel Sh.	Eocene

the sample is diluted with an excess of n-alkane (e.g., pentane, hexane or heptane). Another definition is the fraction that is soluble in benzene and insoluble in n-alkanes (52,53,60-63). These solubility characteristics are a function of both molecular size and polarity; the latter is related to both the NSO heteroatom functional groups and the aromatic/aliphatic carbon ratio.

Tissot pointed out in 1984 (58) that "*asphaltenes can be thought of as small fragments of kerogen, having a comparable structure. Thus kerogen, asphaltenes, resins, and hydrocarbons might be considered to form a continuum, with decreasing size, heteroatom content, and polarity.*" This continuum is a reaction or transformation sequence that proceeds with increasing maturity:

Kerogen   →   Asphaltenes   →   Resins   →   Hydrocarbons

Pyrolysis of kerogens and asphaltenes has demonstrated the nature and logic of this conversion sequence (58,63-64). Relative to elemental composition, initial asphaltenes have a much lower atomic O/C ratio, a slightly lower atomic S/C ratio, and almost the same H/C and N/C ratios as their source kerogens (53).

We should caution that the above concept of the genetic relationship between kerogens and asphaltenes differs from the more historic view that asphaltenes are condensation and/or alteration products of hydrocarbons and resins. Certainly, in some petroleum processing treatments and probably at higher maturation levels in nature, various reactions do form new products with asphaltene solubility characteristics. These new condensation products may be regarded as altered asphaltenes and intermediates in the coke or pyrobitumen formation process (62-64). Contamination of original asphaltenes by subsequently formed or altered products, of course, will result in a less definitive correlation between an asphaltene and its source kerogen.

Presently, the unsatisfactory state of knowledge regarding structure of organic sulfur in kerogens and asphaltenes is also the case for coals. Coal is a special form of kerogen derived largely from higher terrestrial plant organic matter. Advances in analytical methods for solids with respect to sulfur characterization will benefit all fuel sciences. Sulfur K edge X-ray absorption spectroscopy (XANES and XAFS) shows promise (65). This technique applied to petroleum asphaltenes is reported by George et al. (66) in this volume, and is being investigated for quantitative determination of sulfur forms in other complex solid materials such as coals and kerogens.

Sulfur in Natural Gas

Natural gases vary in H_2S content from negligible amounts to over 90 mole percent, but most gases have a negligible H_2S content. Natural gases with high H_2S contents often also contain dissolved elemental sulfur and polysulfides. Gases with more than a few percent H_2S are generally limited to carbonate reservoirs at relatively high temperatures and in strata associated with evaporites. These rocks and associated waters furnish sulfate (from anhydrite) that can be reduced to H_2S by bacteria at low temperatures and by thermochemical sulfate reduction at higher temperatures. The H_2S from thermochemical sulfate reduction can accumulate in high concentrations if the reservoirs contains little reactive iron. Geologic and geochemical controls on the distribution of H_2S in natural gas have been reviewed (<u>67</u>). Several informative papers were given in the symposium dealing with thermochemical sulfate reduction and other aspects of sulfur in natural gas (see abstracts of 197th ACS Meeting), but most of these papers did not become available for this volume. Hopefully, some will be published elsewhere.

Recent Advances in Molecular Structure of OSC

Major advances in analytical chemistry in the last two decades have greatly accelerated the geochemical study of fossil fuels. In particular, high resolution capillary gas chromatography coupled with mass spectrometry (GC-MS) has allowed the determination of molecular structure more precisely, with smaller samples, and with less prior separation and purification than was required previously. As a result, knowledge about petroleum composition has increased greatly. Composition of petroleum is now used much more effectively to infer source-rock facies (reflecting depositional environments), maturation levels (thermal history), migration pathways, and the various secondary alteration processes (biodegradation and water washing). The so-called biomarker molecules retaining structures indicative of their biological precursors are now used routinely in petroleum geochemistry for these applications and inferences. Hydrocarbon biomarkers, especially the steranes, hopanes, isoprenoids and n-alkanes have been the major molecules used for these purposes (<u>68-70</u>).

Advances in petroleum characterization at the molecular structure level by GC-MS methods renewed interest in OSC. Within the past few years, at least one-thousand new and novel OSC that previously were not known to be present in petroleum and bitumens have been reported. Tentative molecular structures inferred from GC-MS and other techniques have been confirmed in many cases by synthesis of authentic reference-compounds. The difficult and time-consuming synthetic work has been crucial in validating many of the novel structures. Another key finding has been that immature bitumens and crude oils (samples that have not received significant thermal stress) differ markedly from the previously known OSC in that they have carbon-skeletons resembling ubiquitous biomarker hydrocarbons (e.g., n-alkanes, isoprenoid alkanes, steranes, and hopanes). This similarity, of course, suggests that the hydrocarbons and OSC have common biogenic precursors.

In this overview we can only illustrate by selected examples the general structural types of OSC that have been identified in recent years and point out major conclusions to be drawn from these studies (<u>71-92</u>). Additional studies representing current trends in OSC research are included in this volume under the heading *Molecular Structure of Sulfur Compounds and Their Geochemical Significance*. These papers largely relate to: (1) early diagenetic reactions in initial sedimentary deposits, (2) immature bitumens and associated kerogens, (3) initial reactions involved in the transformation of kerogen and bitumen to crude oil, and (4) molecular indications of depositional environmental and thermal maturity conditions.

Structures I-XIII in Figure 3 illustrate the new generation of OSC quite different from those previously known (cf. Figures 1 and 2). The beginning of the new era in OSC research was the identification of two series of terpenoid sulfides I and II in petroleum reported by Payzant et al. in 1983 and 1985 (71-72) dealing with Alberta oils and oil sands, and of the C_{35} thiophene-containing hopane III reported by Valisolalao et al. in 1984 (76) from studies of bitumen extracted from an immature black shale. The structure of compound III strongly suggested that it was formed by incorporation of sulfur into bacteriohopanetetrol as shown in Figure 4A. The hopanetetrol is a cell membrane constituent of prokaryotes. This case was the first definitive indication at the molecular level of sulfur incorporation into a specific biogenic precursor molecule.

Additional precursor-product relationships were indicated by the identification of a C_{20} isoprenoid thiophene (IV; Figure 3) by Brassell et al. in 1986 (77) in both recent and ancient deep-sea sediments. This structure strongly suggested an origin by incorporation of reduced inorganic sulfur species into phytol and/or archaebacterial phytenes or their diagenetic products as indicated in Figure 4B. Other C_{20} isoprenoid thiophenes (e.g., V and VI; Figure 3) have been identified subsequently as reported by Sinninghe Damsté et al. in 1986 and 1987 (80,81), Sinninghe Damsté and de Leeuw in 1987 (82) and by Rullkötter et al. in 1988 (83).

Further examples are the C_{25} highly branched isoprenoid thiophenes (e.g., VII; Figure 3) which have been identified as abundant OSC from upwelling areas forming diatomaceous sediments such as in the Monterey Formation of California by Sinninghe Damsté, et al. (87,89). Precursors for these thiophenes are believed to be C_{25} highly-branched isoprenoid alkenes with two to four double bonds as shown in Figure 4C. These isoprenoid alkenes are known to be ubiquitous in young sediments as reported by Robson and Rowland (92) and were recently found to occur in Antarctic sea-ice diatoms by Nichols et al. (93). Therefore, this thiophene series has been suggested as biological markers for diatoms (88).

These examples convincingly demonstrate that specific OSC are formed during the early stages of diagenesis by reactions of reduced sulfur species with specific biogenic substrates. The reactive substrates are proposed to contain either carbon-carbon double bonds or other reactive functional groups that react with either hydrogen sulfide or polysulfides to form the OSC (88). These views are consistent with evidence from sulfur isotopes that H_2S produced by microbial sulfate reduction is the major source of reduced sulfur in sediments

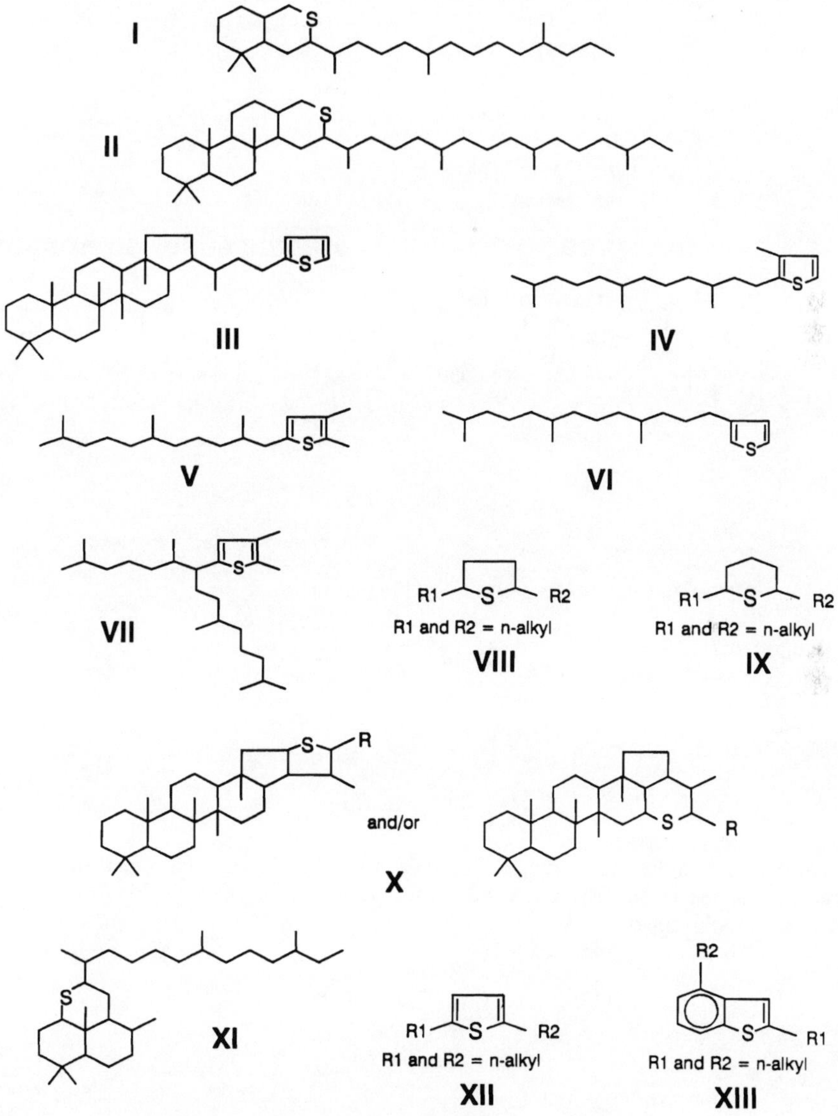

Figure 3. Examples of sulfur compounds with structures related to well known
biological markers.

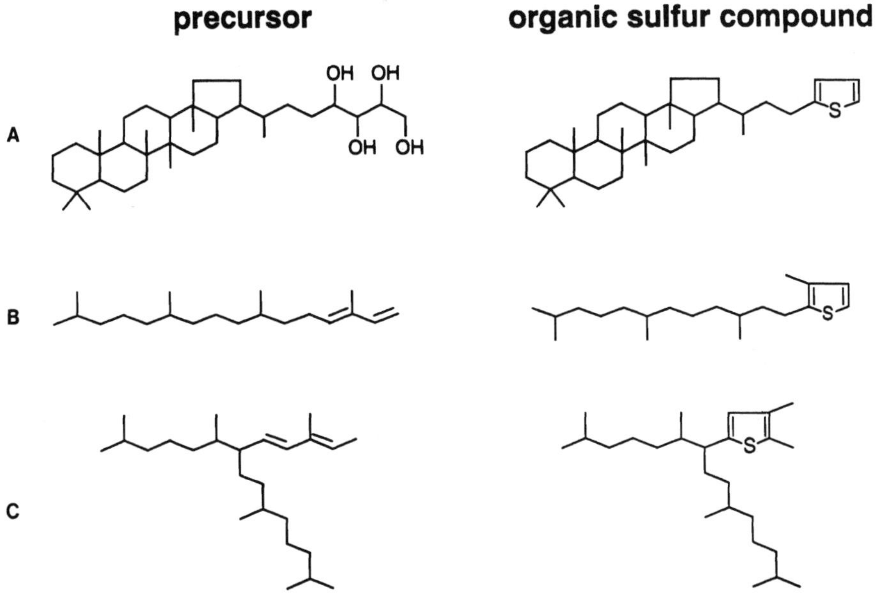

Figure 4. Examples of sulfur compounds and their presumed precursors: A) bacteriohopanetetrol, B) phyta-1,3-diene, and C) a C_{25} highly branched isoprenoid alkadiene.

and crude oils (17-20,22-24,28-29), and they are in agreement with other evidence presented in this volume and elsewhere (53, 99-106).

Despite the overwhelming evidence for the very early chemical incorporation of sulfur into sedimentary organic matter, two alternative explanations for the formation of some of these OSC have been proposed. First, Schmid et al. (90) in 1987 suggested that the direct reaction of n-alkanes with elemental sulfur during early maturation may account for the series of 2,5-dialkylthiolanes (VIII); and 2,6-dialkylthianes (IX) identified in immature crude oils and bitumens. This view was partially supported by laboratory experiments showing reactions of S° with n-alkanes. However, Sinninghe Damsté et al. (86) reported problems in accepting the proposal of Schmid at al. based on comparisons of the distribution patterns of n-alkanes and OCS with linear carbon skeletons in a number of samples. The second alternative suggests a biosynthetic origin for some of the OSC found in petroleum. Cyr et al. (73) proposed a biosynthetic origin for the bicyclic (I), tetracyclic (II) terpenoid sulfides and hopane sulfides (X). In components belonging to these series, the authors noted that the sulfur atom was attached to the second carbon atom of the alkyl side chain of the hydrocarbon analogues. This common structural feature was thought to reflect the site specificity of the biosynthetic pathway whereby sulfur is incorporated into the hydrocarbon framework. Continuing this view, Payzant et al. (74) proposed that the bicyclic- and tricyclicsulfides (I) and (XI) may be accessory pigments in photosynthesis because they have the carbon framework of carotenoids. However, these OSC have never been reported to occur in biota. The reaction of H_2S (HS^-) or H_2S_x (HS_x^-, S_x^{2-}) with reactive functionalized biogenic molecules during deposition or very early diagenesis seems more reasonable than either a direct biochemical origin or reaction of elemental sulfur with n-alkanes.

Origin of Sulfur in Fossil Fuels and Related Materials

Sulfate, the dominant form of sulfur in water and soil where some oxygen is present, is the ultimate source of sulfur in fossil fuels. Normal sea water contains about 28 mmoles/L of sulfate and fresh waters very much less. Higher plants and planktonic algae both assimilate sulfate and convert it into biosynthetic OSC that are part of the organic debris supplied to accumulating sediments. However, the relatively low total-sulfur content in biomass contrasts with the high concentrations of organically-bound sulfur in many sediments and fossil fuels. Biosynthetic OSC are largely the sulfur containing amino acids (cysteine and methionine) in proteins, sulfolipids in membranes, and sulfate esters in cell wall carbohydrates (96-101). Therefore, original OSC in biological debris not only have structures very different from those in fossil fuels, but they also are easily degraded by microbial and enzymatic processes that release sulfur. The released sulfur is largely in H_2S, methyl mercaptan and dimethyl sulfide in young sediments, but some is converted to more oxidized inorganic forms such as elemental sulfur, polysulfides, thiosulfate, polythionates, sulfite and sulfate (2,100). A number of other OSC are found in biological systems but they are quantitatively less important (96). For many years, it has been evident that original biomass is a minor sulfur source for sedimentary organic matter that contains moderate to large amounts of sulfur, but it may furnish a significant fraction of the total sulfur in low-sulfur kerogen and coal.

The major source of sulfur incorporated into pyrite and sedimentary organic matter is the dissimilatory H_2S formed by sulfate-reducing bacteria acting on aqueous sulfate (e.g., sea water in marine environments) either at the sediment-water interface or at shallow depths in organic-rich sediment. Sulfur isotope studies have been critical in establishing this fact (16-21). In some environments with restricted deep-water circulation (silled basins) or in highly productive upwelling areas in oceans, microbial sulfate reduction also may occur in the water column and lead to bottom-waters containing H_2S.

Hydrogen sulfide does not exist alone in natural sediment systems, but is always associated with variable concentrations of partially oxidized species with intermediate valence states between -II (H_2S, HS^-, S^{2-}) and +VI (HSO_4^-, SO_4^{2-}). Sulfate is kinetically inactive at low temperatures and generally is not in equilibrium with other sulfur species. On the other hand, reduced forms of sulfur in general are very reactive and concentrations are controlled by both thermodynamics and kinetics for the various species. The oxidation of microbially produced H_2S can be caused by (1) dissolved oxygen at anoxic/oxic boundaries, (2) by reducible cations such as Fe^{3+} released from minerals and (3) by microbial processes involving biochemical oxidation and reduction reactions in both oxic and anoxic environments. Currently, research directed toward understanding anoxic oxidation of H_2S in biological systems is very active (2,100). The capacity of chemoautotropic bacteria to oxidize hydrogen sulfide as an energy source for CO_2 fixation is well known, however the relatively recent realization that entire biological communities exist in deep ocean sites (using sulfide derived from hydrothermal vents) has stimulated new interest in the ability of biological systems to derive energy by sulfide oxidation in the dark. These studies are producing improved analytical methods for determining the numerous sulfur species (organic and inorganic) in biological systems and in sediments (2,100). Although details of sulfide oxidation mechanisms in anoxic environments are not completely understood, elemental sulfur and polysulfides H_2S_x are ubiquitous in sediments, as are at least traces of other intermediate oxidation state species such as thiosulfate, polythionates, and sulfite. Hydrogen sulfide and elemental sulfur (S^0 valence state) react rapidly and reversibly without biological mediation to form various catenated polysulfides (H_2S_x; x = 2,3,4,5 etc.) and establish the reactive aqueous $H_2S/S^0/H_2S_x$ system in sediments. This system is the major Eh controlling mechanism in reducing sediments (23,102).

Pyrite is a major sink for reduced sulfur from the $H_2S/S^o/H_2S_x$ system to the extent that reactive iron is available. In sediments with limited amounts of reactive iron, reduced sulfur species in the $H_2S/S^o/H_2S_x$ system are available for reaction with the sedimentary organic matter.

Polysulfides are extremely reactive in both oxidation and reduction reactions and readily incorporate sulfur into labile biogenic compounds that have functions such as the keto ($R_2C=O$), hydroxyl ($RCH_2\text{-}OH$), amino ($R_2\text{-}NH$) groups and/or carbon-carbon double bonds ($>C=C<$). Polysulfides also can cause oxidation of almost any functional group to a carboxylic acid or reduction to a hydrocarbon without sulfur remaining in the final organic product (103). LaLonde (104-105) discusses the importance of polysulfides in forming OSC and sulfur-rich kerogen in young sediments at low temperatures. As discussed in the previous section, thiolanes, thianes, thiophenes and organic polysulfides are concluded to be formed very early in this manner. Polysulfides probably also assist the natural polymerization process (kerogen formation) in sulfur-rich environments by forming sulfide and/or polysulfide cross-linkages (perhaps similar to the vulcanization of rubber). Francois (106-107) and others have demonstrated the early incorporation of sulfur into humic acids in modern marine sedimentary environments and stress the probable importance of polysulfides in these reactions. Aizenshtat et al. also reported on the involvement of polysulfides in reactions with recent organic matter in the hypersaline Solar-Lake environment (108).

Further evidence for the addition of H_2S to carbon-carbon double bonds very early in sediments, and further insights into reaction mechanisms, have been reported by Vairavamurthy and Mopper in 1987 and 1989 (109,110). They identified 3-mercaptopropionic acid (3-MPA) as a major thiol in anoxic intertidal marine sediment and demonstrated that the thiol formation could occur by the reaction of HS^- with acrylic acid in sediment water and seawater at ambient temperature: The formation of 3-MPA was hypothesized to occur by a Michael addition mechanism whereby the nucleophile HS^- adds to the activated double bond in the α,β-unsaturated carbonyl system:

$$CH_2=CH\text{-}CO_2^- + HS^- + H_2O \rightarrow HS\text{-}CH_2CH_2\text{-}CO_2^- + OH^-$$

They suggested that this could be a major pathway for the incorporation of sulfur into sedimentary organic matter during early diagenesis.

In the later work (110) they reported the relative rates of additions of both bisulfide (HS^-) and polysulfide (S_4^{2-}) to acrylic acid and to acrylonitrile ($CH_2=CH\text{-}CN$) for a range in pH and ionic strength. Results showed that at equal nucleophile concentrations, the addition of the polysulfide ion was much faster than that of the bisulfide ion. The difference in rate for the two nucleophiles was greater for acrylic acid, which is largely the anion under the pH conditions, than for acrylonitrile which is neutral. Effects of ionic strength on rates led to the suggestion that hypersaline environments also may favor organosulfur formation by the Michael addition mechanism.

Expectations for the Future

Although recent advances have led to a number of important conclusions and insights with respect to sulfur in fossil fuels, many remaining questions provide challenging opportunities for further research.

One area of intrigue relates to preliminary reports (78,86,111) of sulfur cross-linked polymers isolated from low maturity high-sulfur crude oils. Presumably, these materials are still being investigated, and further evaluation of their character and their significance can be expected.

An important opportunity related to petroleum exploration concerns the nature of transformations of OSC occurring between immature oils (or bitumens) and more mature crude oils. Easily identified OSC in immature oils and bitumens are mainly alkylthiolanes, alkylthianes, and alkylthiophenes whereas more mature crude oils (i.e., conventional crude oils) are dominated by rather low-molecular-weight benzo- and dibenzothiophenes. This difference signals potentially useful maturity reaction sequences in the OSC that require further investigation.   Cyclization and aromatization of 2-alkyl- and 2,5-dialkylthiophenes (XII) containing side chains with four or more linear carbons may be a major pathway to the alkylbenzothiophenes as suggested by Perakis (79) and Sinninghe Damsté et al. (84,89). This explanation is supported by the relatively high abundance of 2- and 4-alkyl- and 2,4-dialkylbenzothiophenes (XIII) in crude oils and bitumens (79,84) and the observation that more mature samples contain relatively higher amounts of these benzothiophenes (89). Similar cyclizations may account for benzothiophene moieties inferred to be present in some macromolecules (85,95). Eglinton et al. (94) found an increased abundance of alkylbenzothiophenes over alkylthiophenes in flash pyrolysis of artificially matured kerogens that suggests these reactions may proceed within the macromolecules during diagenesis and maturation.   Of course the entire problem of understanding the differences in composition of OSC between immature and more mature samples concerns not only molecular changes in the early OSC, but also the newly generated product that may form later with increasing maturity.

Further advances in various areas of OSC research can be expected in the next few years and many will be connected with continuing investigations of the mode of occurrence of sulfur in macromolecules (kerogens, asphaltenes and resins).   The difficult problem of understanding sulfur bonding and distribution of sulfur containing functional groups in macromolecules will continue to be given high priority because it is strongly related to the important implication of early oil generation from sulfur-rich kerogens. Determining the nature and abundance of polysulfide functions in kerogens, asphaltenes and resins remains a very important challenge.

Literature Cited

1. Manowitz, B.; Lipfert, F. W. In Geochemistry of Sulfur in Fossil Fuels; Orr, W. L.; White, C. M., Eds.: ACS Symposium Series (this volume); American Chemical Society: Washington, DC, 1990 .
2. Saltzman, E. S.; Cooper, W. J., Eds. Biogenic Sulfur in the Environment ACS Symposium Series 393; American Chemical Society: Washington, DC, 1989; 672 pp.
3. Guenther, A; Lamb, B.; Westberg, H. In Biogenic Sulfur in the Environment; Saltzman, E. S.; Cooper, W. J., Eds.; ACS Symposium Series 393; American Chemical Society: Washington, DC, 1989; 14-30.
4. Martin, R. L.; Grant, J. A. Anal. Chem. 1965, 37, 644-49.
5. Martin, R. L.; Grant, J. A. Anal. Chem. 1965, 37, 649-657.
6. Drushel, H. V. Anal. Chem. 1969, 41, 569-76.
7. Drushel, H. V. ACS Division of Petroleum Chemistry Preprints 1970, 15 (2), C12-42.
8. Demaison, G. J. Amer. Assoc. Petr. Geol. Bull. 1977, 61, 1950-61.
9. Roadifer, R. E. In Exploration for Heavy Crude Oil and Natural Bitumens; Meyer, R. F., Ed.; AAPG Studies in Geology #25; American Association of Petroleum Geologists: Tulsa 1987; 3-23.
10. Baker, E. W.; Lauda, J. W. In Biological Markers in the Sedimentary Environment; Johns, R. B., Ed.; Elsevior 1977; 125-224
11. Filby, R. H.; Van Berkyl, G. J. In Metal Complexes in Fossil Fuels; Filby, R. H.; Branthaver, J. F., Eds.; ACS Symposium Series No. 344; American Chemical Society: Washington, DC, 1987; 2-39.
12. Breger, I. A., Ed. Organic Geochemistry; Macmillon: New York, 1963; 658 pp.
13. Eglinton, G. and Murphy, M. T. J., Eds. Organic Geochemistry Methods and Results; Springer-Verlag: Berlin, 1969; 828 pp.
14. Erdman, J. G., In Fluids in Subsurface Environments, AAPG Memoir 4; Young, A.; Galley, J. E., Eds.; AAPG Press: Tulsa, 1965; 20-52.
15. Silverman, S. R., In Fluids in Subsurface Environments, AAPG Memoir 4; Young, A.; Galley J. E., Eds.,; AAPG Press: Tulsa, 1965; 53-65.
16. Fuex, A. N., Jour. Geochem. Research 1977, 7, 155-188.
17. Thode, H. G. In Fluids in Subsurface Environments, AAPG Memoir 4; Young, A.; Galley, J. E., Eds.; AAPG Press: Tulsa, 1965; 367-377.
18. Thode, H. G.; Monster J.; Dunford, H. B. Amer. Assoc. Petr. Geol. Bull. 1958, 42, 2619-2641.
19. Krouse, H. R. J. Geochem. Exploration 1977, 7, 189-211.
20. Nielsen, H. In Handbook of Geochemistry; Wedpohl, K. H., Ed., Springer-Verlag: Berlin, 1974; vol. II-I, Sec. 16-B.
21. Zobell, C. E. In Organic Geochemistry; Breger, I. A., Ed.; MacMillan: New York, 1963; 579-595.
22. Goldhaber, B. M. and Kaplan, I. R. In The Sea; Goldberg, E. D., Ed.; Wiley: New York, 1974; vol. 5, 569-655.
23. Berner, R. A. Geochim. Cosmochim. Acta 1963, 27, 563
24. Berner, R. A. Geochim. Cosmochim. Acta 1984, 48, 605-615.
25. Tissot, B. P.; Welte, D. H., Petroleum Formation and Occurrence (Second Edition); Springer-Verlag: Berlin, 1984; 699 pp.

26. Hunt, J. M. Petroleum Geochemistry and Geology; W. H. Freeman: San Francisco, 1979; 617 pp.

27. Didyk, B. M., Simonett, B. R. T., Brassell, S. C. and Eglinton, G. Nature 1978, 272, 216-222.

28. Orr, W. L. ACS Div. Petr. Chem. Preprints 1975, 21, 417-421.

29. Orr, W. L. In Oil Sand and Oil Shale Chemistry; Strausz, O. P.; Lown, E. M., Eds.; Verlag Chemie: New York, 1978; 223-243.

30. Dean, R. A.; Whitehead, E. V. In 7th World Petroleum Congress: Panel Discussion 23; Paper 7, 1967.

31. Mehmet, Y. Alberta Sulfur Research Quart. Bull. 1971, 6, 1-17.

32. Gal'pern, G. D. Int. J. Sulfur Chem. 1971, 6, 115

33. Coleman, H. J.; Hopkins R. L.; and Thompson, C. J. Int. J. Sulfur Chem. 1971 (Part B), 6, 41-62.

34. Rall, H. T.; Thompson, C. J.; Colemam, H. J.; Hopkins, R. L. Sulfur Compounds in Crude Oil; U. S. Bureau of Mines. Bulletin 659; US Government Printin Office: Washington, DC, 1972; 187 pp.

35. Thompson, C. J. In Organic Sulfur Chemistry; Freidlina, R., Kh.; Skarova, A. E., Eds.; Pergamon Press, 1981; 189-208.

36. Aksenov, V. S. and Kamyanov, V. F. In Organic Sulfur Chemistry; Freidlina R. Kh. and Skarova, A. E. Eds.; Pergamon Press, 1981; 1-13.

37. Gal'pern, G. D. In The Chemistry of Heterocyclic Compounds Thiophene and Its Derivitives; Gronowitz, S., Ed.; Wiley, 1985; 325-335.

38. Boduszynski, M. M. Energy & Fuel 1987, 1, 2-11.

39. Boduszynski, M. M. Energy & Fuel 1988, 2, 597-613.

40. Tetter, R. M. Mass Spectrometry Reviews 1985, 4, 123-143.

41. Ho, T. Y.; Rogers,M. A.; Drushel, H. V.;. Koons, C. B Amer. Assoc. Petr. Geol. Bull. 1974, 58, 2338-2348.

42. Orr, W. L. Amer. Assoc. Petr. Geol. Bull. 1974, 50, 2295-2318.

43. Radke, M.; Welte D. H.; Willsch, H. Geochim. Cosmochim. Acta 1982, 46, 1-10.

44. Radke, M.; Willsch, H.; Welte, D. H. In Advances in Organic Geochemistry 1985; Leythaeuser, D.; Rullkötter, J., Eds.; Org. Geochem.1986, 10, 51-63

45. Hughes, W. B. In Petroleum Geochemistry and Source Rock Potential of Carbonate Rocks, AAPG Studies in Geology #18; Palacas, J. G., Ed.; AAPG Press: Tulsa, 1984; 181-196.

46. Radke, M. In Advances in Organic Geochemistry ; Brooks, J. and Welte, D. Eds; Academic Press: London, 1987; vol 2, 141-207.

47. Radke, M. Marine Petrol. Geol. 1988, 5, 224-236.

48. Schou, L.; Myhr,M. B. In Advances in Organic Geochemistry 1987; Mattavelli, L.; Novelli, L. Eds; Org. Geochem. 1988, 13, 61-66.

49. Leythaeuser, D.; Radke, M.; Willsch, H. Geochim. Cosmochim. Acta. 1988, 52, 2879-2891.

50. Gransch, J. A.; Posthuma, J. In Advances in Organic Geochemistry 1973; Tissot, B.; Bienner, F. Eds.; Editions Technip: Paris; 1974; 727-739.

51. Durand, B. In Kerogen - Insoluble Organic Matter from Sedimentary Rocks; Durand, B., Ed.; Editions Technip: Paris, 1980; 13-34.

52. Behar, F; Vandenbroucke, M. Org. Geochem. 1987, 11, 15-24.

53. Orr, W. L. In Advances in Organic Geochemistry 1985; Leythaeuser, D; Rullk ötter, J. Eds.; Org. Geochem. 1986, 10, 499-516.
54. Durand, B. and Monin, J. C. In Kerogen - Insoluble Organic Matter from Sedimentary Rocks; Durand, B., Ed.; Editions Technip: Paris, 1980; 113-142.
55. Jones, R. W. In Petroleum Geochemistry and Source Rock Potential of Carbonate Rocks, AAPG Studies in Geology #18; Palacas, J. G., Ed.; AAPG Press: Tulsa, 1984; 163-180.
56. Lewan, M. D. Phil.Trans.R. Soc. Lond. 1985, A315, 123-134.
57. Tannenbaum, E; Aizenshtat, Z. Org. Geochem.1985, 8, 181-192.
58. Tissot, B. P. Amer. Assoc. Petr. Geol. Bull. 1984, 68, 545-563.
59. Tissot, B.; Pelet, R; Ungerer, Ph. Amer. Assoc. Petr. Geol. Bull. 1987, 71, 1445-1466.
60. Speight, J. G.,;Moschopedis, S. E. In Chemistry of Asphaltenes; Bunger, J. W.; Li, N. C., Eds.; ACS Advances in Chemistry Series 195; American Chemical Society: Washington, DC, 1981; 1-15.
61. Speight, J. G. In Polynuclear Aromatic Compounds; Ebert, L. B., Ed., ACS Advances in Chemistry Series 217; American Chemical Society: Washington, DC, 1988; 201-215.
62. Speight, J. G. In Preprints ACS Div. Petr. Chem. Denver 1987, 32, No. 2, 413-418.
63. Pelet, R.; Behar, F.; Monin, J. C. In Advances in Organic Geochemistry 1985; Leythaeuser, D; Rullkötter, J., Eds.; Org. Geochem. 1986, 10, 481-498.
64. Behar, F.; Pelet, R. Energy & Fuel 1988, 2, 259-264.
65. Huffman, G. P.; Huggins, F. E.; Sudipa Mitra; Shah, N.; Pugmire, R. J.; Davis, B; Lytle, F. W.; Greegor, R. B. Energy & Fuel 1989, 3, 200-205.
66. George, G. N.; Gorbaty, M. L.; Kelemen. S. R. In Geochemistry of Sulfur in Fossil Fuels; Orr, W. L.; White, C. M., Eds.: ACS Symposium Series (this volume); American Chemical Society: Washington, DC, 1990.
67. Orr, W. L. In Advances in Organic Geochemistry 1975,; Campos R.; Goni, J., Eds.; Enadisma: Madrid, 1977; 571-597.
68. Seifert, W. K.; Moldowan, J. M. In Biological Markers in the Sedimentary Record; R. B. Johns, Ed.; Elsevier: Amsterdam, 1986; 261-286.
69. Johns, R. B.,Ed. Biological Markers in the Sedimentary Record; Elsevier: Amsterdam, 1986; 364 pp.
70. Yen, T. F.; Moldowan, J. M.. Eds. Geochemical Biomarkers; Harwood Academic: Chur Switzerland, 1988; 438 pp.
71. Payzant, J. D.; Montgomery, D. S.; Strausz, O. P. Tetrahedron Lett. 1983, 24, 651-654.
72. Payzant, J. D.; Montgomery, D. S.; Strausz, O. P. Tetrahedron Lett. 1985, 26, 4175-4178.
73. Cyr, T. D.; Payzant, J. D.; Montgomery, D. S.; Strausz, O. P. Org. Geochem. 1986, 9, 139-143
74. Payzant, J. D.; Montgomery, D. S.; Strausz, O. P. Org. Geochem. 1986, 9, 357-369.

75. Payzant, J. D.; Montgomery, D. S.; Strausz, O. P. AOSTR J. Res. 1988, 4, 117-131.
76. Valisolalao, J.; Perakis, N.; Chappe, B.; Albrecht, P. Tetrahedron Lett. 1984, 25, 1183-1186.
77. Brassell, S. C.; Lewis, C. A.; de Leeuw, J. W.; de Lange, F.; Sinninghe Damsté, J. S. Nature, 1986, 320, 160-162.
78. Schmid, J. C. Marquers biologique soufrés dans les pétroles. Ph. D. dissertation, Univ. of Strasbourg, 1986 263 pp.
79. Perakis, N. Séparation et détection sélective de composés soufrés dans les fractions lourdes des pétroles. Géochimie des benzo[b]thiophènes; Ph. D. dissertation, Univ. of Strasbourg, 1986.
80. Sinninghe Damsté, J. S.; ten Haven, H. L.; de Leeuw, J. W.; Schenck, P. A. In Advances in Organic Geochemistry 1985; Leythaeuser, D.; Rullkötter, J., Eds.: Org. Geochem. 1986, 10, 791-805.
81. Sinninghe Damsté, J. S.; Kock-van Dalen, A. C.;de Leeuw, J. W.; Schenck, P. A. Tetrahedron Lett. 1987, 28, 957-960.
82. Sinninghe Damsté, J. S.; de Leeuw, J. W. Int. J. Environ. Anal. Chem. 1987, 28, 1-19.
83. Rullkötter, J.; Landgrag M.; Disco, U. J. High Res. Chrom. & Chrom. Commun. 1988, 11, 633-638.
84. Sinninghe Damsté, J. S., de Leeuw, J. W.; Kock-van Dalen, A. C.; de Zeeuw, M. A.; de Lange, F.; Rijpstra W. I. C.; Schenck, P. A. Geochim. Cosmochim. Acta 1987, 51, 2369-2391.
85. Sinninghe Damsté, J. S.; Kock-van Dalen, A. C.; de Leeuw, J. W.; Schenck, P. A. J. Chromatog. 1988, 435, 435-452.
86. Sinninghe Damsté, J. S.; Rijpstra, W. I. C.; de Leeuw, J. W.; Schenck, P. A. In Advances in Organic Geochemistry 1987; Novelli, L; Mattavelli, L., Eds.; Org. Geochem. 1988, 13, 593-606.
87. Sinninghe Damsté, J. S.; van Koert, E. R.; Kock-van Dalen, A. C.; de Leeuw, J. W.; Schenck, P. A. Org. Geochem. 1989, 14, 555-567.
88. Sinninghe Damsté, J. S.; Rijpstra, W. I. C.; Kock-van Dalen, A. C.; de Leeuw, J. W.; Schenck, P. A. Geochim. Cosmochim. Acta 1989, 53, 1343-1355.
89. Sinninghe Damsté, J. S.; Rijpstra, W. I. C.; de Leeuw, J. W.; Schenck, P. A. Geochim. Cosmochim. Acta 1989, 53, 1323-1341.
90. Schmid, J. C.; Connan J.; Albrecht, P. Nature 1987, 329, 54-56.
91. Sheng Guoying; Fu Jimo; Brassell, S. C.; Gowar, A. P.; Eglinton, G.; Sinninghe Damsté, J. S.; de Leeuw, J. W.; Schenck, P. A. Geochem. 1987, 6, 115-156.
92. Robson, J. N.; Rowland, S. J. Nature 1986, 324, 561-563.
93. Nichols, P. D.; Volkman, J. K; Palmisano, A. C.; Smith G. A.; White D. C. J. Phycol. 1988, 24, 90-96.
94. Eglinton, T. I.; Philp, R. P.; Rowland, S. J. Org. Geochem. 1988, 12, 33-41.
95. Sinninghe Damsté, J. S.; Eglinton, T. I.; de Leeuw, J. W.; Schenck, P. A. Geochim. Cosmochim. Acta 1989, 53, 873-889.
96. Orr, W. L., In Handbook of Geochemistry; Wedpohl, K. H., Ed., Springer-Verlag: Berlin, 1974; vol. II-I, Sec. 16-L.

97.   Casagrande, D. J.; Siefert, K.; Berschinski, C; Sutton, N. Geochim. Cosmochim. Acta 1977, 41, 161-167

98.   Casagrande, D. J.; Siefert, K. Science 1977, 195, 675-676.

99.   Price, F. T.; Shieh, Y. N. Economic Geology 1979, 74, 1445-1461

100. Vetter, R. D.; Matrai, P. A.; Javor, B.; O'Brien, J. In Biogenic Sulfur in the Environment; Saltzman, E. S.; Cooper, W. J., Eds.; ACS Symposium Series 393; American Chemical Society: Washington, DC, 1989; 243-261.

101. Fischer, U. In Biogenic Sulfur in the Environment; Saltzman, E. S.; Cooper, W. J., Eds.; ACS Symposium Series 393; American Chemical Society: Washington, DC, 1989; 262-279.

102. Boulegue, J; Lord III, C. J.: Church, T. M. Geochim. Cosmochim. Acta 1982, 46, 453-464

103. Pryor, W. A., Mechanisms of Sulfur Reactions; McGraw-Hill: New York, 1962; 241 pp.

104. LaLonde, R. T.; Ferrara, L. M.; Hayes, M. P. Org. Geochem. 1987, 11, 563-571.

105. LaLonde, R. T. In Geochemistry of Sulfur in Fossil Fuels; Orr, W. L.; White, C. M., Eds.: ACS Symposium Series (this volume); American Chemical Society: Washington, DC, 1990; this volume.

106. Francois, R. Geochim. Cosmochim. Acta 1987, 51, 17-27.

107. Francois, R. Limnol. Oceanogr. 1987, 32, 964-972.

108. Aizenshtat, Z.; Stoler, A.; Cohen, Y.; Nielsen, H. In Advances in Organic Geochemistry 1981; Bjorøy, M., Ed.; Wiley: Chichester, 1983; 279-288.

109. Vairavamurthy, A.; Mopper, K. Nature 1987, 329, 623-625.

110. Vairavamurthy, A.; Mopper, K. In Biogenic Sulfur in the Environment; Saltzman, E. S.; Cooper, W. J., Eds.; ACS Symposium Series 393; American Chemical Society: Washington, DC, 1989; 231-242.

111. Mycke, B.; Schmid, J. C.; Albrecht, P. Abstracts of Papers, 19 7th American Chemical Society Meeting; Dallas, 1989; GEOC Abstract 24.

RECEIVED March 5, 1990

Chapter 2

Geochemistry of Sulfur in Coal

C.-L. Chou

Illinois State Geological Survey, 516 East Peabody Drive, Champaign, IL 61820

This paper summarizes our understanding of the geochemistry of sulfur in coal in the following areas: 1) abundance of sulfur in coals of major coal basins in the U.S., 2) distribution of sulfur in coal lithotypes and macerals, 3) characteristics and geochemical significance of sulfur-containing organic compounds, 4) sulfur isotopic studies relating to the sources of sulfur in coal, and 5) sedimentary environments controlling the geochemistry of sulfur in coal. The evidence suggests that the sulfur in low-sulfur coal ($\leq 1\%$ S) is derived primarily from parent plant material. In medium-sulfur (>1 to $<3\%$ S) and high-sulfur ($\geq 3\%$ S) coal, there are two major sources of sulfur: 1) parent plant material, and 2) sulfate in seawater that flooded peat swamps. Compositional variations of sulfur in coal are largely controlled by sedimentary environments during peat accumulation and by postdepositional changes (diagenesis). A model of the formation of pyrite and organic sulfur compounds in high-sulfur coal is presented.

High sulfur content in coal hinders the use of coal resources because sulfur dioxide emissions from utility and industrial boilers are a cause of acid rain. Thus, research into the nature of sulfur in coal is important for improving coal utilization. Geochemical studies of sulfur in coal provide information about the abundance, distribution, and speciation of sulfur in coal. Many of these properties are determined by geological environments and processes of coal formation.

This chapter reviews research on the abundance of sulfur in major coal basins in the U.S., the forms of sulfur in coals, the distribution of sulfur in coal lithotypes and macerals, and the nature of sulfur-containing organic compounds in coal. Next, the origin of sulfur in coal is reviewed based on the evidence from the distribution and speciation of sulfur in peat, and from stratigraphic, isotopic, and trace element data. Finally, the origin of sulfur in coal is explained by a geochemical model.

0097–6156/90/0429–0030$06.75/0

Abundance of Sulfur in Coals of Major Coal Basins in the U.S.

The U.S. has vast coal resources of various geologic age (Carboniferous to Tertiary) and rank (lignite to anthracite). The regions with major coal resources in the conterminous U.S. are divided into six provinces: Eastern, Interior, Rocky Mountains, Northern Great Plains, Gulf, and Pacific Coast. In addition, there are Alaskan coal fields (1). Information concerning geographic area, age, rank, reserve base, and sulfur abundance in coals from major basins in these provinces is summarized in Table I.

The sulfur content of U.S. coals varies considerably but are most commonly within the range of 0.5 to 5% total sulfur. A coal is classified as low-sulfur, medium-sulfur, or high-sulfur on the basis of its total sulfur content (on an as-received basis) of 1% or less, more than 1% and less than 3%, and 3% or more, respectively (1). The Keystone Coal Industrial Manual (2) is a general reference describing the coal seams and coal quality of each coal-producing state.

Forms of Sulfur

The major forms of sulfur in coal are pyritic, organic, and sulfate. Pyritic and organic sulfur generally account for the bulk of sulfur in coal. Elemental sulfur also occurs in coal, but only in trace to minor amounts; it is not determined in routine coal analyses.

Pyritic Sulfur. Pyrite is the predominant sulfide mineral in coal. Other sulfide minerals found include marcasite, pyrrhotite, sphalerite, galena, and chalcopyrite. Pyrite occurs in several forms, most commonly as 1) nodules and partings, 2) disseminated pyrite, having particle sizes ranging from a few millimeters to less than one micrometer, 3) thin, platy pyrite in cleats, 4) pyrite veins, and 5) pyrite in permineralized peat or coal balls. Scanning and transmission electron microscopes reveal irregular polycrystalline forms, framboids, and isolated single pyrite crystals (7,8).

Disseminated pyrite grains are associated mainly with the microlithotypes such as vitrite, bimacerite, trimacerite, and carbominerite (9-11). The size distribution of the various forms of pyrite in coal has been studied by a number of researchers, primarily for the purpose of assessing the removability of pyrite by physical coal cleaning (9-10,12-17).

Organic Sulfur. Organic sulfur is calculated in ASTM method D2492-84 as the difference between total sulfur and the sum of pyritic plus sulfate sulfur (18). Pyritic sulfur is determined on the basis of the amount of iron extracted by nitric acid, and a significant error can be caused by extraction of non-pyritic iron or incomplete dissolution of pyrite (19). As a result, the errors made in the measurement of total sulfur, pyritic sulfur and sulfate sulfur are reflected in the organic sulfur value. In order to avoid these errors, researchers have tried to determine organic sulfur directly. Methods developed include electron probe microanalysis (20-24), scanning electron microscopy with energy-dispersive X-ray analysis (SEM-EDX) (25,26), and transmission electron microscopy (TEM) (27-30). Other methods such as a soft X-ray technique (31), and determination of sulfur volatilized by oxidation under an

Table I. Sulfur abundance in coals of major coal basins in the U.S.

Coal basin	Geographic area*	Rank	Reserve base**	Sulfur abundance
Eastern Province (*Pennsylvanian age*)				
Appalachian Basin	WV, PA, OH, E. KY, AL, VA, TN, MD	Bituminous	98.7	Widely variable, mostly 0.5 to 5% sulfur. Large quantities of low-sulfur coal occur in the Eastern Kentucky Coal Field and the Southern Field of West Virginia.
Pennsylvania Anthracite Region	PA	Anthracite	7.1	0.5-1.5% sulfur.
Interior Province (*Pennsylvanian*)				
Illinois Basin	IL, IN, W. KY	Bituminous	109.6	Mostly high-sulfur coal. Approximately 5% of the total reserve is low- to medium-sulfur coal.
Western Interior Basin	MO, IA, OK, KS, AR	Anthracite Bituminous	0.10 11.1	Generally high-sulfur coal. Some low- to medium-sulfur coal occurs in Arkansas.
Southwestern Region	TX	Bituminous	6.1	Highly variable, 1.1-8.9% sulfur.
Rocky Mountains Province (*Late Cretaceous to Eocene*)				
Eleven basins: Wind River, Green River, San Juan, Bighorn, Hams Fork, Uinta, Denver, Southwestern Utah,	WY, CO, UT, NM, AZ, MT, ID	Anthracite Bituminous Subbituminous Lignite	0.03 22.3 49.1 4.2	Predominantly low-sulfur coal.

Raton Mesa, Black Mesa,
Tertiary lake beds

Northern Great Plains Province (*Tertiary*)

Powder River Basin, Williston Basin, Black Hills, North Central	MT, WY, ND, SD	Bituminous	1.4	Low-sulfur coal. The Big George deposit of the Powder River Basin contains coal with as low as 0.2% sulfur (3). Lignite in the Williston Basin has an average of 0.6% sulfur.
		Subbituminous	126.3	
		Lignite	28.1	

Gulf Province (*Tertiary*)

	TX, AL, AR, LA, MS, TN	Lignite	15.5	Typically 0.7-1.9% sulfur.

Pacific Coast Province (*Eocene*)

	WA, OR, CA	Bituminous	1.0	Low- to medium-sulfur coals.
		Subbituminous	6.6	
		Lignite	0.022	

Alaskan Coal Fields (*Cretaceous to Tertiary*)

	AK	Bituminous	3.9	Generally low-sulfur coal.
		Subbituminous	2.9	

*Abbreviations for state names: AK, Alaska; AL, Alabama; AR, Arkansas; AZ, Arizona; CA, California; CO, Colorado; IA, Iowa; ID, Idaho; IL, Illinois; IN, Indiana; KS, Kansas; KY, Kentucky; LA, Louisiana; MD, Maryland; MO, Missouri; MS, Mississippi; MT, Montana; ND, North Dakota; NM, New Mexico; OH, Ohio; OK, Oklahoma; OR, Oregon; PA, Pennsylvania; SD, South Dakota; TN, Tennessee; TX, Texas; UT, Utah; VA, Virginia; WA, Washington; WV, West Virginia; WY, Wyoming.

**Demonstrated reserve base in billion short tons. Data are taken from a U.S.D.O.E. Energy Information Agency publication (4), Nelson (5), and Merritt (6).

oxygen plasma (32) are being developed.   X-ray absorption fine-structure
(XAFS) spectroscopy has been used to determine the molecular structure of
organic sulfur in coal (33-37) (see the chapter by George et al. in this book).
Analysis of a suite of maceral separates for the X-ray absorption near edge
structure (XANES) within about 20-50 eV of the absorption edge showed
peaks characteristic of compounds containing sulfur atoms bonded in aromatic
systems (37).

Sulfate Sulfur.   Coal contains trace to minor amounts of sulfate sulfur.
Sulfate minerals such as gypsum [$CaSO_4 \cdot 2H_2O$] and barite [$BaSO_4$] are found
in some fresh coals.   The following iron sulfate minerals are weathering
products of pyrite (38-41): szomolnokite [$FeSO_4 \cdot H_2O$], rosenite [$FeSO_4 \cdot 4H_2O$],
melanterite [$FeSO_4 \cdot 7H_2O$], coquimbite [$Fe_2(SO_4)_3 \cdot 9H_2O$], roemerite [$FeSO_4$
$\cdot Fe(SO_4)_3 \cdot 14H_2O$], jarosite (usually a sodium jarosite) [$(Na,K)Fe_3(SO_4)_2(OH)_6$],
and halotrichite [$FeAl_2(SO_4)_4 \cdot 22H_2O$].   The sodium sulfate minerals, mirabilite
[$Na_2SO_4 \cdot 10H_2O$] and thenardite [Na_2SO_4] found in coal refuse (42) are reaction
products of iron sulfates with sodium associated with coal (43).

Elemental Sulfur.   In 1942, Chatterjee (44) reported the presence of elemental
sulfur in weathered Indian coal but not in fresh samples.   He suggested that,
during weathering, pyrite is first oxidized to ferrous and ferric sulfates, and
that then ferric sulfate oxidizes pyrite to elemental sulfur.   The presence of
elemental sulfur in U.S. coals was confirmed recently by Richard et al. (45)
and White and Lee (46).   Duran et al. (47) used extraction and gas
chromatographic analysis to determine elemental sulfur in a suite of U.S.
coals.   They found that elemental sulfur (0.03-0.17%) is present in coal that
has been exposed to the atmosphere, but is absent in pristine samples that
have been processed and sealed under a nitrogen atmosphere.   These data
support Chatterjee's discovery that elemental sulfur in coal is a   weathering
product.
        Yurovskii (48) recognized the importance of elemental sulfur in the
formation of pyritic and organic sulfur compounds and suggested that the
elemental sulfur found in coal today is a primary substance formed during coal
formation.   However, this view is unsupported by recent data (47).
        Beyer et al. (49) found that, during microbial desulfurization, pyritic sulfur
decreases and elemental sulfur increases with time, whereas the organic sulfur
remains unchanged.   They suggested that microbial oxidation of pyrite
produces ferric sulfate [$Fe_2(SO_4)_3$] and that the simultaneous inorganic reaction
of ferric iron with pyrite produces elemental sulfur and ferrous iron, as
follows:

$$2Fe^{3+} + FeS_2 \text{ (pyrite)} \rightarrow 3Fe^{2+} + 2S°$$

In a coal desulfurization study, Narayan et al. (50) were able to extract an
appreciable amount of elemental sulfur (36% of total sulfur) with
perchloroethylene at 120 °C from weathered coal, but not from fresh coal.
Hackley et al. (51) determined the isotopic composition of elemental sulfur
extracted by perchloroethylene and obtained results consistent with the
interpretation that the elemental sulfur originates from the oxidation of pyrite.
        Greer (52) reported that some coals contain significant amounts of

elemental sulfur (up to 0.34%). He pointed out that elemental sulfur, if recorded as organic sulfur, can cause significant error in the determination of organic sulfur by the ASTM method. Stock et al. (53) and the chapter by Stock and Wolny in this book have more information on elemental sulfur in coal.

Distribution of Sulfur in Coal Lithotypes and Macerals

Coal lithotypes are megascopically recognizable components of coal. Several nomenclatures have been used to identify the lithotypes [see Damberger et al. (54) for a review of several common classifications]. Coal lithotypes have fairly distinct maceral compositions, which originate from different combinations of plant materials and evolve along somewhat different paths during coalification.

Data on a 1.73-meter column from the Herrin seam in Jefferson County, Illinois, illustrate the spatial variability of pyritic and organic sulfur contents in lithotypes of a coal seam. This column was collected in an underground mine and examined in the laboratory; lithotypes were hand-picked for analysis. The vertical variation of lithotypes in this column is shown in Figure 1. The hand-picked lithotype separates were analyzed for proximate and ultimate analyses, forms of sulfur, and maceral compositions. The detailed results will be published elsewhere. The relation between organic sulfur and pyritic plus sulfate sulfur in the lithotypes is shown in Figure 2. In vitrain, bright banded coal (BBC), subbright banded coal (SBBC), and dull coal, organic sulfur varies from 0.67 to 2.1%, and the pyritic plus sulfate sulfur from 0.26 to 4.7%. In samples of fusain and mineralized coal, organic sulfur and pyritic plus sulfate sulfur contents display more variation. The ratio of pyritic plus sulfate sulfur to organic sulfur is higher in fusain and mineralized coal than in most vitrain and banded coal, reflecting the occurrence of pyrite in cell voids of fusain and in mineral partings, as well as the deposition of pyrite in permineralized peat (coal balls) (55).

The determination of sulfur in individual macerals (the microscopic constituents of coal) is difficult because of the small size of the macerals and their intimate association with one another in most coals. Analysis of the total sulfur content of individual macerals separated from two coals shows that fusinite had the lowest sulfur content (56). Raymond (57), and more recently Boudou et al. (58), performed direct determinations of organic sulfur, using an electron microprobe. They found that the organic sulfur content of vitrinite approximates the organic sulfur content of the bulk coal. Dyrkacz et al. (59,60) used density gradient centrifugation (DGC) to separate coals into vitrinite, exinite, and inertinite. Tseng et al. (61), using transmission electron microscopy (TEM), showed that sporinite is high in organic sulfur, vitrinite is intermediate in organic sulfur, and inertinite is low in organic sulfur. Organic sulfur is more variable in exinite than in vitrinite and inertinite. Hippo et al. (62) determined the organic sulfur content of individual macerals separated by DGC and confirmed that sporinite has higher organic sulfur contents than other macerals. Results of a recent study on the organic sulfur distribution in coals of the Illinois Basin Coal Sample Program, using SEM-EDX (26), were consistent with these earlier findings. Vitrinite, the predominant maceral in Illinois Basin coals, averages 85% in the Herrin Coal (63) and accounts for about 80-90% of the organic sulfur in bulk Illinois Basin coals.

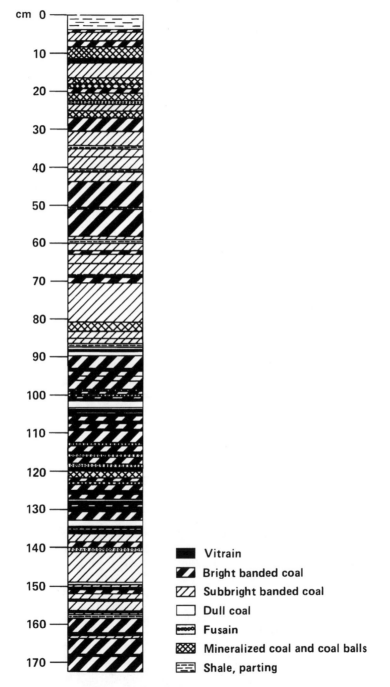

Figure 1.   Coal lithotype variation in a 1.73-meter vertical column of Herrin (No. 6) Coal, Jefferson County, IL.

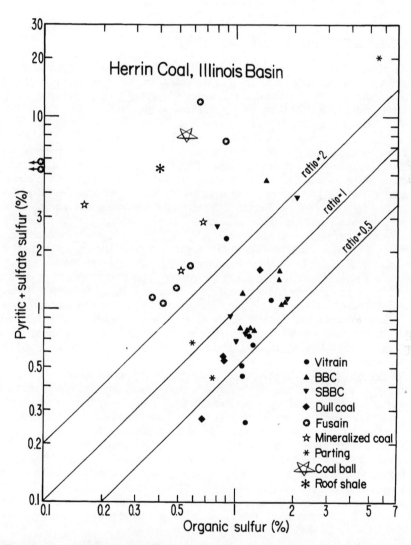

Figure 2.   Variation of pyritic plus sulfate sulfur versus organic sulfur in lithotype samples hand-picked from the column of coal.   Diagonal line indicates the ratio of pyritic plus sulfate sulfur to organic sulfur.

Gondwana coals in the Southern Hemisphere are significantly higher in inertinite content and lower in vitrinite content than Carboniferous coals in the Northern Hemisphere. Among some Permian coals in South Africa, coal with a high vitrinite content has a high sulfur content (64), reflecting vitrinite has a higher sulfur content than inertinite.

Sulfur-Containing Organic Compounds in Coal

Studies of sulfur-containing organic compounds in coal provide information on the origin of organic sulfur and the transformation of sulfur compounds during coal maturation. The distribution and types of organic sulfur compounds in coal have profound implications for desulfurization processes.

The following kinds of organic sulfur compounds are thought to occur in coal: 1) thiols, 2) sulfides and disulfides, and 3) thiophene and its derivatives. Our knowledge about the molecular structure of organic sulfur compounds has been keeping pace with the development of new analytical instrumentation and methods. Methods like the ASTM procedures, SEM-EDX, TEM, and electron probe microanalysis provide excellent data on the forms and amounts of sulfur in coal but no data on the molecular structure of the organosulfur components. On the other hand, methods such as Curie-Point pyrolysis in combination with gas chromatography-mass spectrometry (GC-MS), are capable of identifying organic sulfur species, but provide little information regarding the quantities of these species in coal. Thus, it is necessary to use two or more complementary methods to fully characterize the sulfur-containing organic compounds in coal.

White et al. (65) positively identified a variety of thiophenes in a Bevier seam coal extract. Using low-voltage high-resolution mass spectrometry they were able to observe low concentrations of compounds in the extract that had molecular formulae consistent with those of some thiols, sulfides and disulfides. It should be emphasized however, that no thiol, sulfide or disulfide was positively identified in the extract.

White (66) provides a review of the methods for identification of sulfur-containing polycyclic aromatic hydrocarbons in coal and coal-derived materials. High-resolution glass-capillary gas chromatography (GC) in conjunction with sulfur-selective detectors, and combined GC-MS are standard tools in modern organic geochemistry laboratories. The dual flame ionization and sulfur-specific flame photometric detector (FID/FPD) is frequently used during GC analysis of sulfur compounds. For samples having a low concentration of sulfur-containing polycyclic aromatic compounds (PAC), chemical isolation of target compounds may be essential (67). Nishioka (68) developed ligand-exchange chromatography using silica gel impregnated with $PdCl_2$, for isolation of sulfur-containing PAC which were subsequently analyzed by capillary GC and GC-MS.

Indanethiol and naphthalenethiol were tentatively identified in solvent extracts of a high-volatile bituminous coal from Poland (69). Data from Pyroprobe flash pyrolysis GC-MS and continuous isothermal pyrolysis by Calkins (70) suggested that the organic sulfur in coal is mainly aliphatic and thiophenic and that the relative proportion of aliphatic sulfides and mercaptans to total organic sulfur decreases with increasing rank (60-80% in lignite, 30-50% in high-sulfur bituminous coal, and zero in anthracite). Short-chain dialkyl sulfides, $(C_3H_7)_2S$ and $(C_4H_9)_2S$, were tentatively identified in bituminous coals by pyrolysis-GC-FPD (71). Van Graas et al. (72) used Curie-point pyrolysis-MS and Curie-point pyrolysis-GC-MS to identify several thiophenic compounds in coals ranging from high-volatile bituminous to anthracite. Hayatsu et al. (73) identified benzothiophene and dibenzothiophene in a bituminous coal decomposed by oxidation with $Na_2Cr_2O_7$. Because the decomposition procedure was carried out at a relatively low temperature (250 °C), these thiophenic compounds are described as being indigenous to the coal. Hayatsu et al. (74) determined the relative abundance of aromatic and heteroaromatic compounds in three coals of different rank (lignite, bituminous coal, and anthracite), also by oxidation with $Na_2Cr_2O_7$. The sulfur-containing aromatic compounds (benzothiophene, dibenzothiophene, and benzonaphthothiophene) were found in bituminous coal and anthracite, but not in lignite. Thus, the abundance of various types of organic sulfur compounds in coal may be related to the rank of coal.

White and Lee (46) identified organic sulfur compounds in a benzene extract from a Kentucky high-volatile bituminous coal, using a high-resolution glass-capillary GC combined with a mass spectrometry, and high resolution GC equipped with dual FID/FPD. Tentatively identified were C_2- and C_3-benzothiophenes, methyldibenzothiophenes, and phenanthrothiophene. Oxidation of a vitrinite from high-volatile bituminous coal, Mankham Main, with performic acid in methanol followed by X-ray photoelectron spectroscopy (XPS) yielded data consistent with the presence of sulphides and thiophenes in the original coal (75).

Attar and co-workers (76-78) have developed a method of quantifying the different types of organic sulfur compounds (thiol, sulfide, disulfide, thiophene and its derivatives) in coal, using thermokinetic analysis. Organic sulfur compounds in coal are reduced to H_2S when the coal is heated in the presence of a catalyst at temperatures between <160 °C to 500 °C. The results suggested that 15-30% of the organic sulfur is sulfidic. Attar and co-workers reported that about 30-40% of the organic sulfur in lignite is thiolic and the rest occurs as sulfides and thiophenic compounds. In bituminous coal, 40-60% of the organic sulfur is thiophenic and the rest is bonded in thiols and sulfides. A similar pyrolysis method has been applied to studying the distribution of sulfur functional groups in brown coals by Stanek et al. (79). However, the results from the thermokinetic approach are tentative at best. As pointed out by Given (80, p. 151), the thermokinetic method does not accurately determine the organic sulfur compounds, because it is questionable whether a coal with complex organic structure and mineral impurities will behave in the same way as model polymers with sulfur functional groups.

Ignasiak (81) showed that a third of the sulfur in Raša coal of Yugoslavia is in the form of sulfides, and that the coal contained no thiol. Boudou et al. (58) performed programmed temperature sulfur-reduction experiments on three high-sulfur coals, including a Provence coal (Cretaceous) from France and two Appalachian Basin coals (Pennsylvanian). The H_2S release pattern of the subbituminous Provence coal of fluviolacustrine origin is different from that of the Appalachian Basin coals of bituminous rank which originated in a marine deltaic environment. Thiophenic compounds and sulfur compounds of high molecular weight are less abundant in the Provence coal than in the Appalachian Basin coals.

In summary, considerable progress has been made in recent years in the characterization of sulfur-containing organic compounds in coal. Much of this progress is attributed to the development of GC-MS. However, systematic study of many coal samples from different seams is needed to establish a firm relationship between the nature of organosulfur compounds and either coal rank or depositional environment.

Origin of Sulfur in Coal

Nature of Sulfur in Peat and Formation of Pyrite during Early Diagenesis. The bulk of the sulfur in low-sulfur coal and some of the sulfur in medium- and high-sulfur coal are incorporated at the time of peat accumulation. Most coal pyrite and organic sulfur compounds are formed during early diagenesis of coal. Studies of the nature of sulfur in peats give important clues about the origin of sulfur in coal (82). The sulfur content in peat is closely related to its depositional environment. In general, peats accumulating under the influence of seawater have a higher sulfur content than peats accumulating in a freshwater environment. The sulfur species found in peat include sulfate, hydrogen sulfide, elemental sulfur, ester sulfate, pyritic sulfur, and organic sulfur (83). Pyritic sulfur comprises an average of 14% of the total sulfur in a marine peat, while the fraction of pyritic sulfur in freshwater peats is an order of magnitude lower. The ester sulfate fraction represents 25% of the total sulfur in peats of both marine and nonmarine origin. This form of sulfate is believed to be geochemically stable, and hence is an important source of H_2S that is produced by bacterial reduction of aqueous sulfate (84). The hydrogen sulfide thus generated may react with iron to form iron sulfide species and with organic matter to form organosulfur constituents.

In marine and lacustrine muds, the initial sulfide phase precipitated during early diagenesis is mackinawite ($FeS_{0.9}$) which is subsequently converted to greigite (Fe_3S_4) and pyrite (FeS_2) (85-89). This reaction path leads to the formation of framboidal pyrite (88,90). However, in salt marsh sediments under low pH and low sulfide ion activity conditions, direct precipitation of pyrite by reaction of ferrous iron with elemental sulfur without the formation of iron monosulfides as intermediates has been reported (85-87,89,91,92). This reaction is one possible pathway for the precipitation of pyrite as single crystals (89).

In the freshwater peat swamp, bacterial reduction of organic sulfur in plant tissues may be an important process in the formation of pyrite (93). Altschuler et al. (93) proposed that in the Everglades peat, pyrite precipitates directly by the reaction of HS^- or organic sulfide (produced by reduction of oxysulfur compounds in dissimilatory respiration) with ferrous iron in the degrading tissues. Pyrite formation in low-sulfur coal may be accounted for by this process.

Stratigraphic Evidence Regarding the Origin of Sulfur in Coal. It has long been recognized that the abundance of sulfur in coal is related to the sedimentary environment of the coal-bearing strata. White (94) was among the first to suggest that the high sulfur content of Illinois Basin coal was related to the marine and brackish environment. He based this conclusion on the observation that high-sulfur coals in the Illinois Basin are commonly overlain by marine shale and limestone, but low-sulfur coals in the Appalachian Basin are not. Williams and Keith (95), in their study of the regional distribution of sulfur in the Lower Kittanning coal, confirmed White's hypothesis. They showed that the sulfur content is higher in coal overlain by marine strata than in coal topped by nonmarine rocks.

In the Illinois Basin, the roof rocks of the Herrin Coal have been classified into four types: marine roof, nonmarine gray-shale roof, wedge-type transitional roof, and pod-type transitional roof (96). The sulfur content of the coal is related to the roof type (97): it is highest in coals with a marine roof or with a pod-type transitional roof, intermediate in coal with a wedge-type transitional roof, and lowest in coal with a gray-shale (nonmarine) roof. The sulfur content in coal thus is controlled to a large extent by the sedimentary environment immediately following peat accumulation. Seawater sulfate is the major source of sulfur in high-sulfur coal. It is present as the result of marine incursion into the peat bed.

The variation of sulfur concentration in coal beds has also been studied in major coal fields in Europe and China. For example, the Sarnsbank and Katharina seams of the Ruhr basin, which have roof strata of marine origin, have high sulfur contents (98). Sedimentological and geochemical studies of the Xishan coal basin, Taiyuan, Shanxi Province, China, showed that the medium- and high-sulfur coals in the Late Carboniferous Taiyuan Group were deposited in a tidal flat to lower deltaic plain environment, whereas the Early Permian, mostly low-sulfur coals in the Shanxi Group formed in a deltaic-fluvial environment (99,100). Kreulen (101) reported that the Raša coal, from Istria, Yugoslavia, has a high organic-sulfur content. For a discussion of marine influence on Raša coal, see the chapter by White et al. in this book.

Sulfur Isotope Composition and the Sources of Sulfur in Coal. Sulfur has four stable isotopes of atomic masses 32, 33, 34, and 36. The sulfur isotopic composition of a sample is generally characterized by its $^{34}S/^{32}S$ ratio, expressed in terms of $\delta^{34}S$. It is the permil (‰) deviation in the $^{34}S/^{32}S$ ratio of a sample from a standard (troilite in the Cañon Diablo meteorite) (102).

Sulfur isotopic measurements can shed light on the origin of sulfur in coal. The $^{34}S/^{32}S$ ratio depends on the source of sulfur and the geologic processes involved during coal formation. For example, isotopic compositions are different for the two principal sources of sulfur in coal: 1) the sulfur preserved from the precursor plant material, and 2) the sulfur derived from the bacterial reduction of dissolved sulfate in ambient waters. Plant assimilation of sulfur results in a slight depletion of ^{34}S (4-4.5‰) relative to the $\delta^{34}S$ in the dissolved sulfate source (102,103). In contrast, the dissimilatory bacterial reduction of sulfate results in a large isotopic fractionation; sulfide sulfur can be depleted as much as 60‰ in the heavy isotope (89,104-106).

Earlier work on the isotopic compositions of pyrite, elemental sulfur, and organic sulfur in coals from Japan, Australia, and Germany (107-110) was summarized by Nielsen (102). Smith and Batts (110) showed that organic sulfur in Australian coals ranging in age from Permian to Tertiary has a large isotopic variation (+2.9 to +24‰) in coals with more than 1% sulfur, whereas organic sulfur in low-sulfur coals (less than 1% sulfur) has a narrow isotopic composition between +4.6 and +7.3‰. The relatively uniform isotopic composition of organic sulfur in low-sulfur coal probably reflects the uniform isotopic composition of freshwater sulfate since Permian time (110). A detailed study by Hunt and Smith (111) of the sulfur isotopic variations in low-sulfur coal (≤1% total sulfur, ≤0.2% pyritic sulfur) of Permian basins in Australia showed that both the sulfur content and $\delta^{34}S$ value increased in the downstream direction of the fluviodeltaic system. This trend is parallel with changing sedimentary environments from braided alluvial through fluvial-upper delta plain and lower delta plain to marginal marine conditions. The data indicated an increased availability of sulfate for plant growth, preferred assimilation of ^{32}S during plant growth, and slightly increasing marine influence downstream.

The sulfur isotopic composition of Illinois Basin coals has been studied by Price and Shieh (112), Westgate and Anderson (113), and Liu et al. (114-115). Figure 3 shows the variation of $\delta^{34}S$ in organic sulfur with the organic sulfur content in Illinois Basin coals. Low-sulfur coal (<0.8% organic sulfur) has $\delta^{34}S$ values between +5 and +10, similar to the range for Australian low-sulfur coal studied by Smith and Batts (110) and for low-sulfur Tertiary coals in three intermontane basins in the western U.S. (116). The sulfur isotopic composition in low-sulfur coal is similar to that in dissolved sulfate of average river water and freshwater peat in the Okefenokee swamp, Georgia (117). This suggests that the organic sulfur in most low-sulfur coals originates from the primary plant sulfur (112,113,116). In contrast, the $\delta^{34}S$ value of organic sulfur in high-sulfur coal are more variable and tend to be more negative than those of low-sulfur coal (Figure 3). This implies that a large portion of organic sulfur in high-sulfur coal originates from bacteriogenic sulfide.

The isotopic composition of pyritic sulfur is more variable than that of organic sulfur. Figure 4 compares the sulfur isotopic composition of organic and pyritic sulfur between high- and low-sulfur coals in the Illinois Basin. The $\delta^{34}S$ values of organic sulfur are relatively uniform among the three samples taken at the top, middle, and base of the seam at each mine, and are more negative in high-sulfur than in low-sulfur coal. This suggests that a significant portion of organic sulfur in the former is derived from bacteriogenic sulfide. The apparent organic sulfur isotopic uniformity indicates that isotopes of organic sulfur may have been homogenized during coalification. The contrasting highly variable isotopic composition of pyritic sulfur indicates that there are probably several generations of pyrite. Isolated microenvironments probably exist in peats and precoal matter; these microenvironments give rise to inhomogeneous distribution of sulfate-reducing bacteria and contribute to the complexity of pyrite formation.

Use of Trace Elements in Studying the Origin of Sulfur in Coal. The enrichment of high-sulfur coal relative to low-sulfur coal in certain trace elements provides evidence regarding the origin of sulfur in coal. The Herrin Coal of the Illinois Basin is a good example, because it has been analyzed extensively for many trace elements (118). Most resources of the Herrin seam are high-sulfur coal, but low-sulfur and medium-sulfur coal occurs in restricted areas. A comparison of major and trace element abundances in high-sulfur coal with those in low- to medium-sulfur coal of the Herrin seam indicates that high-sulfur coal is higher in several major and trace elements (including boron, iron, molybdenum, mercury, thallium and uranium) than low-sulfur and medium-sulfur coal (97). Like sulfur, the elements boron, molybdenum, and

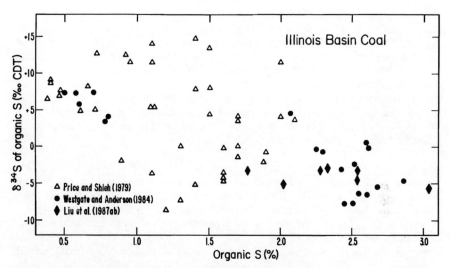

Figure 3. Relationship between $\delta^{34}S$ value and organic sulfur content in Illinois Basin coals. Data are from the literature (112-115).

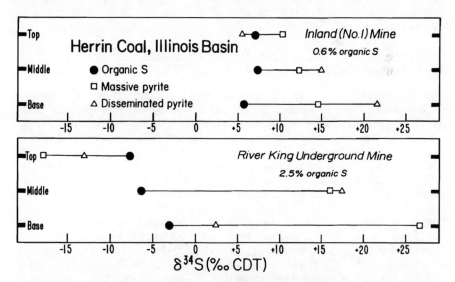

Figure 4. Comparison of isotopic composition of pyritic and organic sulfur in low- and high-sulfur coals in the Illinois Basin. Data are taken form Westgate and Anderson (113). Samples were collected from the top, middle, and base of the Herrin Coal at each mine.

uranium, which are enriched in high-sulfur coal, probably are derived from seawater that flooded the swamp and terminated peat accumulation (97), because concentrations of these elements are much higher in seawater than river water. The other three elements (iron, thallium, and mercury) cannot be derived from seawater, because the concentrations of these elements in seawater are extremely low. These elements have a terrigenous source, and are incorporated in pyrite during diagenesis of peat and coal in a sulfur-rich and reducing environment.

Low-sulfur coal in the western U.S. has a low molybdenum content with an average of 2.1 ppm (118), compared to an average of 10.7 ppm in the Illinois Basin (97). Thus, the molybdenum content appears to be indicative of a sedimentary environment during coal formation.

Basinal Brines as a Source of Sulfur in High-Sulfur Coals. Sulfide minerals, such as pyrite and sphalerite, in coal seams may be deposited from basinal hydrothermal fluids. The occurrence of epigenetic sphalerite in Illinois Basin coals has been described by Hatch et al. (119) and Cobb (120). Whelan et al. (121) studied the isotopic composition of pyrite and sphalerite in coal beds from the Illinois Basin and the Forest City Basin, and suggested that some of the coals were affected by Mississippi Valley-type hydrothermal solutions.

High-sulfur coals of Pennsylvanian age in the Canadian Maritimes Basin contain 5-8% sulfur. However, this sulfur is associated predominantly with freshwater geologic settings. Recent sulfur isotopic results support the hypothesis that the sulfur in these coals is derived from a bedrock evaporite source (122).

Formation of Pyrite and Organic Sulfur Compounds in High-Sulfur Coal: a Geochemical Model

The geochemistry of sulfur in coal is largely controlled by sedimentary environments during peat accumulation, as well as by the geologic conditions during diagenesis following peat formation. In low-sulfur coal, peat accumulates in a nonmarine setting and is never influenced by sulfate-rich seawater: in this case, the sulfur is derived primarily from the parent plant material. In the case of high-sulfur coal, seawater is the primary source of sulfur. A geochemical model of the formation of pyrite and organic sulfur compounds in high-sulfur coal is outlined in Figure 5.

Source of Sulfur. Most of the sulfur in high-sulfur coal arises from the occasional seawater inundations over the coastal swamp during peat accumulation or from the seawater that flooded the peat swamp and terminated peat accumulation. Seawater sulfate diffused freely into the underlying peat and was reduced by microorganisms to H_2S, $S°$, and polysulfides.

Formation of Pyrite. Iron is carried to the peat swamp, before seawater transgression, as ferric oxide and hydroxides adsorbed on fluvial clays (123). During early diagenesis in a reducing environment, ferric iron is reduced to ferrous, which reacts with hydrogen sulfide to form iron monosulfide. If the basic mechanism of pyrite formation is similar to that in   marine sediments

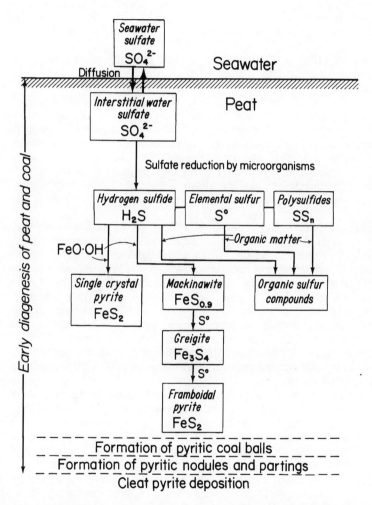

Figure 5.   A model for the origin of pyrite and organic sulfur compounds in high-sulfur coal.

(89), iron monosulfide (FeS) is transformed by reaction with elemental sulfur through several sulfide phases; from mackinawite ($FeS_{0.9}$) through greigite (Fe_3S_4) to framboidal pyrite (FeS_2). Disseminated pyrite single crystals may be formed by direct precipitation of FeS_2. Plant structure, preserved in coal balls, indicates that pyritic coal balls originate late in the peat stage, but before the peat is strongly compacted. Pyrite nodules and partings are formed in the process of transformation of peat to coal. Pyrite in cleats and veins is deposited from migrating solutions, after the compaction of peat into coal.

Formation of Organic Sulfur Compounds. Organic sulfur compounds may have formed during an early stage of the coalification process (humification) when plant debri is decomposed by bacterial activity to a premaceral humic substances. Hydrogen sulfide, elemental sulfur, and polysulfides produced by dissimilatory processes may have reacted with organic matter to form organic sulfur compounds. Much progress has been made recently in understanding the reaction pathways. Mango (124) demonstrated that carbohydrates react with hydrogen sulfide at low temperatures (100 °C to 200 °C) to yield a variety of organosulfur compounds including thiophenes, thiols, sulfides, and sulfones. Casagrande et al. (125) showed that reaction of H_2S with peat to form organosulfur compounds occurs at room temperature in a relatively short time.

Elemental sulfur could also play a role in the formation of organic sulfur compounds in coal (46). Laboratory experiments have demonstrated that the reactions of elemental sulfur with hydrocarbons generates organosulfur compounds, including thiophene derivatives (126-129). Casagrande and Ng (130,131) showed that humic acids from the Okefenokee peat react with elemental sulfur to form organic sulfur compounds, which increase as a function of time. [However, the elemental sulfur generated during early diagenesis should not be confused with the elemental sulfur found in weathered coal today. The former is completely converted to organic and pyritic sulfur during early diagenesis. Pristine coals do not contain elemental sulfur, and any significant amount of elemental sulfur in coal is believed to be a weathering product.]

The reactions between polysulfides and organic compounds have been studied extensively. Aizenshtat and co-workers (132,133) demonstrated the importance of polysulfides in the formation of organosulfur compounds in kerogen and pre-kerogen organic matter. Francois (134) provided experimental data suggesting that inorganic polysulfides (SS_n) react rapidly with organic matter to form sulfur-containing functional groups. Furthermore, LaLonde et al. (135) demonstrated that polysulfide reacts at room temperature with conjugated ene carbonyls to form thiophenes. Thus, recent studies have shown that hydrogen sulfide, elemental sulfur, and polysulfides can react with organic matter to form organic sulfur compounds similar to those believed present in some coals.

Conclusions

Geochemical studies of sulfur in coal provide information on the forms, distribution, and origin of sulfur in coal. The variation of sulfur content in coal is controlled by geologic conditions during coal formation. This is particularly true in the case of the sedimentary environment during peat accumulation and postdepositional changes (early diagenesis). Several lines of evidence (sulfur forms in peat, stratigraphic association, sulfur isotopic geochemistry, trace element abundances) lead to the conclusion that sulfur in plant material is the principal source of sulfur in low-sulfur coal. In medium- and high-sulfur coals, seawater is also a predominant source of sulfur. Seawater sulfate, which diffuses into the peat, is reduced by anaerobic, sulfate-reducing bacteria. Reactions of the reduced sulfur species in interstitial water with iron and organic matter form mineral sulfides (mainly pyrite) and organic sulfur compounds. Speciation of organic sulfur compounds in coal appears to be related to coal rank; thiophenic compounds are reported to be more abundant in bituminous coal and anthracite than in low-rank coals.

Acknowledgments

I thank Wilson Orr and Curt White for the opportunity to participate in the Symposium and Heinz Damberger for supporting my effort. This work benefited from many suggestions and detailed comments by Curt White. Constructive reviews were provided by Rich Cahill, Heinz Damberger, Joyce Frost, Wilson Orr, Curt White, and three anonymous reviewers. This work has been supported in part by the Illinois Department of Energy and Natural Resources through its Coal Development Board and Center for Research on Sulfur in Coal, and the CRSC Coal Industry Committee.

Literature Cited

1. Wood, G. H., Jr.; Kehn, T. M.; Carter, M. D.; Culbertson, W. C. U.S. Geological Survey Circular 891, 1983.
2. Richardson, C. V. (General Manager) Keystone Coal Industry Manual; Maclean Hunter Publishing Co.: Chicago, 1989.
3. Kent, B. H. Geol. Soc. Amer. Special Paper 1986, 210, 105-122.
4. Energy Information Administration Estimation of U.S. Coal Reserves by Coal Type, Heat and Sulfur Content; DOE/EIA-0529; 1989.
5. Nelson, W. J. Intern. Jour. Coal Geol. 1987, 8, 355-365.
6. Merritt, R. D. Geol. Soc. Amer. Special Paper 1986, 210, 173-200.
7. Gluskoter, H. J. Energy Sources 1977, 3, 125-131.
8. Hsieh, K. C.; Wert, C. A. Materials Sci. Eng. 1981, 50, 117-125.
9. Frankie, K. A.; Hower, J. C. Fuel Process. Tech. 1985, 10, 269-283.
10. Frankie, K. A.; Hower, J. C. Intern. Jour. Coal. Geol. 1987, 7, 349-364.
11. Harvey, R. D.; DeMaris, P. J. Org. Geochem. 1987, 2, 343-349.
12. Bomberger, D. R.; Deul, M. Trans. Soc. Mining Eng., AIME 1964, 229, 65-69.
13. McCartney, J. T.; O'Donnell, H. J.; Ergun, S. U.S. Bur. Mines Rept. Investigations 7231; 1969.
14. Greer, R. T. Energy Sources 1978, 4, 23-51.
15. Carlton, R. W. In Processing and Utilization of High Sulfur Coals; Attia, Y. A., Ed.; Elsevier: Amsterdam, 1985; pp 3-17.
16. Kneller, W. A.; Maxwell, G. P. In Processing and Utilization of High Sulfur Coals; Attia, Y. A., Ed.; Elsevier: Amsterdam, 1985; pp 41-65.
17. Moza, A. K.; Neavel, R. C. Fuel 1986, 65, 547-551.

18. American Society for Testing and Materials In Annual Book of ASTM Standards; ASTM: Philadelphia, 1989; Vol. 05.05, pp 266-269.
19. Markuszewski, R. Jour. Coal Quality 1988, 7, 1-4.
20. Sutherland, J. K. Fuel 1975, 54, 132.
21. Solomon, P. R.; Manzione, A. V. Fuel 1977, 56, 393-396.
22. Harris, L. A.; Yust, C. S.; Crouse, R. S. Fuel 1977, 56, 456-457.
23. Raymond, R., Jr.; Gooley, R. Scanning Electron Microscopy 1978, 1, 93-108.
24. Raymond, R., Jr.; Gooley, R. In Analytical Methods for Coal and Coal Products; Karr, C., Ed.; Academic: New York, 1979, vol. III, pp 337-356.
25. Straszheim, W. E.; Greer, R. T.; Markuszewski, R. Fuel 1983, 62, 1070-1075.
26. Demir, I.; Harvey, R. D. Geol. Soc. Amer. Abs. Prog. 1989, 21, A162.
27. Hsieh, K. C.; Wert, C. A. Fuel 1985, 64, 255-262.
28. Hsieh, K. C.; Wert, C. A. Proc. Intern. Conf. Coal Sci. 1985, 826-829.
29. Wert, C. A; Hsieh, K. C.; Tseng, B.-H.; Ge, Y.-P. Fuel, 1987, 66, 914-920.
30. Wert, C.; Ge, Y.; Tseng, B. H.; Hsieh, K. C. Jour. Coal Quality 1988, 7, 118-121.
31. Hurley, R. G.; White, E. W. Anal. Chem. 1974, 46, 2234-2237.
32. Paris, B. In Coal Desulfurization, Chemical and Physical Methods; Wheelock, T. D., Ed.; ACS Symposium series No. 64; American Chemical Society: Washington, DC, 1977; pp 22-31.
33. Spiro, C. L.; Wong, J.; Lytle, F. W.; Greegor, R. B.; Maylotte, D. H.; Lamson, S. H. Science 1984, 226, 48-50.
34. Wong, J.; Spiro, C. L.; Maylotte, D. H.; Lytle, F. W.; Greegor, R. B. In EXAFS and Near Edge Structure III; Hodgson, K. O.; Hedman, B.; Penner-Hahn, J. E., Eds.; Springer-Verlag: Berlin, 1984; pp 362-367.
35. Huffman, G. P.; Huggins, F. E.; Shah, N.; Bhattacharyya, D.; Pugmire, R. J.; Davis, B.; Lytle, F. W.; Greegor, R. B. In Processing and Utilization of High Sulfur Coals II; Chugh, Y. P.; Caudle, R. D., Eds.; Elsevier: New York, 1987, pp 3-12.
36. Huffman, G. P.; Huggins, F. E.; Shah, N.; Bhattacharyya, D.; Pugmire, R. J.; Davis, B.; Lytle, F. W.; Greegor, R. B. ACS Division of Fuel Chemistry Preprints 1988, 33, 200-208.
37. Huffman, G. P.; Huggins, F. E.; Mitra, S.; Shah, N.; Pugmire, R. J.; Davis, B.; Lytle, F. W.; Greegor, R. B. Energy & Fuels 1989, 3, 200-205.
38. Kossenberg, M.; Cook, A. C. Mineral. Mag. 1961, 32, 829-830.
39. Gluskoter, H. J.; Simon, J. A. Illinois State Geological Survey Circular 432, 1968.
40. Ehlers, E. G.; Stiles, D. V. Amer. Mineral. 1965, 50, 1457-1461.
41. Gruner, D.; Hood, W. C. Trans. Illinois State Acad. Sci. 1971, 64, 156-158.
42. Kinoshita, K.; Honda, T.; Tanaka, N. Jour. Mining Inst. Kyushu (Kyushu Kozan Gakkaishi) 1966, 34, 377-380.
43. Gluskoter, H. J.; Shimp, N. F.; Ruch, R. R. In Chemistry of Coal Utilization, Second Supplementary Volume; Elliott, M. A., Ed.; Wiley: New York, 1981; pp 369-424.

44. Chatterjee, N. N. Quart. Jour. Geol. Mining Metall. Soc. India 1942, 14, 1-7.
45. Richard, J. J.; Vick, R. D.; Junk, G. A. Environ. Sci. Tech. 1977, 11, 1084-1086.
46. White, C. M.; Lee, M. L. Geochim. Cosmochim. Acta 1980, 44, 1825-1832.
47. Duran, J. E.; Mahasay, S. R.; Stock, L. M. Fuel 1986, 65, 1167-1168.
48. Yurovskii, A. Z. Sulfur in Coals 1960, English Translation, U.S. Bureau of Mines Technical Translation TT 70-57216, 1974, p 455; available from National Technical Information Service: Springfield, VA.
49. Beyer, M.; Ebner, H. G.; Assenmacher, H.; Frigge, J. Fuel 1987, 66, 551-555.
50. Narayan, R.; Kullerud, G.; Wood, K. V. ACS Division of Fuel Chemistry Preprints 1988, 33, 193-199.
51. Hackley, K. C.; Buchanan, D. H.; Coombs, K.; Chaven, C.; Kruse, C. W. Second Rolduc Symposium on Coal Science, 1989, Fuel Process. Tech. 1990, 24.
52. Greer, R. T. Scanning Electron Microscopy 1979, 1, 477-486.
53. Stock, L. M.; Wolny, R.; Bal, B. Energy & Fuels 1989, 3, 651-661.
54. Damberger, H. H.; Harvey, R. D.; Ruch, R. R.; Thomas, J., Jr. In The Science and Technology of Coal and Coal Utilization; Cooper, B. R.; Ellingson, W. A., Eds.; Plenum Press: New York, 1984; pp 7-45.
55. Love, L. G.; Coleman, M. L.; Curtis, C. D. Trans. Royal Soc. Edinburgh: Earth Sciences 1983, 74, 165-182.
56. Reggel, L.; Wender, I.; Raymond, R. Fuel 1970, 49, 281-286.
57. Raymond, R., Jr. In Coal and Coal Products: Analytical Characterization Techniques; Fuller, E. L., Jr., Ed.; ACS Symposium Series No. 205; American Chemical Society: Washington, DC, 1982, pp 191-203.
58. Boudou, J. P.; Boulègue, J.; Maléchaux, L.; Nip, M.; de Leeuw, J. W.; Boon, J. J. Fuel 1987, 66, 1558-1569.
59. Dyrkacz, G. R.; Horwitz, E. P. Fuel 1982, 61, 3-12.
60. Dyrkacz, G. R.; Bloomquist, A. A.; Ruscic, L.; Horwitz, E. P. In Chemistry and Characterization of Coal Macerals; Winans, R. E.; Crelling, J. C., Eds.; ACS Symposium Series No. 252; American Chemical Society: Washington, DC, 1984; pp 65-77.
61. Tseng, B.-H.; Buckentin, M.; Hsieh, K. C.; Wert, C. A.; Dyrkacz, G. R. Fuel 1986, 65, 385-389.
62. Hippo, E. J.; Crelling, J. C.; Sarvela, D. P.; Mukerjee, J. In Processing and Utilization of High Sulfur Coals II, Chugh, Y. P.; Caudle, R. D., Eds.; Elsevier: New York, 1987; pp 13-22.
63. Harvey, R. D.; Dillon, J. W. Intern. Jour. Coal Geol. 1985, 5, 141-165.
64. Roberts, D. L. Intern. Jour. Coal Geol. 1988, 10, 399-410.
65. White, C. M.; Douglas, L. J.; Perry, M. B.; Schmidt, C. E. Energy & Fuels 1987, 1, 222-226.
66. White, C. M. In Handbook of Polycyclic Aromatic Hydrocarbons; Bjørseth, A., Ed.; Marcel Dekker: New York, 1983; pp 525-616.
67. Lee, M. L.; Willey, C.; Castle, R. N.; White, C. M. In Polynuclear Aromatic Hydrocarbons: Chemistry and Biological Effects; Bjørseth, A.; Dennis, A. J., Ed.; Battelle Press: Columbus, 1980; pp 59-73.
68. Nishioka, M. Energy & Fuels 1988, 2, 214-219.

69. Bodzek, D.; Marzec, A. Fuel 1981, 60, 47-51.
70. Calkins, W. H. Energy & Fuels 1987, 1, 59-64.
71. Chou, M.-I. M.; Lake, M. A.; Griffin, R. A. Jour. Anal. Appl. Pyro. 1988, 13, 199-207.
72. van Graas, G.; de Leeuw, J. W.; Schenck, P. A. Adv. Org. Geochem., Phys. Chem. Earth 1980, 12, 485-494.
73. Hayatsu, R.; Scott, R. G.; Moore, L. P.; Studier, M. H. Nature 1975, 257, 378-380.
74. Hayatsu, R.; Winans, R. E.; Scott, R. G.; Moore, L. P.; Studier, M. H. Fuel 1978, 57, 541-548.
75. Jones, R. B.; McCourt, C. B.; Swift, P. Proceedings International Conference on Coal Science, Düsseldorf, 1981, pp 657-662.
76. Attar, A. In Analytical Methods for Coal and Coal Products; Karr, C., Ed.; Academic: New York, 1979; Vol. III, pp 585-624.
77. Attar, A.; Dupuis, F. In Coal Structure; Gorbaty, M. L.; Ouchi, K., Eds.; ACS Advances in Chemistry Series No. 192; American Chemical Society: Washington, DC, 1981; pp 239-256.
78. Attar, A.; Hendrickson, G. G. In Coal Structure; Meyers, R. A., Ed.; Academic: New York, 1982; pp 131-198.
79. Stanek, J.; Buryan, P.; Macak, J. Acta Mont. 1986, 73, 225-235; Chem. Abstr. 1988, 106, 70070d.
80. Given, P. H. In Coal Science; Gorbaty, M. L.; Larsen, J. W.; Wender, I., Eds.; Academic: Orlando, 1984; Vol. 3, pp 63-252.
81. Ignasiak, B. S.; Fryer, J. F.; Jadernik, P. Fuel 1978, 57, 578-584.
82. Casagrande, D. J. In Coal and Coal-bearing Strata: Recent Advances; Scott, A. C., Ed.; Geol. Soc. Special Publication No. 32, 1987; pp 87-105.
83. Casagrande, D. J.; Siefert, K.; Berschinski, C.; Sutton, N. Geochim. Cosmochim. Acta 1977, 41, 161-167.
84. Casagrande, D.; Siefert, K. Science 1977, 195, 675-676.
85. Berner, R. A. Jour. Geol. 1964, 72, 293-306.
86. Roberts, W. M. B.; Walker, A. L.; Buchanan, A. S. Mineral. Deposita 1969, 4, 18-29.
87. Rickard, D. T. Stockholm Contrib. Geol. 1969, 20, 67-95.
88. Sweeney, R. E.; Kaplan, I. R. Econ. Geol. 1973, 68, 618-634.
89. Goldhaber, M. B.; Kaplan, I. R. In The Sea; Goldberg, E. D., Ed.; Wiley: New York, 1974; vol. 5, pp 569-655.
90. Berner, R. A. Econ. Geol. 1969, 64, 383-384.
91. Love, L. G.; Murray, J. W. Amer. Jour. Sci. 1963, 261, 433-448.
92. Howarth, R. W. Science 1979, 203, 49-51.
93. Altschuler, Z. S.; Schnepfe, M. M.; Silber, C. C.; Simon, F. O. Science 1983, 221, 221-227.
94. White, D. U.S. Bureau of Mines Bulletin 1913, 38, 52-84.
95. Williams, E. G.; Keith, M. L. Econ. Geol. 1963, 58, 720-729.
96. Damberger, H. H.; Nelson, W. J.; Krausse, H.-F. In Proceedings, First Conference on Ground Control Problems in the Illinois Coal Basin; Chugh, Y. P.; Van Besien, A., Eds.; Southern Illinois University, Carbondale; 1980, pp 14-32.
97. Chou, C.-L. Memoir Geol. Soc. China 1984, 6, 269-280.
98. Mackowsky, M.-Th. In Coal and Coal-bearing Strata; Murchison, D.;

Westoll, T. S., Eds.; Oliver and Boyd: Edinburgh, 1968; pp 309-321.

99. Pan, Suixian; Cheng, Baozhou <u>Sedimentary Environments of Taiyuan Xishan Coal Basin</u>; Ministry of Coal Industry Press: Beijing, China, 1987; p 631. (In Chinese, with an English abstract).

100. Yang, Xilu; Pan, Suixian; Cheng, Baozhou <u>Coal Science Technology</u> (Beijing) 1987, <u>6</u>, 16-26. (In Chinese with an English abstract).

101. Kreulen, D. J. W. <u>Fuel</u> 1952, <u>31</u>, 462-467.

102. Nielsen, H. In <u>Handbook of Geochemistry</u>; Wedepohl, K. H., Ed.; Springer-Verlag: Berlin, 1978; Vol. II-2, pp (16-B) 1-40.

103. Chukhrov, F. V.; Ermilova, L. P.; Churikov, V. S.; Nosik, L. P. <u>Org. Geochem.</u> 1980, <u>2</u>, 69-75.

104. Kaplan, I. R.; Emery, K. O.; Rittenberg, S. C. <u>Geochim. Cosmochim. Acta</u> 1963, <u>27</u>, 297-331.

105. Kaplan, I. R.; Rittenberg, S. C. <u>Jour. Gen. Microbiol.</u> 1964, <u>34</u>, 195-212.

106. Chambers, L. A.; Trudinger, P. A. <u>Geomicrobiol.</u> 1979, <u>1</u>, 249-293.

107. Rafter, T. A. In <u>Biogeochemistry of Sulfur Isotopes</u>; Jensen, M. L., Ed.; Proc. National Science Foundation Symposium, Yale University, New Haven; 1962; pp 42-60.

108. Vinogradov, V. I.; Kizilsthein, L. Ja. <u>Litol. Pol. Iskopaiem.</u> 1969, <u>5</u>, 149-151.

109. Čuchrov, F. V.; Ermilova, L. P. <u>Ber. deut. Ges. Geol. Wiss.</u> 1970, <u>15</u>, 255.

110. Smith, J. W.; Batts, B. D. <u>Geochim. Cosmochim. Acta</u> 1974, <u>38</u>, 121-133.

111. Hunt, J. W.; Smith, J. W. <u>Chem. Geol.</u> 1985, <u>58</u>, 137-144.

112. Price, F. T.; Shieh, Y. N. <u>Econ. Geol.</u> 1979, <u>74</u>, 1445-1461.

113. Westgate, L. M.; Anderson, T. F. <u>Intern. Jour. Coal Geol.</u> 1984, <u>4</u>, 1-20.

114. Liu, C. L.; Hackley, K. C.; Coleman, D. D. <u>Fuel</u> 1987a, <u>66</u>, 683-687.

115. Liu, C. L.; Hackley, K. C.; Coleman, D. D.; Kruse, C. W. <u>Fuel Process. Tech.</u> 1987b, <u>15</u>, 377-384.

116. Hackley, K. C.; Anderson, T. F. <u>Geochim. Cosmochim. Acta</u> 1986, <u>50</u>, 1703-1713.

117. Casagrande, D. J.; Price, F. T. <u>Geol. Soc. Amer. Abs. Prog.</u> 1981, <u>13</u>, 423-424.

118. Gluskoter, H. J.; Ruch, R. R.; Miller, W. G.; Cahill, R. A.; Dreher, G. B.; Kuhn, J. K. <u>Illinois State Geological Survey Circular 499</u>, 1977.

119. Hatch, J. R.; Gluskoter, H. J.; Lindahl, P. C. <u>Econ. Geol.</u> 1976, <u>71</u>, 613-624.

120. Cobb, J. C. Ph. D. Thesis, University of Illinois, Urbana, 1981.

121. Whelan, J. F.; Cobb, J. C.; Rye, R.O. <u>Econ. Geol.</u> 1988, <u>83</u>, 990-1007.

122. Gibling, M. R.; Zentilli, M.; McCready, R. G. L. <u>Intern. Jour. Coal Geol.</u> 1989, <u>11</u>, 81-104.

123. Carroll, D. <u>Geochim. Cosmochim. Acta</u> 1958, <u>14</u>, 1-28.

124. Mango, F. D. <u>Geochim. Cosmochim. Acta</u> 1983, <u>47</u>, 1433-1441.

125. Casagrande, D. J.; Idowu, G.; Friedman, A.; Rickert, P.; Siefert, K.; Schlenz, D. <u>Nature</u> 1979, <u>282</u>, 599-600.

126. Martin, T. H.; Hodgson, G. W. <u>Chem. Geol.</u> 1973, <u>12</u>, 189-208.

127. Martin, T. H.; Hodgson, G. W. <u>Chem. Geol.</u> 1977, <u>20</u>, 9-25.

128. de Roo, J.; Hodgson, G. W. <u>Chem. Geol.</u> 1978, <u>22</u>, 71-78.

129. Przewocki, K.; Malinski, E.; Szafranek, J. Chem. Geol. 1984, 47, 347-360.
130. Casagrande, D. J.; Ng, L. Nature 1979, 282, 598-599.
131. Casagrande, D. J. In Proceedings Ninth International Congress on Carboniferous Stratigraphy and Geology, 1979, Compte Rendu; Cross, A. T., Ed.; Southern Illinois University Press: Carbondale, 1985; Vol. 4, pp 299-307.
132. Aizenshtat, Z.; Stoler, A.; Cohen, Y.; Nielsen, H. In Advances in Organic Geochemistry 1981; Bjørøy, M., Ed.; Wiley: Chichester, 1983; pp 279-288.
133. Dinur, D.; Spiro, B.; Aizenshtat, Z. Chem. Geol. 1980, 31, 37-51.
134. Francois, R. Limnol. Oceanogr. 1987, 32, 964-972.
135. LaLonde, R. T.; Ferrara, L. M.; Hayes, M. P. Org. Geochem. 1987, 11, 563-571.

RECEIVED February 28, 1990

Chapter 3

Environmental Aspects of the Combustion of Sulfur-Bearing Fuels

B. Manowitz and F. W. Lipfert

Department of Applied Science, Brookhaven National Laboratory, Upton, NY 11973

This paper describes the origins of sulfur in fossil fuels and the consequences of its release into the environment after combustion, with emphasis on the United States. Typical sulfur contents of fuels are given, together with fuel uses and the resulting air concentrations of sulfur air pollutants. Atmospheric transformation and pollutant removal processes are described, as they affect the pathways of sulfur through the environment. The environmental effects discussed include impacts on human health, degradation of materials, acidification of ecosystems, and effects on vegetation and atmospheric visibility. The paper concludes with a recommendation for the use of risk assessment to assess the need for regulations which may require the removal of sulfur from fuels or their combustion products.

Sulfur species are found in ambient air in most parts of North America and in most industrial countries. Their sources include natural emissions (biogenic and volcanic), smelting of ores and other industrial refining processes, and combustion of sulfur-bearing fuels. This paper will focus on the combustion sources in the United States and some of the effects of their sulfur emissions. The environmental effects of sulfur in the environment have been of interest for many years and much of the information presented here has been drawn from the various conference proceedings and assessment documents that have been published in recent years (1-11). When specific references are not listed in the text, the information represents a consensus from these various sources.

Sources of Sulfur Emissions

Types of Fuels and Their Uses. Fuel combustion is the overwhelming source of energy in the United States, providing heating, electric

0097–6156/90/0429–0053$06.00/0

power, transportation, and energy for manufacturing processes. The
fuels used vary by application, the major categories being natural
gas, gasoline, distillate fuel oil, heavy and residual fuel oils,
coke and coal, in approximate increasing order of their sulfur
contents. This is also roughly the order of increasing difficulty
in fuel handling and combustion, so that the cleaner fuels tend to
be used in the smaller, more widely dispersed facilities, and the
dirtier fuels in larger, centralized facilities (residential
firewood is an exception). The effects of fuel combustion on the
environment (air quality) will depend on the impurities in the fuel
(i.e., sulfur and ash), the efficiency of combustion, the height of
the stack through which combustion gases are discharged, and the
dispersive capacity of the atmosphere.

Typical ranges of sulfur contents for various fuel types and
rates of sulfur emission per million Btu released are given in
Table I, as well as the estimated total national sulfur emissions.
The 18 million metric tons of SO_2 emitted from fuel combustion in
1985 is down about 20% from the peak, which occurred about 1970.
Table II gives the national distribution of residential heating
fuels as of 1983; natural gas is the fuel of choice everywhere
except in the Northeast, where it is evenly divided with oil. Note
that electric heating tends in effect to be heating by means of
coal combustion except in the Northwest, where hydro and nuclear
have larger shares of electric generation.

Table I. Types of Fuels and Associated Sulfur Emissions (1985)

Fuel Type	S Content lb/10^6 Btu	U.S. SO_2 Emission* 10^3 metric tons/yr
Bitum. Coal	1 - 8	14,930
Anthracite	1	38
Lignite	1.2 - 3.0	616
Resid. Oil	0.3 - 3.0	1374
Dist. Oil	0.1 - 0.3	329
Natural Gas	0.0006	71
Gasoline	0.03	252
Diesel Fuel	0 - 0.1	392
Wood	>0.15	26

* Source: Ref. 12

Trends. This picture has changed considerably from that which existed
prior to and during the decade after World War II, the time at which
many adverse environmental effects may have peaked. Prior to the
development of transcontinental natural gas pipelines, (local) coal
and coke were widely used for residential and commercial space
heating, with severe effects on local air quality. In the 1960s,
high-sulfur residual fuel oil was heavily used by commercial and
industrial sources along the Eastern seaboard. Following the passage
of the 1970 Clean Air Act, fuel sulfur limitation regulations were
implemented throughout the U.S. as necessary to meet ambient air

Table II. Distribution of Residential Space Heating Fuels
(% of Housing Units)

Region: housing: fuel type	Northeast total urban		Midwest total urban		South total urban		West total urban	
electric*	7.2	5.5	10.0	7.6	31.0	30.0	21.5	19.0
coal	1.2	0.6	0.3	0.1	1.1	0.1	0.1	--
oil	44.5	44.5	8.1	4.3	8.2	9.4	2.7	2.5
gas	42.5	47.9	77.8	87.0	51.8	59.2	67.3	72.0
wood	3.0	0.5	3.4	0.6	6.7	1.2	5.4	2.0

* The fuels used to generate electricity will vary locally and by
region. Source: Ref. 3

quality standards. In addition, newly constructed sources were
required to emit no more than 1.2 pounds of SO_2 per million Btu
consumed, even if this required installation of gas cleaning
devices ("scrubbers"). In 1977, regulations were further tightened
to limit the use of excessively tall stacks as a means of achieving
acceptable ground-level air quality, to prevent worsening of air
quality in locations with air quality better than the standards,
and to require at least 70% removal of sulfur for large sources.
This last provision resulted in a practical SO_2 emission limit of
0.6 $lb/10^6$ Btu. Sulfur removal devices or scrubbers have their own
secondary environmental impacts by virtue of the need for land
disposal of the sludge.

Sources of Sulfur in Coal and Oil. The major coal beds of eastern
North America are of Pennsylvanian age. During that time, there
was a constantly fluctuating sea level across flat lowlands over
the North American interior. Coal was formed just before the onset
of marine conditions, so that coal swamp forests occurred on broad
lands along or near the sea shore. Thicker sections accumulated on
the more rapidly subsiding Illinois and Forest City basins and in
the Appalachia fireland basin (14).
    Western coals were formed during the Cretacious and Tertiary.
During those time periods, the region was emergent land, so the
coal swamps were in contact with fresh rather than saline waters.
    The first step in the formation of coal is the alteration of
plant material into peat by biochemical (i.e., microbial)
processes. The overall properties and composition of a given peat
depend upon the environmental conditions that existed during the
alteration period after burial. During the biochemical stage,
microorganisms can reduce the sulfate in saline pore water to an
active form of sulfur, e.g., H_2S. H_2S will react with components of
the sediment, forming pyrite when iron is present and forming

organic sulfur compounds with the lignin-cellulose-humic matter present.

Plants with a different biochemistry are found in fresh water environments, as compared to marine environments. They decay to different products and the resultant peat will have a lower sulfur content. However, it is the fresh water contact that provides the general explanation of the low sulfur content of Western U.S. coals, and the saline water contact that provides the general explanation of the high sulfur content of the Eastern coals (Figure 1). Further evidence of this phenomenon is that those Eastern coal beds that have high sulfur content are overlain with marine shale or limestone, whereas the low sulfur coal beds are not (3, 14).

The sulfur content of crude oils varies from a fraction of a percent to more than ten percent. The general explanation that the source of the sulfur arises from biological reactions on marine sulfate holds for oil as well as for coal. However, the situation is more complex because during maturation, migration, and reservoir retention, processes can occur which will alter the distribution of sulfur compounds. Thermal alteration processes taking place at great depths result in gradual desulfurization of oils. The precipitation of asphaltenes by light organic fractions generally results in desulfurization. Biodegradation of oils in the reservoir generally results in a concentration of sulfur compounds in the non-biodegraded fraction, the portion extracted. Thus the sulfur content of fuel oils depends to some extent upon the biogeochemical characteristics of oil-bearing deposits.

Removal of Sulfur from Fossil Fuels. In the case of coal it becomes important to quantify the organic sulfur fraction versus the inorganic, for various grades of coal, and to identify the sulfur compounds in the organic fraction. Inorganic sulfur may often be removed prior to combustion through physical cleaning of the coal, which can also remove a portion of the mineral matter. However, chemical cleaning, a more complex and expensive process, is required to remove the organically bound sulfur. The alternative to removing sulfur prior to combustion is to process the combustion gases through any of a number of chemical processes, often referred to collectively as "scrubbers." These processes add substantially to the cost of the facility and tap off a part of the energy produced, as well as producing a waste stream that must be disposed of.

Removal of sulfur compounds from fuel oils is usually accomplished at the refinery. Although most of the sulfur in oil is in the high boiling fraction, some middle distillates have high sulfur contents. A better understanding of the chemistry of sulfur compounds in oil will also facilitate their efficient removal.

Fate of Sulfur Emissions

It is commonly assumed that all fuel sulfur is oxidized to SO_2 during the combustion process, regardless of its chemical form in the fuel; a small percentage is emitted as sulfuric acid or metal sulfate particles (typically <1% for coal and 1-2% for oil). Note that the older grate combustion processes for coal typically

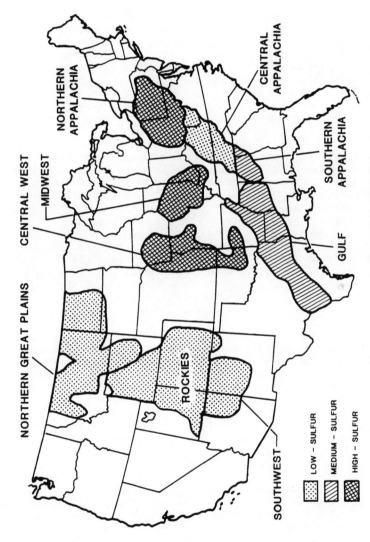

Figure 1. Sulfur content of major coal regions of the United States.

retained some of the fuel sulfur in the ash (up to about 10-20%).
(A further exception is lignite combustion, for which up to 50% of
the fuel sulfur may be retained in the ash, depending on the sodium
content of the fuel (15).)   As the combustion gas plume travels
downwind in the atmosphere, it is subject to various chemical
transformation and removal processes which can have important
bearings on the nature of any adverse effects on the environment.

Atmospheric Transformations. SO_2 can be oxidized to form H_2SO_4 in
either the gas phase through photochemical reactions, or in the
aqueous phase in cloud water (16).   Clouds can then evaporate,
leaving behind sulfate particles.   The photochemical reactions are
fairly slow, 1-3%/hr, but operate during all daylight hours.
Aqueous reactions are faster, but require the presence of clouds.
The relative importance of these two sulfate formation pathways
will vary with season and climatic factors.   The chemical form of
airborne sulfur can play a major role in determining the nature and
rate of any adverse effects, as discussed below.
    Sulfate particles are usually associated with ammonium;
occasionally larger particles involving calcium or other crustal
elements are found.   The acidity of sulfate particles will depend
on their source and the extent of contact with ambient ammonia.   In
the Eastern U.S., the average composition seems to vary between
letovicite in the summer $[(NH_4)_3H(SO_4)_2]$ and ammonium bisulfate in
the fall $[NH_4HSO_4]$(17).   Free H_2SO_4 is found on occasion.
    The low vapor pressure of H_2SO_4 and ammonium sulfates is an
important property, which ensures that they remain as particles
under normal atmospheric conditions.   In contrast, nitric and
hydrochloric acids exist as vapors, and ammonium nitrate tends to
be unstable.   For this reason, the term "acid aerosol," which is
one of the pollutants of current health concern, relates mostly to
acidic sulfates (17).

Atmospheric Removal Mechanisms.   The solubility of SO_2 in water is
central to its removal from the atmosphere (18); this removal is
responsible for the maintenance of a steady concentration state
(rather than a continuous build-up in response to continuous
emissions).   Removal mechanisms are termed "wet" if associated with
hydrometeors; "dry," if otherwise. SO_2 can readily be
removed by dry deposition to vegetation and other moist surfaces,
including the oceans.   However, it dissolves only slightly in
water, according to physical (Henry's Law) solubilities; its
solubility is enhanced considerably by dissolution to form
bisulfite ion, but this solubility is pH-dependent and quite
limited for pH < 5 (16).   Thus the uptake of SO_2 by surface moisture
depends on its buffering capacity (5, 8).   Sea water is highly
buffered and may be the "perfect" absorber of SO_2.   Uptake of SO_2
into cloud water or surface moisture layers may be greatly enhanced
by aqueous-phase oxidation, especially by hydrogen peroxide (H_2O_2)
(16).   In surface moisture, attack of the surface by H_2SO_3 or H_2SO_4
can provide buffering (19).
    Removal of sulfate particles is dominated by wet removal
processes, mostly by dissolution into cloud water concurrent with
cloud formation, followed by deposition in precipitation (16).   Dry

deposition of sulfate particles is limited by the low diffusion coefficient of small aerosol particles (20).

The pH of precipitation that has absorbed SO_2 or sulfates will depend on the other ions present, which can include NO_3^- formed from nitrogen oxide emissions and base cations from various dust sources, as well as HCO_3^- from atmospheric CO_2 and organic acids from natural sources. Precipitation with pH values below 5.6 (the value for equilibrium with atmospheric CO_2) is popularly referred to as "acid rain."

Wet removal processes are further controlled by precipitation types and rates. Dry deposition processes on surfaces are affected by atmospheric transport rates that mix fresh pollutant into the surface boundary layers and by the physical properties of particles. For the Eastern U.S., the approximate annual deposition rates of sulfate can be compared as follows (Table III), considering that deposition flux is the product of a concentration and a velocity of deposition (V_d) (20):

Table III.   Sulfur Fluxes

Process	Typical S Conc. (g/m^3)	Typical V_d (m/sec)	Annual S Flux (g/m^2)
Precip. (aq. conc.)	0.8	3×10^{-8} ($\sim$1 m/yr)	0.8
Dry dep. (gas)	8×10^{-6}	0.005	1.3
Dry dep. (particle)	2×10^{-6}	0.001	0.06

According to these estimates, dry removal processes are slightly more important than wet, and dry deposition of (gaseous) SO_2 far exceeds that of its sulfate transformation products. This is a result of the characteristic SO_4^- particle sizes (typically 0.1-1 μm) being near the minimum for sedimentation and other physical deposition processes. One of the results of this difference in deposition velocities is that sulfur particles will remain airborne and thus travel further from their original sources than SO_2. Considering that in the Eastern U.S., the annual sulfur emission rate is about 5 g/m^2 (as sulfur), which is much larger than the combined deposition fluxes given above, we conclude that a large portion of Eastern U.S. sulfur emissions is transported elsewhere rather than being deposited near their sources (Note that there are strong local variations in this material balance). This conclusion is consistent with pollutant flux estimates over the western Atlantic Ocean (21), and with modeling results (22).

Atmospheric Transport Processes. If an atmospheric constituent is not removed, then it will be transported by the winds, which vary greatly in time and space. The vertical dimension is particularly important, not only because transport speeds increase with height but because the probability of encountering clouds increases with height and the likelihood of being trapped by a ground-based

inversion is decreased. The tallest stacks are used by electric
utilities, sometimes reaching 300 m. Since about 1960, the release
heights of sulfur emissions have increased substantially in the
U.S., as a way of meeting ground level air quality standards. As
of 1985, about 50% of U.S. point source SO_2 emissions came from
stacks taller than 150 m (12). Approximate calculations indicate
that only about 5-10% of tall stack emissions will be deposited
within a radius of about 500 km, depending on precipitation. The
prevailing winds in the Eastern U.S. during times of worst regional
air quality tend to be from the Southwest (23).

It may be instructive to consider the interplay between
emissions from a distributed area source (which could be SO_2 from
urban space heating or biogenic emissions from a swamp, for
example) and dry deposition to that same area. Note that area
sources will deposit relatively more of their emissions than
elevated point sources since their emissions are released close to
the surface. By using air quality dispersion models (24), we have
determined that the annual air concentration ($\mu g/m^3$) in an area
source (either peak or averaged over the area) is proportional to
the emission density (g/m^2sec). The proportionality factor is
approximately 15 sec/m for an area source of kilometer scale or
larger. Note that this factor has the appearance of an inverse
deposition velocity (1/0.067 m/sec), and thus it corresponds to an
"emission velocity." By comparison with typical long term average
dry deposition velocities for sulfur, which rarely exceed 0.01
m/sec, we conclude that dry deposition will never dominate the
budget, i.e., that independent of the strength of an area source, a
substantial fraction of the emissions will travel downwind from the
source. This simple calculation implies that sulfur must have been
distributed throughout the Northeastern U.S. countryside when the
major emissions sources were home heating and local industry,
before the advent of tall stacks.

Ambient Concentration Levels. SO_2 air concentration patterns
downwind of a source burning sulfur-bearing fuel depend on many
factors, including stack height, atmospheric conditions, and the
presence of aerodynamic obstacles. Tall stacks tend to result in
low concentrations (1-20 ppb) when averaged over time periods
exceeding several days, due to the dispersive nature of the
atmosphere. However, under adverse atmospheric conditions (highly
convective conditions which bring the plume to the ground),
concentrations can approach 1 ppm for periods of the order of one
hour. These levels should be contrasted with stack exit gas SO_2
concentrations, which are often of the order of 2000 ppm or more.

Agglomerations of low-level sources (i.e., area sources)
respond differently to atmospheric conditions, in that the highest
concentrations occur during periods of air stagnation. The
emission density is the controlling factor under such conditions,
which for space heating emissions, may be represented by the
product of housing density, heating fuel sulfur content, and
heating demand. During the worst air pollution episodes of many
years ago, SO_2 concentrations sometimes exceeded 1 ppm for several
days, in conjunction with elevated levels of many other air
pollutants.

The clustering of many large fuel-burning facilities with tall stacks (such as in the Ohio Valley, for example) has created a different type of source, with predominantly regional characteristics. The combined impacts of these facilities has raised the long-term regional background SO_2 level to about 10-12 ppb and is a major contributor of sulfate particles in the atmosphere. Short term SO_2 levels are still controlled by the characteristics of individual facilities, since it is rare for the plumes from neighboring plants to coincide.

By way of comparison, the threshold of SO_2 odor detection is about 0.5 ppm (instantaneous basis), and the U.S. national health standards are 140 ppb for 24 hours and 30 ppb for an annual average. These standards are currently being met in most U.S. locations.

Sulfate particle concentrations average 8-10 $\mu g/m^3$ annually in the Eastern U.S.; it is rare for short term (3-24 hour) levels to exceed about 50 $\mu g/m^3$. There are no Federal ambient standards for sulfate particles.

Effects of Airborne Sulfur Compounds

Deposition of sulfur in regions where the soils are deficient in sulfur may be considered a beneficial effect; all of the other effects of sulfur air pollution are considered adverse. They include effects on human health, materials degradation, vegetation, and atmospheric visibility, and acidification of soils, watersheds and freshwaters. Limitations on the scope of this paper will greatly restrict our discussions of the details of these effects; see the References 1-11 for more detailed information.

There are three modes of damages due to sulfur emissions that must be considered, depending on the receptor: damages due to SO_2, due to sulfate particles, and due to acidified precipitation.

Human Health Effects. SO_2 is considered a mild respiratory irritant, affecting primarily the upper respiratory tract ([25]). The evidence for SO_2 health effects comes from animal toxicological experiments, chamber studies of human volunteers, and epidemiological studies. However, in many of the epidemiological studies, it has been difficult to isolate the effects of the various pollutants which tend to occur at the same times and places (for example, SO_2 and particulates) ([26]).

Early tests of human volunteers at SO_2 levels from 1-8 ppm showed a great deal of variability in their tolerance ([27]). Definite but transient effects were shown on pulse and respiration rates. Some people become acclimated to such high levels of SO_2. More recent experiments have shown respiratory sensitivity among some asthmatics at levels as low as 0.25 ppm, particularly while exercising ([28]).

Sulfate particles may penetrate into the deeper airways, depending on their size, the breathing rate, and the amount of hygroscopic growth in transit. Acidic particles can reduce the pH of mucus, which in turn affects the action of the cilia in clearing foreign matter from the lung. This may explain why the combination of sulfur and other particulate matter at high concentrations was

so deadly during the notorious air pollution episodes of many years
ago (26). In addition, recent studies have been interpreted to
support an hypothesis that prolonged breathing of sulfuric acid
particles can result in permanent lung alterations (29). Since
industrial H_2SO_4 exposure levels of 500-1000 $\mu g/m^3$ are commonly
tolerated, it appears that there are marked individual differences
in susceptibility.

Materials Degradation Effects. Virtually all of the effects of SO_2
on materials are associated with its dissolution in surface
moisture as the primary mechanism of deposition (19). Thus an
appropriate environmental index for materials damage should include
not only the average ambient SO_2 concentration but also a measure
of the portion of time that the material surfaces are wet and thus
receptive to SO_2 deposition. Sulfur in the atmosphere has an
additional adverse affect on some materials through acidification
of precipitation; however, for both "dry" deposition of SO_2 and
impact by acid precipitation, the primary damage mode is attack by
dilute acids.
     The materials known to be sensitive to such attack are
primarily those presenting a relatively thin facade of a substance
that reacts readily with dilute acids (especially sulfuric). These
include zinc (galvanized steel), certain paints, unprotected carbon
steel. Copper (bronze) and carbonate stones (marble, limestone,
some sandstones) may be attacked by acids, but their "sensitivity"
will depend on the stock thickness and the intended
service life. In the case of outdoor sculpture, for example, works
of permanent value will be "sensitive" to deposited acids.
     For simple materials such as zinc or $CaCO_3$, research has shown
stoichiometry between deposited sulfur and the base material (30).
Thus in equivalent molar units, the deposition rate of sulfur is
equal to the removal rate of the base material. At present
environmental conditions in the U.S., these rates are low in terms
of expected lifetimes of consumer-oriented components. There are
some exceptions, such as galvanized fence wire, for which SO_2
deposition rates may be 2-3 times higher than on large flat
surfaces (31). For $CaCO_3$, dissolution in (normal) rain can be an
important mode of material loss and acts to remove the more soluble
$CaSO_4$, creating conditions more receptive to additional SO_2
deposition (32).
     In the U.S., the most common construction material in terms of
exposed surface area is comprised of various types of painted
surfaces, especially painted wood, which is found to some extent on
most residential buildings. The sensitivity of paints depends on
their constituents; research has shown that acid-soluble components
such as $CaCO_3$ or ZnO may be selectively leached out from the coating
(33). To our knowledge, the effects of such leaching on coating
service lives have not yet been quantified.

Acidification Effects. Deposition of acidifying substances upon
the landscape can have several different types of adverse effects,
depending on the buffering capacity of the ecosystem. These
acidifying substances include both acid precipitation and dry
deposition of SO_2 and sulfate particles. In non-arid regions,

precipitation ultimately dissolves all of these deposits and
flushes them through the watershed into receiving surface waters,
after which they eventually end up in the oceans.

The adverse effects of this chain of events can include direct
damage to vegetation, acidification of soils, and acidification of
surface waters with attendant impacts on flora and fauna. In the
case of managed (agricultural) soils, soil acidification effects
may be overcome through use of lime. Liming has also been used to
control the pH of lakes.

Many factors control whether a given water body will become
acidified as a result of a given deposition regime. In addition to
the deposition rate and the lake residence or turnover time, these
include the ratio of water surface area to watershed area, the
composition of the lake bottom, the residence time of incident
precipitation en route through the watershed, and the buffering
capacity of the watershed. The presence of organic material can
also be important.

There is considerable debate as to the pH or alkalinity level
for surface waters beyond which significant harm begins, depending
on whether the criteria for "harm" include elements of the
ecosystem other than game fish. For example, Swedish research ($\underline{34}$)
has shown that some crustaceans and plankton may disappear from
waters at pH values less than about 6, some fish species at pH < 5,
and some species at pH < 4.5. Most observers would probably agree
that alkaline lakes (pH > 7) are "healthier" than recently
acidified lakes, in any event. Figure 2 compares the 1985
isopleths of precipitation pH and the percentages of lake area with
pH < 5.0 ($\underline{35}$). The correspondence is not particularly good,
indicating the importance of other factors with regard to
acidification of surface waters. More recent data on stream
chemistry ($\underline{36}$) in the Eastern U.S. shows that only 2.7% of the
aggregate total stream length has zero acid neutralizing capacity
(i.e., already acidified) but 11.7% with less than 50 μeq/L (in
danger of acidification). Some researchers have been able to show
correspondence between surface water sulfate concentration and
precipitation sulfate deposition ($\underline{37}$).

There is currently widespread concern about the potential
effects of air pollution and acid deposition on forest health. In
the U.S., declines have been noted in red spruce, fir, sugar maple,
and hardwoods, for example. The only firm evidence of atmospheric
pollution damage comes from effects of ozone and from nearby
effects of strong point sources such as smelters or power plants;
no hard evidence of forest damage due to regional-scale air
pollution or deposition has been shown. Laboratory experiments
have shown foliar effects on some species at pH levels of 3.5 or
lower, although sometimes a positive impact is seen due to the
fertilization effect of nitrogen deposition in simulated acid rain.
The current consensus thinking attributes the observed forest
declines to a combination of effects, including ozone and possibly
acid deposition, along with increased environmental stress from
other factors including natural phenomena.

Effects on Atmospheric Visibility. In contrast to many of the
adverse effects of sulfur emissions described above, effects on

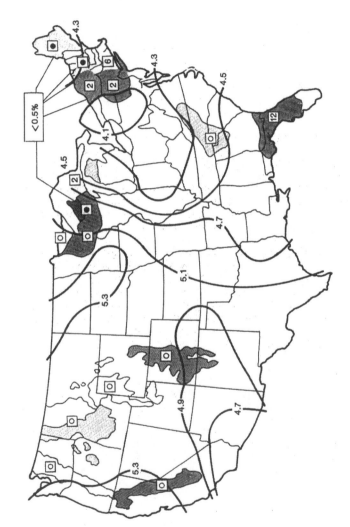

Figure 2. Comparison of 1985 isopleths of precipitation pH and the percentage of lake area with pH<5.0. (Reproduced from ref. 3.)

visibility (visual range) are relatively straightforward, since the effects may be calculated from the principles of physical optics. Of the atmospheric sulfur compounds, only sulfate particles are involved. Atmospheric moisture is a complicating factor since it has its own effects on visibility and also causes the sulfate particles to grow hygroscopically, which in turn changes their optical properties.

Visual range is inversely related to the atmospheric scattering coefficient, which is proportional to the fine-particle mass concentration. The two most important constituents in urban areas are elemental carbon and sulfates.

Atmospheric visibility is a property valued by many people, as has been shown in consumer surveys, in addition to affecting transportation safety and the amount of solar energy reaching the ground. In the East, visibility can be reduced to 5 km or less under hazy, humid conditions (25). In the arid west, it can exceed 200 km under ideal conditions.

Conclusions

The combustion of sulfur-bearing fuels creates effects on health and the environment which are perceived to be deleterious. Stringent regulations have been placed on the use of these fuels; through the years these regulations have been tightened further and this process may well continue, in response to acid deposition issues, for example. Therefore, as a specific conclusion, it becomes increasingly important to characterize the sulfur compounds in fossil fuels in order to facilitate their efficient removal (38, 39).

In a more general sense, however, it is even more important to strengthen the scientific foundations on which societies base their policies for the protection of human health and the environment. The principles of quantitative risk assessment can then provide a rational policy basis, so that well-founded regulations will ultimately ensue, taking into account all of society's needs.

Literature Cited

1. International Electric Research Exchange, Effects of SO_2 and Its Derivatives on Health and Ecology. Vol. 1-Human Health. Vol. 2-Natural Ecosystems, Agriculture, Forestry, and Fisheries. Electric Power Research Institute: Palo Alto, CA, 1981.
2. National Acid Precipitation Assessment Program, Annual Reports, NAPAP: Washington, DC, 1984, 1985, 1986.
3. National Acid Precipitation Assessment Program, Interim Assessment, The Causes and Effects of Acidic Deposition, Vols. I-IV, U.S. Government Printing Office: Washington, DC, Sept. 1987.
4. Sulphur and Its Inorganic Derivatives in the Canadian Environment, NRCC No. 15015, National Research Council Canada: Ottawa, Canada, 1977.
5. Nriagu, J.O., Ed. Sulphur in the Environment; Wiley-Interscience: New York, 1978.

6.   The Regional Implications of Transported Air Pollutants: An
     Assessment of Acidic Deposition and Ozone, Office of Technology
     Assessment: Washington, DC, July 1982.
7.   Acidification Today, Swedish Ministry of Agriculture,
     Environment '82 Committee, 1982.
8.   The Acidic Deposition Phenomenon and Its Effects, Critical
     Assessment Review Papers, Volume I, Atmospheric Sciences,
     Altshuler, A.P., Ed.; Volume II, Effects Sciences, Linthurst,
     R.A., Ed.; EPA-600/8-83-016, U.S. Environmental Protection
     Agency, 1984.
9.   Air Quality Criteria for Particulate Matter and Sulfur Oxides,
     EPA-600/8-82-029, U.S. Environmental Protection Agency:
     Research Triangle Park, NC, 1982.
10. Environmental Health Criteria 8, Sulfur Oxides and Suspended
     Particulate Matter, World Health Organization: Geneva, 1979.
11. Johnson, R.W. and Gordon, G.E., Eds. The Chemistry of Acid
     Rain, Sources and Atmospheric Processes, ACS Symposium Series
     349, American Chemical Society: Washington, DC, 1987.
12. Anthropogenic Emissions Data for the 1985 NAPAP Inventory,
     EPA-600/7-88-022, U.S. Environmental Protection Agency, 1988.
13. U.S. Census Bureau, 1983 Annual Housing Survey, U.S. Dept. of
     Commerce, 1985.
14. Williams, E.G.; Kieth, M.J. Economic Geology 1963, 58, 720-29.
15. Compilation of Air Pollution Factors, Report AP-42, U.S.
     Environmental Protection Agency: Research Triangle Park, NC,
     1977; 3rd ed., pp 1-7.
16. Schwartz, S.E. Science 1989, 243, 753-63.
17. Lipfert, F.W.; Morris, S.C.; Wyzga, R.E. Env. Sci. Tech., in
     press.
18. Schwartz, S.E.; Freiberg, J.E. Atmos. Environ. 1981, 15, 1129-
     144.
19. Lipfert, F.W. J.APCA 1989, 39, 446-52.
20. Committee on Atmospheric Transport and Chemical Transformation
     in Acid Precipitation. Appendix C, Atmospheric Deposition
     Processes; Acid Deposition, Atmospheric Processes in Eastern
     North America; National Academy Press: Washington, DC, 1983.
21. Galloway, J.N.; Church, T.M.; Knap, A.H.; Whelpdale, D.M.;
     Miller, J.M. In The Chemistry of Acid Rain, Sources and
     Atmospheric Processes; Johnson, R.W. and Gordon, G.E., Eds.;
     ACS Symposium Series No. 349; American Chemical Society:
     Washington, DC, 1987; pp 39-57.
22. Kleinman, L.I. Atmos. Environ. 1983, 17(6), 1107-121.
23. Hidy, G.M., In The Chemistry of Acid Rain, Sources and
     Atmospheric Processes; Johnson, R.W.; Gordon, G.E., Eds.; ACS
     Symposium Series No. 349; American Chemical Society,
     Washington, DC, 1987; pp 10-27.
24. Lipfert, F.W.; Dupuis, L.R.; Schaedler, J.S. Methods for
     Mesoscale Modeling for Materials Damage Assessment, Brookhaven
     National Laboratory Report to U.S. Environmental Protection
     Agency, BNL 37508, April 1985.
25. Air Quality Criteria for Particulate Matter and Sulfur Oxides,
     EPA-600/8-82-029, U.S. Environmental Protection Agency:
     Research Triangle Park, NC, 1982.

26. Lipfert, F.W. In <u>Handbook of Environmental Chemistry</u>, Hutzinger, O., Ed.; Springer-Verlag: Heidelberg, 1989.
27. Amdur, M.O.; Melvin, W.W.; Drinker, P. <u>Lancet</u> 1953, <u>ii</u>, 758-59.
28. Sheppard, D.; Wong, W.S.; Uehara, C.F.; Nadel, J.A.; Boushey, H.A. <u>Am. Rev. Resp. Dis</u>. 988, <u>122</u>, 873-78.
29. Lippmann, M. <u>Env. Health Perspectives</u> 1989, <u>79</u>, 203-05.
30. Lipfert, F.W. In <u>Handbook of Environmental Chemistry</u>; Hutzinger, O., Ed.; Springer-Verlag: Heidelberg, 1989; Vol. 4/Part B.
31. Haynie, F.H. In <u>Durability of Building Materials and Components</u>; ASTM STP-691, Amer. Soc. for Testing Materials: Philadelphia, 1980; pp 157-75.
32. Lipfert, F.W. <u>Atmospheric Environment</u> 1989, <u>23</u>, 415-29.
33. Balik, C.M.; Fornes, R.E.; Gilbert, R.D.; Williams, R.S. <u>The Micro-Macro Effects of Acid Deposition on Painted Wood Substrates</u>, draft report to U.S. Environmental Protection Agency, 1989.
34. <u>Acidification Today</u>, Swedish Ministry of Agriculture, Environment '82 Committee, 1982.
35. <u>The Causes and Effects of Acidic Deposition</u>, National Acid Precipitation Assessment Program, Interim Assessment, Vol. I, U.S. Gov't Printing Office: Washington, DC, Sept. 1987.
36. <u>1988 Annual Report</u>, National Acid Precipitation Assessment Program: Washington, DC, 1989.
37. Hendrey, G.R.; Hoogendyk, C.G.; Gmur, N.F. Analysis of Trends in the Chemistry of Surface Waters in the United States, BNL 34956, May 1984.
38. White, C.M.; Lee, M.L. <u>Geochimica Cosmochimica Acta</u> 1980, <u>44</u>, 1825-832.
39. White, C.M.; Douglas, C.J.; Perry, M.B.; Schmidt, C.E. <u>Energy & Fuels</u> 1987, <u>1</u>, 222-26.

RECEIVED March 12, 1990

Chapter 4

Polysulfide Reactions in the Formation of Organosulfur and Other Organic Compounds in the Geosphere

Robert T. LaLonde

Department of Chemistry, State University of New York, College of Environmental Science and Forestry, Syracuse, NY 13210

The environmental conditions for the laboratory simulation of diagenic synthesis are outlined. In particular, formation of five-membered, sulfur-containing heterocycles (S-heterocycles) is considered. Two types of chemical transformations occurring at the low-temperature of diagenesis are reviewed. One of these transformations involves the low-temperature oxidation of ketones and aldehydes by elemental sulfur in the presence of ammonia. The second is a nonoxidative-reductive transformation involving polysulfide dianion which acts on conjugated ene carbonyls. The response of product formation to changes in conditions for both types of reactions is described, but the variation of product structure is emphasized. The implications of laboratory results for the occurrence of S-heterocycles in the geosphere is given in several instances. The presence of heterocycles and their element sources in the geosphere is considered briefly.

The goal of laboratory geochemical simulation is to understand the basis for chemical change in the geosphere. Our approach to simulation has been to define first the limits of the conditions. Thereafter we assess the types of relevant chemical processes occurring under these conditions and choose experimental substrates judged to be consistent with both the chemical processes and the pertinent geochemical product. The selection of substrates will evolve from adopting the molecules possessing the necessary chemical functionality and subsequently choosing molecules more closely resembling those of the natural setting. Thus the initial focus is on chemical process and product.

Since we are interested in the sulfur-carbon chemistry of diagenesis, the relevant geochemical processes are limited to those occurring within the very narrow temperature span of approximately 0 to 60 $^\circ$C. The reactions must occur under hydrous conditions although in media having at least some capacity to accommodate lipophilic source molecules as a result of the presence of substances such as low grade carbohydrates and small carboxylic acids or their salts. If the sulfur source is a form of reduced sulfur from active microbial reduction of sulfate, the pH will range from near neutral

0097–6156/90/0429–0068$06.00/0

to strongly basic, barring a simultaneous acid-forming metabolism or a sulfide precipitating mechanism. As an example of a high pH medium, an ammonia stabilized polysulfide environment at pH 8 has been observed (1). An even more extreme, active sulfate-reducing environment has been reported to have a pH of 9.7 (2).

Obviously since the products of the geochemical processes contain sulfur and carbon, sources for these elements must be considered but the selection of experimental substrates representing them will await prior chemical process analysis. The result of such an analysis that is relevant to the geosynthesis of compounds from polysulfide dianion and conjugated ene carbonyls will be reviewed briefly later.

This paper describes two low-temperature reactions of sulfur and carbon capable of producing five-membered S-heterocycles. Ammonia is involved in one of them, but its role will be considered only incidentally when it influences the reactions of sulfur and carbon sources with each other. Heterocycles are singled out since some of their members, especially the five-membered, sulfur-containing rings, which are relatively stable, lend themselves to analysis and as a consequence are good probes for evaluating the predictions of laboratory simulations.

For purposes of organization, a distinction is made between oxidation-reduction processes and nonoxidative-reduction processes. The former is illustrated by summarizing briefly the work of Asinger and his associates who over the course of twenty-five years investigated the low-temperature oxidative-reductive reactions of ketones and elemental sulfur in the presence of ammonia or amines. This work deviates from hydrous conditions in nearly all cases. Nevertheless it is apropos both in relation to the temperatures employed and the variety of heterocycles produced. It points the way to reinvestigation under hydrous conditions.

Nonoxidative-reductive reactions of sulfur and carbon are illustrated through examples from our ongoing studies of sulfur, in the form of polysulfide, reacting with conjugated ene carbonyls as the carbon source.

A later section of this paper addresses briefly the question of types of S-heterocycles present in recent geological samples. Finally, the presence of appropriate source molecules in the geosphere is treated in the same section.

Low Temperature Oxidation-Reduction

Ketones possessing alpha hydrogen atoms react at room temperature and below with elemental sulfur in the presence of ammonia to give sulfur and nitrogen containing heterocycles as summarized in Figure 1, which shows the products arranged from top to bottom in the order of increasing oxidation state of the five position. These transformations were studied extensively for their synthetic utility by Asinger and associates (3). However, the rationale for Asinger's initial investigation (4) was the low-temperature capture of reactive intermediates whose identification would point the way to the complex series of events involved in the Willgerodt-Kindler reaction (5), a transformation whereby aryl ketones, the original and probably the most frequently employed substrates, are treated with sulfur and ammonia, or amines. The reaction is carried out at temperatures well in excess of 100 °C (6-7). As a consequence the carbonyl migrates to the methyl terminus of the alkyl chain and ultimately manifests itself as a carboxamide or carboxylic acid as shown in Equation 1. However, in contrast to the normally high

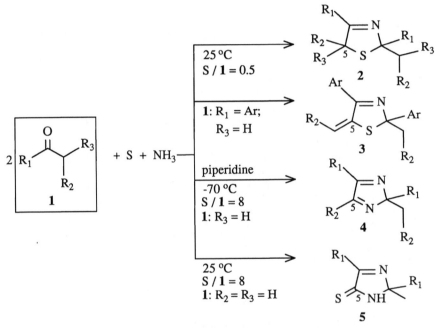

Figure 1. Heterocyclic compounds resulting from the action of elemental sulfur and ammonia on two moles of carbonyl compound **1** possessing alpha-hydrogen. The S/1 ratio refers to the mole ratio of elemental sulfur to carbonyl compound.

$$\text{ArCO(CH}_2)_n\text{CH}_3 \xrightarrow{\text{NH}_3, \text{S}} \text{Ar(CH}_2)_n\text{CONH}_2 \qquad (1)$$

temperatures of the Willgerodt-Kindler reaction, Asinger and associates (4), referring to unpublished work, indicated that the ketimine from 1-butylamine and acetophenone is converted to 2-phenylthioacetomorpholide when the ketimine is treated with sulfur and morpholine in methanol at 20 °C.

As opposed to ketones, the same low-temperature action of sulfur and ammonia on aldehydes received very little attention presumably because the results of the initial exploration discouraged utilization in synthetic chemistry. Again in reference to unpublished work, the product mixture was claimed to contain 3-thiazoline, formed in 10% yield (3). The low-temperature synthesis of aldehyde-generated thiazolines (2, $R_1 = R_2 = H$), as depicted in Figure 1, possibly may have some geochemical interest when viewed in the light of reports that such thiazolines are dehydrogenated to thiazoles by sulfur at 130 °C (8-10).

Asinger's studies demonstrated that product formation is sensitive to the ratio of sulfur to ketone (1), the structure of the ketone, the replacement of ammonia by amines, the temperature and the medium. Room temperature (20-25 °C) reactions in which the ratio of sulfur to ketones is 0.5 favors the formation of 3-thiazoline, 2, as shown in Figure 1. The formation of 5-alkylidene-3-thiazolines, 3, sometimes competes with the formation of 3-thiazolines; such is the case when aryl ketones such as 1-phenylpropan-2-one and 1-phenylbutan-2-one are employed (4). Also the additional presence of hydrogen sulfide promotes the generation of 1,2,4-trithiolanes and 1,2,4,5-tetrathiolanes from ketones and aldehydes at the expense of 3-thiazoline formation (11-12). Increasing the S/ketone ratio to 8 favors the formation of the 3-imidazoline-5-thione (5), a product which has a greater tendency to result from aryl methyl ketones (3).

The synthesis of 3-thiazolines from ketones requires that the latter is treated neat or in dry solvents (3). The presence of water diminishes the yield however. 3-Pentanone with aqueous sodium polysulfide in the presence of ammonia gave an 18% yield (after correction for recovered ketones) of 3-thiazoline but under anhydrous conditions, an 85% yield (13).

An example demonstrating the effect of temperature on product formation is the case of alpha methylene ketones. The latter react in methanol solution at -70 °C with sulfur (S/ketone = 8) in the presence of ammonia and piperidine giving 2H-imidazoles, 4, in high yield (14). However, a diminished yield of this type of product results when the temperature is raised to 0 °C. Interestingly, in the case of 2H-imidazole formation, sulfur is promoting the oxidation of ketone without being incorporated into the product.

The requirement of alpha hydrogen for product formation leads to the reasonable view (4) that an early step is thiation at an alpha carbon resulting from the attack of an enolate, or its ketimine equivalent, on electrophilic sulfur. Subsequent thiolketimine formation, conceivably occurring by the sequence of steps shown in Scheme 1, would lead to the end product 3-thiazoline when the thiolketimine reacts with the second mole of ketone, as shown in the last step. As a consequence of enolization and thiation, one ketonic alpha carbon has become oxidized. The formation of the other heterocycles, 3-5 (Figure 1), requires that the same carbon atom be raised to still higher oxidation states. Perhaps, further oxidation occurs in a manner similar to the first, that is, through enolization followed by additional thiation.

Contrasting the actions of sulfur and hydrogen sulfide on ketones is an instructive conclusion to this section. While sulfur results in the formation of various heterocycles as noted above, hydrogen sulfide in the presence of amines or ammonia at room temperature yields geminal dithiols or thioketones as exemplified in the report of Mayer and co-workers (15).

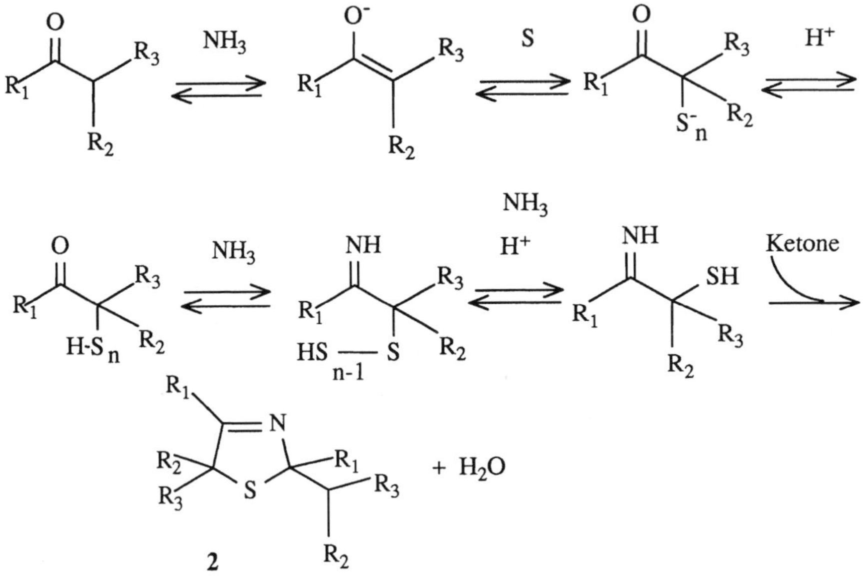

Scheme 1

Low Temperature Nonoxidative-Reductive Processes

As discussed in the previous section, chemical transformation resulted from oxidizing elemental sulfur reacting with reducing carbon in the form of an enolate anion. The initial roles of sulfur and carbon can be regarded as reversed in the reactions discussed in the present section. Sulfur as the polysulfide dianion can be considered as a nucleophile, a relatively electron rich species, which can react with a conjugated ene carbonyl. The ene carbonyl is regarded as the electrophile or electron deficient species. An interesting outcome of sulfur acting in this manner is the formation of covalent sulfur-carbon bonds and carbon-carbon bonds as well.

The potential for forming S-heterocycles from low-temperature processes initiated by nucleophile-electrophile combinations is well known for the action of hydrogen sulfide on enones. Some examples will be presented below. However, an exceptional generation of S-heterocycles was first substantiated in a report of Del Mazza and Reinecke ([16]) who proposed, on the basis of NMR and MS spectral data, that the product obtained in nearly quantitative yield when Fromm and Hubert ([17]) treated chalcone, 6, with sodium polysulfide was 2,4-dibenzoyl-3,5-diphenylthiolane, 7. The Fromm-Hubert experiment had been carried out under anhydrous conditions.

That the chalcone units combined in a head-to-tail manner was conclusively decided by showing that the thiolane could be dehydrogenated by dichlorodicyanoquinone in boiling chlorobenzene to a dibenzoyldiphenylthiophene, 8, whose ^{13}C NMR revealed the presence of two different benzoyl groups ([18]). It followed that each of the two possible modes of head-to-head fusion of chalcone units in a thiolane could be rejected.

The foregoing indication of the special action of alkali polysulfide on a conjugated enone led to an attempt to understand how the initial reaction between polysulfide dianion and chalcone could be such a facile one. The details of the analysis of this problem, based primarily on the frontier molecular orbital (FMO) approach, have been reported ([19]). It will suffice to reiterate here only the conclusions of this analysis: first, the elongation of catenated sulfur in polysulfide dianion will enhance its nucleophilicity relative to sulfide dianion; second, the elongation of a conjugated alkene or ene carbonyl system, through the incorporation of additional conjugated vinyl units, will enhance the reactivity of the conjugated system when it interacts with the type of nucleophile just described; third, the attachment of phenyl groups to the carbonyl carbon or to the alkenic terminus of the conjugated system may further enhance the reactivity of a conjugated ene carbonyl system.

Molecular orbital theory also predicts that a nucleophile of the sulfide type will bond at the carbon terminus of a conjugated ene carbonyl system; that is, the nucleophile will bond with the electrophile in the Michael addition mode of reaction ([20]). Thus, the reaction of polysulfide dianion with an enone represented by a chalcone may proceed initially in such a manner as shown in Scheme 2, which reproduces one of the several pathways

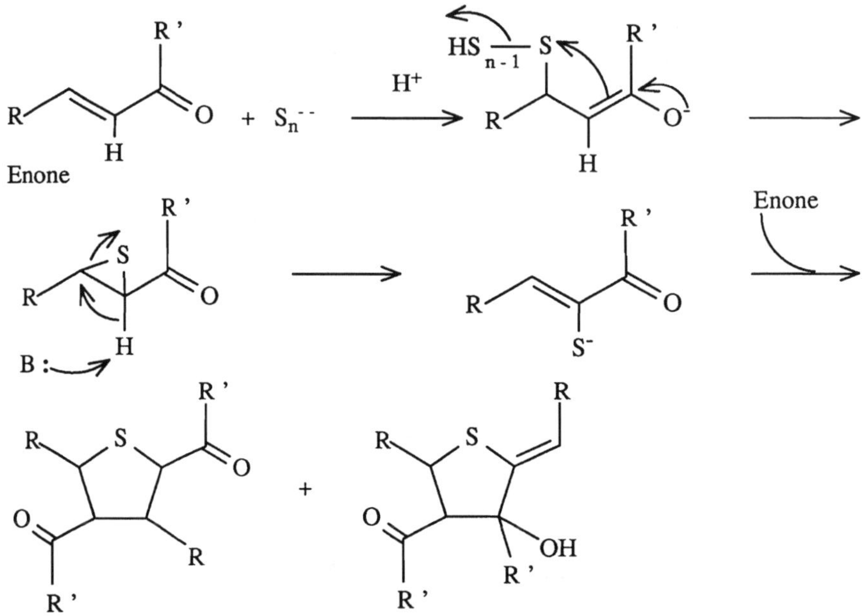

Scheme 2

from chalcones to thiolanes that have been suggested (16,18,21). However, once the first carbon-sulfur bond is made, nucleophilicity would be transferred to the carbon moiety which now bears a negative charge as an enolate anion. Simultaneously, the sulfur atom attached to carbon would become electrophilic. Thereafter, the attack by the carbon terminus of an enolate anion on the carbon-bonded sulfur atom, such as in Scheme 2, could result in thiocycle formation through the displacement of the remainder of the sulfur chain. Therefore, catenated sulfur is regarded not only as a reactive nucleophile in polysulfide dianion but also as an electrophile once one end of the original chain is attached to carbon. The same considerations that apply to the vinylogous extension of catenated electrophilic carbon also apply to extension of electrophilic sulfur as well. Thus extended catenation of sulfur results in both a more reactive nucleophile when the chain is dianionic and in a more reactive electrophile when the chain is neutral (22).

The carbon electrophiles studied in my laboratory were chosen because of the preceding theoretical considerations and in view of a need to facilitate the analysis of the anticipated complex product mixtures. Thus, aromatic substituted ene carbonyls were selected because UV-absorbing aromatic units would be incorporated unchanged into products and consequently would assist the detection and isolation of these products.

Several variously substituted chalcones, 9, underwent conversion to five-membered S-heterocycles under standard conditions consisting of a saturated solution of hydrated sodium polysulfide in 95% ethanol at room temperature (23). Two different chalcones reacted to give four different thiolanes. Two of

9: $R_1 = R_4 = Ar$; $R_2 = R_3 = H$

10: $R_1 = R_4 = Ph$; $R_2 = CH_3$; $R_3 = H$

11: $R_1 = R_4 = Ph$; $R_2 = H$; $R_3 = CH_3$

12: $R_1 = R_2 = R_3 = H$; $R_4 = Ph$

13: $R_1 = CH_3$; $R_2 = R_3 = H$; $R_4 = Ph$

14: $R_1 = Ph$; $R_2 = R_3 = R_4 = H$

15: $R_1 = Ph$; $R_2 = R_3 = H$; $R_4 = CH_3$

16: $R_1 = R_2 = Ph$; $R_3 = R_4 = H$

17: $R_1 = t\text{-}Bu$; $R_2 = R_3 = H$; $R_4 = Ph$

the four resulted from the combination of two molecules of the same chalcone; the other two resulted from each of two different chalcones combining with one another in two ways (24-22). However, some chalcones, among them the α- (10) and β-methyl chalcone (11) whose structures are given below, were unreactive.

Besides chalcones, other types of α,β-unsaturated carbonyls affording five-membered S-heterocycles were cinnamaldehyde, 12, and the α,β-unsaturated methyl ketone 13 and phenyl ketones 14 and 15. However, another phenyl ketone, 16, substituted at the α position or tertiary butyl ketone 17 failed to yield S-heterocycles.

S-Heterocycle Structure

In nearly all cases, the action of alkali polysulfide on a chalcone under standard conditions gave but a single thiolane in which the two chalcone units had combined head-to-tail as depicted in compound 7. Generally, the structures were ascertained through spectral evidence which was consistent for a large series of thiolanes. Proton NMR showed from the magnitude of the three-bond proton-proton coupling constants ($^3J_{H-H}$) that in all cases the hydrogen atoms attached at ring positions 3, 4 and 5 were *trans-anti*. (See structure 7 for the numbering of ring atoms.) Unlike the other protons, the relative configuration of the fourth ring hydrogen attached to position 2 could not be ascertained directly from the value of its coupling with the proton at position 3. However, the *ortho* methoxy substituted chalcone 18, whose structure is shown in Figure 2, gave a mixture of the two diastereomeric thiolanes: the all *trans*, and the *cis, anti, trans*, 19 and 20 respectively. Similarly, *ortho*-methoxychalcone 21 gave a mixture of diasteriomers 22 and 23. The structures 19, 20 and 23 were determined by X-ray crystallography (21). Other experiments demonstrated that the *cis, anti, trans* diastereomers 20 and 23 were formed first but were converted to the corresponding all *trans* diastereomers in time (21).

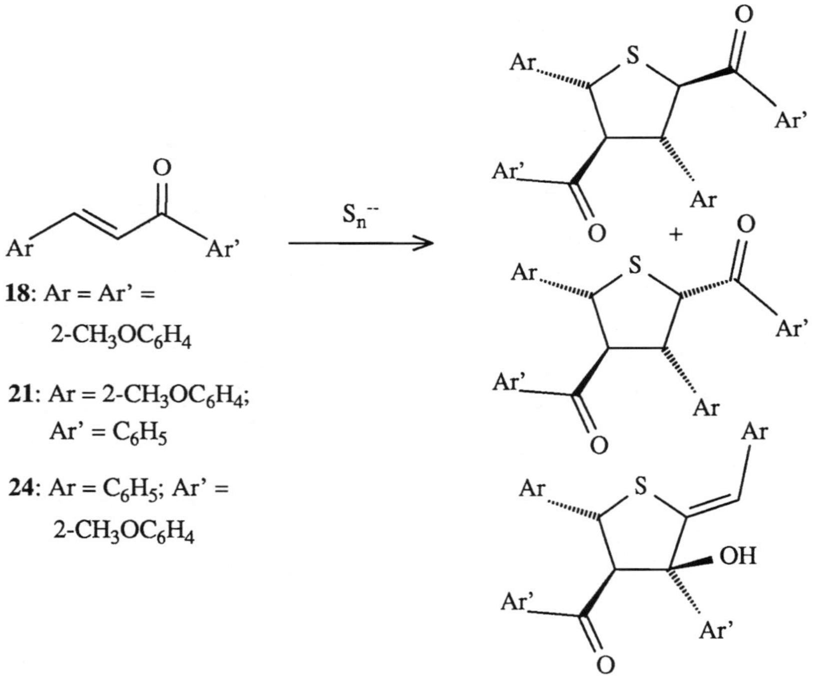

18: Ar = Ar' = 2-$CH_3OC_6H_4$

21: Ar = 2-$CH_3OC_6H_4$; Ar' = C_6H_5

24: Ar = C_6H_5; Ar' = 2-$CH_3OC_6H_4$

19: Ar = Ar' = 2-$CH_3OC_6H_4$

22: Ar = 2-$CH_3OC_6H_4$; Ar' = C_6H_5

25: Ar = C_6H_5; Ar' = 2-$CH_3OC_6H_4$

20: Ar = Ar' = 2-$CH_3OC_6H_4$

23: Ar = 2-$CH_3OC_6H_4$; Ar' = C_6H_5

26: Ar = C_6H_5; Ar' = 2-$CH_3OC_6H_4$

32: Ar = Ar' = C_6H_5

Figure 2. Thiolanes resulting from the head-to-tail (**19**, **20**, **22**, **23** and **25**) and head-to-head (**26** and **32**) combinations of chalcone units.

The head-to-tail combination of conjugated enone units also was evident in the S-heterocyclic products obtained from phenyl vinyl ketone 14 when the latter was treated with alcoholic polysulfide. The product mixture contained the 2,4-dibenzoylthiolane and the corresponding thiophene obtained in 5.8 and 3 % yields respectively.

The action of alkali polysulfide on chalcones does not always result in the exclusive formation of the head-to-tail structural type of thiolanes. Treatment of 2'-methoxychalcone, 24, gave both the usual structural type of thiolane, 25 (as shown in Figure 2), and the head-to-head arrangement of chalcone units as evident in the structure of the stable tertiary alcohol 26, whose structure was determined by X-ray crystallography (24). The same carbon-sulfur skeleton evident in 26 also was found in the thiophenes 27 and 28 as shown in Figure 3. These thiophenes resulted from the previously mentioned combination of two molecules of cinnamaldehyde with sulfur from sodium polysulfide. The formation of ethyl ester 28, presumably by auto-oxidation of the aldehyde under highly alkaline conditions, may be of considerable consequence to the occurrence of carbonyl-free thiophenes in the geosphere since the presence of 28 points to corresponding carboxylic acid formation when aqueous rich solvents replace those which are alcohol rich. The thiophene carboxylic acids, or their salts, could undergo decarboxylation. Thus alkali promoted, Cannizzaro type auto-oxidation-reduction of aldehydes followed by decarboxylation might serve as a deoxygenation mechanism for aldehydic thiophenes. Similarly 29 was obtained from 4-phenylbuten-2-one, as depicted in Figure 3, while thiophenes 30 and 31 resulted from the combination of a single molecule of dieneones with sulfur from sodium polysulfide (19).

The production of thiolanes has been studied with regard to reaction temperature, the composition of the alkali polysulfide and solvent (19,23,25). In the range of 0 - 30 °C, thiolane 7 was the sole product when chalcone was treated in an alcoholic or aqueous alcoholic solvent rich in the alcohol. However, in boiling methanol, at 65 °C, the product consisted not only of thiolane 7, but also dihydrochalcone ($PhCH_2CH_2COPh$). The same treatment of 4,4'-dimethoxychalcone similarly resulted in the formation of 4,4'-dimethoxydihydrochalcone (26) although under standard conditions this chalcone was recovered unchanged. These results help to define the sensitivity of conjugated enones to another well known mode of polysulfide reactivity, namely reduction. With regard to chemical geosynthesis, the result points to another possible polysulfide involvement occurring at a somewhat higher temperature: the conversion of a carbon source to its more highly saturated form.

Concerning the composition of alkali polysulfide, the yield of thiolane 7 suffered none in utilizing alkali polysulfide prepared from hydrated sodium sulfide or lithium or potassium sulfide rather than anhydrous sodium sulfide. When the ratio of sulfur to hydrated sodium sulfide was 8:1 or greater, the thiolane 7 resulted whereas at a ratio of 4:1, or using commercial Na_2 , several components containing both sulfur and carbon were detected in the solid product.

Other results make it clear that solvent can have a striking influence on the nature of the product. The use of aqueous alcoholic solvents rich in alcohol such as 70 - 95% aqueous alcohols, resulted in good conversion of several chalcones to thiolanes typified by the structure 7. However, the use of the solvent 80% water - 20% ethanol resulted in the formation of both thiolane structural types 7 and 32 (Figure 2) although the conversion of

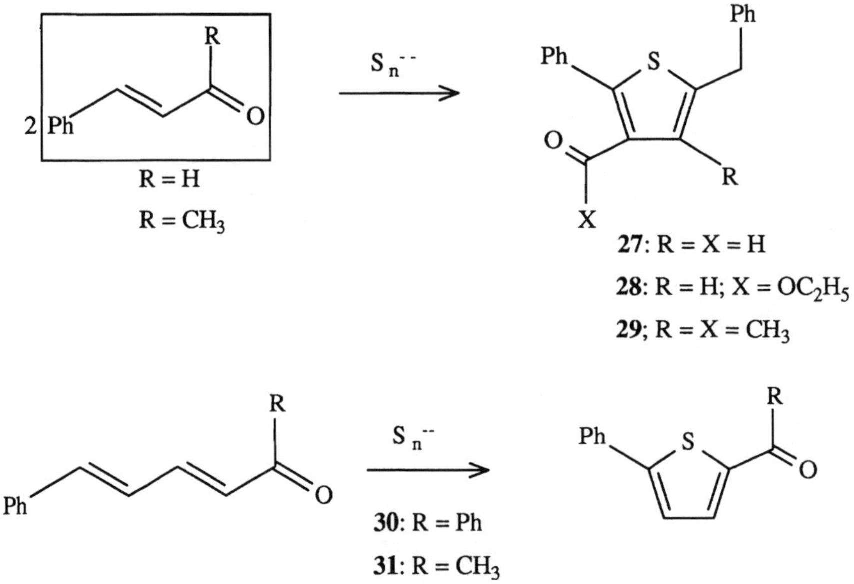

Figure 3. Thiophenes formed from the action of sodium polysulfide on cinnamaldehyde (top) and dienones (bottom).

chalcone to thiolane was somewhat lower than normal. The thiolane **32** was also obtained from chalcone when the solvent was ethylene glycol. Still another altogether different product type, 1,3-diphenyl-3-thioxopropen-1-one (**18**) resulted from the use of the solvent 1,2-dimethoxyethane under anhydrous conditions.

The reactions of conjugated enones with polysulfides are noteworthy in comparison to reactions of the same compounds with hydrogen sulfide in the presence of alkali, ammonia or amines. The action of hydrogen sulfide and ammonia on 4-phenylbutenone, **13**, at room temperature produced the mercaptophenylbutenone **33** and azadithiobicyclo[3.3.1]nonanes **34** and **35** as shown below (27,28). However, treatment of the same enone with sodium sulfide in aqueous ethanol at 0°C resulted in the formation of the tetrahydrothiopyranol **36** (16,28). Similarly, the action of hydrogen sulfide in concentrated aqueous potassium hydroxide on chalcone, **6**, at 0 °C resulted in the formation of diphenylmercaptopropanone **37** (29) while the treatment of the same enone with hydrogen sulfide and ammonia produced the tetrahydrothiopyranol **38**. Sulfides are sometimes the product. Thus hydrogen sulfide bubbled through an ethanolic solution of 2-butenone at room temperature produced the sulfide **39** (16). Therefore by comparison these results furnish ample testimony to the exceptional outcome that polysulfide effects when acting on the same conjugated enones.

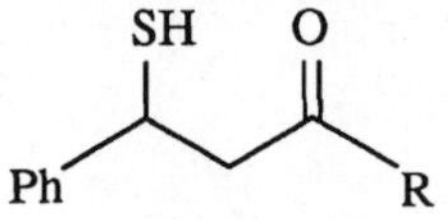

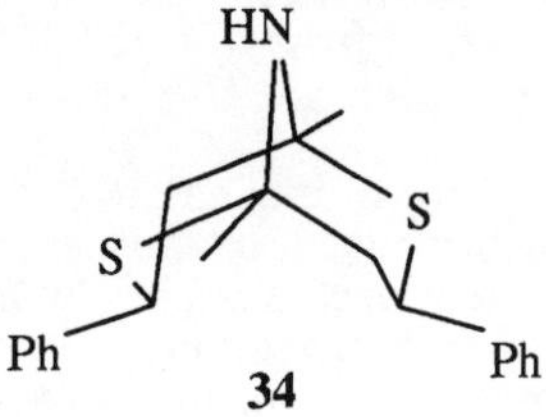

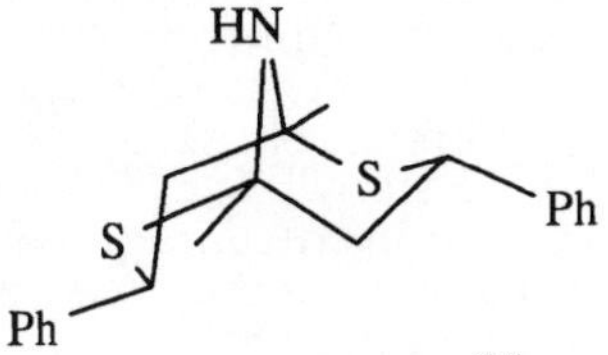

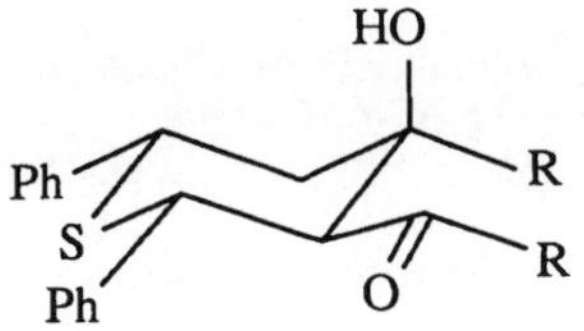

The Presence of S-heterocycles and their Element Sources in the Geosphere

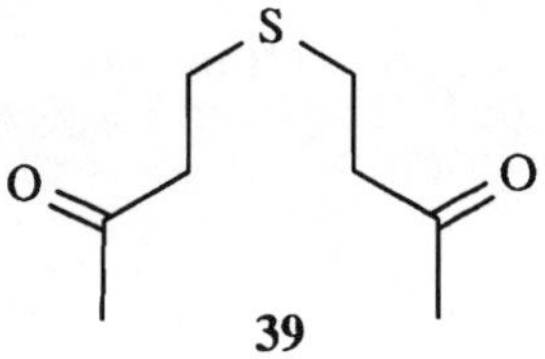

The appearance of thiophenes in fuel sources is reported in an abundance of papers. (See for recent examples: in coal (30); in oil shale liquification fluids (31); in crude oil (32)). However, if attention is restricted primarily to modern geological settings, such as modern sediments, the reports of S-heterocycle occurrence are minuscule in comparison. Whelan and co-workers detected thiophene, 2-methyl- and 3-methylthiophene in a modern marine sediment (33). The occurrence of these compounds was ascribed either to a biological source or to a low-temperature reaction. Brassell and co-workers found an isoprenoidal thiophene in an immature marine sediment (34). Phenyl and alkyl substituted thiazoles have been detected in petroleum (35) although there are no reports of thiazole presence in a modern geological sample. Moreover, the significance of thiazoles as products of the geosphere may be lessened by the caution that they represent primarily the products of biosynthesis rather than geosynthesis. This same caution must be applied to the occurrence of other S-heterocycles as well. However, now that low-temperature processes for the formation of such compounds have been demonstrated in the laboratory, the likelihood that these processes also occur during diagenesis must at least be given consideration.

The question of the source molecule occurrence must be addressed. The coincidence in anaerobic environments of ammonia and reduced divalent sulfur species, which includes hydrogen sulfide, bisulfide anion and sulfide dianion, has been given adequate recognition. (See for example references 36,37.) The coincidence of polysulfide and ammonia has already been noted in an earlier part of this report (1). Polysulfide has been found in tidal and salt marsh sediments as well (38).

Polysulfide requires high pH for stabilization (39). In the course of studying the oxidation of H_2S, Hoffmann (40) detected polysulfide presence as an intermediate in the pH range of 6-9. However the less catenated members of the S_n^{--} series of ions require higher pH for stabilization than highly catenated members. Chivers, in his review of aqueous solution polysulfide equilibria, indicates S_2^{--} is stabilized only at the highest pH levels, S_3^{--} at somewhat lower levels, S_4^{--} at an intermediate level and S_5^{--} at nearly pH 7 or somewhat below (41). According to our FMO analysis, the more highly catenated forms are those which are the more reactive with a given appropriate ene carbonyl (19).

Regarding carbon sources, the chemical process analysis directed attention to conjugated ene carbonyls. However, assessing the natural occurrence of the proximate carbon source has been somewhat more difficult than doing the same for its sulfur counterpart. Seemingly, the carbon molecules biosynthesized in the largest amounts would be good starting candidates for the type of carbon source molecules required for the ene carbonyl-polysulfide route leading to stable S-heterocycles. Carotenoids and lignin have been suggested as carbon sources. The oceanic production of carotenoids in modern times was estimated at nearly 10^8 t/year, and continental production very nearly the same. The production of lignin was estimated to be of the order of 10^9 t/year (19). Chiefly through facile degradation in oxidative environments, the resulting ene carbonyl products could be an allochthonous, proximate carbon source.

Carbohydrates, whose continental and marine productions have been estimated at nearly 10^{11} and 10^{10} t/year respectively (42), have qualified for recognition as a possible carbon source. Indeed, the treatment of carbohydrate with hydrogen sulfide at a temperature as low as 100 °C is reported to result in S-heterocycle formation (43). Recently, glucose has been treated with hydrogen sulfide at 40 °C then pyrolyzed in the course of product analysis. Although S-heterocycles were found in the pyrolysate (44), the conditions to which the carbohydrates were subjected may result in their conversion to the proximate carbon source, which only then is sufficiently reactive enough to combine with hydrogen sulfide. Thus, the role assumed of carbohydrates as proximate carbon source molecules in low-temperature reactions is presently not readily appraised.

Conclusion

The laboratory demonstrations of low-temperature routes to S-heterocycles through oxidation and electrophile-nucleophile combination point the way to previously overlooked geosynthetic routes to S-heterocycles. These routes assume significance when considered together with the occurrence of polysulfide, sulfur and ammonia environments as well as the occurrence of several S-heterocycles. The polysulfide route to S-heterocycles, which also involves conjugated ene carbonyls as the proximate carbon source molecules, would seem more plausible presently because highly abundant biotic sources of carbon molecules, namely carotenoids and lignin, are readily conceived. The oxidation route to S-heterocycles has plausibility too with regard to temperature but requires further study, especially in connection with the course of the reaction in aqueous or hydroxylic solvents and with aldehydes as a proximate carbon source. Moreover, the natural source of carbon for this latter reaction type must be given further evaluation.

Acknowledgments

The author acknowledges the contributions of the many undergraduate chemistry majors, who through special problems, senior research or summer research participation have assisted his study of polysulfide chemistry. Generous support for their work was provided by the Sterling-Winthrop Research Institute. The author also acknowledges the receipt of a NYS/UUP Experienced Faculty Travel Award in support of his participation in the symposium on Geochemistry of Sulfur in Fossil Fuels.

Literature Cited

1. Aizenshtat, Z.; Stoler, A.; Cohen, Y.; Nielsen, H. Advances in Organic Geochemistry; Bjoroy, J., Ed.; Wiley: New York, 1981, pp 279-288.
2. Smith, R. L.; Oremland, R. S. Limnol. Oceanogr. 1987, 32 794-803.
3. Asinger, F.; Offermanns, H. Angew. Chem. Interna. Edit. 1967, 6, 907-919.
4. Asinger, F.; Schaefer, W.; Halcour, K.; Saus, A.; Triem, H. Agnew Chem.Internat. Edit. 1964, 3, 19-28.
5. Pryor, W. A. Mechanisms of Sulfur Reactions; McGraw-Hill: New York, 1962; pp 127-138.
6. Carmack, M.; Spielman, M. A. In Organic Reactions, Vol. 3: Adams, R., Ed.; Wiley: New York; 1946, pp 83-107.
7. Brown, E. V. Synthesis 1975, 358-375.
8. Asinger, F.; Thiel, M.; Pallas, E. Justus Liebigs Ann. Chem. 1957, 603 37-49.
9. Thiel, M.; Asinger, F.; Fedtke, M. Justus Liebigs Ann. Chem. 1958, 615, 77-84.
10. Asinger, F.; Schroeder, L.; Hoffmann, S. Justus Liebigs Ann. Chem. 1961, 648, 83-95.
11. Asinger, F.; Thiel, M.; Lipfert, G.; Plessmann, R. E.; Mennig, J. Agnew. Chem. 1958, 70, 372.
12. Asinger, F.; Thiel, M.; Lipfert, G. Justus Liebigs Ann. Chem. 1959, 627, 195-212.
13. Asinger, F.; Thiel, M.; Schroeder, L. Justus Liebigs Ann. Chem. 1957, 610, 49-56.
14. Asinger, F.; Leuchtenberger, W. Justus Liebigs Ann. Chem. 1974, 1183-1189.

15. Mayer, R.; Hiller, G.; Nitzschke, M.; Jentzsch, J. Angew. Chem. Internat. Edit. 1963, 2, 370-373.
16. Del Mazza, D.; Reinecke, M. G. J. Org. Chem. 1981, 46, 128-134.
17. Fromm, E.; Hubert, E. Justus Liebigs Ann. Chem. 1912, 394, 301-309.
18. LaLonde, R. T. J. Chem. Soc. Chem. Commun. 1982, 401.
19. LaLonde, R. T.; Ferrara, L. M.; Hayes, M. P. Org. Geochem. 1987, 11, 563-571.
20. Radunz, H.-E. Kontakte 1977, 3-10.
21. LaLonde, R. T.; Codacovi, LM.; Cun-heng, H.; Cang fu, X.; Clardy, J.; Krishnan, B. S. J. Org. Chem. 1986, 51 4899-4905.
22. Meyer, B.; Peter, L.; Spitzer, K. Homoatomic Rings, Chains and Macromolecules of Main-Group Elements; Rheingold, A. L., Ed.; Elseview: Amsterdam, 1977, pp 477-497.
23. LaLonde, R. T.; Horenstein, B. A.; Schwendler, K.; Fritz, R. C.; Florence, R. A. J. Org. Chem. 1983, 48, 4049-4052.
24. LaLonde, R. T.; Florence, R. A.; Horenstein, B. A.; Fritz, R. C.; Silveira, L.; Clardy, J.; Krishnan, B. S. J. Org. Chem. 1985, 50, 85-91.
25. Reinecke, M. G.; Morton, D. W.; Del Mazza, D. Synthesis, 1983, 160-161.
26. Fritz, R. C.; LaLonde, R. T. unpublished observations.
27. Fromm, E.; Haas, F. Justus Liebigs Ann. Chem. 1912, 394, 291-300.
28. Forward, G. C.; Whiting, D. A. J. Chem. Soc. C, 1969, 1647-1652.
29. Tanaka, H.; Yokoyama, A. Chem. Pharm. Bull. 1960, 8, 275-279.
30. Herod, A. A.; Smith, C. A. Fuel 1985, 64, 281-283.
31. Braekman-Danheux, C. J. Anal. Appl. Pyrol. 1985, 7, 315-322.
32. Arpino, P. J.; Ignatiadis, I.; DeRycke, G. J. Chromatogr. 1987, 390, 329-348.
33. Whelan, J. K.; Hunt, J. M.; Berman, J. Geochim. Cosmochim. Acta 1980, 44, 1767-1785.
34. Brassell, S. C.; Lewis, C. A.; deLeeuw, J. W.; DeLange, F.; Sinninghe Damste, J. S. Nature 1986, 320, 160-162.
35. Aksenov, V. S.; Kamyonov, V. F. In Organic Sulfur Chemistry; Freidina, R. K.; Skorova, A. E. Eds,; Pergamon: Oxford, 1981, pp 1-13.
36. Richards, F. A.; Anderson, J. J.; Cline, J. D. Limnol. Oceanogr. 1971, 16, 43-50.
37. Cline, J. D.; Richards, F. A. Limnol. Oceanogr. 1972, 17, 885-900.
38. Luther, G. W. III; Giblin, A. E.; Varsolona, R. Limnol. Oceanogr. 1985, 30, 727-736.
39. Schwarzenbach, G.; Fischer, A. Helv. Chim. Acta 1960, 43, 1365-1390.
40. Hoffman, M. R. Environ. Sci. Technol. 1977, 11, 61-66.
41. Chivers, T. In Homaotomic Rings, Chains and Macromolecules of Main Group Elements; Rheingold, A. L. Ed.; Elsevier: Amsterdam, 1977, pp 499-537.
42. Salisbury, F. B.; Ross, C. W. Plant Physiology, 3rd Ed.,; Wadsworth Publishing Co.: Belmont, CA 1985, p 217.
43. Mango, F. D. Geochim. Cosmochim. Acta, 1983, 47, 1433-1441.
44. Suzuki, N.; Philip, R. P. in press.

RECEIVED March 5, 1990

Chapter 5

Isolation of Sulfur Compounds from Petroleum

O. P. Strausz, E. M. Lown, and J. D. Payzant

Department of Chemistry, University of Alberta, Edmonton, Alberta T6G 2G2, Canada

Analytical procedures for the isolation of the sulfide and thiophene classes of compounds from petroleum are described. The methods are based on the selective oxidation of first the sulfides to sulfoxides, followed by silica gel separation of the sulfoxides and their reduction back to sulfides. The thiophenes are then separated from the sulfide-free oil by selective oxidation to sulfones, followed by silica gel separation of the sulfones from the oil and their reduction back to thiophenes.

The analytical chemistry of sulfur compounds in petroleum is attracting attention owing to the global increase in the market share of sulfur-rich heavy crudes, the statutory limitations of sulfur content of refined fuels and the increasingly recognized wealth of biogeochemical information contained in the sulfur compounds, reflecting the origin, biodegradation, thermal and water washing history of the petroleum. In spite of the large number of individual sulfur compounds identified in petroleum (1–20), sulfur occurs in the form of only a few functional groups, mainly as thiophenes and cyclic sulfides, and in some crudes as disulfides and mercaptans as well. The detailed characterization of the sulfur compounds in petroleums requires methods for the selective isolation of the sulfide and thiophene classes of compounds for further analysis. Recently, a non-destructive, sulfur K edge X-ray spectroscopic method has been reported for the determination of the quantities of sulfides and thiophenes in asphaltenes (21).

Methods based on distillation or mercuric chloride adduct formation (1–5) are tedious and applicable only to the isolation of select, individual compounds. The methods of choice would, instead, isolate the various classes of sulfur-containing compounds from petroleum in high yield and purity.

An earlier procedure reported from this laboratory for the isolation of sulfides was based on the selective oxidation of the sulfides to the sulfoxides with photoexcited singlet oxygen followed by silica gel chromatographic separation of the sulfoxides (9). Reduction of the isolated sulfoxides back to the sulfides and a subsequent chromatographic purification resulted in the isolation of the sulfides as a pale yellow oil. However, when the sulfur is in a five-membered ring, the photooxidation may lead to side reactions (22). Therefore, other oxidation methods had to be explored.

For the isolation of thiophenes from petroleum and other fossil fuels, a number of methods have been reported, including chromatography on silver nitrate (23) or palladium chloride-impregnated silica gel (24), and oxidation to the more polar sulfones followed by chromatographic separation of the sulfones, their reduction back to the thiophenes and chromatographic separation of the thiophenes (25). The

0097–6156/90/0429–0083$06.00/0

oxidation procedure employed 30% H_2O_2/acetic acid/benzene (16 hour reflux). This, however, in many cases, leads to oxidation of the aromatic ring, resulting in low recoveries of the thiophenes (26,27). Many reagents and conditions have been used for the oxidation of thiophenes to their sulfones (28–30), of which m-chloroperbenzoic acid under neutral anhydrous conditions (31) appears to be the most suitable.

In this paper we describe selective oxidation procedures to isolate cyclic sulfides on the one hand, and thiophenes, on the other, from petroleums.

Method for Separating the Sulfides

The overall scheme for the separation of the thiophenes and sulfides from the maltene fraction (n-pentane-soluble fraction) of bitumen is outlined in Figure 1. The method is based on the large difference in polarity between the highly polar sulfoxides and the mildly polar sulfides, and the ease and high yields with which they can be interconverted. Oxidation of the maltene fraction by (n-C_4H_9)$_4$NIO$_4$ converts aliphatic sulfides to sulfoxides while leaving the thiophenes intact. This selective oxidation of sulfides to sulfoxides is central to the method. Subsequent chromatography on silica gel separates the highly polar sulfoxides and other polar compounds from the maltene. The sulfoxides may then be converted back to sulfides by reduction with LiAlH$_4$. A subsequent chromatographic step separates the low-polarity sulfides from the other polar substances. It is important that the oxidation stops at the sulfoxide rather than continuing to the sulfone since six-membered ring saturated sulfones are difficult to reduce back to sulfides (32). Many oxidizing reagents will convert sulfides to sulfoxides (33), however, most of them will convert sulfoxides to sulfones when used in excess. In the isolation of sulfides from petroleum, it is necessary to employ an excess of reagent since the quantity of sulfides is unknown, as is the amount of oxidant which may be consumed in other reactions with the petroleum.

Tetrabutylammonium periodate in the refluxing mixed solvent toluene/methanol (5/1 by volume) cleanly oxidizes the sulfides to sulfoxides. When three equivalents of tetrabutylammonium periodate were used, the test sulfides 2-dodecyl thiolane and 2-dodecyl thiane were >99% converted after a 20-minute reflux to the corresponding sulfoxides (98%) and sulfones (2%). After 60 minutes the yield of sulfone had increased to 10%. The reaction is highly dependent on the exact conditions employed. Small deviations from the recommended conditions may produce unacceptable results (34). Sulfones of saturated sulfides are difficult to reduce (32) and thus their generation should be minimized for maximum recovery of the sulfides.

Model reactions showed that 2-dodecyl-5-methyl thiophene, benzo[b]thiophene, 2-n-butyl benzo[b]thiophene and dibenzothiophene were unaffected by these reaction conditions after a 4-hour reflux. For the LiAlH$_4$ reduction step the solvent dioxane (b.p. 100°C) is preferred to either diethyl ether or tetrahydrofuran since preferential reduction of certain sulfides in the latter solvents has been observed (8). The results of the sulfide analyses for several petroleums from Alberta are summarized in Table I.

There is considerable variation in the sulfide content in these petroleums, ranging from 16% for Peace River, a bitumen, to 0.2% for Pembina, a conventional oil. The sulfide GC-FID chromatograms of some selected samples are shown in Figure 2, which shows considerable variation between the samples. For example, Bellshill Lake contains substantial quantities of monocyclic sulfides possessing a linear (n-alkane) carbon framework and these appear as partially-resolved clusters of peaks on the bottom chromatogram of Figure 2 (19,35).

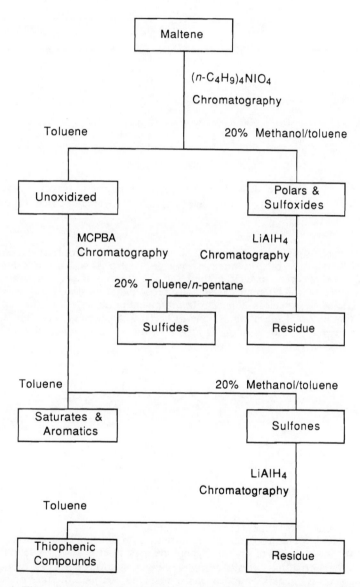

Figure 1. Flow diagram for the separation of the sulfide and thiophene classes of compounds from the maltene fraction of petroleum. (Reproduced from Reference 34. Copyright 1989, American Chemical Society.)

Table I. Sulfide and Thiophene Contents of Some Alberta Petroleums

Sample	Depth (m)	Sufides %[a]	Thiophenes
Syncrude	0	6.9 ± 0.9[b]	6.4 ± 0.5[c]
Bellshill Lake	920	3.1	5.5
Lloydminster	670	2.6	5.7
Peace River	558	16.0	7.6
Wolf Lake	500	3.6	2.9
Suncor coker feed	0	n.d.	2.7
Cold Lake	500	n.d.	2.4
Pembina	1080	0.2	0.91
Leduc	1580	0.3	1.1

[a]As percent of the maltene. [b]Average of three determinations. Errors are twice the standard deviation. [c]Average of four determinations. Errors are twice the standard deviation.

These monocyclic sulfides are degraded by certain microorganisms (36) and thus they are not present in the other samples shown in Figure 2, which are extensively biodegraded. Much of the material in these samples consists of a complex mixture of polycyclic sulfides which manifests itself as a broad underlying hump in the GC-FID chromatograms. Nevertheless, a number of prominent peaks appear on the upper two traces of Figure 2 and these are assigned (9, 10) to a series of bicyclic (B_n) and tetracyclic (T_n) terpenoid sulfides, as shown below. These terpenoid sulfides are resistant to biodegradation and are thus in

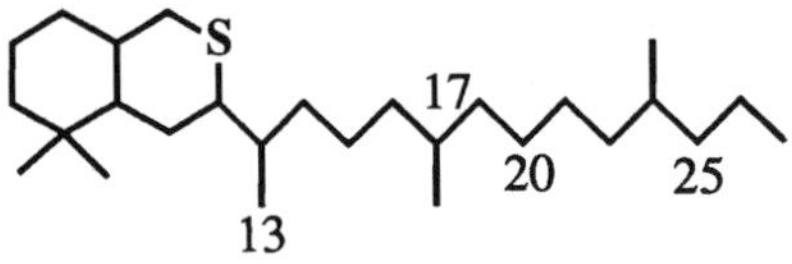

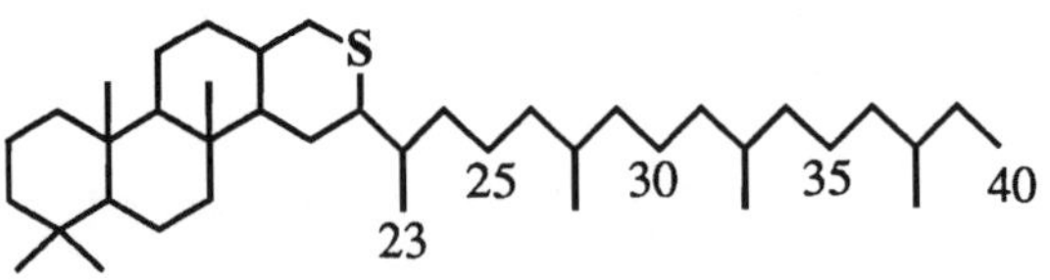

high relative abundance in these biodegraded samples.

The GC-FID chromatogram of the sulfides from Bellshill Lake petroleum is shown in the lowest panel of Figure 2 where the most intense peak is due to the C_{13} bicyclic terpenoid sulfide. Peaks corresponding to various series of monocyclic sulfides appear as clusters on top of the unresolved complex mixture of sulfides. These monocyclic sulfides are mainly complex mixtures of homologous series of isomeric thiolanes and thianes which have an n-alkane carbon framework. Generalized structures are shown below. Recent experiments have shown that the thiolanes and thianes derived from n-alkanes will interconvert under simulated geological conditions, with the sulfur atom migrating along the linear chain (35).

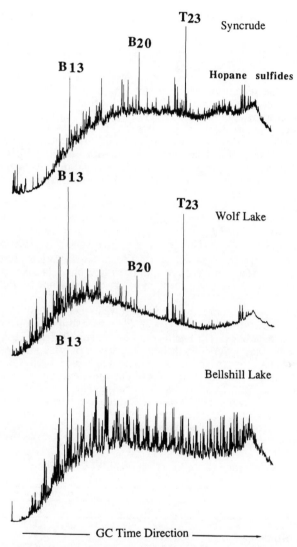

Figure 2. GC-FID chromatograms for the sulfide fractions from different Alberta petroleums. The peaks labeled B_{13} and B_{20} correspond to the bicyclic terpenoid sulfides with 13 and 20 carbons, respectively. The peak labeled T_{23} corresponds to the tetracyclic terpenoid sulfide with 23 carbons and peaks due to the hopane sulfides are indicated at the end of the chromatograms. The clusters of peaks spaced one carbon apart on the Bellshill Lake trace correspond mainly to complex mixtures of isomeric monocyclic sulfides possessing a linear carbon framework. These sulfides have been removed by biodegradation from the upper two samples. For more complete peak identification see References 9, 10 and 35. (Reproduced from Reference 34. Copyright 1989, American Chemical Society.)

Method for Separating the Thiophenes

The method for the separation of the thiophenes from petroleum is schematically illustrated in Figure 1. After removal of the sulfides from the maltene by oxidation to sulfoxides and chromatography, the remaining unoxidized material is subjected to another oxidation step designed to convert the thiophenes to their sulfones. The resulting sulfones are highly polar and can be readily separated from the mixture by chromatography. The sulfones are then reduced with LiAlH$_4$ which converts them back to the low-polarity thiophenes which are separated from the mixture by subsequent chromatography. This method is similiar in principle to the procedure for the isolation of the sulfides.

A variety of methods have been described for the isolation of thiophenes from fossil fuels (13,23–25). An oxidation-reduction procedure very similiar to the present method has been described (25), the difference being in the reagent used for the oxidation step. In the method of Willey *et al.* (25), a 30% H$_2$O$_2$/acetic acid/benzene mixture under reflux for 16 hours was used. This is satisfactory for some thiophenes, but in other cases further oxidation to quinones and phenols takes place and thus recoveries were often poor since these substances could not be reduced back to thiophenes (26,27).

In an attempt to minimize overoxidation we explored the oxidation system *m*-chloroperbenzoic acid/NaHCO$_3$/CH$_2$Cl$_2$ (34) with three thiophenes and three polyaromatic hydrocarbons. The results are summarized in Table II, where it is seen that the thiophenes are converted to their sulfones after only 30 minutes reaction time and the polyaromatic hydrocarbons are either unaffected by the oxidation or are oxidized much more slowly. The sulfones of the thiophenes listed in Table II are not oxidized further under these conditions. The thiophene content of Syncrude maltene was found to be 6.4% by the present method while the recovery was only 4.2% using the method of Willey *et al.* (25). Increasing the time of the oxidation reaction in the present procedure from 20 to 60 minutes had only a minor (±10%) effect on the yield of isolated thiophenes.

Table II. Rate of Oxidation of Various Compounds by *m*-Chloroperbenzoic Acid[a]

Compound	% Remaining after			
	0 min.	15 min.	30 min.	120 min.
Benzo[b]thiophene	100	16	<1	<1
2-*n*-Butylbenzo[b]thiophene	100	<1	<1	<1
Dibenzothiophene	100	2	<1	<1
Phenanthrene	100	100	100	100
Anthracene	100	83	64	33
Pyrene	100	92	86	81

[a]Reaction details are described in Reference 34.

Oxidation by *m*-chloroperbenzoic acid is catalyzed by the presence of acids. The peracid itself is too weak to catalyze the reaction, but its reduction product *m*-chlorobenzoic acid is a considerably stronger acid and will catalyze the reaction. The addition of NaHCO$_3$ (other bases such as Na$_2$CO$_3$ and Na$_2$HPO$_4$ were also found to be satisfactory) neutralizes the *m*-chlorobenzoic acid as it is formed and thus controls the oxidation. It should be noted that *m*-chloroperbenzoic acid is a considerably more vigorous oxidant in the presence of water and in fact many aromatic hydrocarbons were quickly attacked when the oxidation was conducted in a two-phase H_2O/CH_2Cl_2 system. The presence of water in the reaction system must therefore be carefully avoided in order to prevent overoxidation.

The present method, however, does not recover parent thiophenes. Although thiophenes which possess alkyl substitutents in the 2- and 5- positions are oxidized to their corresponding sulfones in fair yield under these conditions ($\underline{31}$), other complex reactions may occur during the oxidation of less substituted thiophenes ($\underline{28}$). Moreover, it was found that the sulfone of 2-dodecyl 5-methyl thiophene was not reduced by $LiAlH_4$ in ether. On the other hand, analysis of Athabasca maltene by field ionization mass spectrometry failed to reveal the presence of any thiophenes although benzothiophenes and dibenzothiophenes were abundant ($\underline{37}$). Thiophenes have been isolated from the pyrolysis oil of Athabasca asphaltene by ligand exchange chromatography using 10% $AgNO_3$ on silica gel ($\underline{13}$).

Another limitation of the method is that the reduction of the sulfones with $LiAlH_4$ occasionally does not regenerate the starting thiophene, as mentioned above. For example, benzo[b]thiophene sulfone is reduced with $LiAlH_4$ in ether to the corresponding 2,3-dihydro compound ($\underline{38,39}$). However, when 2-*n*-butyl benzo[b]thiophene sulfone was reduced with $LiAlH_4$ in ether, the starting thiophene was regenerated. Dibenzothiophene sulfone was also reduced to dibenzothiophene under these conditions.

The present oxidation reaction proceeds under mild conditions and is easy and quick to perform. In addition, the method is quite reproducible and yields samples of thiophenes which are relatively free of impurities such as aromatic hydrocarbons and sulfides. The large difference in polarity between the sulfones and the remaining low-polarity material ensures that the separation is little affected by the length and number of alkyl substitutents on the aromatic ring systems and is thus relatively forgiving to minor changes in the activity of the chromatographic adsorbants. Previously-described methods for the separation of thiophenes, based on ligand exchange chromatography ($\underline{13,23}$–$\underline{27}$), involve rather exacting chromatography since the difference in chromatographic behaviour between the thiophenes and the various aromatic hydrocarbons with which they usually occur is slight.

The results of the analyses for the thiophenes for several petroleums are summarized in Table I. The thiophene contents vary from 0.3% to 7.6%. The GC-FID chromatograms for a few samples are displayed in Figure 3, where the peaks corresponding to dibenzothiophene and the isomeric monomethyldibenzothiophenes are indicated. Peak assignments are based on GC/MS determination of the molecular weight and the previously-assigned retention order of the isomeric monomethyldibenzothiophenes ($\underline{26}$). The numbering system of dibenzothiophene is as follows:

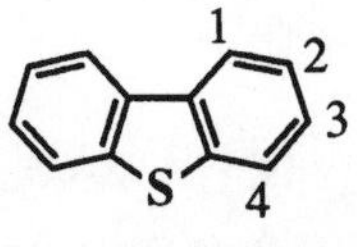

As may be seen from Figure 3, there is considerable difference in the general appearance of these GC-FID traces from the various petroleums. In the Syncrude sample the quantities of dibenzothiophene and monomethyldibenzothiophenes are low and most of the dibenzothiophenes have a $\geq C_2$ alkyl side chain(s) attached to the ring system. This contrasts sharply with the Bellshill Lake sample where dibenzothiophene, mono- and dimethyldibenzothiophenes dominate the mixture. With

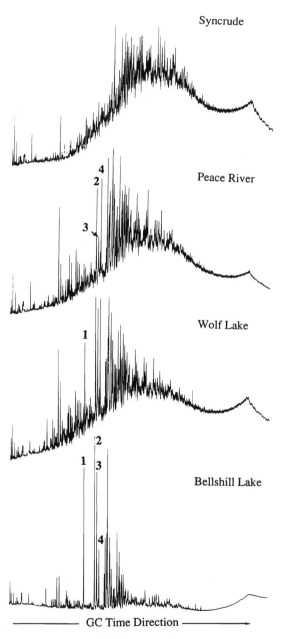

Figure 3. GC-FID chromatograms for thiophene compound fractions from Alberta petroleums. The peak labels are as follows: 1. dibenzothiophene, 2. 4-methyldibenzothiophene, 3. 2- and 3-methyldibenzothiophene and 4. 1-methyldibenzothiophene. Samples are arranged in order of their depth of burial. Note the shift toward lower molecular weight compounds and a reduction in the amount of the unresolved complex mixture with increasing depth of burial. (Reproduced from Reference 34. Copyright 1989, American Chemical Society.)

increasing thermal maturation, much of the unresolved complex mixture of polyalkylated dibenzothiophenes is converted into a limited number dibenzothiophenes with short side chains. Benzothiophenes are represented by the complex series of peaks which elute prior to dibenzothiophene in the chromatograms of Figure 3. As may be seen, the abundance of the benzothiophenes relative to the dibenzothiophenes declines with increasing depth of burial.   Also, the abundance of 1-methyldibenzothiophene relative to the 2-, 3- and 4-methyldibenzothiophene isomers declines with increasing depth of burial.

Acknowledgment

The financial support of the Alberta Oil Sands Technology and Research Authority, the Natural Sciences and Engineering Research Council of Canada and Imperial Oil Ltd. is gratefully acknowledged.

Literature Cited
1.   Smith, H.M. U.S. Bur. Mines Bull. 1968, 642, 1–136.
2.   Rall, H.T.; Thompson, C.J.; Coleman, H.J.; Hopkins, R.L. U.S. Bur. Mines Bull. 1972, 659, 1-187.
3.   Gal'pern, G.C. Int. J. Sulfur Chem. 1971, B6, 115–130.
4.   Gal'pern, G.C. Russian Chem. Rev. 1976, 45, 701–720.
5.   Gal'pern, G.C. In The Chemistry of Heterocyclic Compounds. Thiophene and its Derivatives;Vol. 1; Gronowitz, S., Ed.; J. Wiley & Sons; 1985; pp 325-351.
6.   Orr, W.L. In Oil Sand and Oil Shale Chemistry; Strausz, O.P., Lown, E.M., Eds.; Verlag Chemie; 1978; pp 223–243.
7.   Orr, W.L. Org. Geochem. 1986, 10, 499–516.
8.   Payzant, J.D.; Montgomery, D.S.; Strausz, O.P. Tetrahedron Lett. 1983, 24, 651–654.
9.   Payzant, J.D.; Montgomery, D.S.; Strausz, O.P. Org. Geochem. 1986, 9, 357–369.
10.   Payzant, J.D.; Cyr, T.D.; Montgomery, D.S.; Strausz, O.P. Tetrahedron Lett. 1985, 26, 4175–4178.
11.   Payzant, J.D.; Cyr, T.D.; Montgomery, D.S.; Strausz, O.P. In Geochemical Biomarkers; Yen T.F.; Moldowan J.M., Eds.; Harwood Academic; 1988; pp 133–147.
12.   Cyr, T.D.; Payzant, J.D.; Montgomery, D.S.; Strausz, O.P. Org. Geochem. 1986, 9, 139–43.
13.   Payzant, J.D.; Montgomery, D.S.; Strausz, O.P. AOSTRA J. Res. 1988, 4, 117–131.
14.   Valisolalao, J.; Perakis, N.; Chappe, B.; Albrecht, P. Tetrahedron Lett. 1984, 25, 1183–1186.
15.   Brassell, S.C.; Lewis, C.A.; de Leeuw, J.W.; de Lange, F.; Sinninghe Damsté, J.S. Nature 1986, 320, 160–162.
16.   Sinninghe Damsté, J.S.; ten Haven, H.L.; de Leeuw, J.W.; Schenck, P.A. Org. Geochem. 1986, 10, 791–805.
17.   Sinninghe Damsté, J.S.; de Leeuw, J.W.; Kock-van Dalan, A.C.; de Zeeuw, M.A.; de Lange, F.; Rijpstra, W.I.C.; Schenck, P.A. Geochimica et Cosmochimica Acta 1987, 51, 2369–2391.
18.   de Leeuw, J.W. In Organic Marine Geochemistry; Sohn, M.L., Ed.; American Chemical Society: Washington; 1986; pp 33–61.
19.   Schmid, J.C.; Connan, J.; Albrecht, P. Nature 1987, 329, 54–56.
20.   Strausz, O.P.; Lown, E.M.; Payzant, J.D. "The Nature and Geochemistry of Sulfur-Containing Compounds in Alberta Petroleums", this volume.
21.   George, G.N.; Gorbaty, M.L. J. Am. Chem. Soc. 1989, 111, 3182–3186.

22. Takata, T.; Ishibashi, K.; Ando, W. Tetrahedron Lett. 1985, 26, 4609–4612.
23. Joyce, W.F.; Uden, P.C. Anal. Chem. 1983, 55, 540–543.
24. Nishioka, M.; Campbell, R.M.; Lee, M.L.; Castle, R.N. Fuel 1986, 65, 270–273.
25. Willey, C.; Iwao, M.; Castle, R.N.; Lee, M.L. Anal. Chem. 1981, 53, 400–407.
26. Kong, R.C.; Lee, M.L.; Iwao, M.; Tominaga, Y.; Pratap, R.; Thompson, R.D.; Castle, R.N. Fuel 1984, 63, 702–708.
27. Nishioka, M.; Lee, M.L.; Castle, R.N. Fuel 1986, 65, 390–396.
28. Raasch, M.S. In The Chemistry of Heterocyclic Compounds. Thiophene and its Derivatives; Vol. 1; Gronowitz, S., Ed.; J. Wiley & Sons; 1985; pp 571–627.
29. Schank, K. In The Chemistry of Sulphones and Sulphoxides; Patai, S.; Rappoport, Z.; Stirling, C., Eds.; J. Wiley & Sons; 1988; pp 165–232.
30. Block, E. In Reactions of Organosulfur Compounds; Academic press; 1978; p 16.
31. Mukherjee, D.; Dunn, L.C.; Houk, K.N. J. Am. Chem. Soc. 1979, 101, 251–252.
32. Weber, W.P.; Stromquist, P.; Ito, T.I. Tetrahedron Lett. 1974, 30, 2595–2598.
33. Madesclaire, M. Tetrahedron, 1986, 42, 5459–5495.
34. Payzant, J.D.; Mojelsky, T.M.; Strausz, O.P. Energy and Fuels 1989, 3, 449–454.
35. Payzant, J.D.; McIntyre, D.D.; Mojelsky, T.M.; Torres, M.; Montgomery, D.S.; Strausz, O.P. Org. Geochem. 1989, 14, 461–473.
36. Fedorak, P.M.; Payzant, J.D.; Montgomery, D.S.; Westlake, D.W.S. Appl. Environ. Microbiol. 1988, 54, 1243–1248.
37. Payzant, J.D.; Hogg, A.M.; Montgomery, D.S.; Strausz, O.P. AOSTRA J. Res. 1985, 1, 183–203.
38. Bordwell, F.G.; McKellin, W.H. J. Am. Chem. Soc. 1951, 73, 2251–2253.
39. Lee, M.L.; Willey, C.; Castle, R.N.; White, C.M. In Polynuclear Aromatic Hydrocarbons: Chemistry and Biological Effects, Bjørseth A., Dennis A.J., Eds.; Battelle; 1980; pp 59–73.

RECEIVED April 17, 1990

Chapter 6

Microbial Metabolism of Organosulfur Compounds in Petroleum

Phillip M. Fedorak

Department of Microbiology, University of Alberta, Edmonton, Alberta T6G 2E9, Canada

The microbial metabolism of a large number of hydrocarbons has been thoroughly studied and described in the literature. In contrast, much less is known about the metabolism of organosulfur compounds found in petroleum. These investigations have been hampered by the commercial unavailability of most of the sulfur compounds of interest. Some of the first studies on the biodegradation of sulfur compounds focused on their fate and removal from oil-contaminated environments. Another major area of study is the microbial process of "biodesulfurization" which has been suggested as a means of selectively removing sulfur compounds from petroleum prior to refining. Information gathered from these two areas of research provide the basis of the present knowledge of metabolism of organosulfur compounds in petroleum. This information is reviewed with emphasis on the metabolism of dibenzothiophenes and n-alkyl tetrahydrothiophenes (n-alkyl thiolanes).

Microorganisms are capable of metabolizing an extremely wide variety of organic compounds. These include naturally occurring compounds derived from biochemical syntheses and many synthetic chemicals which have been released into the environment over the last few decades. In many cases, as demonstrated experimentally, the complete degradation of an organic compound leading to release of much of the carbon as carbon dioxide, a process known as mineralization, can be accomplished by an individual microbial species - i.e. a pure culture. In other cases, a number of different types of microorganisms - i.e. a mixed culture or consortium - must work together to mineralize an organic compound. In the latter situation, the metabolic endproducts of one microbe serve as substrates for other microbes in the population. Some organic compounds will not serve as a carbon or energy sources. However, these may be transformed in the presence of other organics which serve as primary energy source. This phenomenon is known as co-metabolism.

To demonstrate microbial metabolism of a specific compound, the compound of interest (substrate) is added to a microbial culture and incubated under suitable conditions. At various times, the culture and suitable abiotic controls are analyzed to determine whether there has been a decrease in the amount of the substrate due to

0097–6156/90/0429–0093$06.00/0

microbiological activity. If this occurred, the culture may be further examined to detect and identify intermediates and endproducts.

Heterotrophic microorganisms obtain energy from the oxidation of organic compounds. To grow on these substrates, there must be a reducible compound or ion, known as a terminal electron acceptor, available for the ultimate disposal of the electrons from these oxidations. Molecular oxygen serves as the terminal electron acceptor under aerobic conditions. In the absence of oxygen (anaerobic conditions), nitrate, sulfate, carbon dioxide or ferric iron may serve as terminal electron acceptors for specific microorganisms or microbial populations. The energy yield from aerobic oxidation of organic compounds is much greater than that from anaerobic oxidations. Therefore the aerobic cultures metabolize substrates and grow much faster than anaerobic cultures. Hence most studies of the metabolism of compounds found in petroleum have used aerobic culture conditions.

The amount of information on the microbial metabolism of hydrocarbons is much greater than the amount of information on the metabolism of organosulfur compounds (OSC). As work is progressing with studies on the latter group of compounds, results are showing that there are many similarities between the biodegradation pathways of OSC and those of hydrocarbons with similar structures. Therefore, included below is a short section describing some aspects of hydrocarbon metabolism.

Metabolism of Hydrocarbons

Aerobic hydrocarbon metabolism has been well studied and reviewed by an number of authors (1-6). Two general pathways are shown in Figures 1 and 2. The initial oxidation of alkanes commonly occurs on the terminal carbon yielding an alcohol (Figure 1). This is further oxidized to an aldehyde and then to a carboxylic acid which undergoes a series of β-oxidations. Each β-oxidation yields acetic acid and a carboxylic acid with two fewer carbon atoms than the previous acid.

The aerobic metabolism of aromatic hydrocarbons by bacteria leads to ring cleavage and mineralization. In contrast, fungi generally hydroxylate aromatic hydrocarbons without further degradation (2, 4). Figure 2 shows the initial oxidation of benzene by aerobic bacteria producing a 1,2-dihydroxy aromatic compound (e.g. catechol) which is a key intermediate for subsequent ring cleavage by either ortho or meta fission. Diols are common intermediates in the degradation of polycyclic aromatics. For example, 1,2-dihydroxynaphthalene, 3,4-dihydroxyphenanthrene and 1,2-dihydroxyanthracene are intermediates in the bacterial metabolism of naphthalene, phenanthrene, and anthracene, respectively.

Until quite recently, hydrocarbons were thought to be resistant to microbial metabolism under anaerobic conditions. Although there has been no unambiguous demonstration that alkanes and alicyclic hydrocarbons can be metabolized in the absence of molecular oxygen, there have been numerous recent reports of aromatic hydrocarbon degradation under anaerobic conditions. Much of this work has been summarized by Grbic´-Galic´ (7, 8). In brief, benzene, toluene, ethylbenzene, styrene, o-xylene and naphthalene have been shown to be mineralized by methanogenic consortia using carbon dioxide as the terminal electron acceptor. Using medium containing ^{18}O-labeled water, Vogel and Grbic´-Galic´ (9) showed that the initial oxidations of benzene and toluene gave phenol and p-cresol, respectively, with the oxygen originating from water. In addition, there have been recent reports on the anaerobic metabolism of several aromatics compounds under denitrifying conditions where nitrate serves as the terminal electron acceptor. The biodegradation of benzene (10), toluene, m- and p-xylenes (11), naphthalene and acenaphthene (12) has been demonstrated under these conditions.

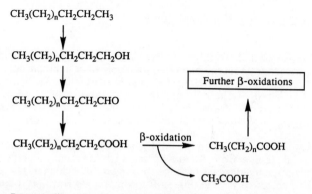

Figure 1. Intermediates in the aerobic microbial metabolism of *n*-alkanes by a terminal oxidation and subsequent β-oxidations.

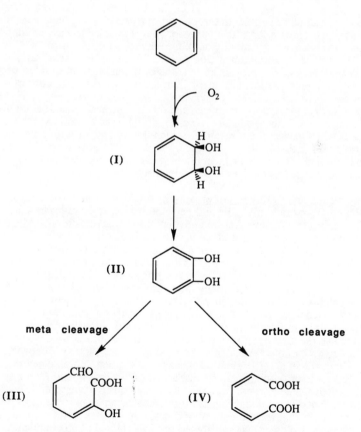

Figure 2. Intermediates in the aerobic bacterial metabolism of benzene. (I): *cis*-benzene dihydrodiol, (II): catechol, (III): 2-hydroxymuconic semialdehyde, (IV): *cis, cis-* muconic acid.

Metabolism of Organosulfur Compounds

Petroleum contains a wide variety of OSC (13-17). However, in comparison to the amount of knowledge about the microbial metabolism of hydrocarbons, much less is known about the metabolism of OSC found in petroleum. These investigations have been hampered by the commercial unavailability of most of the sulfur compounds of interest. Because only a small number of OSC are available to workers studying microbial metabolism, few biodegradation pathways have been elucidated.

Some of the first studies on the biodegradation of sulfur compounds focused on their fate and removal from oil-contaminated environments. Another major area of study is the microbial process of "biodesulfurization" which has been suggested as a means of selectively removing sulfur compounds from petroleum prior to refining. The biodesulfurization strategy has been to have the microbes oxidize the OSC yielding water-soluble products which can be separated from the petroleum. The subject has been recently reviewed by Foght, J.M.; Fedorak, P.M.; Gray, M.R.; Westlake, D.W.S. (Microbial desulfurization of liquid fossil fuels. In: Microbial Mineral Recovery, Ehrlich, H.L. Ed., in press). In addition, comparisons of the composition of petroleum samples from reservoirs that have been subjected to biodegradation with those that have not been subjected to biodegradation have indicated the relative susceptibility of OSC to microbial metabolism.

This article reviews information gathered from these areas of research which provides the bulk of our present knowledge of metabolism of OSC in petroleum. The order of presentation is that of increasing experimental control over the microbial activities. First, observations of the persistence of OSC in petroleum-contaminated environments and their occurrence in petroleum reservoirs will be presented. In these situations, we exert little or no control over the microbial activities which occur. Next, results from laboratory studies with crude petroleum or petroleum fractions will be presented. Finally, laboratory work with pure OSC will be summarized. Using the latter controlled conditions, metabolic intermediates and pathways for a few OSC have been elucidated.

Environmental and Reservoir Observations. Evidence from the chemical analyses of samples from crude oil spills and petroleum reservoirs suggests that the OSC vary in susceptibility to metabolism under environmental conditions. Metabolism is slow and selective, and the observed variability is due to the microbial population, ambient conditions and physical/chemical properties of the petroleum.

Samples from two major oil spills that occurred in the marine environment in the late 1970's were analyzed with specific reference to the sulfur heterocycles in the aromatic fraction. The Brittany coast of France heavily contaminated by 66 million gallons of oil from the *Amoco Cadiz* wreckage, and the open waters of the Gulf of Mexico received 140 million gallons of crude oil from the Ixtoc 1 blowout. A detailed chemical study of the fate of petroleum from the *Amoco Cadiz* spill revealed that the mixture of residual petroleum in the littoral zone was enriched in complex hydrocarbon components and aromatic OSC such as dibenzothiophenes (18). Gundlach et al. (19) reported that the alkyldibenzothiophenes persisted as major aromatic molecular markers three years after the *Amoco Cadiz* spill. This enrichment appeared to result from selective metabolism of susceptible compounds, suggesting that dibenzothiophenes are more recalcitrant than some other petroleum components. In contrast, enrichment of dibenzothiophenes in the mousse generated from the Ixtoc 1 blowout was attributed to physical and photochemical processes (20) with no evidence for microbial degradation playing a role in the enrichment. Because of their persistence in the environment, Friocourt et al. (21) has suggested the use of dibenzothiophenes as markers of oil pollution.

The biodegradation of crude oils in their reservoirs is well documented (14, 22). It occurs in the presence of meteoric water which supplies dissolved oxygen and nutrients including phosphate and fixed nitrogen. Microenvironments may exist in which aerobic and anaerobic activities occur in close proximity so that intermediates of aerobic metabolism may become substrates for anaerobic bacteria. In reservoirs, microbes are most active at the oil-water interface and at temperatures between about 20° and 60 to 75°C.

Deroo et al. (23) observed that the decreasing n-paraffinic and isoprenoid contents of western Canadian crude oils were paralleled by decreasing thiophenic content of aromatic fractions. They attributed these losses to *in situ* biodegradation and water washing of reservoir oils. Westlake (24) examined the saturates, aromatics and sulfur heterocycles of three Kumak crude oils by capillary gas chromatography (GC). These oils from the Canadian Beaufort Basin came from depths of 4439, 7048 and 7566 ft (25). The shallow oil had undergone extensive biodegradation and contained no n-alkanes (24). The oil from the middle depth had completely lost the n-alkanes up to C_{16} and partially lost n-alkanes up to C_{20}. The reservoir temperature at 7566 ft was 67°C (25) and the oil showed no evidence of biodegradation. GC analyses with a sulfur-specific detector showed that the two deepest samples contained a variety of C_1- and C_2-benzothiophenes, dibenzothiophene, C_1-, C_2- and C_3-dibenzothiophenes (24). However, in the shallow oil, there was a marked decrease in the amounts of alkylbenzothiophenes present and dibenzothiophene was absent.

Williams et al. (26) studied a number of South Texas oils of a common genetic type which exhibited a broad range of *in situ* biodegradation. Although the benzothiophenes and dibenzothiophenes were present in extremely low concentrations, the authors clearly observed that these compounds were susceptible to biodegradation. Their chromatograms showed that the unsubstituted parent compounds, benzothiophene and dibenzothiophene, were more readily degraded than most of the corresponding alkyl substituted compounds and that in general, the C_1-benzothiophenes were removed to a greater extent than the C_2-benzothiophenes.

A series of n-alkyl substituted tetrahydrothiophenes (thiolanes) and thiacyclohexanes (thianes) shown in Figure 3 have been isolated and identified from non-biodegraded petroleums (27). These n-alkyl monocyclic sulfides were not found in petroleums which had been subjected to biodegradation in their reservoirs. These observations suggested that this group of compounds is biodegradable.

Laboratory studies with petroleum and petroleum fractions. The first report of microbial oxidation of sulfur heterocycles was published by Walker et al. (28). Using computerized mass spectrometry (MS) to analyze the residual oil extracted from aerobically-grown mixed cultures, they observed losses of 40% of the dibenzothiophenes and 50% of the naphthobenzothiophenes. They concluded that the sulfur-containing aromatics were approximately twice as resistant to microbial degradation as their hydrocarbon analogues.

The degradation of sulfur heterocycles in Prudhoe Bay crude oil by aerobic enrichment cultures of marine and soil microorganisms was studies by Fedorak and Westlake (29, 30). Figure 4 shows typical results obtained when the aromatic fraction of the oil was analyzed by capillary GC with a sulfur-specific detector. The mixed microbial populations in these enrichments were able to degrade the n-alkanes, pristane, phytane and a number of polycyclic aromatics in this oil (31, 32). There were some similarities between the biodegradability of the sulfur heterocycles in Figure 4 and that of the aromatic hydrocarbons reported by Fedorak and Westlake (31, 32). For example, although cultures without N and P addition were able to remove a number of the sulfur heterocycles from the oil, the addition

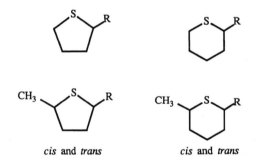

cis and trans   cis and trans

Figure 3. Structures of n-alkylmonocyclic sulfides found in the sulfide fraction of Bellshill Lake crude oil. R=n-alkyl C_{10} to at least C_{30}.

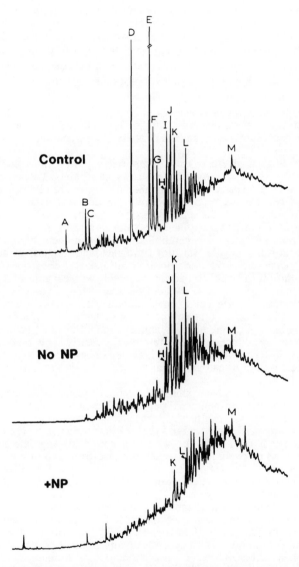

Figure 4. GC analysis of the sulfur heterocycles in the aromatic fraction of Prudhoe Bay crude oil using a sulfur-specific detector. The oil was extracted from cultures inoculated with greenhouse soil and incubated aerobically for 27 days at 20°C. Control = autoclaved soil in mineral medium; No NP = soil in mineral medium without N and P added; +NP = soil in mineral medium with N and P added. Peak identification: A = C_2-benzothiophene; B, C= C_3-benzothiophenes; D = dibenzothiophene (DBT); E, F, G = C_1- DBTs; H, I, J, K = C_2-DBTs; L = C_3- DBT; M = benzonaphthothiophene. (Reproduced with permission from Ref. 30. Copyright 1984 Kluwer Academic Publishers.)

of N and P increased the variety and amounts of these compounds which were removed. Also, the resistance toward microbial attack increased with the increased number of alkyl carbons in the sulfur-containing compound. For the benzothiophene series, the C_2-compound was attacked before the C_3-compounds. For the dibenzothiophene series, dibenzothiophene was attacked before the C_1-substituted compounds which were attacked before the C_2-compounds which were attacked before the C_3-compound (30). The same trend was observed for the alkyl substituted naphthalenes and phenanthrenes (31, 32). This order of removal of sulfur heterocycles in these laboratory studies is consistent with the loss of these compounds in petroleums which have undergone biodegradation in their reservoirs (24, 26).

Recently Foght and Westlake (33) reported a pure culture study in which *Pseudomonas* sp. HL7b was grown aerobically on Prudhoe Bay crude oil as its sole carbon source. This bacterium was unable to grow on the aliphatic compounds or dibenzothiophene. However, a number of aromatic hydrocarbons supported its growth. In addition, many of the sulfur heterocycles were removed from the oil by HL7b yielding a GC profile similar the the bottom chromatogram in Figure 4.

To determine if microbial metabolism may have been the reason for the absence of *n*-alkyl monocyclic sulfides (thiolanes and thianes) in biodegraded petroleums (27), the sulfide fraction from Bellshill Lake oil was isolated and added to a pure culture of an aerobic bacterium which was capable of growing on 2-*n*-dodecyltetrahydrothiophene (2-*n*-dodecylthiolane) (34). Figure 5 compares the chromatograms of the extracted oil from the culture after 14 days of incubation with that of the original sulfide fraction and that of the thiourea non-adduct of the sulfide fraction. In the latter case, the *n*-alkyl monocyclic sulfides had been chemically removed by thiourea adduction (27). These results clearly show that the bacterial culture was able to degrade the mixture of *n*-alkyl monocyclic sulfides isolated from a crude oil yielding a chromatogram that was essentially the same as that of the thiourea nonadduct.

Eckart and co-workers have published a series of papers on laboratory studies of biodesulfurization of petroleum and petroleum fractions. The ability of various aerobic mixed cultures to desulfurize Romashkino crude oil (1.69 wt.% S) was addressed by Eckart et al. (35). After 5 days of incubation at 30°C in sulfur-free mineral medium with oil as sole source of carbon and sulfur, approximately 55% of the total sulfur was recovered in the aqueous phase from two of the most active cultures. In another study, gas oil (1.2 to 2 wt.% S), vacuum distillates (1.8 to 2 wt.% S) and fuel oil (up to 4 wt.% S) were used as sole carbon and sulfur sources for the oil-degrading microorganisms (36). The addition of an emulsifying agent was required to enhance desulfurization. Sulfur removals of up to 20% from the gas oil, 5% from the vacuum distillates, and 25% from the fuel oil were observed after 5 to 7 days of incubation. In a later study (37), approximately 30% of the sulfur was removed from fuel-D-oil by a mixed population of bacteria. The removal of benzothiophene, dibenzothiophene and naphthobenzothiophene was shown by high resolution MS analysis. Hydrocarbon degradation was observed in each of these studies. For example, in the latter study with fuel-D-oil , the decreases in the *n*-alkane and aromatic content were 59% and 14%, respectively.

Eckart et al. (38) investigated the biodesulfurization of Romashkino crude oil under anaerobic conditions. Mixed cultures, in which *Desulfovibrio* spp. were the predominant organisms, were grown with lactate as a source of carbon and energy under conditions where sulfate served as the terminal electron acceptor. When the pH of the medium was controlled at 7.2, desulfurization of 26 to 40% was observed after 48 h. Emulsifying agents were not required in these cultures.

Studies with Individual Organosulfur Compounds

The results in Figure 4 clearly show that a number of alkyl benzothiophenes and dibenzothiophenes can be metabolized (or co-metabolized) by microorganisms and the biodesulfurization studies suggest that other OSC can undergo biodegradation. Yet, the metabolism of only a few OSC has been studied in detail.

Thiophenes. Although several attempts have been made to demonstrate the aerobic degradation of thiophene, none have succeeded (39-41). However, the anaerobic metabolism of thiophene was reported by Kurita et al. (42). Hydrogen sulfide was released from thiophene when bacterial cultures, obtained from oil sludges, were grown with polypeptone. The fate of the carbon atoms was not determined but they could not be found as lower hydrocarbons (C_1 to C_4).

Several carboxythiophenes are metabolized by aerobic microorganisms (39-41, 43) but these will not be discussed because they are not commonly found in petroleums. Sagardía et al. (43) reported the metabolism of 2-methylthiophene and 3-methylthiophene by a pure culture of *Pseudomonas aeruginosa*. They observed a loss of these substrates from the medium with a concurrent consumption of molecular oxygen. However, no intermediates were identified.

Fedorak et al. (34) studied the aerobic metabolism of n-alkyl tetrahydrothiophenes (thiolanes) which have been found in non-biodegraded petroleums. Two representative compounds, 2-n-dodecyltetrahydrothiophene (DTHT) and 2-n-undecyltetrahydrothiophene (UTHT) were synthesized and fed to pure cultures of n-alkane-degrading bacteria. Figures 6 and 7 show the structures of the parent compounds and the metabolites found in the cultures. In both cases, trace amounts of the corresponding sulfones and sulfoxides were detected. However, the major products were carboxylic acids which resulted from the initial attack on the alkyl side chain as illustrated in Figure 1. This mode of attack was the same as observed for n-dodecylbenzene metabolism by bacteria (44) and fungi (45). Five β-oxidations of the side chain of DTHT would yield 2-tetrahydrothiopheneacetic acid (Figure 6). Similarly, four and five β-oxidations of the side chain of UTHT would yield 3-(2-tetrahydrothiophenyl)propanoic and 2-tetrahydrothiophenecarboxylic acids, respectively (Figure 7). Although the fate of the carboxylic acid-containing metabolites was not unequivocally determined, they were assumed to have undergone ring cleavage because total molar amounts of tetrahydrothiophene-containing compounds recovered from the cultures were always much less than the amount of substrate added.

Benzothiophene. There have been several reports on the aerobic co-metabolism of benzothiophene (43, 46, 47). Bohonos et al. (46) identifed some benzothiophene metabolites by GC-MS and the structures of these are shown in Figure 8. Although the only compounds found were oxidized on the thiophene ring, they could not exclude the possibility of oxidation of the benzene ring. Finnerty et al. (48) found that benzothiophene was transformed to unidentified water-soluble products by a dibenzothiophene-oxidizing bacterium.

Benzothiophene has been shown to be susceptible to microbial metabolism under anaerobic conditions. Kurita et al. (42) observed the release of hydrogen sulfide from this substrate but they did not determine the fate of the carbon atoms. In studies to determine the extent of biodesulfurization of OSC, Köhler et al. (49) added benzothiophene dissolved in paraffin oil to cultures of *Desulfovibrio* sp. After 6 days of incubation, the amounts of desulfurization ranged from 3.1 to 9.2%.

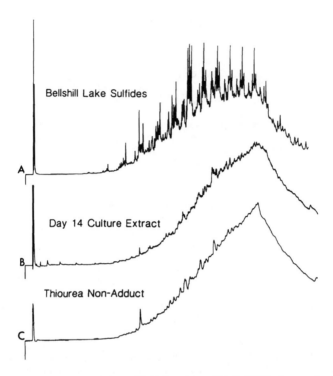

Figure 5.   GC analysis of the sulfide fraction of Bellshill Lake crude oil using a sulfur-specific detector. (A) Unaltered sulfides; (B) residue from a bacterial culture incubated aerobically for 14 days with the sulfide fraction; (C) thiourea nonadduct of sulfide fraction. (Reproduced with permission from Ref. 34. Copyright 1988 American Society for Microbiology.)

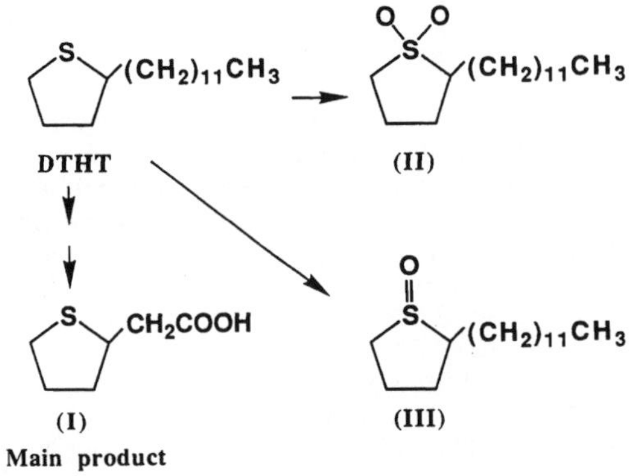

Figure 6. Metabolites found in aerobic bacterial cultures growing on 2-*n*-dodecyltetrahydrothiophene (DTHT). (I): 2-tetrahydrothiopheneacetic acid, (II): DTHT-sulfone, (III): DTHT- sulfoxide.

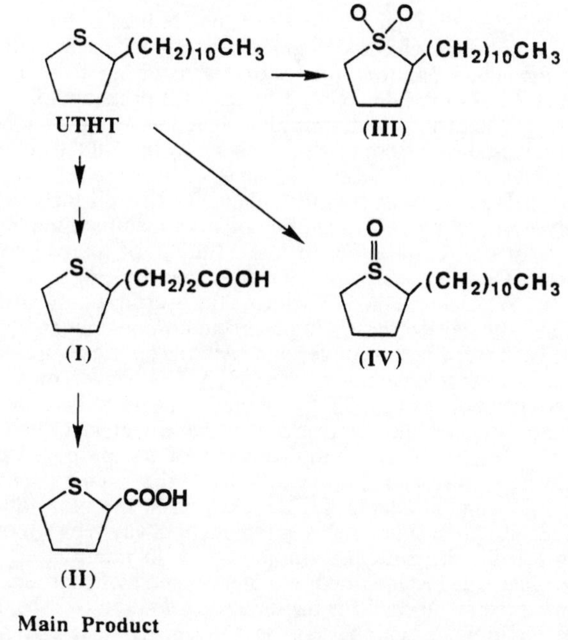

Figure 7. Metabolites found in aerobic bacterial cultures growing on 2-*n*-undecyltetrahydrothiophene (UTHT) (I): 3-(2-tetrahydrothiophenyl)propanoic acid, (II): 2-tetrahydrothiophenecarboxylic acids (III): UTHT-sulfone, (IV): UTHT-sulfoxide.

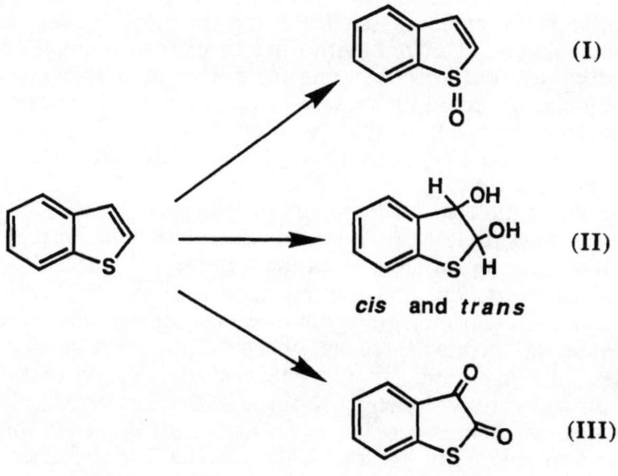

Figure 8. Oxidation products found in aerobic cultures growing on naphthalene and co-metabolizing benzothiophene. (I): benzothiphene-sulfoxide, (II): 2,3-dihydrobenzothiophene-2,3-diol   (III): 2,3-dihydrobenzothiophene-2, 3-dione.

Grbic'-Galic' (8) identified a number of metabolites in an aquifer-derived methanogenic consortium which could degrade benzothiophene and Figure 9 shows some of the cyclic intermediates. Intermediates resulting from ring cleavage included 2-methyl-2-hexanol, 3-hexenol, 2-methyl-1,2-propanediol, thiopropionic acid, hexanoic acid, 2-hexenoic acid, propionic acid and acetic acid. Methane and carbon dioxide were produced from benzothiophene by this mixed culture.

Dibenzothiophenes. Because of its commercial availability, dibenzothiophene is the most extensively used compound in studies of organosulfur metabolism. It has been used as a model compound in the studies of petroleum and coal biodesulfurization. These topics have been reviewed by Foght, J.M.; Fedorak, P.M.; Gray, M.R.; Westlake, D.W.S. (Microbial desulfurization of liquid fossil fuels. In: Microbial Mineral Recovery, in press) and Monticello and Finnerty (50).

There have been a few reports of aerobic bacterial species which will grow on dibenzothiophene as their sole carbon source (51, 52). However, most studies have focused on its co-metabolism (33, 53, 54) which appears to be a more common phenomenon. Intermediates of the biodegradation have been identified in a number of studies (55-57). Figure 10 is a simplification of the proposed pathway for dibenzothiophene metabolism showing the two endproducts: 3-hydroxy-2-formylbenzothiophene (HFBT), which is the more abundant, and dibenzothiophene-sulfoxide. [For a more complete pathway see (56) or (58)]. This pathway shows 1,2-dihydroxydibenzothiophene as an intermediate and this is analogous to the diols found as intermediates of aromatic hydrocarbons.

The enzyme systems needed for bacterial degradation of dibenzothiophene usually require an inducer which may be the substrate itself, naphthalene, anthracene or salicylate (59, 60). However, Pseudomonas sp. HL7b co-metabolizes dibenzothiophene constitutively (33). That is, a glucose-grown suspension of HL7b cells immediately started to oxidize dibenzothiophene when it was added to the medium and yielded HFBT among other metabolites.

None of the pure cultures that produced HFBT have been shown to further metabolize this compound. Bohonos et al. (46) found two further oxidation products, 3-hydroxybenzothiophene and 2,3-dihydrobenzothiophene-2,3-dione in aerobic mixed cultures co-metabolizing dibenzothiophene. Recently, Mormile and Atlas (61) inoculated portions of the filter-sterilized supernatant from a dibenzothiophene-degrading culture with soil and sediment samples and observed the loss of HFBT using a spectroscopic method. Under their aerobic growth conditions, they also observed the release of carbon dioxide from these cultures indicating that these products of dibenzothiophene degradation can be further oxidized. In addition, they observed carbon dioxide production from dibenzothiophene-sulfoxide.

Working with a mutated bacterial strain, Isbister et al. (62) demonstrated a novel mechanism of aerobic oxidation of dibenzothiophene which involved the specific excision of the sulfur atom from the molecule (Figure 11). Studies with ^{35}S-labeled dibenzothiophene showed the release of the radioactivity into the aqueous phase and ion chromatography showed the appearence of sulfate. There was no radioactive carbon dioxide released when this microorganism was incubated with ^{14}C-labeled dibenzothiophene. GC-MS analysis showed that the oxidation product was 2,2'-dihydroxybiphenyl. Kargi and Robinson (52) also report the release of sulfate from dibenzothiophene. This OSC served as the sole carbon and sulfur source in their cultures of the aerobic thermophile Sulfolobus acidocaldarius.

A few investigations have considered the anaerobic metabolism of dibenzothiophene. In biodesulfurization studies, Köhler et al. (49) dissolved this

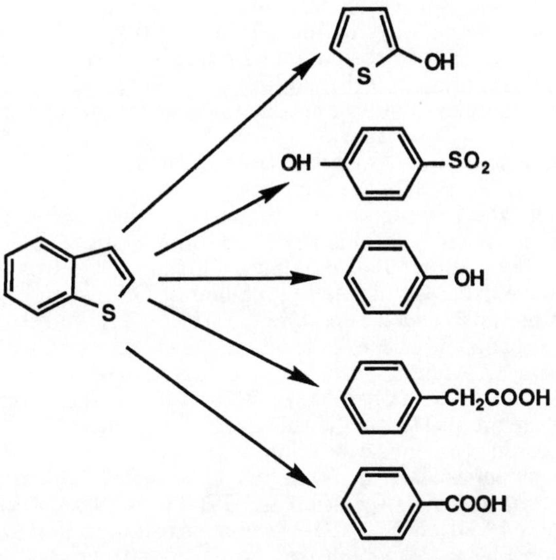

(I) (II) (III) (IV) (V)

Figure 9. Some aromatic intermediates of benzothiophene metabolism found in an anaerobic methanogenic consortium. (I): 2-hydroxythiophene, (II): *p*-hydroxybenzenesulfonic acid, (III): phenol, (IV): phenylacetic acid, (V): benzoic acid.

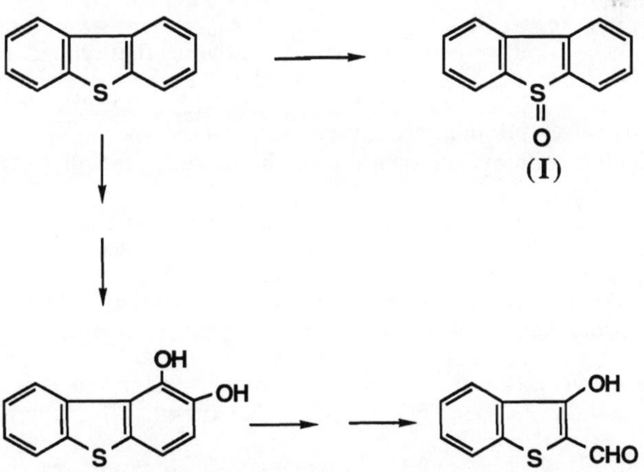

(I) (II) (III)

Figure 10. Simplification of the proposed pathways from dibenzothiophene co-metabolism by pure cultures of aerobic bacteria. (I): dibenzothiophene-sulfoxide, (II): 1,2-dihydroxydibenzothiophene, (III): 3-hydroxy-2-formylbenzothiophene.

compound in paraffin oil and added this solution to cultures of *Desulfovibrio* spp. Under conditions where sulfate was the terminal electron acceptor, they observed 15.9% desulfurization after 6 days of incubation. Maka et al. (63) enriched dibenzothiophene-metabolizing mixed cultures from soil and sewage sludge. They observed 60 to 80% removal of dibenzothiophene within 2 weeks at 37°C. However, neither the identities of the intermedates nor the terminal electon acceptor were given.

Although studies of biodegraded petroleum have indicated that alkyl dibenzothiophenes can be metabolized by microorganisms, this has not been demonstrated without the presence of oil. Some preliminary investigations on 4-methyldibenzothiophene co-metabolism by *Pseudomonas* strain HL7b have been done (Fedorak and Payzant, unpublished data). This substrate was synthesized from dibenzothiophene using the metalation method of Gilman and Jacoby (64). Based on GC analysis, this yielded a mixture of 71% 4-methyldibenzothiophene and 29% unreacted dibenzothiophene. This mixture was incubated with the HL7b cells in phosphate buffer. After acidification, the supernatant was extracted with methylene chloride and this was back extracted with 5% sodium bicarbonate solution to separate the products from the substrates. Subsequent acidification and extraction of the sodium bicarbonate solution with methylene chloride gave two sulfur-containing compounds when analyzed by GC (Figure 12). Compound 1 had the same retention time and mass spectrum as HFBT which had previously been isolated from a culture of HL7b (33). CG-MS analysis showed that Compound 2 gave a molecular ion m/z = 192 which was 14 mass units greater than HFBT. These results are consistent with Compound 2 being methyl-HFBT. Studies on the bacterial metabolism of methyl-substituted aromatic hydrocarbons such as 1-methyl-naphthalene (65), 2-methylnaphthalene (66), and 3-methyl- and 4-methyl-biphenyls (67) have shown that the unsubstituted ring is preferentially oxidized.

Based on the peak areas from the GC analysis, HFBT accounted for 58% of the two products recovered and methyl-HFBT accounted for 42% (Fedorak and Payzant, unpublished data). This proportion was quite different than that of the added substrates suggesting that the methyl-substituted compound was less susceptible to the biotransformation or that both the methylated ring and the nonmethylated benzene ring of 4-methyldibenzothiophene were susceptible to cleavage yielding a mixture of HFBT and methyl-HFBT for this one substrate. None-the-less, this preliminary study suggested that microbial metabolism of isomers of C_1-dibenzothiophene should yield the isomers of methyl-HFBT.

Dibenzylsulfide. Figure 13 shows the identified products of aerobic and anaerobic metabolism of dibenzylsulfide. Babienzien et al. (68) incubated this substrate with a mixed population of aerobic bacteria and detected several water-soluble organic products. However, only benzylmercaptoacetic acid was identified. Köhler et al. (49) incubated dibenzylsulfide with a selected strain of *Desulfovibrio* utilizing molecular hydrogen under anaerobic conditions where sulfate served as the terminal electron acceptor. Toluene and benzylmercaptan were identified as metabolites. In addition, ^{35}S-sulfide was found in cultures incubated with ^{35}S-dibenzylsulfide.

Other Organosulfur Compounds. There have been reports of the microbial metabolism of other OSC. However, few of these studies have given the identities of intermediates or organic endproducts of the OSC. For example, aerobic cultures have been reported to remove sulfur from phenyl sulfide (62). Thioxanthene and thianthrene were transformed to water-soluble products by a dibenzothiophene-oxidizing bacterium (48). In addition, thianthrene and thioxanthene served as sole carbon sources for the aerobic thermophile *S. acidocaldarius* (69) which released sulfate from these compounds.

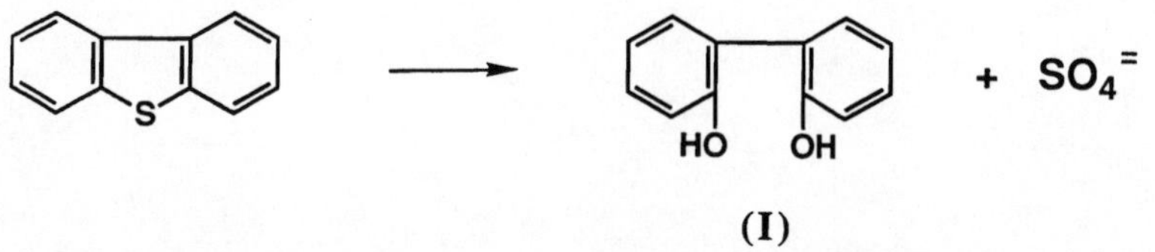

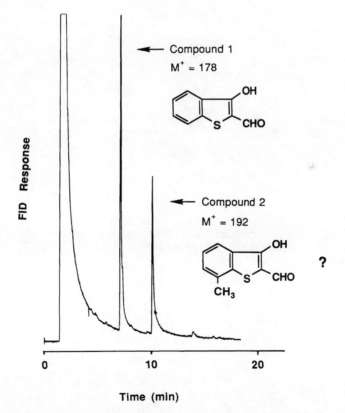

(I)

Figure 11. Specific excision of the sulfur atom from dibenzothiophene by an aerobic bacterium. (I): 2,2'-dihydroxybiphenyl.

Figure 12. Products formed by *Pseudomonas* HL7b co-metabolising a mixture of 4-methyldibenzothiophene and dibenzothiophene.

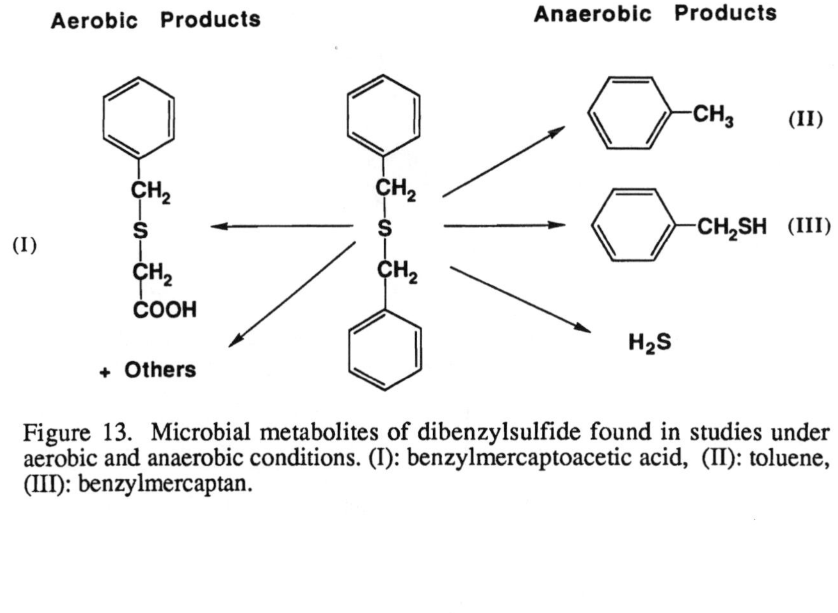

Figure 13. Microbial metabolites of dibenzylsulfide found in studies under aerobic and anaerobic conditions. (I): benzylmercaptoacetic acid, (II): toluene, (III): benzylmercaptan.

Some volatile compounds such as methanethiol, dimethylsulfide and dimethyldisulfide have been shown to yield methane when they were added to anaerobic cultures derived from aquatic sediments (70, 71). Kiene et al. (72) showed that methane bacteria and sulfate-reducers competed for dimethyldisulfide when it was added at low concentrations to anaerobic aquatic sediments. They also isolated a methanogen that metabolized dimethyldisulfide to methane and carbon dioxide (72). Recently Oremland et al (73) detected trace amounts of ethane released from anoxic sediment slurries. This could be stimulated by the addition of ethanethiol or diethylsulfide and inhibited by the addition of bromoethanesulfonic acid which specifically inhibits methane bacteria. These results indicated that methane bacteria co-metabolized these two OSC.

In anaerobic studies to determine the extent of biodesulfurization of several OSC, Köhler et al. (49) dissolved dibenzyldisulfide, butylsulfide and octylsulfide individually in paraffin oil and added these solutions to cultures of *Desulfovibrio* sp. After 6 days of incubation, there were small amounts (3.5 to 8.6%) of desulfurization.

Conclusions and Challenges

As is evident from the focus of this symposium, there is a great interest in sulfur compounds in petroleum. In addition, there is no doubt that microbial metabolism has played a major role in the quantity and quality of the World's petroleum. Yet there is relatively little information on the microbial metabolism of OSC found in petroleum. This paper provided a thorough review of the literature on the topic but it discussed only 20 pure OSC. I am not aware of work which has been done with any other compounds. In their studies of several crude oils, Rall et al. (13) isolated and identified 176 organic sulfur compounds which were grouped into 13 classes. Many more sulfur compounds have been identified since their work. Thus there is much more to be learned about microbial transformations of these compounds.

Further progress in this field will require the collaboration between organic chemists and microbiologists. The chemists will have to synthesize commercially unavailable compounds in sufficient quantities so that the microbiologists can feed these to microbial cultures and look for transformation products. When products are found, the skills of the chemists will be needed again to synthesize the proposed metabolites to allow for their unequivocal identification.

If this collaboration is established, predictably the first studies that will be undertaken will work with aerobic cultures because they are easy to handle and the rates of metabolism or co-metabolism are relatively fast. However, as shown by Grbic´-Galic´ (8), benzothiophene is susceptible to biodegradation under anaerobic methanogenic conditions. What other OSC can be metabolized in this manner? Now that the anaerobic metabolism of some aromatic hydrocarbons with nitrate serving as the terminal electron acceptor has been demonstrated (11, 12), can the same be shown for OSC? Some of the anaerobic biodesulfurization studies using sulfate as the terminal electron acceptor have indicated that organic sulfur compounds can be converted to water-soluble products (38, 49). These observations open a whole new area for further investigations.

The topic of microbial metabolism of OSC in petroleum has been of interest for over three decades. In fact, U.S. patents were issued in the early 1950's for methods of biodesulfurization (74, 75). Although there has been progress in understanding the microbial metabolism of OSC, this area of research is still in its infancy. With the appropriate collaboration between organic chemists and microbiologists, there are many new facts to uncover.

Acknowledgments

I thank Dr. Dunja Grbic´-Galic´ for reviewing this manuscript and for providing preprints of her recent review articles. I also thank Dr. John Payzant for preparing 4-methyldibenzothiophene.

Literature Cited

1. Britton, L.N. In Microbial Degradation of Organic Compounds; Gibson, D.T., Ed.; Marcel Dekker Inc., New York, 1984; p 89.
2. Gibson, D.T.; Subramanian, V. In Microbial Degradation of Organic Compounds, Gibson, D.T. Ed.; Marcel Dekker Inc., New York, 1984; p 181.
3. Trudgill, P.W. In Microbial Degradation of Organic Compounds; Gibson, D.T., Ed.; Marcel Dekker Inc., New York, 1984; p 131.
4. Cerniglia, C.E. In Petroleum Microbiology; Atlas, R.M., Ed.; Macmillan Publishing Co., New York, 1984; p 99.
5. Perry, J.J. In Petroleum Microbiology. Atlas, R.M.; Ed.; Macmillan Publishing Co., New York, 1984; p 61.
6. Singer, M.E.; Finnerty, W.R. In Petroleum Microbiology. Atlas, R.M.; Ed.; Macmillan Publishing Co., New York, 1984; p 1.
7. Grbic´-Galic´, D. Dev. Ind. Microbiol. 1989, 30, 237-253.
8. Grbic´-Galic´, D. In Soil Biochem.; Bollag, J.M.; Stotzky, G. Eds.; Marcel Dekker Inc., New York, 1990; Vol. 8, p 117.
9. Vogel, T.M; Grbic´-Galic´, D. Appl. Environ. Microbiol. 1986, 52, 200-202.
10. Major, D.W.; Mayfield, C.I.; Barker, J.F. Ground Water 1988, 26, 8-14.
11. Zeyer, J.; Kuhn, E.P.; Schwarzenbach, R.P. Appl. Environ. Microbiol. 1986, 52, 944-947.
12. Mihelcic, J.R.; Luthy, R.G. Appl. Environ. Microbiol. 1988, 54, 1182-1187.
13. Rall, H.T.; Thompson, C.J.; Coleman, H.J.; Hopkins, R.L. Sulfur Compounds in Crude Oil. U.S. Department of the Interior, Bureau of Mines Bulletin 659. 1972.
14. Tissot, B.P.; Welte, D.H. Petroleum Formation and Occurrence. 2nd ed., Springer-Verlag, New York. 1984.
15. Payzant, J.D.; Montgomery, D.S.; Strausz, O.P. Org. Geochem. 1986, 9, 357-369.
16. Cyr, T.D.; Payzant, J.D.; Montgomery, D.S.; Strausz, O.P. Org. Geochem. 1986, 9, 139-143.
17. Strausz, O.P. Proc. 4th UNITAR/UNDP Conf. on Heavy Crude and Tar Sands. Alberta Oil Sands Technology and Research Authority, Edmonton, Canada, 1989; Vol. 2, p 607.
18. Atlas, R.M.; Boehm, P.D.; Calder, J.A. Estuarine Coastal Shelf Sci. 1981, 12, 589-608.
19. Gundlach, E.R.; Boehm, P.D.; Marchand, M.; Atlas, R.M.; Ward, D.M.; Wolfe, D.A. Science 1983, 221, 122-129.
20. Patton, J.S.; Rigler, M.W.; Boehm, P.D.; Fiest, D.L. Nature 1981, 290, 235-238.
21. Friocourt, M.P.; Berthou, F.; Picart, D. Toxicol. Environ. Chem. 1982, 5, 205-215.
22. Connan, J. In Adv. Pet. Geochem.; Brooks, J.; Welte, D.H., Ed.; Academic Press, London, 1984; Vol. 1, p 299.

23. Deroo, G.; Tissot, B.; McCrossan, R.G.; Der, F. In Oil Sands, Fuel of the Future; Hills, L.V. Ed.; Can. Soc. Pet. Geol. Mem., 1974, 3, 148-167.
24. Westlake, D.W.S. In Proc. 1982 Intl. Conf. on Microbial Enhancement of Oil Recovery, Donaldson E.C.; Clark J.B. Eds.; Bartlesville Energy Technology Center, Bartlesville, Oklahoma. 1983, p 102-111
25. Burns, B.J.; Hogarth, J.T.C.; Milner, C.W.D. Bull. Can. Pet. Geol. 1975, 23, 295-303.
26. Williams, J.A.; Bjorøy, M.; Dolcater, D.L.; Winters, J.C. Org. Geochem. 1986, 10, 451-461.
27. Payzant; J.D., McIntyre, D.D.; Mojelsky, T.W.; Torres, M.; Montgomery, D.S.; Strausz, O.P. Org. Geochem. 1989, 14, 461-473.
28. Walker, J.D.; Colwell, R.R.; Petrakis, L. Can. J. Microbiol. 1975, 21, 1760-1767.
29. Fedorak, P.M.; Westlake, D.W.S. Can. J. Microbiol. 1983, 29, 291-296.
30. Fedorak, P.M.; Westlake, D.W.S. Water, Air, Soil Pollut. 1984, 21, 255-230.
31. Fedorak, P.M.; Westlake, D.W.S. Can. J. Microbiol. 1981, 27, 432-443.
32. Fedorak, P.M.; Westlake, D.W.S. Water, Air, Soil Pollut. 1981, 21, 255-230.
33. Foght, J.M; Westlake, D.W.S. Can. J. Microbiol. 1988, 34, 1135-1141.
34. Fedorak, P.M.; Payzant, J.D.; Montgomery, D.S.; Westlake, D.W.S. Appl. Environ. Microbiol. 1988, 54, 1243-1248.
35. Eckart, V.; Hieke, W.; Bauch, J.; Gentzsch, H. Zentralbl. Bakeriol. II. 1980, 135, 674-681.
36. Eckart, V.; Hieke, W.; Bauch, J.; Bohlmann, D. Zentralbl. Bakeriol. II. 1981, 136, 152-160.
37. Eckart, V.; Hieke, W.; Bauch, J.; Gentzsch, H. Zentralbl. Mikrobiol. 1982, 137, 270-279.
38. Eckart, V.; Köhler, M.; Hieke, W. Zentralbl. Mikrobiol. 1986, 141, 291-300.
39. Amphlett, M.J.; Callely, A.G. Biochem. J. 1969, 112, 12p.
40. Cripps, R.E. Biochem. J. 1973, 134, 353-366.
41. Kanagawa, T.; Kelly, D.P. Microb. Ecol. 1987, 13, 47-57.
42. Kurita, S.; Endo, T.; Nakamura, H.; Yagi, T.; Tamiya, N. J. Gen. Appl. Microbiol. 1971, 17, 185-198.
43. Sagardía, F.; Rigau, J.J.; Martínez-Lahoz, A.; Fuentes, F.; López, C.; Flores, W. Appl. Microbiol. 1975, 29, 722-725.
44. Sariaslani, F.S.; Harper, D.B.; Higgins, I.J. Biochem. J. 1974, 140, 31-45.
45. Fedorak, P.M.; Westlake, D.W.S. Appl. Environ. Microbiol. 1986, 51, 435-437.
46. Bohonos, N.; Chou, T.-W.; Spanggord, R.J. Jpn. J. Antibiot. 1977, 30(suppl), 275-285.
47. Fuentes, F.A. Proc. 84th Ann. Mtg. Am. Soc. Microbiol., St. Louis, Missouri, 1984; Abst.# N26.
48. Finnerty, W.R.; Shockley, K.; Attaway, H. In Microbial Enhanced Oil Recovery; Zajic, J.E., Cooper, D.C., Jack, T.R. and Kosaric, N. Eds.; PennWell Publishing Company, Tulsa, Oklahoma, 1983; p 83.
49. Köhler, M.; Genz, I.-L.; Schicht, B.; Eckart, V. Zentralbl. Mikrobiol. 1984, 139, 239-247.
50. Monticello, D.J.; Finnerty, W.R. Ann. Rev. Microbiol. 1985, 39, 371-389.
51. Malik, K.A.; Claus, D. 5th Intl. Fermentation Symp., Berlin, 1976; Abst. #23.03.
52. Kargi, F.; Robinson, J.M. Biotechnol. Bioeng. 1984, 26, 687-690.

53. Hou, C.T.; Laskin, A.I. Dev. Ind. Microbiol. 1976, 17, 351-362.
54. Kodama, K. Agric. Biol. Chem. 1977, 41, 1305-1306.
55. Kodama, K.; Nakatani, S.; Umehara, K.; Shimizu, K.; Minoda, Y.; Yamada, K. Agric. Biol. Chem. 1970, 34, 1320-1324.
56. Kodama, K.; Umehara, K.; Shimizu, K.; Nakatani, S.; Minoda, Y.; Yamada, K. Agric. Biol. Chem. 1973, 37, 45-50.
57. Laborde, A.L.; Gibson, D.T. Appl. Environ. Microbiol. 1977, 34, 783-790.
58. Ensley, B.D., Jr. In Microbial Degradation of Organic Compounds; Gibson, D.T. Ed.; Marcel Dekker Inc., New York, 1984; p 309.
59. Kodama, K. Agric. Biol. Chem. 1977, 41,1193-1196.
60. Monticello, D.J.; Bakker, D.; Finnerty, W.R. Appl. Environ. Microbiol. 1985, 49, 756-760.
61. Mormile, M.R.; Atlas, R.M. Appl. Environ. Microbiol. 1988, 54, 3183-3184.
62. Isbister, J.D.; Wyza, R.; Lippold, J.; DeSouza, A.; Anspach, G. In Reducing Risks from Environmental Chemicals through Biotechnology; Omenn, G. S., Ed.; Plenum Press, New York, 1988; p 281.
63. Maka, A.; McKinley, V.L.; Conrad, J.R.; Fannin, K.F. Proc. 87th Ann. Mtg. Am. Soc. Microbiol., Atlanta, Georgia, 1987; Abst. # O-54.
64. Gilman, H.; Jacoby, A.L. J. Org. Chem. 1938, 3, 108-119.
65. Dean-Raymond, D.; Bartha, R. Dev. Ind. Microbiol. 1975, 16, 97-110.
66. Williams, P.A.; Catterall, F.A.; Murray, K. J. Bacteriol. 1975, 124, 679-685
67. Fedorak, P.M.; Westlake, D.W.S. Can. J. Microbiol. 1983, 29, 497-503.
68. Babenzien, H.-D.; Genz, I.; Köhler, M. Z. Allg. Mikrobiol. 1979, 19, 527-533.
69. Kargi, F. Biotechnol. Lett. 1987, 9, 478-482.
70. Zinder, S.H.; Brock, T.D. Nature 1978, 273, 226-228.
71. Kiene, R.P.; Capone, D.G. Microb. Ecol. 1988, 15, 275-291.
72. Kiene, R.P.; Oremland R.S.; Catena, A.; Miller, L.G.; Capone, D.G. Appl. Environ. Microbiol. 1986, 52, 1037-1045.
73. Oremland, R.S.; Whiticar, M.J.; Strohmaier, F.E.; Kiene, R.P. Geochim. Cosmochim. Acta 1988, 52, 1895-1904.
74. Strawinski, R.J. U.S. Patent 2 521 761, 1950.
75. Strawinski, R.J. U.S. Patent 2 574 070, 1951.

RECEIVED March 5, 1990

STUDIES OF DEPOSITIONAL ENVIRONMENTS

Chapter 7

Geochemistry of Organic and Inorganic Sulfur in Ancient and Modern Lacustrine Environments

Case Studies of Freshwater and Saline Lakes

Michele L. Tuttle, Cynthia A. Rice, and Martin B. Goldhaber

U.S. Geological Survey, MS 916, Box 25046, Denver Federal Center, Denver, CO 80225

Abundances of sulfur species (monosulfide, disulfide, and organosulfur) and their isotopic compositions were used to determine sulfur geochemistry in sediment from two freshwater lakes and three saline lakes, and in two Paleogene lacustrine oil shales. Concentrations of reactants (SO_4^{2-}, organic matter, and iron) as well as their reactivity are controls on the extent of sulfate reduction, sulfide-mineral formation, and sulfidization of organic matter. In freshwater lakes containing low sulfate concentrations and in the freshwater oil shale of the Rundle Formation, sulfate availability limits the amount of sulfide-mineral formation and the mineral isotopic values are near those of the lake sulfate. Iron and organic-carbon availability limit sulfide-mineral formation in relatively short-lived, high sulfate lakes and the isotopic composition of these minerals is generally depleted in ^{34}S relative to the initial sulfate. In high-pH saline lakes, the rate of iron sulfidization is significantly decreased. In lakes undergoing rapid fluctuation in lake level, diagenetic processes such as H_2S diffusion complicate the sulfur geochemistry. In very long-lived lakes such as those that deposited the Green River Formation oil shale, the isotopic composition of sulfide minerals is enriched in ^{34}S relative to the original sulfate entering the lakes. The sulfate reservoir in these long-

lived lakes evolves to ^{34}S-enriched values.
Formation of organosulfur in saline lakes
occurs predominantly from the sulfidization of
organic matter by bacteriogenic H_2S. The
sulfur found in lacustrine shale oil may
evolve from this bacteriogenic sulfur.

Sulfur geochemistry is recognized as a powerful approach
to understanding the depositional and diagenetic histories
of sedimentary environments. Sulfur is ubiquitous in
nature and involved in both abiotic and biotic processes
(Figure 1). Also, sulfur transformations are sensitive to
both pH and redox conditions. Using interpretive results
from sulfur studies in modern lakes, our research seeks to
reconstruct the sulfur geochemistry during deposition and
diagenesis of lacustrine oil shales.

Total sulfur concentrations and sulfur isotopic
compositions are the two most common parameters used to
interpret sulfur geochemistry in modern and ancient
sediments. Since total sulfur includes inputs from
multiple processes in the sedimentary sulfur cycle,
important information regarding individual processes may
be lost. A more fruitful approach would be to distinguish
separate residences of sulfur and to analyze their
individual isotopic compositions. We have extensively
employed the additional insights that this latter approach
provides. In keeping with the theme of this symposium, we
focus our investigation on organosulfur (sulfur bound in
organic matter). However, in order to understand
processes controlling the incorporation of sulfur into
organic matter, we consider data from all types of
lacustrine sulfur.

First we review controls on the amount and isotopic
composition of various forms of sulfur in lacustrine
environments. Next, we summarize the diverse behavior of
sulfur in sediment from two freshwater environments; in
sediment from three modern, productive, saline lakes; and
in oil shales deposited in freshwater and saline
lacustrine environments. Lastly, our results are
integrated in order to produce models that 1) predict the
extent of formation and isotopic composition of sulfide
minerals in response to major controls on sulfur
geochemisty; and 2) show the formational pathway of
organosulfur in lacustrine oil shale and its derivative
oil.

Controls on Sedimentary Sulfur Chemistry. The
variability of sulfur geochemistry in lacustrine
environments is due, in part, to the large concentration

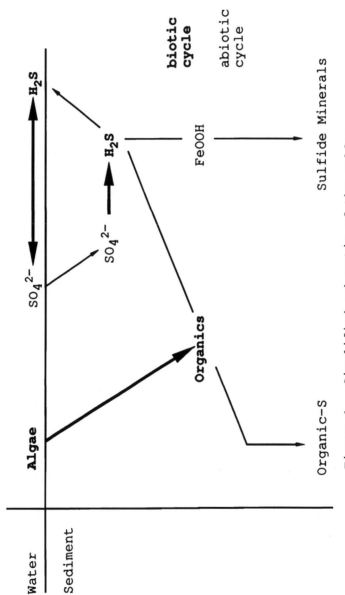

Figure 1. Simplified schematic of the sulfur geochemistry in productive, saline lakes.

ranges of reactants involved in key reactions of the
sedimentary sulfur cycle--the bacterially mediated
reduction of sulfate and the production of iron-sulfide
minerals. The sulfate-reduction reaction in its
simplified form is (1):

$$2CH_2O + SO_4^{2-} \longrightarrow H_2S + 2HCO_3^-, \tag{1}$$

where CH_2O is a generic carbohydrate molecule used
bacterially as an energy source with sulfate as the
electron acceptor.

The amount of organic matter (CH_2O) in lake sediment
is directly related to both lake productivity and
preservation during deposition and early diagenesis.
Organic productivity in lakes is highly variable, as low
as 0.6 g C m^{-2} yr^{-1} in oligotrophic tundra lakes to as
high as 640 g C m^{-2} yr^{-1} in eutrophic equatorial lakes
(2). Preservation of organic matter in lake sediment is
related to the length of time the organic matter is in
contact with oxygenated waters. In lakes with oxygenated
bottom water, aerobic mineralization produces refractory
organic components incapable of yielding the fermentative
degradation products utilizable by sulfate-reducing
bacteria. If anoxic conditions are established quickly,
either in the bottom water or just below the sediment-
water interface, aerobic mineralization is minimal and
metabolizable organic matter supports extensive bacterial
sulfate reduction (3-4).

The other major reactant in Equation 1 is sulfate
(SO_4^{2-}). Sulfate concentrations are highly variable in
lake waters, from 3 x 10^{-5} mol/L in soft-water lakes in
crystalline-rock drainage basins to 1.6 mol/L in
hypersaline lakes (2). In productive, freshwater lakes,
sulfate reduction typically goes nearly to completion (5).
As sulfate concentrations increase, amounts of organic
matter eventually become insufficient for complete sulfate
reduction to occur. This is the case in "normal" marine
sediment where a linear relation between total reduced
sulfur and organic-carbon concentrations is observed.
Sea-water sulfate concentration is 0.028 mol/L and the
ratio of total reduced sulfur to organic-carbon
concentrations (often referred to as S/C) in marine
sediment is 0.33 (6).

The amount of reduced sulfur in freshwater
lacustrine sediment and in most marine sediment is a
function of the availability of the limiting reactant
during sulfate reduction--whether sulfate or organic
matter. This simple two end-member model must frequently
be modified for saline lacustrine sediment and for some
marine sediment in order to reflect the capacity of the

sediment to remove H_2S. Incorporation of H_2S by the
sediment is dominantly controlled by the reaction of iron
and reduced sulfur species as shown by the following
reactions (modified from equations in 7-8):

$$FeOOH + H^+ + 3/2S^{2-} \rightarrow FeS + 1/16S_8 + 2OH^- \quad (2)$$
$$FeS + 1/8S_8 \rightarrow FeS_2. \quad (3)$$

The elemental sulfur (S_8) in Equation 3 is generally
dissolved as a polysulfide ion. In iron-poor sediment or
sediment in which iron resides largely in refractory
minerals, only small amounts of sulfide minerals form.
Excess H_2S in these sediments slowly reacts with organic
matter to form organosulfur.

   In summary, there are three master variables
controlling sulfide-mineral formation in marine and
lacustrine environments. In marine sediment,
metabolizable organic matter limits sulfide-mineral
formation except in carbonate-rich sediment or euxinic
(H_2S-bearing) basins where iron may be limiting. Sulfide-
mineral formation in freshwater sediment is generally
sulfate limited. In saline lakes containing high amounts
of dissolved sulfate, we expect sulfide-mineral formation
to be limited by either organic matter or by iron. The
results summarized in this paper confirm our expectation
and identify key processes controlling these variables.

Controls on Sedimentary Sulfur Isotopy. Processes
controlling lacustrine sulfur geochemistry (Figure 1) are
recorded not only by the amounts of mineralogical sulfur
and organosulfur, but also by their isotopic composition.
The sulfur isotopic composition ($\delta^{34}S$) of a sulfur phase
is determined by measuring its $^{34}S/^{32}S$ (R) and comparing
the ratio to that of the Cañon Diablo troilite (CDT), the
standard most often used for sulfur:

$$\delta^{34}S \text{ in } \permil = [(R_{sample} - R_{standard}) / R_{standard}] * 1000. \quad (4)$$

Sulfur isotope systematics in sedimentary environments are
reviewed in several excellent references (1, 9-11), and
are only briefly discussed in this paper.
The predominant organisms in highly productive lakes are
generally algae and bacteria. These organisms contain low
levels of naturally occurring organosulfur formed by
assimilating sulfate, reducing it to sulfide, and using
the sulfide in production of compounds such as amino
acids. This process is termed assimilatory reduction of
sulfate. Assimilated sulfur has an isotopic composition
similar to the dissolved sulfate in the lake (11). In
contrast, dissimilatory sulfate reduction occurs when

bacteria use sulfate as an electron acceptor during
decomposition of organic matter. This process involves a
substantial fractionation of sulfur isotopes with the
product, H_2S, being depleted in ^{34}S relative to the
reactant, sulfate. The systematics of sulfur isotopic
evolution during dissimilatory reduction are depicted in a
Rayleigh fractionation plot (Figure 2). The plot shows
the evolution of $\delta^{34}S$ of the residual sulfate,
instantaneously produced H_2S, and accumulated sulfide
reservoirs as a function of the extent of reduction of the
initial sulfate reservoir. The curves on the plot were
calculated using Rayleigh fractionation equations (9), an
initial sulfate isotopic composition of
$\delta^{34}S$ = 10‰--typical for sulfate in rivers and lakes
(9,12), and an instantaneous fractionation value ($\Delta_{SO4-H2S}$)
of 30‰--the vertical distance between the "sulfate" and
"instantaneous H_2S" curves. The choice of 30‰ is
intermediate; within the lacustrine range of 10‰ in some
freshwater lakes (9) to 60‰ in some saline lakes (13).
For discussion of factors controlling the magnitude of
$\Delta_{SO4-H2S}$ see (1,10). As the sulfate reservoir is depleted
(moving from left to right on the abscissa in Figure 2),
both the sulfate and instantaneously produced H_2S become
progressively enriched in ^{34}S. Given a system in which
only partial reduction of dissolved sulfate occurs (less
than 20% for the case modeled in Figure 2), the
accumulated H_2S will be depleted in ^{34}S relative to the
original sulfate reservoir by a value approaching $\Delta_{SO4-H2S}$.
Marine sediments in contact with ocean-water sulfate
exhibit similar isotope systematics. In a system with a
sulfate reservoir that is appreciably reduced as in many
low sulfate, freshwater lakes, the accumulated H_2S will
approach the initial isotopic composition of the sulfate
reservoir. Subsequent abiotic sulfidization reactions
involving H_2S have relatively small isotope fractionations
so that the solid phase products dominantly reflect the
biotic sulfate-reductive processes.

Case Studies. The discussion of individual studies in
this paper are intended as brief summaries of important
results from a variety of ancient and modern lake
sediments discussed in other papers (13-16). These
studies include two Paleogene lacustrine oil shales--the
Green River Formation (Colorado, Utah, and Wyoming) and
the Rundle Formation (Queensland, Australia). The
locations of these formations are shown in Figure 3, and
key characteristics of the deposits are compared in
Table I. Also included are results from studies of three
modern productive saline lakes (Soap Lake, Washington;
Great Salt Lake, Utah; and Walker Lake, Nevada) and two

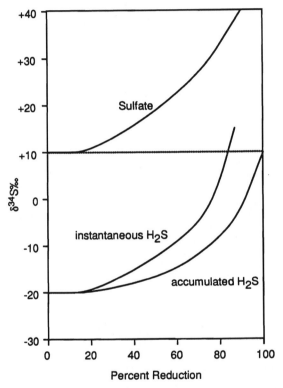

Figure 2. Rayleigh fractionation curves of $\delta^{34}S$ as a function of the percent of the initial sulfate reduced.

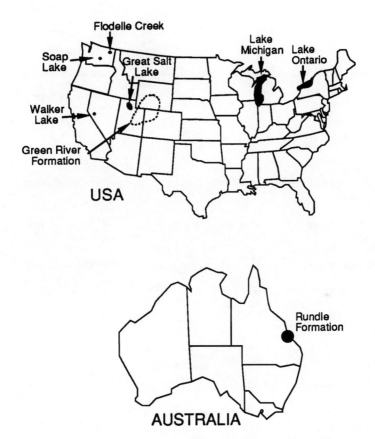

Figure 3. Locations of Flodelle Creek, Washington; Soap Lake, Washington; Lake Michigan; Lake Ontario; Great Salt Lake, Utah; Walker Lake, Nevada; approximate areal extent of Green River Formation, Utah, Colorado, Wyoming; and approximate areal extent of Rundle Formation, Queensland, Australia.

freshwater environments (the Great Lakes and a spring pool
in a peat bog located near Flodelle Creek, Washington).
The morphometric and geochemical characteristics of the
modern lakes are given in Table II and their locations
shown in Figure 3.

Table I. Comparison of characteristics of oil-shale
deposits of the Green River and Rundle Formations

Characteristic	Green River Formation	Rundle Formation
Age	Paleogene	Paleogene
Environment of deposition	Lacustrine	Lacustrine
Classification	Lamosite (algal)	Lamosite (algal)
Avg oil yield (L/tonne)	126	105
Area of deposit[a] (km^2)	4500	45
Thickness[a] (m)	up to 640	40–350
Barrels of oil equiv.[a]	1.2×10^{13}	2.7×10^9
Data source	(17)	(18)

[a]Green River data for Piceance basin only.

Sampling and Analytical Methods

Sediment cores were obtained with a gravity, a piston, or
a hand-driven Livingston corer. All samples except those
from Walker Lake were collected under a nitrogen
atmosphere and pore water extracted by centrifuging each
core sample. Samples of the water column were taken with
a Van Dorn sampler. All samples were kept frozen until
analyzed. Twenty-five sediment samples were collected
from Soap Lake--nine from above the chemocline (core
length of 0.28 meters) and 16 from below the chemocline
(core length of 0.71 meters). Sixty-four sediment samples
from Great Salt Lake were collected from two cores--29 for
sulfur and related element chemistry (core length 3.7

Table II. Morphometric and geochemical characteristics of Flodelle Creek spring pool (FC), Washington; Lake Michigan (GL); Walker Lake (WL), Nevada; Great Salt Lake (GSL), Utah; and Soap Lake (SL), Washington. n.d., not detected. --, no data. FC data ([19, 20]); GL data ([21]); WL data ([22-24]); GSL data ([25, 26]); SL data ([27-29]).

Feature	FC	GL	WL	GSL	SL
Area (km^2)	0.000005	57,800	150	4,250	3.7
Maximum depth (m)	1	281	35	10	28
Elevation (m)	1036	820	1,210	1,280	330
Mean ann. precip. (m)	0.45	0.83	0.10	0.43	0.20
Mean ann. evap. (m)			1.2	1.5	1.1
Productivity (g C m^{-2} yr^{-1})	<100	130	--	145	360
Brine type	--	--	Na-Cl-CO$_3$	Na-Mg-Cl	Na-CO$_3$-SO$_4$
TDS (ppm) surface	87	160	10,500	254,000	26,200
bottom water					144,400
pH	7.4	8.3	9.4	7.4	9.8
Sulfate (mol/L) surface	0.000028	0.00020	0.028	0.20	.03
bottom water					.34
H$_2$S (mol/L)	n.d.	n.d.	n.d.	0.001	0.12

meters) and 35 sediment samples for sulfur chemistry and isotopy (core length 5.2 meters). Thirty-six sediment samples were collected from Walker Lake (core length of 144 meters), 11 sediment samples from Flodelle Creek peat bog (core length of 2.9 meters), and 14 sediment samples from Lake Michigan (core length of 6.0 meters).

Samples from the Green River Formation were obtained from each of the three depositional basins. In order to minimize lake-margin effects, depocenter cores were sampled using a sampling procedure designed to ensure maximum coverage given the limited number of samples analyzed. One hundred fifteen samples were collected--35 samples from the Greater Green River basin, Wyoming; 41 samples from the Piceance basin, Colorado; and 39 samples from the Uinta basin, Utah. The ten Rundle samples are from outcrop; the samples showed no obvious signs of weathering such as oxidation of iron-sulfide minerals. Generally, two samples were collected from each of the oil-shale-bearing members of the Rundle Formation. Oil-shale samples were ground to pass through a -115-mesh screen (130 μm mean diameter).

Total sulfur was determined using a commercial induction-furnace/infrared-detection system. Dissolved sulfur (sulfate and hydrogen sulfide) and solid sulfur species (monosulfide, FeS; disulfide, FeS_2; sulfate; and organosulfur, S_{org}) were separated and analyzed by an HCl/Cr^{2+}/Eschka-fusion scheme shown schematically in Figure 4 (30). Sulfur contained in FeS and FeS_2 are collectively referred to as S_{min}. The isotopic composition of the separated sulfur forms was determined on the products of the separation scheme by conversion to SO_2 and analysis by mass spectrometry. Organic-carbon (organic-C) concentrations were determined on acid-treated sediment samples using a commercial induction-furnace/thermal conductivity cell system. In oil-shale samples and lacustrine sediment known to contain predominantly algal-derived organic matter, conversion from organic-C concentrations to organic-matter concentrations was made using a factor of 1.27. This conversion factor is an average ($s = 0.05$) determined from data on Green River, Type I kerogen (31). Organic matter in the spring-pool sediment was measured directly by a loss-on-ignition technique. "Reactive" iron (sum of HCl-soluble iron and iron in FeS_2) is believed to reflect iron available for sulfidization (32). By analyzing the filtrate from the first separation step of the sulfur-separation scheme, HCl-soluble iron was determined using atomic absorption spectroscopy (AAS). Total iron was determined by lithium-metaborate fusion (33) and AAS.

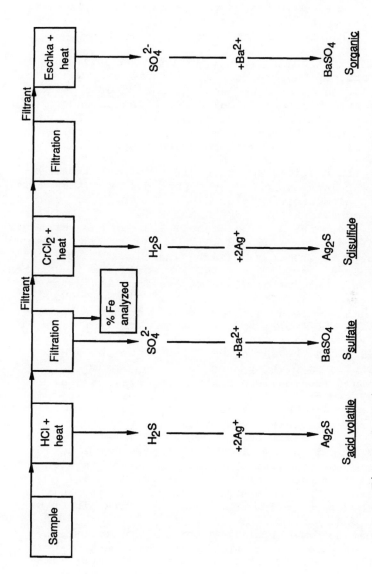

Figure 4. Diagrammatic representation of the method designed to collect and analyze forms of sulfur.

Results and Discussion

The chemical and isotopic data are presented in Table III
as average values and standard deviations. The Rundle
data were better represented by geometric means and
deviations. All chemical data are on a carbonate-free
(CF), dry-weight basis. The modern-lake studies, lowest
sulfate to highest sulfate, are discussed first.

Freshwater Environment. We evaluated the sulfur
geochemistry of two freshwater environments. One
environment is an area of local ground-water upwelling
within a peat bog near Flodelle Creek, northeastern
Washington. The spring pool is small (Table II) and
subsurface deposits grade upward from coarse-grained to
granular sand, to fine-grained organic-C-rich units
interfingering with sandy units, to woody peat and
organic-C-rich silt/clay units (19). Of the lakes
studied, the spring pool is the most dilute (Table II)
with a total dissolved solid (TDS) concentration only 0.3%
that of seawater with 34,800 ppm TDS (34). Primary
productivity rates have not been measured; however, Wetzel
(2) predicts rates <100 g C m^{-2} yr^{-1} for bog environments.
   Pyrite is the only sulfide-mineral detected in the
Flodelle Creek spring pool, and its formation is limited
by the small amounts of dissolved sulfate in the spring-
pool water, 2.8 x 10^{-5} mol/L. The $\delta^{34}S$ value of the
sulfate has not been measured, but is probably between
0 and 10‰, the range for most sulfur found in granitic
rocks (11) such as those in the Flodelle Creek drainage
basin (19). In the spring-pool sediment, the average S_{min}
concentration is 0.04 wt% and the $\delta^{34}S_{min}$ value is +0.7‰.
These values are consistent with complete, or near
complete, reduction of a small pore-water sulfate
reservoir to H_2S that reacts with iron in the sediment to
form sulfide minerals. The average S_{org} concentration is
0.16 wt% (0.7 wt% of the organic matter) and the $\delta^{34}S_{org}$
value is +2.8‰. The concentration and isotopic
composition of sulfur in the organic matter is within the
range predicted for assimilatory sulfur in this
environment. The sedimentary organosulfur probably is
inherited directly from the biota that lived in or near
the peat bog.
   Great Lakes Michigan and Ontario constitute the
second freshwater environment (data from Lake Ontario are
from 21). The Great Lakes are large freshwater lakes
(Table II) formed from glacial scouring during the retreat
of the Wisconsin ice sheet. They have a salinity about
0.5% that of seawater. Primary productivity in the Great

Lakes is relatively low compared to rates in the saline lakes (Table II).

The dissolved sulfate concentration is 2.0 x 10^{-4} mol/L in Lake Michigan and 3.0 x 10^{-4} mol/L in Lake Ontario; the $\delta^{34}S$ value of the sulfate in Lake Ontario is ≈ +7‰. In the upper 1 meter of near-shore Lake Michigan sediment, the average S_{min} concentration is 0.17 wt% and the average $\delta^{34}S_{min}$ value is +10.1‰; the average S_{org} concentration is 0.01 wt% and the average $\delta^{34}S_{org}$ value is +10.0‰. In the upper 50 centimeters of Lake Ontario sediment, total-sulfur concentrations average 0.15 wt% and the average $\delta^{34}S_{total}$ value is +12.8‰ ($\delta^{34}S_{total}$ calculated from data in 21). Presumably 52% of the sulfur resides in monosulfides, 32% in pyrite and/or elemental sulfur, and 16% in organic matter (21). As in Flodelle Creek spring-pool sediment, low sulfate concentrations and near complete sulfate reduction in Great Lake sediment produce small amounts of sulfide minerals enriched in ^{34}S. Most of the organosulfur is probably assimilatory sulfur. The sedimentary sulfur in the Great Lakes is enriched in ^{34}S by up to 6‰ relative to the lake sulfate. Nriagu and Coker (21) propose that this ^{34}S enrichment is in part due to preferential loss from the lake sediment of H_2S depleted in ^{34}S.

Sediments deposited in Flodelle Creek spring pool and the Great Lakes have similar and relatively uncomplicated sulfur geochemistry that is controlled by two processes. These processes are the assimilation of sulfur into living biota and its subsequent deposition as organosulfur when the organism dies, and the complete reduction of the pore-water sulfate to H_2S that forms sulfide minerals. Low dissolved sulfate concentrations limit the amount of sulfide minerals formed. The $\delta^{34}S$ value of most of the S_{min} is essentially the same as the dissolved sulfate. The possible exceptions are minerals formed in sediment from which some ^{34}S-depleted H_2S had diffused.

Walker Lake. Walker Lake, located in west-central Nevada, is a remnant of pluvial Lake Lahontan, which, at its high stand, had a depth of 280 meters (24). Walker Lake is the terminus of the Walker River. Although its salinity is only 30% that of seawater, its sulfate concentration is identical to that of seawater (0.028 mol/L). The Walker Lake samples represent 300,000 years of depositional history (Yang, unpublished report). Preliminary interpretations of paleontological and mineralogical data (Benson and Spencer, written communication) suggest that lake levels have fluctuated.

Table III. Chemical and isotopic averages and standard deviations (±) for sediment samples from Flodelle Creek, Washington; Lake Michigan (Goldhaber and Nicholson, unpublished data); Lake Ontario (21); Walker Lake, Nevada (13); Great Salt Lake, Utah (13); Soap Lake, Washington (13); Green River Formation, USA (13); and Rundle Formation, Australia (unpublished data). Rundle chemical data are reported as the geometric mean and geometric deviation ({}). All chemical concentrations are reported on a dry weight basis and a carbonate-free basis. Sulfur isotope values are parts per mil relative to Cañon Diablo troilite. tot, total; dis, disulfide; min, iron-sulfide minerals; org, organo; rct, reactive; nd, not determined; (-), too few samples to calculate standard deviation; --, no data. *%Fetotal for Great Salt Lake

	% Stot	% Sdis	% Smin	% Sorg	% Ferct*	% Corg	δ^{34}Sdis‰	δ^{34}Sorg‰
FRESHWATER								
Flodelle Creek	.24±.22	.04±.04	.04±.04	.16±.14	1.6±.27	20±12	.73±2.8	2.8±3.6
Lake Michigan	.12±.09	.07±.09	.09±.09	.02±.01	nd	nd	nd	nd
Lake Ontario	.14 (-)	nd	nd	nd	nd	nd	13 (-)	nd
SALINE								
Walker Lake	.94±.56	.38±.35	.69±.43	.05±.03	3.6±.74	1.2±.88	-15±17	-4.4±12
Great Salt Lake								
Stage 1 (saline)	.50±.12	.31±.14	.31±.14	.20±.04	1.7±.31	2.5±1.3	-21±3.1	nd
Stage 2 (saline)	.44±.23	.27±.21	.27±.21	.16±.08	1.8±.39	3.5±.55	-19 (-)	nd
Stage 3 (fresh)	.90±.21	.87±.25	.87±.25	.06±.08	2.4±.75	1.2±.57	-15±8.3	nd
Soap Lake								
Noneuxinic	.08±.02	.04±.02	.04±.02	.04±.004	2.2±.28	4.0±1.1	-29±9.7	-20±4.5
Euxinic	.20±.08	.08±.07	.12±.07	.09±.02	1.6±.87	9.6±4.1	-20±5.4	-14±2.7

Table III. Continued.

	% Stot	% Sdis	% Smin	% Sorg	% Ferct	% Corg	δ³⁴Sdis‰	δ³⁴Sorg‰
GREEN RIVER FM.								
Greater Green River basin								
Stage 1	2.4±1.2	1.8±.71	1.9±.72	.24±.09	2.7±.41	9.4±4.1	14±8.7	18±6.4
Stage 2	.71±.57	.44±.46	.52±.46	.05±.07	2.2±1.0	3.8±4.9	26±8.1	27±7.0
Stage 3	1.4±.51	.98±.52	.99±.51	.21±.12	2.0±.76	7.6±4.1	18±11	19±10
Piceance basin								
Stage 1	1.9±1.0	1.5±.95	1.6±.94	.19±.08	2.8±1.0	13±8.1	17±8.9	17±8.0
Stage 2	1.7±.23	.86±.54	.89±.52	.39±.50	2.4±1.1	15±7.5	29±5.0	28±4.0
Stage 3	1.0±.65	.78±.52	.78±.52	.25±.28	2.5±.78	16±12	35±12	37±11
Uinta basin								
Stage 1	.68±.62	.69±.39	.69±.39	.03±.03	2.3±.92	1.8±2.8	14±9.2	14±4.6
Stage 2	.66±.85	.32±.31	.47±.51	.05±.03	2.6±1.1	5.6±5.1	28±8.7	26±9.2
Stage 3	.76±.69	.43±.32	.45±.31	.17±.26	1.9±.62	14±11	28±9.5	30±7.3
RUNDLE FM.	.50(4.2)	.13(5.4)	.17(5.2)	.12(4.0)	3.1(3.3)	7.1(5.5)	7.9±10	8.8±5.0

Walker Lake sediment contains variable amounts of S_{min}, 0.10 to 2.2 wt%, and $\delta^{34}S_{min}$ values ranging from -42‰ to +25‰. On the average, 44% of the sulfur in sulfide minerals resides in monosulfides and 56% resides in pyrite. Both minerals coexist in many samples and have similar $\delta^{34}S$ values. The exact reason for monosulfide preservation in this sediment is not known; however, Berner (35) attributes monosulfide preservation in Black Sea sediment to insufficient elemental sulfur (polysulfides) to completely convert FeS to FeS$_2$ (Equation 3). Both sulfide minerals in sediment from Walker Lake are typically depleted in ^{34}S with a $\delta^{34}S_{min}$ average of -15‰. At a few depths, however, $\delta^{34}S$ values are similar or enriched in ^{34}S relative to the sulfate in the modern lake ($\delta^{34}S \approx +10‰$; calculated from pore-water sulfate data).

The S_{min} concentration in Walker Lake is related to the $\delta^{34}S_{min}$ values (Figure 5a). This relation can be reconciled with the Rayleigh equations; a hypothetical field of values bounded by Rayleigh fractionation curves of accumulated H$_2$S is shaded in Figure 5a (curves calculated assuming a $\Delta_{SO_4-H_2S} = 55‰$ and 100% extent of reduction produces 1.3 wt% S_{min}). This observation suggests that the amount and isotopic composition of sulfide minerals produced is a function of the extent of sulfate reduction. This in turn is generally dependent on the size of the initial sulfate reservoir in permeable sediment, or on pore-water sulfate concentrations in impermeable sediment. The sediment containing abundant sulfide minerals enriched in ^{34}S (upper right hand corner of Figure 5a) are probably impermeable and reduction of sulfate in pore water proceeds to completion. The boxed data in Figures 5a and 5b are for samples from the upper 12 meters of sediment and are discussed below.

Organic-C concentrations in sediment from Walker Lake average 1.2 wt% (Table III) and are low relative to most values in the other lakes studied. Yet, except in the upper 12 meters of sediment (boxed data in Figure 5b), S_{min} concentrations are an order of magnitude higher than predicted for equivalent organic-C contents in "normal" marine sediment, S/C of 0.33 (6). In Black Sea sediment, S/C values >0.33 are attributed to euxinic depositional conditions (36). Although Walker Lake is seasonally stratified, mineralogical and ostracode data (Spencer and Forester, unpublished reports) do not support long periods of euxinic bottom water. An alternative hypothesis explaining the consistently high S/C values in most of the samples, is that sedimentary organic matter from Walker Lake has a greater reactivity than that in marine sediment. Greater reactivity results in a larger

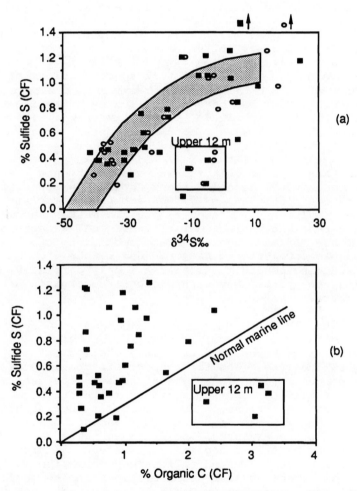

Figure 5. Scatter plots of carbonate-free data from Walker Lake: (a) $\delta^{34}S$ versus weight % sulfur in FeS minerals (circles) and FeS_2 minerals (squares); (b) weight % organic carbon versus weight % sulfide sulfur and "normal" marine line.

percentage of the organic matter being utilized by
sulfate-reducing bacteria. Eventually all the reactive
organic matter is consumed and sulfate reduction ceases.
Samples in the upper 12 meters of sediment have relatively
high organic-C contents and low S_{min} contents, possibly
reflecting ongoing sulfate reduction and sulfide-mineral
formation.

The organic matter in Walker Lake sediment contains
an average of 3.3 wt% sulfur--almost three times the
amount in most living organisms (1.2 wt%; $\underline{1}$) and has a
$\delta^{34}S_{org}$ averaging -4.4‰. The high concentrations of the
^{34}S-depleted sulfur in the organic matter indicate that
the organic matter has been sulfidized by bacteriogenic
H_2S. If the organic matter is extensively biodegraded as
suggested above, it likely contains abundant functional
groups reactive to H_2S.

The sulfur geochemistry in Walker Lake is more
complex than that in the freshwater environments. Because
of its reactivity, organic matter apparently supports more
sulfate-reduction than equivalent amounts in marine
sediment. Ultimately, however, organic matter becomes the
limiting reactant during sulfate reduction, except in
impermeable sediment where sulfate reduction in pore water
proceeds to completion. Biogenic H_2S reacts with abundant
iron producing monosulfides that are only partially
converted to pyrite, and with biodegraded organic matter
forming organosulfur.

Great Salt Lake. During the past 25,000 years, Great
Salt Lake has varied from the deep (300 meters),
freshwater Lake Bonneville to a very shallow (<8 meters),
highly saline lake. The first-order expectation for
sulfur concentrations in a lake alternating between fresh
water (low sulfate) and saline water (high sulfate) are:
1) when concentrations of sulfate are low during high
stands (fresh water), low total-sulfur concentrations in
the sediment are expected; and 2) when sulfate is
abundant during saline deposition (present salinity is 7
times that of seawater and sulfate concentration is 0.20
mol/L), high total-sulfur concentrations in the sediment
are expected. Averages in Table III show that just the
opposite occurs--sediment deposited in Great Salt Lake
during freshwater conditions contains twice as much total
sulfur per unit weight as that from saline lake stages,
and much more S_{min} than reported for "normal" marine
sediment with equivalent amounts of organic matter (Figure
6a).

Pyrite was the only sulfide mineral observed in
sediment from Great Salt Lake. The S_{min} concentrations
are linearly dependent on total-iron concentrations

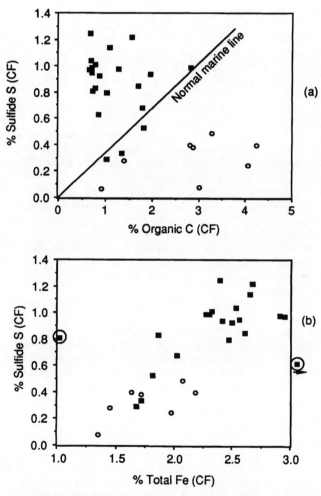

Figure 6. Scatter plots of carbonate-free data from Great Salt Lake samples (squares, freshwater stage; circles, saline stage): (a) weight % sulfide sulfur versus weight % organic carbon; (b) weight % sulfide sulfur versus weight % total iron.

(Figure 6b); however, the processes producing the linear
trend differ between high and low stands. During the
Bonneville high stand, the sulfur cycle would have been
similar to that in the freshwater lakes studied; all the
sulfate in the pore water was likely reduced and
quantitatively removed by reaction with detrital iron
(total iron is 2.4 wt%) to form low abundances of
^{34}S-enriched sulfide minerals. Additional iron was
undoubtedly still available for sulfidization.   During
subsequent saline stages, chemical sedimentation of $CaCO_3$
dominated, resulting in clay-poor and, consequently, iron-
poor sediment (total iron is 1.7 wt%). Hydrogen sulfide
in this low-iron sediment was continually removed by
reaction with iron until all available iron was
sulfidized, at which time excess pore-water H_2S began to
accumulate. This excess H_2S eventually diffused along
concentration gradients into underlying freshwater
sediment where it sulfidized the remaining available iron.
The diffusing H_2S would have been depleted in ^{34}S, as
$\delta^{34}S_{min}$ in the saline sediment is -21‰, and, when combined
with the primary freshwater ^{34}S-enriched sulfide, produced
a much more ^{34}S-depleted isotopic signature (average
$\delta^{34}S_{min}$ value of -15‰) than predicted for freshwater lake
sediment. On the average, 15% of the total iron is
sulfidized in the saline stage of Great Salt Lake and 33%
in the freshwater stage. Not only was iron more abundant
in the freshwater stage, it was apparently also more
easily sulfidized.
     The sulfur content of organic matter in sediment
from Great Salt Lake's saline stage and the uppermost part
of the freshwater stage averages 7 wt%. These high
concentrations indicate that a large proportion of the
organosulfur in these sediments formed from sulfidization
of organic matter by bacteriogenic H_2S. In sediment from
the mid- to lower-freshwater stage, the sulfur
concentrations are much lower, generally 1 wt%.
Sulfidization occurred to a much lesser degree in this
freshwater sediment than in that near the saline-
freshwater sediment interface.
     The Great Salt Lake data give us an opportunity to
study sulfur geochemistry in a lake undergoing dramatic
lake-level fluctuations. Originally, the sulfur
geochemistry in the Bonneville high stand was probably
similar to that in the freshwater lakes--formation of
assimilatory organosulfur and small amounts of
^{34}S-enriched sulfide minerals. Additional sulfide
minerals formed as ^{34}S-depleted H_2S diffused from
overlying saline sediment and reacted with the remaining
available iron. Diffusion was controlled by gradients
established when the production of H_2S in the saline

sediment exceeded the capacity of iron to form sulfide minerals. Much of the organosulfur in the saline sediment was formed from the reaction of bacteriogenic H_2S and organic matter.

Soap Lake. Soap Lake, the terminal lake of the Grand Coulee (Washington) is a saline, high pH, sodium-bicarbonate lake with high primary productivity (Table II). It contains two depositional environments separated at a water depth of 20 meters by a chemocline. Above the chemocline, waters are seasonally oxic and have a salinity about 80% that of seawater and a sulfate concentration of about 0.03 mol/L. Sediment that accumulated in this setting is termed "noneuxinic". Conditions below the chemocline are permanently euxinic with a salinity six times seawater and a sulfate concentration of 0.34 mol/L. In response to temperature changes, ephemeral $NaSO_4 \cdot 10H_2O$ crystals form in sediment near the chemocline. Sediment from below the chemocline is termed "euxinic".

In the noneuxinic sediment, about 6% of the pore-water sulfate ($\delta^{34}S = +13‰$ at the sediment-water interface) has been reduced to H_2S ($\delta^{34}S = -44‰$) at a depth of 10 centimeters below the sediment-water interface. The pore-water H_2S concentration is initially 7.5×10^{-4} mol/L and increases with depth to 4×10^{-3} mol/L indicating that sulfate reduction is occurring throughout the core. The isotope fractionation is very large ($\Delta_{SO_4-H_2S} \approx 60‰$).

In euxinic sediment, the isotopic composition of the pore-water sulfate ($\delta^{34}S \approx +25‰$) and H_2S ($\delta^{34}S \approx -26‰$) is enriched in ^{34}S relative to the values in noneuxinic pore water because reduction is more extensive (35% of the sulfate is reduced versus 6% in noneuxinic pore water). The H_2S concentration and isotopic composition profiles in the euxinic sediment are constant with depth indicating that no significant reduction is occurring in the sediment. The high H_2S concentration (110 mmol/L) may have an inhibitory effect on the sulfate-reducing bacteria.

In light of the high pore-water H_2S concentrations, the very low concentrations of total sulfur that average 0.08 wt% in noneuxinic sediment and 0.20 wt% in euxinic sediment, as well as very low S_{min} concentrations (about 1/2 those for total sulfur), are extremely surprising. These anomalies are a result of lack of iron availability for sulfidization. The noneuxinic sediment contains between 1.3 and 2.4 wt% reactive iron, but only 1% to 3% of it is sulfidized. The euxinic sediment contains a larger range of reactive iron concentrations (0.58 to 3.0 wt%) -- 5 to 11% of which is sulfidized. Even though

these percentages of sulfidized iron are small, the S_{min}
concentrations are still linearly dependent on the amount
of reactive iron in the sediment (Figure 7). The low S_{min}
concentrations and the accumulation of pore-water and
monimolimnion (layer below the chemocline) H_2S evidently
result from the kinetic inhibition of iron sulfidization
by high pore-water pH, a hypothesis that is consistent
with experimental results from our laboratory showing the
dramatic decrease in iron-sulfidization rates with
increasing pH (Stanton and Goldhaber, written
communication).

Iron monosulfides comprise about 20% of the iron
sulfide minerals in the noneuxinic sediment and about 50%
in the euxinic sediment. Their greater preservation in
the euxinic sediment, as in sediment from Walker Lake, is
probably a result of insufficient elemental sulfur
formation in this extremely reducing environment.
Seasonally aerobic conditions at the sediment-water
interface of the noneuxinic sediment would promote
elemental sulfur formation from the oxidation of H_2S that
accumulates in pore water.

Although iron-sulfide formation is inhibited in
sediment from Soap Lake, the large amounts of H_2S that
accumulate in the pore water do promote progressive
sulfidization of organic matter as shown in the depth
plots of S_{org} concentrations and $\delta^{34}S_{org}$ (Figures 8a and
8b). Living algae within the lake water column likely
have a sulfur concentration similar to that reported for
other algae (1.2 dry wt% [1]); and, assuming the algae
assimilate sulfate from the surface layer of the lake, the
$\delta^{34}S_{org}$ will be the same as the surface-water sulfate
(+10‰). Based on these assumed values, the organic
matter at the sediment-water interface has already lost
assimilatory sulfur because S_{org} concentrations in organic
matter are only 0.54% in noneuxinic sediment and 0.59% in
euxinic sediment. As shown by the isotopic composition of
the S_{org} in the shallowest samples (-13‰ in noneuxinic
sediment and -12‰ in euxinic sediment), at least 50% of
this S_{org} is bacteriogenic H_2S. S_{org} increases with depth
and its isotopic composition continues to decrease,
indicating that progressive sulfidization of the organic
matter with bacteriogenic H_2S occurs. The formation of
organosulfur occurs at a shallower depth in the noneuxinic
core than in the euxinic core (Figures 8a and 8b),
possibly because degraded organic matter contains more
functional groups to react with the H_2S, or because
depositional rates are slower above the chemocline than
below the chemocline. In both sedimentary environments,
the formation of organosulfur appears to cease near the
bottom of the core (Figure 8b). The S_{org} may never reach

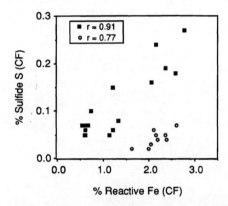

Figure 7. Weight % sulfide sulfur versus weight % reactive iron (both on a carbonate-free basis) in euxinic sediment (squares) and noneuxinic sediment (circles) from Soap Lake.

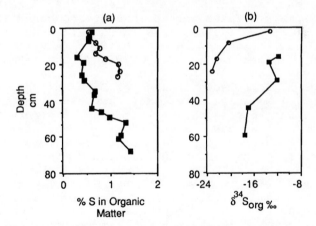

Figure 8. Depth plots of (a) sulfur content of organic matter and (b) $\delta^{34}S$ of organosulfur in euxinic sediment (squares) and noneuxinic sediment (circles) from Soap Lake.

the isotopic composition of the pore-water H_2S (-26‰ in the euxinic core and -44‰ in the noneuxinic core) because some assimilatory sulfur persists in the structure of the organic matter (Figure 8b).

The sulfur geochemistry in Soap Lake is the most complex of any for the lakes studied. High sulfate concentrations and high productivity result in large H_2S production. Less than 3% of the reactive iron in the noneuxinic sediment and less than 11% of the reactive iron in the euxinic sediment have reacted with the H_2S to form sulfide minerals, probably because high pH kinetically inhibits the reaction. As a result, the H_2S accumulates in pore water and in the monimolimnion. Monosulfides are better preserved in the sediment below the chemocline probably because the very reducing conditions prevent their conversion to pyrite. Pore-water H_2S reacts with the organic matter progressively forming more organosulfur with depth.

Green River Formation Oil Shale. The Green River Formation was deposited in two ancient lakes--Lake Uinta to the south and Lake Gosiute to the north. Lake Uinta occupied two depositional basins, Uinta basin (Utah) and Piceance basin (Colorado). Lake Gosiute occupied the Greater Green River basin (Wyoming) and was probably shallower and, at times, more oxidizing than Lake Uinta. These Green River lakes were permanent and bound in many areas by broad mud flats that were submerged during high stands (37-40). The entire body of Green River data discussed below are from three phases of the lakes collectively spanning 6 my of deposition; the phases are lake stage 1, relatively freshwater to brackish-water deposition; lake stage 2, saline deposition; and lake stage 3, transgressive-phase deposition. In Lake Gosiute, deposition in lake stage 3 was in relatively fresh water, but in the Piceance and Uinta basins, deposition was in a stratified lake that maintained a saline monimolimnion at least during the early part of the stage (13).

The average S_{min} concentration of 1.0 wt% for all samples from the Piceance basin is the highest of the modern or ancient sediments studied. The average S_{min} concentration of 0.52 wt% for all samples from the Uinta basin is about one-half of that from the Piceance basin (1.0 wt%), and data for the Greater Green River basin, averaging 0.82 wt%, are intermediate. "Reactive" iron concentrations in all three basins are fairly low (lake-stage averages range from 1.9 to 2.8 wt%), but an average of 40% is sulfidized, accounting for the generally abundant disulfides (pyrite and minor marcasite; FeS_2) and minor pyrrhotite (Fe_xS; $x \leq 1$). As observed in Soap Lake, iron sulfidization may have initially been inhibited by the high pH's required for the Green River lakes to have deposited sodium-carbonate mineral phases. If so, inhibiting conditions were eventually overcome during diagenesis as abundant sulfide-minerals formed.

Sulfate reduction in the ancient Green River lakes is not adequately modeled by the simple Rayleigh fractionation equations. The Green River sedimentary sulfide $\delta^{34}S$ values average +25‰ and are as high as +49‰. These values are more enriched in ^{34}S than any conceivable inflow sulfate whose isotopic composition was likely between 2‰ and 11‰, the range of values for sulfate in rivers flowing through Utah today (41). The extreme ^{34}S-enrichment and the abundance of Green River sulfide minerals suggests a two phase evolution of the sulfate reservoir. First, a large proportion of the reservoir was reduced (near right side of Figure 2), producing sulfides with high $\delta^{34}S$ values. This extreme reduction of what must have been a relatively large sulfate reservoir was driven by organic-matter concentrations up to 50 wt%. Reduction was never complete, however, and a small, very ^{34}S-enriched residual sulfate reservoir was necessarily present. Because the depositional rate within these lakes was very slow (0.08 millimeters rock/year), sulfate reduction took place at the sediment-water interface in contact with the monimolimnion or, alternatively within the water column itself. Sulfate continued to enter the lake, but was never added in sufficient enough quantities to completely obscure the ^{34}S-enriched signal of the residual sulfate. Second, this ^{34}S-enriched-sulfate reservoir evolved over hundreds-of-thousands, even millions, of years. Bacterial reduction continued to produce H_2S that was depleted in ^{34}S relative to the sulfate reservoir, and was extremely enriched in ^{34}S relative to any conceivable source providing sulfate into the lake. The extent of sulfur isotope evolution to ^{34}S-enriched $\delta^{34}S$ values is dramatically illustrated in the $\delta^{34}S_{min}$ depth profile for stage 2 Lake Uinta in the Uinta and Piceance basins (Figures 9a and 9b). These profiles imply that even during saline stage 2, Lake Uinta maintained a stable water column in which mixing was minimized, sulfate reduction occurred, and the H_2S/SO_4 was very high. The reversal of the long-term trend towards more ^{34}S-depleted values at the top of each of these Green River profiles coincides with the beginning of the stage 3 transgression during which the oil-shale-rich Mahogany zone was deposited. Likely, a large increase in the input of sulfate during this transgression temporarily disrupted the isotopic evolutionary trend.

The $\delta^{34}S$ depth profile for the Greater Green River
basin stage 2 samples (Figure 9c) shows little or no long-
term evolutionary trend towards ^{34}S-enriched values.
Instead, the $\delta^{34}S$ oscillates between generally very ^{34}S-
enriched values around +40‰ and ^{34}S-depleted spikes with
$\delta^{34}S$ values of +8‰. This behavior suggests that the
water-column sulfur reservoir became enriched in ^{34}S in a
much shorter period of time than it did in Lake Uinta,
perhaps reflecting a much smaller aqueous sulfur
concentration than in Lake Uinta. The isotopic spikes to
^{34}S-depleted $\delta^{34}S$ values in the Greater Green River basin
stage 2 profile correlate with volcanic-ash units. If the
aqueous sulfur reservoir was small during this stage, the
lake would have been isotopically sensitive to inputs of
ash-related sulfur, $\delta^{34}S$ of 0‰ to +6‰ (42). This
process may also explain some of the scatter in the
isotopic data from the other two basins, although their
sulfur reservoir was large enough to continue its long-
term evolutionary trend. Because discrete ash layers are
generally not preserved in our Lake Uinta cores, this
hypothesis is difficult to test.
      The average amounts of sulfur in the Green
River organic matter are slightly higher than
expected in living algae and range from 1.4 wt% in
samples from the Uinta basin to 2.5 wt% in samples
from the Greater Green River basin. The isotopic
composition of the organosulfur is generally
identical to that of the coexisting sulfide mineral,
supporting formation of organosulfur by reaction of
organic matter and bacteriogenic H_2S.
      In summary, sulfate reduction was nearly complete in
the ancient Green River lakes and relatively large amounts
of H_2S were produced. Formation of sulfide minerals may
have been initially inhibited because of high pore-water
pH's as in sediment from Soap Lake, but eventually,
sulfide minerals did form. Most of the organosulfur was
likely from sulfidization of organic matter by
bacteriogenic H_2S. Stability of the lakes over millions
of years resulted in an unusual evolution of sulfur
isotopes to extremely ^{34}S-enriched compositions. This
evolution was periodically interrupted during input of
^{34}S-depleted volcanic ash, and during transgressive stages
when there was an increased inflow of ^{34}S-depleted
sulfate.

Rundle Formation Oil Shale. The Rundle oil shale was
deposited as a 300 meter thick sedimentary section in a
freshwater, ephemeral lake; depositional cycles in the
Rundle are described by Coshell (43). A typical cycle
began with periods of multiple transgressive-regressive

events, resulting in reworking of lake sediment. The beginning of each transgressive event is marked by a layer of transported coaly plant material with gastropods. Gradually, the transgressive-regressive events give way to a permanent-water stage depositing laminated oil shale; the lake at this maximum depth was still relatively shallow, around 7 meters (45). Following the permanent-water stage, the lake sediment abruptly dried, followed by a long period of subaerial exposure before starting a new depositional cycle. Rocks deposited during individual cycles are about 5 meters thick (44). Our samples from the Rundle Formation were collected from the deeper-water lake stages.

The Rundle samples contain relatively low concentrations of total sulfur (geometric mean of 0.5 wt%) compared to the Green River samples. The average concentration of sulfur residing in sulfide minerals, 0.17 wt%, is almost as low as that in the euxinic sediment from Soap Lake, 0.12 wt%. However, in contrast to euxinic sediment from Soap Lake, the $\delta^{34}S_{min}$ values are relatively enriched in ^{34}S ($\delta^{34}S_{min}$ average = +8‰) and within the range +5 to +15‰ typical of sulfate isotopic compositions in freshwater lakes (12). Iron has not limited sulfide mineral formation because the geometric average for reactive iron is 3.1 wt% and abundant siderite is present in some samples. The sedimentary sulfide $\delta^{34}S$ values and relatively high organic-C contents (up to 50 wt%), suggest that sulfate was completely reduced and sulfide-mineral formation was sulfate limited. In contrast to Lakes Uinta and Gosiute, the ephemeral nature of the Rundle lake prevented the long-term evolution of the sulfur reservoir to values enriched in ^{34}S relative to inflow sulfate.

The isotopic composition of organosulfur in the Rundle samples cannot be used to discriminate between assimilated sulfur and sulfur incorporated into organic matter by sulfidization. Its elevated concentration, 2.3 wt% of the organic matter, however, suggests that organosulfur was, at least in part, formed by sulfidization processes.

Sulfur Models for Lacustrine Environments

Depositional and Diagenetic Behavior of Sulfur. The key controls on sulfur behavior in freshwater-lake, saline-lake, and marine sediments include the concentration of reactants for bacteriogenic H_2S formation and sulfide-mineral formation. Table IV illustrates the relative importance of the independent reactants (organic matter, dissolved sulfate, and iron) in each of the lacustrine environments studied. We have shown that in saline lacustrine environments additional factors other than simple abundance of these reactants may be important in determining their effective reactivity. In Soap Lake, high pore-water pH may be an important kinetic control on the sulfidization of iron. The Great Salt Lake data identify the major role that diffusional processes can play in the sulfur story. The Walker Lake data show that, for equivalent amounts of organic-C, more sulfide minerals form in lake sediment than in marine sediment when both are organic-matter limited.

Table IV. Relative importance of abundances of dissolved sulfate (SO_4^{2-}), sedimentary organic matter (OM), and sedimentary iron (Fe) in controlling the sulfur geochemistry in Flodelle Creek (FC), Great Lakes (GL), Walker Lake (WL), Great Salt Lake (GSL), Soap Lake (SL) sediments; and Green River (GRF) and Rundle (RF) Formations. XXX, greatest; X, least

	SO_4^{2-}	Organic Matter	Iron	other
FC	XXX			
GL	XXX			
WL		XX		OM reactivity
GSL			XX	diffusion
SL			XX	high pH
GRF	X		X(?)	high pH isotope evolution
RF	XXX			

The record of the extent of the sulfate-reduction and iron-sulfidization reactions are reflected in the amount of sedimentary sulfide minerals formed and their isotopic composition. Figure 10 schematically shows the relation between these two variables for: 1) low-sulfate lakes where sulfate availability limits sulfide-mineral formation; 2) sulfate enriched lakes where iron and organic-C limit sulfide-mineral formation; and 3) very long-lived lakes where the sulfur reservoir evolves isotopically to $\delta^{34}S_{min}$ values more enriched in ^{34}S than sulfate entering the lake. Rayleigh fractionation curves on Figure 10 are based on $\Delta_{SO_4-H_2S} = 10‰$ and complete sulfate reduction producing 0.15 wt% S_{min} in freshwater lakes; and $\Delta_{SO_4-H_2S} = 30‰$ in saline lakes. Complete sulfate reduction in saline lakes only occurs during progressive diagenesis in a closed sediment, thus the amount of S_{min} formed is difficult to predict.

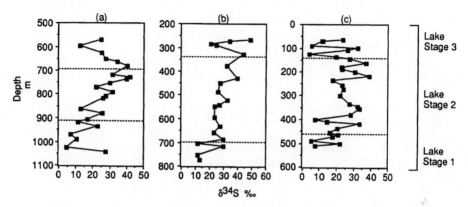

Figure 9. Depth plots of $\delta^{34}S_{min}$ in samples from the Uinta basin (a), Piceance basin sediment (b), and Greater Green River basin (c).

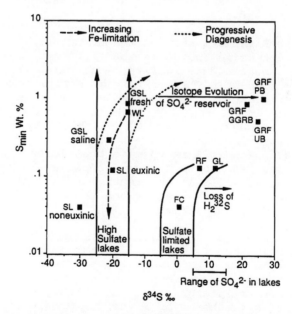

Figure 10.   Schematic showing the predicted Rayleigh behavior of weight % S_{min} and $\delta^{34}S_{min}$ in low-sulfate lakes, high-sulfate lakes with sediment in contact with lake water, and high sulfate lakes with sediment in contact with pore water and closed to lake water.

In a low-sulfate lake (lower Rayleigh fractionation curve in Figure 10), the isotopic composition of sulfur in sulfide minerals is fixed at a $\delta^{34}S$ value near the composition of the sulfate in the lake by essentially complete reduction within the sediment column. The S_{min} concentrations reflect the lake's low sulfate concentration. Preferential loss of ^{34}S-depleted H_2S from the sediment moves the $\delta^{34}S_{min}$ values to the right.

With increasing sulfate concentrations, complete sulfate reduction becomes increasingly difficult unless, during diagenesis, the pore water is isolated from the lake water and enough sulfate, organic matter, and iron are available to form sulfide minerals. The isotopic composition of the S_{min} becomes systematically more negative relative to the initial sulfate, and sulfide-mineral formation is limited by organic-C availability as in Walker Lake or iron availability as in Soap Lake and Great Salt Lake. In Soap Lake's noneuxinic sediment, the S_{min} has more negative $\delta^{34}S$ values than predicted from the "high-sulfate" curve, because of the large fractionation factor in the sediment relative to that assumed to calculate the curve.

In long-lived lakes such as proposed for deposition of the Green River oil shale, organic matter is sufficient to support extensive (but less than 100%) sulfate reduction. The residual sulfate reservoir becomes extremely enriched in ^{34}S and, given sufficient stability of the lake, $\delta^{34}S_{min}$ values isotopically evolve to values more enriched in ^{34}S than the initial sulfate composition. The lower S_{min} abundance in the Green River oil shale from the Uinta basin indicates that the sulfur reservoir in this sub-basin was smaller than in the Piceance basin, but was still large enough to allow isotopic evolution to occur. Green River lakes apparently had sulfate concentrations intermediate between those in the freshwater and saline lakes.

<u>Quantitative Model of Lacustrine Organosulfur</u>. Data for organosulfur in Soap Lake point out the importance of dissimilatory sulfur (H_2S) relative to assimilatory sulfur with respect to accumulation of sulfur in organic matter. These data have helped quantify the pathway of formation of organosulfur in the Green River and Rundle oil shale and shale oil. Figure 11 schematically shows the quantitative evolution of sulfur in lacustrine organic matter from assimilation in living algae in the lake to production of oil by retorting or catagenesis.

Algae, presumably with 1.2 wt% sulfur, is the major source of organic matter in sediment from Soap Lake. At the sediment-water interface, assimilated sulfur is

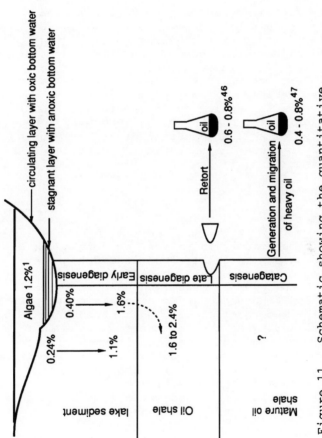

Figure 11. Schematic showing the quantitative evolution of sulfur in lacustrine organic matter. Data from 1,46-47.

preferentially lost and secondary sulfidization of the organic matter by H_2S accounts for up to 50% of the organosulfur. The organic-C concentrations in sediment deposited in the anoxic, euxinic bottom water (9.6%) is over twice that in sediment deposited in seasonally oxic waters (4.0%), reflecting the preservation of organic matter deposited in permanently anoxic waters. The dashed arrow between concentrations of organic matter deposited in anoxic waters and in the oil shale indicates that the rich oil shales were probably deposited in permanently anoxic waters. Oil products from retorting and catagenesis contain between 11 and 15% of the organosulfur in the oil-shale source rock assuming an average oil yield of 58 L/tonne (Mahogany zone of the Piceance basin, this study) and specific gravity of 0.89 (calculated from data in 47). The source of this organosulfur in the lacustrine oil has been debated. It may be indigenous to the bitumen fraction of the oil shale, or it may result from sulfidization of the bitumen by H_2S produced during decomposition of sulfide minerals or kerogen during oil generation. Because the isotopic composition of the various sulfur forms in the oil shale are similar to those in the oil (41), differentiating between these two pathways is difficult.

Conclusions

The sulfur geochemistry in saline, productive lakes is complex and does not readily lend itself to interpretation by traditional methods based on marine or freshwater-lake studies. The problem is magnified when working with ancient saline lake sediment as diagenetic overprinting may erase any record of depositional processes. Given a basic understanding of the key controls on sulfur geochemistry in the sedimentary environment, we can test certain hypotheses about the behavior of sulfur in modern lacustrine environments. The results enable us to understand much about the deposition and diagenesis of lacustrine, organic-C-rich sediments such as those forming the Green River and Rundle oil-shale deposits.

Literature Cited

1. Goldhaber, M. B.; Kaplan, J. R. In The Sea; Goldberg, D., Ed.; John Wiley & Sons: New York, 1974; Vol. 5, pp 569-655.
2. Wetzel, R. G. Limnology; Saunders College Publishing: Philadelphia, 1983; 767 p.

3. Pfennig, N.; Widdel, F. Royal Soc. London Philos. Trans. Series B 1982, 298, 433-441.
4. Westrich, J. T.; Berner, R. A. Limnnol. Oceanogr. 1984, 29, 236-249.
5. Berner, R. A.; Raiswell, R. Geochim. Cosmochim. Acta 1983, 47, 855-862.
6. Sweeney, R. E. Ph.D. Thesis, University of California, Los Angeles, 1972.
7. Rickard, D. T. Am. Jour. Sci. 1974, 274, 941-952.
8. Rickard, D. T. Am. Jour. Sci. 1975, 275, 636-652.
9. Nakai, N.; Jensen, M. L. Geochim. Cosmochim. Acta 1964, 28, 1893-1912.
10. Chambers, L. A.; Trudinger, P. A. Geomicrobiology J. 1979, 1, 249-293.
11. Nielsen,H. In Lectures in Isotope Geology; Jäger, E.; Hunziker, J. C., Eds.; Springer-Verlag: Berlin, 1979; pp 283-312.
12. Holser, W. T.; Kaplan, I. R. Chem. Geol. 1966, 1, pp 93-135.
13. Tuttle, M. L. Ph.D. Thesis, Colorado School of Mines, Golden, 1988.
14. Goldhaber, M. B.; Tuttle, M. L.; Baedecker, M. J. Geol. Soc. Am. Annual Mtg., 1984, abstract no. 33625.
15. Tuttle, M. L.; Goldhaber, M. B. Geol. Soc. Am. Annual Mtg., 1986, abstract no. 108714.
16. Tuttle, M.L.; Goldhaber, M. B.; Rice, C. A. Geol. Soc. Am. Annual Mtg., 1987, abstract no. 130630.
17. Donnell, J. R. Oil and Gas J. 1980, 78, pp 218-224.
18. Rowley, P. G.; Brown, T. In Oil Shale--the Environmental Challenges II; Petersen, K. K., Ed.; Colorado School of Mines: Golden, 1982; pp 1-28.
19. Johnson, S. Y.; Otton, J. K.; Macke, D. L. Geol. Soc. Am. Bull. 1987, 98, pp 77-85.
20. Zielinski, R. A.; Otton, J. K.; Wanty, R. B.; Pierson, C. T.; Chem. Geol. 1987, 62, pp 263-289.
21. Nriagu, J. O.; Coker, R. D. Limnol. Oceanogr. 1976, 21, pp 485-489.
22. Benson, L. V. Quat. Research 1981, 16, pp 390-403.
23. Benson, L. V.; Spencer, R. J. U.S. Geol. Survey Open-File Report 83-740 1983, 53 p.
24. Benson, L. V.; Mifflin, M. D. U.S. Geol. Survey Water Res. Inv. 85-4262 1986, 14 p.
25. Eugster, H. P.; Hardie, L. A. In Lakes--chemistry, geology, physics; Lerman, A.; Ed.; Springer-Verlag: New York, 1978; pp 237-294.
26. Utah Dept. of Natural Resources Bull. 116; Gwynn, J. W.; Ed.; 1980; 400 p.
27. Anderson, G. C. Limnol. Oceanogr. 1958, 3, 259-270.

28. Edmondson, W. T. In _Limnology in North America_;
     Frey, D. G.; Ed.; Univ. of Wisconsin Press:
     Madison, 1963; pp 371-392.
29. Friedman, I.; Redfield, A. C. _Water Res. Research_
     1971, _7_, 874-898.
30. Tuttle, M. L.; Goldhaber, M. B.; Williamson, D. L.;
     Talanta 1986, _33_, 953-961.
31. Tissot, B. P.; Welte, D. H. _Petroleum Formation and
     Occurrence_, Second Edition; Springer-Verlag:
     Berlin, 1984; 699 p.
32. Berner, R. A. _Geochim. Cosmochim. Acta_ 1984, _48_,
     605-615.
33. Ingamells, C. O. _Anal. Chim. Acta_ 1970, _52_, 323-334.
34. Broecker, W. S.; Peng, T. _Tracers in the Sea_;
     Lamont-Doherty Geol. Observatory Pub.: New York,
     1982; 690 p.
35. Berner, R. A. In _The Black Sea--geology, chemistry,
     and biology_; Degens, E. T.; Ross, D. A.; Eds.; Am.
     Assoc. Pet. Geol. Memoir 20: Tulsa, 1974; pp 524-
     531.
36. Leventhal, J. S. _Geochim. Cosmochim. Acta_ 1983, _47_,
     133-137.
37. Eugster, H. P.; Surdam, R. C. _Am. Assoc. Pet. Geol.
     Bull._ 1973. _84_, 1115-1120.
38. Ryder, R. T.; Fouch, T. D.; Elison, J. H. _Geol. Soc.
     Am. Bull._ 1976, _87_, 496-512.
39. Johnson, R. C. _Geology_ 1981, _9_, 55-62.
40. Johnson, R. C. In _Cenozoic Paleogeography of Western
     United States_; Flores, R. M.; Kaplan, S. S.; Eds.;
     Rocky Mountain Paleogeography Symposium 3; Society
     of Economic Paleontologists and Mineralogists:
     Denver, 1985; pp 247-276.
41. Mauger, R. L. _24th Int'l. Geol. Congress_, 1972, pp
     19-27.
42. Rye, R. O.; Luhr, J. F.; Wasserman, M. D. _J. Volc.
     Geothermal Research_ 1984, _23_, 109-123.
43. Coshell, L. _Proc. 1st Australian Workshop on Oil
     Shale_, 1983, pp 25-30.
44. Patterson, J. H. _Chem. Geol._ 1988, _68_, 207-219.
45. Crisp, P. T.; Ellis, J.; Hutton, A. C.; Korth, J.;
     Martin, F. A.; Saxby, J. D. _Australian oil Shales: a
     compendium of geological and chemical data_, CSIRO
     Institute of Energy and Earth Resources: North Ride
     NSW, Australia, 1987.
46. Ingram, L. L.; Ellis, J.; Crisp, P. T.; Cook, A. C.
     Chem. Geol. 1983, _38_, pp 185-212.
47. Peer, E. L. _Proc. 1st Int'l. Conf. on Future of
     Heavy   Crude   Oils and Tar Sands_, 1979, pp
     651-654.

RECEIVED March 16, 1990

Chapter 8

Characterization of Organic Matter in Sulfur-Rich Lacustrine Sediments of Miocene Age (Nördlinger Ries, Southern Germany)

Jürgen Rullkötter, Ralf Littke, and Rainer G. Schaefer

Institute of Petroleum and Organic Geochemistry (ICH–5) at the Research Center (KFA) Jülich, D–5170 Jülich 1, Federal Republic of Germany

A sequence of lacustrine sediments in the Nördlinger Ries, a large impact crater in southern Germany, contains a highly bituminous "laminite" series with up to 25% total organic carbon and 5% sulfur. Deposition under saline conditions is indicated by the presence of dolomite, calcite and (rare) gypsum as well as a complex mixture of organic sulfur compounds (OSC) in the extractable organic matter portion which is remarkably large considering the low thermal maturity. The organic facies in the laminite is variable. This is reflected by bulk properties of the organic matter, the composition of organic particles as well as the distribution of various biological marker compound classes. Correlation of bulk, microscopic and molecular data is not straightforward and is impeded by the fact that the structures and/or the geochemical significance of a number of abundant OSC are presently unknown.

Previous studies of core samples from research well Nördlingen-1973 revealed unusually high bitumen contents in the lacustrine sediments of the Nördlinger Ries (1-2). This was corroborated by "oil" shows observed during later exploration drilling in this area. The high bitumen contents were in contrast to the assumed mild temperature history of the sediments. Temperatures usually required for the thermal generation of large quantities of bitumen (3) have never been reached in the Nördlinger Ries sediments (1). Present bottom-hole temperatures, although not measured, are certainly below 30°C, and they are unlikely to have been much higher in the past as indicated by vitrinite reflectance values below 0.35% in all samples studied (this work and 1). Early generation of asphaltic material is known, however, from sulfur-rich (marine) carbonate sediments and evaporite deposits (3-5). It was the objective of this study to characterize the organic matter in the most oil-prone, highly bituminous "laminite" series of the sediments in the Nördlinger Ries in terms of their depositional environment and possible vertical and lateral organofacies variations.

0097–6156/90/0429–0149$06.25/0

Geological Background

The Nördlinger Ries is a circular shaped Miocene sedimentary basin 20 km in diameter and located about 100 km northwest of Munich in southern Germany. It was formed by meteorite impact into the Late Jurassic carbonate sediments of the Schwäbische Alb mountain range (6) about 15 Ma ago (7). The extended material formed a rim around the crater which prevented freshwater supply except by rainfall. The climate in southern Germany was semi-arid during the Miocene (8). The basement of the crater consists of suevite (meteorite impact brecciae) containing high-pressure modifications of quartz.

Post-impact sediments were penetrated by research well Nördlingen-1973 near the center of the basin and described in detail by Jankowski (8). They consist of a "basal unit" overlying the suevite, a 140 m thick laminite series, 60 m of marl and a clay layer at the top. Bituminous sediments mainly occur in the laminite series (Figure 1). In the Nördlingen-1973 well, in which the laminite series was recovered between 256 and 111 m depth, Jankowski (8) distinguished four subunits: a basal clinoptilolite subunit very rich in organic matter (256-244 m), an analcime subunit with low bitumen concentration (244-195 m), a bituminous subunit (195-145 m) and a diatomaceous subunit with lower bitumen content (145-111 m). Deposition of these sediments was estimated to have extended over a period of 0.3 to 2 Ma (8).

Mineralogically the sediments of the laminite series consist of carbonates (mainly dolomite and calcite), various clay minerals, zeolites, opal, quartz and rare gypsum (8). The occurrence of gypsum, based on our X-ray diffraction (XRD) data, is restricted to the marl unit. Deposition, according to Jankowski (8), occurred in a periodically evaporitic, stagnant lake. The high bitumen concentrations most probably were responsible for the preservation of unusual minerals such as Mg-rich calcites (up to 25% Mg; 9) and bituminous smectites (Müller, G., University of Heidelberg, personal communication, 1989).

Samples

The 27 samples investigated in this study are from four exploration wells of BEB Erdgas und Erdöl GmbH which were drilled in the Nördlinger Ries in the early 1980s. Well NR-10 is in the center of the Ries basin close to research well Nördlingen-1973. In both wells, facies and thickness of the lithologic subunits are very similar to each other (Figure 1). Bituminous sediments in well NR-10 mainly occur between 250 and 236 m and between 180 and 150 m depth. The former sequence corresponds to the clinoptilolite subunit and the latter to the bituminous subunit of Jankowski (8). Wells NR-20 and NR-30 are also in the central basin within a few kilometers from well Nördlingen-1973. Well NR-40 is located at the margin of the basin. In wells NR-20 and NR-30 bituminous sediments occur in the entire lower half of the lacustrine sediments, in particular close to the base of the laminite series. This base in NR-20 is 30 m deeper and in NR-30 30 m shallower than in NR-10. In well NR-40 bituminous sediments were encountered near the base of the sediment sequence (90 - 100 m; Figure 1).

Samples were selectively taken from the most bitumen-rich layers in the four wells and thus are not representative of the entire sediment sequence in the Nördlinger Ries.

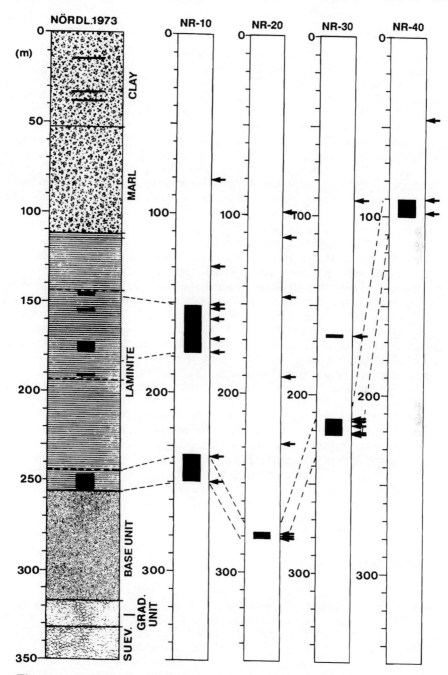

Figure 1. Rough correlation of lithologic units in wells NR-10 through NR-40 with those in research well Nördlingen-1973. Black intervals indicate sections particularly rich in organic matter. Sampling is shown by arrows.

Experimental Methods

For microscopic studies, whole rock samples were embedded in a resin. The resultant block was ground flat and polished perpendicular to the bedding of the sediment. Maceral abundances were determined using a point counting procedure first in fluorescence mode and subsequently in reflected white light on a Zeiss Axiophot microscope at 600x magnification. More details of the preparation and counting procedures were described by Littke et al. (10).

Sediments were dried at 50°C and ground prior to extraction which was performed using a modified flow-blending technique (11) with dichloromethane as solvent. Total extracts were separated by medium pressure liquid chromatography into saturated hydrocarbons (including olefins), aromatic hydrocarbons (including sulfur aromatics and saturated sulfides) and heterocomponents (12). The saturated hydrocarbon fractions, dissolved in *n*-heptane, were analyzed by gas chromatography (GC); conditions: 25 m x 0.32 mm i.d. fused silica capillary coated with CP Sil 8 silicone (0.19 μm film thickness). The oven temperature was programmed from 80°C to 300°C at 4°C/min. The split ratio was 1:100. Squalane was used as internal standard. For the analysis of the aromatic hydrocarbon fractions (in *n*-hexane) the following GC conditions were used: 15 m x 0.25 mm i.d. fused silica capillary coated with immobilized poly-methylsiloxane (0.1 μm film thickness). The oven temperature was programmed from 110°C to 350°C at 3°C/min. The split ratio was 1:25. Samples were injected using the programmed temperature vaporization technique (KAS System, Gerstel). Data evaluation was achieved by a chromatography data system (Multichrom, VG Laboratory Systems) connected to the gas chromatographs (Hewlett Packard models 5710 and 5890, respectively).

Compound identifications were made by combined gas chromatography-mass spectrometry (GC-MS) based on relative retention times and mass spectral interpretations. The instrument used was a Finnigan 5100 computerized GC-MS system equipped with a 50 m x 0.32 mm i.d. fused silica capillary coated with CP Sil 8 CB (0.25 μm film thickness). Helium was used as carrier gas and the temperature program was as follows: 110°C (2 min)- 3°C/min - 320°C.

In some cases the presence of sulfur in certain constituents of the aromatic hydrocarbon fractions was confirmed by GC with a sulfur-specific detector (FPD); for conditions see (13).

Bulk Sediment Properties

Total Organic Carbon Contents (TOC). In the 27 samples investigated for this study the TOC values vary between 1.9 and 25.5% (Table I). The four samples from the marl series (shallowest sample in each well; Figure 1) have TOC values in the lower part of this range (1.9 - 2.9%). The bituminous subunit of the laminite series in well NR-10 is particularly rich in organic matter.

Volume percentages of organic matter identifiable under the microscope only correlate well with total organic carbon contents if the unstructured groundmass is included in the volume calculation. From a relationship between TOC values, average densities of the rock matrix and the organic matter we deduce that in most samples the microscopic kerogen evaluation accounted for at least 80% of

Table I: Information on sample origin, carbonate content and bulk organic matter properties of sediments from the bituminous laminite series of the Nördlinger Ries (for abbreviations see text). Carbonate content is expressed as $CaCO_3$. The residue not recoverable from the liquid chromatography column accounts for the remainder of 100%. n.d. = not determined.

Well	Depth (m)	$CaCO_3$ (%)	TOC (%)	S_{tot} (%)	Macerals (%) terrigenous	alginite	groundmass	Extract (mg/g TOC)	Fractions (%) saturated HC	"aromatic HC"	hetero-compounds	HI (mg HC/g TOC)
NR-10	81.5	4.6	2.5	2.7	19.6	80.4	0	16.5	16.5	3.9	57.2	269
NR-10	130.0	22.2	7.2	1.9	4.1	95.9	0	37.8	4.1	4.1	86.8	476
NR-10	151.5	0	12.3	1.7	2.5	97.5	0	132.4	3.7	2.2	66.1	470
NR-10	153.8	0	15.0	1.4	1.5	3.2	95.3	65.6	5.6	0.9	69.6	823
NR-10	159.5	3.3	11.9	1.8	1.6	98.4	0	112.9	7.0	2.1	73.8	474
NR-10	170.5	1.7	25.5	1.0	0.6	2.7	96.6	45.3	6.8	1.2	74.4	920
NR-10	178.0	12.4	4.5	1.3	0	100.0	0	45.1	11.9	1.8	65.3	470
NR-10	236.5	16.6	3.5	2.8	2.5	36.0	61.5	78.8	12.4	3.3	65.0	465
NR-10	250.0	7.5	14.4	3.0	0	6.1	93.9	81.5	11.3	4.7	65.6	863
NR-20	98.1	3.5	2.9	1.5	23.7	76.3	0	12.2	21.7	5.1	n.d.	329
NR-20	112.5	0.5	4.2	1.8	51.8	48.2	0	185.0	2.3	0.7	26.8	340
NR-20	145.7	17.9	1.7	1.4	6.2	93.8	0	30.4	24.0	0.5	76.4	177
NR-20	90.5	15.5	7.7	1.3	2.4	23.8	73.8	56.6	6.6	0.2	50.3	620
NR-20	227.5	28.0	3.4	0.5	0	100.0	0	70.6	11.3	1.4	68.1	564
NR-20	278.5	4.8	4.5	4.8	12.4	27.0	60.6	62.6	6.9	2.8	35.8	400
NR-20	279.1	0	5.1	3.9	27.5	72.0	0	67.9	3.7	2.8	28.1	410
NR-20	280.0	32.4	4.0	3.3	5.6	16.9	77.5	159.8	5.4	3.2	24.3	492
NR-30	92.7	22.8	2.1	4.8	8.5	41.5	50.0	18.3	20.1	1.3	64.9	324
NR-30	167.8	15.0	7.8	2.0	5.1	80.9	14.0	65.1	9.2	2.1	69.2	539
NR-30	214.0	33.4	3.7	1.5	n.d.	n.d.	n.d.	51.6	7.3	5.8	52.4	617
NR-30	215.1	5.8	10.3	3.5	1.3	27.9	70.8	116.7	6.4	7.0	52.8	626
NR-30	218.0	38.4	5.8	2.3	0	63.0	37.0	45.5	6.4	5.0	42.0	683
NR-30	221.4	26.8	2.8	2.7	4.1	95.9	0	72.9	8.3	4.0	25.1	515
NR-30	222.9	0	8.4	4.3	8.1	30.2	61.7	59.6	4.3	6.5	33.1	480
NR-40	46.8	26.3	1.9	1.2	0	100.0	0	67.6	10.1	3.9	67.8	325
NR-40	91.3	23.0	4.6	2.2	2.9	61.9	35.2	89.6	11.1	5.4	71.2	584
NR-40	98.4	15.0	7.7	3.2	5.0	24.1	70.9	56.6	6.8	3.9	64.3	594

the total organic matter present. The organic matter is distributed homogeneously within the sediment, *i.e.*, there is rarely a distinct layering of organic and inorganic matter as observed, *e.g.*, in the Messel Oil Shale, a freshwater-lacustrine deposit of Eocene age (14). On the other hand, the type of the organic matter in the Nördlinger Ries sediments was found to differ vertically over short distances so that several organic facies zones could be distinguished in some of the polished blocks studied under the microscope.

Mineralogy. The most important mineral constituents were determined by X-ray diffraction for five samples (four of them from NR-10). Quartz and opal (mainly CT) are important components in all five samples. In a marl (NR-10, 81.5 m), calcite, kaolinite, pyrite, gypsum and illite/smectite mixed-layer minerals occur in addition to those.

Two sediments from the bituminous subunit (NR-10, 151 and 170 m) are almost free of carbonate. Opal and quartz are dominant in these samples and are accompanied by mixed-layer clay minerals, kaolinite and illite. In the most organic-carbon-rich sediment (NR-10, 170.5 m; 25.5% C_{org}) opal is present as amorphous opal A and not as opal CT like in the other sample. Furthermore, the carbonate signals in the X-ray diffractogram are extremely broad. This indicates that the minerals in this sediment are diagenetically less altered than those in the other samples studied.

The sediment from 236 m depth in well NR-10 is from the clinoptilolite subunit and its main mineral constituents are dolomite, quartz, analcime and opal, together with some clay minerals and pyrite, as well as a little siderite and calcite. A similar mineral composition was determined for a marl sample from 46 m depth in well NR-40 which contains additional gypsum, more calcite and less analcime, however.

Sulfur Content. Total sulfur contents of the sediments are highly variable between 0.5 and 5% (Table I). Total sulfur includes gypsum, but because this is rare and apparently restricted to the marl unit, its contribution is considered to be of low significance. There was no indication of the presence of free sulfur in the XRD traces. Overall, the sediments from well NR-10 contain slightly less sulfur than those from the other three wells. The highest sulfur concentrations in all four wells occur at the base of the laminite series (Figure 2). Although some pyrite has been detected by X-ray diffraction, most of the sulfur appears to be bound in the organic matter. Sulfur and iron determinations on kerogen concentrates (which usually occlude most of the pyrite) of four of the most organic-matter-rich sediments revealed high sulfur but very low iron concentrations (Table II). The low iron and thus pyrite contents are due to the limited iron content of the carbonate-rich sediments in the rim of the impact crater which was the main source of mineral matter deposited in the crater basin. The iron contents in the four kerogen samples differ greatly, *i.e.*, they vary by a factor of five.

In a similar way, organic sulfur contents are highly variable. In the isolated kerogens, S_{org}/C_{org} ratios differ by a factor of three. Only two of them would be termed Type II-S kerogens in a strict sense (atomic $S_{org}/C_{org} > 0.04$; 15). Both are from the basal part of the laminite series (*cf.* Figure 2), and although further kerogen isolations were not made, a rough calculation assuming low iron

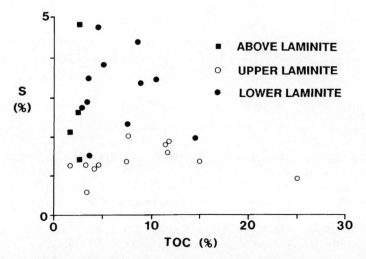

Figure 2. Total sulfur content (wt.% in rock) versus total organic carbon content (TOC) for the marl section above the laminite (squares) and in the upper (open circles) and lower laminite sections (closed circles).

contents throughout indicates that Type II-S kerogens only occur in the lower laminite whereas all of the upper laminite samples formally are regular Type II kerogens (Figure 2). Inspection of the other analytical data (*e.g.*, extract yields) shows, however, that this distinction may be artificial and misleading in the case of the Nördlinger Ries sediments. For example, based on atomic H/C ratios all of the samples in Table II would be Type I kerogens according to Tissot and Welte (3). In general, the atomic S_{org}/C_{org} ratios of most of these sediments, particularly those of the lower laminite series, are more typical of marine sediments. The high availability of sulfate in the former Nördlinger Ries lake is a less common feature of lacustrine depositional settings.

Table II: Carbon, sulfur, nitrogen and iron elemental data of kerogens (containing pyrite) isolated from Nördlinger Ries core samples. H/C, N/C and S_{org}/C_{org} are atomic ratios.

Well	Depth (m)	TOC (%)	H/C	N/C	S_{tot} (%)	Fe (%)	S_{pyrite} (%)	S_{org}	S_{org}/C_{org}
NR-10	151.5	49.5	1.66	0.03	4.50	0.78	0.90	3.60	0.027
NR-10	170.5	57.8	1.85	0.01	3.93	0.24	0.28	3.65	0.024
NR-10	250.0	50.5	1.86	0.02	7.43	1.28	1.47	5.96	0.044
NR-30	222.9	43.2	1.82	0.02	9.30	0.53	0.61	8.69	0.075

Kerogen Composition. The bulk of the organic matter in the Nördlinger Ries sediments consists of brightly fluorescing alginite (Table I). Most of the individual alginite particles are smaller than 20 μm in diameter (alginite B; 16) and thin-walled (<2 μm). In a few samples, the alginites form layers that are derived from either planktonic algal blooms or benthic algal mats. Although a great variety of alginites in terms of size, shape and fluorescence was found in different samples, only a few of them are large and specific enough to be related to extant species. For example, Botryococcus alginite was observed in only a single sample (NR-10, 130.0 m).

Exceptional are some of the most organic-matter-rich sediments which mainly contain a brightly fluorescing, almost homogeneous groundmass around large mineral grains instead of well-preserved and identifiable alginite particles. The organic nature of this groundmass becomes evident from a comparison of TOC values and volume percentages of macerals. The high TOC values can only be explained if the fluorescing groundmass consists almost entirely of organic matter. Occasional alginite within the groundmass indicates that it is derived from algae which were diagenetically reworked and homogenized (structurally degraded) by bacteria. The deposition of large amounts of extremely small, *i.e.*, submicroscopic planktonic organisms together with a few larger algae as precursor material of the groundmass seems to be less likely but cannot be completely ruled out. On the other hand, there is no correlation between the hopane abundances in the hydrocarbon fractions and the percent groundmass, but if the hopanoid content is indeed a true quantitative indicator of bacterial activity a major part of this information may as well be hidden in the polar extract fractions and in the kerogen.

Organic particles from higher land plants (sporinite, vitrinite, inertinite) are subordinate in the Nördlinger Ries sediments. Their concentration is higher in the uppermost, less bituminous part of the sequence than in the central and basal parts (Table I).

There is a tendency of increasing hydrogen indices (HI) from Rock-Eval pyrolysis with increasing TOC values (Table I) but not a strict correlation of these two parameters. Only three samples have HI values in excess of 800 mg hydrocarbons/g TOC. All of them have TOC values higher than 15%, and these hydrogen-rich kerogens, which would be termed Type I according to the nomenclature of Tissot et al. (17) consist almost entirely of a microscopically structureless groundmass.

In most of the sediments the hydrogen indices vary between 450 and 650 mg HC/g TOC (Table I), although the TOC values range from about 3 to 12%. The kerogens of some consist of alginite, but in others is a mixture of alginite and structureless groundmass. There seems to be no systematic variations between the maceral, HI and TOC relationships within this group of samples.

Sediments with HI values below 400 mg HC/g TOC mainly comprise marl samples. In general, all these relatively hydrogen-lean samples contain a significant proportion of terrigenous organic particles (18.3% terrigenous material on average in samples with HI < 400 and 4.4% in samples with HI > 400).

Hydrocarbon Distributions

The extract yields of the Nördlinger Ries sediments, normalized to organic carbon, differ greatly and, with the exception of the marl samples, are much higher than would have been expected for immature organic matter at the shallow depth setting (Table I). Such anomalies are known for organic matter in some carbonates (4, 18-20), evaporites (5) and some diatomites (15), and are related to the high sulfur content of the organic matter in sediments deposited under strongly reducing conditions and with a limited supply of iron in the presence of sufficient amounts of sulfate (Type II-S kerogens).

Saturated plus olefinic hydrocarbons (collectively called saturated hydrocarbons for simplicity here) account for about 5-15% of the extractable organic matter in most of the samples (higher values in the marls, Table I). There is a rough tendency of decreasing relative proportions of saturated hydrocarbons with increasing organic carbon content; *i.e.*, heterocomponents and highly polar asphaltic components are particularly abundant relative to the hydrocarbon fractions in the sediments with the highest bitumen contents. The "aromatic hydrocarbons" in almost all samples are only a small fraction of the total hydrocarbons in the extract.

The saturated hydrocarbon distributions of the marl samples are dominated by long-chain *n*-alkanes of higher land plant origin (21) with a strong odd-over-even carbon number predominance. Hopanoid hydrocarbons are the next most abundant constituents, but other hydrocarbons particularly abundant in the laminite samples described hereafter are also clearly recognizable.

Four types of saturated and olefinic hydrocarbons are characteristic of nearly all organic-matter-rich laminite sediments (Figures 3a-c). This is independent of the presence of alginite as the dominant maceral or a predominance of the

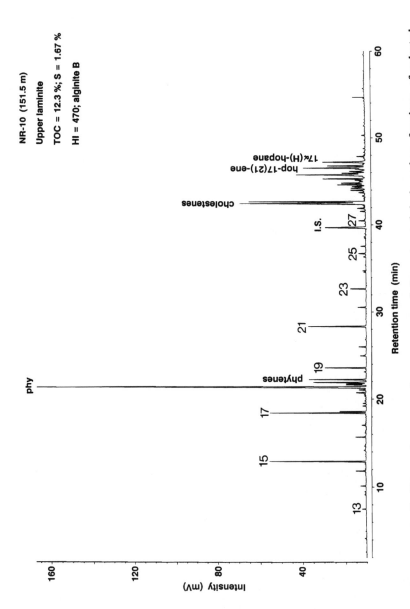

Figures 3a–d. Capillary column gas chromatograms of the saturated hydrocarbon fractions of selected samples from the laminite series of the Nördlinger Ries. *n*-Alkanes are indicated by number of carbon atoms; phy = phytane, G = gammacerane, G′ = gammacerane-2-ene, I.S. = internal standard (squalane).

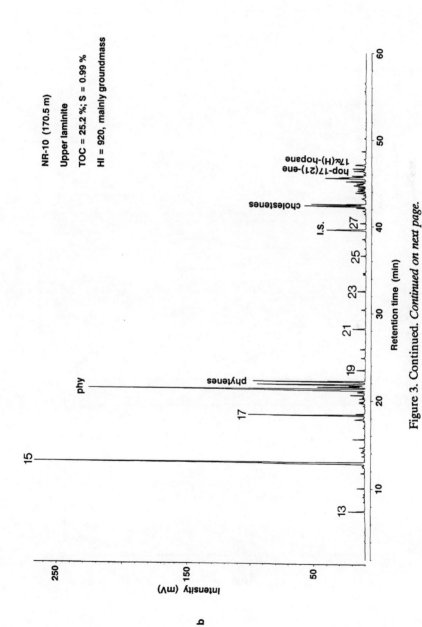

Figure 3. Continued. *Continued on next page.*

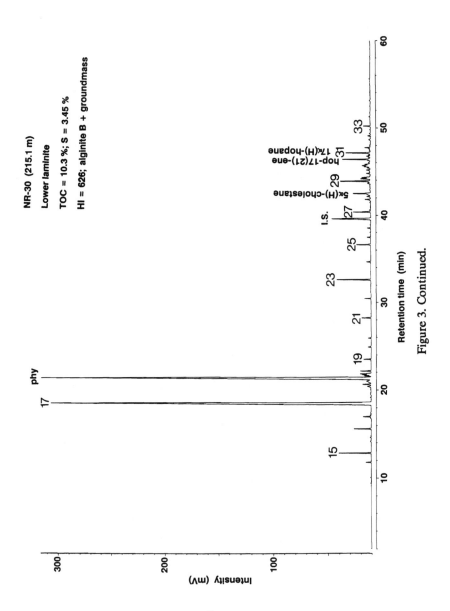

Figure 3. Continued.

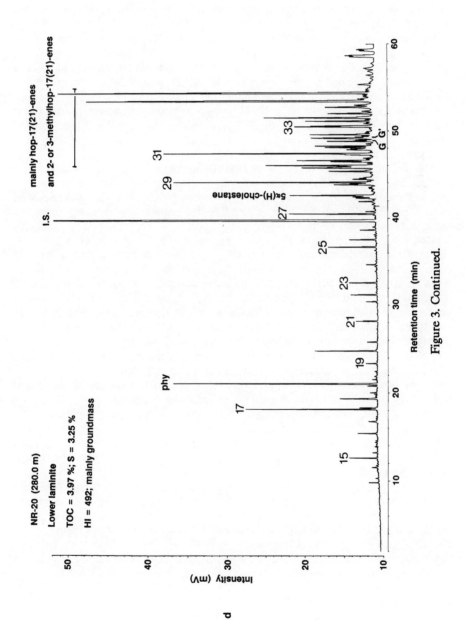

Figure 3. Continued.

microscopically structureless organic groundmass. First, phytane and phytenes are particularly abundant in most samples (Figure 3a). Phytane/pristane concentration ratios are greater than 10 in all cases and indicate an enhanced salinity in the water column during the time of sediment deposition (22). Thus, the high phytane concentrations may be related to archaebacterial lipids as well as chlorophyll. Secondly, the next most abundant saturated hydrocarbons, or the predominant compounds in some cases, are the C_{15} and/or C_{17} n-alkanes (Figures 3b,c) which are presumably derived from higher planktonic algae (e.g., green algae; 23). The relative concentrations of these two straight-chain alkanes differ strongly, but there is no obvious systematics in these variations. Other n-alkanes, although mostly much less abundant, ranged from C_{19} to C_{27} in carbon number with a maximum at $n\text{-}C_{21}H_{44}$ and a strong odd-over-even carbon number predominance. Their origin, particularly that of the lower carbon number species, is less obvious (planktonic, bacterial, or [bacterially] shortened wax alkanes?).

Thirdly, in the polycyclic saturated hydrocarbon range, the dominant components usually are the isomeric pair cholest-4-ene and cholest-5-ene (Figures 3a,b) accompanied by smaller amounts of C_{28} and C_{29} sterenes and steradienes. Finally, hop-17(21)-ene and $17\alpha(H)$-hopane are the only abundant hopanoids of bacterial origin (24). The absence of $17\beta(H)$-hopanes and the limited range of hopanoids is surprising considering the low level of maturity (presence of phytenes and cholestenes), and in this respect the hopanoid distribution is different from various deep sea sediments at a similar diagenetic stage (e.g., 25-27). Equally unusual, in some samples the sterenes are almost absent and replaced by $5\alpha(H)$-cholestane (Figures 3c,d). The depth difference is only a few tens of meters which excludes a simple thermal effect and indicates non-thermal processes to be responsible for the relatively early occurrence of these stable saturated hydrocarbons. Phytenes and cholestenes, for example, are absent or of low abundance in the samples with the higher sulfur contents. This may be the result of preferential or more complete removal of the olefins, as a result of the addition of H_2S, and the relative enrichment of the saturated hydrocarbons in the saturated + olefinic hydrocarbon fractions.

The sediment from 280.0 m of well NR-20 is an exception with respect to the saturated hydrocarbon distributions within the laminite series. Here, a complete series (C_{27}, C_{29}-C_{35}) of hop-17(21)-enes and 2β- or 3β-methylhop-17(21)-enes (as approximately 1:1 mixtures) are the most significant constituents (Figure 3d). Both series exactly coelute on the GC column used, and the C_{35} members (C_{36} for the methylated series) are by far the most abundant homologs (the two largest peaks, apart from the internal standard in Figure 3d represent the respective 22S- and 22R-epimers). There is no microscopic or other (e.g., mineralogical) evidence to explain the exceptional character of the hydrocarbon distribution of this particular sediment sample. The relatively high abundance of gammacerane and gammacerene in this sample has been reported by ten Haven et al. (28). These compounds were not found in the other samples studied.

Organic Sulfur Compounds (OSC)

The "aromatic hydrocarbon" fractions of the laminite sediments from the Nördlinger Ries mainly contain OSC (thiophenes, thiolanes and unidentified sulfur compounds). Pure aromatic hydrocarbons are present only in minor concentrations. The types and distributions of OSC are highly variable. Three examples are shown in Figures 4a-c.

C_{20} isoprenoid thiophenes are always the most abundant single sulfur-bearing compounds (Figures 4a-c). They occur as two major isomers with the 3-methyl-2-(3,7,11-trimethyldodecyl)thiophene dominating over the 3-(4,8,12-trimethyltridecyl)thiophene (Figures 4a,b). Occasionally, they are accompanied by the corresponding thiolanes (Figure 4b). Higher and lower isoprenoid thiophene and thiolane (pseudo)homologs occur in distribution patterns differing from sample to sample. For example, the lower laminite sample from well NR-30 (215.1 m) mainly contains C_{20} to C_{25} isoprenoid thiophenes (Figure 4b). A corresponding lower laminite sample from well NR-10 (250.0 m) displays a complex mixture in the C_{16} to C_{20} carbon number range (Figure 4c). n-Alkylthiophenes are less abundant, 2-tetradecylthiophene usually being the most significant member in this series (Figures 4a,b). Exceptional are the C_{33} thiolanes and thiophenes in the sample from 250.0 m of well NR-10 (Figure 4c), which probably also have an unbranched carbon skeleton. The same is true for the C_{37} and C_{38} mid-chain thiolanes (29) in the same sample (Figure 4c). While the latter are probably related to the long-chain unsaturated ketones known to occur in Prymnesiophytae algae (30), there is no clue as to the biogenic precursor of the C_{33} thiophenes and thiolanes.

The upper laminite sample from NR-10 (151.5 m), besides the 2-tetradecylthiophene, contains an isomer with an identical mass spectrum which according to the slightly shorter retention time may have a methyl-branched carbon skeleton (Figure 4a). Small amounts of a thienylhopane are present in all samples studied, but sulfur-bearing steroids could not be detected despite the abundance of sterenes in the saturated hydrocarbon fractions.

A variety of unidentified sulfur species in the "aromatic hydrocarbon" fractions of the Nördlinger Ries sediments are present in addition to the known OSC. The sample from 151.5 m in well NR-10 contains a series of isomers and homologs with very simple mass spectra. Molecular ions are at m/z 440, 454, 468 or 482 and the base peaks either at m/z 124 or at m/z 138 (Figure 4a); there are no further significant fragments in the spectra. A complex mixture of isomers with a molecular weight of 538 occurs in the sample from 215.1 m depth of well NR-30 (Figure 4b). Fragment peaks, *e.g.*, at m/z 280, 313 and 341 may indicate that these compounds are long-chain molecules with a sulfur moiety in the middle of the chain. In each case, these unknown OSC are abundant components of the "aromatic hydrocarbon" fraction. They may contain significant geochemical information which will be released only when the precise structures have been elucidated.

Concluding Discussion

The results of the geochemical and petrographic characterization of the organic matter in the Nördlinger Ries sediments should be considered preliminary. At this stage, they provide a coherent picture of the gross depositional environment and explain the occurrence of the high concentrations of free bitumen. On the other hand, details of the variations of several geochemical parameters and of the molecular signatures in the extractable material still need further clarification.

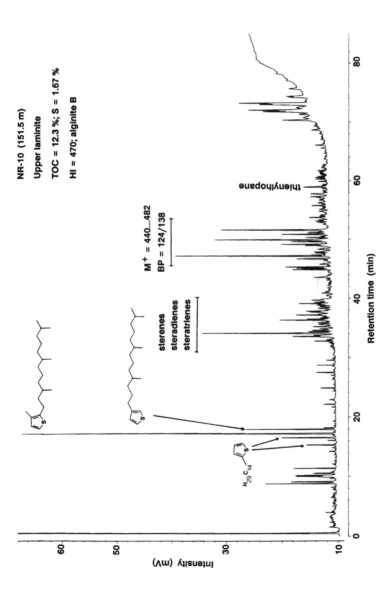

Figures 4a–c. Capillary column gas chromatograms of the "aromatic hydrocarbon" fraction of selected samples from the Nördlinger Ries. M⁺ = molecular ion, BP = base peak in the corresponding mass spectra. The sample from 151.5 m of well NR-10 contains steroid olefins due to incomplete liquid chromatography separation.

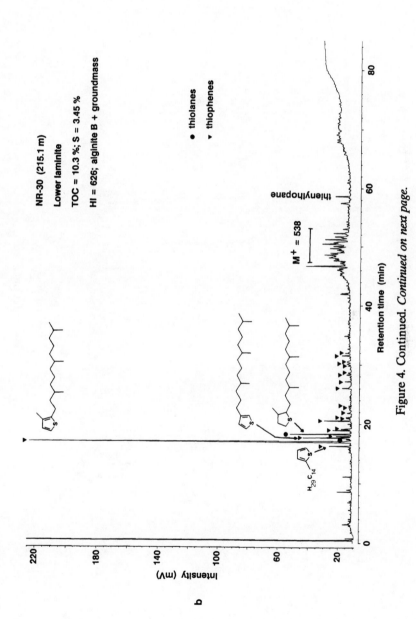

Figure 4. Continued. *Continued on next page.*

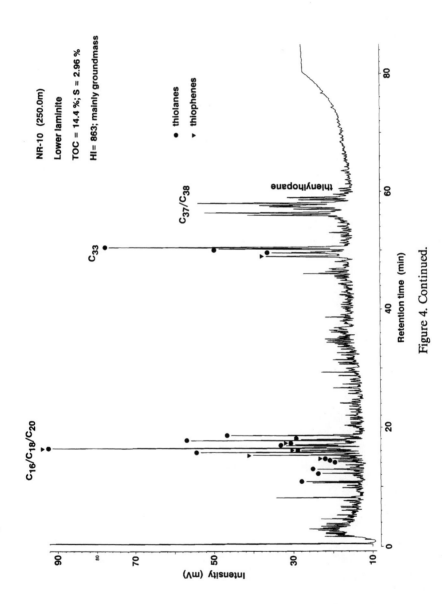

Figure 4. Continued.

Small algae (alginite B) are the main contributors of the sedimentary organic matter in the Nördlinger Ries; bacterial biomass and higher plant debris are subordinate components. In those samples which contain a structureless groundmass as the main microscopically recognizable maceral type, supposed to be due to bacterial reworking of the algal material, the dominant molecular signal is still that of the algae in most cases. How this can have raised the hydrogen index is not clear.

High sulfur contents of the sediments, mainly bound to the organic matter, indicate that strongly reducing conditions prevailed probably due to salinity stratification of the water column and oxygen depletion in the lower part of it. Although the occurrence of carbonate minerals and (occasionally) gypsum indicates that the former Nördlinger Ries lake was sometimes evaporitic, the molecular signals pointing to the bioproduction of higher algae (C_{15} and C_{17} n-alkanes) may indicate that algal blooms occurred in the surface waters during times of decreased salinity, *e.g.*, after freshwater supply from rainfalls. This delicate balance between a generally slightly evaporitic system and occasional freshwater supply may also explain some small-scale organofacies variations visible in kerogen analysis under the microscope.

The high bitumen concentrations are a consequence of the high sulfur content of the organic matter. Early diagenetic sulfur incorporation into organic matter in (partly) calcareous or evaporitic sediments evidently leads to a kerogen network less strongly condensed by stable carbon-carbon bonds than in typical clastic sediments. Furthermore, carbon-sulfur and sulfur-sulfur bonds apparently may be cleaved relatively early as diagenesis progresses, and this will form highly polar and asphaltic bitumen at low levels of thermal maturity.

The assessment of organofacies by the analysis of molecular indicators so far was restricted to the saturated and aromatic hydrocarbon fractions, and it should be kept in mind that both fractions together in many cases do not represent more than 20% of the total bitumen. Conclusions should be drawn cautiously from this limited molecular data set although no major discrepancies occurred in comparison to the organic petrography results. It became clear, however, that the information gained from the organic sulfur compounds in the "aromatic hydrocarbon" fractions was partly complementary to that in the saturated hydrocarbon fractions. This shows again, that the early diagenetic reactions of inorganic sulfur species with sedimentary organic matter can be highly selective. It has been observed before (31-32) that this can lead to a quenching effect and that even more information may be hidden in high-molecular-weight organic sulfur compounds not yet analyzed in the case of the Nördlinger Ries sediments.

Acknowledgments

We are grateful to BEB Erdgas und Erdöl GmbH (Hannover) for core samples of the Nördlinger Ries sediments and for permission to publish the results. We thank Dr. M. Radke for providing facilities for extraction and liquid chromatography. Analyses of sulfur-bearing species in the aromatic hydrocarbon fractions were performed by H. Willsch. We obtained technical support from W. Benders, U. Disko, R. Harms, B. Kammer, F.J. Keller, W. Laumer, A. Fischer, O. Schmitz, H. Schnitzler and S. de Waal. Discussions with J.S. Sinninghe

Damsté (Technical University Delft) and H.L. ten Haven were stimulating and helpful as were the careful reviews of J. Gormly, Th. McCulloh and W.L. Orr (all Mobil Oil, Dallas) and an anonymous referee. The manuscript was typed by Mrs. B. Schmitz.

Literature Cited

1. Hollerbach, A.; Hufnagel, H.; Wehner, H. Geol. Bavarica 1977, 75, 139-53.
2. Wolf, M. Geol. Bavarica 1977, 75, 127-38.
3. Tissot, B.P.; Welte, D.H. Petroleum Formation and Occurrence; Springer-Verlag: Heidelberg, 1984.
4. Gehman, H.M., Jr. Geochim. Cosmochim. Acta 1962, 26, 885-97.
5. Evans, R.; Kirkland, D.W. In Evaporites and Hydrocarbons; Schreiber, B.C., Ed.; Columbia University Press: New York, 1988; pp. 256-99.
6. Pohl, J.; Gall, H. Geol. Bavarica 1977, 76, 159-75.
7. Wagner, G.A. Geol. Bavarica 1977, 75, 349-54.
8. Jankowski, B. Bochumer geol. u. geotechn. Arb. 1981, 6, 315 pp.
9. Förstner, U.; Rothe, P. Geol. Bavarica 1977, 75, 49-58.
10. Littke, R.; Baker, D.R.; Leythaeuser, D. Org. Geochem. 1988, 13 , 549-59.
11. Radke, M.; Sittardt, H.G.; Welte, D.H. Anal. Chem. 1978, 50, 663-65.
12. Radke, M.; Willsch, H.; Welte, D.H. Anal. Chem. 1980, 52, 406-11.
13. Rullkötter, J.; Orr, W.L. In Geochemistry of Sulfur in Fossil Fuels; Orr, W.L.; White, C.M., Eds.; American Chemical Society: Washington; this volume.
14. Jankowski, B.; Littke, R. Geowissenschaften in unserer Zeit 1986, 4, 73-80.
15. Orr, W.L. Org. Geochem. 1986, 10, 499-516.
16. McKirdy, D.M.; Kantsler, A.J. Aust. Petr. Explor. Assoc. (APEA) J. 1980, 20, 68-86.
17. Tissot, B.; Durand, B.; Espitalié, J.; Combaz, A. Amer. Assoc. Petr. Geol. Bull. 1974, 58, 499-506.
18. Tannenbaum, E.; Aizenshtat, Z. Org. Geochem. 1985, 8, 181-92.
19. Palacas, J.G. Petroleum Geochemistry and Source Rock Potential of Carbonate Rocks, AAPG Studies in Geology No. 18; Amer. Assoc. Petr. Geol.: Tulsa, 1984; 208 pp.
20. Jones, R.W. In Petroleum Geochemistry and Source Rock Potential of Carbonate Rocks, Palacas, J.G., Ed., AAPG Studies in Geology No. 18, Amer. Assoc. Petr. Geol.: Tulsa, 1984; pp. 163-80.
21. Eglinton, G.; Hamilton, R.J. Science 1967, 156, 1322-35.
22. ten Haven, H.L.; de Leeuw, J.W.; Rullkötter, J.; Sinninghe Damsté, J.S. Nature 1987, 330, 641-43.
23. Blumer, M.; Guillard, R.R.L.; Chare, D. Mar. Biol. 1971, 6, 183-89.
24. Ourisson, G.; Albrecht, P.; Rohmer, M. Pure Appl. Chem. 1979, 51, 709-29.
25. Stein, R.; Rullkötter, J.; Littke, R.; Schaefer, R.G.; Welte, D.H. Proc. ODP, Sci. Res. 1988, 103, 567-85.
26. Rullkötter, J.; Mukhopadhyay, P.K.; Welte, D.H. Init. Repts. DSDP 1987, 93, 1163-76.
27. Rullkötter, J.; Mukhopadhyay, P.K.; Schaefer, R.G.; Welte, D.H. Init. Repts. DSDP 1984, 79, 775-806.

28. ten Haven, H.L.; Rohmer, M.; Rullkötter, J.; Bisseret, P. Geochim. Cosmochim. Acta 1989, 53, 3073-79.
29. Sinninghe Damsté, J.S.; Rijpstra, W.I.C.; Kock-van-Dalen, A.C.; de Leeuw, J.W.; Schenck, P.A. Geochim. Cosmochim. Acta 1989, 53, 1343-55.
30. Volkman, J.K.; Eglinton, G.; Corner, E.D.S.; Sargent, J.R. Advances in Organic Geochemistry - 1979 1980, pp. 219-27.
31. Schmid, J.C. Ph.D. Thesis, Université de Strasbourg, 1986.
32. Sinninghe Damsté, J.S.; Rijpstra, W.I.C.; de Leeuw, J.W.; Schenck, P.A. Org. Geochem. 1988, 13, 593-606.

RECEIVED March 5, 1990

Chapter 9

Incorporation of Sulfur into Recent Organic Matter in a Carbonate Environment (Abu Dhabi, United Arab Emirates)

F. Kenig and A. Y. Huc

Institut Français du Pétrole, B.P. 311,92506, Rueil Malmaison, Cedex, France

In the carbonate, hypersaline, Holocene environment of Abu Dhabi (U.A.E.) microbial mats, <u>Avicennia</u> mangrove soils and lagoonal seaweeds constitute specific sources for three different types of organic matter that are incorporated into the sediment. Elemental analysis of "kerogens" isolated from living plants, surface and buried sediments shows that the organic sulfur content is low in living tissues and increases significantly with depth. This observation can be explained by incorporation of inorganic sulfur species, made available by bacterial sulfate reduction, into the organic matter during early diagenesis.

Organic sulfur compounds, including mainly thiophenic compounds, have been recognized in extracts and pyrolysates of kerogens. The distribution of the thiophenic compounds differs significantly among the types of organic matter considered, including those from microbial mats and mangrove paleosoils.

The distribution patterns of the thiophenic compounds, which can be related to the distribution of saturated and unsaturated compounds, is clearly controlled by the type of molecular precursors in which inorganic sulfur is incorporated.

Environmental conditions are usually considered to be the main factors controlling for sulfur content of sedimentary organic matter (1). The favourable sedimentary conditions proposed for incorporation of sulfur into organic matter include organic matter richness, anaerobic conditions, and iron depleted environments. Under anaerobic conditions, if sulfate is available, bacterial sulfate-reduction is promoted and the resulting inorganic sulfur species are able to react with and become incorporated into organic matter.

Iron availability in a sulfate-reducing environment leads to the formation of pyrite and, consequently, to the consumption of inorganic sulfur species (2) preventing its incorporation into the organic matter. However, in iron-starved environments, such as carbonate environments protected from sources of terrigenous sediments, there is insufficient iron to bring about appreciable pyrite formation (2). Part of the inorganic sulfur species produced by bacterial sulfate reduction is introduced into pyrite, and part remains available for incorporation into organic

0097–6156/90/0429–0170$06.00/0

matter. Crude oils generated by carbonate sequences are usually richer in sulfur than crude oils generated by siliciclastic sequences (3).

The recent sedimentary environment of Abu Dhabi, located in the south-western portion of the Arabian Gulf (Figure 1), is a favourable setting for studying the incorporation of sulfur into sedimentary organic matter. The reasons for this are the following: (a) at least 90 weight% of recent deposits in this area are carbonates (4); (b) a desertic hinterland and a long distance to regional siliciclastic sources, including the Shott Al Arab and the Zagros chain, limits the contribution of iron to that contained in sea water (Figure 1); (c) local concentrations of decomposing organic matter allow development of anoxic niches and bacterial sulfate-reduction. Morever the *in situ* accumulation of specific biological communities --microbial mats, Avicennia mangrove, and seaweeds-- makes possible a direct comparison of the sulfur status between the living precursors and the derived sedimentary organic matter.

Geological Setting

The recent lagoon-sabkha system of Abu Dhabi is the result of a Holocene regression that began 4000-5000 years B.P. (5) and which is probably still active. An open lagoon has been developed behind a barrier islands system (Figure 2). This lagoon has been and is still being infilled by intertidal sedimentary accretion of aragonitic sediments of chemical and biogenic origin. Progradation has resulted in the formation of a saline supratidal plain called a sabkha (Figure 3). The sabkha is a locus of carbonate diagenesis (aragonite to calcite and dolomite), of primary precipitation of the evaporite minerals (gypsum and anhydrite), and of their subsequent secondary growth. With an average salinity of 60 g/l in the lagoon and with a salinity of surface water reaching 280 g/l in sabkha channels, Abu Dhabi is defined as a hypersaline environment.

Sedimentary organic matter originates from three main types of biological precursors growing in this system (Figure 2). The lagoonal seaweeds (Halodule) grow in the infratidal zone. Avicennia , black mangrove, grows in the intertidal zone. The Avicennia mangrove community is currently exposed to adverse conditions, mainly to high salinity. Therefore its distribution is not as widespread as when the lagoon was more open. As a result of its low productivity, the recent mangrove is unable to produce a soil equivalent to the relatively organic carbon rich paleosoils that have been observed buried in the sabkha (an accumulation of roots and leaves). Instead it produces a soil characterized by sparse root-traces and trunks. The buried paleosoil can be divided into two parts: a "peat" composed mostly of leaves and roots, and an underlying zone of lower intertidal sediments which contain much less organic carbon and which is colonized only by the root system.

The microbial mats, generally refered to as algal mats, grow in the upper intertidal zone. They are essentially formed of cyanobacteria which are progressively colonizing the intertidal zone. These micro-organisms form a massive organic fabric. Periodic storms cover this organic film with inorganic sediments, which in turn are recolonized by cyanobacteria. This repetitive process results ultimately in a stack of layers of organic carbon rich algal layers sandwiched between inorganic storm layers. The microbial mat forms an "algal belt" at the periphery of the lagoon (6), where it can be commonly followed laterally for more than twenty kilometers in some areas (Figure 2). The width of this belt can reach two kilometers because of the high tidal range (up to 1.5 m) in this very flat region. The microbial mats we studied represent only 30% to 60 % of the total belt (7). They can be refered to as the "smooth mat" of Kinsman and Park (1976) or as the "polygon mat" of Kendall and Skipwith (1968). The other 40% to 70% are mostly thin algal films unlikely to be preserved in the sediment record (7).

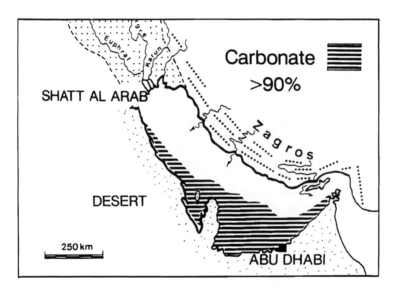

Figure 1. Location map of carbonate sedimentation in the Arabian Gulf (adapted from B.H. PURSER, 1980).

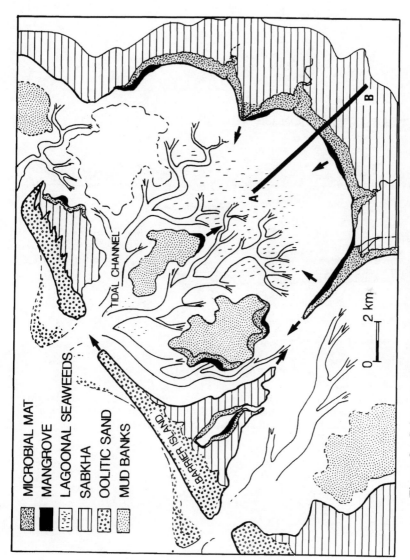

Figure 2. Schematic map of a part of the recent sedimentary system of Abu Dhabi showing the location of living communities, suppliers of sedimentary organic matter. Arrows on the scheme represent progradation direction.

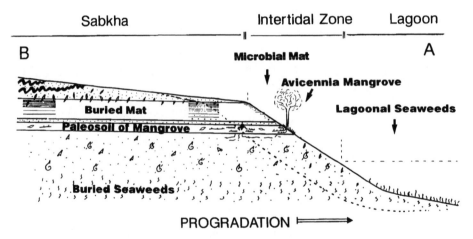

SCHEMATIC SECTION OF ABU DHABI LAGOON COAST LINE

Figure 3. Schematic section through the intertidal zone and the sabkha, showing the position of organic matter precursors and their location in the sabkha after burial. The theoretical position of this section is displayed on fig.2.

These three types of biological precursors are incorporated into the sediment where they grow. Transport mechanisms do not seem to have influence on the accumulation of the organic matter in this system. As the lagoon is infilled by intertidal sedimentary accretion, these organic facies are progressively "buried" and then evolve in a sedimentary environment where mineral diagenesis is active: the sabkha (Figure 3). An organo-sedimentary, infratidal to supratidal, sequence of up to 1.8 m is fossilized. The major unit in the sequence is the intertidal facies including the microbial mat and the paleosoil of mangrove which have been recognized over wide surfaces. A seven kilometers long channel (Mussafah channel) dredged landward perpendicularly to the intertidal zone and cutting across the sabkha gave us the opportunity to study the lateral extension of the system. A microbial mat stretching over seven kilometers without interruption has been identified. Its thickness can reach 60 cm with a maximum total organic carbon (TOC) content of 2.5%. A paleosoil of mangrove stretching landward over two kilometers has been identified. It has a maximum thickness of 30 cm and a TOC of up to 10%. The lagoonal seaweed bearing sediments were also observed four kilometers inland within the sabkha. They have a maximum TOC of 0.8%.

The organic matter incorporated into the sediment is subjected to sulfate-reduction. According to field observation, bacterial sulfate reduction is probably already active in microbial mats beneath the few millimeters corresponding to the living mat. Decimetric-sized lumps filled with H_2S have been observed at the surface of one algal mat in a tidal channel. This situation occurs where the permanent presence of water has prevented the formation of dessication polygons and has allowed trapping of H_2S beneath the impermeable living mat. An odor of hydrogen sulfide is a permanent feature of surface lagoonal sediments reflecting the activity of sulfate reducing bacteria. Even the poorly developed recent mangrove soil is affected by bacterial sulfate-reduction.

Sampling

Microbial mats. Five samples of surficial microbial mats, currently accumulating in a sabkha channel, were selected along a distance of 1400 m. They correspond to a sequence that begins at the seaward part of the channel where the mat section is dominated by layers of carbonate mud up to 1 cm thick interbedded with thin organic layers with a thickness inferior to 0.5 mm. Along this sequence, the relative importance of the carbonate layers decreases progressively landward to a point where the microbial mat forms a massive bed of organic matter containing gypsum crystals, subordinate carbonate sediment and exhibiting accumulation of H_2S in surficial lumps. The TOC of this sequence ranges from 1.6% seaward to 4.5% landward. Its thickness varies between 20 cm and 30 cm. The five samples of modern microbial mats are quite representative of the different environments in which the mat, which is likely to be buried and preserved, is growing. The sample of modern microbial mat used for extraction and pyrolysis of the kerogen, corresponds to the massive mat zone exhibiting H_2S in surficial lumps. It has a TOC of 4.5%.

Buried mat samples were collected from the sabkha , in pits or along artificial channels. They were selected to correspond to the modern mat samples in terms of depositional environments. The buried mat sample used for detailed geochemical studies has been collected from the sabkha two kilometers inland at a depth of 60 cm. Its sedimentary structure is well preserved and it contains 2% TOC.

Mangrove soils. The sample of mangrove paleosoils used for extraction and pyrolysis was collected, from the sabkha, one kilometers inland at a depth of 50 cm. It corresponds to the upper part of the soil (30 cm thick), made from the accumulation of leaves and roots. The TOC of this sample is 8.2 %. Selective diagenesis of organic tissues observed in thin sections of the soil leads to a relative concentration of cuticles. Other samples of paleosoil studied were collected in the same area and apparently accumulated in a similar environment. Samples of living Avicennia were collected. Roots and leaves were treated separately.

Lagoonal muds. Two surface samples of seaweed bearing sediments (TOC 1.5% and 2%) were collected, in the infratidal zone of the main lagoon in an area covered with Halodule. Four buried lagoonal sediments containing seaweed debris were selected: two samples were recovered by push-coring from a depth of 70 cms in the lagoon itself (TOC 0.9% and O.6%), and two others collected from the sabkha, 3 km and 2.5 km inland, at a depth of 1.50 m (TOC 0.9% and 1%). A sample of living Halodule was also collected from the lagoon.

Experimental

In the field the samples were stored on ice in a cool box and transfered into a deep freezer after a few hours. They were kept frozen until they were freeze-dried in the I.F.P laboratory.

Extraction and extract separation. The freeze dried samples were ground and extracted with chloroform for one hour at 55°C. Free sulfur was removed by percolating the extracts over an activated copper column. Resulting extracts were separated using thin layer chromatography (Merck precoated T.L.C. Silica-gel 60f-254) with cyclohexane as the eluting solvent. Three fractions were obtained: an immobile polar fraction, an "intermediate" fraction, and a saturated/unsaturated hydrocarbon fraction.

Gas chromatography and gas chromatography-mass spectrometry. Gas chromatography of the "intermediate" fraction was carried out with a Varian 6000 instrument equipped with a flame ionization detector (FID) and a sulfur selective flame photometric detector (FPD)(8). The oven temperature was programmed from 60° to 100° at 10°C/min and from 100° to 300° at 3°C/min. The column used was a JW, DB5, 30 m, φ: 0.32, having a 0.25 μ film thickness, split just before the FID and FPD detectors. Gas chromatography of the saturated/unsaturated fraction was carried out on a Varian 2000 instrument equipped with a FID detector. The oven temperature was programmed from 50°C to 110°C at 10°C/min and from 110° to 380°C at 3°/min. The column was a Chrompack (length 25 metres, φ: 0.32) with a film thickness of 0.12 μ.

   Gas Chromatography-Mass Spectrometry (GC-MS) analysis was carried out on a Varian instrument coupled to a Nermag R10-10 quadrupole mass spectrometer. Separation was achieved on a JW DB1 (60 m, φ: 0.32, film thickness 0.25 μ) capillary column.

Kerogens. Bitumen free samples were demineralized in order to obtain a kerogen. A first decarbonation was made using HCl. The decarbonated samples were then treated with HCl (6N) and HF (40%) for further demineralization. The samples were then re-extracted with chloroform in order to eliminate elemental sulfur. The kerogens obtained were studied by elemental analysis (C,H,O,N,S,Fe). In order to determine the content of organic sulfur, a stoichiometric calculation was made, iron present in the kerogen was assumed to be exclusively in pyritic form, hence sulfur necessary to satisfy the stoichiometric balance of pyrite (FeS_2) was subtracted from the total sulfur. The sulfur remaining after calculation was assumed to be organically bound sulfur (9).

Pyrolysis. Open pyrolysis was performed using a "minifour" apparatus (10). Heating temperature was programmed from 130°C to 550°C at 60°C/min and 5 min at 550°C.

The pyrolysates were separated using the same methods as for the extracts. GC and GC-MS were carried out with the same instruments as for the extracts.

Results

Analysis of Kerogen.

Isolated kerogens of the five modern microbial mats sampled in the sabkha channel were studied by elemental analysis (Table I, Figure 4A). Their average S/C atomic ratio is 0.015. The S/C values of these modern mats vary from 0.010 (in the seaward part of the channel) to 0.019 (in the landward part of the channel, in the massive algal mat). Unfortunately this ratio cannot be compared with living organic matter. The living part of the microbial mat represents only the upper few millimeters of the modern mat and was not specially sampled in the field. The average S/C atomic ratio of the six buried microbial mats studied is 0.027. Their S/C values vary between 0.019 and 0.037 (Table I, Figure 4A). These values are significantly higher than those in modern microbial mats. The sulfur content of the polar fraction of extracts is only slightly lower than the sulfur content of kerogens. The polar fraction of the extracts of both a modern and a buried mat contain 2.1% and 3.5% of sulfur (S/C: 0.011 and 0.019) respectively, whereas the kerogens of the same samples have slightly lower sulfur content, 2.7% and 3.8%, respectively (S/C: 0.017 and 0.022).

Table I. Elemental analysis of "kerogens" of living plant , surface sediments and buried sediments from Abu Dhabi recent sedimentary system

	C*	H*	O*	N*	Sorg*	Sorg/C**	H/C**	O/C**	PYRITE[+]
Modern	60.10	6.30	24.23	7.65	1.72	0.010	1.26	0.30	0.43
Microbial	60.97	6.94	23.04	7.13	1.91	0.012	1.37	0.28	0.60
Mat	61.76	6.71	22.84	6.36	2.33	0.014	1.30	0.28	1.11
	61.94	1.46	20.83	6.92	2.79	0.017	1.46	0.25	0.41
	61.14	7.73	20.67	7.34	3.13	0.019	1.51	0.25	0.43
Buried	66.37	6.60	20.34	3.24	3.46	0.019	1.19	0.23	5.94
Microbial	66.01	6.51	20.60	3.08	3.80	0.022	1.18	0.23	4.74
Mat	65.02	6.65	20.89	2.87	4.67	0.027	1.21	0.24	5.53
	65.30	7.21	20.18	2.60	4.71	0.027	1.32	0.23	4.18
	60.18	8.00	20.65	5.70	5.46	0.034	1.59	0.25	3.90
	60.49	6.98	20.49	6.07	5.96	0.037	1.38	0.25	8.44
Avicennia	62.64	7.98	28.49	0.89	0	0	1.52	0.34	0
living plant	57.12	6.24	36.09	0.55	0	0	1.31	0.47	0
Paleosoil	63.98	6.84	22.75	2.09	4.34	0.025	1.28	0.26	12.0
of	63.03	6.26	24.48	1.38	4.85	0.029	1.19	0.29	1.54
Mangrove	60.62	6.18	26.55	1.96	4.68	0.029	1.22	0.33	1.09
	61.96	7.02	23.50	2.60	4.92	0.030	1.36	0.28	10.20
Seaweeds	45.98	5.34	46.47	1.24	0.97	0.008	1.39	0.76	0
Surf. Sed.	64.35	7.26	23.48	2.53	2.37	0.014	1.35	0.27	6.49
+Seaweeds	63.23	7.58	23.34	2.66	3.20	0.019	1.43	0.27	9.30
Buried Sed	64.35	7.10	23.58	2.43	3.15	0.019	1.34	0.28	11.29
+ Seaweeds	63.28	6.56	24.84	1.99	3.33	0.020	1.24	0.29	10.16
	64.60	7.44	20.67	2.28	4.95	0.029	1.38	0.24	12.69
	60.68	7.32	22.36	4.26	5.38	0.033	1.45	0.27	6.75

* Calculated on the basis of dry and ash free organic matter   ** atomic ratio
+ Weight % of the isolated kerogen

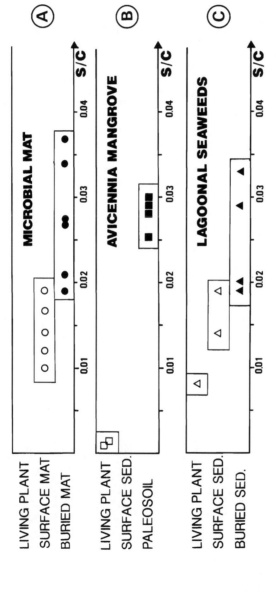

Figure 4. S/C atomic ratios from kerogens of living plants, modern sediments, and buried sediments of the three main types of organic matter encountered in the Abu Dhabi recent sedimentary system, (A) the microbial mats, (B) the Avicennia mangrove, (C) the lagoonal seaweeds.

Elemental analysis of kerogens prepared from living plants of <u>Avicennia</u> trees revealed a near absence of sulfur in leaves, trunks, and branches (Table I, Figure 4B). There is no trace of organic sulfur compounds in the "intermediate" fraction of the extracts. No modern soil of <u>Avicennia</u> mangrove has been studied because of its limited development. In the four samples of paleosoils or buried roots analysed S/C atomic ratios average 0.028 with a maximum of 0.030 (Figure 4B).

The "kerogen" of a living lagoonal seaweed <u>Halodule</u> contains only a very low amount of sulfur (Table I, Figure 4C). The extract of the same sample does not contain detectable organic sulfur compounds. On the other hand, the kerogens of two surface sediments covered by seaweed patches contain organic sulfur. Their respective S/C atomic ratios are 0.014 and 0.019. The two samples collected at a depth of 70 cm in the lagoon have S/C values of 0.019 and 0.020. The two other samples, which are buried in the sabkha and still contain recognizable remains of the seaweed <u>Halodule</u>, have S/C values of 0.029 and 0.033.

<u>Pyrolysates and Extracts.</u>

<u>The Microbial Mat.</u> The "intermediate" fraction of the extracts and pyrolysates of the kerogens of a modern and buried mat were analysed by GC with coupled FID and FPD and GC-MS. In the extract of the modern mat, two isomers of a C_{20} thiophenic isoprenoid (compounds I and II) already reported in the literature (<u>11-12</u>) are the most abundant organic sulfur compounds (Figure 5B). The "intermediate" fraction of the pyrolysate of the corresponding sample, is also dominated by the same two isomers of thiophenic isoprenoids that exhibit a similar internal ratio as in the extract. The presence of a third isomer was detected in smaller quantity (compound III)(<u>13</u>). Other thiophenic compounds are present in the sample and were tentatively identified using mass spectra data, GC retention time, and literature (<u>14-16</u>). They are similar to the compounds found in the "intermediate" fraction of the pyrolysate of the buried mat (Figure 5D).

The "intermediate" fraction separated from the pyrolysate of the buried mat is also dominated by the compounds I and II which show the same relative concentration ratio as in the related extract and as in the modern mat extract and pyrolysate. The FPD trace (Figure 5D) displays numerous other sulfur compounds, among them n-alkylthiophenes ranging from C_{13} to C_{30}, peaking between C_{14} and C_{21}, and other isoprenoids, branched and mid-chain thiophenes in the C_{13}-C_{20} range (Table II).

Table II. Some of the sulfur compounds found in the
microbial mat pyrolysate (Figure 5D)

I	3-methyl-2-(3,7,11-trimethyldodecyl)thiophene
II	3-(4,8,12-trimethyltridecyl)thiophene
III	2,3-dimethyl-5-(2,6,10-trimethylundecyl)thiophene
VI	isoprenoid thiophene C18
V	branched C19 thiophene

The distribution of n-alkanes, isoprenoids and other branched hydrocarbons in the saturated/unsaturated hydrocarbons fraction of the modern microbial mat ranges mostly between C_{14} and C_{21}, in extracts and pyrolysates (Figure 5A and 6C). It is similar to the distribution of the hydrocarbons described in the "top mat" of the Gavish sabkha in Israël (<u>17</u>). The main differences are the presence, in the extract of

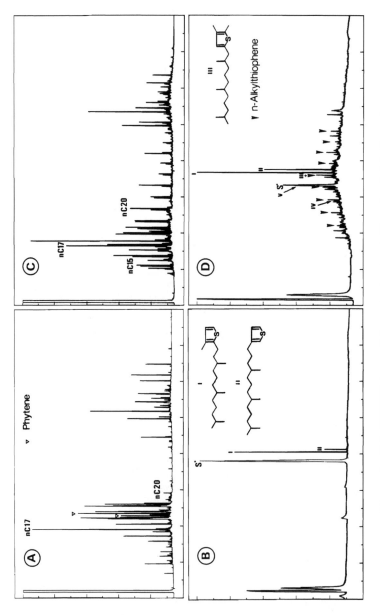

Figure 5. (A) GC FID trace of the extract saturated/unsaturated hydrocarbons fraction of a modern microbial mat. (B) GC FPD trace of the extract "intermediate" fraction of a surface microbial mat. (C) GC FID trace of the pyrolysate saturated/unsaturated hydrocarbons fraction of a modern microbial mat. (D) GC FPD trace of the "intermediate" fraction of a buried microbial mat. Labelled compounds are listed in Table II.

the Abu Dhabi mat, of the highly branched C_{20} isoprenoid and the important concentration of the phytene isomers.

The Paleosoil of Mangrove. The GC trace of the saturated/unsaturated hydrocarbons of the extract from the buried soil of mangrove is dominated by odd n-alkanes probably derived from cuticular waxes (Figure 6A). The "intermediate" fraction of the same extract is poor in sulfur compounds (Figure 6B): only a C_{29} sterane thiol previously identified in the Rozel Point oil (18), and by a C_{30} n-alkylthiophene. The former compound might be related to the C_{29} sterene dominantly present in the saturated/unsaturated fraction (Figure 6A).

Saturated and unsaturated hydrocarbons (Figure 6C) of the pyrolysate are largely dominated by straight-chain hydrocarbons ranging from C_{14} to C_{32}. There is a drastic drop of concentration of saturated and unsaturated hydrocarbons after the C_{30} monoenes elutes. The C_{28}, C_{29}, C_{30} olefins are dominant with their diolefinic counterparts. The FPD trace of the "intermediate" fraction of the pyrolysate (Figure 6D) is dominated by a C_{30} n-alkylthiophene and by a C_{30} n-alkylthiolane-thiol tentatively identified from mass spectra. The n-alkylthiophenes dominate throughout the fraction with a distribution ranging from C_{13} to C_{32}. The C_{28}-C_{30} are the most prominent thiophenes. The presumed C_{28}-C_{30} thiolane-thiols show the same distribution as the C_{28}-C_{30} thiophenes.

The intensity of the thiolanes-thiols C_{28}-C_{30} on the FPD trace is exagerated because of the two sulfur atoms carried by thiolane-thiols and because of the quadratic response of the FPD detector. The mass-spectra of the C_{30} thiolane-thiol, displays the fragments m/z: 55 (100%), m/z: 87 (62%) the typical thiolane ring, a fragment m/z: 451 (17%) which corresponds to the C_{30} thiolane, and a molecular ion M^+:484 (60%) which could be result of the adjonction of a thiol (S-H) to the C_{30} thiolane. The molecules eluting in doublets with the thiolane-thiols are not identified. The C_{20} thiophenic isoprenoids are also present, with the compound I dominating.

Discussion

Kerogens and extracts from living plants, including leaves and roots of <u>Avicennia</u> and <u>Halodule</u> contain a very small amount of organic sulfur. Biosynthesized sulfur found in organisms is generally associated with the most labile classes of compounds, including proteins, and is likely to be eliminated during acid treatment (19). Therefore, the influence of such biosynthesized sulfur compounds is probably negligible in the sulfur content of the kerogens. Kerogens of buried lagoonal seaweed-bearing sediments is richer in organic sulfur in comparison to kerogens of surface sediments and living <u>Halodule</u> plants. The same is true for buried soils of mangrove relative to the samples of living <u>Avicennia</u> plants and for the kerogens of buried microbial mats relative to their modern mat equivalents. The increase of sulfur content in sedimentary organic matter strongly suggests that sulfur was incorporated into the organic matter during early diagenesis.

The S/C atomic ratios of kerogens of modern and buried microbial mats covers a large range of values. The lowest S/C values were found for the modern microbial mat with thick carbonate-sand layers collected near the lagoon. Landward, as the carbonate layers get thinner and more muddy, the S/C values increase progressively, until they reach the maximum value which is found in the modern samples, in the massive mat with lumps filled with H_2S. This could be explained, if we consider the aptitude of this thick mineral layers to allow water circulation, and so a better oxygenation of the system. On the contrary the massive mat with the much thinner mineral layers seems to form a much closer system as suggested by the occurrence of lumps filled with H_2S.

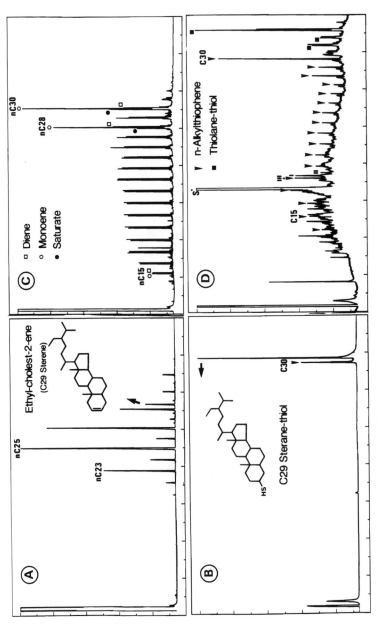

Figure 6. (A) GC FID trace of the extract saturated/unsaturated hydrocarbons fraction of a paleosoil of mangrove. (B) GC FPD trace of the extract of the "intermediate" fraction of a paleosoil of mangrove. (C) GC FID trace of the pyrolysate saturated/unsaturated hydrocarbons fraction of the paleosoil of mangrove. (D) GC FPD trace of the pyrolysate "intermediate" fraction of the paleosoil of mangrove. Labelled compounds are listed in Table II.

Sedimentological evidence, position in the sequence, and comparison with modern mats enables correlation between the sulfur content of buried mats and their depositional environment. For a similar type of organic matter different environments of sedimentation and early burial seem to result in different sulfur enrichment.

The incorporation of sulfur in organic matter at the early stage of diagenesis is supported by the presence in the extracts and kerogen pyrolysates of organic sulfur compounds. In the pyrolysates as well as in the extracts of modern and buried microbial mats, the organic sulfur compounds are present in significant amounts. They are dominated by the C_{20} isoprenoid isomers and especially by compound I. The abundance of phytene isomers in the extract fraction can be related to this dominance since a common precursor exhibiting a similar carbon skeleton can be evoked (11,14). The distribution of the organic sulfur compounds, mostly between C_{14} and C_{21}, which matches the distribution of saturated and unsaturated, branched and isoprenoid hydrocarbons in the extract and pyrolysate of the modern mats, can also be rationalized by invoking the same functionalized precursors (20).

The pyrolysate of the kerogen is the only fraction from the mangrove paleosoil which contains significant amounts of organic sulfur. The typical distribution of straight-chain hydrocarbons in the saturated/unsaturated hydrocarbons fraction in the pyrolysate of the paleosoil of mangrove could be the result of cleavage during pyrolysis of a macromolecule which can tentatively be assigned to a "cuticular resistant aliphatic biopolymer" (21-22). This type of macromolecule which is not hydrolyzable and not extractable is a part of the kerogen. The relationship between the distribution patterns of the straight chain hydrocarbons and the n-alkylthiophenes and n-alkylthiolane thiols of the "intermediate" fraction of the paleosoil of mangrove pyrolysate suggests the possibility of incorporation of sulfur into such a geopolymer. It has been shown that addition of H_2S to double bounds occurs in recent sediments and leads to the formation of thiolanes (23). Thus the presence in the saturated/unsaturated hydrocarbons fraction of C_{28}-C_{30} dienes (Figure 6C), could explain the formation of the dominant C_{28}-C_{30} thiolanes and the dominance of C_{28}-C_{30} thiophenes (Figure 6D).The n-alkanes distribution of the extract (Figure 6A) which may derived from cuticular waxes is not printed in the distribution of sulfur compounds in the extract and in the pyrolysate. These molecules were not able to accept an inorganic sulfur atom to generate an organic sulfur compound. On the contrary the C_{29} sterane-thiol in the extract can be tentatively related to the C_{29} sterene. The nonextractable character of the geopolymer in which sulfur is suspected of being incorporated, would explain the low content of organic sulfur compounds in the extract.

The distribution of thiophenes in the microbial mat and in the paleosoil of mangrove are noticeably different. In the microbial mat, isoprenoid, branched, midchain and n-alkylthiophenes occur mainly between C_{14} and C_{21} and are dominated by the C_{20} thiophenic isoprenoids. In the paleosoil of mangrove, if some of the isoprenoids and branched thiophenes are present, the sulfur compounds are dominated by the C_{28}-C_{30} n-alkylthiophenes and by a presumed C_{30} n-alkylthiolane thiol, and in the extract by a C_{29} sterane-thiol. These differences are likely to be a reflection of the type of molecules in which sulfur might have been incorporated.

As already assumed the intramolecular incorporation of sulfur into the organic matter is not a random process (20).The distribution of the sulfur compounds will depend on the chemical capacity of the molecules present during early diagenesis in each kind of organic matter to accept an inorganic sulfur atoms. These two different types of organic matter enable the intramolecular incorporation of sulfur in different molecules, and consequently result in different distribution patterns of organic sulfur compounds.

Thiolanes are usually considered to be intermediates in the formation of thiophenes (20-22). Their presence in these modern sediments was expected.

However, with the exception of the supposed thiolane thiols present in the mangrove pyrolysate, no thiolanes were recognized. At this time we have no explanation for their absence, except that thiolanes are fragile molecules. Hence, their destruction or aromatization to thiophenes during sample treatment is a possibility.

Acknowledgments

We thank authorities of Abu Dhabi for their cooperation. We thank TOTAL CFP and TOTAL ABK for funding the field trip to Abu Dhabi and local support, and especially Dr. J.L. Oudin and Dr. R. Boichard for their interest in the subject, and advice. We also thank Professor B. H. Purser, Dr. J.C. Plaziat, and Dr. F. Baltzer of Orsay PARIS XI University, for their help in the field and considerable advice in the sedimentological studies. We aknowledge the help of Mrs Da Silva who carried on the GC-MS analysis, Mrs Fabre, and Mrs Bernon for their permanent help. This manuscript benefit from critical rewiew by Dr. H. L. ten Haven and Dr. D. W. Kirkland.

Literature Cited.

1. Gransh, J. A.; Posthuma, J. In Advances in Organic Geochemistry,1973, Tissot, B.; Bienner, F., Eds.; Technip: Paris, 1974, pp 727-739.

2. Berner, R. A. Geochim. Cosmochim. Acta 1984, 48, 605-615.

3. Tissot, B. P.; Welte, D. H. Petroleum Formation and Occurrence 2nd Ed, Springer: Berlin, 1984, p 669.

4. Purser, B. H. Sedimentation et diagenese des carbonates neritiques recents 1980; Technip: Paris, tome 2.

5. Evans, G.; Schmidt, V.; Bush, P.; Nelson, H. Sedimentology 1969, 12, 145-159.

6. Kendall, C. G. St. C.; Skipwith, Bt P. A. d'E. Journal of Sedimentary Petrology 1968, 38, 4, 1040-1058.

7. Kinsmann, D. J. J.; Park, R. K. In Stromatolites; Walter, M. H., Ed; Elsevier Amsterdam, Development in Sedimentology 1976, 20, pp 421-433.

8. Castex, H.; Roucache, J.; Boulet, R. Revue de l'Institut Francais du Petrole 1974, 29, 1, 3-38.

9. Durand, B.; Monin, J.C. In Kerogen; Durand, B. Ed; Technip Paris, 1980, pp 113-142.

10. Vandenbroucke, M.; Behar, F. In Lacustrine Petroleum Source Rocks; Fleet, A. J.; Kelts, K.; Talbot, M. R., Eds; Blackwell: Oxford, Geological Society Special Publication 1988, 40, pp 91-101.

11. Brassell, S. C.; Lewis, C. A.; de Leeuw J. W.; de Lange, F.; Sinninghe Damste, J. S. Nature 1986, 320, 160-162.

12. Rullkotter, J.; Landgraff, M.; Disko, U. J. High Res. Chrom & Chrom. Commun. 1988, 11, 633-638.

13. Sinninghe Damste, J. S.; Kock-van Dalen, A. C., de Leeuw J. W.; Schenck, P. A. Tetrahedron Lett. 1987, 28, 957-960.

14. Sinninghe Damste, J. S.; ten Haven, H.L.; de Leeuw, J. W.; Schenck, P. A. In Advances In Organic Geochemistry, 1985; Leythaeuser, D.; Rullkotter, J., Eds.; Pergamon: Oxford, Org. Geochem. 1986, 10, pp 791-805.

15. Sinninghe Damste, J. S.; de Leeuw, J. W.; Kock-van Dallen A. C.; de Zeeuw M. A.; de Lange F.; Rijpstra W. I. C.; Schenck, P. A. Geochim. Cosmochim. Acta 1987, 51, 2369-2391.

16. Sinninghe Damste, J. S.; Rijpstra W. I. C.; de Leeuw, J. W.; Schenck P. A. Geochim. Cosmochim. Acta 1989, 53, 1323-1341

17. de Leeuw, J. W.; Sinninghe Damste, J. S.; Klok, J.; Schenck, P. A., Boon, J. J. In Hypersaline Ecosystems; Friedman G. M.; Krumbein, W. E., Eds.; Springer Verlag: Berlin, Ecological Studies 1985, 53, 350-367.

18. Schmid, J.C. Ph.D. Thesis, University Louis Pasteur, Strasbourg, 1986.

19. Francois, R. Geochim. Cosmochim. Acta 1987, 51, 17-27.

20. Sinninghe Damste, J. S.; Rijpstra, W. I. C.; Kock Van Dalen A.C.; de Leeuw, J. W.; Schenk P.A. Geochim. Cosmochim. Acta 1989, 53, 1343-1355.

21. Nip, M; Tegelaar, E. W.; Brinkhuis, H.; de Leeuw J.W.; Schenck, P.A.; Holloway P.J. In Advances in Organic Geochemistry; 1985, Leythaeuser D.; Rullkotter J., Eds.; Pergamon: Oxford, Org. Geochem. 1986, 10, pp 769-778.

22. Tegelaar, E. W.; de Leeuw, J. W.; Largeau, C.; Derenne, S.; Schulten, H. R.; Muller, R.; Boon, J. J.; Nip, M.; Sprenkels, J. C. M. J. of Anal. and Appl. Pyr. 1989, 15, 29-54.

23. Vairavamurthy, A.; Mopper, K. Nature 1987, 329, 623-625.

RECEIVED March 5, 1990

Chapter 10

Sulfur and Pyrite in Precursors for Coal and Associated Rocks

A Reconnaissance Study of Three Modern Sites

A. M. Bailey[1], J. F. Sherrill[2], J. H. Blackson[3], and E. C. Kosters[4]

[1]Geology Department, University of Southwestern Louisiana, Box 44530, Lafayette, LA 70504
[2]Ebasco Services, Inc., Greensboro, NC 27407
[3]Dow Chemical Company, Midland, MI 48667
[4]Department of Geology, University of Utrecht, P.O. Box 80, 021, 3508, TA, Utrecht, Netherlands

Investigations have been undertaken on sulfur in recent sediments from known settings in south Louisiana to determine controlling factors for sulfur in coals and associated shales. These sediments range in organic content from peats to clays and have been subjected to different degrees of marine incursion. Results from chemical studies show a general increase of total sulfur with both marine incursion and organic matter. Partitioning of this sulfur varies with organic content. In clays most of the sulfur is pyritic, while in organic-rich samples organic sulfur is generally dominant, or present in concentrations approximately equal to pyritic sulfur. Results from petrographic studies show that pyrite occurs as individual crystals in all samples examined and as framboidal aggregates. In addition, rounded clouds of fine particles with diameters of <1 micron occur. Microanalyses of these microcrystallites and larger individual crystals show S/Fe ratios of around 2 in most grains examined. An exception occurs at one site where oxidation has resulted in S/Fe ratios of <2.

The distribution of sulfur in fossil fuels and associated shales is controlled to a large extent by factors operating in sediments from which these rocks formed. Key studies of sulfur in modern sediments include Casagrande (1), Casagrande et al. (2), Cohen, Spackman and Dolsen (3), Buston and Lowe (4), Altschuler (5), Kaplan et al. (6), Berner (7), Davison, Lishman and Hilton (8) and others. An excellent review of the literature on sulfur and iron in natural aqueous systems is given by Morse et al. (9). These investigators demonstrate the importance of marine influence and organic matter on sulfur incorporation in peats and clastic sediments, but tend not to treat the effects of both variables over large ranges of concentrations. The Mississippi River Delta Plain contains vast quantities of recent sediments with almost continuous ranges of organic matter. Included in these sediments are clays, defined by the Louisiana Geological Survey as having <15% organic matter, mucks with 15-75% organic matter and peats with >75% organic matter. These deposits, which were generally freshwater originally, have been recently subjected to different degrees of marine incursion. Sediments are saturated with waters of various salinities to depths of greater than several meters and are incorporating secondary sulfur to varying degrees. Hence, they offer a particulary good setting in which to study the incorporation of sulfur into sediments with a wide range of organic matter.

0097–6156/90/0429–0186$06.00/0

In addition, relatively little petrographic work, such as that done by Cohen, Spackman and Dolsen (3) on Florida peats, has been attempted on Mississippi River Delta Plain sediments. This paper describes results of a reconnaissance study undertaken to examine, both chemically and petrographically, the occurrence of sulfur in Mississippi Delta Plain sediments.

Figure 1 provides a framework for a summary discussion of representative studies of sulfur in recent sediments. Here, corners A, B, C, and D represent freshwater peats, marine peats, marine sediments, and lacustrine sediments, respectively. Peat studies, which deal with material falling in region A-B of Figure 1, include those by Casagrande (1), Casagrande et al.(2), Cohen, Spackman and Dolsen (3), Buston and Lowe (4), and Altschuler et al. (5). While a few data on clays are given by some of these authors, the studies are focused on peats that generally contain more than 75% organic matter, which corresponds to approximately 50% C. They have been formed in waters with a range of chloride salinities from 0 ppt (freshwater) to 19 ppt (marine).

Average total sulfur concentrations in these deposits, which are primarily from the Okefenokee and Everglades swamp-marsh complexes, range from less than 1 percent for freshwater peats to more than 5% for marine peats (1,2,5). In addition, Cohen, Spackman and Dolsen (3) determined averages of 3.5% total sulfur for Everglades brackish peats and 3.7% for freshwater peat overlain by brackish peats, but did not find higher values for marine peats. Bustin and Lowe (4) list an average of 2.86% total sulfur for brackish sedge peats from the Fraser River Delta, British Columbia.

Organic sulfur is the dominant form in peats described in these studies. Pyrite, however, is abundant in brackish and marine peats, occurring in void spaces in or between plant debris (3). In a study of pyrite formation in freshwater peats, Altschuler et al. (5) determined parallel decline in ester sulfate with increases in pyrite as depth increased and concluded that pyrite formed at the expense of organic sulfur. In general, framboidal morphology is present at all salinities. Altschuler et al. (5) and Lowe and Bustin (10) found monosulfides to be minor in peats.

Representative studies of sulfur in clastic sediment, located in region C-D of Figure 1, include studies by Kaplan et al. (6), Berner (7), and Davison, Lishman, and Hilton (8) and generally deal with material that has much less organic matter than sediments described in peat studies. From the cited references, organic carbon contents for freshwater lacustrine sediments from the English Lake District range from 4 to 16% (8), while those for California marine basins are generally less than 7% (6). As with peat studies, the observed clastic sediments were deposited under waters with different salinities, from freshwater to marine.

Total sulfur concentrations in freshwater lacustrine sediments described in these studies may reach about 1 % (8), but are generally less than 0.4%. Marine sediments in California generally have average total sulfur concentrations of less than 1%, but are usually greater than 0.6% (6). Values for pyritic sulfur of 0.31 to 1.98% are given for coastal and marine sediments of Long Island Sound (7).

In these clastic sediments the dominant form of sulfur is pyritic, while organic sulfur is usually present only in trace amounts. For this reason, much work on sulfur in these sediments focuses on pyrite formation and its crystallization has been studied in detail by Berner (11), Sweeney and Kaplan (12), Rickard (13), Rickard (14) and others. Under saline and hypersaline conditions precipitation of monosulfides may be the initial step. Sulfur is then added to these precipitates, converting them to pyrite. Laboratory studies indicate that if griegite is present in the original precipitate, sulfurization may produce framboidal aggregates (12). Conversion may depend on chemical factors such as H_2S concentrations (9). In contrast, in conditions that are undersaturated with respect to monosulfides, but supersaturated with respect to pyrite, pyrite may form directly and rapidly from solution (Goldhaber and Kaplan (15), Rickard (14), Howarth (16), Luther et al. (17)).

In summary, these studies point toward important roles for organic matter and salinity in determining sulfur behavior, but are not focused on intermediate sediments and, with a few exceptions (3), are not petrographic in nature. A reconnaissance investigation was undertaken to ascertain possible influences on sulfur incorporation in Mississippi Delta Plain sediments and to determine forms of sulfur in these sediments.

Methods

Site Selection. Work by Kosters (18) and Kosters et al. (19) served as a guide for selection of 3 sites used in the study. These sites have different estimated salinities as indicated by vegetation and proximity to the present coast, and they contain sediments with a range of organic content, from peat to clay, as indicated by examination of available cross sections. The sites are located near Avery Island, in the Barataria Basin, and near Gueydan, Louisiana (Figure 2).

Avery Island deposits are described by Kosters and Bailey (20) as a blanket of peats and mucks that formed on the relatively slowly subsiding Holocene Maringouin/Teche delta lobe. Clastic influx ended when the river shifted and peats started accumulating at this time. Subsequent submergence has caused the accumulation of organic-poor sediments. Thus, deposits at the bottom and top are organic-poor, while deposits in the center of the section are true peats. Subsidence and/or elevation of sea level brought about marine incursion so that the area is now a brackish marsh. Abundant cypress debris in the peat beds indicates that they represent freshwater swamps. A cross section of this site is shown in Figure 3.

Barataria Basin deposits are described by Kosters et al. (19) as lobe deposits located between the present Mississippi River and Bayou LaFourche. They consist of clays, mucks and true peats. True peat horizons are located in the top 1 meter of the section and developed during times of reduced clastic input from adjacent distributaries. Salinities of this site are relatively low, averaging 0.23 ppt in surface water and 1.11 ppt in porewater, because the site is about 80 km inland from the Gulf of Mexico, but are increasing with time. A cross section of the site is shown in Figure 4.

Gueydan deposits are described by Kosters and Bailey (20) as mucks and peats that accumulated in a channel on the Pleistocene surface. Organic matter is supplied primarily by debris from ingrowing vegetation, while mineral matter is washed in from surrounding areas. The vegetation indicates that the deposit is essentially freshwater (18). A cross section of the site is shown in Figure 5.

Sampling. Two sets of vibracores were collected at each site. A previously collected set was used, along with other cores in each area, to determine stratigraphic and sedimentologic relationships. Color, texture, and structure were used to establish different lithofacies, which were later verified or modified using x-ray radiography. Ash and total sulfur contents were then determined for channel samples of these units.

A second set of vibracores, as well as associated open water samples were collected for more detailed work. Lithologic units were determined employing the criteria used for the first set and channel samples taken for chemical work. In addition, 1 x 2 cm rectangles for petrographic work were cut from the center of each lithostratigraphic unit less than about 25 cm in length. Two samples were taken in thicker units, one at the top and one at the bottom of the unit. Cores were stored in sealed tubes in a walk-in cooler operated at 5° C.

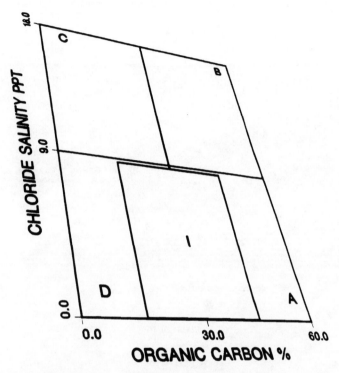

Figure 1. Framework for main environments examined in studies of sulfur in recent sediments. Freshwater peats, marine peats, marine sediments, and lacustrine sediments occupy corners A, B, C, and D, respectively. Sediments in this study range from A to D and a large proportion are in region I.

Figure 2. Location map for the three main sites of this study.

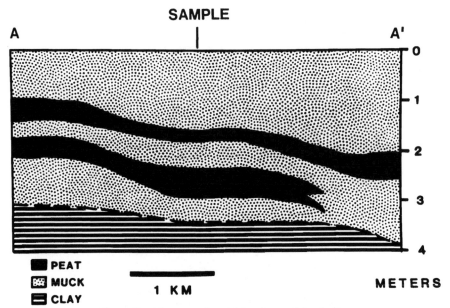

Figure 3.   Cross section of the Avery Island site.

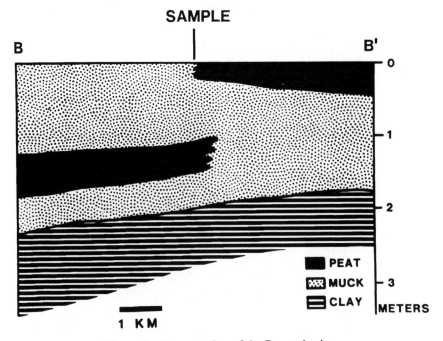

Figure 4.   Cross section of the Barataria site.

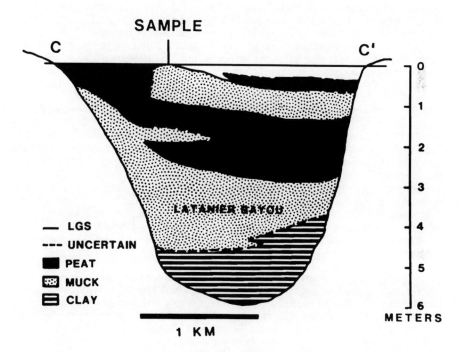

Figure 5.   Cross section of the Gueydan site.
(LGS indicates boundary established by Louisiana Geological Survey).

Chemical Analysis. More detailed chemical and petrographic analyses were carried out on samples from the second set of cores. Initially 80 g of the mixed channel sample were placed in a gas pressured cell and pore fluid was expressed using nitrogen at 68.95 bars. Pore fluids were then analyzed for pH, sulfate, and chloride, while the sediment was dried at 100°C and ground to a fine powder using a ball mill. Oxidation of minor sulfur present in monosulfides probably occurred during drying, resulting in slightly higher sulfate sulfur values. Since microscopic examination indicated that monosulfides are minor (see **Results** section), this increase is thought to be negligible. Analysis of these powders gave concentrations of total organic matter, total sulfur, total iron and forms of sulfur. Total organic matter was determined by ashing the dried sample at 555°C in a muffle furnace for 2 hours (21). Total sulfur was determined by two methods. One set of samples was sent to Dickinson Laboratories, El Paso, Texas, where sulfur was determined using a Leco sulfur analyzer. These results were verified by the authors using methods developed by Spielholtz and Diehl (22) and McGowan and Markuszewski (23) in which samples were wet oxidized with perchloric acid. Total sulfur as sulfate was determined in the resulting solutions by photometric measurement, while total iron was determined by atomic absorption spectrophotometry. Errors in total sulfur and iron are less than $\pm 0.1\%$, whereas errors in organic matter determination are around $\pm 1\%$.

Sequential extractions of the initial fine powder provided estimates of the 3 major forms of sulfur. Extraction with 0.1 N LiCl solution and photometric measurement of the resulting $BaSO_4$ suspension after addition of $BaCl_2$ resulted in the determination of sulfate sulfur. The extracted sample was treated using the method of Begheijn, Van Breeman and Velthorst (24) to remove nonpyritic iron and then extracted with 2N HNO_3 to release pyritic Fe. Following this, pyritic Fe was determined using atomic absorption spectrophotometry. Stoichiometry was used to determine pyritic S based on pyritic Fe. Organic sulfur was determined by subtracting the sum of sulfate and pyritic sulfur from total sulfur. Artificial mixtures of pyrite were used to test the extraction procedures and it was found that the method gave an average recovery efficiency of about 86 $\pm 2\%$. Lord (25) reported an efficiency of 92.7% ± 1.1 using a similar method.

Petrographic Work. Petrographic studies required preparation of thin sections. Briefly, rectangular blocks of material were treated with ethanol to remove water and then infiltered with hard grade LR White resin, a low viscosity resin developed for biological work. Following this, the block was mounted on a petrographic slide and a polished thin section produced (see Bailey and Blackson (26) and Blackson and Bailey (27)). A total of 57 sections were produced in this way, at least 1 from each thinner lithofacies unit and 2 from thicker lithofacies units.

These thin sections were examined using transmitted and reflected light microscopy. Further, small portions were removed from thin sections, and then microtomed on an ultramicrotome with a diamond knife to give 80 nanometer slices which were supported on carbon coated collodion copper grids. These were examined using a JEOL 1200EX transmission electron microscope operated at 120 KV accelerating voltage and fitted with a Link AN10000 analyzer with an ultrathin window detector to allow detection of oxygen. FeS_2 and FeS standards obtained from Aesar Johnson Matthey, Inc were used for quantitation.

Results

Chloride, sulfate, and pH analyses of pore solutions from the three sites are shown in Figure 6. Figure 7 depicts organic matter, total sulfur and total iron for the same cores and Figure 8 shows data on forms of sulfur.

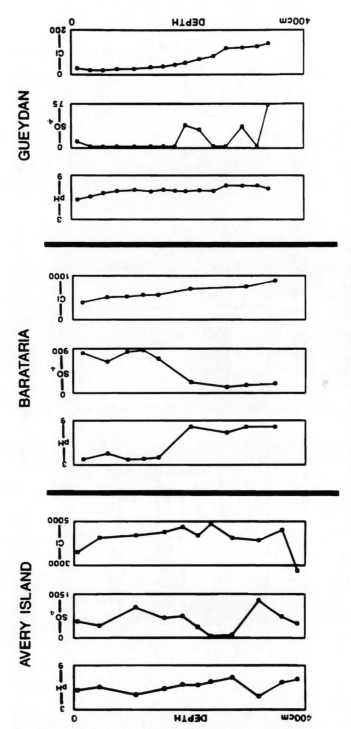

Figure 6. pH, sulfate and chloride concentrations in porewaters.

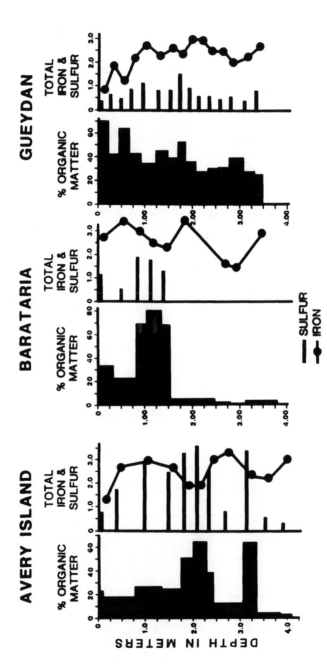

Figure 7. Organic matter, total sulfur and total iron in solids (expressed on a dry weight basis).

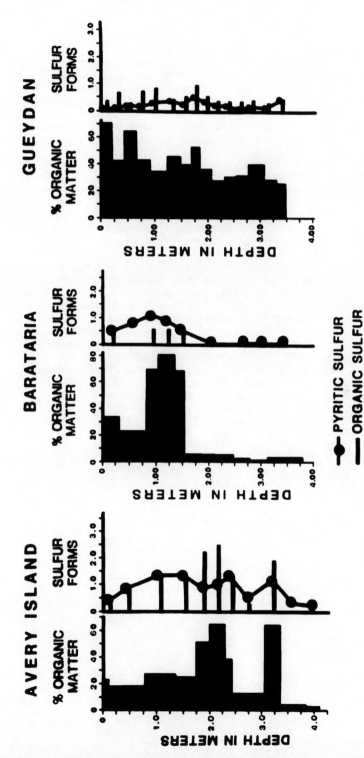

Figure 8. Organic matter and sulfur forms (expressed on a dry weight basis). Sulfate concentrations are minor and are not shown.

Petrographic data consist of light photomicrographs, electron micrographs, x-ray microanalyses, and observations on size and distribution of pyrite crystals. Light microscopy was carried out on all 57 sections, while electron microscopy and x-ray microanalysis were carried out on two sections, one from the Avery site (depth 210 cm) and one from the Barataria site (depth 106 cm). Micrographs and x-ray microanalysis of iron sulfides are shown in Figures 9 and 10, respectively.

Octahedral pyrite crystals occur in all examined thin sections, including those within 5 cm from the surface, and range in size from 0.5 to 10 microns. Larger crystals tend to be found at greater depth, but small crystals occur throughout and overall concentrations of pyrite do not appear to increase much over the depth intervals of the cores.

Irregular aggregates of crystals (A, Figure 9) and framboidal aggregates of crystals (B, Figure 9) are found in addition to individual crystals . Crystals in these aggregates range in size from 0.2 to 4 microns, whereas the aggregates range in size from 10 to 44 microns. These aggregates are found in the top 5 cm of sediment at the Avery Island and Barataria sites, while the uppermost occurrence of framboidal pyrite is at 31 cm at the Gueydan site. Framboidal aggregates are generally scarce in the upper half of the Gueydan core. Some aggregates were observed in all lithologies and at all three sites.

In addition to normal framboidal aggregates in which crystals are generally larger than approximately 1 micron and in which the volume of the body is filled primarily by pyrite, there are rounded bodies comprised of small crystals that range in size from submicroscopic under the light microscope to 1 micron in diameter and which contain transparent material between the crystals (C and D, Figure 9). These bodies are 10-250 microns in diameter and are generally larger than associated normal framboids. These are of lesser abundance compared to coarser material. Crystals within the aggregates are often regulary arranged.

X-ray microanalyses of these small crystals show two different results. At the Barataria site, diminished S/Fe ratios relative to pyrite (A, Figure 10) occur in the outer portions of these fine grains (E, Figure 9). A prominant O peak is also seen. Interiors of these grains show S/Fe ratios around 2 and little O. A more extensive examination of fine grains at the Avery site (F, Figure 9) also showed S/Fe ratios of about 2 and no O (B, Figure 10) for all grains examined.

Discussion

Results from surface and porewater analyses were used to gauge present chloride salinities, an indication of marine influence (Table I).

Table I. Chloride Salinities of Examined Sites.

	Open Water	Average Pore Water
Avery Island	2.46 ppt	7.73 ppt
Barataria	0.23 ppt	1.11 ppt
Gueydan	0.05 ppt	0.11 ppt

These data verify that the Gueydan site is freshwater, Barataria intermediate, and Avery Island brackish. It will be noted that pore water salinities are greater than associated open waters. Evapotranspiration may produce such increases in

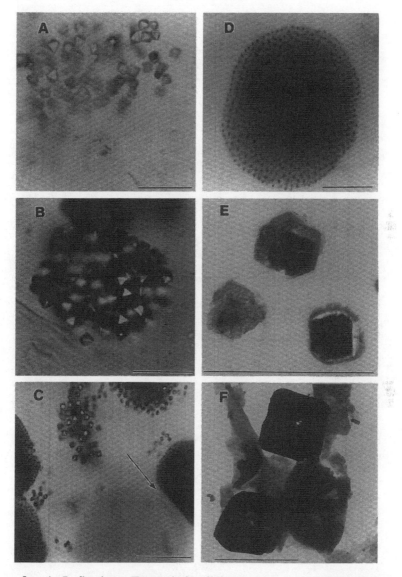

Figure 9. A. Reflection - Transmission light micrograph showing aggregate of pyrite crystals (Bar = 10 microns). B. Reflection - Transmission light micrograph showing framboidal aggregate of pyrite crystals (Bar = 10 microns). C. Reflection - Transmission light micrograph showing clouds of pyrite microcrystallites. Note separation by clear layer indicated by the arrow (Bar = 10 microns). D. Reflection - Transmission light micrograph showing ordering in clouds of microcrystallites (Bar = 10 microns). E. Transmission electron micrograph of microtomed section through a cloud of Barataria microcrystallites. Note the alteration around the grains (Bar = 1 micron). F. Transmission electron micrograph showing Avery Island microcrystallites (Bar = 1 micron).

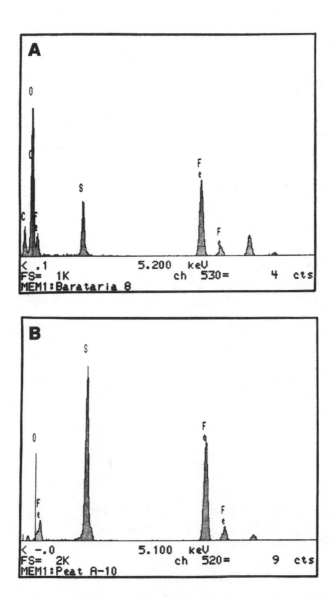

Figure 10.   A. X-ray spectrum of the margin of a microcrystallite in a diffuse cloud from the Barataria site. B. X-ray spectrum of a microcrystallite in a diffuse cloud from the Avery Island site.

concentration (King et al. (28)). In addition, there are temporal variations in pore water salinities in the upper 0.5 meters due to seasonal variations in admixtures of freshwater (29). The average salinity, however, appears to be approximately equal to values below 0.5 meters (29) and these values do not appear to be strongly influenced by this seasonal variation. Hence, average pore water salinities are considered good measures of average marine influence.

Total organic matter data from both earlier and later cores show that samples range from peats to clays (Figure 8). From these and the salinity data, it can be concluded that most of these samples fall in region I of Figure 1.

To investigate relationships between salinity, organic matter, and total sulfur, a plot was constructed showing total sulfur as a function of salinity and organic matter (Figure 11). Data include those from the 2 cores taken at each of the three sites. In addition, information from a more saline core taken farther to the south in the Barataria Basin and data from Frazier and Osanik (30) and Brupbacher et al. (31) were included. The results show that there is a general increase of total sulfur with both organic matter and salinity where there is at least slight marine influence. Values for total sulfur (on a dry weight basis) range from 0.1% to more than 3%.

Similar results can be obtained through a simple multiple linear regression in which total sulfur is regressed on chloride salinity and organic matter and in which the regression passes through the origin. The resulting regression plane, shown in Figure 11 shows a positive correlation between total sulfur, and salinity and organic matter. An F value of about 264 is given for the model along with a squared multiple correlation coefficient of 0.88. Results also indicate only minor interaction between salinity and organic matter. In the low-organic part of the diagram (organic matter<0.2) these direct proportionalities result from the general reaction

$$2 \ CH_2O + SO_4^{-2} \ \text{------>} \ H_2S + 2HCO_3^- \ \text{(Berner (32))}$$

where CH_2O is a simplified general formula for organic matter, SO_4^{-2} is dissolved sulfate, and H_2S may react with available iron to form iron sulfides. Since total sulfur increases with both variables it appears that, in these particular clays, both organic matter and sulfate control incorporation of sulfur. Of the 2 variables, organic matter appears to be most critical because total sulfur increases fairly rapidly with organic matter in this part of the diagram and because a close association between pyrite and the limited number of organic fragments can be observed microscopically.

Isolation of the system from continued addition of sulfate and continued reaction will cause an inverse relationship to develop between organic carbon and sulfur which depends on sulfate concentrations in trapped porewaters. High C/S ratios will be characteristic of freshwater sites and low C/S ratios characteristic of saline sites (Berner and Raiswell (33)).

In mucks and peats (organic matter >0.2) sulfate is still an inhibiting factor so that total sulfur still goes up with increasing salinity. Production of H_2S, however, is probabably no longer inhibited by reactive organic mattter. With more organic matter, fixation by reaction with organic matter, in addition to pyrite formation, may control total sulfur incorporation. Data on forms of sulfur (next section) indicate appreciable organic sulfur in these sediments, and work by Casagrande (1) and Casagrande et al. (2) describes reactions for organic incorporation of sulfur. An exception to the general increase of total sulfur with organic matter occurs at the low-salinity site in more organic-rich samples. Here total sulfur does not increase with increasing organic matter because no more sulfate is available.

Examination of the data on forms of sulfur appears to substantiate processes implied by the data on total sulfur. With some exceptions, organic sulfur is more important, or of approximately equal importance in organic-rich samples, whereas

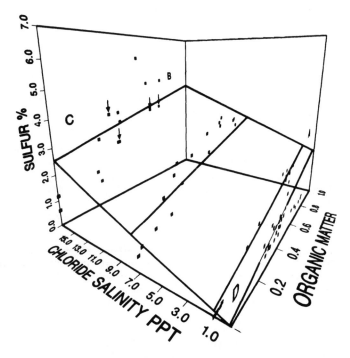

Figure 11.   Total sulfur (dry weight basis) as a function of chloride salinity and organic matter. Arrows indicate points taken from Frazier and Osanik (30). Salinity of the most saline site (line CB) was estimated using data from Brupbacher, Sedberry, and Willis (31). The plane A-B-C-D represents the simple multiple linear regression of total sulfur on chloride salinity and organic matter and the 4 lines in the plane the regression values for total sulfur at the 4 sites.

pyritic sulfur is dominant in organic-poor samples (Figure 8). Pearson correlation coefficients also show a relationship between organic matter and organic sulfur, though the relationship with salinity is stronger. In general, for a given salinity, sulfur partitions itself between organic and iron sulfide phases based on the availability of iron and organic matter. In this sense, the samples appear to occupy the expected intermediate position between clastic sediments and peats.

Petrographic work is difficult to correlate with bulk chemical analyses due to the sporadic distribution of pyrite. However, petrographic observations offer clues revealing processes involved in iron sulfide formation.

For example, individual crystals of pyrite are thought by numerous workers (Goldhaber and Kaplan (15), Rickard (13), Howarth (16), Luther et al. (17)) to form by rapid direct precipitation from solution and where the solution is undersaturated with respect to monosulfides but supersaturated with respect to pyrite. The presence of small individual crystals in all samples of this study, even in samples from close to the surface, is consistent with rapid formation of this type of pyrite.

In contrast, formation of framboidal pyrite may be a slower and more complex process involving a greigite precursor which forms from mackinawite in the presence of a small amount of oxygen (Berner (11), Sweeney and Kaplan (12)). Original monosulfide precipitation requires supersaturation with respect to these phases. The presence of framboidal pyrite in close proximity to single crystals in these sediments appears to indicate spatially varying microenvironments and/or changes with time and superposition of the results. To some extent, the individual crystals may represent earlier conditions when concentrations of sulfate were lower. The general scarcity of framboidal pyrite in the upper half of the Gueydan section, where sulfate is almost absent (Figure 6) supports this model. Framboidal pyrite is more common in the lower part of the section where sulfate increases, and is abundant at Barataria and Avery Island where sulfate is abundant (Figure 6).

The significance of loose, rounded aggregates of finer crystals described in the first paragraph of the **Results** section is harder to determine. These are not common, but some can be found in all three sections associated with organic-rich horizons. From the morphology they may represent framboids in an early state of formation. As noted, if this is the case, some monosulfides might be expected. X-ray microanalyses of microcrystallites in the Barataria Section do show variable S/Fe ratios. However, significant O peaks in the x-ray spectrum and low pH's and relatively high sulfate concentrations in the pore solutions for the upper part of the Barataria section (Figure 6) point to oxidation of these fine-grained bodies during seasonal fluctuations of water levels at this locality. Centers of grains show S/Fe ratios around 2. Such oxidation has been described by Luther et al. (17) and Luther and Church (34) for New England and Delaware marshes, respectively. The Avery Island data show S/Fe ratios around 2 and no O peak (Figure 10). Apparently, ratios of around 2 are more normal for these grains. If sulfurization occured, it must have been fairly rapid. Morse and Cornwell (35) also note a general scarcity of monosulfides in framboids in their SEM investigations of marine sediments. Similar submicroscopic iron sulfides have been observed in coals (36).

Summary

There is a general positive correlation between both sulfate from marine incursion and organic matter, and total sulfur. This results from the importance of both organic matter and sulfate in H_2S production and from the fact that most of the sulfur is trapped by organic matter and by pyrite formation in these sediments. Total sulfur is partitioned primarily between pyritic and organic forms. Pyritic sulfur is dominant in clays, whereas organic sulfur is of greater or approximately equal importance in organic-rich samples. This pyrite appears to be forming both directly as individual crystals, and by rapid sulfurization of monosulfides or direct precipitation of pyrite in early framboidal aggregates.

Acknowledgments

The authors wish to thank the Louisiana Geological Survey for aid at several stages of the study, and the McIlhenny Company for help during the collection of samples near Avery Island. The authors also wish to thank Dow Chemical Company for use of facilities and the USL Foundation for partial financial support. Part of this work represents a thesis done by one of the authors (J. S.). Reviews by Dr. D. J. Casagrande, Dr. R. Markuszewski, Dr. W. E. Straszheim, and three anonymous reviewers resulted in significant improvements. The final responsibility for any errors rests with the authors.

Literature Cited

1. Casagrande, D. J. In Coal Bearing Strata: Recent Advances; Scott, Ed; Geological Society: London, 1987, pp 87-105.
2. Casagrande, D. J.; Idowu, G.; Friedman, A.; Rickert, P.; Siefert, K.; Schlenz, D. Nature 1979, 282, 599-600.
3. Cohen, A. D.; Spackman, W.; Dolsen, P. International Journal of Coal Geology 1983, 4, 73-96.
4. Bustin, R. M.; Lowe, L. E. Journal of the Geological Society, London 1987, 144, 435-450.
5. Altschuler, Z. A.; Schnopfe, M. M.; Silber, C. C.; Simon, F. O. Science 1983, 221, 221-227
6. Kaplan, I. R.; Emery, K. O.; Rittenberg, S. C. Geochim. et Cosmochim. Acta 1963, 27, 297-331.
7. Berner, R. A. American Journal of Science 1970, 268, 1-23.
8. Davison, W.; Lishman, J. P.; Hilton, J. Geochim. et Cosmochim. Acta 1985, 49, 1615-1620.
9. Morse, J. W.; Millero, F. J.; Cornwell, J. C.; Rickard, D. Earth-Science Reviews 1987, 24, 1-42.
10. Lowe, L. E.; Bustin, R. M. Canadian Journal of Soil Science 1985, 65, 531-541.
11. Berner, R. A. Journal of Geology 1964, 72, 293-306.
12. Sweeney, R. E.; Kaplan, I. R. Economic Geology 1973, 68, 618-634.
13. Rickard, D. T. American Journal of Science 1975, 275, 636-652.
14. Rickard, D. T. American Journal of Science 1974, 274, 941-952.
15. Goldhaber, M. B.; Kaplan. I. R. In The Sea Vol. 5; Goldberg, Ed.; Wiley: New York, 1974, pp 569-655.
16. Howarth, R. W. Science 1979, 203, 49-51.
17. Luther, G. W., III; Giblin, A.; Howarth, R. W.; Ryans, R. A. Geochim. et Cosmochim. Acta 1982, 46, 2665-2669.
18. Kosters, E. C. Louisiana Peat Resources. Dept. of Natural Resources, Louisiana Geological Survey, Baton Rouge, LA 1983, 63 pp.
19. Kosters, E. C.; Chumra, G. L.; Bailey, A. Journal of the Geological Society, London 1987, 144, 432-434.
20. Kosters, E. C.; Bailey, A. Gulf Coast Assoc. Geol. Soc. Trans. 1983, 33, 311-325.
21. ASTM. Annual Book of ASTM Standards: Part 26; Gaseous Fuels; Coal and Coke; Atmospheric Analysis. ASTM, Philadelphia, 1974, pp 507-511.
22. Spielholtz, G. I.; Diehl, H. Talanta 1966, 13, 991-1002.
23. McGowan, C. W.; Maruszewski, R. Fuel 1988, 67, 1091-1095.

24. Begheijn, L. Th.; Van Breeman, N.; Velthorst, E. J. Commun. in Soil Science and Plant Analysis 1978, 9, 873-882.
25. Lord, C. J. J. Sed. Petrology 1982, 52, 664-666.
26. Bailey, A.; Blackson, J. Scanning Electron Microscopy/1984/IV, 1984, 1475-1481.
27. Blackson, J. H.; Bailey, A. EMSA Bulletin 1985, 15, 115-117.
28. King G. M.; Klug, M. L.; Wiegert, R.; Chalmers, A. G. Science 1982, 218, 61-63.
29. Feijtel, T. C. Ph.D. Thesis, Louisiana State University, Baton Rouge, 1986.
30. Frazier, D. E.; Osanik, A. In Environments of Coal Deposition; E. C. Dapples, Hopkins, M. E. Eds.; Spec. Paper Geol. Soc. America 114: GSA: Boulder, 1969; pp 63-85.
31. Brupbacher, R. H.; Sedberry, J. E.; Willis, W. M. Louisiana Agriculture Experiment Station Bulletin, No. 672, 1973, p. 34.
32. Berner, R. A. Geochim. et Cosmochim. Acta 1984, 48, 605-615.
33. Berner, R. A; Raiswell, R. Geochim. et Cosmochim. Acta 1983, 47, 855-862.
34. Luther, G. W., III; Church T. M. Marine Chemistry 1988, 23, 295-309.
35. Morse, J. W.; Cornwell, J. C. Marine Chemistry 1987, 22, 55-69.
36. Greer, R. T. Colloidal and Interface Science 1976, 4, 411-423.

RECEIVED March 5, 1990

Chapter 11

Formation of Iron Sulfides in Modern Salt Marsh Sediments (Wallops Island, Virginia)

T. S. White[1], J. L. Morrison[1,2], and L. R. Kump[1]

[1]Department of Geosciences, Pennsylvania State University,
University Park, PA 16802
[2]Materials Research Laboratory, Pennsylvania State University,
University Park, PA 16802

Iron sulfide abundances and sediment pore fluid
chemistries varied considerably between three closely
spaced but environmentally distinct cores taken from
evaporative marsh panne, low marsh, and tidal creek
subenvironments within a Spartina salt marsh. From
these cores, three depositional settings were discerned
in the subsurface: marsh, tidal flat, and bay. The
formation of iron sulfides in these sediments was found
to be rapid, with pyrite occurring as the dominant
sulfide phase. Extant marsh chemistries have an
overprinting effect on the underlying sediments
producing relative high amounts of pyrite directly
beneath the marsh sediments. Sulfide grain size and
morphologies varied nonsystematically between cores.

Wallops Island, Virginia, is a Holocene transgressive barrier island
system off the eastern coast of the Delmarva Peninsula (1).
Chincoteague and Assateague Islands lie to the north of Wallops
Island, separated by Chincoteague Inlet which provides an inlet for
open marine waters into Chincoteague Bay and an outlet for lagoonal
and estuarine waters from the bay (Figure 1). The surface water
drainage divide of the Delmarva Peninsula is very close to the ocean,
therefore most of the precipitation runoff is directed to Chesapeake
Bay. As a consequence, the bays between the Delmarva Peninsula and
the offshore barrier islands are sediment starved with the majority
of the sediments derived from adjacent eroding marshes and through
distribution of tidal delta and washover sediments.
    A typical barrier-island system can be divided into three
depositional environments (barrier beach, lagoonal/bay, and tidal
channel-delta complex) each of which contains a number of smaller
subenvironments (2). This work focused on a typical bay-fill
sequence in which bay sediments are overlain by tidal flat deposits

0097–6156/90/0429–0204$06.00/0

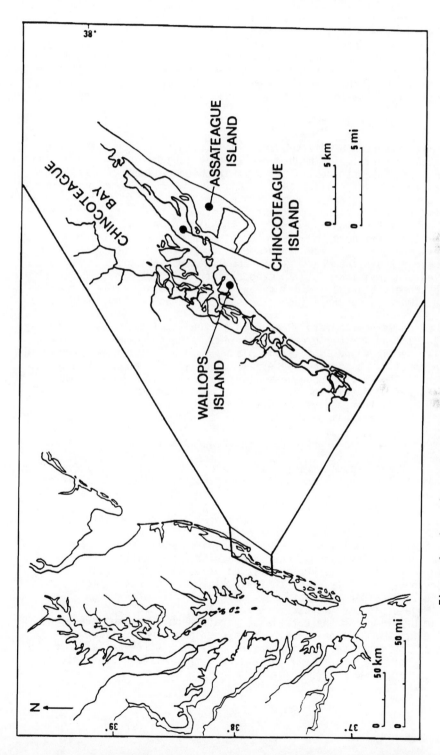

Figure 1. Location map of Wallops and Chincoteague Islands, Virginia [taken from Dade (1)].

which in turn are overlain by tidal flat deposits which in turn are overlain by a modern Spartina marsh.

The main objective of this research effort was to characterize diagenetic transformations among the various sulfide phases and analyze pore fluid chemistry with respect to depth. In addition, it also examined if the overlying organic-rich Spartina marsh sediments affected the formation of iron sulfides in the underlying sediments.

Sample Collection and Processing

The salt marsh can be divided into two distinct environments: high and low marsh (3). The high marsh is only subject to tidal flushing during the highest tides. This study focused on the low marsh where tidal flushing is regular.

Cordgrass present throughout the marsh is Spartina alterniflora. Two forms of this plant are identifiable. The tall form occurs along the banks of the tidal creek and is tolerant of normal marine salinities; the short form is tolerant of higher salinities and occurs throughout the low marsh where such water may be encountered.

Three subenvironments within the low marsh (Figure 2) were sampled and analyzed: (1) an evaporative or marsh panne, (2) the low marsh proper, and (3) a tidal creek. Sampling was performed (during March, 1988) using three inch aluminum tubing and a vibracoring unit (4).

Cores were immediately returned to the lab facilities of the Wallops Island Marine Science Consortium. The cores were cut into smaller sections and immediately placed into a nitrogen atmosphere in glove boxes. Sub-sections were taken from various depths along the length of the core. Each sub-section was split; one half of the split was immediately stored under an inert nitrogen atmosphere and frozen for further analyses. One quarter was processed for sedimentological analysis and solid phase chemistry. The remaining quarter was centrifuged to obtain pore fluids for analysis. Sample preparation was performed under a nitrogen atmosphere in the glove boxes.

The frozen samples were returned to The Pennsylvania State University for preparation and petrographic analysis. Preparation involved drying the sample at 105°C in a nitrogen atmosphere. The samples were pulverized to -16 mesh using a mortar and pestal, mixed with a Hexacol epolite resin and hardener, and then centrifuged to ensure complete mixing with the sediment. After curing, the resins were cut longitudinally and placed in one inch stainless steel molds where additional resin was poured over the cured resin. The one inch pellets were then polished and viewed using reflected light microscopy. A twenty point reticule was used to better facilitate recording occurrences.

Megascopic and microscopic analyses of the sediments provided the information for the correlated stratigraphic columns presented in Figure 3. Marsh, tidal flat, and bay sediments were discerned from each of the three cores. These sediments were differentiated on the basis of grain size, fossil content, and the relative abundance of Spartina root and rhizome debris.

Figure 2. Evaporative panne, marsh proper, and tidal creek subenvironments. In Figure 2a, an inundated evaporative panne is shown in the foreground with a *Spartina* marsh in the background. Figure 2b shows an exposed evaporative panne during low tide and the vibracoring unit. The tidal creek with high form *Spartina alterniflora* growing along its perimeter, and the short form growing towards the interior of the marsh is visible in Figure 2c.

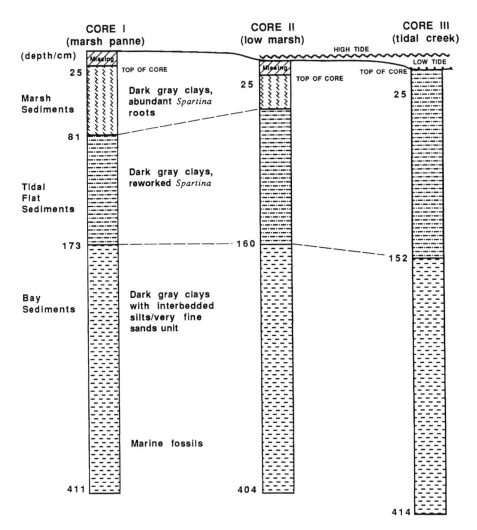

Figure 3. Stratigraphic columns of the cored subenvironments.

Analytical Methods

Pore waters were analyzed for pH, alkalinity, salinity, and sulfate concentration following centrifugation. pH was measured with a Ross combination electrode standardized with commercial NIST traceable buffers. Alkalinity determinations were made by a single addition of 0.01 N HCL (5). Salinity was determined with a refractometer, and with an accuracy of +/- 0.5 o/oo. Sulfate was determined turbidimetrically (6).

Solid phase chemical analyses included determination of total organic C content, and the distribution of S between iron monosulfides (acid-volatile sulfur or AVS) and pyrite (the difference between total reducible sulfur and AVS). Total organic carbon was measured coulometrically following combustion at 1050°C (7). Acid-volatile sulfur and total reducible sulfur analyses followed the procedure of Canfield et al. (8). A microbiological assay of the abundance of sulfate reducing bacteria was performed according to (9).

Results

Pore fluid chemistries of the various samples are summarized in Figure 4. The distribution of salinities (Figure 4A) within each core is fairly regular but distinct from each of the other cores. The tidal creek core (core 3) has pore water salinities closest to that of normal seawater, perhaps reflecting a constant supply of seawater in the tidal creek. The evaporative marsh panne (core 1) has salinities greater than normal sea water. Marine waters may reach the evaporative pannes via shallow subsurface hydrologic communication with the tidal creeks and through surface flooding during high tides (10). Standing water in the evaporative panne is subject to evaporation which concentrates the sea water salts. The "plumbing" which allows for elevated salinity at depth is not fully understood. Fluctuations in near-surface salinities do occur seasonally (11). The highest alkalinity pore fluids correspond to the most saline waters (Figure 4B); core 2 most closely matches seawater with the upper portion of the other curves trending towards it.

In Figure 4C, the pH of the pore fluids is shown to be in the range of 7-7.8 in all three cores. Low marsh and tidal creek sediments (cores 2 and 3, respectively) show a gradual increase in pH with depth while the evaporative panne shows a decrease in pH with depth followed by an increase.

The results of dissolved sulfate concentrations are summarized in Figure 4D. Sulfate concentrations are highest in the top of the evaporative panne (core 1) sediments. Sediments of the evaporative panne and tidal creek (cores 1 and 3 respectively) do not reproduce the anticipated closed system behavior for sediments undergoing sulfate reduction, i.e. progressive decrease with depth (12). Instead, initial depletions in the top meter are followed by sulfate increases deeper in the cores. The marsh core (core 2) does not contain detectable amounts of dissolved sulfate at any depth.

A microbiological assay of the abundance of sulfate reducing bacteria was performed to determine the distribution of colony

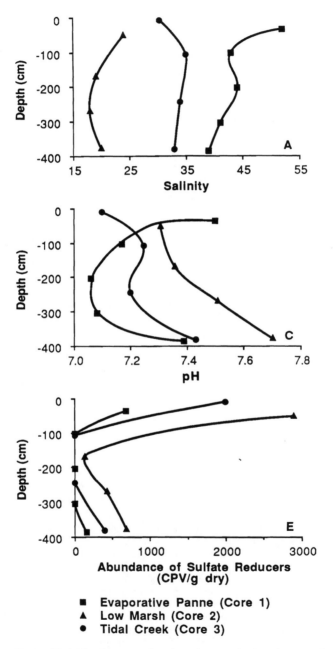

Figure 4.   Pore fluid/sediment chemistries and abundance of sulfate reducers from the various cores.

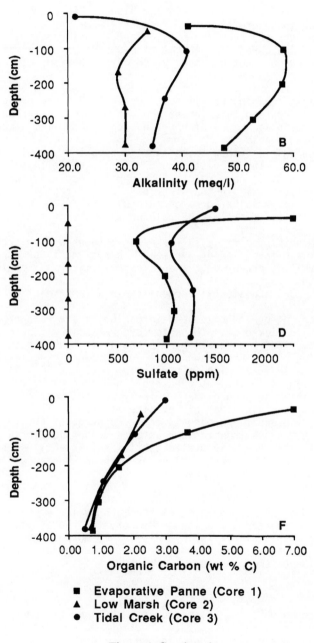

Figure 4. Continued.

forming units (CFU) with respect to depth (Figure 4E). The sediments of the marsh core (core 2) contain the highest CFU counts. All three cores show an abundance of sulfate reducers near the top of the core trending towards zero near the middle, and then increasing towards the bottom of the core. In general, the abundance of sulfate reducers is low, but not unexpectly, so given the time of sampling (March).

Organic carbon profiles (Figure 4F) in all three subenvironments show the same decreasing profile with depth (13). The evaporative panne (core 1) contain the highest amounts of organic carbon which corresponds to an abundance of Spartina roots and filamentous algae visible in hand specimen.

Iron monosulfides show an antithetic relationship with pyritic sulfur (Table I). The highest amounts of acid volatile sulfides (which remained small compared to S in pyrite) occur directly beneath the marsh sediments within the upper portions of the tidal flat deposits. The high value for iron monosulfides in the upper part of the tidal creek sediments may be related to high rates of sulfate reduction at these depths.

High amounts of pyritic sulfur occur throughout our cores suggesting rapid formation of pyrite, either directly or through an iron - monosulfide precursor. The relatively minor AVS content of these sediments suggest that if AVS acts as a precursor, its existence is as a transient intermediate in a fast reaction to pyrite.

A variety of pyrite sizes and morphologies are recognized (Figure 5). In Figure 5A, SEM photomicrographs show fine-grained framboidal pyrite. Framboidal pyrite and an irregular mass of inequigranular pyrite are shown in Figure 5B. An irregular shaped mass of pyrite in inertinite is presented in Figure 5C. All of these morphologies and associations presented in Figure 5 are very common in the stratigraphic record, particularly in coal-bearing strata.

A record was maintained of the size of each pyrite occurrence. Core 1 shows a gradual increase in the percentage of pyrite particles less than six microns in size whereas core 3 shows a decrease in the percentage of the same size pyrite particles (Table I). Core 2 shows no consistent pattern. The tidal creek with an abundance of sulfate in a reducing environment appears to produce more pyrite of smaller sizes. On the marsh panne, larger pyrite particles are formed in this hypersaline, high alkalinity environment.

A record of morphology classes for each pyrite occurrence was kept during petrographic analyses. Monocrystalline pyrite includes euhedral and subhedral pyrite crystals. This morphology class is always more prevalent than framboidal pyrite except at the top of core 1 and the bottom of core 3.

Discussion

Organic Carbon and Pyrite Relationships. Organic carbon enters most marine sediments as detritus at the sediment-water interface and then goes progressive decomposition with burial. However, in salt marshes, where the subsurface production of organic matter (in the form of roots and rhizomes) can be as great as eight times the above-ground production (14), a major addition of organic matter

Table I. Sulfide characterization data.

Sample (depth/cm)	Environment	Acid Volatile Sulfides (mg S/g dry sediment)	Pyritic Sulfur (mg S/g dry sediment)	Monocrystalline* Pyrite	Framboidal* Pyrite	Grain Size (μm)*		
						0-6	6-12	>12
I -36	Marsh	0.060	4.540	0.0	14.1	60.6	26.7	12.7
I -102	Tidal Flat	0.100	18.230	19.2	9.2	70.6	18.4	11.0
I -203	Bay	0.001	18.130	11.9	10.2	78.0	15.3	6.7
I -305	Bay	0.020	6.280	20.0	0	80.0	16.3	3.7
I -386	Bay	0.030	9.140	16.4	7.6	81.0	15.1	3.9
II -51	Marsh/Tidal Flat	0.140	8.310	12.4	5.7	80.7	14.9	4.4
II -168	Tidal Flat	0.060	13.660	5.0	4.3	80.6	12.9	6.5
II -269	Bay	0.040	5.910	2.9	0	89.1	7.9	3.0
II -376	Bay	0.000	8.450	8.5	4.2	74.5	14.9	10.6
III -10	Tidal Flat	0.200	18.400	4.6	3.1	88.5	10.7	0.8
III -107	Tidal Flat	0.001	23.000	11.5	6.2	77.0	16.8	6.2
III -244	Bay	0.030	10.370	14.3	4.8	78.1	12.4	9.5
III -381	Bay	0.005	7.100	3.0	9.1	66.7	15.2	18.1

Evaporative panne subenvironment (core 1) = I 36-386
Low marsh subenvironment (core 2) = II 51-376
Tidal creek subenvironment (core 3) = III 10-381

*Percentage of the observed pyrite.

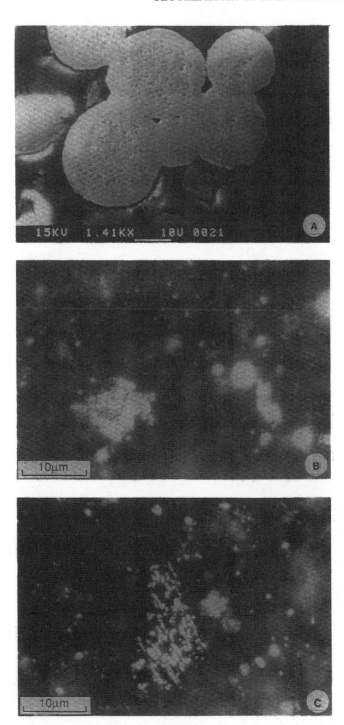

Figure 5.   Photomicrographs of Iron Sulfides.

occurs at shallow depths in the sediments. If this material is digestible, it should stimulate the activity of anaerobic sulfate reducing bacteria in this largely anoxic environment. Thus one might expect an association between pyrite (the mineralogical end-product of sulfate reduction) and the remains of Spartina roots and rhizomes.

The iron sulfides found intimately associated with fresh plant tissues in near surface sediments were in most cases classified as framboidal. In addition, irregular pyrite morphologies were identified deeper in the cores, infilling organic tissues that would be classified by coal petrographers as inertinite.

The pyrite found at depth associated with inertinite probably originated near the surface. It is believed that inertinite (as the remains of vascular plants) does not form a good substrate for sulfate-reducing bacteria (15). Thus, extensive and continued formation of pyrite at depth in these sediments is unlikely.

This conclusion is made despite the apparent increase in sulfate-reducing bacteria and sulfate concentrations at depth in these sediments. This increase in sulfate concentration is coincident with a decrease in alkalinity, and thus might represent the oxidation of pyrite at depth:

$$FeS_2 + 7/2\ O_2 + 2\ HCO_3^- \longrightarrow Fe^{2+} + 2\ SO_4^{2-} + 2\ CO_2 + H_2O \qquad (1)$$

The supply of O_2 would thus imply deep circulation of oxidizing pore fluids in these sediments. The continued decrease of organic carbon might be tied to this excess supply of oxidants at depth.

Diagenetic Overprinting. Pye (14) noted that the nature of authigenic iron minerals in salt marsh sediments is dependent on accretion rates, initial reactive iron content, marsh vegetation, and fluctuations in sulfate-reducing bacterial activity. Given that the Chincoteague Bay marshes are eroding, accretion rates appear to be negligible; the sediments may be subject to exhumation. Therefore, the zone of sulfate reduction will not migrate through the sediment column since the sediment-water interface is not being elevated. In fact, the zone of sulfate reduction may be maintained at a given depth for a considerable period of time, and may remigrate through underlying sediments which have already been subject to these processes. The aforementioned occurrence of pyrite infilling inertinite could be a manifestation of such a stage of pyrite formation. However, a more metabolizable organic material is necessary for bacterial consumption and sulfate reduction. This energy source is not readily available in these sediments. Thus, pyrite associated with inertinite was probably emplaced near the surface.

The superposition of marsh sediments on older bay and tidal flat sediments has likely increased their pyrite content. Environmental overprinting has been reported in both recent sediments (16) and in the rock record (17).

Pyrite Morphology. Pyrite formed during early diagenesis in marsh sediments has essentially two modes of occurrence, as small single crystals and as framboids (18). The processes that lead to one form or the other remain unclear. Euhedra may form by direct precipitation where pore waters are supersaturated with respect to pyrite but undersaturated with respect to mackinawite or greigite, or by the reaction between mackinawite and elemental sulfur. Framboids may also develop inorganically during the reaction of iron monosulfides with elemental sulfur. Giblin and Howarth (19) observed pyrite occurring predominantly as small single crystals and framboids. They believed that framboids result from iron monosulfide precursors whereas the small single crystals arise when pyrite precipitates directly.

In all three cores of our study, the framboidal pyrite shows an overall gradual decrease with depth followed by an increase at the bottom (Table I). This trend may be related to the same trend visible in the plots of abundance of sulfate reducers and iron monosulfides. The top of core 3 is the only outlier to the decreasing then increasing trend in percentage framboids. This outlier corresponds to the most reducing Eh value, the lowest alkalinity, the highest amount of iron monosulfides, and the second highest amounts of sulfate, sulfate reducers, and pyrite obtained from any of the cores. Apparently, framboids do not form as readily in the tidal creek as compared to the marsh panne (core 1). The small amount of AVS in the upper portion of core 1 may be a function of uptake through framboid formation; core 1 has the most framboids of any of the cores.

Cores 1 and 3 show an antithetic relationship between framboidal and monocrystalline pyrite whereas core 2 has the same decreasing then increasing with depth profile as observed for framboids. The genetic relationship between these pyrite morphologies is evident. Where AVS content is high and pyritic sulfur content is low, the monocrystalline morphologies are more prevalent than framboids. However, where framboid content is high, monocrystalline pyrite and AVS content are low, whereas pyritic sulfur content is high. Framboid formation through iron monosulfide conversion to pyrite appears to boost the total pyrite content of the sediments, presumably beyond levels attained through direct pyrite precipitation.

Conclusions

The pore water profiles, solid phase chemistry, and pyrite petrography in three closely spaced cores from salt marsh subenvironments show complex trends and rapid variations.

Marsh overprinting on the underlying sediments produced relatively high amounts of pyritic sulfur at shallow depths (100 to 203 cm).

The iron sulfides were very fine-grained with the 0 to 6 micron size fraction most common. The reducing environment of the tidal creek produced more pyrite of smaller grain sizes, whereas larger pyrite particles were formed in the hypersaline, high alkalinity sediments of the marsh panne. Monocrystalline pyrite morphologies were more prevalent than framboidal pyrite where AVS-content was relatively high and pyritic sulfur content was low. However, where framboid content was high, monocrystalline pyrite and AVS-content was low, and pyritic sulfur content was high. Framboids appear to form more readily in the marsh panne as compared to the tidal creek.

Framboid formation may boost the total pyrite content of the sediments beyond levels attained through direct precipitation.

Acknowledgments

The authors acknowledge the efforts of the members of Geosc. 597B during the Spring semester of 1988 which provided the majority of the geochemical data used throughout this publication, and the Department of Geosciences at The Pennsylvania State University which provided the necessary logistical support for this research endeavor. The National Aeronautical and Space Administration (NASA) provided accessibility to Wallops Island. We also thank two anonymous reviewers for their helpful suggestions.

Literature Cited

1. Dade, W.B. M.S. Thesis, The Pennsylvania State University, Pennsylvania, 1983.
2. Reinson, G.E. In Facies Models; Walker, R.G., Ed.; Geol. Assoc. Canada, 1984; pp 119-140.
3. Leatherman, S.P. Barrier Island Handbook; University of Maryland, 1982; p 109.
4. Sanders, J.E.; Imbrie, J. Geol. Soc. Amer. Bull. 1963, 74, 1287-92.
5. Strickland, J.D.H.; Parsons, T.R. A Practical Handbook of Seawater Analysis; Strickland, J.D.H.; Parsons, T.R., Eds.; Alger Press Ltd., 1977; pp. 11-34.
6. Standard Methods for the Examination of Waste and Wastewater; Port City Press, MD, 1985, 16th edition.
7. Ramirez-Rojas, A.J. Ph.D. Thesis, The Pennsylvania State University, Pennsylvania, 1988.
8. Canfield, D.E.; Raiswell, R.; Westrich, J.T.; Reaves, C.M.; Berner, R.A. Chem. Geol. 1986, 54, 149-55.
9. Postgate, J.R. The Sulphate-Reducing Bacteria; Cambridge University Press, New York, 1984.
10. Nuttle, W.K.; Hemond, H.F. Global Biogeochem. Cycles. 1988, 2, 91-114.
11. Casey, W.H.; Guber, A.; Bursey, C.; Olsen, C.R. EOS Trans. Amer. Geoph. Union 1986, 67, 1305.
12. Berner, R.A. Geochim. Cosmochim. Acta. 1984, 48, 605-15.
13. Bluth, V.S. M.S. Thesis, The Pennsylvania State University, Pennsylvania, 1989.
14. Pye, K. Nature 1981, 294, 650-52.
15. Lyons, W.B.; Gaudette, H.E. Org. Geochem. 1979, 1, 151-55.
16. Cohen, A.D.; Spackman, W.; Dolsen, P. International Journal of Coal Geology 1984, 4, 73-96.
17. Williams, E.G.; Keith, M.L. Economic Geology 1963, 58, 720-729.
18. Raiswell, R. Am. Jour. Sci. 1982, 282, 1244-1263.
19. Giblin, A.E.; Howarth, R.W. Limnol. Oceanogr. 1984, 29, 47-63.

RECEIVED April 11, 1990

CHARACTERIZATION OF SULFUR IN FOSSIL FUEL MATERIALS

Chapter 12

Sulfur K-Edge X-ray Absorption Spectroscopy of Petroleum Asphaltenes and Model Compounds

Graham N. George, Martin L. Gorbaty, and Simon R. Kelemen

Corporate Research Laboratories, Exxon Research and Engineering Company, Annandale, NJ 08801

The utility of sulfur K-edge X-ray absorption spectroscopy for the determination and quantification of sulfur forms in nonvolatile hydrocarbons has been investigated. X-ray Absorption Near Edge Structure (XANES) spectra were obtained for a selected group of model compounds, for several petroleum asphaltene samples and for Rasa coal. For the model compounds the sulfur XANES was found to vary widely from compound to compound, and to provide a fingerprint for the form of sulfur involved. The use of third derivatives of the spectra enabled discrimination of mixtures of sulfide and thiophenic model compounds, and allowed approximate quantification of the amount of each component in the mixtures, in the asphaltene samples and the coal. These results represent the first demonstration that nonvolatile sulfide and thiophenic sulfur forms can be distinguished and approximately quantified by direct measurement.

One gap in our knowledge of the chemistry of heavy hydrocarbons and coal concerns the chemical forms and quantities of organically bound sulfur in these materials. At best, current knowledge is qualitative, and is based almost entirely upon characterization of volatile products. For coals, total organic sulfur is quantified by the difference between total and pyritic sulfur analyses. The different forms of sulfur have been assigned by a variety of methods including mass spectroscopy of extractable materials and of liquid products, methyl iodide derivatization, catalytic decomposition, and oxidative techniques (1). More recently, methods have been reported in using a two step chemical modification, in conjunction with ^{13}CNMR spectroscopy to determine the chemical forms of sulfur in non-volatile petroleum materials (2). However, to date, no available method is completely adequate for directly determining both the forms and the amounts of organically bound sulfur in native coals and petroleum materials.

Several papers in this book and in the recent literature (3) discuss use of pyrolysis techniques coupled with gas chromatography and mass spectrometry to determine forms of organically bound sulfur, but these methods introduce an uncertainty due to the possible interconversion of these sulfur forms during the heating step. For example, it has been shown that when benzyl sulfide was heated to 290°C, tetraphenyl thiophene, hydrogen sulfide and stilbene were produced (4). Coupled with heat and mass transport limitation considerations, particularly for viscous liquids and solids, it is not unreasonable to question whether at least some of the thiophenic forms observed by these techniques were produced during the analysis and may not have been present in the original sample.

One way to overcome this uncertainty is by direct measurement. The technique used must be element specific (in this case, for sulfur), environment sensitive, and must be able to observe the entire sample. X-ray Absorption Spectroscopy (XAS) is one such method. This report investigates the applications of X-ray absorption near edge structure (XANES) spectroscopy for the purpose of speciating and quantifying the forms of organic sulfur in solids and nonvolatile liquids. In earlier work Hussain et al. (5), Spiro et al. (6), and later Huffman et al. (7,8) demonstrated the potential of sulfur X-ray absorption spectroscopy for the qualitative determination of sulfur forms in coals; however, they made no attempt at quantification. Sulfur K-edge X-ray absorption spectroscopy has also recently found applications in characterizing sulfur in several different biological systems (9-12).

The approach taken in this work was to examine the sulfur X-ray Absorption Near Edge Structure (XANES) spectra of organic compounds containing sulfur functionalities representative of those believed to be present in coals and heavy petroleum. The compounds investigated included mercaptans, sulfides, disulfides and thiophenes. Compounds containing more oxidized forms of sulfur were also investigated, both for the sake of completeness and in order to explore the possibility of using sulfur X-ray absorption spectroscopy in conjunction with oxidative techniques for quantifying organically bound sulfur forms. Spectra were also obtained for three different petroleum asphaltene samples and for Rasa coal. The latter sample was chosen for this study because it has an unusually high amount of organically bound sulfur, and an unusually low level of pyritic sulfur (13,14).

Experimental Section

Materials. All model compounds used in this study were obtained from Aldrich Chemical Company and were used without further purification. The asphaltene samples were prepared from petroleum residua by precipitation from n-heptane following the procedure of Corbett (15). A sample of Rasa coal was generously provided by Dr. Curt White.

Sample Preparation. Solid samples were finely powdered and dusted onto mylar tape. In order to minimize thickness effects some model compound samples were diluted with boron nitride. Liquid samples were prepared as thin films sandwiched between two pieces of 6.3μm thick polypropylene film.

Data Collection. Data were collected on samples at room temperature at the Stanford Synchrotron Radiation Laboratory on beam line VI-2 with the storage ring SPEAR running in dedicated mode (30-90 mA at 3 GeV). The 54 pole wiggler was operated in undulator mode with magnetic fields of 0.14-0.15T, with a platinum coated focusing mirror and using a Si(111) double crystal monochromator. In order to ensure the maximum possible resolution, the aperture upstream of the mirror was reduced until the XANES spectrum of a sodium thiosulfate standard showed no further sharpening with subsequent reduction. Under these conditions we estimate an upper limit for the resolution of approximately 0.5 eV. In order to minimize atmospheric attenuation of the X-ray beam, the experimental apparatus was enclosed in a bag containing helium gas. X-ray absorption was monitored as the X-ray fluorescence excitation spectrum using a Stern-Heald-Lytle detector (16) or by measuring the total electron yield. The energy scale was calibrated with reference to the lowest energy sulfur K-edge absorption peak of a sodium thiosulfate standard, assumed to be 2469.2 eV (17). Although the absolute value for the energy calibration contains some uncertainty, the relative accuracy of the energy scale proved to be reproducible to within less than 0.1 eV. Higher derivatives of XANES spectra were calculated using a cubic polynomial fit over a 0.4 eV range for each data point. In all cases spectra were normalized to the height of the edge jump which was obtained by extrapolating the EXAFS spline function to the edge. All spectra presented are averages of 2-4 thirty minute scans. Under these conditions, spectra with acceptable signal to noise ratios could be obtained from samples containing less than 1 mole % sulfur.

Results And Discussion

Model Compound XANES. Figure 1 compares the sulfur K edge XANES of a number of model compounds. The XANES can be seen to vary widely from compound to compound. The first inflection energies of all compounds examined are tabulated in Table I. As might be anticipated (10,18), the first inflections of compounds with more oxidized sulfur are notably higher in energy than for those with reduced sulfur. The total span in energy is quite large, being some 12.4 eV between thiohemianthraquinone and potassium sulfate. Casual inspection of Figure 1 might suggest that the XANES from sulfur in different environments can be similar as, for example, with dibenzylsulfide and dibenzothiophene. However, examination of Table I reveals that the edge of dibenzothiophene is displaced from that of dibenzylsulfide, the first inflection energy being some 0.6 eV higher for the former compound. Much of the structure observed in XANES spectra corresponds to transitions to bound states, although other phenomena such as shake-up and shake-down satellites, and continuum resonances (18) also contribute. Bound state transitions commonly follow simple dipole selection rules, and an intense feature at a K edge would thus be expected to correspond to a transition to a level possessing significant p-orbital character.

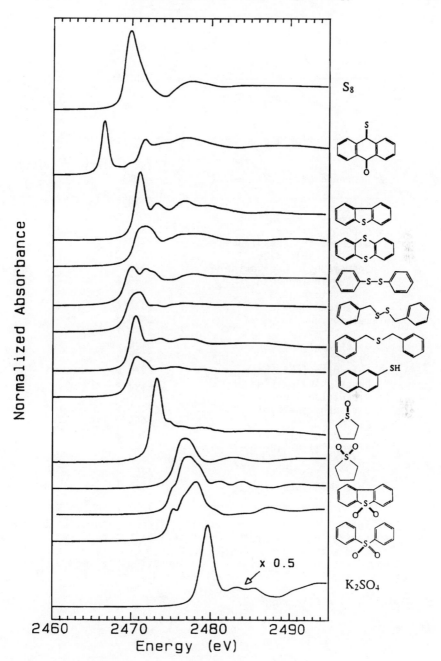

Figure 1. Sulfur K-edge XANES spectra and formulae of selected model compounds illustrating the diversity of sulfur XANES spectra. All data have been normalized to the height of the edge jump. Because of the intensity of the spectrum of K_2SO_4 the vertical scale has been expanded by a factor of two.

TABLE I.   Sulfur K edge First Inflection Energies (b)

Compound	First Inflection (a) (eV)
K$_2$SO$_4$	2478.5
tetramethylenesulfone	2475.6
diphenylsulfone	2474.8
dibenzothiophenesulfone	2474.7
dimethyl sulfoxide	2472.8
tetramethylenesulfoxide	2472.4
2-methylthiophene	2470.6
benzothiophene	2470.4
dibenzothiophene	2470.4
methionine	2470.3
thianthrene	2470.2
dioctylsulfide	2470.1
cystine	2470.1
tetramethylthiophene	2470.0
benzylphenylsulfide	2469.9
dibenzylsulfide	2469.8
2-naphthalenethiol	2469.8
cysteine	2469.2
diphenyldisulfide	2469.2
dibenzyldisulfide	2469.1
sulfur	2469.1
iron pyrite	2468.4
thiohemianthraquinone	2466.1

a.   First inflection points were obtained from the position of the lowest energy maximum of the first derivative, and are considered accurate to better than 0.1eV. Energy standardization is relative to the lowest energy peak of Na$_2$S$_2$O$_3$ as discussed in the Experimental section.

b.   Modified from Reference 19.

Although a detailed analysis of the XANES spectra in Figure 1 is outside the scope of this paper, it is apparent that the spectra can readily be used as a fingerprint for the electronic nature of organically bound sulfur.

Thus the sulfur XANES spectra of compounds with similar sulfur electronic environments were found to be similar. For example, dibenzothiophene and benzothiophene were found to give similar sulfur XANES spectra, whereas dibenzothiophene and thianthrene, whose sulfur atoms are in significantly different environments, exhibit dissimilar sulfur XANES. It is a simple matter to distinguish between the forms of sulfur in the pure compounds investigated by simple examination of the sulfur XANES spectra.

XANES Of Mixtures Of Model Compounds And Of Asphaltenes. To address the question of whether sulfur XANES spectra can be used to distinguish and quantify different sulfur forms in mixtures, spectra were collected of mixtures of dibenzothiophene and dibenzylsulfide. Illustrative sulfur XANES spectra for the pure compounds and a 1:1 molar mixture are shown in Figure 2, together with the corresponding third derivative spectra. Although the absorption line shapes on the left side of Figure 2 all appear very similar, significant differences are revealed by the third derivatives. Because of this it proved possible to identify each compound in the presence of the other. Additionally by simply measuring the heights of the third derivative features at 2469.8 eV and 2470.8 eV relative to the baseline, an approximate estimate of the amounts of each component can be obtained. A calibration plot for this purpose is illustrated in Figure 3.

Sulfur XANES spectra, and the corresponding third derivative spectra of three asphaltene samples and Rasa coal are shown in Figure 4. Even casual inspection of the figure shows that, while the absorption lineshapes on the left side of Figure 4 look quite similar, the third derivatives show clear differences. By comparison with the model compound data, and in agreement with previously reported EXAFS results (19), we conclude that the organic sulfur bound in asphaltene 1 consists almost entirely of dibenzothiophenic forms. The sulfurs in samples 2 and 3, however, appear to be mixtures of thiophenic and sulfide forms (note the absence of the sulfide peak at 2469.8 eV in the spectrum of sample 1, and the appearance of this feature in the other two). This conclusion is consistent with pyrolysis data for these asphaltene samples, which yield larger quantities of hydrogen sulfide from samples 2 and 3 relative to 1.

Assuming that the composition of the sulfur forms in the asphaltene samples and the Rasa coal is approximated by the simple two component mixture of dibenzothiophene and dibenzylsulfide models, an estimate of the relative molar quantities of sulfide and thiophenic forms can be obtained as described above and from Figure 3. These approximate values are listed in Table II.

This simple analysis indicates that sample 3 contains about 50% sulfide forms and 50% thiophenic forms (ie. about a 1:1 molar mixture), while sample 2 contains 43% sulfide and 37% thiophenic. In the latter case, the totals do not add to 100%. For X-ray fluorescence data this method of quantification is susceptible to errors resulting from distortion of the XANES spectra by thickness

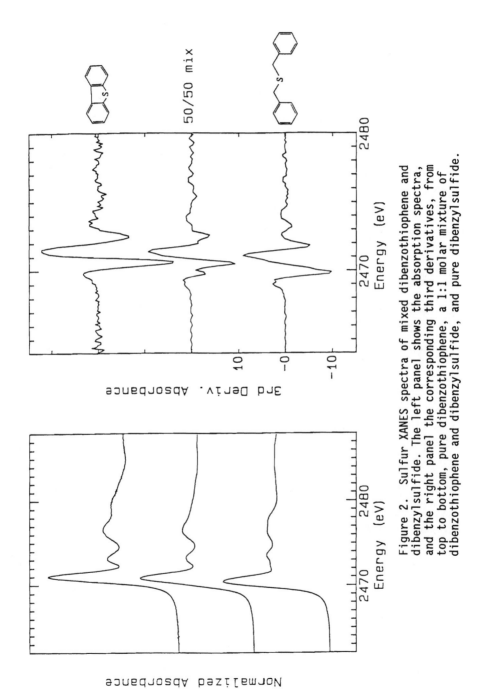

Figure 2. Sulfur XANES spectra of mixed dibenzothiophene and dibenzylsulfide. The left panel shows the absorption spectra, and the right panel the corresponding third derivatives, from top to bottom, pure dibenzothiophene, a 1:1 molar mixture of dibenzothiophene and dibenzylsulfide, and pure dibenzylsulfide.

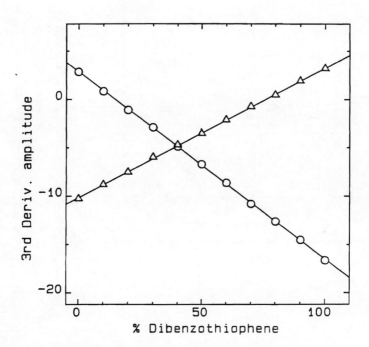

Figure 3. Calibration plot for quantification of dibenzothiophene and dibenzylsulfide mixtures, from mixtures of the pure compounds. Points denoted as "Δ" indicate the third derivative amplitude at 2469.8 eV, and those as "O" the third derivative amplitude at 2470.4 eV.

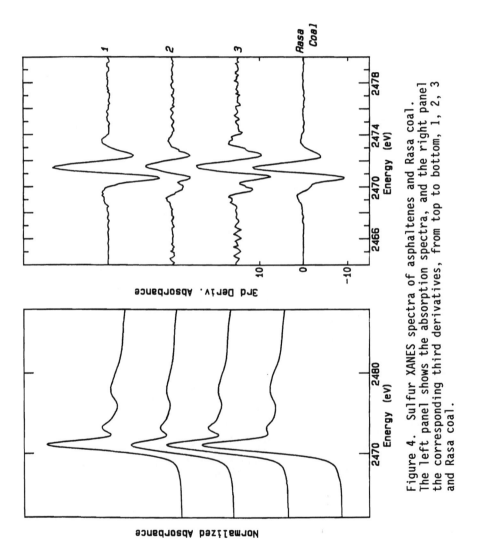

Figure 4. Sulfur XANES spectra of asphaltenes and Rasa coal. The left panel shows the absorption spectra, and the right panel the corresponding third derivatives, from top to bottom, 1, 2, 3 and Rasa coal.

Table II. Organic Sulfur Forms Approximately Quantified by XANES

Sample	% Sulfide(± 10)[a]	% Thiophenic(± 10)[a]
Asphaltene 1	0	100
Asphaltene 2	43(54)	37(46)
Asphaltene 3	50	50
Rasa Coal	30	70

[a]Approximate uncertainty based on measurements on mixtures of model compounds

effects. For accurate quantification, care must be taken, especially for model compounds, by diluting the samples sufficiently to ensure that minimal thickness effects occur. Such thickness effects may be indicated for sample 2 by the low values for both sulfur forms determined above, although because of the dilute nature of this sample we consider this unlikely. A more probable explanation is that a range of slightly different sulfur types, of both sulfide and thiophenic forms, causes a broadening of the features of the XANES spectrum. In agreement with this the structure of the third derivative spectrum of sample 2 does appear to be broadened relative to that of the spectrum of sample 3 in Figure 4. Based on this assumption, the data were normalized to 100%, giving a rough approximation of the amounts of sulfide and thiophenic sulfur, which are shown in parentheses in the table.

The analysis indicates that 30% of the organic sulfur is sulfide and 70% is thiophenic in the Rasa coal sample. Some confirmation of these values comes from the work of Kavcic (20), who showed that about 75% of the sulfur in this coal was not reactive toward methyl iodide; this lack of reactivity was attributed to the sulfur being bound in ring structures.

Despite the overall similarity between the XANES spectrum of the mixed model compounds and the asphaltene and coal samples, there are some significant differences in the relative energies of the various features. One possible explanation for these differences is that subtle electronic differences at the sulfur are reflected in the sulfur XANES. Such differences might be expected to arise from extended ring structures or from substituents. This will be a subject for further study.

Conclusions

This work has demonstrated for the first time that organically bound sulfide and thiophenic sulfur forms can be distinguished and in some manner quantified directly in model compound mixtures and in petroleum asphaltenes and coal. The use of the third derivative XANES spectra was the critical factor in allowing this analysis. The tentative quantitative determinations of sulfur forms appear to be consistent with the chemical behavior of the asphaltene and coal samples. Further work is in progress to extend these techniques to other nonvolatile and solid hydrocarbon materials.

Acknowledgments

X-ray absorption spectra were recorded at the Stanford Synchrotron Radiation Laboratory, which is funded by the Department of Energy, under contract DE-AC03-82ER-13000, Office of Basic Energy Sciences, Division of Chemical Sciences and the National Institutes of Health, Biotechnology Resource Program, Division of Research Resources. The writers wish to thank Dr. Curt White for providing a sample of Rasa coal.

Literature Cited

1. Van Krevelen, D.W. Coal; Elsevier: Amsterdam, 1961; p 171.
2. Rose, K.D; Francisco, M.A. J. Am. Chem. Soc. 1988, 110, 637-638.
3. Calkins, W. H. Energy & Fuels 1987, 1, 59-64.
4. Fromm, E.; Achert, O. Ber. 1903, 36, 534-546.
5. Hussain, Z.; Umbach, E.; Shirley, D.A.; Stohr, J.; Feldhaus, J. Nucl. Instrum. Methods 1982, 195, 115-131.
6. Spiro, C.L.; Wong, J.; Lytle, F.W.; Greegor, R.B.; Maylotte, D.H.; Lamson, S.H. Science 1984, 226, 48-50.
7. Huffman, G.P.; Huggins, F.E.; Shah, N.; Bhattacharyya, D.; Pugmire, R.J.; Davis, B.; Lytle, F.W.; Greegor, R.B. In Processing and Utilization of High Sulfur Coals II; Chugh, Y. P.; Caudle, R. D., Eds.; Elsevier: Amsterdam, 1987, pp 3-12.
8. Huffman, G.P.; Huggins, F.E.; Mitra, S.; Shah, N.; Pugmire, R.J.; Davis, B.; Lytle, F.W.; Greegor, R.B. Energy & Fuels 1989, 3, 200-205.
9. Hedman, B.; Frank, P.; Penner-Hahn, J.E.; Roe, A.L.; Hodgson, K.O.; Carlson, R.M.K.; Brown, G.; Cerino, J.; Hettel, R.; Troxel, T.; Winick, H.; Yang, J. Nucl. Instrum. Methods Phys. Res. Sect. 1986, A246, 797-803.
10. Frank, P.; Hedman, B.; Carlson, R.M.K.; Tyson, T.; Roe, A.L.; Hodgson, K.O. Biochemistry 1987, 26, 4975-4979.
11. George, G.N.; Byrd, J.; Winge, D.R. J. Biol. Chem. 1988, 263, 8199-8203.
12. Hedmann, B.; Frank, P.; Gheller, S.F.; Roe, A.L.; Newton, W.E.; Hodgson, K.O. J. Am. Chem. Soc. 1988, 110, 3798-3805.
13. Given, P. H. Prog. Energy Combust. Sci. 1984, 10, 149-154.
14. Ignasiak, B.S.; Fryer, J. F; Jadernik, P. FUEL 1978, 57, 578-584.
15. Corbett, L.W. Anal. Chem. 1969, 41, 576-579.
16. Lytle, F. W.; Greegor, R. B.; Sandstrom, D. R.; Marques, E. C.; Wong, J.; Spiro, C. L.; Huffman, G. P.; Huggins, F. E. Nucl. Instrum. Meth. 1984, 226, 542-548.
17. Sekyama, H.; Kosugi, N.; Kuroda, H.; Ohta, T. Bull. Chem. Soc. Jpn. 1986, 59, 575-579.
18. Bianconi, A. In X-ray Absorption; Koningsberger, D. C.; Prins, R. Eds.; John Wiley & Sons: New York, 1988; p 573-662.
19. George, G. N; Gorbaty, M. L. J. Am. Chem. Soc. 1989, 111, 3182-3186.
20. Kavcic, R. Bull Sci. Conseil Acad. RPF, Yugoslav 1954, 2, 12.

RECEIVED March 5, 1990

Chapter 13

Direct Determination of Total Organic Sulfur in Coal

J. T. Riley, G. M. Ruba[1], and C. C. Lee

Center for Coal Science, Department of Chemistry, Western Kentucky University, Bowling Green, KY 42101

A method for the direct determination of organic sulfur in coal was developed. Samples of -60 mesh (250 μm) coal were extracted with boiling 2 M HNO_3, which removes essentially all mineral sulfur. After washing and drying, the extracted samples were analyzed for moisture, ash, and total sulfur. The moisture and ash-free (maf) sulfur values for eight bituminous and subbituminous coals obtained by this method were almost identical to the maf sulfur values for deep-cleaned samples of the coals. The deep-cleaned samples were prepared by float/sink separation of -60 mesh coal in 1.30 specific gravity media, followed by milling the float coal to particle sizes less than 10 μm and subsequent float/sink-centrifugation cleaning. The maf organic sulfur values determined were generally less than those obtained using ASTM Method D 2492. The precision for the new method was much better than that obtained with the ASTM method.

The types of sulfur in coal, as well as their distribution and reactivity, have a profound impact on the efficiency of desulfurization processes. For practicality, coal sulfur forms are commonly classified as sulfate sulfur, pyritic sulfur, and organic sulfur. Pyritic and organic sulfur account for almost all the sulfur in coals. Sulfate sulfur is usually much less than 0.1% in freshly mined coals, and increases as the coals are exposed to the atmosphere or "weather".

The commonly accepted method for determining coal sulfur forms is ASTM Method D 2492 (1). In this method, the sulfate sulfur is determined by extracting -60 mesh (250 μm) coal with hot dilute HCl and determining the sulfate sulfur gravimetrically as $BaSO_4$. Pyritic sulfur is determined by extracting the HCl-treated coal with hot 2 M HNO_3 for 30 minutes, or overnight with 2 M HNO_3 at room temperature, and determining the iron present in the extract. The pyritic sulfur in the coal is calculated from the iron content, assuming that all the extracted iron came from FeS_2 in the coal.

[1]Current address: Merrell Dow Research, 9550 N. Cionsville Road, Indianapolis, IN 46268

0097–6156/90/0429–0231$06.00/0

Organic sulfur is then calculated as the difference between the total sulfur and the sum of the sulfate and pyritic sulfur.

It is generally accepted that there are several forms of mineral sulfur, other than sulfate and pyritic, present in coals. Mossbauer studies of inorganic sulfur forms in coal have shown the presence of pyrrhotite (FeS) (2), and the presence of other sulfides such as sphalerite (ZnS), and chalcopyrite (CuFeS$_2$) have been reported by other workers (3). Several investigators have reported as much as 1% sphalerite in some northwestern Illinois coals (4-6). The presence of elemental sulfur has also been reported (7-9). The failure to account for these mineral sulfur forms results in high values for organic sulfur using the ASTM method.

The direct determination of the organic sulfur in coal has been the goal of researchers for many years. Kuhn and coworkers (10) developed a method for the direct determination of organic sulfur in coal by first removing the sulfate sulfur and nonpyritic iron by extraction with dilute HCl, followed by extraction with LiAlH$_4$ in tetrahydrofuran to remove pyrite, before determining total sulfur in the residue. They assumed that all mineral sulfur was removed in the two extractions, leaving only that sulfur associated with the hydrocarbons in the coal. The determined organic sulfur values for nine coals were generally 0.2-0.3% lower than the calculated ASTM organic sulfur values.

Microprobe analysis of sulfur in coal offers a method of direct determination of organic sulfur concentrations in coal. The microprobe uses a finely focused electron beam which strikes a carefully selected area in a coal particle to produce x-rays characteristic of the elements present (11,12). Pretreatment of the coal with hydrochloric acid to remove nonpyritic iron and sulfate sulfur is necessary for best results. Use of accurate calibration standards allows reasonably good agreement between organic sulfur values calculated by ASTM Method D 2492 and those obtained by this method (12). Use of a scanning electron microscope (SEM) in conjunction with an energy dispersive x-ray spectrometer (EDX) allows better identification of lithotypes in prepared coal particles and thus a better analysis for organic sulfur (13-16). SEM-EDX is a relatively fast technique for organic sulfur analysis, but has received little attention as a reliable method. Reasons for this lack of attention may be the cost of the instrumentation and skills needed for performing the analysis. One study reported analysis for 9 coals with organic sulfur values ranging from 1.27 to 3.25% (dry, mineral-matter-free basis) with an average deviation of 0.35% (17.8% relative) from the ASTM organic sulfur values (13). Some of this deviation may be due to the lack of a direct method for organic sulfur analysis that could serve as a reference method.

A procedure for the direct determination of the sulfate, sulfide, pyritic, and organic sulfur in a single sample of coal has been reported by McGowan and Markuszewski (17). The method uses various strengths of perchloric acid as the selective oxidizing agent. The results obtained for the analysis of three coals were comparable to ASTM results and the relative standard deviations for nine sulfate, four pyritic, six organic, and four total recovered sulfur determinations were 2.7, 3.4, 2.4, and 2.4%, respectively.

This paper reports a practical and reliable method for the determination of organic sulfur in coal. This new method involves extraction of -60 mesh coal with dilute nitric acid (2 M) followed by the determination of moisture, ash, and sulfur in the extract residue. Sulfur values are reported on a dry, ash-free basis.

Experimental

Eight coal samples were used in this study. Characterization data and information about the rank and origin of the eight coals are given in Table I. Whole seam channel samples of the Kentucky and Tennessee coals were collected and prepared according to ASTM Method D 2013. The subbituminous coal from the Wyodak seam (#86039) was a run-of-mine sample, and two medium volatile bituminous coals were obtained from the Pennsylvania State University Coal Sample Bank (#86040 = PSOC 1138 and #86041 = PSOC 1195). All coal samples were subjected to standard analysis by ASTM methods, or methods with equivalent or better precision, as follows: proximate analysis using the LECO MAC-400 moisture, ash, and volatile matter analyzer; ultimate analysis using the LECO CHN-600 carbon, hydrogen, and nitrogen analyzer and the LECO SC-132 sulfur analyzer. The forms of sulfur data given in Table II represent the averages of 4 analyses (duplicate analysis on two different days) by ASTM Method D 2492.

The analysis for organic sulfur was performed by using 6.0 g of -60 mesh (250 μm) coal with 120 ml of 2 M HNO_3 in a 250 ml beaker or Erlenmeyer flask. The mixture was heated, with constant stirring, to boiling and boiled gently for 30 minutes. The mixture was then filtered while hot through two pieces of Whatman #1 filter paper, and the extracted coal washed with several portions of hot deionized water. The total washings were 400-500 ml per sample. The extracted coal was dried in a recirculating nitrogen atmosphere at 110°C for 3 hours (18). The coal was then analyzed for moisture, ash, and sulfur before reporting the sulfur values on a moisture and ash-free (maf) basis.

Clean coal samples of the eight coals were obtained using a two-step procedure (18). In the first step 500 g portions of -60 mesh coal were cleaned using 4 liters of 1.30 specific gravity $ZnCl_2$ solutions in a float/sink apparatus (19). The coal was slowly mixed with the $ZnCl_2$ solution to insure thorough wetting of the coal surface. The mixture was allowed to set overnight to allow the float and sink portions to separate and the float and sink portions were then filtered using two sheets of Whatman #1 filter paper in Buchner funnels. The float coal was then washed repeatedly on the filter with deionized water until the filtrate did not give a positive test for the chloride ion. The coal was then dried in a recirculating nitrogen atmosphere at 110°C for 3 hours.

The particle size of the precleaned coals was reduced to mean diameters near 10 μm by milling in a stirred-ball slurry attritor mill. Slurries containing 20% by weight coal dispersed in deionized water were milled for 2 minutes using 6.3 mm diameter stainless steel media (140 cc slurry/kg media) and an agitator shaft speed of 290 rpm (18). The slurries were then filtered and dried in a recirculating nitrogen atmosphere at 110°C for 3 hours.

Float/sink separations were carried out on the milled products using aqueous solutions of zinc chloride (1.30 specific gravity). Twenty gram samples of the dried coals were thoroughly mixed with 150 ml of aqueous $ZnCl_2$ before centrifuging at 1700 rpm for 45 minutes with a model K, size 2, centrifuge

Table I. Characterization of Coals[a]

I.D.[b]	85098	85099	86024	86025
Seam	WKY #11	WKY #12	WKY #10	WKY #9
State[c]	KY	KY	KY	KY
Mine	Sinclair	Sinclair	Gibraltar	Gibraltar
% Moisture[d]	4.3	5.6	4.8	4.1
% Ash	19.0	15.9	23.0	15.5
% Vol Matter	34.7	33.0	31.6	35.3
% Carbon	63.2	66.3	60.3	66.7
% Hydrogen	4.4	4.4	4.2	4.6
% Nitrogen	1.3	1.5	1.4	1.4
% Sulfur	6.0	4.0	4.5	4.6
% Oxygen (by diff.)	6.2	7.9	6.7	7.2
Btu/lb	11,830	12,050	11,010	12,350
Apparent rank[e]	hvAb	hvBb	hvBb	hvAb

[a] Adapted from reference 20.
[b] Accession number, Center for Coal Science.
[c] Coals are from Muhlenberg County.
[d] Moisture is as-determined; all other analyses are reported on a dry basis.
[e] Using as determined moisture.

Table I. Characterization of Coals (cont.)

I.D.[a]	86038	86039	86040	86041
Seam	ETNA	Wyodak	U.Kittanning	L.Kittanning
County/State	Marion/TN	Campbell/WY	Clearfield/PA	Cambria/PA
Mine	Sand Mtn.	Jacobs Ranch	Penn #4	Dean #1
% Moisture[b]	1.5	22.5	1.3	1.8
% Ash	9.3	9.3	10.2	11.3
% Vol Matter	24.9	43.5	21.0	24.8
% Carbon	80.2	53.8	81.3	75.9
% Hydrogen	4.8	4.6	4.7	4.7
% Nitrogen	1.5	1.0	1.6	1.4
% Sulfur	1.2	0.8	0.7	2.3
% Oxygen (by diff.)	3.2	30.6	1.6	4.4
Btu/lb	14,170	11,570	13,850	13,250
Apparent rank[c]	mvb	sub B	mvb	mvb

[a] Accession number, Center for Coal Science.
[b] Moisture is as-determined; all other analyses are reported on a dry basis.
[c] Using as determined moisture.

Table II. Forms of Sulfur Data for Coals[a]

Coal	Total Sulfur (wt.% ± s)[b]	Pyritic Sulfur (wt. % ± s)	Sulfate Sulfur (wt. % ± s)	Organic Sulfur (wt. % ± s)
85098	7.35 ± 0.087	3.51 ± 0.14	0.68 ± 0.007	3.16 ± 0.160
85099	4.78 ± 0.075	1.92 ± 0.089	0.61 ± 0.012	2.25 ± 0.120
86024	5.78 ± 0.035	3.16 ± 0.22	0.05 ± 0.010	2.57 ± 0.220
86025	5.38 ± 0.079	2.05 ± 0.13	0.60 ± 0.006	2.73 ± 0.150
86038	1.28 ± 0.032	0.74 ± 0.04	0.06 ± 0.005	0.48 ± 0.051
86039	0.84 ± 0.014	0.18 ± 0.01	0.08 ± 0.002	0.58 ± 0.017
86040	0.72 ± 0.010	0.12 ± 0.01	0.02 ± 0.002	0.58 ± 0.014
86041	2.58 ± 0.022	0.91 ± 0.09	0.41 ± 0.008	1.26 ± 0.093

[a] All data are reported on a moisture and ash-free basis.
[b] The value of s is one standard deviation for four determinations.

from International Centrifuge Co. The float and sink fractions were separated, filtered, washed repeatedly with deionized water and dried for 3 hours at 110°C in a nitrogen atmosphere.

Results and Discussion

Analytical values for the eight coals after treatment with 2 M HNO_3 are given in Table III. The reported values are the averages of four determinations (duplicate determinations on different days). A comparison of the dry ash values for the HNO_3-extracted residues described in Table III to the dry ash values for the raw coals described in Table I reflects the reduction in mineral matter caused by extraction of the raw coals with 2 M HNO_3. Carbonates, sulfates, and other minerals dissolve in the acid solution used to extract pyrite.

Table III. Analytical Values for Eight Coals After Extraction With 2 M HNO_3[a]

Coal No.	% Moisture	% Ash (dry)	% Sulfur ± s[b] (moisture and ash-free)
85098	2.84	11.76	2.22 ± 0.080
85099	3.19	10.29	1.33 ± 0.106
86024	3.31	15.47	1.24 ± 0.054
86025	3.12	8.89	1.84 ± 0.085
86038	1.40	6.56	0.50 ± 0.021
86039	0.38	3.92	0.44 ± 0.015
86040	0.46	9.11	0.54 ± 0.005
86041	0.31	9.05	0.87 ± 0.024

[a] Adapted from reference 20.
[b] The value of s is one standard deviation for 4-6 determinations.

Table IV lists the moisture and ash-free sulfur values for the eight coals before and after they were physically cleaned to reduce the mineral matter. The moisture and dry ash values for the cleaned coals are also listed. One can see from the results of the cleaning that the sulfur contents of the eight coals were substantially reduced.

Table IV. Sulfur Values   for Eight Coals Before
and After Physical Cleaning[a]

| | | Cleaned Coal | | |
Coal No.	% Sulfur Raw Coal	% Moisture	% Ash (dry)	% Sulfur
85098	7.35	2.08	1.50	2.17
85099	4.78	1.23	1.80	1.40
86024	5.78	5.38	2.45	1.14
86025	5.38	3.06	1.20	1.89
86038	1.28	0.46	2.35	0.58
86039	0.84	0.15	6.79	0.56
86040	0.72	0.94	1.62	0.59
86041	2.58	1.16	2.20	0.81

[a]   All sulfur values are reported on a moisture and ash-free basis. Each set of values is for the product from a single cleaning experiment. Adapted from reference 20.

Three sets of sulfur data are given in Table V. The organic sulfur values determined by ASTM Method D 2492, which are the averages of duplicate runs on two different days, are reported for the eight coals. Sulfur values determined by the direct extraction method and clean coal sulfur values are reported for comparison. One can see from the data there is very good agreement between the moisture and ash-free sulfur values obtained for the physically cleaned coals and those obtained by the HNO_3 extraction method. The mean difference between the two sets of sulfur values is +/- 0.07% (absolute) with the clean coal sulfur values being slightly higher by an average of 0.02%. Considering that maf sulfur values are calculated using three determined parameters (moisture, ash, and sulfur) the agreement between the two sets of sulfur values is excellent.

Upon examining the maf organic sulfur results obtained by ASTM D 2492 and the direct organic sulfur determination given in Table V one can see there is poor agreement between the two sets of data. The large differences between the results from the two methods can be explained by the inability of the D 2492 method to account for sulfur forms other than sulfate and pyritic. The sulfur in any mineral sulfide such as FeS, ZnS, or PbS, and any elemental sulfur is not analyzed separately, but is included in the total sulfur. Consequently, the presence of sulfur in these forms results in the calculation of high organic sulfur values. Any nonstoichiometric pyrite, which would have

Table V. Summary of Organic Sulfur Data[a]

Coal	ASTM D 2492 % Org. Sulfur ± s (by Difference)	Direct Extraction Method % Org. Sulfur ± s	Clean Coal Sulfur %	ASTM D 2492 Minus Direct Ext. Method
85098	3.16 ± 0.16	2.22 ± 0.080	2.17	0.94
85099	2.25 ± 0.12	1.33 ± 0.106	1.40	0.92
86024	2.57 ± 0.22	1.24 ± 0.054	1.14	1.33
86025	2.73 ± 0.15	1.84 ± 0.085	1.89	0.89
86038	0.48 ± 0.051	0.50 ± 0.021	0.58	-0.02
86039	0.58 ± 0.017	0.44 ± 0.015	0.56	0.14
86040	0.58 ± 0.014	0.54 ± 0.005	0.59	0.04
86041	1.26 ± 0.093	0.87 ± 0.024	0.81	0.39

[a] All sulfur values are reported on a moisture and ash-free basis. The value of s represents one standard deviation for 4-6 determinations. Adapted from reference 20.

something other than a 2:1 sulfur:iron ratio, would also lead to an incorrect value for the pyritic sulfur.

The discrepancy between the ASTM organic sulfur values for coals and the sulfur that remains in "deep-cleaned" coals has been noted elsewhere (Simmons, F.J., Otisca Industries, Ltd., Syracuse, NY, personal communication, 1988.). The ASTM organic sulfur values obtained with -60 mesh raw coal samples are usually greater than the total sulfur values in deep-cleaned ultrafine coal used in slurries for fuels. It is proposed that the HNO_3-extraction method for organic sulfur can be used to determine an accurate value for the sulfur content in deep-cleaned coals.

To answer the obvious question of whether or not the physical cleaning of the coal would preferentially remove coal components enriched in organic sulfur, as well as the mineral sulfur forms, an experiment was conducted on the sink portions of four coals. These samples were the portions of coal that sank in the 1.30 specific gravity media in the process of physically cleaning the four coals. These mineral matter enriched portions were extracted with 2 M HNO_3 (20 ml per g of coal), filtered, washed with deionized water, dried in a nitrogen atmosphere, and the dried residue analyzed for moisture, ash, and total sulfur. The results of quadruplicate analyses of the sink portions are given in Table VI. The moisture and ash-free sulfur values for the four sink portions are almost identical to those obtained for the analysis of the -60 mesh raw coals (Tables III and V). This indicates the organic sulfur present in the coals is not extracted by 2 M HNO_3 and is essentially the same as the sulfur remaining in the coal after an efficient cleaning process to remove the mineral matter.

During extraction with 2 M HNO_3, the coals incorporate nitro groups into their organic structure. Elemental analysis of the residues from the HNO_3-extraction shows an increase in nitrogen and oxygen and a decrease in hydrogen in the extracted coals. Table VII gives the moisture and ash-free hydrogen/carbon, oxygen/carbon, nitrogen/carbon, and sulfur/carbon ratios for the deep-cleaned and HNO_3-extracted coals used.

Table VI. Sulfur Values for HNO_3
Extracted Sink Portions From
Float/Sink Separations[a]

Coal No.	% Sulfur (moisture and ash-free basis)
85098	2.23
85099	1.32
86024	1.22
86025	1.80

[a] Adapted from reference 20.

Table VII. Elemental Ratios of Cleaned and Extracted Coals

Sample[a]	H/C	O/C	N/C	S/C
85039 Cld.	1.060	0.218	0.0149	0.00295
Ext.	0.739	0.393	0.0645	0.00288
85099 Cld.	0.871	0.139	0.0222	0.00742
Ext.	0.690	0.224	0.0583	0.00741
86024 Cld.	0.848	0.127	0.0226	0.00562
Ext.	0.728	0.221	0.0585	0.00648
85098 Cld.	0.926	0.142	0.0195	0.0105
Ext.	0.826	0.145	0.0342	0.0111
86025 Cld.	0.923	0.117	0.0218	0.0113
Ext.	0.786	0.133	0.0359	0.0115
86041 Cld.	0.730	0.0644	0.0159	0.00348
Ext.	0.642	0.0987	0.0291	0.00350
86038 Cld.	0.737	0.0577	0.0170	0.00254
Ext.	0.657	0.0899	0.0264	0.00223
86040 Cld.	0.670	0.0358	0.0139	0.00249
Ext.	0.692	0.0979	0.0299	0.00249

[a] Cld. = deep-cleaned coals.
   Ext. = coals extracted with 2 M HNO_3.

One can see from the data in Table VII that the H/C ratios generally decrease upon extraction with 2 M HNO_3. The coals are arranged in order of increasing rank and the rate of the H/C ratio decrease is lowest in the higher rank coals. The moisture and ash-free O/C and N/C ratios increase upon extraction with HNO_3 and the rate of increase is lowest in the higher rank coals. However, the moisture and ash-free S/C ratio remains essentially constant for the coals upon extraction with HNO_3, regardless of rank. This is an indication that extraction with 2 M HNO_3 does add nitro groups to the coal, replacing hydrogens, but there is essentially no change in the organic sulfur content of the coal.

A recent seven laboratory round robin analysis of the pyrite in nine coals using ASTM Method D 2492 showed a mean relative standard deviation (RSD) of 7.4% for the within-laboratory determinations (Janke, L., Canmet Energy Research Laboratory, Ottawa, Ontario, Canada, personal communication, 1986.). Since organic sulfur in coal is calculated as the difference between the total sulfur and the sum of the sulfate and pyritic sulfur, the determination of organic sulfur using ASTM Method D 2492 cannot be expected to yield a within-laboratory RSD much different from the 7.4% found in the seven laboratory round robin analysis of pyritic sulfur.

In our study the mean percent RSD for the determination of organic sulfur in the eight coals using ASTM Method D 2492 was 6.0%. Duplicate analyses on two different days (4 analyses total) of each of the coals yielded this level of precision. The mean percent RSD for the HNO_3-extraction method of determining organic sulfur in the eight coals was 3.6%. This measure of precision was based on 4-6 analyses of each of the eight coals. Thus, it can be stated that the HNO_3-extraction method for determining total organic sulfur in coal appears to give better precision than the current ASTM method.

Conclusions

A method for the direct determination of the organic sulfur in coal using extraction with hot nitric acid was developed. The method yields total organic sulfur values that are much closer to the sulfur values found in deep cleaned coals and with greater precision than the organic sulfur values obtained by ASTM Method D 2492.

Acknowledgments

The authors gratefully acknowledge partial support of this work through funding from the United States Department of Energy under contract DE-FG22-85PC80514. Support provided by the Robinson Professorship from the Ogden Foundation at Western Kentucky University is greatly appreciated.

Literature Cited

1. "Test Method for the Forms of Sulfur in Coal," Method D 2492, Annual Book of ASTM Standards, Vol. 5.05, Amer. Soc. for Testing and Materials, Philadelphia, PA, 1988.
2. Huffman, G. P.; Huggins, F. E. Fuel 1978, 57, 592.
3. Gluskoter, H.J. Energy Sources, Crane, Russak & Co., Inc.: 1977; Vol. 3(2), pp 125-131.
4. Hatch, J. R.; Gluskoter, H. J.; Lindahl, P. C. Econ. Geol. 1976, 71(3), 613.
5. Ruch, R.R.; Gluskoter, H.J.; Shimp, N.F. "Occurrence and Distribution of Potentially Volatile Trace Elements in Coal"; Environmental Geology Note 72; Illinois State Geological Survey: Urbana, IL, 1974.
6. Miller, W. G. M.S. Thesis, University of Illinois, Urbana, IL, 1974.
7. Berteloot, J. Ann. Soc. Geol. Nord 1947, 67, 195; Chem. Abstr. 1950, 44, 818a.

8. Yurovskii, A. Z., <u>Sulfur in Coals</u>, English translation available from the U.S. Department of Commerce, National Technical Information Service, Springfield, VA 22161, U.S.A., (Ref. No. TT 70-57216).
9. White, C.M.; Lee, M.L. <u>Geochim. Cosmochim. Acta</u> 1980, <u>44</u>, 1825-1832.
10. Kuhn, J.K.; Kohlenberger, L.B.; Shimp, N.F. "Comparison of Oxidation and Reduction Methods in the Determination of Forms of Sulfur in Coal"; Environmental Geology Note 66; Illinois State Geological Survey: Urbana, IL, 1973.
11. Sutherland, J. K. <u>Fuel</u> 1975, <u>74</u>, 132.
12. Raymond, R. T. In <u>Coal and Coal Products: Analytical Characterization Techniques</u>; Fuller, E.L., Jr., Ed.; ACS Symposium Series No. 205, American Chemical Survey: Washington, DC, 1982, pp 191-203.
13. Straszheim, W. E.; Greer, R. T.; Markuszewski, R. <u>Fuel</u> 1983, <u>62</u>, 1070.
14. Maijgren, B.; Hubner, W.; Norrgard, K.; Sundvell, S. <u>Fuel</u> 1983, <u>62</u>, 1075.
15. Timmer, J. M.; van der Burgh, N. <u>Fuel</u> 1984, <u>63</u>, 1645.
16. Clark, C. P.; Freeman, G. B.; Hower J. C. <u>Scanning Electron Microsc.</u> 1984, <u>2</u>, 537.
17. McGowan, C. W.; Markuszewski, R. <u>Fuel</u> 1988, <u>67</u>, 1091.
18. Lloyd, W. G.; Riley, J. T.; Kuehn, K. W.; Kuehn, D. W. "Chemistry and Reactivity of Micronized Coal," Final Report, USDOE Contract No. DE-FG22-85-PC 80514, February, 1988.
19. "Test Method for Determining the Washability Characteristics of Coal," Method D 4371, <u>Annual Book of ASTM Standards</u>, Vol. 5.05, Amer. Soc. for Testing and Materials, Philadelphia, PA, 1988.
20. Riley, J.T.; Ruba, G.M. <u>Fuel</u> 1989, <u>68</u>, 1594.

RECEIVED March 14, 1990

Chapter 14

Elemental Sulfur in Bituminous Coals

Leon M. Stock[1] and Ryszard Wolny

Department of Chemistry, University of Chicago, Chicago, IL 60637

Many coals contain elemental sulfur. It is clearly a ubiquitous constituent of bituminous coals that have been exposed to the atmosphere. In contrast, three pristine samples obtained from the Premium Sample Program at the Argonne National Laboratory are free of elemental sulfur within the detection limit of four sensitive analytical methods. Exposure of the pristine coals to the atmosphere leads to the production of elemental sulfur. These observations strongly suggest that sulfur is not a natural constituent of coal, but rather is produced after exposure to the atmosphere by chemical or bacteriological action.

The presence of elemental sulfur in coal has been under discussion for a long time. In 1942, Chatterjee presented arguments to support the view that elemental sulfur was not a natural constituent of coal, but rather was formed by the air oxidation of inorganic sulfur compounds (1,2). In 1960, Yurovskii suggested that elemental sulfur could be a natural constituent of coal (3). He pointed out that this substance could be formed as a by-product during the conversion of iron sulfate to iron pyrite (3). More recently, work has been undertaken in several laboratories to elaborate these issues. This contribution is focused on the results for three pristine bituminous coals from the Premium Coal Sample Program of the Argonne National Laboratory (APCSP) and three other coals that have been exposed to the atmosphere. These substances were obtained from the Illinois Basin Coal Sample Program (IBCSP). The compositions of these coals are set forth in Tables I and II for convenient comparison.

Analysis

Sulfur is a reactive oxidizing agent. Hence, some precautions need to be observed in its removal from coal and in its quantitative measurement. Recent contributions have discussed several different methods for its quantitative analysis (4-9). The results for two coal samples, one a pristine sample of Illinois No. 6 coal, APCSP-3, and the other an exposed sample of Illinois No. 6 coal, IBCSP-1, are summarized in Table III.

---

[1]Current address: Chemistry Division, Argonne National Laboratory, Building 200, Argonne, IL 60439–4831

0097–6156/90/0429–0241$06.00/0

All of the investigators have observed that the amount of sulfur that can be extracted from the pristine coal is below the detection limit. Conservatively, these observations mean that there is less than 0.001% sulfur in the pristine coal samples.

The results for exposed samples of Illinois No. 6 coal indicate that 0.11 ± 0.06 wt% elemental sulfur is present. All the methods except the bacteriological method, which appears to underestimate the sulfur content, provide reasonably consistent results. While not highly precise, the four physicochemical methods provide accurate information, and suggest that 1 to 5% of total sulfur in the exposed coals from this Basin is in the elemental form. It is pertinent to note that the spectrophotometric method of analysis that has been developed by Buchanan and Chaven does not involve gas chromatography and is especially suitable for the analysis of the concentration of small amounts of sulfur in extracts (8).

Data for Coals

The results that have been obtained for three pristine coals and six other samples are summarized in Table IV.

Table I. Composition of Coals[a]

Sample	Content, wt%, dry basis				
	C	H	N	O	Ash
Upper Freeport mv, bituminous APCSP-1	74.23	4.08	1.35	4.72	13.03
Wyodak subbituminous APCSP-2	68.43	4.88	1.02	16.90	8.77
Illinois No. 6 hv, bituminous APCSP-3	65.65	4.23	1.16	10.11	15.48
Illinois No. 6 hv, bituminous IBCSP-1	69.20	5.12	1.28	9.49	10.5
Illinois No. 2 hv, bituminous IBCSP-2	73.92	5.29	1.52	9.02	6.9
Illinois No. 6 hv, bituminous IBCSP-5	63.60	4.56	1.19	7.55	18.5

[a]The compositions were provided by Karl Vorres at Argonne National Laboratory and Carl Kruse at Illinois State Geologic Survey.

Table II. Sulfur Content of Coals[a]

Sample	Content, wt%, dry basis		
	Sulfur Pyrite	Sulfur Organic	Sulfur Sulfate
Upper Freeport mv, bituminous APCSP-1	1.77	0.54	0.01
Wyodak subbituminous APCSP-2	0.17	0.43	0.03
Illinois No. 6 hv, bituminous APCSP-3	2.81	2.01	0.01
Illinois No. 6 hv, bituminous IBCSP-1	1.27	3.00	0.05
Illinois No. 2 hv, bituminous IBCSP-2	2.29	0.94	0.06
Illinois No. 6 hv, bituminous IBCSP-5	2.57	1.94	0.00

[a]The compositions were provided by Karl Vorres at Argonne National Laboratory and Carl Kruse at Illinois State Geologic Survey.

Table III. Elemental Sulfur for Two Illinois Basin Coal Samples

Analytical Method	Elemental Sulfur, wt%		Reference
	pristine, ASCSP-3	exposed, IBCSP-1	
GC-Hall Detector	0.00	0.07	(6)
GC-MS	0.00	0.10	This work
GC-MS	0.00	0.23	(7)
Spectroscopic	0.00	0.13	(8)
Bacteriological		0.02	(9)

Table IV. Elemental Sulfur in Coal

Sample[a]	Extraction solvent	Elemental sulfur, wt%	Reference
Pristine Samples from Argonne National Laboratory Program			
Upper Freeport[b]	hexane	0.00	(6)
Wyodak[b]	hexane	0.00	(6)
Illinois No. 6[b]	hexane	0.00	(6)
Illinois No. 6[b]	tetrachloroethylene	0.00	This work, (7, 8)
Exposed Samples from Illinois Basin Program			
Illinois No. 6 (IBCSP-1)	ethanol	0.07	(6)
Illinois No. 2	ethanol	0.15	(6)
Illinois No. 2	hexane	0.16	(6)
Illinois No. 2	acetone	0.13	(6)
Illinois No. 2	benzene	0.17	(6)
Illinois No. 6[b] (IBCSP-5)	tetrachloroethylene	0.0006	(8)
Illinois No. 6	tetrachloroethylene	0.11	(8)
Exposed Samples from Other Sources			
Iowa	cyclohexane	0.21	(4)
Wyoming, Smith	hexane	0.002	(6)
Indiana Bog[c]	tetrachloroethylene	1.2	(7)

[a]Please see Tables I and II for the composition of the coals obtained from the Argonne National Laboratory and the Illinois Basin Sample Programs.
[b]Special precautions were taken with these samples to avoid exposure to the atmosphere.
[c]This sample contains 11.6% total sulfur.

Pristine Coals

A variety of solvents including hexane, acetone, tetrahydrofuran, toluene and perchloroethylene have been used in these studies together with four analytical strategies. No sulfur has been detected. This feature is well illustrated in Figure 1, which displays the contrasting results for the pristine Wyodak sample, APCSP-2, and another exposed sample of this low sulfur coal. Elemental sulfur at the 0.002% level is easily detected in the exposed sample, none is observed in the pristine sample.

Non-pristine Coals

The samples from the Illinois Basin Coal Sample Program contain 0.1 to 0.2 wt% elemental sulfur. The only exception was reported by Narayan and coworkers who recently reported that IBCSP-2 contained about 1% elemental sulfur (7). Further work led to the realization that they had inadvertently

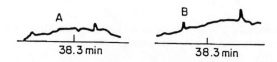

Pristine sample from Wyodak Seam, Undetectable Sulfur

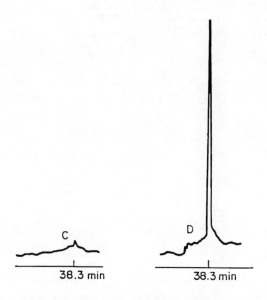

Exposed sample from Wyodak Seam, 0.002% Sulfur

Figure 1. Chromatographic observations for pristine and exposed samples of low sulfur Wyodak coal. Elemental sulfur emerges at 38.3 minutes under the conditions of the analysis. A. Unconcentrated extract of APCSP-2. B. The extract of APCSP-2 after 20-fold concentration. C. Unconcentrated extract of exposed Wyodak coal. D. The same extract after 20-fold concentration. (Adapted with permission from ref. 6. Copyright 1985 Fuel.)

analyzed an unusual sample of material that contained more than 11 wt% total sulfur (Narayan, R., private communication, 1988). Thus, there are no major discrepancies.

The results for other samples from other sources provide results that are comparable to the results for the samples from the Illinois Basin program.

Intentional Exposure

In the course of our work, we exposed several pristine coals to the atmosphere. The results are presented in Table V.

Table V. Elemental Sulfur Content of Pristine Coals after
Intentional Exposure to the Atmosphere

Coal Sample	Elemental Sulfur, wt %, dry basis		
	Initial	49 days	9 months
Upper Freeport mv, bituminous APCSP-1	BDL	5.1×10^{-5}	4.1×10^{-4}
Wyodak subbituminous APCSP-2	BDL	1.4×10^{-3}	1.0×10^{-3}
Illinois No. 6 hv, bituminous APCSP-3	BDL	1.7×10^{-4}	4.7×10^{-3}
Illinois No. 6 hv, bituminous APCSP-3 Humid atmosphere	BDL	2.0×10^{-3}	2.8×10^{-1}

[a]The designation BDL indicates that the quantity of elemental
sulfur in the original sample was below the detection limit.

It is evident that detectable quantities of sulfur form in the pristine coals in less than two months. The rate of formation of sulfur appeared to be enhanced significantly when the coal was suspended above water in a closed desiccator. Clearly, a broad array of factors such as particle size, air currents and so forth will influence the rate of sulfur production, but there is no doubt that it is a facile process.

Elemental Sulfur Origins in Coal

Although somewhat beyond the scope of our work, it is pertinent to note that sulfur can be produced by two very well recognized pathways. One involves the conversion of pyrite to sulfur by water and oxygen. Bergholm thoroughly investigated this reaction in 1955 (10,11). The other involves the production of elemental sulfur from pyrite by bacteriological action (9). This process has been examined in many laboratories in recent years because it offers an opportunity for coal desulfurization (9, 12). A third somewhat more speculative interpretation for the appearance of elemental sulfur in coal has been advanced by Narayan and his associates (7). They proposed that naturally occurring organic polysulfides may decompose to form elemental sulfur during extraction procedures. In point of fact, organic polysulfides have been detected in sedimentary materials (13) and complex polythionates are known end products of the bacterial decomposition of inorganic sulfides (14, 15). Although experiments specifically designed for the detection of these substances in coals have not been carried out, some logical limits can be placed on their abundances in coal. Kuhn and his associates developed and tested a new method for the determination of the pyrite content of coals via reduction with lithium aluminum hydride (16). They reported that this method provided information that was well within experimental error of the information provided by the oxidative ASTM procedure (14). Careful review of their results, however, reveals that the lithium aluminum hydride reduction method provided pyrite content data that were consistently lower than the data that were provided by the oxidative method. This finding is exactly opposite to the observation that would be expected if the coals contained polysulfides because the reduction method would count polysulfide sulfur as pyrite whereas the oxidative method would count polysulfide sulfur as organic sulfur. The results for the representative group of coals that were examined by Kuhn and coworkers suggest that the polysulfide content of bituminous coals is probably quite small.

The notion that naturally occurring organic polysulfides in coal decompose to form elemental sulfur has also been tested in another way. Buchanan and his associates have shown that the $^{32}S/^{34}S$ ratios of the elemental sulfur and the pyrite in another Illinois Basin Coal Sample Program coal are similar and different from the $^{32}S/^{34}S$ ratio for the organic material in the same coal (Buchanan, D., private communication, 1989). This result infers that pyrite is the source of elemental sulfur. Thus, we conclude that oxidative chemical and bacteriological processes convert pyrite to elemental sulfur when pristine coals are exposed to the atmosphere.

Acknowledgment

It is a pleasure to acknowledge the support of the Center for Research for Sulfur in Coal.

Literature Cited

1. Chatterjee, N. N. Quart. J. Geol. Mining. Met. Soc. India 1942, 14, 1-8.
2. Chatterjee, N. N. Chem. Abs. 1947, 41, 1410g.
3. Yurovskii, A. Z. Sulfur in Coals, English translation available from the U.S. Department of Commerce, National Technical Information Service, Springfield, VA 22161, USA (REF. No. TT 70-57216), 1960.

4.   Richard, J. J.; Vick, R. D.; Junk, G. A. Environ. Sci. Technol. 1977, 11,
     1084.
5.   White, C. M.; Lee, M. L. Geochimica et Cosmochimica Acta 1980, 44,
     1825-1832.
6.   Duran, J. E.; Raymahasay, S.; Stock, L. M. Fuel 1985, 65, 1167-1168.
7.   Narayan, R.; Kullerud, G.; Woods, K. V. Prepr. Pap. Am. Chem. Soc.,
     Div. Fuel Chem. 1988, 33(1), 193-197.
8.   Buchanan, D.; Chavan, D. Proceedings: Fourteenth Annual EPRI Fuel
     Science Conference; Electric Power Research Institute: Palo Alto, CA,
     1989.
9.   Schicho, R. N.; Brown, S. H.; Kelly, R.M.; Olson, G. J.; Prepr. Pap.
     Am. Chem. Soc., Div. Fuel Chem. 1988, 34(4), 554-560.
10. Bergholm, A. Jernkontorets Annaler Arg. 1955, 139, 531-549.
11. Bergholm, A. Chem. Abs. 1955, 49, 15408e.
12. Couch, G. R. Biotechnology and Coal, IEA Coal Research, London,
     1987, Chapter 4.
13. Sinninghe Damste, J. S.; deLeeuw, J. W. Abstracts Fourteenth
     International Meeting on Organic Geochemistry, Paris, 1989, Keynote
     Paper 10.
14. Steudel, R.; Holdt, G.; Gobel, T.; Hazen, W. Angew. Chem. Int. Ed.
     Engl. 1987, 26, 151-153.
15. Fisher, U. Biogenic Sulfur in the Environment, Saltman, E.S. and
     Cooper, W. J. Eds.; Am. Chem. Soc. Symposium Series 1989, 393, 262-
     279.
16. Kuhn, J. K.; Kohlenberger, L. B.; Shimp, N. F. Environ. Geol. Notes
     (Ill. State Geol. Surv.) 1973, 66.

RECEIVED March 13, 1990

Chapter 15

Multiple-Heteroatom-Containing Sulfur Compounds in a High Sulfur Coal

Randall E. Winans and Paul H. Neill

Chemistry Division, Argonne National Laboratory, 9700 South Cass Avenue, Argonne, IL 60439

Flash vacuum pyrolysis of a high sulfur coal has been combined with high resolution mass spectrometry yielding information on aromatic sulfur compounds containing an additional heteroatom. Sulfur emission from coal utilization is a critical problem and in order to devise efficient methods for removing organic sulfur, it is important to know what types of molecules contain sulfur. A high sulfur Illinois No. 6 bituminous coal (Argonne Premium Coal Sample No. 3) was pyrolyzed on a platinum grid using a quartz probe inserted into a modified all glass heated inlet system and the products characterized by high resolution mass spectrometry (HRMS). A significant number of products were observed which contained both sulfur and an additional heteroatom. In some cases two additional heteroatoms where observed. These results are compared to those found in coal extracts and liquefaction products.

The nature of the organic sulfur compounds in coal is not known in detail. However, the development of methods for removing this sulfur would benefit from this information. The objective of this study is to examine the nature of sulfur compounds containing more than one heteroatom in a coal. The compounds that are being characterized were released by rapid vacuum pyrolysis of a high sulfur Illinois No. 6 coal from the Argonne Premium Coal Sample Bank (#3). The analytical pyrolysis approach was taken to help reduce secondary thermal reactions which occur in atmospheric, long residence time pyrolysis and liquefaction reactions. Other sources of materials for specific compound characterization include extracts and selective oxidative degradation products.

Two recent papers examining coal extracts showed little evidence for multiple heteroatom species in the small molecules ([1-2]). The extracts of an abnormally high sulfur coal from Yugoslavia were found to contain species with several sulfurs per molecule ([3]). HRMS analysis of extracts from the Argonne Premium Coal Sample used in this study showed evidence for species containing both SO and SN ([4-5]). Specific compounds containing sulfur and

0097–6156/90/0429–0249$06.00/0

another heteroatom have been identified in liquids derived from solvent refined coal processes (2,6-9). Heteroatom species in a high temperature bituminous coal carbonization product have been investigated using HRMS (10).

Analytical pyrolysis has been used extensively for studying coals, macerals separated from coals (11-13) and kerogens (14). Three modes are typically used: PyGCMS, PyMS (low resolution) and PyHRMS. The last method was used in this study with both low voltage and 70 eV electron impact ionization. Because of the very close precise masses of C_3 and SH_4 a resolving power of approximately 60,000 at MW=200 is necessary to separate hydrocarbons from sulfur containing species. Although it is easier to assign structures from GCMS data, the HRMS approach allows one to examine higher molecular weight and more polar species. In addition, chemical modification can be used to alter the yields of various types of compounds. For example, the volatility of aromatics with hydroxyl groups can be reduced by the exchange of the acidic protons with potassium ions. By comparing products from the exchanged and unaltered sample, the contributions from the hydroxyl groups can be assessed. Recently, an excellent description of the use of HRMS to study coal derived liquids was published (15). Molecular formulae data and possible structures produced in this study will be compared to the data derived from extracts, carbonization and liquefaction products.

EXPERIMENTAL

The composition of the coal sample used in this study is given in Table I. This coal's contact with oxygen has been minimized in that all processing was done in a nitrogen atmosphere and the samples were stored in flame sealed glass ampules (16). The potassium exchanged coal sample was prepared by stirring 5 gms of the fresh coal in a 10% KOH methanol solution for 72 hrs. A demineralized sample was prepared by treating the fresh coal with aqueous HCl and HF under a nitrogen atmosphere (17). Bulk pyrite was further removed from the demineralized sample which had been ground to a 1 micron particle size by a sink-float method in aqueous CsCl solution using a non-ionic surfactant to disperse the fine particles. The higher density minerals such as pyrite sink and the float contains the organic portion of the coal.

The samples were pyrolyzed in two modes. In the batch technique, 10 mg of sample were heated at 450°C under vacuum (10^{-6} torr) and tars collected at room temperature. The potassium salts of these tars were prepared by exchange with 10% KOH in THF. In the second method the samples were flash pyrolysed at 650°C on a platinum mesh using a quartz probe inserted into an all glass inlet system designed in this laboratory.

The HRMS data were collected on a Kratos MS-50 operating with an EI source at low eV with resolving power of 80,000 (batch tars) and at 70 eV with 50,000 resolution (in situ pyrolysis). The data was transferred to another computer system and analyzed using a program which we have developed. Initially, empirical formulae were assigned and deviations between calculated and observed masses determined. Next, peaks were sorted by heteroatom content, and finally by hydrogen deficiency (HD = rings

Table I. Analyses of the Coal and Treated Coals

	% C (daf)	Atoms/100C				% Ash (dry)	% Pyrite (dry)
		H	N	S	O		
Fresh	77.7	77.0	1.5	1.2	13.0	15.5	2.8
Acid Demineralized	75.0	83.2	1.6	1.0	13.5	7.7	~2.8
Float							<1.0

+ unsaturation). Combinations of heteroatoms which were normally considered include:

None	S	N
O	S_2	N_2
O_2	SO	NO
O_3	SO_2	NO_2
O_4	SO_3	
	SN	
	SNO	

Normally, over 95% of total ions could be assigned probable formulae. Additional combinations did not explain any more of the data. Finally, the data was evaluated for usual patterns of ion peaks in any particular group. Typically, a series separated by 14 mass units(methylene) was seen in each group with a certain hydrogen deficiency and heteroatom content.

RESULTS AND DISCUSSION

The results of the LVHRMS analysis on the vacuum pyrolysis tars are quite surprising in that 89% of the total ion intensity contained at least one heteroatom and that 38% contained at least one sulfur. These data are illustrated in Figure 1 which displays the distribution of various heteroatom combinations as a function of the hydrogen deficiency. The oxygen and multiple oxygen species have been summed together and the same was done for nitrogen species which did not contain sulfur. The sulfur species are shown individually and one can readily observe that SO_x, where x=1-3, species dominate. Overall, it is very apparent that this sample is dominated by aromatic oxygen containing molecules. For example, the base peak has an empirical formula of $C_{20}H_{18}O$.

It has been shown that high resolution and precise mass measurements are important in the MS characterization of very complex mixtures (9-11,15). This is demonstrated in Table II where the six ion peaks found at nominal mass $= 268$ and their assigned formulae are shown. Again notice the diversity of heteroatom species for just this one nominal mass. Also, note that most of these peaks are quite abundant even at this relatively high mass. Five peaks agree very well between the calculated and observed mass. This is quite typical with the deviation being approximately one millimass unit (mmu) or less. The peak assignment for the fifth is a poor fit and in this case the data are examined to determine if the assignment fits into a series. For this set of LVHRMS data good assignments were made out to the highest mass at $m/z = 600.4410(C_{43}H_{56}N_2)$.

A select list of species containing sulfur plus an additional heteroatom is given in Table III. These groups of compounds have been observed in coal derived products from various sources. To be able to make reasonable comparisons only data from bituminous coals are used. Possible structures for the formulae given in Table III are shown in Scheme I. It should be emphasized that further information is necessary to positively assign structures. The viability of these structural assignments will be discussed in the next three sections.

Table II. Peaks Found at Nominal m/z = 268 for LVHRMS Data from 450°C Vacuum Tar with Assigned Formulae

Peak No.	Obs. m/z	Relative Abundance	Formula	Calc. m/z	Dev. (mmu)
1	268.0393	10.5	$C_{18}H_6NO_2$	268.0398	-0.5
			$C_{16}H_{12}S_2$	268.0380	1.3
			$C_{19}H_8S$	268.0347	4.6
2	268.0889	3.6	$C_{20}H_{12}O$	268.0888	0.1
			$C_{17}H_{16}SO$	268.0922	-3.3
3	268.1102	33.1	$C_{17}H_{16}O_3$	268.1100	0.2
			$C_{20}H_{14}N$	268.1126	-2.4
4	268.1264	23.0	$C_{21}H_{16}$	268.1252	1.2
			$C_{18}H_{20}S$	268.1286	-2.2
			$C_{17}H_{18}NO_2$	268.1338	2.4
5	268.1362	8.4	$C_{15}H_{24}S_2$	268.1320	4.2
6	268.1659	2.3	$C_{15}H_{24}O_4$	268.1674	1.5
			$C_{18}H_{22}NO$	268.1701	-4.2

Table III. Comparison of Formulae Identified from Extracts, Liquefaction Products, Carbonization Products, or PyMS for Structures Containing Sulfur Plus One Additional Heteroatom

Base Formula	HD	Extract (1,5-6)	Liquefaction (21)	Carbonization Product (10)	PyMS (this work)
$C_{12}H_{10}S_2$ (I)	8	+(1)	-	-	-
$C_{12}H_8S_2$ (II)	9	-	+	-	+
$C_{14}H_8S_2$ (III)	11	-	+	-	-
$C_{11}H_7SN$ (IV)	9	-	+	-	-
$C_{12}H_9SN$ (V)	9	+(5)	+	-	+
$C_{12}H_8SO$ (VII)	9	+(5-6)	+	+	+
$C_{16}H_{10}SO$ (VIII)	12	-	-	-	+

- not observed, + observed

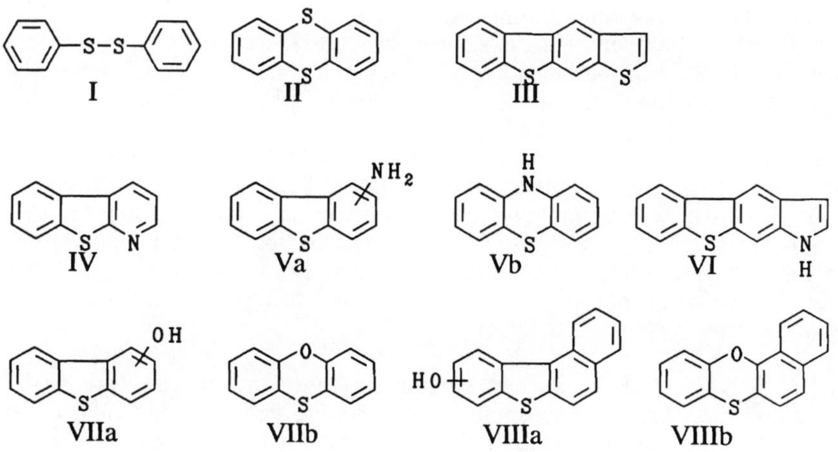

I II III
IV Va Vb VI
VIIa VIIb VIIIa VIIIb

Scheme I.

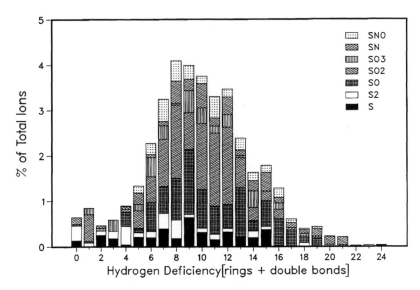

Figure 1. Distribution of sulfur containing ions from LVHRMS of the 450°C vacuum tar.

SULFUR-SULFUR. As can be seen in Figure 1, the contribution from ions with two sulfurs is quite small. No evidence for disulfides is seen, although molecular formulae consistent with their presence have been observed in extracts (1). It is possible that they could have formed during the workup from thiophenols. However, there is no conclusive evidence for the existence of aromatic thiols in bituminous coals. They are most likely highly reactive, especially in the presence of the large number of phenolic compounds which are found in coals (18). However, there is a series in the vacuum pyrolysis tar which could possibly be alkylated thianthrenes(II). There is a report that thianthrene has been identified from oxidation by chlorine of a lignite coal (19). Thianthrenes will most likely be converted to dibenzothiophenes upon long contact time thermolysis such as in a liquefaction process. They are not observed in products from such reactions. In addition, in a model compound study it was found that thianthrene is converted to dibenzothiophene during pyrolysis from 550 to 900°C (20).

SULFUR-NITROGEN. Species containing sulfur and nitrogen are slightly more abundant than those containing two sulfurs per molecule. There are species with the empirical formula (IV) in the pyrolysis tar based on the carbon number, but higher alkylated species can not be distinguished from (V). Very small amounts of benzothienopyridine have been positively identified in liquefaction products (21). Species with formula (V) were seen in the PyMS as well as in extracts and liquefaction products. Aminodibenzothiophenes(Va) were identified in a very careful analysis of an SRC heavy distillate (21).

It seems likely that aromatic amines which are found in liquefaction products have been produced by a combination of thermolysis and hydrogenation. There is no evidence for aromatic amines in coals from either selective oxidation degradations (22) or from direct X-ray Photoelectron Spectroscopy measurements (23). Oxidations would produce very stable nitroaromatics which are not seen. Another possible structure for this formula is phenoxazine(Vb). Such a molecule would not survive high temperature combined with long reaction times. Although annelated thiophene with a pyrrole(VI) would appear to be a likely structure in coal, there is no evidence for its existence in any of the coal derived materials.

SULFUR-OXYGEN. From Figure 1 it can be seen that the greatest number of two heteroatom species with one sulfur is found in this category. The series of peaks containing S,O for HD = 5-15 are shown in Figure 2. The most

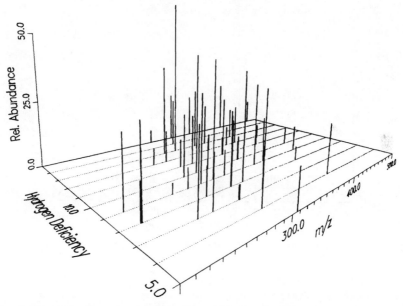

Figure 2. Peaks of ions containing S and O from LVHRMS of the vacuum tar.

abundant group of ions is at HD = 9 which corresponds to formula (VII) in Table III. Hydroxydibenzothiophenes(VIIa) have been observed in liquefaction products (8). The amount of the 1-hydroxy isomer has been semiquantified to have a concentration of only 3.4 μg/g. Also, compounds with this formula have been observed by HRMS in high-temperature coal carbonization products (10). We propose that phenoxathiin(VIIb) is another possible structure. It is easily formed from diaryl ethers and sulfur, so that one could visualize this reaction during the early stages of coalification.

Although it is impossible to distinguish between the two structures (VIIa and VIIb) from the HRMS data, the samples can be chemically modified to facilitate differentiation. Siskin and Aczel (24) showed that the volatility of a mixture of phenols can be dramatically reduced by exchanging the acidic proton with potassium. We have shown that the same approach works for vacuum pyrolysis of coals and coal macerals (17). The distribution of ions which contain both sulfur and oxygen is shown in Figure 3 for the raw coal and the potassium exchanged coal. At HD = 9 the yields are essentially the same. This result is consistent with the occurrence of structure (VIIb) in this coal. However, it is possible but unlikely that all of the hydroxydibenzothiophene originally existed as aryl ethers and the hydroxy form has been formed during the pyrolysis.

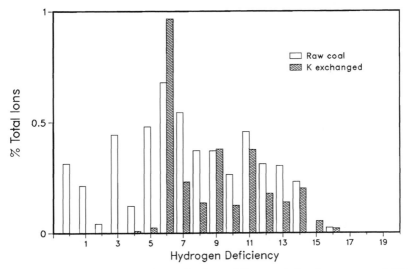

Figure 3. Comparison of ion intensity from SO containing fragments of 450°C tars.

The data shown in Figure 3 were acquired at 70 eV to investigate the possible occurrence of more saturated species which are discriminated against by the low eV, very high resolution experiment. At high resolution the spectrometer is scanned at 1000 sec/decade starting at the high mass end and thus more volatile species are not observed. Note that there is a significant amount of ions observed for HD ≤ 6 for the raw coal. In contrast, very little is seen for the exchanged coal in this region except for at HD = 6. It is unclear why this has occurred.

SULFUR-SEVERAL ADDITIONAL HETEROATOMS. From the data presented in Figure 1 it is apparent that the most abundant sulfur species includes two oxygens. It is unlikely that these are sulfones since this coal has not been exposed to oxygen and the pyrolysis is carried out under a vacuum. In addition, sulfones have not been observed in characterization of extracts or liquefaction products where there can be prolonged exposure to oxygen. The empirical formulae for the major groups are given in Table IV. One of the more abundant groups (HD = 8) has been identified in carbonization products (10). The ions observed for the most abundant group are shown in Figure 4. Note that the high mass peaks are the most intense which is a little unusual for a strict methylene series.

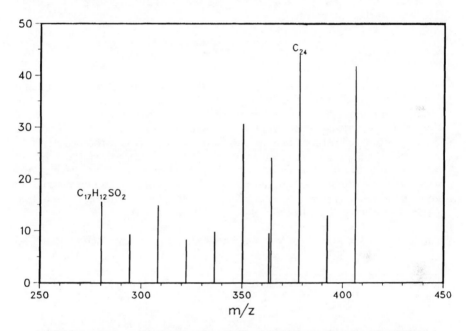

Figure 4. Selected ions containing SO_2 and HD = 12 from LVHRMS.

This effect might be due to the combination of several different types of molecules in the series including general structures A-B or A-B-C where A, B and C are various types of heteroaromatics or functionalized aromatics. Such combinations may account for many for the species listed in Table IV. In this table the base formula for each homologous series is given.

Although less abundant there is also a significant number of ions which have four heteroatoms, one sulfur and three oxygens. The main series of peaks are at HD = 9 and 14. The first group is quite typical while the second starts at a relatively high carbon count which may be an example of a combination of aromatics as previously described. Finally, a small number of ions were observed to have one of each heteroatom. Structural assignments would be very speculative at this time.

CONTRIBUTIONS FROM PYRITE. This coal has a significant amount of pyrite which may produce sulfur containing organic molecules during pyrolysis.

To investigate this possibility we have pyrolysed a sample of the coal which has most of the pyrite removed by a float-sink process using aqueous $CsCl_2$ and a surfactant to disperse the coal which is a technique we have used to concentrate coal macerals(25). The coal is first demineralized with HCl/HF and then ground to a very fine particle size($\simeq 1$ micron) prior to the float-sink step. The final ash content is less than one percent. The results, which are shown in Figure 5, suggest that the bulk pyrite is probably not contributing to the formation of organic polycyclic aromatic sulfur compounds. However, it is apparent that the pyrite is contributing to the formation of organic sulfides (HD <6) during the pyrolysis.

CONCLUSIONS

For this particular bituminous coal from the Herrin Seam, it is apparent that there are a significant number of sulfur heteroaromatics which contain additional heteroatoms. Oxygen is the major heteroatom found in these species which is not surprising considering the oxygen content of this coal. There are 13 oxygens per 100 carbons and approximately 69 aromatic carbons per 100 in this coal. If one assumes that 11 of the oxygens are associated with an aromatic structure in some form such as hydroxyl or ether, then there is one oxygen for every 6-member aromatic ring. This is probably an overestimation but it does demonstrate that one should expect these multiple heteroatom species to be indigenous to this coal.

Table IV. Peaks from Low Voltage PyHRMS Which Have Been Identified as Having Sulfur Plus More than One Additional Heteroatom

Base Formula	HD	Maximum Number of Methyls/Methylenes	Number of Peaks in a Series	%Total Ionization
$C_{14}H_{16}SO_2$	7	7	7	1.03
$C_{14}H_{14}SO_2$†	8	8	9	1.61
$C_{13}H_{10}SO_2$	9	9	6	0.81
$C_{15}H_{12}SO_2$	10	15	11	1.45
$C_{14}H_8SO_2$	11	12	11	1.60
$C_{17}H_{12}SO_2$	12	9	12	1.71
$C_{17}H_{10}SO_2$	13	11	10	0.66
$C_{14}H_{12}SO_3$	9	10	5	0.48
$C_{22}H_{18}SO_3$	14	6	4	0.36
$C_{13}H_{13}SNO$	8	9	7	0.45
$C_{17}H_{19}SNO$	9	6	5	0.30

† Compounds with this formula were observed in carbonization products. The data for this formula were taken from ref. 10.

In addition, there is evidence that a class of compounds including thianthrene(II), phenoxathiin(VIIb) and phenoxazine(Vb) may be present in this coal. These compounds are thermally quite reactive and would probably not survive extended pyrolysis or liquefaction conditions. Instead, they will convert into the more stable dibenzothiophenes. This would suggest that chemical or biological removal of these compounds could be more efficient than a thermal process. Work is in progress to better characterize these structures by among other approaches HRMS/MS.

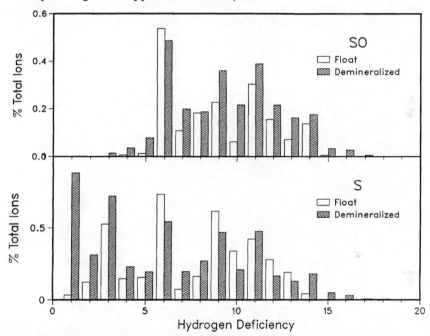

Figure 5. Comparison of ion intensities with and without bulk pyrite for selected species.

ACKNOWLEDGMENTS

This work was performed under the auspices of the Office of Basic Energy Sciences, Division of Chemical Sciences, U.S. Department of Energy, under contract no. W-31-109-ENG-38.

LITERATURE CITED

1.   White, C. M.; Douglas, L. J.; Perry, M. B.; Schmidt, C. E. Energy Fuels 1987, 1, 222-226.
2.   Nishioka, M. Energy Fuels 1988, 2, 214-219.

3.   White, C. M., this volume.
4.   Winans, R. E.; McBeth, R. L.; Young, J. E. Processing and Utilization of High Sulfur Coals III (in press).
5.   Neill, P. H.; Xia, Y. J.; Winans, R. E. Preprints, Div. Fuel Chem. ACS 1989, 34(3), 745-751.
6.   Nishioka, M.; Lee, M. L.; Castle, R. N. Fuel 1986, 65, 390-396.
7.   Later, D. W.; Lee, M. L.; Bartle, K. D.; Kong, R. C.; Vassilaros, D. L. Anal. Chem. 1981, 53, 1612-1620.
8.   Nishioka, M.; Lee, M. L.; Kudo, H.; Muchiri, D. R.; Baldwin, L. J.; Pakray, S.; Stuart, J. G.; Castle, R. N. Anal. Chem. 1985, 57, 1327-1330.
9.   Aczel, T. Rev. Anal. Chem. 1972, 1, 226-261.
10.   Schultz, J. L.; Kessler, T.; Friedel, R. A.; Sharkey, A. G. Fuel 1972, 51, 242-246.
11.   Winans, R. E.; Hayatsu, R.; Scott, R. G.; McBeth, R. L. In "Chemistry and Characterization of Coal Macerals"; Winans, R. E.; Crelling, J. C., Eds.; ACS Symposium Series No. 252, American Chemical Society:Washington, D.C., 1984; pp. 137-155.
12.   Winans, R. E.; Scott, R. G.; Neill, P. H.; Dyrkacz, G. R.; Hayatsu, R. Fuel Proc. Tech. 1986, 12, 77-88.
13.   Nip, M.; DeLeeuw, J. W.; Schenck, P. A., Geochim. Cosmochim. Acta 1988, 52, 637-648.
14.   Sinninghe Damste, J. S.; Koch-van Dalen, A. C.; DeLeeuw, J. W.; Schenk, P. A. J. Chromatogr. 1988, 435, 435-452.
15.   Schmidt, C. E.; Sprecher, R. F.; Batts, B. D. Anal. Chem. 1987, 59, 2027-2033.
16.   Vorres, K. S.; Janikowski, S. K., Preprints, Div. Fuel Chem. ACS 1987, 32(1), 492-499.
17.   Winans, R. E.; Scott, R. G.; Neill, P. H.; Dyrkacz, G. R.; McBeth, R. L.; Hayatsu R. Proceedings, 1985 International Conference on Coal Science 1985, pp. 687-690.
18.   Oae, S., In "Organic Sulfur Chemistry"; Bernardi, F.; Csizmadia, I. G.; Mangini, A., Eds.; Elsevier:Amsterdam, 1985; pp. 1-67.
19.   Roy, M. M. Brennst. Chemie 1965, 46, 407-411.
20.   Aitken, J.; Heeps, T.; Steedman, W. Fuel 1968, 47, 353-357.
21.   Nishioka, M.; Campbell, R. M.; West, W. R.; Smith, P. A.; Booth, G. M.; Lee, M. L.; Kudo, H.; Castle, R. N. Anal. Chem. 1985, 57, 1868-71.
22.   Hayatsu, R.; Scott, R. G.; Winans, R. E. In "Oxidation in Organic Chemistry, Part D"; Trahanovsky, W. S., Ed.; Academic:New York, 1982; pp. 279-354.
23.   Jones, R. B.; McCourt, C. B.; Swift, P. Proceedings, 1981 International Conference on Coal Science, 1981, pp. 657-662.
24.   Siskin, M.; Aczel, T. Proceedings, 1981 International Conference on Coal Science 1981, pp. 651-656.
25.   Dyrkacz, G. R.; Bloomquist, C. A. A.; Horwitz, E. P. Sep. Sci. Technol. 1981, 16, 1571-1588.

RECEIVED April 13, 1990

Chapter 16

Organosulfur Constituents in Rasa Coal

Curt M. White[1], L. J. Douglas[1], R. R. Anderson[1], C. E. Schmidt[1], and R. J. Gray[2]

[1]Pittsburgh Energy Technology Center, U.S. Department of Energy, P.O. Box 10940, Pittsburgh, PA 15236
[2]Ralph Gray Services, 303 Drexel Drive, Monroeville, PA 15146

The petrographic, proximate, and ultimate analyses of Rasa coal are reported, along with information on the vitrinite reflectance, carbon aromaticity, sulfur and carbon isotopic abundances, ash analysis and forms of sulfur. A Rasa coal containing 11.79% sulfur was extracted with a pyridine/toluene azeotrope and the extract was analyzed by low-voltage, high-resolution mass spectrometry. The molecular ions of 1440 individual compounds were detected, and using this precise mass information, molecular formulae were assigned. The average deviation between the measured and theoretical masses was 0.0003 amu. Many homologous series of organosulfur compounds were detected. Some families of organosulfur compounds contained two or three sulfur atoms per molecule. The degree of alkylation of many homologous series maximized at either three, four, or five alkyl carbons.

Rasa coal has generated substantial interest over the years because of its high organic sulfur content. Information concerning the nature and distribution of organic sulfur moieties in coal is desirable for the design and evaluation of coal desulfurization processes. A major impediment to the character-

0097–6156/90/0429–0261$07.50/0
© 1990 American Chemical Society

ization of organosulfur constituents in lower organic sulfur coals is the high hydrocarbon background, which interferes with the spectral observation of organosulfur compounds. Rasa coal does not suffer significantly from this problem. Thus, for the scientist interested in characterizing organic sulfur components in coal, Rasa coal is a good starting point. The organosulfur constituents in Rasa coal may be the same as, or similar to, those of other coals. Characterization of the organosulfur compounds in Rasa coal is expected to be a guide to the analysis of organosulfur compounds in coals having much lower organic sulfur contents.

Rasa coal occurs in several seams on the Istrian Peninsula, the lowest seam being Cretaceous. Hamrla has described the geologic conditions and surroundings in which the 10 seams that constitute Rasa coals were formed (1-3). According to Hamrla, the coal-bearing strata are lacustrine and brackish facies of the Upper Paleocene. The lowest coal seam occurs on the erosion surface of the Upper Cretaceous. Apparently, the coal forming plants grew under lacustrine and/or brackish conditions. The resulting plant debris was periodically inundated with seawater. The roofs of the coal beds are mainly marine limestone and flysch. In the region where the Rasa coals occur, there was no vigorous folding or faulting (1-3). After studying the maceral content, Hamrla concluded that the Rasa coals were formed under completely anaerobic conditions.

Kreulen reported that Rasa coal was a humic coal and contained no sapropelites, waxes, or resins (4). He noted that Rasa coal exhibits dual character, i.e., it exhibits both low- and high-rank characteristics. Chemical tests indicated that a small amount of the sulfur present is in side chains, and a large amount occurs in ring structures. Using a statistical structural analysis method based on density, van Krevelen computed that 59% of the carbon in Rasa coal is aromatic (4).

Kavcic treated several samples of Rasa coal, hydrogenated Rasa coal, and the residue left after acetophenone extraction of Rasa coal with methyliodide to differentiate between the types of sulfur present. Results indicated that about 75% of the sulfur was thiophenic and 25% was reactive with methyliodide (5). Kavcic later investigated the ratio of $^{32}S/^{34}S$ in a Rasa coal and found it to be 22.9 ($\delta^{34}S_{CDT}$ = -30.1) (6).

Kavcics' results on the forms of organic sulfur were confirmed by Ignasiak et al. (7), who determined that about one third of the sulfur in Rasa coal is in thioether links. This result was determined using two independent techniques. Ignasiak et al. also concluded that mercaptanic sulfur was absent from Rasa coal because unspecified chemical tests failed to reveal its presence.

Until recently, little was known with certainty about the chemical identity of organosulfur compounds in coal. White and coworkers characterized some organosulfur constituents extractable from both Homestead, Kentucky, coal (8) and Bevier seam coal (9). Boudou et al. have studied the organosulfur components in the Currie point pyrolysis products from Provence, Muskingum, and Meigs coals by GC-MS (10). Many organosulfur compounds were tentatively identified. Nishioka published an excellent paper describing the characterization of organosulfur constituents extracted from a Rock Springs No. 7 coal taken from Sweetwater, Wyoming (11). Even though significant progress has been made on the characterization of organosulfur constituents from coal, almost nothing is known with certainty concerning the nonthiophenic constituents in coal. One purpose of the present manuscript is to begin to define the possible nature of nonthiophenic sulfur moeities in coal.

Experimental

A single lump, approximately 2 kg, of Rasa coal was obtained from M. Eckert-Maksic of the Rudjer Boskovic Institute of Zagreb, Yugoslavia. The lump was randomly selected, and no special storage conditions were used. The coal was ground in air until it passed through a minus 60-mesh screen.

Petrographic studies were conducted using the air-dried pulverized coal mixed with a nonreactive epoxy binder. Air drying was performed at ambient temperature. The ASTM method specifies minus 20 mesh (minus 850µm). The Rasa coal sample was finer. The mixture, which contained about 18 to 25 percent plastic, was briquetted in a one-inch-internal-diameter cylindrical mold using pressures up to 5000 psi. The mold contained top and bottom plugs. The briquette was ground and polished for microscopic examination according to ASTM D2797-85 (12).   One pellet of Rasa coal was prepared using a Buehler Automet polishing device. The coal macerals were defined according to ASTM D2796-88 (13). Use of brand names facilitate understanding and does not necessarily imply endorsement by the U.S. Department of Energy.

A Leitz Ortholux microscope was used to determine the maceral content of Rasa coal. The microscope was fitted with an oil- immersion 60X fluoride objective and 10X high eyepoint oculars to give an effective magnification of 720 diameters. A point count system of analysis was used for the maceral determination. Four points were identified per field, and a total of 1000 counts were determined for Rasa coal. This system employs a point count stage and an ocular graticule. The volume percent of macerals was calculated according to ASTM D2799-86 (14).

Elemental analysis of the minus 60-mesh coal was performed using ASTM methods D3177, D3178, and D3179 (15-17). Moisture and ash were determined by ASTM methods D3173 and D3174 (18,19), volatile matter by ASTM method D3175 (20), and forms of sulfur by ASTM method D2492 (21). Elemental analysis of the ash was performed using ASTM method D3682 (22). Carbon aromaticity was determined using ^{13}C NMR CP-MAS procedures described elsewhere (23). X-ray powder diffraction analysis of the mineral matter in the whole coal was performed using a Rigaku powder diffractometer. The diffraction lines were compared with those in the Powder Diffraction File (24). The $\delta^{34}S_{CDT}$ and $\delta^{13}C_{PDB}$ measurements were made in duplicate at Global Geochemistry in Canoga Park, California.

The mean maximum reflectance of vitrinite was determined in oil (ca. 1.517 index of refraction) and in 546-nm wavelength green light. A photomultiplier photometer was used to measure vitrinite reflectance. Two glass standards were used to calibrate the equipment. A Leitz MPV2 microscope photometer was used. The reflectance values were determined as designated in ASTM D2798-85 (25). Twenty-five vitrinite reflectance values were determined for the Rasa coal.

An aliquot of the ground coal (142.9 g) was mixed with an equal weight of Celite and extracted for five days in a Soxhlet using the pyridine/toluene azeotrope (22:78) having a boiling point of ~110°C. The ground coal was mixed with Celite in order to prevent chanelling of the solvent through the coal in the Soxhlet thimble. The pyridine/toluene azeotrope was used because it boils at a slightly lower temperature than pure pyridine. After five days, the extract was filtered through a 10-micron Teflon filter. The filtrate was subjected to rotary evaporation to remove bulk solvent, followed by further evaporation of solvent using a stream of flowing nitrogen over the warmed (50°C) extract. When the extract weight loss slowed, the extract was placed in a vacuum oven at 50°C. After two weeks, the extract reached nearly constant weight, yielding 32.9 g (23.0%). Elemental analyses of the extract were performed at Huffman Laboratories.

The low-voltage, high-resolution mass spectra (LVHRMS) were obtained on a Kratos MS-50 high-resolution mass spectrometer interfaced to a Kratos DS-55 data system. The Rasa coal extract was vaporized directly into the ion source by a direct-insertion probe. Spectra were recorded at probe temperatures of 100°C, 200°C, 300°C, and 350°C. The instrument was scanned from 700 to 60 amu at a scan

rate of 1 mass decade/300 s. Multiple scans were
acquired. The mass spectrometer was operated at a
dynamic resolution of one part in 25,000. The
ionizing voltage was kept at 11.5 eV. Details concerning the low-voltage, high-resolution mass spectrometric procedure and data reduction routines are
presented elsewhere (26). A complete description of
the direct-insertion probe technique, and an
auxiliary computation method to improve the inter-
polation of sample peak masses, have also been
reported (27).

Results and Discussion

Petrographic examination of the Rasa coal shows it to
be predominately vitrinite and/or bituminite (90+%)
with some liptinite (2-3%) and very little inertinite
(2-3%). The vitrinite occurs as subangular to
rounded particles surrounded by an eucollinitic-to-
resinous material. The eucolinite groundmass has no
specific morphology and fits the description for the
maceral bituminite. There are some fungal spores and
a small amount of micrinite, macrinite, and inerto-
detrinite. The coal has an abundance of very small
pyrite crystals relative to North American coals.
Many of these crystals appear to be euhedral
(octahedrons), and most of the pyrite is highly disseminated and less than 5 microns. Microscopic
examination revealed calcite, much of which exhibits
crystal twinning. There is also some dolomite and a
small amount of siderite. X-ray diffraction analysis
of Rasa coal indicated that calcite and dolomite were
the major mineral constituents.

The results of the proximate and ultimate analysis
of Rasa coal appear in Table I. The results of the
forms of sulfur analysis and carbon aromaticity are
also included in Table I. The total sulfur on an as-
received basis was 10.77 weight percent, and includes
0.02 weight percent sulfate, 0.30 weight percent
pyrite, and 10.45 weight percent organic sulfur.
Organic sulfur is calculated as the difference

between total sulfur and the sum of the pyritic and sulfate sulfur. Table II indicates that the ash contained 5.23 weight percent Fe_2O_3, on an as-received basis. Thus, there is more iron in the ash than is accounted for by the pyrite. Therefore, the pyrite value reported in Table I is probably low. If all the Fe in the ash was present as pyrite, as is probably the case, then the coal would have contained 0.63 weight percent pyrite on an as-received basis. Because the pyrite was highly disseminated and present in particles of 5 microns or less, it may have been incompletely leached from the coal during nitric acid treatment, resulting in a low pyrite value.

TABLE I. Properties of Rasa Coal

	As-Received Basis	Moisture-Free Basis	Moisture- and Ash-Free Basis
Moisture	0.68	N/A	N/A
Volatile Matter	48.00	48.33	52.55
Fixed Carbon	43.34	43.63	47.45
Ash	7.98	8.04	N/A
Hydrogen	4.34	4.79	5.21
Carbon	73.29	73.79	80.23
Nitrogen	1.12	1.13	1.23
Sulfur	10.77	10.84	11.79
Oxygen*	2.01	1.42	1.54
Sulfur Forms			
Sulfate	0.02	0.02	0.02
Pyritic	0.30	0.30	0.33
Organic	10.45	10.52	11.44
f_a by ^{13}C NMR CP-MAS	0.65	N/A	N/A

*By difference.

The carbon aromaticity of Rasa coal was 0.65, which is not significantly different from van Krevelen's finding that 59% of the carbon in Rasa coal was aromatic (4). Rasa coal is difficult to classify according to rank. Some literature articles refer to it as a lignite. The high sulfur content of Rasa coal tends to skew the quantitative results for the other elements. The vitrinite reflectance values range from 0.60% to 0.77%. Rasa vitrinite has a mean maximum reflectance in green light of 546 nm and in oil of 0.682%. These values are over twice those reported for lignites, and the mean maximum reflectance is similar to that of many Illinois and Ohio high-volatile B or C rank bituminous coals. Rasa coal is high in volatile matter as expected for a lignite. This is due in small part to the large amounts of calcite and dolomite (carbonates) in Rasa coal. Carbonates yield carbon dioxide when heated, giving an erroneously high volatile matter value, making the coal appear to be of lower rank. Based on the vitrinite reflectance value and the carbon value, Rasa coal appears to be a relatively mature coal.

The major elements present in Rasa coal ash were determined. The results are in Table II. The major elements present in the ash, Ca and Mg, are consistent with the X-ray diffraction findings of calcite and dolomite. The weight percent values in Table II do not and should not add to 100%. The sum of all the values reaches 100% when the weight percent of the minor elements are included.

Rasa coal is exceptional because of its high sulfur content, most of which appears to be organic sulfur. The high organic sulfur content is characteristic of marine-influenced bituminous coals. Rasa coal appears to have been formed in a high-pH marine environment where bacteria thrived. This hypothesis is supported by the presence of substantial amounts of both calcite and dolomite in Rasa coal. The calcium- and magnesium-rich environment where Rasa coal formed is expected to have been alkaline. The

TABLE II. Rasa Coal Ash Analysis

Major Elements in Ash	Wt.% of Element Oxide in Ash, As-Received Basis
SiO_2	5.63
Al_2O_3	7.07
Fe_2O_3	5.23
TiO_2	0.28
CaO	33.03
MgO	10.15
Na_2O	1.00
K_2O	0.42
SO_3	37.16

alkaline marine environment, having relatively high concentrations of Group II ions, fostered bacterial growth in which products of bacterial action contributed to coal formation and sulfur fixation. This scenario fits well with the description provided by Hamrla (1-3) concerning the environmental conditions that existed on the Istrian Peninsula during coalification.

The presence of large amounts of organic sulfur and small amounts of pyrite in Rasa coal is also consistent with the early influence of a marine environment that was low in iron and high in sulfate. The abnormally high sulfur content of Rasa coal suggests a source of sulfur other than that originally in the plants, that became incorporated in the vegetable debris that formed the coal. Plants do not typically contain enough sulfur to account for the very high values found in Rasa coal. The sulfur in Rasa coal is present as a result of bacterial reduction of marine sulfate and subsequent incorporation of the reduced sulfur into the organic matrix. The low pyrite content of the Rasa coal is consistent with minimal input of terrestrial iron by fresh water. The observed facts are consistent with the early depositional environment of Rasa coal being primarily marine and alkaline.

An alkaline marine environment high in H_2S and HS⁻ is favorable for incorporation of sulfur into organic matter; HS⁻ is an aggressive nucleophile. Such conditions also would have been favorable for the formation of polysulfides and elemental sulfur. All of these species, either alone or in combination, are expected to have played a role in the incorporation of sulfur into the vegetable debris that ultimately formed Rasa coal. Many of these species are known to be reactive with hydrocarbons at mild temperatures. Elemental sulfur reacts with hydrocarbons to form organosulfur compounds, including thiophenes at mild temperatures (28,29). Polysulfides react with conjugated ene carbonyls at room temperature to form thiophenes and other sulfur heterocycles (30).

The isotopic abundances of ^{13}C and ^{34}S were measured. Determination of the $\delta^{34}S_{CDT}$ can provide insight into the origin of sulfur in coal. Therefore, the $\delta^{13}C_{PDB}$ and $\delta^{34}S_{CDT}$ were each measured in duplicate. The $\delta^{13}C_{PDB}$ duplicate values were -23.98, and -24.02, while the $\delta^{34}S_{CDT}$ duplicate values were +7.8 and +7.9. The $\delta^{34}S$ values are relative to Canyon Diablo triolite (CDT), while the $\delta^{13}C$ values are relative to Peedee beleminite (PDB). The $\delta^{13}C_{PDB}$ values are typical of those found for many coals. The $\delta^{34}S_{CDT}$ values are also typical of many coals and are consistent with the sulfur in Rasa coal having a marine origin. The $\delta^{34}S_{CDT}$ of seawater sulfate has ranged from about +10 to +23 over the last 250 million years (31). Thus, Rasa coal is depleted in ^{34}S relative to seawater sulfate. A fractionation of at least 14 per mil usually occurs during bacterial sulfate reduction if the available sulfate source approaches infinity, as it does in the ocean. In 1957, Kavcic reported the isotopic abundance of ^{34}S in Rasa coal as a $^{32}S/^{34}S$ ratio of 22.9 ± 0.04 ($\delta^{34}S_{CDT}$ value of -30.1) (6), which is substantially different from the values reported here. The reason for the large discrepancy is not understood. Nevertheless, both sets of measurements show the sulfur in Rasa coal to be depleted in ^{34}S relative to seawater sulfate and thus consistent with the biological reduction of marine sulfate and subsequent incorporation of the reduced sulfur into organic matter that ultimately became Rasa coal.

The elemental analysis of the Rasa coal extract appears in Table III, along with the values for the coal. The elemental analyses of the coal and of the extract are similar. This is consistent with the extract being representative of the organic matter in the coal. The nitrogen content of the extract is slightly higher than that of the coal and is due to a small amount of pyridine remaining in the extract. Pyridine was observed by mass spectrometric analysis of the extract. The oxygen content of the extract is higher than the coal. This could be due to oxidation of the extract or preferential extraction of oxygen-containing compounds from the coal. Alternately, the apparent difference in oxygen content of the coal and extract may simply be due to the way the oxygen determinations were made. The oxygen content of the coal was determined by difference, and thus reflects any errors made during measurement of the other elements. Oxygen in the extract was measured by direct methods. The elemental values for C, H, O, N, and S were then normalized to 100%.

TABLE III.   Elemental Analysis - Moisture-and Ash-Free

	Coal	Extract
C	80.23	78.08
H	5.21	5.77
O	1.54*	3.56
N	1.23	1.65
S	11.79	10.94

*By difference

The extract was analyzed by low-voltage, high-resolution mass spectrometry (LVHRMS). Low ionizing voltages were used to minimize fragmentation. Generally, under low-voltage conditions, the ions detected are molecular ions. High resolution was used to separate ions having the same nominal mass. Table IV displays an octet of peaks detected at m/e 288 that were resolved from one another. The assigned molecular formulas, % of base peak (the largest peak), and the calculated and measured masses are also included. Under low-resolution conditions, these ions would merge into one unresolved peak, providing little or no information. As seen in Table IV, the Rasa coal contains compounds that possess 2 and 3 sulfur atoms per molecule. The resolution required to achieve base-line separation of these ions from their neighbors is shown in Table V. Even using high-resolution techniques, some ambiguities exist that require interpretation. The peak at m/e 288.0979 could be assigned to $C_{20}H_{16}S$, $C_{17}H_{20}S_2$, or $C_{23}H_{12}$. In this case, it is not possible to make a clear assignment to a single molecular formula. Assignment of the peak at m/e 288.0979 to the two sulfur containing formulae is preferred to the hydrocarbon because of the closeness of the precise mass measurement (0.0006 error for $C_{20}H_{16}S$, 0.0027 for $C_{17}H_{20}S_2$, and 0.0040 for $C_{23}H_{12}$). Even though the measured mass is closer to that of $C_{20}H_{16}S$, the $C_{17}H_{20}S_2$ formula cannot be eliminated because lower molecular weight homologs of $C_{17}H_{20}S_2$ were clearly present in the sample. Further, the assignment of the peak at m/e 288.0979 in part to $C_{17}H_{20}S_2$ is supored by observation a ^{34}S peak. The peak at m/e 288.0979 is best assigned to both sulfur-containing molecular formulae. The peak at m/e 288.1899 could be assigned to either $C_{19}H_{28}S$ or $C_{22}H_{24}$ or both. Using a dynamic resolution of one part in 25,000, these two ions are not base-line separated, and thus it is difficult to distinguish between these two possibilities. It should be emphasized, however, that base-line resolution of peaks is not always necessary.

TABLE IV. Multiplet at m/e 288

Measured Mass	Calculated Mass	Molecular Formula	% of Base Peak
288.1899	288.1911	$C_{19}H_{28}S$	16.53
	288.1878	$C_{22}H_{24}$	
288.1514	288.1514	$C_{21}H_{20}O$	6.92
288.1357	288.1350	$C_{18}H_{22}O^{34}S$	0.81
288.1113	288.1150	$C_{20}H_{16}O_2$	2.38
288.0979	288.0973	$C_{20}H_{16}S$	28.67
	288.1006	$C_{17}H_{20}S_2$	
	288.0939	$C_{23}H_{12}$	
288.0781	288.0774	$C_{20}H_{14}{}^{34}S$	5.81
288.0606	288.0609	$C_{19}H_{12}OS$	17.84
288.0085	288.0101	$C_{15}H_{12}S_3$	1.33

TABLE V. Approximate Resolution Required to Achieve Base-Line Separation of the Ions Observed at m/e 288 and Listed in TABLE IV

Doublet	Resolution Required
$C_{19}H_{28}S$, $C_{22}H_{24}$	87,270
$C_{19}H_{28}S$, $C_{21}H_{20}O$	7,250
$C_{21}H_{20}O$, $C_{18}H_{22}O^{34}S$	17,560
$C_{18}H_{22}O^{34}S$, $C_{20}H_{16}O_2$	14,400
$C_{20}H_{16}O_2$, $C_{20}H_{16}S$	16,270
$C_{20}H_{16}S$, $C_{17}H_{20}S_2$	87,270
$C_{20}H_{16}S$, $C_{23}H_{12}$	84,700
$C_{17}H_{20}S_2$, $C_{23}H_{12}$	42,990
$C_{20}H_{16}S$, $C_{20}H_{14}{}^{34}S$	14,470
$C_{20}H_{14}{}^{34}S$, $C_{19}H_{12}OS$	17,450
$C_{19}H_{12}OS$, $C_{15}H_{12}S_3$	5,670

The use of LVHRMS to determine organosulfur constituents in a coal containing a significantly lower amount of organic sulfur than that in Rasa coal is much more difficult and less reliable because the large hydrocarbon background interferes with observation of the organosulfur compounds. Furthermore, the organosulfur compounds are usually present in such low concentrations that observation of a ^{34}S isotope peak is rare. In most coals, the multiplet at m/e 288 would have contained large hydrocarbon ions for $C_{22}H_{24}$ and $C_{23}H_{12}$ at precise masses of 288.1878 and 288.0939, respectively. If present in significant amounts, these ions would have obscured the sulfur-containing compounds. The $C_{22}H_{24}$ would have interfered with $C_{19}H_{28}S$, and the $C_{23}H_{12}$ would have interfered with both $C_{20}H_{16}S$ and $C_{17}H_{20}S_2$. The resolution required to achieve base-line separation of the $C_{22}H_{24}$, $C_{19}H_{28}S$ doublet is 87,000, while the resolution needed to achieve base-line separation of the $C_{23}H_{12}$, $C_{20}H_{16}S$ doublet is 84,700. In fact, based on the results obtained, the presence of these hydrocarbon ions cannot be ruled out. However, if they were present in significant amounts, experience has shown that the measured mass would have been closer to that of the hydrocarbon ions rather than the sulfur-containing ions.

Table VI contains a portion of the low-voltage, high-resolution mass spectrometric information obtained during analysis of the Rasa coal extract (127 of the 1440 peaks observed). The structures shown in Table VI are included to show some of the kinds of possible types of organosulfur compounds that could be present. These structures are consistent with the observed molecular formulae. The structures drawn have not been identified in the coal extract. Many other structures are consistent with the observed molecular formulae. In fact, the thiols are particularly suspicious because Ignasiak et al. failed to find thiols in Rasa coal (7).

Throughout Table VI, it is clear that the mass measurements are made with extremely high accuracy.

TABLE VI. PARTIAL LVHRMS DATA FROM RASA COAL EXTRACT

Possible Compound Type	Number of Alkyl Carbons	Measured M.W.	Calculated M.W.	% of Base Peak	Molecular Formula	³⁴S
	1	124.0352	124.0347	4.00	C_7H_8S	NO
	2	138.0503	138.0503	2.62	$C_8H_{10}S$	NO
	4	166.0802	166.0816	4.54	$C_{10}H_{14}S$	NO
	5	180.0957	180.0973	2.95	$C_{11}H_{16}S$	NO
	1	148.0346	148.0346	5.24	C_9H_8S	NO
	2	162.0500	162.0503	16.15	$C_{10}H_{10}S$	NO
	3	176.0652	176.0657	15.86	$C_{11}H_{12}S$	NO
	4	190.0816	190.0816	15.19	$C_{12}H_{14}S$	NO
	5	204.0971	204.0973	16.90	$C_{13}H_{16}S$	NO
	6	218.1115	218.1130	20.36	$C_{14}H_{18}S$	NO
	0	184.0347	184.0347	5.99	$C_{12}H_8S$	NO
	1	198.0509	198.0503	23.81	$C_{13}H_{10}S$	NO
	2	212.0666	212.0659	51.94	$C_{14}H_{12}S$	YES
	3	226.0813	226.0816	74.55	$C_{15}H_{14}S$	YES
	4	240.0968	240.0973	55.36	$C_{16}H_{16}S$	YES
	5	254.1125	254.1129	55.52	$C_{17}H_{18}S$	YES
	6	268.1282	268.1285	49.25	$C_{18}H_{20}S$	YES
	7	282.1435	282.1442	52.80	$C_{19}H_{22}S$	YES
	8	296.1590	296.1598	33.57	$C_{20}H_{24}S$	NO
	9	310.1744	310.1755	28.39	$C_{21}H_{26}S$	YES
	10	324.1917	324.1912	16.98	$C_{22}H_{28}S$	NO
	11	338.2056	338.2069	22.16	$C_{23}H_{30}S$	NO
	12	352.2199	352.2224	8.14	$C_{24}H_{32}S$	NO
	13	366.2375	366.2369	11.81	$C_{25}H_{34}S$	NO
	14	380.2536	380.2537	10.90	$C_{26}H_{36}S$	NO
	15	394.2704	394.2694	2.20	$C_{27}H_{38}S$	NO

Continued on next page.

TABLE VI. CONTINUED

Possible Compound Type	Number of Alkyl Carbons	Measured M.W.	Calculated M.W.	% of Base Peak	Molecular Formula	^{34}S
	0	186.0506	186.0503	4.10	$C_{12}H_{10}S$	NO
	1	200.0657	200.0659	5.90	$C_{13}H_{12}S$	NO
	2	214.0816	214.0816	6.19	$C_{14}H_{14}S$	NO
	3	228.0961	228.0973	19.35	$C_{15}H_{16}S$	NO
	4	242.1138	242.1129	10.77	$C_{16}H_{18}S$	YES
	1	174.0508	174.0503	3.32	$C_{11}H_{10}S$	NO
	2	188.0661	188.0659	4.15	$C_{12}H_{12}S$	NO
	3	202.0797	202.0816	9.96	$C_{13}H_{14}S$	NO
	4	216.0970	216.0973	11.71	$C_{14}H_{16}S$	YES
	5	230.1100	230.1129	56.56	$C_{15}H_{18}S$	YES
	6	244.1261	244.1285	45.95	$C_{16}H_{20}S$	YES
	7	258.1445	258.1442	16.41	$C_{17}H_{22}S$	YES
	8	272.1586	272.1599	30.50	$C_{18}H_{24}S$	YES
	9	286.1726	286.1755	23.68	$C_{19}H_{26}S$	NO
	0	240.0067	240.0067	42.45	$C_{14}H_8S_2$	YES
	1	254.0225	254.0225	75.53	$C_{15}H_{10}S_2$	YES
	2	268.0380	268.0380	100.00	$C_{16}H_{12}S_2$	YES
	3	282.0531	282.0537	88.26	$C_{17}H_{14}S_2$	YES
	4	296.0716	296.0693	32.61	$C_{18}H_{16}S_2$	YES
	5	310.0851	310.0851	40.83	$C_{19}H_{18}S_2$	YES
	6	324.1015	324.1006	26.01	$C_{20}H_{20}S_2$	YES
	7	338.1157	338.1162	26.69	$C_{21}H_{22}S_2$	NO
	8	352.1297	352.1319	10.79	$C_{22}H_{24}S_2$	NO
	9	366.1478	366.1476	6.10	$C_{23}H_{26}S_2$	NO
	10	380.1622	380.1633	8.64	$C_{24}H_{28}S_2$	NO

n					
0	208.0352	208.0357	3.70	$C_{14}H_8S$	NO
1	222.0487	222.0503	5.56	$C_{15}H_{10}S$	NO
2	236.0661	236.0660	32.94	$C_{16}H_{12}S$	YES
3	250.0812	250.0816	63.72	$C_{17}H_{14}S$	YES
4	264.0976	264.0973	33.72	$C_{18}H_{16}S$	NO
5	278.1124	278.1129	34.68	$C_{19}H_{18}S$	NO
6	292.1286	292.1286	31.50	$C_{20}H_{20}S$	YES

n					
0	210.0498	210.0450	2.31	$C_{14}H_{10}S$	YES
1	224.0674	224.0660	6.89	$C_{15}H_{12}S$	YES
2	238.0817	238.0816	25.23	$C_{16}H_{14}S$	YES
3	252.0959	252.0972	35.99	$C_{17}H_{16}S$	YES
4	266.1124	266.1129	45.73	$C_{18}H_{18}S$	YES
5	280.1283	280.1286	42.59	$C_{19}H_{20}S$	YES
6	294.1443	294.1443	32.38	$C_{20}H_{22}S$	NO
7	308.1595	308.1598	20.88	$C_{21}H_{24}S$	NO
8	322.1761	322.1755	13.04	$C_{22}H_{26}S$	NO
9	336.1906	336.1912	15.23	$C_{23}H_{28}S$	NO
10	350.2040	350.2068	4.25	$C_{24}H_{30}S$	NO

n					
0	242.0205	242.0224	4.44	$C_{14}H_{10}S_2$	NO
1	256.0364	256.0380	5.12	$C_{15}H_{12}S_2$	NO
2	270.0524	270.0537	11.26	$C_{16}H_{14}S_2$	NO
3	284.0669	284.0693	21.32	$C_{17}H_{16}S_2$	NO
4	298.0830	298.0850	26.79	$C_{18}H_{18}S_2$	YES
5	312.1010	312.1014	26.13	$C_{19}H_{20}S_2$	YES
6	326.1153	326.1163	17.37	$C_{20}H_{22}S_2$	YES
7	340.1352	340.1353	6.09	$C_{21}H_{24}S_2$	NO

Continued on next page.

TABLE VI. CONTINUED

Possible Compound Type	Number of Alkyl Carbons	Measured M.W.	Calculated M.W.	% of Base Peak	Molecular Formula	^{34}S
	0	216.0076	216.0068	1.31	$C_{12}H_8S_2$	NO
	1	230.0228	230.0224	4.08	$C_{13}H_{10}S_2$	NO
	2	244.0377	244.0380	3.14	$C_{14}H_{12}S_2$	NO
	3	258.0542	258.0537	7.72	$C_{15}H_{14}S_2$	NO
	0	218.0221	218.0224	4.47	$C_{12}H_{10}S_2$	NO
	1	232.0379	232.0381	12.00	$C_{13}H_{12}S_2$	NO
	2	246.0530	246.0537	16.18	$C_{14}H_{14}S_2$	YES
	3	260.0680	260.0693	24.29	$C_{15}H_{16}S_2$	YES
	4	274.0840	274.0849	48.00	$C_{16}H_{18}S_2$	YES
	5	288.0979	288.1006	28.67	$C_{17}H_{20}S_2$	YES
	6	302.1140	302.1163	24.24	$C_{18}H_{22}S_2$	YES
	7	316.1337	316.1354	34.79	$C_{19}H_{24}S_2$	NO
	8	330.1466	330.1476	25.30	$C_{20}H_{26}S_2$	YES
	1	213.0625	213.0612	3.40	$C_{13}H_{11}NS$	NO
	2	227.0766	227.0769	12.39	$C_{14}H_{13}NS$	NO
	3	241.0932	241.0925	13.40	$C_{15}H_{15}NS$	NO
	4	255.1081	255.1082	7.07	$C_{16}H_{17}NS$	NO
	0	215.0417	215.0429	9.22	$C_{12}H_9NOS$	NO
	1	229.0565	229.0561	15.40	$C_{13}H_{11}NOS$	NO
	2	243.0719	243.0717	10.22	$C_{14}H_{13}NOS$	NO
	3	257.0882	257.0875	8.32	$C_{15}H_{15}NOS$	NO
	4	271.1010	271.1031	4.82	$C_{16}H_{17}NOS$	NO
	5	285.1178	285.1187	3.19	$C_{17}H_{19}NOS$	NO
	6	299.1354	299.1344	2.76	$C_{18}H_{21}NOS$	NO

R					
0	224.0276	224.0296	1.18	$C_{14}H_8SO$	NO
1	238.0472	238.0452	6.97	$C_{15}H_{10}SO$	NO
2	252.0607	252.0609	12.62	$C_{16}H_{12}SO$	NO
3	266.0764	266.0766	22.42	$C_{17}H_{14}SO$	YES
4	280.0922	280.0922	23.94	$C_{18}H_{16}SO$	YES
5	294.1069	294.1079	18.91	$C_{19}H_{18}SO$	NO
6	308.1226	308.1235	15.11	$C_{20}H_{20}SO$	NO
7	322.1418	322.1444	5.43	$C_{21}H_{22}SO$	NO
0	234.0512	234.0503	49.06	$C_{16}H_{10}S$	YES
1	248.0659	248.0659	92.77	$C_{17}H_{12}S$	YES
2	262.0819	262.0816	76.73	$C_{18}H_{14}S$	YES
3	276.0976	276.0972	69.16	$C_{19}H_{16}S$	YES
4	290.1133	290.1129	41.47	$C_{20}H_{18}S$	YES
5	304.1282	304.1285	32.23	$C_{21}H_{20}S$	YES
6	318.1446	318.1442	33.60	$C_{22}H_{22}S$	NO
7	332.1609	332.1598	18.04	$C_{23}H_{24}S$	NO
8	346.1747	346.1755	11.97	$C_{24}H_{26}S$	NO
9	360.1909	360.1912	8.64	$C_{25}H_{28}S$	NO
0	250.0431	250.0453	2.41	$C_{16}H_{10}SO$	NO
1	264.0606	264.0609	5.12	$C_{17}H_{12}SO$	NO
2	278.0770	278.0775	5.83	$C_{18}H_{14}SO$	YES
3	292.0933	292.0922	7.43	$C_{19}H_{16}SO$	YES
4	306.1075	306.1078	7.94	$C_{20}H_{18}SO$	NO
5	320.1232	320.1232	7.17	$C_{21}H_{20}SO$	NO
6	334.1404	334.1392	5.10	$C_{22}H_{22}SO$	NO
7	348.1528	348.1548	2.76	$C_{23}H_{24}SO$	NO

On average, the deviation between the measured and
calculated masses was 0.0003 amu. Further support
for the assigned formulae comes from the fact that an
entire homologous series was observed where the mass
of each homolog was measured with high accuracy. In
many cases, final confirmation of the assigned
molecular formula results from observation of the ^{34}S
isotope peak two amu higher at the correct precise
mass and approximately correct intensity. A ^{34}S
isotope peak was not observed for each sulfur-containing ion. Failure to observe a ^{34}S isotope peak
was due to either low intensity of the isotope peak
or to the presence of an interfering ion.

A summary of the LVHRMS results obtained on the
Rasa coal extract appears in Table VII. This summary
was obtained by taking data from eight individual
scans acquired at different temperatures and adding
them together. The data are summarized according to
the method of Aczel et al. (32). Table VII indicates
that 24 different homologous series containing one
sulfur atom were observed. Among these monosulfur-
containing families of compounds, 181 different
homologs (not including positional isomers) were
observed that contained from 6 to 36 carbons. The
range in Z is also listed in Table VII, where Z is
defined as C_NH_{2N-Z} and is a measure of hydrogen
deficiency. Nearly 50% of the molecular formulae
detected by LVHRMS contained at least one sulfur
atom. In all, 219 distinct structural series were
determined in the Rasa coal extract by LVHRMS; 1440
individual molecular formulae were identified.

Commonly, the most abundant homolog of each series
possessed three, four, or five alkyl carbons. The
observation of several homologous series of alkylated
thiophenes and other organosulfur compounds is
similar to that of Sinninghe Damste' et al. (33).
They observed multiple homologous series of
thiophenes in pyrolysates of sulfur-rich kerogens.

Many of the families of compounds found in
Table VI appear to be related to one another. For
example, the sample contains families of compounds

TABLE VII.   Summary of Compound Types Determined by LVHRMS in a
Pyridine/Toluene Extract From Rasa Coal

Formula Class	Relative Mole Percent of Those Compounds That Ionized	Number of Homologous Series	Z Number Range	Number of Individual Homologs Observed	Carbon Number Range
Monosulfur compounds	31.6	24	4-50	181	6-36
Hydrocarbons	31.1	30	0-58	328	5-42
Disulfur compounds	10.9	18	10-44	136	9-32
Monoxygenates	8.8	24	6-52	178	6-39
Nitrogenates	4.0	24	5-55	138	6-39
Nitrogen-oxygen compounds	3.8	20	9-45	110	10-37
Sulfur-oxygen compounds	3.6	19	10-48	92	9-35
Dioxygenates	2.4	19	14-50	114	14-38
Disulfur oxygen compounds	1.6	10	20-38	67	15-27
Trioxygenates	0.9	12	10-34	47	14-23
Trisulfur compounds	0.7	9	22-38	22	17-26
Dioxygen-sulfur compounds	0.6	10	20-42	27	16-31
Summary	100.0	219	0-58	1440	5-42

having formulae of $C_{12}H_{10}S$ and $C_{12}H_8S$ (dibenzo-thiophenes). The dibenzothiophenes could have been formed by internal cyclization and dehydrogenation of the $C_{12}H_{10}S$ family of compounds. Several other examples of one homologous series being formed from another by dehydrogenation are possible. Dehydro-genation of organic compounds is promoted by elemental sulfur.

The sulfur content of the extract was so high that families of compounds containing two and three dif-ferent heteroatoms per molecule were observed. Bodzek and Marzec have observed compounds from coal extracts that contained both nitrogen and sulfur, both nitrogen and oxygen, and both sulfur and oxygen (34). The data in Table VI show that the Rasa coal contained families of compounds that contained both nitrogen and sulfur, both sulfur and oxygen, and sulfur, nitrogen, and oxygen. It is rare to observe a homologous series of compounds from coal that con-tains three different heteroatoms.

The Rasa coal appeared to be susceptible to oxidation. The organosulfur compounds in the coal appeared to be oxidized when exposed to air. Thus, for many homologous series of compounds that con-tained only sulfur, there was a corresponding series that contained both sulfur and oxygen. The formation of sulfur oxygen bonding was observed by electron spectroscopy for chemical analysis (ESCA) when the Rasa coal was exposed to oxygen (Baltrus, J., Pittsburgh Energy Technology Center, Personal Communication, 1989).

No specific compounds have yet been identified. Although the structures drawn in Table VI are con-sistent with the observed molecular formulae, at present no other evidence suggests that they are valid. The thiol functionalities are particularly suspicious. Nevertheless, the molecular formulae assigned are correct. This investigation represents a starting point for more detailed study of the molecular components present in Rasa and other coals, and has defined some possibilities concerning the

nature of organosulfur compounds in coal. Lastly, it
is possible, but not likely, that the sulfur compounds in the coal were thermally altered either during the 110ºC extraction or during volatilization
into the mass spectrometer.

Acknowledgment

The authors are indebted to M. Eckert-Maksic for providing a sample of Rasa coal, to Joseph Malli, Jr.,
for operating the high-resolution mass spectrometer,
to Francis McCown for performing the Soxhlet extractions, to Richard F. Sprecher for performing the 13C
NMR, to Sidney Pollack for obtaining the X-ray diffraction results, and to the Coal Analysis Branch of
PETC for performing the ultimate and proximate
analysis of the Rasa coal.

Literature Cited

1. Hamrla, M. Geologija 1955, 3, 181-197.
2. Hamrla, M. Geologija 1959, 5, 180-264.
3. Hamrla, M. Rudarsko-Metalurski Zbornik 1960, 3,
   203-216.
4. Kreulen, D. J.W. Fuel 1952, 31, 462-467.
5. Kavcic, R. Bulletin Scientifique, Yugoslavie
   1954, 49, 5809.
6. Kavcic, R. Slovenskega, Kemijskeg a Drustva 1958,
   5, 7-11.
7. Ignasiak, B.S.; Fryer, J.F.; Jadernik, P. Fuel
   1978, 57, 578- 584.
8. White, C.M.; Lee, M.L. Geochim. Cosmochim. Acta
   1980, 44, 1825-1832.
9. White, C.M.; Douglas, L.J.; Perry, M.B.; Schmidt,
   C.E. Energy & Fuels 1987, 1, 222-226.
10. Boudou, J.P.; Boulegue, J.; Malechaux, L.; Nip,
    M.; de Leeuw, J.W.; Boon, J.J. Fuel 1987, 66,
    1558-1569.
11. Nishioka, M. Energy & Fuels 1988, 2, 214-219.
12. Standard Test Method of Preparing Coal Samples
    for Microscopical Analysis by Reflected Light. In

1987 Annual Book of ASTM Standards, Petroleum Products, Lubricants, and Fossil Fuels, Section 5, Volume 05.05, D2797-85, ASTM: Philadelphia, 1987; pp 361-366.

13. Standard Definitions of Terms Relating to Megascopic Description of Coal and Coal Seams and Microscopical Description and Analysis of Coal. In 1987 Annual Book of ASTM Standards, Petroleum Products, Lubricants, and Fossil Fuels, Section 5, Volume 05.05, D2796-82, ASTM: Philadelphia, 1987; pp 357-360.

14. Standard Method for Microscopical Determination of Volume Percent of Physical Components of Coal. In 1987 Annual Book of ASTM Standards, Petroleum Products, Lubricants, and Fossil Fuels, Section 5, Volume 05.05, D2799-86, ASTM: Philadelphia, 1987; pp 371-372.

15. Standard Test Methods for Total Sulfur in the Analysis Sample of Coal and Coke. In 1987 Annual Book of ASTM Standards, Petroleum Products, Lubricants, and Fossil Fuels, Section 5, Volume 05.05, D3177-84, ASTM: Philadelphia, 1987; pp 397-401.

16. Standard Test Methods for Carbon and Hydrogen in the Analysis Sample of Coal and Coke. In 1987 Annual Book of ASTM Standards, Petroleum Products, Lubricants, and Fossil Fuels, Section 5, Volume 05.05, D3178-84, ASTM: Philadelphia, 1987; pp 402-407.

17. Standard Test Method for Nitrogen in the Analysis Sample of Coal and Coke. In 1987 Annual Book of ASTM Standards, Petroleum Products, Lubricants, and Fossil Fuels, Section 5, Volume 05.05, D3179-84, ASTM: Philadelphia, 1987; pp 408-414.

18. Standard Test Method for Moisture in the Analysis Sample of Coal and Coke. In 1987 Annual Book of ASTM Standards, Petroleum Products, Lubricants, and Fossil Fuels, Section 5, Volume 05.05, D3173-87, ASTM: Philadelphia, 1987; pp 382-384.

19. Standard Test Method for Ash in the Analysis Sample of Coal and Coke From Coal. In 1987 Annual

Book of ASTM Standards, Petroleum Products, Lubricants, and Fossil Fuels, Section 5, Volume 05.05, D3174-82, ASTM: Philadelphia, 1987; pp 385-388.

20. Standard Test Method for Volatile Matter in the Analysis Sample of Coal and Coke. In 1987 Annual Book of ASTM Standards, Petroleum Products, Lubricants, and Fossil Fuels, Section 5, Volume 05.05, D3175-82, ASTM: Philadelphia, 1987; pp 389-391.

21. Standard Test Method for Forms of Sulfur in Coal. In 1987 Annual Book of ASTM Standards, Petroleum Products, Lubricants, and Fossil Fuels, Section 5, Volume 05.05, D2492-84, ASTM: Philadelphia, 1987; pp 336-340.

22. Standard Test Method for Major and Minor Elements in Coal and Coke Ash by Atomic Absorption. In 1987 Annual Book of ASTM Standards, Petroleum Products, Lubricants, and Fossil Fuels, Section 5, Volume 05.05, D3682-87, ASTM: Philadelphia, 1987; pp 443-449.

23. Maciel, G.; Bartuska, V.J.; Miknis, F.P. Fuel 1979, 58, 391- 394.

24. Powder Diffraction File, 1986, International Centre for Diffraction Data, 1601 Park Lane, Swarthmore, 1986.

25. Standard Method for Microscopical Determination of the Reflectance of the Organic Components in a Polished Specimen of Coal. In 1987 Annual Book of ASTM Standards, Petroleum Products, Lubricants, and Fossil Fuels, Section 5, Volume 05.05, D2798-85, ASTM: Philadelphia, 1987; pp 367-370.

26. Schmidt, C.E.; Sprecher, R.F.; Batts, B.D. Anal. Chem. 1987, 59, 2027-2033.

27. Schmidt, C.E.; Sprecher, R.F. In Novel Techniques in Fossil Fuel Mass Spectrometry; Ashe, T.R.; Wood, K.V., Eds.; ASTM STP 1019; ASTM: Philadelphia, 1989; pp 116-132.

28. Weitkamp, A.W. J. Amer. Chem. Soc. 1959, 81, 3430-3434.

29. White, C.M.; Douglas, L.J.; Schmidt, C.E.; Hackett, M. Energy and Fuels 1988, 2, 220-223.

30. Lalonde, R.T.; Ferrara, L.M.; Hayes, M.P. Org. Geochem. 1987, 11, 563-571.

31. Claypool, G.E.; Holser, W.T.; Kaplan, I.R.; Sakai, H.; Zak, T. Chem. Geol. 1980, 28, 199-260.

32. Aczel, T.; Williams, R.B.; Chamberlain, N.F.; Lumpkin, H.E. In Advances in Chemistry Series; American Chemical Society: Washington, DC, 1981; Vol. 195, pp 237-251.

33. Sinninghe Damste' J.S.; Kock-van Dalen, A.C.; de Leeuw, J.W.; Schenck, P.A. J. Chromatogr. 1988, 435, 435-452.

34. Bodzek, D.; Marzec, A. Fuel 1981, 60, 47-51.

RECEIVED March 5, 1990

Chapter 17

Distribution of Organic-Sulfur-Containing Structures in High Organic Sulfur Coals

R. J. Torres-Ordoñez, W. H. Calkins, and M. T. Klein

Center for Catalytic Science and Technology, Department of Chemical Engineering, University of Delaware, Newark, DE 19716

The distribution of organic sulfur-containing structures in two coals of high organic sulfur and low pyritic sulfur content was investigated using continuous isothermal flash pyrolysis experiments at 750 to 960°C. The conditions of atmospheric pressure and low contact times (<1 sec) minimized secondary reactions. The volatile sulfur-containing gas products (i.e., H_2S, COS, and CS_2) were measured as a function of pyrolysis temperature and compared to the products from the pyrolysis of a range of model sulfur compounds. The results suggest that the transformation of sulfur in coal goes from the more labile pendent sulfur structures to the more stable heterocyclic sulfur structures as coalification proceeds.

Coal sulfur is conventionally divided into organic and inorganic portions, and the distribution between these two sulfur forms is quite variable. Inorganic sulfur may further be classified into (1) sulfates, which are usually present in very low amounts, and may therefore be neglected, and (2) sulfides, which are predominantly iron pyrites (although other metal sulfides are also present). Although the forms of inorganic coal sulfur are well characterized and directly measurable by standard methods of analysis, the organic sulfur forms are not as well understood. While much effort has been directed towards the determination of the nature of these organic sulfur structures, only indirect methods have so far proven promising (1-3), with the possible exception of XAFS spectroscopy (4)(11).

If sulfur-containing bituminous coal is pyrolyzed in a pyroprobe, two types of sulfur-containing products are observed: (1) low molecular weight compounds such as H_2S, COS, and CS_2; and (2) heterocyclic sulfur-containing compounds such as substituted and unsubstituted thiophenes, benzothiophenes, dibenzothiophenes. The low molecular weight products most likely evolve by cleavage of pendent sulfur-containing side chains on the aromatic ring clusters. The heterocyclic sulfur compounds, on the other hand, apparently arise from cleavage of carbon-carbon bonds which attach the heterocyclic sulfur structures to the coal.

0097–6156/90/0429–0287$06.00/0

In a previous study (1), compounds modeling the reactive sulfur functional groups in coals were pyrolyzed in a fluidized sand bed under isothermal conditions at a range of selected temperatures and short contact times of 0.5-1.0 sec. The results indicated that under these conditions pendent mercaptan and thioether structures pyrolyze to form H_2S, COS, and CS_2 in a narrow temperature range significantly lower than the pyrolysis temperature range for the heterosulfur compounds such as thiophene, benzothiophene and dibenzothiophene. Moreover, the aliphatic mercaptan and thioether groups pyrolyze at lower temperatures (700-850°C) than do the aromatic mercaptan and thioether groups (>900°C), which in turn pyrolyze at lower temperatures than the heterocyclic sulfur structures (>950°C). Apparently, under very high temperature conditions, the heterocyclic compounds are also broken down to H_2S, COS, and CS_2.

The significance of this model compound information is that molecular classes pyrolyse in fairly narrow temperature ranges characteristic of their chemical structures and substitutents. Moreover, these ranges are sufficiently separated in temperature, which allows discrimination among the classes and provides a tool for estimating the types of sulfur functional groups present in coal. The present technique, therefore, involves comparison of the pyrolysis products from the coal (as function of temperature) with those obtained from the sulfur model compounds under the same pyrolysis conditions. (It must be borne in mind, however, that activating groups may alter the pyrolysis ranges of sulfur groups somewhat.)

The previous application of this technique to coal was aimed at sulfur functions (1). Coals of various ranks have been pyrolyzed under the same conditions as the sulfur model compounds (1). This work showed that low-rank coals containing organic sulfur produce largely the low molecular weight sulfur products at lower pyrolysis temperatures, whereas the higher-rank coals produce much lower proportions of these low molecular weight sulfur compounds under the same conditions. This suggests that the transformation of sulfur during coalification is from the more labile pendent sulfur structures to the more stable heterocyclic sulfur structures.

Most of the above observations have been made on high-sulfur coals containing both organic and pyritic sulfur. The pyritic sulfur, which usually occurs in amounts comparable to the organic, produces H_2S, COS and CS_2, as does some of the organic sulfur (10). The pyrolysis results therefore require correction for the contribution from pyritic sulfur. This likely introduces error in the analysis. Moreover, pyrite may play some role in the formation of the organic sulfur structures in the coal. For example, Yurovskii (5) suggests that pyrite masses in coals are surrounded by organic sulfur-rich layers, whose concentration decreases with distance from the pyrite mass center. He attributes the higher organic sulfur concentration near the pyrite to the elemental sulfur that may have been liberated during the formation of the pyrite core. While there are indications that this may not be the case (6,9), it seems desirable to investigate the pyrolysis products (as a function of temperature) of high-sulfur coals which contain little or no pyrite.

This motivates the present work. Two coals of high organic sulfur content and low pyritic sulfur content were chosen for study: a lignite (Spanish Mequinenza) and a high volatile A bituminous coal (New Zealand coal from the Charming Creek mine).

EXPERIMENTAL

The two high-organic low-pyrite sulfur coals used in this study are ground using a SPEX mill in a nitrogen atmosphere, sieved, and then vacuum-dried at 104-107°C for 20 hours. Only 40 grams of the New Zealand coal are available so that preliminary coarse grinding is performed on the SPEX mill and subsequent grinding is done with a mortar and pestle.

Table I shows the ultimate and sulfur forms analyses for these coals (performed by Huffman Laboratories in Golden, CO) on the indicated size fractions. The experiments on the Spanish lignite were conducted with the 74-105 μm and the 105-149 μm fractions; those on the New Zealand coal were conducted with the 74-105 μm fraction. The pyrolysis experiments were conducted at 750-960°C in the continuous flow pyrolyser unit shown in Fig. 1. The coal particles are entrained into the nitrogen stream in the coal feeder and carried over into the fluidized sand bed at various flash pyrolysis temperatures. The coal feeder is suspended from a Mettler balance and the change in its weight with time is monitored by a Linseis recorder. The coal feeder used requires 4 to 5 grams of either of this size fraction for uniform feeding at rates of 0.01-0.04 gram/min. Approximately 1-2 grams of the coal are fed during an experiment, and material recovery is fairly good (~85-92%). Upon entering the pyrolyser, the coal is rapidly heated to the bed temperature at a rate of ~10^5 °C/sec. The high heating

Table I: Analysis for High Organic Sulfur (Low Pyrite) Coals

	% Dry Basis		
	Spanish Mequinenza[1]	New Zealand Charming Creek[2]	
Particle size, μm	74-105	105-149	<63
Ash	14.90	2.79	2.99
Total S	10.73	5.55	5.68
Pyritic S	0.60	0.06	0.07
Sulfate S	0.09	0.03	0.03
Organic S	10.04	5.46	5.58
Carbon	58.34	76.37	76.37
Hydrogen	4.57	5.20	5.16
Nitrogen	0.77	1.02	0.99
Oxygen (by diff.)	10.69	9.07	8.81

[1] See reference (7) for hydrodesulfurization studies on this coal.

[2] See reference (8) for chemical desulfurization studies on this coal.

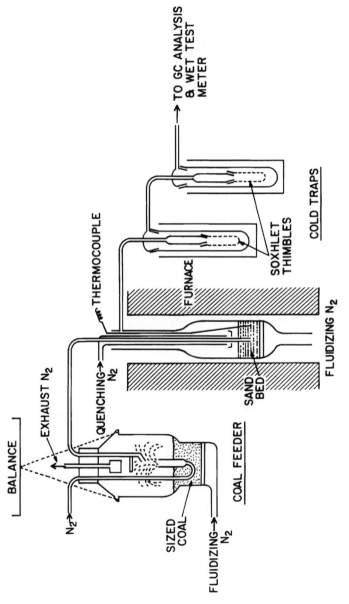

Figure 1. Fluidized Bed Pyrolyzer

rates were necessary to allow exposure of the sample at the desired temperature with a minimum of extra exposure in the heating process. Thermocouple wells in the reactor allow close monitoring of the temperature in the reactor bed and in the region above it (which is usually quenched with cold nitrogen to ~100-130°C below the reactor bed temperature). The residence time is calculated based on the fluidizing gas velocity, assuming that the "free volume" (i.e.; the volume of the expanded bed minus the volume of the sand) is fully utilized. At the temperature, total reactor gas flow rates, and sand bed volumes used, the residence time was about 0.5-1.0 sec. A typical operation began by washing the sand in 10% HNO_3 and distilled water to remove impurities, such as iron, which may act as catalysts, and then calcined at 850° C for at least 12 hours to remove any sulfides and carbonates. The coal feed is then begun and pyrolysis products then exit the pyrolyser to a set of two cold traps fitted with cellulosic thimble filters maintained at 0° C. The outlet gas temperature after the first trap is 30-34° C. Much of the light char formed is entrained in the exit gas and carried into these traps, with most of it in the first trap.

Gases and tars are analyzed. A side stream after the first trap can be vented to a Panametrics hygrometer for measurement of water content. The glass surfaces of the traps are rinsed with methylene chloride to recover the condensed tars; the tars trapped in the cellulosic filters are recovered by Soxhlet extraction in methylene chloride for 24 hours, and subsequently analyzed in an HP5880A gas chromatograph. The final gases exiting the last trap are collected in any of nine 0.5 cm^3 stainless steel sampling loops for analysis in a Perkin-Elmer Sigma 3B gas chromatograph for sulfur gases, and in any of fifteen similar loops for analysis in a Sigma 1B GC for carbon oxides and hydrocarbon gases. The yields of sulfur gases were calculated as % Sulfur based on the original coal pyrolyzed.

H_2S has previously been shown to adsorb on the glass reactor and ancillary equipment ([1]). To avoid errors resulting from this, a very small amount of H_2S in N_2 (500 ppm at 0.5 l/min) is added to the N_2 reactor flow of 1.50-1.80 l/min for one to two hours prior to the actual pyrolysis experiment to allow the quartz reactor and the glass traps to equilibrate with the H_2S. GC analysis of the reactor effluent with the H_2S indicated the preferential adsorption of the H_2S on the reactor system. Similar analysis of the effluent after the H_2S conditioning was turned off did not show any H_2S desorption.

RESULTS AND DISCUSSION

Figure 2 shows the results of the pyrolysis experiments conducted with the Spanish lignite at 750-960°C at residence times of 0.52-0.72 sec. It is seen that under the pyrolysis conditions used, 60 - 70% of the sulfur in this coal appears in the gaseous products as H_2S, COS, and CS_2. As in the previous sulfur study ([1]), the principal sulfur gaseous product at all temperatures is H_2S, with some CS_2 formed at T >840°C. The CS_2 is apparently formed at the expense of the H_2S, by any of several reactions: H_2S may react with the carbon of the coal and/or the methane evolved in the pyrolysis of the coal to form CS_2. A small amount of COS is detected at all temperatures; trace amounts of SO_2 are also detected. Moreover, the total sulfur yield appears to reach a maximum about 900°C. The decrease in sulfur volatilization as pyrolysis temperature is increased above 900°C is attributed to sulfur retention in the char due to the reaction of H_2S with coke or char to form more stable thiophenic structures ([2]). GC/MS analysis of the tars (diluted to 10 ml) from the pyrolysis at 750 and 850°C did not reveal any sulfur-containing structures. Tars from the pyrolysis at 900 and 950°C, however, contain dibenzothiophene.

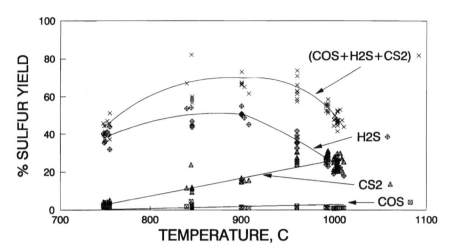

Figure 2.   Sulfur Yield From Spanish Lignite

An estimate of the distribution of organic sulfur types in this Spanish lignite may be made using the results of the sulfur model compound study performed previously (1). This assumes that all sulfur compounds obtained on pyrolysis at 775 - 900° C or lower arise from aliphatic mercaptan or sulfide, and that the sulfur compounds evolved over 900°C arise from aromatic, nonheterocyclic mercaptans or sulfides. Those organic sulfur compounds stable at temperatures up to 950°C are assumed to be heterocyclic, mainly thiophenic structures. Applying these criteria to the Mequinenza lignite, the following approximate organic sulfur distribution may be inferred: 67% aliphatic sulfides and/or mercaptan, essentially 0% aromatic sulfides and 33% heterocyclics (by difference). Since the aliphatic sulfur structures tend to convert to heterocyclic structures on thermal exposure, the aliphatic sulfur content reported may be lower than actual. This inferred distribution reinforces the trend noted in the previous sulfur study (1), i.e., low-rank coals have high proportions of aliphatic sulfur.
Figure 3 shows results from the pyrolysis of the New Zealand bituminous coal at 850, 900 and 930°C at residence times of 0.54-0.75 sec. The inferred approximate organic sulfur distribution is as follows: 26% aliphatic sulfides and mercaptans, 6% aromatic sulfides and mercaptans, and 68% of the more stable thiophenic sulfur structures. These results should be regarded as preliminary since this coal is difficult to feed because it is highly caking. Also, only a small sample was available, so the results shown represent only a few measurements compared to the Spanish lignite. Nevertheless, the results clearly indicate that this higher-rank coal has a lower proportion of labile sulfur structures, and presumably a higher proportion of heterocyclic sulfur than does the lower rank Spanish lignite. This is consistent with the view that coalification converts the organic sulfur components into the more stable heterocyclic form.
Until recently, there have been only a few reports of aliphatic sulfur structures in coal (1,12). These results together with the experiments of Gorbaty et al. (11), give further support for the presence of labile (presumably aliphatic) sulfur moieties in high-organic sulfur-containing coals. In the previous sulfur study (1), one bituminous coal was analyzed for organic sulfur forms both by the

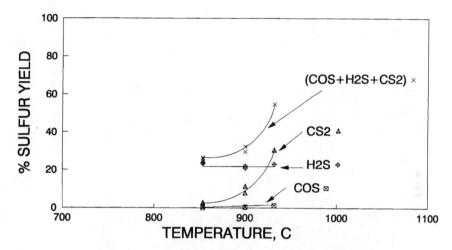

Figure 3.   Sulfur Yield From New Zealand Coal

flash pyrolysis technique described herein and by the Attar technique (2). The latter technique gave a much higher heterocyclic sulfur content than the flash pyrolysis technique.

Table II shows the estimated organic sulfur distribution for the two high-organic sulfur coals investigated in this study, along with the distributions for eight coals of various ranks investigated in a previous study (1). Heating value is shown as a rough measure of coal rank. It is seen that the present results fit in fairly well with the previous results for the coals where the pyritic sulfur concentration was more comparable to the organic sulfur concentration. This suggests that pyrites probably do not influence the formation of organic sulfur structures in coal.

SUMMARY

Comparison of the sulfur-containing products from the flash pyrolysis of a low- and a higher-rank high organic sulfur coal containing little or no pyrite indicates that the low rank coal contains a larger proportion of labile sulfur groups than the higher-rank coal. This is consistent with previous studies of sulfur-containing coals containing a larger proportion of pyritic sulfur. Model compound studies suggest that the labile sulfur structures may be aliphatic, thioethers, mercaptans or disulfides. The more-stable sulfur structures are mainly heterocyclic. This suggests that labile groups convert to the more stable heterocyclic sulfur structures as coalification proceeds. Since similar sulfur distributions were observed for high-pyrite containing coals, it appears probable that the pyrite does not influence the formation of the organic sulfur components of the coal.

Table II: Organic Sulfur Distribution in Various Coals[1]

Coal	Total Sulfur (wt %)	Organic Sulfur (wt %)	Aliphatic Sulfides & Mercaptans [3]	Aromatic Sulfides & Mercaptans	Thiophenic	Heating value (BTU/lb)	Pyrite (wt %)
			% of Organic Sulfur				
Alcoa Texas Lignite	1.30	0.73	82	<1	18	10,555	0.57
Spanish Lignite	*10.73*	*10.04*	*67*	*0*	*33*	*N. D.[2]*	*0.60*
Emery (Utah)	1.19	0.57	61	1-2	39	12,710	0.62
Pittsburgh 8 Shoemaker	4.00	1.36	53	<1	46	12,991	2.64
Pittsburgh 8 R & F	4.85	1.63	49	1-2	49	12,156	3.22
Pittsburgh 8 McElroy	5.47	1.58	49	<2	49	13,098	3.89
Ohio 9 Egypt Valley	2.49	1.46	44	<1	55	13,175	1.03
Illinois 6 Burning Star	3.24	2.04	39	2	59	12,165	1.20
New Zealand HVA	*5.55*	*5.46*	*26*	*6*	*68*	*14,260*	*0.06*
Anthracite Lehigh Valley	0.90	0.60	0	0	100	14,245	0.30

(1) Entries in italics are from this work; the remaining entries are from reference (1).
(2) Not determined.
(3) The labile (presumed to be aliphatic) portion should probably be considered a minimum, as the pyrolysis may convert some aliphatic to hetrocyclic sulfur.

ACKNOWLEDGMENTS

This research was sponsored in part by Amoco Oil Corporation. The research was also financed in part by the State of Delaware as authorized by the State budget Act of Fiscal Year 1988. We would like to acknowledge the help of Prof. Harold H. Schobert of the Pennsylvania State University for providing a sample of the Spanish lignite, and the help of Drs. Richard Markuszewski and Glenn A. Norton of Iowa State University for providing the high-sulfur New Zealand coal sample.

LITERATURE CITED

(1)   Calkins, W. H., *Energy and Fuels* **1987**, 1, 59-64.

(2)   Attar, A., DOE Report DOE,.PC/30145-T1 (DE84007770).

(3)   LaCount, R. B.; et. al., *New Approaches in Coal Chemistry, EDS;* ACS Symposium Series 169; American Chemical Society: Washington, DC, 1981; pp 415-426.

(4)   Huffman, G. P.; Huggins, F. E.; Mitra, S.; Shah, N; Pugmire, R. J.; Davis, B.; Lyttle, F. W.; Greegor, R. B., *Energy and Fuels* **1989**, 3, 200-205.

(5)   Yurovskii, A. Z., *Sulfur in Coal;* Institute of Mineral Fuels, Academy of Sciences of the USSR,: Moscow, Translated and Published for the U.S. Bureau of Mines, 1960.

(6)   Raymond, R., *Proc. Int. Coal Sci.* ,**198**, pp. 857-862.

(7)   Garcia-Suarez, A. B.; Schobert, H. H., *ACS Div. of Fuel Chem. Preprints,* **1988,** vol. 33, pp. 241-246.

(8)   Norton, G. A.; Mroch, D. R.; Chriswell, C. D.; Markuszewski, R., *Proceedings of the 2nd International Conference on Processing and Utilization of High Sulfur Coals,* 1987, pp. 213-223.

(9)   Ge, P.; Wert, C. A., this volume.

(10)   Krouse, H. R.; Yonge, C. J., this volume.

(11)   Gorbaty, M. L.; George, G. N., this volume.

(12)   Ignasiak, B.S.; Fryer, J. F.; Jadernik, P., *Fuel* 1978, 57, 578-584.

RECEIVED February 28, 1990

Chapter 18

Characterization of Organic Sulfur Compounds in Coals and Coal Macerals

Stephen R. Palmer[1], Edwin J. Hippo[2], Michael A. Kruge[1], and John C. Crelling[1]

[1]Department of Geology, Southern Illinois University, Carbondale, IL 62901
[2]Department of Mechanical Engineering and Energy Processes, Southern Illinois University, Carbondale, IL 62901

Peroxyacetic acid oxidation has been used to investigate the type and distribution of organic sulfur species in samples of vitrinite, sporinite and inertinite, separated from the Herrin No.6 and an Indiana No.5 coal seam. It was established that organic sulfur species were selectively preserved during oxidation and their analysis lead to some of the first sulfur-33 NMR spectra obtained for coal. The effects of maceral separation processes on model compounds were also studied. Results from our studies support the following conclusions: 1). Different macerals have different distributions and types of organic sulfur species. 2). Organic sulfur compounds in coal occur at the ends of macromolecular structures. 3). Maceral separation techniques do not affect organic sulfur species in coal. 4). Maceral separation is essential for the chemical characterization of coal. 5). GC-MS and sulfur-33 NMR data agree.

The combustion of sulfur-containing coals leads to the environmentally unacceptable problems associated with acid rain(1). Although current coal cleaning technologies can remove most of the inorganic sulfur from coal (2), no technology is presently in use for the effective removal of organic sulfur. The failure of organic desulfurization processes is due in part to a lack of information regarding the types of organic sulfur present in coal. The optimum approach therefore, would seem to be the initial characterization of the organic sulfur forms in coal followed by the design of appropriate desulfurization technologies.

There are methods for the direct determination of organic sulfur in coal (3,4), but details regarding individual molecular structures are much harder to obtain. Much of the research designed to obtain this information has concentrated on the analysis of coal extracts (5,6) and pyrolysis products (7,8). Unfortunately the sulfur species from pyrolysis processes may be highly modified and usually account for only a very small percentage of the total sulfur in the coal.

Extrapolating the data from these analyses to characterize the whole coal can be very misleading.

Other approaches have been used to characterize the organic sulfur in coal. Programmed temperature reduction (PTR) (9-14) as used by Attar is one such method as is programmed temperature oxidation (PTO) (15-17). Both methods rely on differences in reactivity of the different sulfur species. However, both procedures involve high temperatures and under such conditions transformation between sulfur species is possible. This reaction is highly probable if pyrite or elemental sulfur are present in the sample (18-20).

Further techniques for the determination of molecular structure of organic sulfur species in coal include Curie point pyrolysis (21) and X-ray absorption fine structure (XAFS) spectroscopy (22). In addition there are techniques which provide an overall view of the relationships between sulfur, metals and coal petrology. These include optical microscopy/electron probe microanalysis (10), X-ray photoelectron spectroscopy (XPS) (23) and scanning electron microscopy-energy dispersive X-ray analysis (SEM-EDAX) (10). Although the above mentioned methods provide some information regarding organic sulfur species in coal, and some advances have been made, a routine, low-temperature, unambiguous method for organic sulfur characterization in coal is yet to be found.

In this study we have investigated selective oxidation as a potential organic sulfur characterization approach. In particular we have used the peroxyacetic acid oxidation procedure. Although this selective oxidant has received some attention in the study of lignin (24) and humic acid structures (25), its application to the study of coal has been limited to only a few instances (26,27) with very little information about organic sulfur species being reported.

Peroxyacetic acid oxidation is similar to the peroxytrifluoroacetic acid (Deno) oxidation (28). These peroxide systems are reported to selectively oxidize the aromatic portions of molecules while leaving aliphatic portions intact (29). Peroxyacetic acid will oxidize aromatic units to phenolic units via hydroxylation. These phenolic moieties will oxidize rapidly to ortho and para quinones, the latter of which are unstable are undergo ring fission to form diene carboxylic acids (30).

The selective oxidation of aromatic portions of molecules was demonstrated using model compounds such as toluene, ethylbenzene, n-propylbenzene and iso-propylbenzene (28). The major oxidation products from these compounds were acetic, propionic, butyric and isobutyric acids respectively. In each case the carboxylic acid group marks the position that the aromatic unit used to occupy.

Although the selective oxidation of coal has been extensively studied (31-33), surprisingly little has been reported about sulfur species in the oxidation products. Even less is known about the distribution of organic sulfur species between different coal macerals despite the fact that this information is important for the development of any future desulfurization technology.

In view of these shortcomings we have combined the need to characterize organic forms of sulfur with the recent progress obtained in the separation of coal into its single maceral fractions (34). This affords an opportunity to compare the sulfur chemistries of individual macerals with that of their parent coals.

Experimental

Sample Preparation.   Herrin No.6 (Illinois No.6) and Indiana No.5
(Illinois No.5) coals obtained from the Illinois Geological Survey
Sample Bank were used in this study. The whole coals were split into
four fractions each of which was placed in a sealed, 5-gallon drum.
The fourth fraction was ground to minus 200 mesh and then introduced
into a nitrogen gas powered (100 psi) Sturtevant fluid energy mill.
In this device the coal particle size is reduced to the micron level
by impaction between coal particles themselves and with the
impaction chamber walls.   Proximate and elemental data for these
micronized coals are reported in Table I.

Table I. Proximate and Elemental Data

	Herrin No.6	Indiana No.5
Moisture	14.7	10.4
Volatile Matter	40.5	39.6
Fixed Carbon	49.5	51.4
H-T Ash	10.5	9.0
Carbon	69.20	71.73
Hydrogen	5.12	4.85
Nitrogen	1.28	1.71
Oxygen	9.49	8.93
Sulfate Sulfur	0.05	0.01
Pyritic Sulfur	1.27	1.85
Organic Sulfur	3.00	1.91
Total Sulfur	4.32	3.76
Total Chlorine	0.12	0.02

All percentages are reported on a moisture free basis except for
moisture.

Aliquots of the micronized coal were treated with HCl and HF to
remove carbonate and silicate mineral matter, and aliquots of the
micronized, acid treated coal were floated in a 1.67 specific gravity
solution of CsCl to eliminate pyrite. Some of the floated samples
were then separated by density gradient centrifugation (DGC) into
sporinite, vitrinite and inertinite maceral fractions (34).
      Soxhlet extractions were performed on each of the coal and
maceral samples. The micronized, acid treated and floated samples
were extracted successively with hexane, toluene and finally THF. The
maceral fractions were extracted with THF only. All extracts and
extraction residues were isolated, weighed and the distribution of
sulfur between them established. In addition, the extracts were
examined using FTIR, proton and carbon-13 NMR spectroscopy, GLC-
FID/FPD and GC-MS.

Determination of the Effects of Separation Processes on Organic Sulfur Forms in Model Compounds. Three substituted dibenzothiophenes were subjected to the coal preparation and maceral separation processes. The model compounds used are shown below.

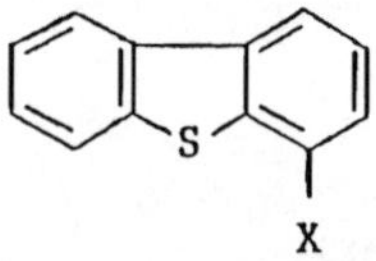

1. $X = CH_2SH$

2. $X = SCH_3$

3. $X = SC_6H_5CH_3$

Each compound was soaked for 1 hr in 38% (wt/v) HCl followed by soaking in 55% (wt/v) HF for 1 hr. Each was then rinsed exhaustively with water and dried under vacuum at room temperature. In a separate experiment each of the model compounds were exposed to a solution of cesium chloride. After sufficient exposure each compound was recovered and washed with copious quantities of distilled water. The percentage recovery, the melting point and spectroscopic data such as FTIR , proton and carbon-13 NMR were recorded.

Mild Oxidation of Coal and Macerals. Two grams of extracted coal was placed in a 3-necked 250mL r.b.flask. Fifty mL of glacial acetic acid was then added followed by 20 mL of 30% (wt/v) hydrogen peroxide. The flask was then heated under reflux and after 1 hr a further 40 mL of hydrogen peroxide was added dropwise so as to maintain the reaction. The reaction was allowed to reflux for a total of 24 h. Maceral oxidations were performed on a 0.5g scale. Other oxidations were performed using 5g of coal with, in some cases, only 40 mL of hydrogen peroxide. These experiments were conducted to generate samples for total sulfur analysis and to study the partial oxidation, dissolution and desulfurization of coal. Aliquots of the soluble oxidation products were retained for analysis including total sulfur, FTIR and NMR analysis. Another aliquot was methylated using the diazomethane method prior to GC-MS and GLC-FID/FPD analysis.

Instrumentation. Proton, carbon-13 and sulfur-33 nuclei were observed at 300, 75 and 23 MHz respectively, using a Varian VXR 300 MHz NMR spectrometer. Proton spectra were recorded using a pulse width of 12us and a pulse delay of 20s. Typically 10 transients were obtained. Carbon-13 spectra were obtained using a pulse width of 4.3us and a pulse delay of 5s. The number of transients recorded varied from 500 - 10,000 depending on sample concentration. Sulfur-33 spectra were obtained using a pulse width of 65us and a pulse delay of 5ms. Typically between 400,000 and 500,000 transients were collected. Extracts were dissolved in either deuterated chloroform or deuterated THF and chemical shifts measured against internal TMS. Oxidation products were dissolved in acetone/deuterium oxide (9:1) and chemical shifts measured against internal TMS for proton and carbon-13 spectra and external ammonium sulfate for sulfur-33 spectra. GC-MS analysis was performed on a Hewlett Packard 5970B MSD fitted with a 30m OV-1 column. Two temperature programs were used. Firstly the oven temperature was ramped from 100 - 300°C at

3°C/min and secondly a ramp from 100 - 300°C at 20°C/min was used. Injector and detector temperatures were maintained at 270°C and 300°C respectively. GLC-FID/FPD analysis was performed on a Varian 3400 gas chromatograph using the same chromatographic conditions as used in the GC-MS runs. FTIR spectroscopy was performed on a Nicolet instrument using the thin film on sodium chloride plates technique for liquids and the potassium bromide pellet technique for solids.

Results and Discussion

Treatment of Model Compounds. As anticipated, each of the sulfur-containing model compounds were recovered quantitatively after their exposure to the micronization, acid treatment and floatation processes used in coal and maceral preparation. In all cases the proton and carbon-13 NMR spectra, the FTIR spectra and the melting points of the recovered materials matched those of the starting materials, indicating that these sulfur species remain unchanged during the processing.

Solvent Extraction. The yields of extracts and extraction residues obtained by successive hexane, toluene and THF extraction of the floated coal samples, plus the percent of the total of the organic sulfur that each extract contained are shown in Table II.

Table II. Extraction yields and sulfur distributions

	Herrin No.6		Indiana No.5	
Extract Type	Yield Wt%	%Tot.O.S.	Yield Wt%	%Tot.O.S.
Hexane	0.5	2.0	0.6	6.6
Toluene	2.5	2.9	1.2	4.6
THF	7.1	4.2	14.9	14.2
Extract Residue	89.9	90.9	83.3	74.6

It is clear from Table II that most of the organic sulfur remains in the insoluble coal matrix and cannot be extracted. The Indiana No.5 coal has a higher quantity of extractable organic sulfur than the Herrin No.6.

Both the extractability and the organic sulfur distribution varies between the two coals. It is interesting that the hexane extracts from both coals and the toluene extract of the Indiana No 5. coal, contain a disproportionately high percentage of the total organic sulfur. Thus, there is organic sulfur enrichment in these extracts and a consequential depletion of the organic sulfur in the extraction residues. This would appear to be good criteria for a desulfurization process but   the low extractabilities obtained with these solvents prevents this from being a viable process.

Characterization of Sulfur Compounds in Extracts.   Each of the extracts and extraction residues were examined by FTIR spectroscopy. The hexane and toluene extracts gave spectra dominated by aliphatic features while the THF extracts had more aromatic and polar functional group characteristics. The extraction residues gave FTIR spectra dominated by aromatic and polar functional group characteristics with minor aliphatic features. All spectra contained peaks that could have arisen from organosulfur groups but the regions occupied by these peaks overlap with those occupied by other non-sulfur functional groups. Hence assignment of sulfur structures was not possible. This demonstrates the futility of trying to identify a minor constituent of a complex substance such as coal using non-selective techniques.

The same type of information was obtained from proton and carbon-13 NMR. Once again there are absorptions that may be due to carbon and hydrogen bonded to sulfur but contributions from other non-sulfur containing structures are highly likely and hence a firm assignment cannot be made.

It is obvious from the FTIR and NMR analyses of these extracts that in order to positively identify organosulfur structures we need an analytical technique that is sulfur selective. That is, a technique that responds to sulfur uniquely. One such technique, applicable to the problem in hand, is GLC-FID/FPD where the flame photometric detector is set in the sulfur selective mode.

Aliquots of hexane, toluene and THF extracts were mixed together proportionately and the resulting combined extract separated by open-column liquid chromatography to give a saturate, an aromatic, a polar 1 and a polar 2 fraction. The saturate fractions from both coals did not contain any sulfur compounds detectable in our GC-FID/FPD system. However, the aromatic fractions from both coals contained a complex series of sulfur compounds. Subsequent GC-MS analysis of these fractions using the selected ion mode (SIM) identified these compounds as a series of benzothiophenes having one, two or three methyl group substituents and a series of dibenzothiophenes with one through four methyl group substituents. Dibenzothiophene itself was also detected but benzothiophene was not. Despite the inherent peculiarities of the FPD detector (non-linearality, compound dependance and quenching), comparison of the FPD traces with the composite SIM traces afforded a very good correlation, illustrating the fact that the composite SIM method did not miss major sulfur compounds.

GC analysis of underivatized polar fractions did not reveal any volatile sulfur compounds. However, once these fractions were methylated with diazomethane, a number of sulfur compounds were detected. (Presumably, the diazomethane methylated either carboxylic acid, phenolic, thiophenolic, sulfonic acid or even alcohol or thiol groups and thereby increased their parent molecules volatility). These additional sulfur compounds are currently under investigation in our laboratories and the results of these studies will be reported later.

Comparison of the extracts taken from the isolated macerals show them all to be very similar, not only to each other , but to the extracts taken from the unfractionated coal. This indicates that the composition of the extract does not vary with maceral type and suggests the extract is probably free to migrate throughout the coal matrix and disperse evenly among the various coal components.

Oxidation of Coal and Coal Macerals. Extraction residues were oxidized as outlined in the experimental section. Two oxidation procedures were adopted:- i). excess oxidant oxidation and ii). oxidant starved oxidation (partial oxidation).

i) Excess oxidant oxidation:- Under the conditions employed the extraction residues were oxidized to soluble products leaving very little residual matter. Typically the percentage of the coal dissolved was around 90-95% although some figures as high as 99% were recorded. The exceptions to this were the sporinite and inertinite macerals, both of which were more resistant to this oxidative dissolution, with only 50-60% dissolving. This no doubt reflects differences in the properties and chemical structures of these macerals. It is also important to note that the yields of oxidation products were higher for floated coal samples vs the micronized and the acid treated samples. This suggests that minerals influence the oxidation of the organic portion of the coal, perhaps catalyzing the production of carbon dioxide.

In the larger scale reaction (5g sample size) it was established that the soluble oxidation products had considerably enhanced sulfur contents when compared with the unoxidized samples. This is explained by the oxidation of carbon in the coals to carbon dioxide and the subsequent concentration of the organic sulfur that remained. It was also established that virtually all of the organic sulfur in the extracted and unoxidized samples could be accounted for by the organic sulfur found in the soluble oxidation products, indicating that very little was lost during the reaction and work-up procedures.

It was a concern to us that some of the organic sulfur may have been oxidized to sulfate. To check for this we tested the soluble oxidation products for sulfate using barium chloride solution. Both the micronized and the acid treated coal samples tested positive for sulfate. This is not surprising since each contains pyrite that would be oxidized to sulfate during the reaction. However, the floated coal samples and the macerals (which have very little residual pyrite <0.1%) tested negative for sulfate indicating that 'over-oxidation' of organic sulfur to sulfate did not occur. This preservation and indeed concentration of the organic sulfur is due to the nature of the oxidant used. It is reported that under strong acidic conditions the peroxyacetic acid becomes protonated and then dissociates to form the hydroxyl cation (35). The hydroxyl cation is a strong electrophile and as such attacks centers of negative charge. For this reason structures bearing a high electron density, for example S, N, O and some aromatic units will be attacked faster than others. In the case of oxidation at sulfur , sulfoxides, sulfones and sulfonic acids would result. These oxidized sulfur forms are strong electron withdrawing groups and would deactivate their parent structures towards further electrophilic attack by the hydroxyl cation. This is our main reason for using peroxyacetic acid for the oxidation of the coal samples. The preservation and concentration of the organic sulfur compounds is essential for their eventual identification by chromatographic and spectroscopic techniques.

ii). Oxidant starved oxidation:- In order to study the early stages in the dissolution and desulfurization of coal using the peroxyacetic acid technique, a sample of the extraction residue from the Herrin No. 6 floated coal was oxidized using insufficient oxidant to dissolve the whole sample. After consumption of the oxidant the products were separated into a soluble fraction and an insoluble fraction by filtration. The insoluble fraction was then extracted with dilute aqueous sodium hydroxide to give a humic acid fraction and a final insoluble residue. The yields of each fraction and their sulfur contents are given in Table III.

Table III. Sulfur contents of the reaction products from the oxidant starved oxidation of Herrin No.6 floated coal

Product fraction	Yield	%Yield	Total S (Wt%)	% of Total S
Soluble products	2.72g	52.0	5.54	86
Humic acids	1.50g	28.7	0.63	6
Insoluble residue	1.00g	19.3	1.33	8

As we can see from Table III, 86% of the sulfur contained in all of the oxidation products resides in the soluble oxidation products even though this product fraction only represents 52% by weight of the combined oxidation products. This indicates that organic sulfur compounds have been preferentially depolymerized from the coal matrix and that this oxidative procedure may be a possible desulfurization process.

To study the desulfurization of the coal as a function of its dissolution in more detail we devised another series of reactions where one gram portions of the extraction residue were reacted with the oxidant for periods of time ranging from 5 minutes to 24 hours. Each oxidation product was then filtered, the quantities of soluble and undissolved coal measured and the amount of sulfur remaining undissolved determined.

It was found that by dissolving only 4% of the coal extraction residue we can remove 25% of its organic sulfur. Similarly by dissolving 20% of the coal extraction residue we can remove 50% of its organic sulfur. It should be stressed that the coal used in this set of reactions had been floated and had no or very little pyrite and therefore the sulfur that was being dissolved must have been organic. The fact that preferential dissolution of sulfur compounds occurs in the early stages of oxidation suggests that a significant proportion of the organic sulfur is more reactive towards the oxidant than the bulk of the coal. This enhanced reactivity of some sulfur units over others may be due to the chemical nature of the sulfur functionalities concerned, for example, disulfides would oxidize and cleave easily under the conditions employed, or it may simply be due to a disproportionately high concentration of sulfur species on the surface of coal particles. These surface organosulfur units would naturally dissolve first because they are in immediate contact with the oxidant.

FTIR and NMR Analysis of Oxidation Products. As expected the FTIR spectra of the soluble oxidation products are dominated by the strong absorptions due to the hydroxyl and carbonyl groups of carboxylic acids. There are absorptions in the spectra which occur in the regions expected for sulfones (1350-1310 and 1160-1120 cm-1) and sulfonic acids (1420-1330 and 1200-1145 cm-1), but these are probably due to non-sulfur containing functional groups, especially carbon-oxygen bonds which are presumably present in much higher concentrations.

In addition each soluble oxidation product was examined by proton and carbon-13 NMR. Each spectra can be split up into three regions; an aliphatic region, an aromatic region and a carbonyl region. If we compare the NMR spectra for a THF extract before and after oxidation we can see that the oxidized sample is much less aromatic than before oxidation. This would suggest that the peroxyacetic acid oxidation is similar to that of the peroxytrifluoroaectic acid or "Deno" oxidation (28,29) in which preferential preservation of aliphatic units is observed.

As with the NMR spectra taken of the unoxidized extracts there are resonances that can be assigned to organic sulfur compounds. However, these resonances are probably due to the more abundant non-sulfur containing species which absorb in the same regions. Thus it is very difficult to characterize organosulfur compounds using proton and carbon-13 NMR alone. This is why the sulfur selective technique sulfur-33 NMR has such great potential for the characterization of organosulfur groups in coal.

Sulfur-33 NMR.   Although sulfur-33 NMR spectroscopy is non-destructive, low temperature, sulfur selective and examines the sulfur atoms directly, there are a number of problems that make routine sulfur-33 NMR of coal very difficult. To begin with the active isotope, sulfur-33, has a low natural abundance (only 0.75%). This coupled with its relatively low sensitivity makes the NMR signals hard to detect. In addition, the sulfur-33 nucleus has a quadrupole moment. This leads to considerable line broadening especially when there is an unsymmetrical electric field gradient surrounding the nucleus. Other problems that are encountered result from the low concentration of sulfur in the coal and from coals inherent insolubility. Fortunately some of these problems can be overcome. For instance, certain oxidation techniques can render coal soluble and at the same time convert its sulfur species to various sulfones and sulfonic acids which have relatively symmetrical electric field gradients about the sulfur atom. Also, since quadrupolar nuclei relax very fast, many of the problems associated with low sensitivity can be overcome in theory, by simply using rapid pulse repetition rates and long accumulation times. However, sulfur NMR still has a long way to come before it can be routinely applied to structural determinations.

Many of the oxidation products analysed by sulfur-33 NMR gave poor signal to noise ratios and little information was obtained. However the oxidation products derived from the floated Herrin No.6 coal did produce organic sulfur signals at -6.8 ppm relative to external aqueous ammonium sulfate (See Figure 1). Since this coal sample had been extracted with THF and floated prior to oxidation, the sulfur signal observed could not be due to sulfate derived from oxidized pyrite nor could it be from oxidized organosulfur compounds present in the extractable component of this coal. This peak must therefore come from organosulfur groups attached to the THF insoluble matrix of coal that is soublized upon oxidation. The peak at -6.8 ppm can be attributed to dialkyl sulfones, aryl sulfones, aryl alkyl sulfones, dibenzothiophene sulfones and sulfonic acids. We believe this peak is due to sulfonic acids since these have been detected as prominent sulfur compounds in the oxidation products by GLC-FID/FPD and GC-MS analysis (See next section). Although this peak can be attributed to a number of structures it must be pointed out that sulfur NMR is in its infancy and that this spectra represents a significant advancement in sulfur NMR of coal products. The potential of sulfur NMR in the analysis of sulfur in coals and other fossil fuels was realized by Retcofsky (36,37). His pioneering work on model compounds led to some of the first sulfur NMR spectra of organic sulfur species. Subsequently the sulfur NMR spectra of many organic compounds, predominantly sulfones, have been obtained (38-40). More recently some sulfur NMR spectra of fuel related products have been published (41,42). It should be pointed out that the spectrum presented in Figure 1 is derived from a coal product that contains 90% of the total organic sulfur in the original coal. (The remaining 10% of the organic sulfur was removed by solvent extraction prior to oxidation). The lines or bars associated with each sulfur compound in Figure 1 represent the chemical shift range over which those sulfur nuclei are reported to resonate (38-40). The broken lines represent chemical shift ranges that are at present uncertain because in the authors opinion insufficient numbers of model compounds of that type have been studied.

GLC-FID/FPD and GC-MS Analysis of Oxidation Products.   To obtain information regarding the molecular structure of the organic sulfur compounds present in samples, each of the soluble oxidation products were methylated using the diazomethane method and then analysed, initially by both dual FID/FPD gas chromatography, and by GC-MS. Representative FID and FPD traces obtained from the Herrin No.6 samples are shown in Figures 2 and 3 respectively, while those for the Indiana No.5 samples are shown in Figures 4 and 5 respectively.

It is clear from these FID traces that significant differences exist between all of the samples analysed. Not only do the different macerals from the same coal exhibit differences, but so do similar macerals from different coals. Thus, not all vitrinites are the same, not all sporinites are the same and not all inertinites are the same. This is especially clear when we compare the two inertinite FID traces. The Indiana No.5 inertinite contains significant amounts of both low and high retention time products which are absent in the Herrin No.6 inertinite. Although the absence of the lower boiling constituents may be due to their evaporation in the work-up procedures, we believe this not to be the case since an identical work-up procedure was used for all samples.

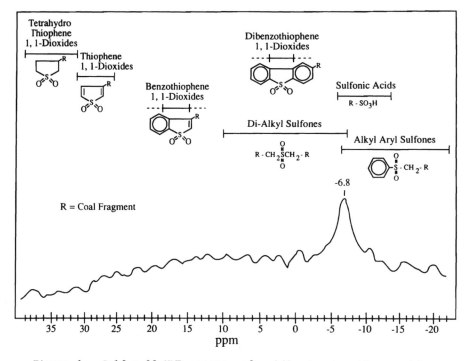

Figure 1.   Sulfur-33 NMR spectra of oxidized extraction residue from floated Herrin No.6 coal.

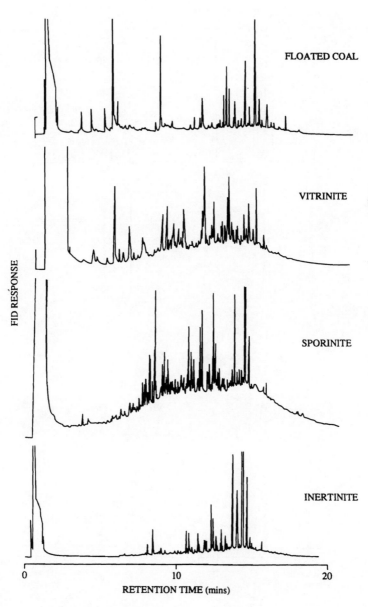

Figure 2. GLC-FID chromatograms of oxidized Herrin No.6 floated coal and maceral samples.

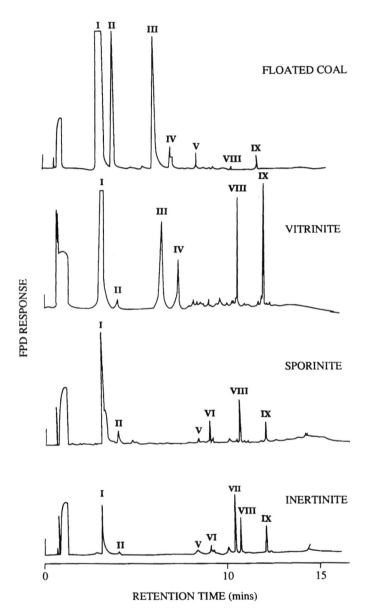

Figure 3. GLC-FPD chromatograms of oxidized Herrin No.6 floated coal and maceral samples.

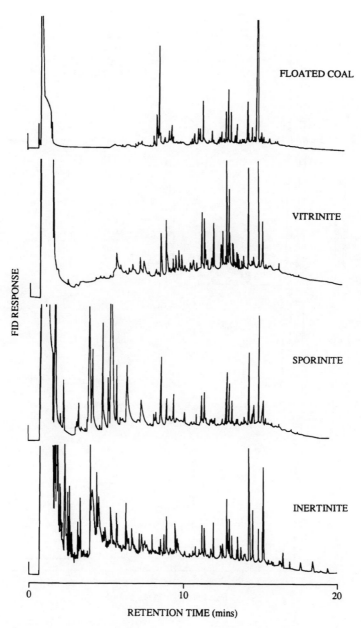

Figure 4.   GLC-FID chromatograms of oxidized Indiana No.5 floated coal and maceral samples.

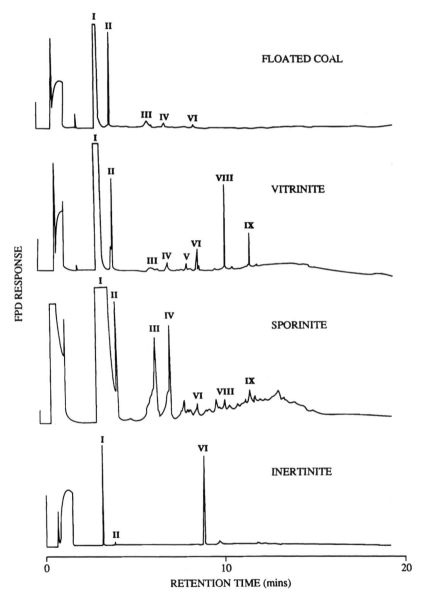

Figure 5.   GLC-FPD chromatograms of oxidized Indiana No.5 floated coal and maceral samples.

Although many of the oxidation products are common to all of the samples analysed, their distribution varies considerably from sample to sample. In addition, there are some oxidation products that appear exclusively in the FID traces of some samples. For instance, there are compounds in sporinite and inertinite samples which do not appear in the FID trace obtained for their parent floated coal. The absence of these compounds in the FID traces of the floated coals is explained by the presence of the more abundant maceral vitrinite, the oxidation products of which either swamp or dilute those from the lesser macerals, making their detection very difficult. Here we see how maceral separation is important for the characterization, not only of the individual macerals themselves, but of the whole coal. Observation of sulfur constituents that are unique to minor macerals components may be difficult to detect during the analysis of a whole coal, but are easily observed during analysis of individual macerals.

The same conclusions can be made by looking at the FPD traces. Once again we see that there are considerable differences between all of the samples analysed. It is clear that a significant portion of the total peak area in the FPD chromatograms is attributable to only nine major sulfur-bearing components. The sulfur chemistry of these samples therefore, as revealed by peroxyacetic acid oxidation, appears to be much less complex than extract and pyrolysis data may lead us to believe. The nine significant sulfur species have been detected in a variety of oxidation products derived from these samples. This includes the oxidation products derived from the partial oxidation of these samples. In no instances were any additional sulfur compounds detected. In other experiments oxidation products were separated by liquid chromatography into GC volatile and GC involatile fractions. The involatile fractions were then re-oxidized and the resulting volatile products analysed for additional sulfur compounds. None were found. The FPD traces that were obtained were very similar to those obtained for the original volatile oxidation products, indicating that the material initially too involatile to pass through the GC column was simply an undegraded version of that which could.

Once again we see sulfur compounds present in some macerals but not in others, and indeed sulfur compounds in macerals which appear to be absent in the floated (unfractionated) coal. For instance, sulfur compounds III and IV contribute significantly to the FPD traces of oxidized Herrin No.6 floated and vitrinite samples and the Indiana No.5 sporinite sample, but are very weak or apparently absent in the remaining samples. Also sulfur compound number VII appears in the Herrin No.6 inertinite sample but no other. We attribute the fact that certain sulfur compounds appear in the sporinite and inertinite chromatograms but not in the chromatograms of their parent floated coals to the presence of a large excess of vitrinite in the floated coal samples.

Without retention time data for authentic standards, all that GLC-FID/FPD analysis can tell us is the number and distribution of the sulfur compounds in the various samples. It cannot tell us what the individual sulfur compounds are. To obtained this information we turn to GC-MS analysis.

Employing exactly the same chromatographic conditions each sample was analysed using GC-MS instrumentation. Using the retention time information obtained from the dual FID/FPD gas chromatography, most of the sulfur compounds could be located on the GC-MS TIC and their mass spectra were obtained. Based on the molecular weight and fragmentation pattern information that was forthcoming, structural assignments for sulfur compounds I through IX were made where possible. These are shown in Table IV. The assignment of structure to compounds I, II and IV was verified using authentic compounds. No authentic compounds VIII and IX were available and consequently their structural assignment must be considered tentative. The presence of the M+2 peak from the sulfur-34 isotope in the mass spectra, helped to confirm the fact that sulfur was present in these molecules.

Table IV. Sulfur compounds identified in oxidation products by GC-MS and GLC-FID/FPD analysis

Sulfur compound	Identification
I	Methylsulfonic acid
II	Ethylsulfonic acid
III	Unknown
IV	Benzenesulfonic acid
V	Unknown
VI	Unknown
VII	Unknown
VIII	Carboxytrimethylbenzene sulfonic acid (T)
IX	Carboxytrimethyl dibenzothiophene-1,1-dioxide (T)

T = tentative assignment

Sulfur compounds I, II and IV, methylsulfonic acid, ethylsulfonic acid and benzenesulfonic acid could have been derived from a number of precursors in the parent coal samples. Firstly, mercaptans and thiols will form sulfonic acids when oxidized with peroxides. However, the presence of methylmercaptan, ethylmercaptan and benzenethiol in an exhaustively extracted coal sample is highly unlikely and we believe that the sulfonic acids did not arise from these compounds. In addition, there is a possibility that these sulfonic acids may have come from pendant or terminal thioether groups.

A more feasible route for the formation of these sulfonic acids would be via the oxidation of disulfides. Under peroxide oxidation the disulfide bond is ruptured giving rise to two sulfonic acid groups. Thus the presence of methyldisulfide, ethyldisulfide and benzenedisulfide units in these coals is suggested. The presence of disulfides may explain the desulfurization results given earlier where 25% of the organic sulfur can be removed by dissolving only 4% of the coal.

Sulfur compound VIII, a carboxytrimethylbenzene sulfonic acid, could have come from an aryl disulfide, a thiol or could be derived from the further oxidation of compound IX, a carboxytrimethyldibenzothiophene-1,1-dioxide. This latter possibility is indicated by the lower concentration of compound IX relative to compound VIII in the oxidation products of coals containing mineral matter. Once again the catalytic effect of the mineral component of coal is indicated.

Another significant point to note about these sulfur compounds is that they all , with the exception of compound VIII, only have one acidic group (whether it be carboxylic or sulfonic). Since the acid group marks the position at which the molecule was bonded into the coal structure, we can conclude that these compounds were only bonded into the coal matrix by one bond, and hence the sulfur compounds must occur as terminal or pendant groups on macromolecular structures. If a sulfur compound was in the middle of a chain or an aromatic cluster for instance, then they would exhibit two, three or more acidic groups depending on their degree of bonding to the chain or cluster. Many of the oxidation products exhibit these features but none of them contain sulfur. These other compounds include short chain dicarboxylic acids and a number of di-, tri-, tetra- and penta carboxylic acids of benzene. Methyl and hydroxyl derivatives of these benzene carboxylic acids have also been identified.

Several of the sulfur compounds remain to be identified. When this is accomplished we will be able to investigate the terminal sulfur group theory in more detail.

Conclusions

The results of this study support the following conclusions:

1.   Coal preparation processes such as micronization, acid treatment, floatation and maceral separation do not affect a series of representative sulfur-containing model compounds. We believe this conclusion can be extrapolated to include those sulfur structures present in coal.

2.   Different macerals have different distributions and types of organosulfur compounds. The sulfur chemistry varies not only between macerals from the same coal, but between similar macerals from different coals.

3.   Only through maceral separation can some of the sulfur compounds be detected. Maceral separation is therefore to be considered essential not only for the characterization of individual macerals but the characterization of whole coals as well.

4.   There are only a limited number of major sulfur structures present in the oxidation products of the extraction residue of coal. These are dominated by sulfonic acids.

5. The majority of the sulfur compounds detected to date are bonded to the coal matrix by only one bond. We believe these sulfur compounds are terminal or pendant groups on macromolecular coal structures.

6. Sulfur NMR spectra have been obtained that represent the organic sulfur derived from the extraction residue of coal. This NMR data agrees very well with the GC-MS data.

Acknowledgments

This work was supported in part by grants from the Illinois Coal Development Board through the Center for Research on Sulfur in Coal, the U.S. Department of Energy through the Coal Technology Laboratory at Southern Illinois University, and the Electric Power Research Institute.

Literature Cited

1. Record, F. A.; Bebenick, D. V.; Kindya, R. J. Acid Rain Information Book; Noyes Data Corp.: Park Ridge, NJ, 1982.
2. Meyers, R. A.; Hart, W. D.; McClanathan, L. C. Coal Process Technol 1981, 7, 89-93.
3. McGowan, C. W.; Markuszewski, R. Fuel 1988, 67, 1091-95.
4. McGowan, C. W.; Markuszewski, R. Fuel Process. Technol. 1987, 17, 29.
5. White, C. M.; Douglas, L. J.; Perry, M. B.; Schmidt, C. E. Energy & Fuels 1987, 1, 222-26.
6. Nishioka, M. Energy & Fuels 1988, 2, 214-219.
7. Calkins, W. H. Energy & Fuels 1987, 1, 59.
8. Philp, R. P.; Bakel, A. Energy & Fuels 1988, 2, 59-64.
9. Attar, A. In Analytical Methods for Coal and Coal Products; Karr, C., Ed.; Academic Press: New York, 1979.
10. Boudou, J. P.; Boulegue, J.; Malechjaux, L.; Nip, M.; de Leeuw, J. W.; Boon, J. J. Fuel 1987, 66, 1558.
11. Attar, A. Fuel 1978, 57, 201.
12. Messenger, L.; Attar, A. Fuel 1979, 58, 655.
13. Attar, A.; Dupuis, F. In Coal Science. Gorbaty, M. L.; Ouchi, K., Eds.; ACS Symposium Series No. 192; American Chemical Society: Washington DC 1981; p239.
14. Attar, A.; Hendrickson, G. T. In Coal Structure; Meyers, R. A., Ed.; Academic Press, N.Y., 1984; Chapter 5.
15. LaCount, R. B.; Anderson, R. R.; Friedman, S.; Blaustein, B. D. Fuel 1987, 66, 909-913.
16. LaCount, R. B.; Anderson, R. R.; Friedman, S.; Blaustein, B. D. Am. Chem. Soc. Div. Fuel Chem., Prepr. 1986, 31, 70.
17. LaCount, R. B.; Gapen, D. K.; King, W. P.; Dell, D. A; Simpson, F. W.; Helms, C. A. In New Approaches in Coal Chemistry; ACS Symposium Series No. 169; American Chemical Society: Washington,DC, 1981; p415.
18. White, C. M.; Douglas, L. J.; Schmidt, C. E.; Hackett, M. Energy & Fuels 1988, 2, 220-223.
19. DeRoo, J.; Hogdson, G. W. Chem. Geol. 1978, 22 71-78.

20. Przewocki, K.; Malinski, E.; Szafranek, J. Chem. Geol. 1984, 47, 347-360.
21. Tromp, P. J. J.; Moulijn, J. A.; Boon, J. J. Fuel 1986, 65, 960.
22. Huffman, G. P.; Huggins, F. E.; Mitra, S.; Shah, N.; Pugmire, R. J.; Davis, B.; Lytle, F. W.; Greegor, R. B. Energy & Fuels 1989, 3, 200-205.
23. Dutta, S. N.; Dowerah, D.; Frost, D. C. Fuel 1983, 62, 840.
24. Chang, H. M.; Allan, G. G. In Lignins; Sarkanen, K. V and Ludwig, C. H., Eds.; Wiley: New York, 1971, p433.
25. Schnitzer, M.; Skinner, S. I. M. Can. J. Chem. 1974, 52, 1072.
26. Hayatsu, R.; Winans, R. E.; Scott, R. G.; Moore, L. P.; Studier, M. H. Fuel 1978, 57, 541.
27. Hayatsu, R.; Winans, R. E.; Scott, R. G.; Moore, L. P.; Studier, M. H. In Organic Chemistry of Coal; Larsen, J. W., Ed.; ACS Symposium Series No. 71; American Chemical Society: Washington, DC, 1978, p108.
28. Deno, N. C.; Greigger, B. A.; Stroud, S. G. Fuel 1978, 57, 455.
29. Deno, N. C.; Greigger, B. A.; Rakitsky, W. G.; Smith, K. A.; Minard, R. D. Fuel 1980, 59, 699.
30. Plesnicar, B. In Oxidation in Organic Chemistry. Part C; Trahanovsky, W. S., Ed.; Academic Press: New York, 1978, Chapter III.
31. Hayatsu, R.; Scott, R. G.; Winans, R. E. In Oxidation in Organic Chemistry. Part D; Trahanovsky, W. S., Ed.; Academic Press: New York, 1982, Chapter IV, 279-354.
32. Lowry, H.H. Chemistry of Coal Utilization. Suppl. Vol.; Wiley: New York, 1963.
33. Bearse, A. E.; Cox, J. L.; Hillman, M. Production of Chemicals by Oxidation of Coal. Battelle Energy Program Rep., Columbus, Ohio, 1975.
34. Dyrkacz, G. R.; Horwitz, E. P. Fuel 1982, 61, 3-12.
35. Derbyshire, D. H. Nature 1950, 165, 401-2.
36. Retcofsky, H. L.; Friedel, R. A. Applied Spectroscopy 1970, 24, 379-80.
37. Retcofsky, H. L.; Friedel, R. A. J. Am. Chem. Soc. 1972, 94, 6579-84.
38. Faure, R.; Vincent, E. J.; Ruiz, J. M.; Lena, L. Org. Mag. Res. 1981, 15, 401-3.
39. Harris, D. L.; Evans, S. A. J. Org. Chem. 1982, 47, 3355.
40. Annunziata, R.; Barbarella, G. Org. Mag. Res. 1984, 22, 250-4.
41. Ngassoum, M. B.; Faure, R.; Ruiz, J. M.; Lena, L.; Vincent, E. J.; Neff, B. Fuel 1986, 65, 142.
42. McIntyre, D. D.; Strausz, O. P. Magnetic Resonance in Chemistry 1987, 25, 36.

RECEIVED February 28, 1990

Chapter 19

Spatial Variation of Organic Sulfur in Coal

Elizabeth Ge[1] and Charles Wert

Department of Materials Science and Engineering, University of Illinois at Urbana–Champaign, Urbana, IL 61801

Transmission electron microscopy has been used to determine the concentration of organic sulfur in coal. Because the electron beam can be focused to a fine spot on the coal specimen, the variation of organic sulfur in the macerals can be measured over distances as short as 1 μm or less. Thus, spatial variation can be determined within a particular maceral and across maceral boundaries in thinned coal foils. The excellent spatial resolution has also permitted us to measure the organic sulfur concentration in close proximity to sulfides; we find that the organic sulfur concentration is constant over the maceral to within 1 μm of the pyrite.

Techniques of transmission electron microscopy have proved valuable in many areas of solid state science. Use of electron diffraction permits identification of crystal types, determination of unit cell sizes and characterization of crystal defects in the phases. Measurement of Energy Dispersive X-ray (EDS) line intensity allows calculation of the elemental composition of the phases. It is difficult to overestimate the value of such applications to metallic alloys, ceramic materials and electron-device alloys (1-4). Applications to coal and other fuels are far fewer, but the studies also show promise, both in characterization of mineral phases and in determination of organic constituents (5-9). This paper reports measurements on a particular feature of coal, the spatial variation of the organic sulfur concentration.

Method

The method depends on measurement of the X-ray radiation from atoms when a solid is irradiated with electrons. X-rays of particular energy are emitted-the energies depend on transitions between specific atomic states. The most prominent X-ray line for sulfur has energy of about 2300 eV, corresponding to a wavelength of about 5.4 Angstroms. That radiation may be detected and the intensity determined by crystal detectors, as in the electron microprobe, wavelength dispersive spectroscopy, or by detectors which discriminate in energy, energy dispersive microscopy, used in scanning electron microscopy or

[1]Current address: Shenyang College of Architectural Engineering, Shenyang, China

0097–6156/90/0429–0316$06.00/0

in transmission electron microscopy. Fortunately, no other elements in coal provide interferences with the S-line, so corrections from that source are unnecessary.

The apparatus used in electron microscopy is shown in Figure 1. The beam of electrons is focused to a fine spot on a thinned coal foil or on particles of powdered coal supported on a thin carbon film. The X-ray radiation from the sulfur (and other impurities) is detected and counted by a detector and associated multi-channel analyzer. The background radiation from the coal matrix is also counted for a predetermined period of time. The area irradiated by the electron beam is typically 0.1 μm^2. Since very little dispersion of the electrons occurs in the foil (or powder particle) the volume from which the X-rays come is typically less than 0.1 cubic μm. Accurate calculation of the organic sulfur concentration depends on accurate determination of this volume. Such determination is difficult for solids, in general, but for matrices of light elements, such as coal, a special technique eases the problem. The mass of material present in the volume is proportional to the Bremsstrahlung radiation--the background counts far away in energy from any characteristic line of elements present in the specimen. Thus the volume can be determined from the X-ray measurement. The accuracy of this determination is within a few %, depending on how reliably the organic composition of the coal is known. A sketch of a typical spectrum is shown in Figure 2.

Such spectra have long been known in X-ray measurements, but their simple use for quantitative calculation of elements present in metallic alloys and ceramic systems has been difficult--interpretation of the Bremsstrahlung radiation almost demands prior knowledge of the alloy composition, the very feature sought. This difficulty is greatly reduced for specimens of light elements such as coal. A simple expression can be written for the concentration of an impurity element, such as S, in a matrix of light elements (in this case, mainly carbon, oxygen and hydrogen).

$$c(S) = A \; C_S/C_b \qquad (1)$$

In this equation, $c(S)$ is the concentration of the sulfur, C_S is the area under the X-ray line for S and C_b the Bremsstrahlung counts over a specific energy range, see the dark areas in Figure 2. For C_S we use the area at half-maximum and for C_b the area under the curve from 10 to 18 keV. The parameter A is a constant which takes into account the geometry of the microscope, the counting efficiency of the detectors and certain characteristics of the matrix. It can be evaluated using known sulfur compounds--we typically use pyrite, a thiophene, a sulfone or films of pure S. For our system, a value of A of 1.5 gives the organic sulfur concentration in wt% for a bituminous coal. The value of A changes only by a few % for lignite or for anthracite (10). Thus the technique permits reliable determination of organic sulfur concentration for any coal.

The basis for the simple expression of Equation 1 is found in many articles on quantitative analytical-electron-microscopy (11). The original application to light matrices, though, was made by Hall for biologically important materials (12,13). Understanding of details of calculation of parameters making up the constant A can be obtained from his work.

Applications

Three applications of this technique are described in the following sections: (1) measurement of variation of $c(S)$ across macerals, (2) determination of the

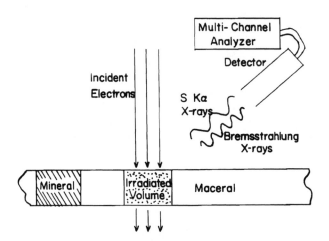

Figure 1.   Sketch of apparatus.

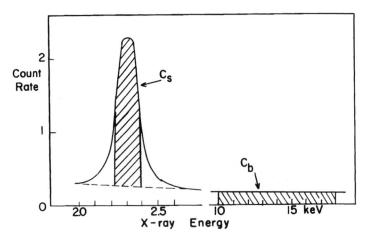

Figure 2.   Typical energy dispersive X-ray spectrum.

average value of c(S) for whole coal and (3) variation of c(S) in the vicinity of pyrite.

Spatial Variation of Organic Sulfur. The excellent spatial resolution of focused electron beams offers the possibility of examining variation of organic sulfur within macerals. The electron microprobe and scanning electron microscope allow resolution of a few microns (14-16). The transmission electron microscope allows even better resolution (less than 1 μm) because the thin foils and powders produce less electron scattering. We have used this capability to measure the distribution of S in a number of coals.

The first such study was reported by Hsieh and Wert (10). Measurements on a thinned vitrinite maceral of an Illinois #6 coal at a spacing of about 0.5 μm showed a variation of about ±5% of the average value of about 2.6 wt%. Other (unreported) measurements using the TEM technique gave about the same results for this coal. Later Tseng et al. measured a traverse across a sporinite maceral embedded in a large vitrinite maceral (17). We have made another such measurement; it is shown in Figure 3. Some 68 measurements were made along a traverse of about 50 mm. In this maceral the average in the vitrinite is about 3.5% with local variation about ±5%. The average in the sporinite maceral is about 5%, again with a variation of about ±5%. Importantly, the jump at the maceral boundary is sharp, a micron or two.

Solid solutions in metallic alloys are normally compositionally very uniform; random variations of 5% would be unusual. Also, a two phase layer normally is found between two solid solutions. The sulfur distribution in coal seems not to behave this way. Apparently, the distribution pattern established at some early stage of coal formation is frozen-in and the organic sulfur is bound so tightly to its hydrocarbon sites that it cannot diffuse until the temperature of the coal is raised to 400°C or above (18).

The Average Organic Sulfur Concentration. We have used TEM measurements to determine the average organic sulfur concentration in whole coals. Such determination requires that a series of measurements be made over all the maceral types. Use of foils, as described in the previous section, is difficult and expensive, since many foils would be required to give assurance that all maceral types were sampled. Consequently, we have found it more convenient to use finely ground powders. Fine coal powder is further ground in a mortar and pestle (with a little alcohol) to produce micron size powders. A drop of the alcohol is transferred to a carbon film supported on a carbon mesh so that a thin sprinkling of ultra-fine powder is deposited grain by grain. We have found that the average of 50 individual measurements gives a good value for the organic sulfur concentration.

A typical measurement is shown in Figure 4 for Illinois #6 coal. The range of individual measurements is from 0.5 to 4.5 wt% S with an average around 2.5 wt%. This average is close to the value of 2.4 wt% determined by the ASTM standard technique. Wert and his colleagues have reported several such comparisons in previous publications (10,17,20). We have just made 21 new TEM measurements and have collected data from the literature for additional SEM-type measurements (either Sem or Electron Microprobe) (21,22). For all of these coals, ASTM values were also available (19). The entire group is plotted in Figure 5. Remarkably good agreement exists for these coals between the electron-optical technique and the ASTM standard method. (The full set of data exists in a Lotus 123 file in format wk.1. The interested reader may obtain

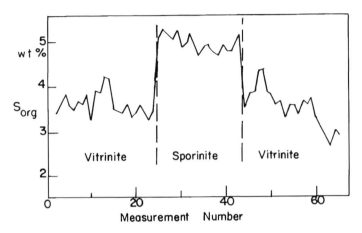

Figure 3.   Variation of S_{org} across maceral types.

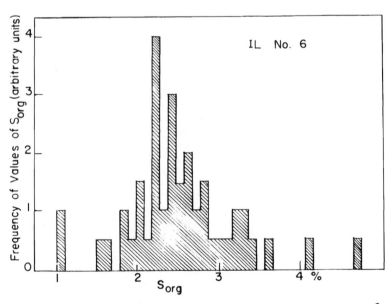

Figure 4.   Distribution of organic sulfur within volumes of about 1 μm^3 as a function of sulfur content. Illinois #6.

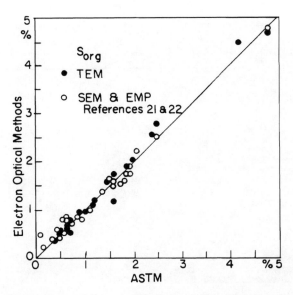

Figure 5. Comparison of measured values of S_{org} between the ASTM and electron-optical techniques. Not all 56 points can be differentiated because some are coincident.

this file by sending a diskette to CAW. It identifies the coal, the investigator, the two values and the reference for the original data. A printout of part of the file is listed in Appendix I.)

Organic Sulfur in the Vicinity of a Pyrite Particle. Distribution of organic sulfur is not uniform through a coal, even within a particular maceral. But the variations about the average value in a particular maceral are small (fractionally) and seemingly random. But the question may be asked as to the "constancy" of the organic sulfur content in the immediate vicinity of a pyrite particle.

Yurovskii claimed (in 1960) that "pyrite concretions in coals are surrounded by coal layers richer in organic sulfur which decreases as the distance from the center of the pyrite concretion increases" (23). He found the enrichment to be enormous, up to 25 times the average value.

His measurements were made on "bulk" coal fragments extracted from the vicinity of pyrite particles. Electron-optical methods, though, can make such measurements in situ and have better spatial resolution than his bulk methods allowed.

We have made such measurements using transmission electron microscopy on two coals--an Illinois #5 and an Illinois #2. The transmission electron micrograph showing pyrite framboids in an otherwise "clean" maceral of a specimen of Illinois #5 coal is shown in Figure 6. The organic sulfur was measured along 3 directions, as shown. The measurements are shown in Figure 7. (Note that the plots are offset along the y-axis for clarity of presentation). No increase in organic sulfur occurs to within 1 µm of the framboids. Then an increase occurs in the signal for sulfur--but the X-ray line for Fe also increases. This may indicate that part of the S-signal has its origin in pyritic-sulfur.

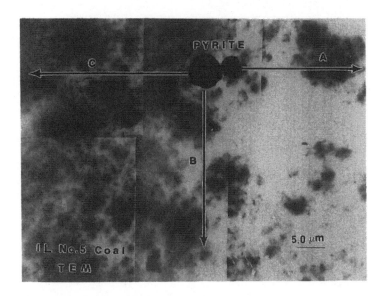

Figure 6.   View of maceral containing pyrite showing traces along which S_{org} was measured. Illinois #5.

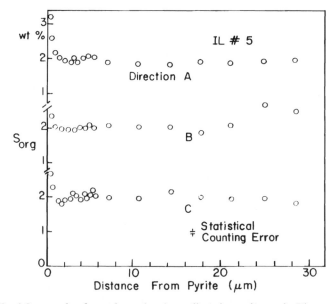

Figure 7.   Measured values along the three directions shown in Figure 6. The measurements are offset in the y-coordinate for ease of interpretation.

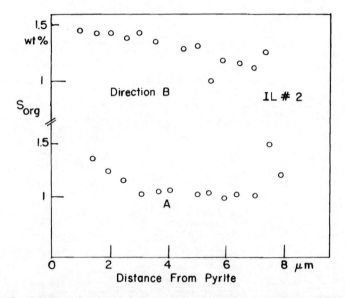

Figure 8.   Measured values along two directions in a specimen of Illinois #2. The measurements are offset in the y-coordinate for ease of interpretation.

A similar measurement for a specimen of Illinois #2 shows the same result, Figure 8. No measurable increase in organic sulfur occurs to within 1 μm of the pyrite.

One must conclude, therefore, that there is no gradient in organic S concentration in the vicinity of the pyrite at distances greater than 1μm. Consequently, the pyrite and adjacent maceral are in a two-phase state with no intervening phase. Since the activity of S in pyrite has a specific value (at a given temperature) independent of the coal, and since the organic sulfur concentration varies among the coals, the sulfur in the maceral and the sulfur in pyrite must not be in thermodynamic equilibrium.

Summary

Transmission electron microscopy has been used to measure the spatial variation of organic sulfur within and between macerals. The spatial resolution of the measurements is superior to any other reported technique.

APPENDIX I. Comparison of Electron Optical and ASTM Values of Organic Sulfur for 56 Coals

COAL	PERSON	TEM %S	ASTM %S	REFERENCE	COAL	PERSON	TEM %S	ASTM %S	REFERENCE
subB	RAYMOND	0.22	0.16	21	SHAN Xi-1	GE	1.09	1.16	NEW
BLIND-C	GE	0.37	0.37	NEW	SHAN Xi-2	GE	1.16	1.58	NEW
hvAb-5	RAYMOND	0.39	0.35	21	CRCCIO7PRC	Ge	1.19	1.19	NEW
WYODAK-A	GE	0.46	0.47	NEW	KCER-9127	CLARK	1.35	1.35	22
PSOC-1020	Clark	0.48	0.09	22	KCER-9127-V	CLARK	1.35	1.35	22
Fu Xin	Ge	0.49	0.49	NEW	PSOC-1018	Clark	1.50	1.59	22
hvAb-2	RAYMOND	0.52	0.57	21	KCER-9419	CLARK	1.52	1.72	22
CRCII1PRC	Ge	0.52	0.49	NEW	KCER-9419-V	CLARK	1.52	1.72	22
BEULAH	GE	0.53	0.70	NEW	PSOC-1019	Clark	1.58	1.44	22
PSOC-1015	Clark	0.55	0.54	22	RM-181	Clark	1.59	1.81	22
POCAHONTAS	GE	0.57	0.50	NEW	Rm-184	Clark	1.61	1.58	22
PSOC-1014	Clark	0.58	0.42	22	RM-181-A	Ge	1.62	1.48	NEW
TANG SHAN	Ge	0.61	0.59	NEW	RM-184-A	Ge	1.72	1.58	NEW
hvAb-3	RAYMOND	0.61	0.63	21	LECO	Clark	1.75	1.86	22
SRC-I	Clark	0.62	0.65	22	KCER-9121-V	CLARK	1.76	1.93	22
STOCKTON	Ge	0.64	0.65	NEW	KCER-9121	CLARK	1.76	1.93	22
SRC-I-RES	Ge	0.65	0.65	NEW	LECA-A	Ge	1.89	1.86	NEW
hvAb-4	RAYMOND	0.66	0.63	21	KCER-9418-V	CLARK	1.90	1.91	22
U-FREEPORT	GE	0.70	0.74	NEW	KCER-9418	CLARK	1.90	1.91	22
PSOC-1012	Clark	0.77	0.94	22	Ill#5	Ge	2.02	1.97	NEW
hvAb-1	RAYMOND	0.78	0.64	21	KCER-9335	CLARK	2.21	2.05	22
CRCII5PRC	Ge	0.79	0.71	NEW	KCER-9335-V	CLARK	2.21	2.05	22
KCER-8055	CLARK	0.81	0.55	22	Rm-183	Clark	2.52	2.48	22
KCER-8055-V	CLARK	0.81	0.59	22	ILL#6	HSIEH	2.54	2.38	10
PSOC-1013	Clark	0.86	0.84	22	RM-183-A	Ge	2.78	2.48	NEW
PITTS-8	GE	0.96	0.89	NEW	PSOC664	Hsieh	4.50	4.20	10
subC	RAYMOND	0.96	1.01	21	RM-182-A	Ge	4.70	4.80	NEW
PaANTHRACITE	Hsieh	0.99	1.10	10	RM-182	Clark	4.78	4.80	22

Acknowledgments

We acknowledge support of the Ivan Racheff Memorial Fund of the University of Illinois.

Literature Cited

1. Dollar, M.; Bernstein, I.M.; Daeubler, M.; Thompson, A.W. Met. Trans. 1989, 20A, 447-451.
2. Baker, I.; Huang, B.; Schulson, E.M. Acta Met. 1988, 36, 493-9.
3. Bruemmer, S.M.; Fluhr, C.B.; Beggs, D.V.; Wert, C.A.; Fraser, H.L. Met. Trans. 1980, 11A, 693-9.
4. Viswanadham, R.K.; Wert, C.A. J. Less Common Metals 1976, 48, 135-150.
5. Wert, C.A.; Hsieh, K.C. Scanning Electron Microscopy 1983, III, 1123-1136.
6. Harris, L.A.; Yust, C. S. Fuel 1976, 55, 233-6.
7. Hsieh, K.C.; Wert, C.A. Mat. Sci. and Eng. 1981, 50, 117-125.
8. Allen, R.M. ; VanderSande, J. Fuel 1984, 63, 24-29.
9. Lauf, R.J. Amer. Cer. Soc. Bull. 1982, 61, 487-90.
10. Hsieh, K.C.; Wert, C.A. Fuel 1985, 64, 255-262.
11. Goldstein, J.I. In Introduction to Analytical Microscopy; Hren, J.J.; Goldstein, J.I.; Joy, D.C., Eds.; Plenum Press: NY, 1979; pp 83-120.
12. Hall, T.A.; Anderson, H.C.; Appleton, T. J. of Microscopy 1973, 99, part 2 , 177-182.
13. Hall, T.A. J. of Microscopy 1979, 117, 146-163.
14. Raymond, R. Jr. Amer. Chem. Soc. Symp. Series 1982, 205, pp 191-203.
15. Solomon, P.R.; Manzoine, A.V. Fuel 1977, 56, 393-6.
16. Greer, R.T. In Coal Desulfurization; Wheelock, Thomas, D., ed.; Amer. Chem. Soc. Symp. Series, No. 64; American Chemical Society: Washington, DC 1977; pp 3-15.
17. Tseng, B.H.; Buckentin, M.; Hsieh, K.C.; Wert, C.A.; Dyrkacz, G.R Fuel 1986, 65, 385-9.
18. Tseng, B.H.; Ge, Y-P; Hsieh, K.C.; Wert. C.A. In Processing and Utilization of High Sulfur Coals II, Chugh, Y.P.; Caudle, R.D., Eds.; Elsevier; Amsterdam, NY, 1987, pp 33-40.
19. Standard Test Method for Forms of Sulfur in Coal, Annual Book of ASTM Standards, ASTM D2442, 1983, 347.
20. Wert, C.A.; Hsieh, K.C.; Buckentin, M.; Tseng, B.H. Scanning Electron Microscopy 1988, 88, , 83-96.
21. Raymond, R.; Hagan, R.C. SEM II 1982, 619-627.
22. Clark, C.P.; Freeman, G.B.; Hower, J.C. SEM II 1984, 537-547
23. Yurovskii, A.Z. Sulfur in Coals, First published in USSR, 1960. Translated by the Indian National Scientific Documentation Center, New Delhi. Published in 1974 by the U.S. Bureau of Mines and the NSF. NTIS Document TT-70-57216, 116-120.

RECEIVED March 14, 1990

Chapter 20

Characterization of Organosulfur Compounds in Oklahoma Coals by Pyrolysis—Gas Chromatography

Allen J. Bakel, R. Paul Philp, and A. Galvez-Sinibaldi

School of Geology and Geophysics, University of Oklahoma, Norman, OK 73019

Extracted coals from eastern Oklahoma were analyzed using pyrolysis-gas chromatography and a flame photometric detector (FPD) to characterize the organosulfur compounds produced by pyrolysis of coals. All coals from the Croweburg seam with calorific values below 13,000 BTU (Table I) were shown to produce similar distributions of organosulfur compounds. The ratio of dibenzothiophenes to thiophenes produced by pyrolysis was shown to be proportional to the calorific value of the coal.

A great deal of recent organic geochemical research has been devoted to the study of organosulfur compounds in organic rich geological samples. Many organosulfur compounds have been identified in crude oils ([1],[2]), rock extracts ([3]), oil asphaltenes ([4]) and kerogen pyrolysates ([5]). The distributions of organosulfur compounds in geochemical samples have been shown to be affected by maturity ([1],[6]) and depositional environment of the original organic matter ([4],[7]). Special attention has been given to the occurrence of organosulfur compounds in samples thought to be derived from organic matter deposited in hypersaline environments. The Rozel Point oil ([7]) and extracts from a marl deposited in a Messinian

0097–6156/90/0429–0326$06.00/0

evaporitic basin in the northern Appennines of Italy (3) have been found to contain a variety of thiophenes (structure i in Appendix I), thiolanes (ii) (tetrahydrothiophenes), thianes (iii) and benzothiophenes (iv). The distributions of organosulfur compounds in the pyrolysates of oil asphaltenes have been shown to resemble those of the aromatic fraction of the oil from which they were isolated (8), and vary according to the depositional environment of the original organic material. Philp and Bakel (4) reported that pyrolysates of asphaltenes isolated from limestone-sourced oils contained high concentrations of benzothiophenes and dibenzothiophenes (v) relative to thiophenes, while asphaltenes from shale-sourced oils produced more thiophenes than benzo- and dibenzothiophenes. Pyrolysis of a variety of kerogens has been shown to yield primarily thiophenes, with smaller amounts of benzothiophenes, dibenzothiophenes (4) and thiolanes (5).

Characterization and identification of organosulfur compounds in various coal derived liquids including hydrogenates (9), coal tars, and coal liquid vacuum residues (10) have been studied in many laboratories. These systems have been studied in order to understand the potential pollution problems created if the liquids are to be used as fuels. Organosulfur compounds identified in coal products include: benzothiophene, hydroxy-, and alkylbenzothiophenes, dibenzothiophene, alkyldibenzothiophenes, benzonaphthothiophenes (vi), phenanthrothiophenes (vii), naphthothiophenes (viii), triphenylenothiophenes (ix) and benzophenanthrothiophenes (x) (see structures in Appendix I). Coal extracts have been found to contain benzo- and dibenzothiophenes (11,12). A variety of techniques, including flash pyrolysis have been used to describe the form of sulfur in the coal matrix (13). Flash pyrolysis of extracted coals has been shown to yield thiophenes, benzothiophenes and dibenzothiophenes (4).

Coals are generated by burial of peats, which are formed in swamps and marshes. Geographic and tectonic requirements for formation of thick peat deposits include slow subsidence, protection of the swamp from erosion by marine inundation or fluvial flood waters and a restricted supply of fluvial sediment. These requirements are met on coastal plains where most Recent peat swamps are located. Microbial activity is capable of degrading great quantities of the plant material which is the ultimate source of peat. Therefore, conditions which inhibit microbial activity are favorable for the accumulation of large peat deposits. The conditions in peat swamps which inhibit aerobic microbial degradation and thus favor accumulation of peat such as low pH and Eh also favor anaerobic sulfate reduction (14).

Organic sulfur in peat is derived from two sources, sulfur from the original plant material, and inorganic sulfur compounds which react with organic material present in peat. Peat is known to contain more sulfur than is present in the original plant material, suggesting an external source of sulfur such as sulfate-rich waters, hydrogen sulfide (H_2S) from the metabolic activity of sulfate reducing bacteria (15) or elemental sulfur (16). H_2S, a metabolic product of anaerobic sulfate reducing bacteria present in peat swamps, is known to react with organic matter in such a way as to permit incorporation of sulfur into the structure of complex organic molecules (17) and has therefore been proposed as an important agent in sulfur enrichment during early diagenesis of peat. In laboratory studies, Casagrande et al. (16) found that peats from the Okefenokee Swamp became enriched in sulfur when exposed to H_2S at room temperature in a matter of days. A large enrichment was observed in the humic acid fraction, which is thought to be a particularly important precursor of coal macerals.

Peats can accumulate in environments of varying salinity. Peats deposited under particularly saline conditions tend to have high sulfur contents, as do peats covered by marine sediments during marine transgressions. The source of the sulfur in these marine-influenced peats is thought to be the sulfate ion in the marine waters, which is reduced to hydrogen sulfide by sulfate reducing bacteria present in anaerobic sediments. Much of the sulfide reacts with iron to form iron sulfides, while excess hydrogen sulfide reacts with organic material such as carbohydrates or humic acids. In support of this model, a correlation has been observed between high concentrations of pyrite and high concentrations of organic sulfur in peats and coals (14).

In order to determine the origin of sulfur in coals, Hackley and Anderson (18) studied the sulfur isotopic composition of various coals from the western United States. Coals from the Powder River Basin, Wyoming were found to have anomalously low $\delta^{34}S$ values (-18.7 to 3.9 parts per mil). The unusual $\delta^{34}S$ values of these coals suggest that original plant material and isotopically light secondary sulfur from bacterial activity were the major contributors to the sulfur content of these coals.

The Hanna Basin coals from Wyoming had higher $\delta^{34}S$ values than coals from the Powder River Basin. The pattern of sulfur isotopic composition found in coals from the Hanna Basin revealed that coals near the top of the coal seam contained more of the heavy sulfur isotope than coals stratigraphically lower in the coal seam. The sulfur isotope pattern observed in the Hanna Basin coals suggests that sulfur was introduced in sulfate-rich water from the top of the coal seam.

Many coal pyrolysis studies have reported only bulk results, such as the amounts of char, water, and NH_3 + H_2S + H_2 produced ([19]). More detailed analytical pyrolysis has been performed on isolated coal macerals. It has been found that the distribution of phenols in the pyrolysates of vitrinite are affected by the rank, or maturity of the coal ([20]). Saiz-Jimenez and de Leeuw ([21]) proposed that certain pyrolysis products of lignin (guaiacyl and syringyl compounds) are of potential use in determination of source material of coals and lignins. Philp et al. ([22]) found that brown coals with similar chemical and petrographic properties (lithotype) yielded similar pyrolysis products, and that the hydrocarbon distribution in the pyrolysis products changed with increasing maturity. Pyrolysis products of several components of the coal (cuticle, resin and soft brown coal woods) were compared to the pyrolysis products of lignins. A number of pyrolysis products of these coal components, including syringyl derivatives and several phenolic compounds, were found to be useful in differentiating angiosperm from gymnosperm sources for the coals. Philp and Bakel ([4]) examined organosulfur compounds in the pyrolysates of extracted Pennsylvanian coals from eastern Oklahoma and identified thiophenes, benzothiophenes, dibenzothiophenes and benzonaphthothiophenes.

The coal reserves in eastern Oklahoma define the southern edge of the interior coal province of North America. Oklahoma coals can be divided into two categories (Figure 1), the Arkoma basin coals, which occur in Coal, Pittsburg, Haskell, Latimer, Sequoyah and Le Flore counties at the southern extremity of the interior coal province, and the northeastern Oklahoma shelf coals, which occur from the Arkoma basin, north to the Kansas border. The coals found in the northeastern shelf area tend to have higher sulfur content (>3%) than the coals from the Arkoma Basin (<2%). Most of the sulfur found in the Arkoma Basin coals is in the form of pyrite (FeS_2). While there are at least 22 recognized

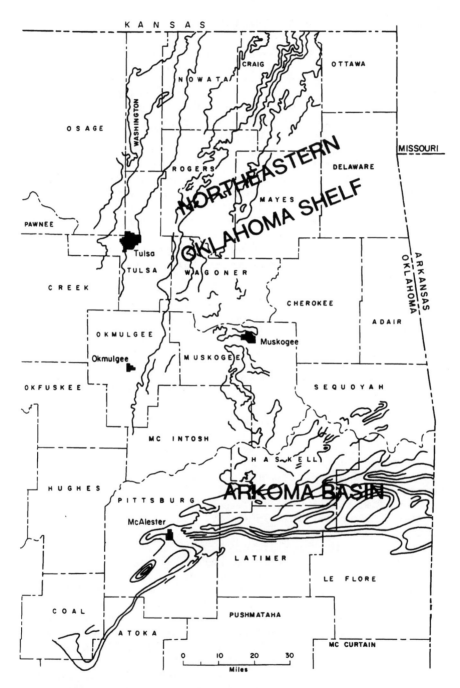

Figure 1. Map showing the coal-producing area of
Oklahoma with outcrop pattern of coals.
(Reproduced with permission from reference 25).

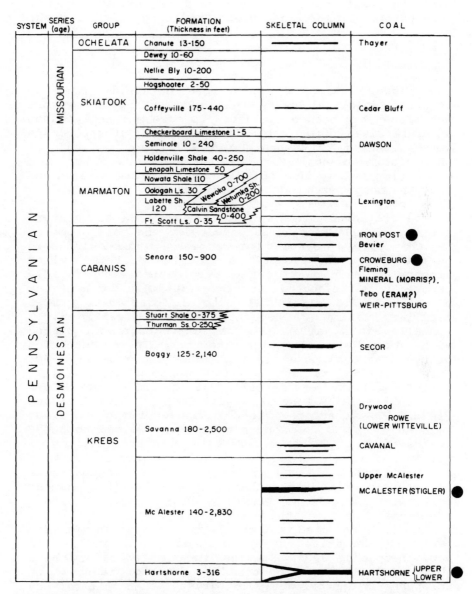

Figure 2. Geologic column of northeastern Oklahoma showing coal beds. Coals included in this study are marked (●). (Reproduced with permission from reference 25).

Desmoinesian coal seams in eastern Oklahoma (Figure 2),
this study will examine only five different coal seams:
Hartshorne, Stigler, McAlester, Croweburg and Iron Post.

Experimental

The eastern Oklahoma coals were obtained from the
Oklahoma Geological Survey Organic Petrography Laboratory
where they had been crushed to -20 mesh and analysed on
an as received basis (Table I). The whole, crushed coals
were extracted for 48 hours with a 1:1 (v/v) mixture of
chloroform/methanol using a soxhlet apparatus. Exhaustive
extraction ensures that soluble compounds are removed and
that the compounds produced during pyrolysis are derived
from the organic matrix of the coal.

The extracted coals were pyrolysed at a temperature
setting of 800°C for 20 seconds using the coil probe of
the Chemical Data Systems Pyroprobe 122 system. It should
be noted that an attempt was made to pyrolyze similar
amounts of each sample. The pyrolysates were rapidly
removed from the heated interface (300°C) under a stream
of helium (1.8 ml/minute) onto a fused silica DB-5
capillary column (J&W Scientific, 30m x 0.32mm i.d.)
installed in a Varian 3300 gas chromatograph equipped
with both a flame ionization detector (FID) for detection
of hydrocarbons and a flame photometric detector (FPD)
for detection of organosulfur compounds. A column
effluent splitter was installed at the end of the column
to allow for simultaneous acquisition of data from the
FID and FPD detectors. The column oven temperature was
held at   -25°C for four minutes, and raised to 300°C
at the rate of 4°C/minute. Tentative identifications of
the FPD peaks (Figure 3) were made by comparison with
chromatograms published by Hughes (1), and comparisons
of retention times of authentic standards (thiophenes,
benzothiophene and dibenzothiophene) analyzed under
identical chromatographic conditions (Table II).

It should be noted that the flame photometric detector
is more sensitive to thiophenes than it is to
benzothiophenes (23). Studies in our laboratory have
shown that the flame photometric detector response to
thiophene is approximately 25% greater than that for the
corresponding quantity of benzothiophene. Hence a
consequence of this non-linearity of response is that the
thiophenes are not as quantitatively dominant as
suggested by some of the FPD pyrograms.

Results

Hartshorne. The Hartshorne coal occurs in the Arkoma Basin, and in some areas ·can be split into upper and lower strata, separated by shales and sandstones. The average total sulfur content of the Hartshorne coal is 1.8%. Higher calorific values of a coal can be correlated with a higher abundance of dibenzothiophenes relative to thiophenes in the FPD pyrograms (Figure 4). It is interesting to note that the total sulfur contents of the two Hartshorne coals shown in Figure 4 are very similar (Table I), suggesting that rank is more important than total sulfur content in determining the distribution of organosulfur compounds in the pyrolysis products from these coals.

McAlester. The distribution of organosulfur compounds in the pyrogram of McAlester coals is dependent on their rank as was the case for the Hartshorne coals (Figure 5). The McAlester coal of lowest rank (#497) also has the highest total sulfur content (Table I). The McAlester coal of lower rank (#497) yields higher concentrations of all organosulfur compounds than coals from the same seam but of higher rank (#499).

Stigler. The Stigler coal is stratigraphically equivalent to the McAlester coal, and typically occurs at shallower depths than the McAlester. Stigler coals of lower rank (#508) yield a broad range of organosulfur compounds upon pyrolysis, including thiophenes, benzothiophenes and dibenzothiophenes and the organosulfur compound distribution of coal #508 is dominated by thiophenes (Figure 6). In contrast, pyrolysates of a higher rank coal (#452) contain a mixture of organosulfur compounds dominated by dibenzothiophenes (Figure 6).

Iron Post. The Iron Post coal is a high sulfur coal (Table I) from the northeastern Oklahoma shelf, which is associated with pyritic shales. FPD chromatograms of the pyrolysis products of two Iron Post coals of similar rank show that the distributions of organosulfur compounds produced by pyrolysis of these two coals are quite similar (Figure 7). This supports the idea of a relationship between rank and organosulfur compound distribution in the pyrolysates of coals.

Table 1.   General Data For Coals

SAMPLE #	FORMATION	COUNTY	TWN/RNG	CALOR.VAL. (BTU) (D-2015)[@@]	%R$_o$	%MOISTURE[*] (D-3302)[@@] (D-3173)	%ASH (D-3174)[@@]
419	Hartshorne	Latimer	T5N R17E	13448	NA	6.4	3.6
512	Hartshorne	Haskell	T9N R21E	14176	NA	3.0	6.5
497	McAlester	Latimer	T6N R21E	12486	NA	4.5	14.3
499	McAlester	Latimer	T6N R21E	13544	0.85	3.0	8.5
452	Stigler	Haskell	T9N R22E	14592	1.25	2.3	4.4
508	Stigler	Haskell	T8N R21E	10120	NA	9.6	21.5
466	Iron Post	Craig	T26N R18E	13412	NA	3.9	5.5
515	Iron Post	Rogers	T24N R17E	13384	NA	4.0	7.2
227	Croweburg	Wagoner	T16N R15E	8709	0.57	3.7	32.3
517	Croweburg	Rogers	T24N R17E	13167	NA	7.6	3.5
397	Croweburg	Okfuskee	T10N R12E	9150	NA	3.0	31.0
395	Croweburg	Okfuskee	T10N R12E	11723	NA	2.4	16.9

* - determined at two different particle sizes.

** - determined as difference of total weight and the sum of %moisture, %ash, and %volatile.

@ - determined as difference of total sulfur and the sum of %S-pyrite and %S-sulfate.

@@ - ASTM method used to determine these values. Method taken from ref. 26.

Examined in This Study

%VOL.MATT. (D-3175)[@@]	%FIXED C[**]	%S-total (D-3177)[@@]	%S-organic[@]	%S-pyrite (D-2492)[@@]	%S-sulfate (D-2492)[@@]	%EXTRACT
38.5	51.5	2.2	0.80	1.40	0.03	2.42
24.2	66.3	2.4	0.40	1.91	0.04	0.23
30.0	51.2	5.2	0.70	4.48	0.02	NA
33.9	54.6	2.6	0.60	1.95	0.01	0.43
25.5	67.8	0.7	0.50	0.17	<0.01	0.17
26.4	42.5	5.5	1.10	2.89	1.55	4.96
44.7	45.9	4.8	2.10	2.63	0.05	2.80
43.9	44.9	4.0	1.87	2.13	<0.01	0.80
32.7	31.3	7.6	1.14	5.41	1.05	NA
32.1	56.8	0.5	0.43	0.02	0.05	1.52
30.0	36.0	6.6	1.00	5.37	0.28	NA
40.5	40.2	3.6	1.30	2.25	0.08	NA

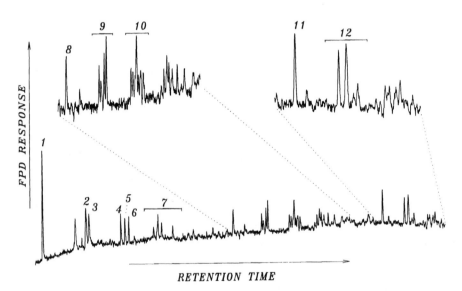

Figure 3. A typical FPD pyrogram of an Oklahoma coal
    with tentative identifications of organosulfur
    compounds. Peak numbers refer to compounds listed
    in Table II. (GC conditions are given in the text
    for all of the chromatograms shown in Figures 3–
    9).

Table II.   Organosulfur Compounds Identified in the Pyrolysates of
Coals from Eastern Oklahoma

---

1.   Thiophene*

2.   2-Methylthiophene*

3.   3-Methylthiophene*

4.   2-Ethylthiophene*

5.   2,5-Dimethylthiophene*

6.   2,3-Dimethylthiophene

7. C_3 Thiophenes

8.   Benzothiophene*

9. C_1 Benzothiophenes

10. C_2 Benzothiophenes

11.   Dibenzothiophene*

12. C_1 Dibenzothiophenes

---

* - Identified by comparison of retention times to authentic standards, others identified by comparison to published chromatograms.

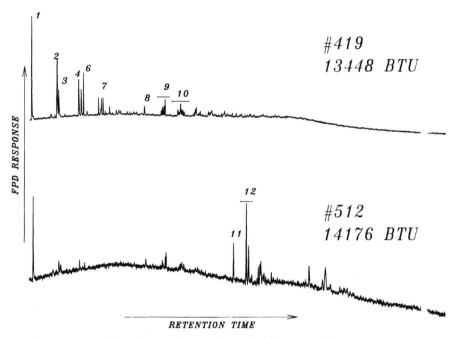

Figure 4. FPD chromatograms of the pyrolysates of two
        Hartshorne coals.

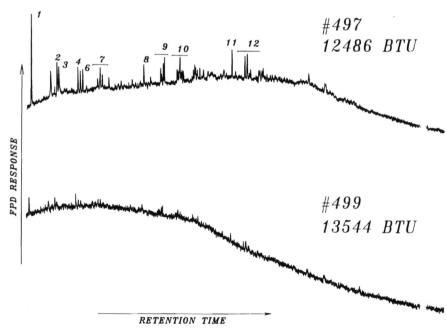

Figure 5.   FPD chromataograms of the pyrolysates of
        two Hartshorne coals.

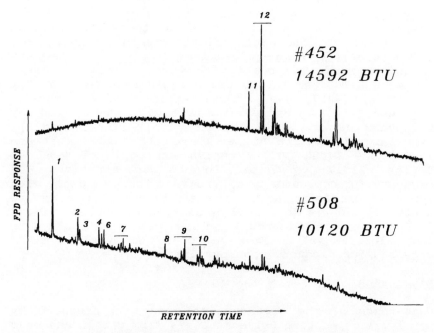

Figure 6. FPD chromatograms of the pyrolysates of two Stigler coals.

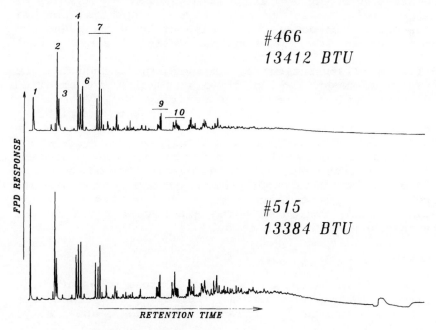

Figure 7. FPD chromatograms of the pyrolysates of two Iron Post coals.

Croweburg. The Croweburg coal has a lower sulfur content than the other coals found on the northeastern Oklahoma shelf (Table I). Coals with calorific values ranging from 8,700 to 13,000 BTU were examined in order to show any maturity trend. The coals with lower calorific values (between 8,700 and 12,000 BTU) (equivalent to R_o of 0.4 to 0.65%) (#227, #397, #395) yield a wide range of organosulfur compounds upon pyrolysis, dominated by thiophene and its alkyl derivatives (Figures 8 and 9). Croweburg coals with higher calorific values (between 12,000 and 14,000 BTU) (equivalent to R_o of 0.65 to 0.75%) (#517) produce lower concentrations of organosulfur compounds upon pyrolysis than the coals with lower calorific values and the organosulfur compound distributions are dominated by dibenzothiophenes (Figure 8). Data shown in Figures 8 and 9 suggest that coals with low calorific values from one coal seam yield similar organosulfur products upon pyrolysis.

Discussion

Comparison of the organosulfur compounds produced by pyrolysis of eastern Oklahoma coals from within one seam shows that lower rank (lower calorific value) coals consistently yield high concentrations of thiophenes relative to benzo- and dibenzothiophenes. Pyrolysates of higher rank coals contain lower concentrations of organosulfur compounds than coals of lower rank and are dominated by dibenzothiophenes. Coals of similar rank (e.g. #517 and #512) from different seams yield markedly different distributions of organosulfur compounds upon pyrolysis, suggesting that the distribution is controlled by source material as well as maturity.

It has been proposed that sulfur-carbon bonds break at lower thermal stress than carbon-carbon bonds in sulfur-rich kerogens (24). Scission of sulfur-carbon bonds provides one explanation for the observed depletion of organosulfur compounds in the pyrolysates of high rank coals. The sulfur groups present in low rank coals are broken from the coal matrix to produce low molecular weight sulfur compounds which are not measured by the techniques used in this study.

Conclusions

- Pyrolysis of Oklahoma coals with a broad range of total sulfur content produce a variety of organosulfur compounds including parent compounds and alkyl

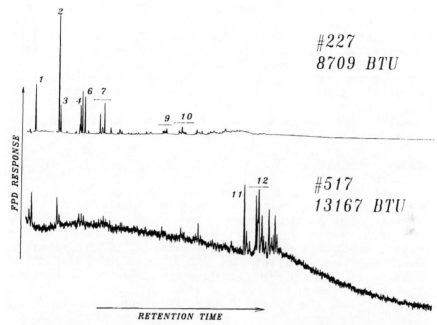

Figure 8. FPD chromatograms of the pyrolysates of two Croweburg coals.

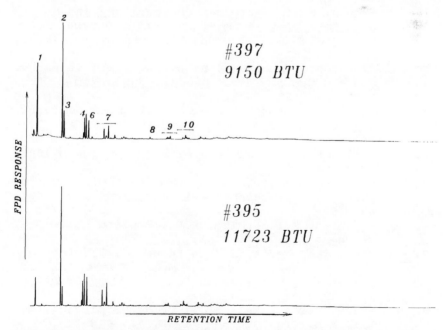

Figure 9. FPD chromatograms of the pyrolysates of two Croweburg coals.

derivatives of thiophenes, benzothiophenes and
dibenzothiophenes.
- Low rank coals from one coal seam (Croweburg) produce
pryrolysates which contain strikingly similar
distributions of organosulfur compounds.
- Comparison of the organosulfur compounds produced by
pyrolysis of coals with various ranks shows that high
rank coals produce higher concentrations of
dibenzothiophenes relative to thiophenes than lower rank
coals from the same coal seam.

Acknowledgments

The authors wish to thank Brian Cardott of the Oklahoma
Geological Survey and Dr. Lloyd M. Wenger for their
contributions to this study.

This work was supported with financial support from
Texaco, Mobil, NSF Grant No. EAR-8608820, DOE Hydrocarbon
Research Program Grant No. DE-FG-22-87FE 61146, the
Oklahoma Mining and Minerals Resource Research Institute
with funds from the Bureau of Mines Allotment Grant No.
G-1164140, and the donors of the Petroleum Research Fund,
administered by the American Chemical Society.

Literature Cited

1.   Hughes, W. In Petroleum Geochemistry and Source
        Rock Potential of Carbonate Rocks; Palacas J.G.,
        Ed.; AAPG Stud. Geol. No. 18; AAPG: Tulsa, Okla.,
        1984; pp 181-196.
2.   Sinninghe Damste, J.; Rijpstra, I.C.; de Leeuw,
        J.W.; Schenck, P.A. Geochim. Cosmochim. Acta
        1989, 53, 1323-1341.
3.   Sinninghe Damste, J.; Haven, H. ten; de Leeuw,
        J.; Schenck, P.A. Org. Geochem. 1986, 10, 791-
        805.
4.   Philp, R.P.; Bakel, A.J. Energy and Fuels 1988,
        2, 59-64.
5.   Sinninghe Damste, J.; Kock-van Dalen, A.C.; de
        Leeuw, J.W.; Schenck, P.A. J. Chrom. 1988, 435,
        435-452.
6.   Eglinton, T.I.; Philp, R.P.; Rowland, S.J. Org.
        Geochem. 1988, 12, 33-41.
7.   Sinninghe Damste, J.; de Leeuw, J.W.,; Kock-van
        Dalen, A.C.; de Zeeuw, M.A.; Lange, F. de;
        Rijpstra, I.C.; Schenck P.A. Geochim. Cosmochim.
        Acta 1987, 51, 2369-2391.
8.   Philp, R.P.; Bakel, A.J.; Galvez-Sinibaldi, A.;
        Lin, L.H. Org. Geochem. 1988, 13, 915-926.

9. Braekman-Danheux C. J. Anal. Appl. Pyrol. 1985, 7, 315-322.

10. Nishioka, M.; Lee, M.L.; Castle R. Fuel 1986, 65, 390-396.

11. White, C.M.; Lee, M.L. Geochim. Cosmochim. Acta 1980, 44, 1825-1832.

12. White, C.M.; Douglas, L.J.; Perry, M.B.; Schmidt, C.E. Energy and Fuels 1987, 1, 222-226.

13. Boudou, J.P.; Boulgue, J.; Malchaux, L.; Nip, M.; de Leeuw, J.W.; Boon, J.J. Fuel 1987, 66, 1558-1569.

14. Stach, E.; Mackowsky, M-Th.; Teichmuller, M.; Taylor, G.H.; Chandra, D.; Teichmuller, R. Stach's Textbook of Coal Petrology. Gebruder Borntraeger: Berlin, 1975; pp 5-17.

15. Casagrande, D.; Ng. L. Nature 1979, 282, 598-599.

16. Casagrande, D.; Idowu, G.; Friedman, A.; Rickert, P.; Siefert, K.; Schlenz, D. Nature 1979, 282, 599-600.

17. Moers, M.E.C.; de Leeuw, J.W.; Cox, H.C.; Schenck, P.A. Org. Geochem. 1988, 13, 1087-1091.

18. Hackley, K.; Anderson, T.; Geochim. Cosmochim. Acta 1986, 50, 1703-1713.

19. Chakrabartty, S.; du Plessis, M. Evaluation of Alberta Plains Coals for Pyrolysis and Liquefaction Processes, Coal Report 85-1, Alberta Research Council, Energy Resources Division, Coal Research Department: Devon, Alberta, Canada 1985; pp 1-24.

20. Sentfle, J.; Larter, S.; Bromley, B.; Brown, J. Org. Geochem. 1986, 9, 345-350.

21. Saiz-Jimenez, C.; de Leeuw, J.W. Org. Geochem. 1986, 10, 869-876.

22. Philp, R.P.; Gilbert, T.; Russell, N. Austr. Coal Geol. 1982, 4, 228-243.

23. Dressler, M. Selective Gas Chromatographic Detectors; Elsevier: Amsterdam, 1986, Chapter 7.

24. Orr, W.L. Org. Geochem. 1986, 10, 499-516.

25. Friedman S.A. Investigation of the Coal Reserves in the Ozarks Section of Oklahoma and Their Potential Uses. Final Report to the Ozark Regional Commission; Oklahoma Geological Survey: Norman, Oklahoma, 1974; pp 1-30.

26. Annual Book of ASTM Standards: Gaseous Fuels; Coal and Coke, American Society for Testing and Materials, 1980, Sec. 5. V. 505, pp 444.

RECEIVED March 5, 1990

Appendix I. Structures of Organosulfur Compounds

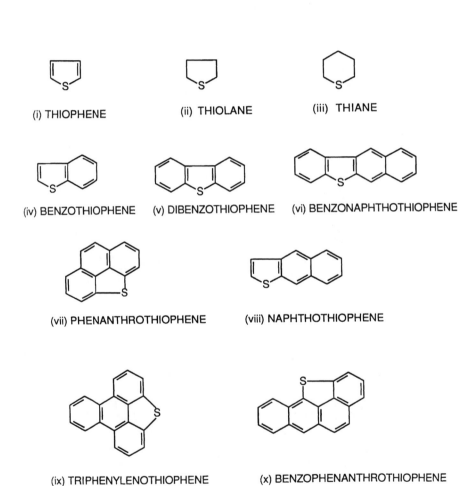

(i) THIOPHENE

(ii) THIOLANE

(iii) THIANE

(iv) BENZOTHIOPHENE

(v) DIBENZOTHIOPHENE

(vi) BENZONAPHTHOTHIOPHENE

(vii) PHENANTHROTHIOPHENE

(viii) NAPHTHOTHIOPHENE

(ix) TRIPHENYLENOTHIOPHENE

(x) BENZOPHENANTHROTHIOPHENE

Chapter 21

Coal Desulfurization by Programmed-Temperature Pyrolysis and Oxidation

Jean-Paul Boudou

Université Pierre et Marie Curie, Centre National de la Recherche Scientifique UA 04 0196, Laboratoire de Géochemie et Métallogénie, T16-25, 5° étage, 4, Place Jussieu, 75252 Paris Cedex 05, France

We are investigating the processes of coal desulfurization using mass spectrometry for analysis of gases evolved during programmed temperature pyrolysis (PTP) and programmed temperature oxidation (PTO). The results show that the sulfur evolution patterns and the intensity of sulfur emission are influenced by experimental conditions (particle size, atmosphere composition, thermal gradient), coal pretreatment (acid washing, organic solvent extraction, low temperature ashing), and coal rank. The emission patterns are independent of the coal organic and pyritic sulfur contents. The data provided by the PTP and PTO techniques are in agreement with those obtained with other techniques such as flash pyrolysis-gas chromatography.

The reduction of sulfur emissions during the world-wide utilization of fossil fuels is a major environmental concern. Lowering the cost and improving the efficiency of emission control techniques associated with coal combustion demand a fundamental understanding of coal desulfurization processes. This understanding is based upon an accurate description of sulfur behaviour during thermal degradation processes, and more generally, on a chemical characterization of sulfur contained in fossil fuels.

Programmed temperature heating of fossil fuels enables us to discriminate among the successive events which occur during faster thermal degradation. Programmed temperature oxidation (PTO) gives gas release patterns which reflect the molecular composition of the sample (1-4). PTO is different from programmed temperature hydrodesulfurization (5), programmed temperature reduction (PTR) (6-7), programmed temperature pyrolysis (PTP) (8-9), and flash pyrolysis-GC-MS (10-11). Pyrolytic and oxidative approaches have been used sequentially (12-13) as routine test methods for coal, kerogen and heavy oil characterization.

In this work, consideration has been given to the significance of the peaks and shoulders in the gas profiles from PTP and PTO to provide insight into the processes of coal desulfurization.

Experimental

The light gaseous species containing C,H,O,N, and S emitted during temperature programmed heating of coal in a stream of argon or air at atmospheric pressure were continuously monitored using mass spectrometry. A detailed description of the method has been given elsewhere (14). Gas evolution profiles were obtained by heating (25°C/min) a sample (< 2 mg) of powdered material in a flowing gas stream (4 to 5 ml/min at atmospheric pressure). The flowing gas stream was either argon, or air, or a mixture of argon and oxygen. The evolved gases were analyzed using a quadrupole mass spectrometer (Leybold Quadrex 100). The signal amplitude of a gas detected by mass spectrometry was proportional to the corresponding evolution rate. The gas profiles of a given micro-sample were similar between duplicate measurements, but frequent calibrations of the signal were necessary to correct a shifting of the coefficient of response due to a slow degasing of the connection fittings between the MS analysis chamber, the pressure converter and the high vacuum pump. Calibration was done using residual water and a reference sample every five to ten experiments.

The MS technique is the most appropriate method for on-line gas analysis because of its low cost, easy data handling, and simple maintenance. It offers high selectivity, fast and simultaneous detection of a number of gases, and the ability to quickly analyze very small samples. M/z 02, 15, 16, 18, 28, 34 (hydrogen sulfide), 44, 55, 57, 64, 76, 78, 84, 85, and 94 were routinely detected during PTP. M/z 02, 18, 28, 30, 44, 60, 64 (sulfur dioxide), and 80 were routinely detected during PTO.

Py-GC experiments were performed using a modified SGE pyrolysis inlet, interfaced either with a Chrompack gas chromatograph (model 437S) equipped with a flame ionization detector and a double flame photometric detector or with a Delsi gas chromatograph (model DI300) interfaced with a Delsi mass spectrometer (model R10 10). The control of the Chrompack GC and the data acquisition were done with a PCI-Chrompack program and a PC computer.

The commercial version of the SGE pyrojector is a continuous-mode pyrolyzer designed to be used with existing injectors fitted to standard gas chromatographs. Pyrolysis gases are pushed through a special needle into the injector by a pyrolysis carrier gas enabling control of the residence time of the products in the hot zone of the pyrolyzer. This pyrolysis system has several disadvantages : secondary reactions are likely in the pyrolyzing zone, and the transfer of the pyrolysis products to the column is complicated by a cold zone between the pyrolysis inlet and the injector. In our pyrolyser a 1/4", 1/8" (o.d., i.d) quartz furnace liner went through an electrical winding micro-furnace and the heated jacket of the GC injector. The micro-furnace was adjusted at the top of the heated jacket of the GC injector to avoid any cold spot. A 1/8" split was fitted at the top of the transfer quartz line in the GC furnace. The split flow-rate was controlled either with a micro needle valve or with a mass flow regulator (Alphagaz). Helium was used as the gas chromatographic mobile phase at a column head pressure of 0.9 to 2.5 bar, a split-vent flow of 15-30 ml/min and a average linear velocity of 24 cm/s. Gas chromatographic separations were performed using fused silica columns of either 50 or 25 m lengths, an internal diameter of 0.25 mm and a 0.1 to 0.25 μm film of a non-polar stationary phase (BP1, BP5 or CPSIL8). The column was operated isothermally at 50°C for 4 minutes and then temperature programmed at 5°C/min to 285°C and held for 9 minutes. A small powdered sample (size : 0.1 to 0.6 mg, granulometry < 80μm) was bored out with the sharp needle tip (70 mm long) of a SGE solids injectors -with its flat tipped plunger taken back - then injected into the pyrolyzer by pushing the plunger through the needle onto an alumina wool plug in the center of the furnace which was heated isothermally using a Minicor or a Microcor III P temperature controller. The hot zone was 15 mm. The transfer temperature from the hot zone to the top of the GC column was about 300°C. The identification of the products was performed by GC retention times and by MS of pure reference compounds.

Table I describes the coal samples studied. The coals examined came from very different geographical areas including French coal samples from the Deutch SBN coal bank and from CERCHAR, and New Zealand coal samples from the N.Z. Geological Survey (16). In the Provence coalfield, as in most New Zealand coalfields, and most coalfields over the world, seams having a strong marine influence are high in sulfur, and sulfur content decreases with increasing stratigraphic depth below marine beds (16). These observations suggest that freshwater peat deposits incorporated sulfur by downward diffusion of marine sulfide after marine transgression.

The low sulfur coal samples studied ranged from peat to anthracite, while the high organic sulfur coals having a low pyritic sulfur content ranged from lignite to medium volatile bituminous. Coal samples were crushed in an agate mortar and sieved through a wire-cloth sieve having square openings of 80 μm. All coal samples were preserved in closed glass bottles maintained in a freezer at about -25 °C.

Table I. Selected properties of coal samples

Origin	Rank	$\overline{R}_s(\%)$	VM % daf	Tmax (15)	Ash	C	N	S	Pyr.S
					% dry basis				
Chatam Island, N.Z., ENMIN-GD503	Peat	nd.	58.94	395	8.69	63.24	0.34	1.05	0.32
Charleston, N.Z., No 43 (16)	LigA/ Sub C	0.40	53.60	409	2.00	67.03	0.51	6.20	0.26
Provence, FR, SBN-136FR45	Sub B	0.48	54.83	415	19.67	59.63	1.45	4.96	1.04
Illinois No 6, SBN-511US43	hvCb	0.47	47.43	414	10.20	69.95	1.40	4.10	1.20
Vouters, FR, SBN-140FR36	hvAb	0.84	37.17	438	5.41	77.82	0.89	0.85	0.26
Webb/Baynes, Buller field, NZ	hvAb	0.82	39.10 (a)	443	0.89	82.67	0.90	4.19	0.51
Rockies, Buller field,NZ	mvb	1.25	33.40 (b)	455	1.50	nd.	0.62	6.96	0.53
Méricourt, FR, SBN-138FR17	lvb	1.33	16.70	477	29.25	62.53	0.92	0.62	0.19
Escarpelles, FR, SBN-139FR08	sa	2.29	9.03	563	12.65	79.88	0.93	0.72	0.11
La Mure, FR, CERCHAR-870382	an	5.15	2.27	nd	6.98	89.25	0.84	0.46	0.08

(a) Volatile matter of low sulfur coal from same seam : about 35 %
(b) Volatile matter of low sulfur coal from same seam : about 27 %
NOTE: Tmax parameter from Rock-Eval analysis as defined by Espitalié et al. (*15*).

Blocks of pure pyrite embedded in rock from a hydrothermal area were hand-picked under an optical microscope. The pyrite grains were crushed, sieved and preserved in the same way as the coal. We choose mineral pyrite rather than coal pyrite because it is easier to obtain. The behaviour differences between different pyrites are due to particle size effects or matrix effects, but not to pyrite itself, which is a well defined chemical and mineralogical species.

The pyrolysis properties of four model compounds were examined. Their molecular structures are shown in Table II. The three and four-ring molecules, containing a benzo [b] thiophenic unit were synthesized by Dr. Cagniant (L.S.C.O., Metz University). A polycyclohexanesulfide is an aliphatic sulfide synthesized by Dr. N. Spassky (Laboratory of Macromolecular Chemistry, Paris VI University). Polymeric aromatic sulfide was represented by a polybenzosulfide provided by Philips Petroleum.

Table II. Peak temperature of sulfur containing model compounds during programmed temperature heating (25°C/min from 250 to 875°C)

Sulfur models	Peak temperature of hydrogen sulfide during PTP	Peak temperature of sulfur dioxide during PTO
	300	300
	no peak	480
	no peak	525
	no peak	525-600

Results and Discussion

Particle Size.

Figure 1 shows the effect of particle size on the evolution of m/z 02, 15, 18, 34 (hydrogen sulfide), 57 (alkanes) and 64 during PTP of the Provence coal. Figure 2 shows the evolution of m/z 18, 44, 64 (sulfur dioxide) during PTO of the Provence coal. The differences in maceral content due to sieving were avoided by passing the largest whole coal fraction through smaller sieves. The effect of particle size on the observed sulfur gas evolution profile was particularly pronounced when the particle size was greater than 80 μm. The effect of an increase of particle size on the PTP and PTO profiles was an increase in the amount of sulfur gas evolved. During PTP as particle size increased the first H_2S peak was enhanced and the amplitude of the second peak was considerably decreased with a simultaneous increase of the m/z 64 evolution (elemental sulfur). Large grains of pyrite tended to decompose much less and appeared to not compete for hydrogen. Hydrogen radical is necessary for the quenching of organic radicals (m/z 57, m/z 15) and sulfur radicals resulting from the decomposition of organic sulfur groups. In the case of the PTO experiments increased particle size decreased gas formation (water, CO_2, SO_2) due to slower penetration of oxygen into

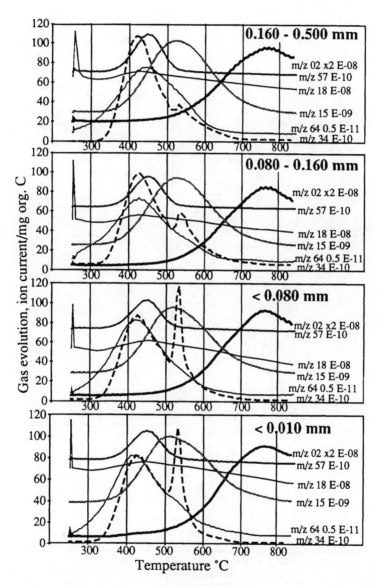

Figure 1. Effect of particle size on PTP of Provence coal. Ion current is multiplied by a factor on the right hand side of the figure.

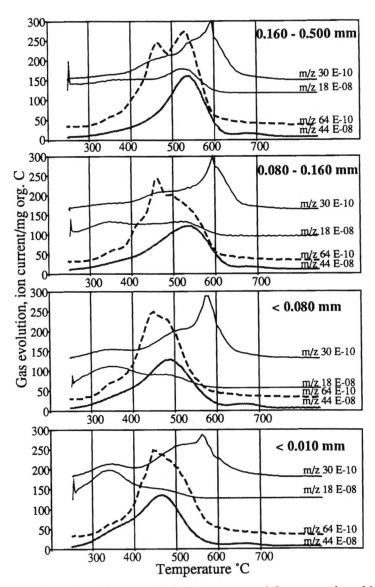

Figure 2. Effect of particle size on PTO of Provence coal. Ion current is multiplied by a factor shown on the right hand side of the figure.

the coal particles, as shown, in Figure 3. The slower penetration of oxygen into the coal causes an increase in CO formation. Increased CO formation can result from oxygen starvation or an increase of the thermal gradient (Figure 4.).

Acid Washing.

Provence coal contains about 10% calcite. Stirring cold dilute HCl with the coal increased sulfur evolution during PTP and PTO (Figure 5). This effect is due to calcite removal. During PTP and PTO calcite starts to decompose into CaO and CO_2 at 400°C. CaO traps sulfur gases (17-21). Other trapping reactions may also be involved, as indicated by results obtained on acid washed coal samples that contained little or no carbonates. Acid washing of sediments from tidal flats and salt marshes of Connecticut that are primarily organic rich clay free of carbonates (22), showed a surprising enhancement of sulfur evolution. These observations suggest that agents other than calcite can be involved in SO_2 fixation during slow heating. A number of acid-washable impurities or mineral additives for SO_2 sorbents, such as NaCl, $CaCl_2$, KCl, Na_2CO_3, Fe_2O_3, coal ash, aqueous solution of $FeSO_3$, $FeSO_4$, and SiO_2, are known to enhance oxidative sorbency of SO_2 (23). H_2S and SO_2 could also be trapped by exchangeable cations linked with clays and with organic polar groups in the organic matrix or by the solid itself (clays, organic polar groups, char (24). Polar groups could be more readily involved in crosslinking reactions when they have been acid-washed, since slow pyrolysis of acid salts produces more CO_2 and less water than the free acid (25-26). Exchangeable cations can also accelerate the air gasification of coal (27-29). A simple acid washing, as well as an HF/HCl demineralization, of Illinois No 6 coal, which contains miscellaneous silicates but little or no calcite (m/z 44 chromatogram from PTP is flat), produced a removal of exchangeable metals in the coal which either catalyze the oxidation of carbon, sulfur and nitrogen, or trap sulfur containing gases, as seen in Figure 6.

Addition of Pure Mineral Pyrite.

Addition of pure mineral pyrite (particle size <40μm) to the Provence coal enhanced the second H_2S peak in the PTP profile and decreased the evolution of other hydrogen containing gases (Figure 5). During PTP, elemental sulfur was emitted from pure pyrite in the same temperature interval that the second pyritic H_2S peak was emitted from coal-pyrite mixtures during PTP. Pyrite-pyrrhotite conversion gives pyrrhotite and atomic sulfur. Then nascent sulfur is converted to H_2S by reaction with weak C-H bonds in the coal leading to a slight depletion of CH_4 evolution and to sulfur chemically bound to carbon (30-34).

Mixing pure pyrite with the Provence coal enhanced the SO_2 peak in the sulfur PTO profile (Figure 5.). The central part of the peak results from a local increase of the thermal ramp due to the strongly exothermic carbon and pyrite oxidation. Paulik (2) showed that the oxidation of pure pyrite is a very heterogeneous process which gave rise either to a large exothermic peak or to a series of numerous needle peaks. The pattern observed is thought to be dependent on the surface energies of the pyrite particles which may vary greatly. Active sites on pyrite particles may be created by lattice distortions, foreign lattice elements incorporated in the lattice, chemisorption of gases and oxidation of surface lattice elements. These heterogeneous exothermic events disturb the reaction temperature control necessary to differentiate between SO_2 generated from organic sulfur and from pyritic sulfur (35-37). The maximum sulfur dioxide evolution from PTO of both pure pyrite and pyrite free organic matter occurs at higher temperature (> 500°C) than in coal-pyrite mixtures. Examination of various kerogen-rock couples, containing the same pyrite, often showed a peak at higher

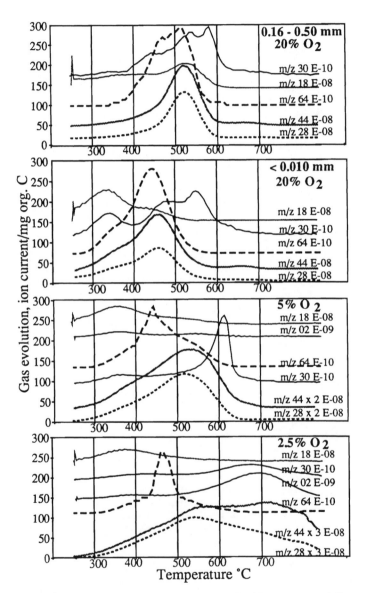

Figure 3. Effect of atmosphere composition on PTO of Provence coal. Ion current is multiplied by a factor shown on the right hand side of the figure.

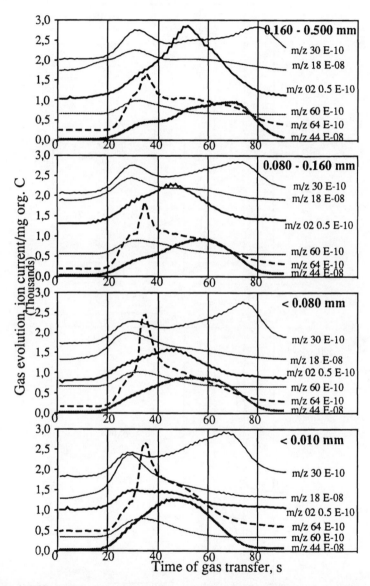

Figure 4. Effect of particle size on gas evolution during rapid heating of Provence coal. The sample was introduced in the oven heated isothermally at 700°C. The flowing gas stream was air. The time of gas tranfer is the time of gas formation and tranfer from the oven to the detector. Ion current is multiplied by a factor shown on the right hand side of the figure.

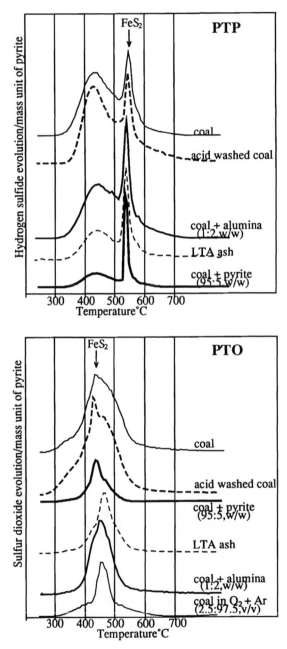

Figure 5. Effect of treatments on sulfur gas evolution during PTP and PTO of Provence coal.

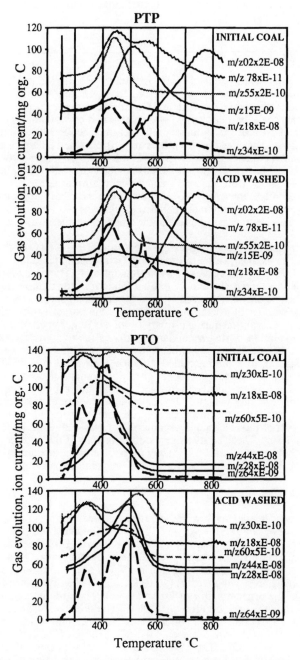

Figure 6. Effect of acid washing on PTP and PTO of Illinois No 6 coal. Ion current is multiplied by a factor shown on the right hand side of the figure.

temperature for rock having a low organic content than for the corresponding kerogen enriched in organic matter.

Organic Solvent Extraction.

Provence coal was acid washed then extracted by pyridine-ethylenediamine (50:50,v/v) at room temperature in an ultrasonic bath. Organic solvent was removed by freeze drying for several days. The residue of freeze drying was a black powder like the initial raw coal. Extraction yield was determined by weighing the extract. The residue was about 50% of the initial sample and only about 30% if extraction was performed on nonacid washed coal (White et al. (38) extracted 23% of Rasa coal with a pyridine/toluene mixture). As Kuczynski and Andrzejak (39) have shown, treatment of coal with acid results in a significant increase in the extractable organic matter. The organic matter extracted from the acid-treated coal is similar to the raw coal and the residue. More recently Larsen (40) showed that little bond cleavage occurs during the demineralization procedure, the mass balance of pyridine extraction of coal is much improved, pyridine retention is reduced and the amount of material extracted increases significantly.

Figure 7 compares the ion chromatograms from PTP and PTO of the initial coal and of its extract as recovered after freeze drying (without any particle size calibration by crushing and sieve analysis). M/z 15 evolution during PTP and m/z 30 during PTO proved that little solvent remained in the extract. The evolution of molecular hydrogen during PTP, which is only slightly sensitive to particle size (surface to volume ratio), indicated that the composition of the initial coal and of the extract is somewhat different. The early hydrogen evolution during primary pyrolysis (cracking/distillation) and the low hydrogen evolution during secondary pyrolysis (coke formation above about 600°C) revealed that the extract either has a lower average molecular weight than the original coal, leading to a preferential distillation over cracking -which consumes donatable hydrogen and gives some coke, or a high content of donatable hydrogen, leading to a release of molecular hydrogen during the primary pyrolysis and a quenching of free radicals which otherwise would give some coke and molecular hydrogen during secondary pyrolysis, or both. In comparison with the initial organic sulfur, a small part of the extractable organic sulfur was converted by PTP into H_2S and by PTO into SO_2. The PTP hydrogen sulfide profile of the extract produced an additional peak between 650 and 700 °C showing a different composition of the extract leading to a different behaviour of sulfur during programmed heating.

Low Temperature Ashing.

Provence coal was treated by low temperature plasma ashing (LTA) in order to selectively remove organic sulfur (41-43). The as received coal sample (crushed to -3 mm) was ashed during seven hours in a Technics Plasma GmbH Plasma-Processor 200 E. Low temperature ashing partially removed the organic sulfur without oxidation of pyritic sulfur. The sulfur content of this"ash" was 5.05 % weight (dry), and the pyrite content was 2.35 % weight (dry) (46.6 % of the total sulfur instead of 20.9% in the initial coal). The organic sulfur of LTA "ash" expressed in percent of the initial coal was 1.22%, and the pyritic sulfur was 1.06 %. The selective loss of organic sulfur was 2.7% of the initial coal, which contains 3.92% of organic sulfur, and there was no loss of pyritic sulfur. Figure 5 presents the SO_2 profile of the LTA "ash" sample and the Provence coal profile. The SO_2 profile of LTA "ash" showed an increase of the central part of the profile due to the selective removal of organic sulfur which occur on the shoulders before and after the pyrite decomposition. Figure 8 shows that the SO_2 curve difference between the original Provence coal and the low temperature ash

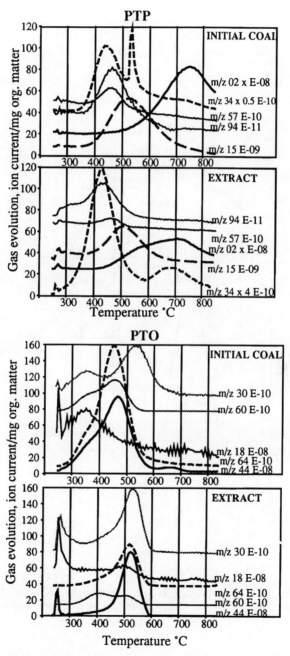

Figure 7. Comparison of Provence coal with its extract (pyridine + ethylenediamine, 50:50, v/v). Ion current is multiplied by a factor shown on the right hand side of the figure.

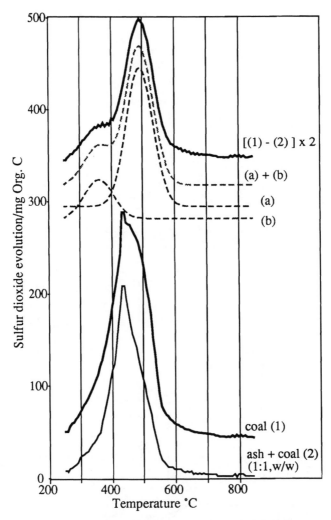

Figure 8. Fitting of two Gaussian curves with the curve difference between the sulfur dioxide evolution from PTO of the Provence coal and of the Low Temperature Ash derived from it.

derived from it -related to the initial total sulfur- showed two overlapping peaks of organic sulfur which could be fitted in terms of two Gaussian peaks. The first peak was small and centered around 350°C. The second was larger and centered at about 500°C, i.e. the peak temperature of CO_2. Therefore, LTA would selectively remove more thermally stable sulfur. LTA is different from PTO because LTA oxidation occurs without any pyrolysis. PTO oxidation produces first a pyrolysis which may lead to formation of a char resistant to oxidation. The selective removal of thermally stable sulfur by LTA is in agreement with MacPhee's work (44) which showed an increase in the ratio of aliphatic to aromatic carbon with time of LTA and leads to the conclusion that excited oxygen reacted with aromatic structures of coal more than with aliphatic ones. Huntington's work (45) showed that polynucelar aromatic compounds can be extraordinarily reactive toward autoxidation.

Coal Sulfur Content and Coal Rank.

Figure 9 shows the evolution of H_2S from PTP and of SO_2 from PTO, expressed with respect to the sulfur content of various coals ordered by increasing rank from peat to anthracite. The H_2S peak temperature, corresponding to the decomposition of organic sulfur in the PTP profile, increased from about 400°C for the peat sample to about 490°C for the hvAb and mvb coal. This increase of peak temperature synchronised with a rapidly decreased gaseous evolution. As seen in Figure 10, the changes shown by PTP parallels both the decreased amount of thiophenic compounds produced by flash pyrolysis and their increased molecular size. A detailed qualitative analyses of these multiring aromatic compounds occurring at high rank has recently been done by Lee's group (46).

Contrary to H_2S, the amount of SO_2 evolution remains constant with coal maturation. The peak temperature of SO_2 rises from about 350°C for the peat sample to about 650°C for the anthracite sample. This temperature increase with coal maturation could be due to a loss of inflammable volatile matter which accelerates the char oxidation, and a relative enrichment of the char in condensed aromatic nuclei more resistant to pyrolytic breakdown, as seen by PTP and Py-GC, but also to a reduction of the size of micropores during coalification which hampers oxygen penetration into the solid matrix.

Conclusions

The effects of experimental conditions on sulfur gas formation during coal programmed temperature pyrolysis (PTP) or oxidation (PTO) were investigated using mass spectrometry to analyse evolved gases. The results showed that experimental conditions dramatically influence the sulfur evolution patterns during PTP and PTO. The physical kinetics of sulfur gas emission during PTP or PTO do not reflect the chemical kinetics of gas formation when the coal particle size is large, when there is a high thermal gradient, or when there is a low molecular oxygen content during oxidation. Acid washing may remove carbonates and indigeneous exchangeable metals which trap hydrogen sulfide or sulfur dioxide. Exchangeable metals and minerals, such as alumina during pyrolysis, may in return catalyze the desulfurization thus changing the observed sulfur gas emission profile. The selective removal of organic sulfur, by solvent extraction or by low temperature ashing, showed some important effects on the decomposition of the organic and pyritic sulfur during pyrolysis or oxidation.

The comparison of the PTP and PTO sulfur profiles of coal samples having low or high sulfur contents and ranging from peat to anthracite showed that the technique provides a plain fingerprint of the change of sulfur composition. However, the results are in agreement with those from more accurate analytical techniques of sulfur

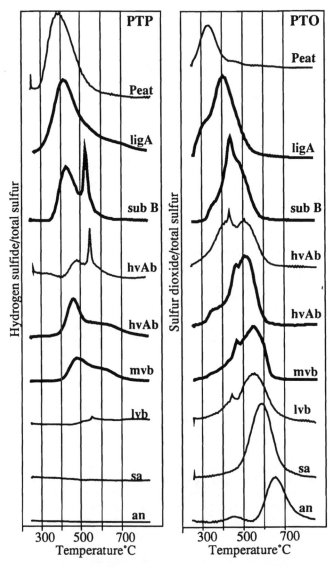

Figure 9. Effect of coal rank on sulfur evolution. ASTM D388 abbreviations of the ranks. Normal line : low sulfur coal, bold line : high sulfur coal.

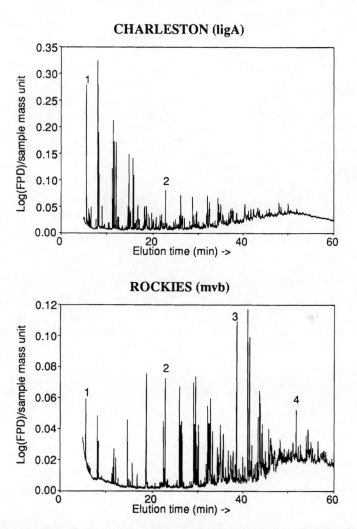

Figure 10. Py-GC-FPD traces showing the effect of maturation on the distribution of sulfur compounds released by flash pyrolysis at 600 °C.
(1 : thiophene, 2: benzothiophene, 3: dibenzothiophene, 4 : naphtobenzothiophene)

characterization used for low or average rank coals, such as flash pyrolysis-gas chromatography-mass spectrometry. The PTO technique offers unmatched capability of sulfur chemical characterization of semianthracite and anthracite coal.

Further developments of the work include a more accurate study of the mechanisms of desulfurization processes using instrumental improvements. This will enable an easy quantitation of gas yield and a thermochemical approach of elemental processes. We also have been using model polymers to better study the interactions of pyrite and sulfur with the organic matrix during coal pyrolysis, oxidation and combustion (34) and to examine more accurately the specific role of organic sulfur in thermal degradation processes.

Acknowledgments

We acknowledge the review of this paper by the editors. We are grateful to Dr. P. Suggate (N.Z. Geological Survey) for providing the New Zealand coal samples and also for comments on the paper. We appreciate the coal samples from the Dutch Centre for Coal Specimen, S.B.N. and from CERCHAR (France). We thank Dr. Cagniant (L.S.C.O., Metz University), Dr. N. Spassky (Laboratory of Macromolecular Chemistry, Paris VI University) and Philips Petroleum for the sulfur reference compounds.

This research was supported by SRETIE (Air-Programme "Métrologie et Réduction des Polluants à l'Emission"), CNRS (GRECO "Hydroconversion et Pyrolyse du Charbon") and IFP (Contrat N° 11642 and N° 12452).

Literature Cited

1. Chantret, F. Bull. Soc. Fr. Mineral. Crist. 1976, 92, 462-467.
2. Paulik, F.; Paulik J.; Arnold, M. J. Thermal Anal. 1982, 25, 327-340.
3. Lacount, R.B.; Gapen, D.K.; King, W.P.; Dell, D.A.; Simpson, F.W.; Helms, C.A. In New Approaches in Coal Chemistry; Blaustein, B.D.; Bockrath, B.C.; Friedman, S., Eds., ACS Symp. Series 169, American Chemical Society : Washington, DC, 1981; pp 415-426.
4. LaCount, R.B.; Anderson, R.R.; Friedman S.; Blaustein B.D. Fuel 1987, 66, 909-913.
5. Yergey, A.L.; Lampe, F.W.; Vestal, M.L.; Day, A.G.; Fergusson, G.J.; Johnston, W.H.; Snyderman, J.S.; Essenhigh, R.H.; Hudson J.E. Ind. Eng. Chem. Process Des. Develop. 1974, 13, 233-240.
6. Attar, A. In Analytical methods for coal and coal products; Karr, C., Ed.; Acad. Press New York, 1979; pp 585-624.
7. Attar, A.; Villoria, R.A.; Verona, D.F.; Parisi, S. ACS Fuel Div. Chem., American Chemical Society : Washington, DC, 1985; pp 1213-1221.
8. Madec, M.; Espitalié, J. J. Anal. Appl. Pyrolysis 1985, 8, 201-219.
9. Krouse, H.R.; Yonge, C.J. In Geochemistry of sulfur in fossil fuels. Symposium series Book; Orr, W.L.; White, C.M., Eds.; American Chemical Society : Washington, DC, 1990.
10. Boudou, J.P.; Boulègue, J.; Maléchaux, L.; Nip, M.; de Leeuw, J.W.; Boon, J.J. Fuel 1987, 66, 1558-1569.
11. Eglinton, T.I.; Sinninghe Damsté, J.S.; Kohnen, M.E.L.; de Leeuw, J.W.; Larter, S.R. In Geochemistry of sulfur in fossil fuels. Symposium series Book; L. Orr, W. L.; White,C.M., Eds.; American Chemical Society : Washington, DC, 1990.
12. Whelan, J.K.; Solomon, P.R.; Deshpande, G.V.; Carangelo, R.M. Energy & Fuels 1989, 2, 65-73.

13. Le Perchec, P.; Thomas, M.; Fixari, B.; Bigois, M. In Analytical chemistry of heavy oils/resids. Preprints Symp. Div. Petr. Chem. American Chemical Society : Washington, DC, 1989; Vol. 34, pp 261-267.
14. Boudou, J.P.; Espitalié, J.; Marquis, F. New methodologies for coal characterization; Charcosset, H. , Ed.; Elsevier : Amsterdam, 1990.
15. Espitalié, J.; Marquis, F.; Barsony, I. In Analytical Pyrolysis; Voorhees, K.J., Ed.; Butterworth and Co, Pub., 1984; pp 276-304.
16. Suggate, R.P New Zealand Department of Scientific and Industrial Research Bulletin 1959, 134, 1-113.
17. Jüntgen, H.; van Heek, K.H. Fortschritte der Chemischen Forschung 1970, 13, 601-699.
18. Attar, A. Fuel 1978, 57, 201-211.
19. Furfari, S.; Cyprès, R. Fuel 1982, 61, 453-459.
20. Erten, M.H. In Coal Science and Technology. Processing and utilization of high sulfur coals; Attia, A., Ed.; Elsevier : Amsterdam, 1985; Vol. 9, pp 451-466.
21. Borgward, R.H.; Bruce, K.R.; Blake, J. Ind. Eng. Chem. Res. 1987, 26, 1993-1998.
22. Canfield, D.C.; Raiswell, R.; Westrich, J.T.; Reaves, C.; Berner, R.A. Chemical Geology 1986, 54, 149-155.
23. Badin, E.J. In Coal Science and Technology; Anderson, L.L., Ed.; Elsevier : Amsterdam, 1984; Vol. 6, pp 40-78.
24. Panagiotidis, T.; Richter, E.; Jüntgen, H. Carbon 1988, 26, 89-95.
25. Schafer, H.N.S. Fuel 1979, 58, 6667-679.
26. Schafer H.N.S. Fuel 1980, 59, 295-304.
27. Hengel, T.D.; Walker, P.L. Fuel 1984, 63, 1214-1220.
28. Charcosset, H.; Tournayan, L.; Nickel, B.; Jeunet, A. In Symposium on the fundamentals of gasification; Winans, R.E.; Jenkins, R.G.; Vorres, K.S.; Stephens, H.P., Eds.; ACS Div. Fuel Chem., American Chemical Society : Washington, DC, 1989; Vol. 34, pp 29A-E.
29. Zhang, Z.G.; Kyotani, T.; Tomita, A. Energy & Fuels 1989, 3, 566-571.
30. Cernic-Simic, S. Fuel 1962, 41, 141-151.
31. Huang, E.Y.K.; Pulsifer, A.H. In Coal desulfurization. Chemical and physical methods. ACS Symposium series; Wheelock, T.D., Ed.; American Chemical Society : Washington, DC, 1977; Vol. 64, pp 290-303.
32. White, C.M.; Douglas, L.J.; Schmidt, C.E. Energy & Fuels 1988, 2, 220-223.
33. Shahab, Y.A.; Siddiq, A.A. Carbon 1988, 26, 801-802.
34. Boudou, J.P.; Roudot, P.; Aune, J.P. (in preparation).
35. Warne, S.St.J.; Bloodworth, A.J.; Morgan, D.J. Thermochim. Acta 1985, 93, 745-748.
36. Schouten, J.C.; Blomaert, F.Y.; Hakvoort, G.; van den Bleek, C.M. In Intern. Conf. Coal Sci. J.A. Moulijn, J.A.; Nater, K.A.; Chermin, H.A.G., Eds; Elsevier : Amsterdam, 1987; pp 837-840.
37. Schouten, J.C.; Hakvoort, G.; Valkenburg, P.J.M.; van den Bleek, C.M. Thermochim. Acta 1987, 114, 171-178.
38. White, C.M.; Douglas, L.J.; Anderson, R.R.; Scmidt, C.E.; Gray, R.J. In Geochemistry of sulfur in fossil fuels. ACS Symposium series book; Orr, W.L.; White, C.M., Eds.; American Chemical Society : Washington, DC, 1990.
39. Kuczynski, W.; Andrzejak, A. Fuel 1961, 40, 203-206.
40. Larsen, J.W.; Pan, C.S.; Shawver, S. Energy & Fuels 1989, 3, 557-561.
41. Gluskotter H. Fuel 1965, 44, 285-291.

42. Paris, B. In <u>Coal desulfurization. Chemical and Physical Methods. ACS</u>
     <u>Symposium series</u>; Wheelock, T.D., Ed.; American Chemical Society :
     Washington, DC, 1977; Vol. 64, pp 22-34.
43. Guilianelli, J.L.; Wiliamson, D.L. <u>Fuel</u> 1982, <u>61</u>, 1267-1272.
44. MacPhee, J.A.; Nandi, B.N. <u>Fuel</u> 1981, <u>60</u>, 169-170.
45. Huntington, J.G.; Mayo, F.R.; Kirshen, N.A. <u>Fuel</u> 1979, <u>58</u>, 31-36.
46. Skelton, R.J.; Chang, Jr. H.C.K.; Farnsworth, P.B.; Markides, K.E.; Lee, M.L.
     <u>Anal. Chem.</u> 1989, <u>61</u>, 2292-2298.

RECEIVED February 28, 1990

MOLECULAR STRUCTURE OF SULFUR COMPOUNDS AND THEIR GEOCHEMICAL SIGNIFICANCE

Chapter 22

Nature and Geochemistry of Sulfur-Containing Compounds in Alberta Petroleums

O. P. Strausz, E. M. Lown, and J. D. Payzant

Department of Chemistry, University of Alberta, Edmonton, Alberta T6G 2G2, Canada

Many new and novel sulfur compounds have been identified in petroleum, the main components being two classes of cyclic sulfides, the terpenoid-derived ones with two through six rings, and the five- and six-membered ring monocyclic *normal* alkane-derived varieties. They all possess alkyl side chains forming homologous series, the distribution of which is thought to reflect the thermal maturity, biodegradation, water washing and redox state of the oil.

Long progressions of alkyl benzo- and dibenzothiophenes have been detected in the aromatic fraction of Athabasca oil sand bitumen, along with some higher aromatic thiophenes.

Oil sand asphaltenes pyrolysis oil also contain *n*-alkane-derived cyclic sulfides along with *n*-alkane-derived thiophenes, benzo- and dibenzothiophenes, showing that the precursor oil was of *n*-alkanoic origin. Approximately 25% of the sulfur in these asphaltenes is in the form of sulfides.

Sulfur in petroleum occurs in many different molecular environments but the bulk of it is usually present as either alkyl substituted five- and six-membered cyclic sulfides or thiophenes and condensed aromatic thiophenes, especially dibenzothiophenes (1). The latter type of compounds are generally more abundant than the former ones. The sulfur content has an influence on the physical properties of the oil, such as viscosity and gravity, determining its commercial value and a broad correlation has been known to exist between sulfur content and API gravity (2). The sulfur is not homogeneously distributed in the oil, but instead the sulfur content varies from fraction to fraction. Higher-boiling fractions generally contain more sulfur, heteroatoms N, O and metals than the lower-boiling fractions and the sulfur content increases in the direction: saturates < aromatics < polars < resins < asphaltene. In the highest molecular weight portions of the aromatics, resins and asphaltene the sulfur compounds are built into and form part of the polymeric framework of the large molecules. In spite of the long list of individual organosulfur compounds detected and isolated from petroleum to date, it was only recently, in the 1980s, that homologous series of sulfur compounds with clear-cut biomarker character have been discovered and their biogeochemical significance recognized. The first of these compounds, reported in 1983 (3), were bicyclic and tetracyclic terpenoid sulfides and sulfoxides, followed by series of publications on other sulfur compounds by

0097–6156/90/0429–0366$08.75/0
© 1990 American Chemical Society

ourselves (4–11) and the research groups in Strasbourg, France (12–14) and Delft, Holland (15–19).

Intensive studies on the sulfur-rich bitumens and heavy oils of the immense Northern Alberta oil sand, carbonate and associated heavy oil deposits since the early 1980s not only led to the discovery of homologous series of novel cyclic terpenoid sulfides but also to the recognition of characteristic structural differences that may exist among sulfur compounds present in the oil, depending on the fraction of the oil from which the sulfur compounds were isolated. All the heavy oils studied from Alberta were biodegraded; on the other hand, most of the other oils from Alberta (Leduc, Bellshill Lake, etc.) and elsewhere (Prudhoe Bay, Enniskillen, Ontario, etc.) were non-biodegraded but nevertheless were also found to contain these cyclic terpenoid sulfides which now appear to be ubiquitous components of oils containing sulfides.

Detection of Cyclic Terpenoid Sulfides and Sulfoxides

The IR spectrum of the most polar fraction of deasphaltened Alberta bitumens obtained from silica gel column chromatography (3) revealed the presence of significant amounts of aliphatic sulfoxides, Figure 1. Employing a combination of electron impact and field ionization mass spectrometry, the latter producing only parent molecular ions, it was possible to detect the presence of a multitude of naphthenic sulfoxides (3) by exact mass measurements, Table I. The data revealed the presence of cyclic, and only cyclic, sulfoxides in these oils ranging in ring number

Table I. Mass Measurements of Sulfoxides

Formula	Measured	Difference x 10^{-4} [a]	z [b]
$C_{10}H_{18}SO$	186.1074	-4	-2
$C_{10}H_{20}SO$	188.1232	-3	0
$C_{11}H_{20}SO$	200.1230	-5	-2
$C_{13}H_{24}SO$	228.1545	-3	-2
$C_{14}H_{28}SO$	244.1864	+3	0
$C_{15}H_{28}SO$	256.1865	+5	-2
$C_{17}H_{30}SO$	282.2012	-6	-4
$C_{17}H_{32}SO$	284.2181	+8	-2
$C_{18}H_{32}SO$	296.2170	-4	-4
$C_{19}H_{36}SO$	312.2494	+7	-2
$C_{23}H_{40}SO$	364.2807	+7	-6

[a]Obs. – calc. [b]Value of z in $C_nH_{2n+z}SO$.

from one to six and in molecular weight from *ca* 200 to 600. The field ionization mass spectrum (FIMS) of the sulfoxides is reproduced in Figure 2 and plots of relative abundances versus carbon number for each of the ring 1 to ring 6 series are shown in Figure 3 (4).

In further studies it was also found that the aromatic fraction of these bitumens as obtained from silica gel chromatography also contained naphthenic sulfides (5) with molecular weight and ring number distribution similar to those of the sulfoxides. Relative abundances *versus* carbon number as obtained from the field ionization mass spectrum for each ring number series are shown in Figure 4. It is clear from these and

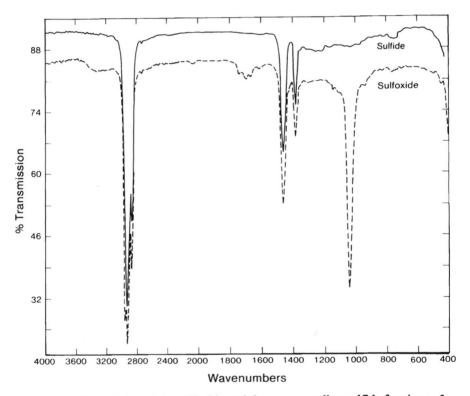

Figure 1. IR spectrum of the sulfoxide and the corresponding sulfide fractions of
deasphaltened Athabasca bitumen. (Reproduced with permission from Ref. 3.
Copyright 1983 Pergamon Press Ltd.)

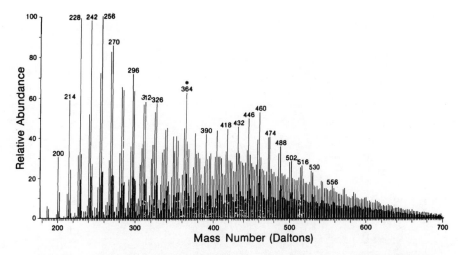

Figure 2. FIMS of the sulfoxide fraction of deasphaltened Athabasca bitumen. Note the prominent peak at 364 daltons, $C_{23}H_{40}SO$. (Reproduced with permission from Ref. 4. Copyright 1985, Alberta Oil Sands Technology and Research Authority.)

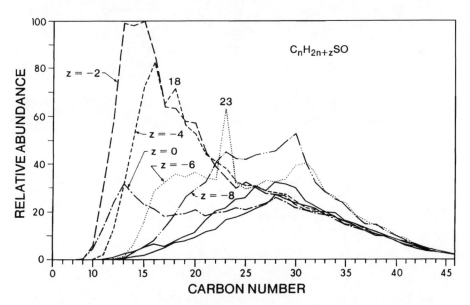

Figure 3. z-Plots for the sulfoxide fraction of deasphaltened Athabasca bitumen as derived from FIMS (Figure 2). The $z = -2$, -4 and -6 traces have maxima at $n = 13$, 18 and 23, respectively. These substances differ by five carbon atoms (an isoprene unit) and one ring. (Reproduced with permission from Ref. 4. Copyright 1985, Alberta Oil Sands Technology and Research Authority.)

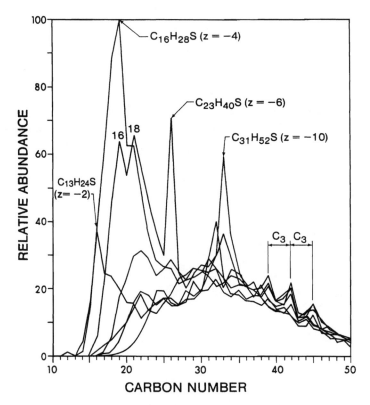

Figure 4. z-Plots for the homologous series of saturated sulfides as derived from FIMS of the aromatic fraction of deasphaltened Athabasca bitumen. The elemental formulae of the prominent peaks were confirmed by high-resolution EIMS. (Reproduced with permission from Ref. 5. Copyright 1985, Alberta Oil Sands Technology and Research Authority.)

the previous Figures that each of the sulfide and sulfoxide series with a given ring number shows characteristic maxima in their distribution curve and that the maxima for the sulfide series and for the sulfoxide series nearly coincide.

The sulfides and sulfoxides can be interconverted. Oxidation of the sulfides with either photochemically-generated singlet oxygen or preferably with tetrabutylammonium periodate (6) selectively oxidizes the sulfides to the sulfoxides and reduction of the sulfoxides with LiAlH$_4$ converts them to the sulfides. (For these and other analytical methods, *cf.* Reference 7) Reduction of the sulfoxides with LiAlH$_4$ showed that the resultant sulfides were identical to the native sulfides isolated from the aromatic fraction of the oil.

In subsequent experiments the sulfides and sulfoxides were separated together from the oil as follows. The maltene was oxidized with either photochemically-generated singlet oxygen or n-Bu$_4$NIO$_4$ in order to convert the sulfides to sulfoxides. From the oxidized maltene, the most polar fraction containing the sulfoxides along with the carboxylic acids was separated by chromatography, reduced with LiAlH$_4$ and the resultant sulfide fraction obtained by chromatographic separation as a pale yellow oil. A capillary GC trace of the sulfide fraction thus prepared from Athabasca oil sand bitumen is shown in Figure 5. This GC trace, which is typical of the biodegraded oils of Northern Alberta, is somewhat reminiscent of the GC trace of the saturate fraction of the oil featuring a broad complex mixture with sharp superimposed peaks due to some of the cyclic terpenoid components. A similar pattern is observed for the hydrocarbons obtained from the hydrodesulfurization of the sulfides.

The sulfoxides comprise several percent of the sulfides in most samples studied. In one Athabasca sample taken from 30 m below the surface and handled under an argon atmosphere, the sulfoxide content was about 2% of the sulfides. These sulfoxides were probably formed in a slow, partial, *in-situ* oxidation of the sulfides by water-dissolved air.

Structure of Bicyclic Terpenoid Sulfides

The mass spectra from capillary GC/MS gave, for the elemental formulae of the B series of peaks (Figure 5), $C_nH_{2n-2}S$, showing that they are bicyclic molecules. All members of the series exhibited a base peak of m/z 183 corresponding to $C_{11}H_{19}S$. This suggests that the rest of the carbon is present as an alkyl side chain, C_nH_{2n+1}, attached to the bicyclic nucleus.

Reduction of the bicyclic sulfides with Raney nickel yielded a homologous series of monocyclic terpanes as illustrated by the example of the C_{13} member (3,8):

Lastly, the NMR spectrum of B$_{13}$ obtained on an ~60% enriched sample (9), Figure 6, clearly shows the presence of the two geminal ring methyls and one terminal chain methyl. On this basis the molecular structure of compound B$_{13}$ has been identified as 1α(H), 6ß(H)-2,2-dimethyl-9-ethylbicyclo-[4,4,0]-8-thiadecane:

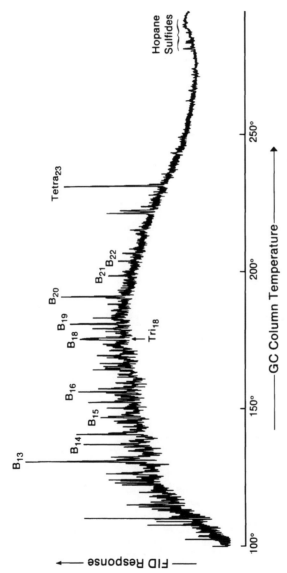

Figure 5. Capillary GC trace of the sulfide fraction from Athabasca maltene. Peaks labelled B_n correspond to bicyclic terpenoid sulfides with n carbons. The tricyclic terpenoid C_{18} sulfide (Tri_{18}) is not resolved from the B_{18} sulfide. Peaks corresponding to the tetracyclic C_{23} sulfide ($Tetra_{23}$) and the hexacyclic sulfides are indicated. (Reproduced with permission from Ref. 10. Copyright 1986, Pergamon Journals Ltd.)

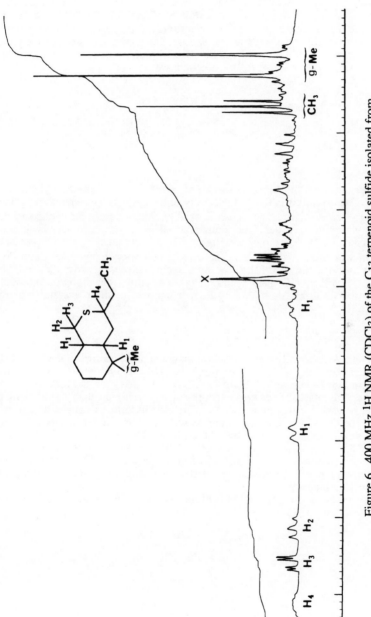

Figure 6. 400 MHz ^{1}H NMR (CDCl$_3$) of the C$_{13}$ terpenoid sulfide isolated from Lloydminster heavy oil. X indicates impurity. The integration curves suggest that the sample is 55–60% pure. (Reproduced with permission from Ref. 9. Copyright 1988, Harwood Academic Publishers.)

The carbon range of these bicyclic terpenoid sulfides lies in the 12–27 region corresponding to C_1–C_{16} alkyl side chains, as can be seen from the m/z 183 single ion recording capillary GC/MS (SIR–GC/MS) spectrum, Figure 7, and other data to be presented later. It is also evident from Figure 7 that each member with a fixed carbon number in the series is present in at least two isomeric forms, the dominant form being assigned to the 9α(H) series while the structures of the other isomers remain unclarified. The variation of relative retention times and signal intensities of the peaks on the GC column shows that the side chains are isoprenoid in nature and thus, the structure of the C_{27} member of the main isomeric series is

The structure of the bicyclic sulfide was further confirmed by deuterium labelling (9). The method is based on the acidity of the hydrogens attached to carbons α to the SO_2 groups of sulfones which can be readily obtained from the cyclic sulfides by oxidation with hydrogen peroxide/acetic acid. The dicyclic sulfone was treated with n-BuLi/CD_3SOCD_3 at room temperature for 16 h (D_2O quench) and converted to the sulfone in which the hydrogens on the α carbons are exchanged for deuterium atoms. From the structure established for the bicyclic sulfide, this number should be three. Indeed, capillary GC mass spectra, Figure 8, show that the experimentally observed number is three, thus lending further support to the correctness of the derived structure.

Structure of Tricyclic Terpenoid Sulfides

This class of sulfides normally represents a minor constituent relative to the bicyclic or tetracyclic terpenoid sulfide (v.i.) components (10). However, in one instance involving a middle Devonian oil (Rainbow, N.W. Alberta) the tricyclic terpenoid complement of the total sulfides was exceptional, much higher than the bicyclic or tetracyclic terpenoid sulfide complements. The carbon number distribution of this series covers the C_{17}–C_{31} range and again, as with the bicyclic terpenoid series, the presence of some isomers is in evidence in the SIR–GC/MS of the m/z 251 base peak of the series, Figure 9. The mass spectrum of the C_{18} member features a strong parent ion and a somewhat weaker m/z 251 ion, while all the spectra of the higher homologues show the m/z 251 peak to be the base peak and the molecular ion peak to be of much weaker intensity.

Raney nickel reduction yields a homologous series of bicyclic terpenoid hydrocarbons of basic structure

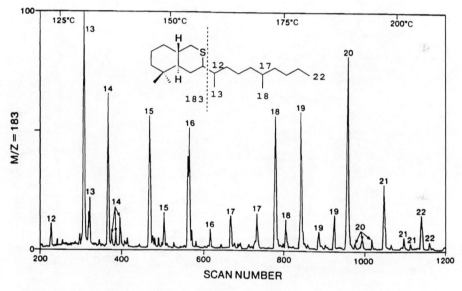

Figure 7. Partial *m/z* = 183 SIR–GC/MS chromatogram showing the distribution of bicyclic terpenoid sulfides in the C_{12}–C_{22} range in Athabasca maltene.

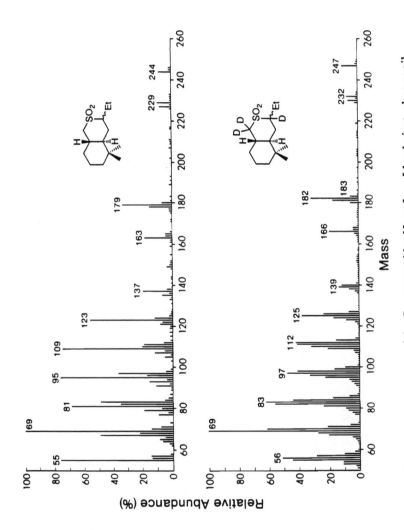

Figure 8. Mass spectra of the C_{13} terpenoid sulfone from Lloydminster heavy oil (upper panel) and the corresponding deuterated sulfone (lower panel) showing the incorporation of three atoms of deuterium. (Reproduced with permission from Ref. 9. Copyright 1988, Harwood Academic Publishers.)

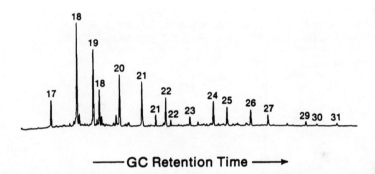

Figure 9. Variation in the carbon number distribution pattern of the tricyclic sulfides in a Cerro Negro sample as shown by the $m/z = 251$ fragmentogram from the SIR–GC/MS experiment. Note the intensity distribution of the peaks. Minima occur at C_{17}, C_{23} and C_{28}. (Reproduced with permission from Ref. 10. Copyright 1986, Pergamon Journals Ltd.)

which has been detected in the saturate fraction of the oil (20). The intense m/z 251 base peak shows that, as in the case of the bicyclic sulfide, the sulfur in this series as well is attached to the second carbon atom of the R side chain. Oxidation of the sulfides to the sulfones followed by deuterium exchange led to the incorporation of two atoms of deuterium, indicating that the second valence of sulfur is attached to a carbon atom which has one hydrogen atom attached to it. This information taken together leads to the consideration of two possible structures for the series:

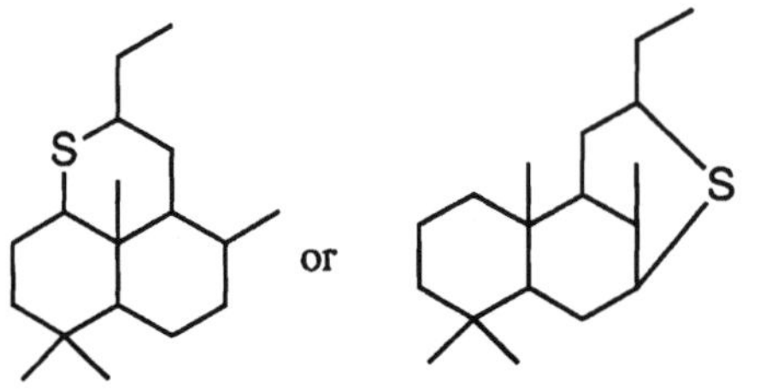

or

The mass spectrometric fragmentation pattern of the C_{18} compound showing a strong m/z 155 peak appears to favor the first structure. Therefore, the tentative structure we assign to the C_{31} *(v.i.)* member of the tricyclic terpenoid sulfides is:

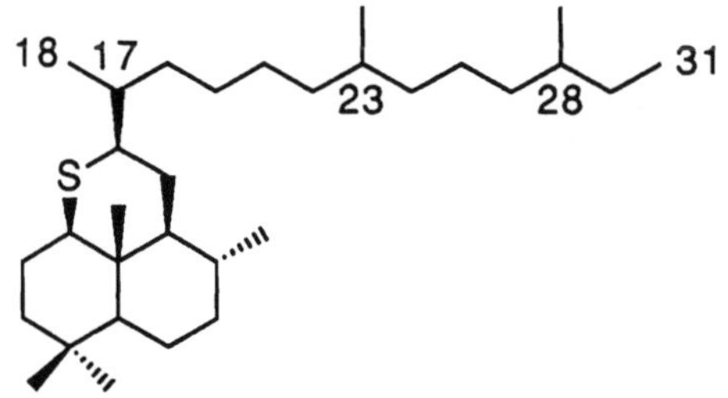

Tetracyclic Terpenoid Sulfides

The structure of this series, as in the previous cases, was established from mass spectral and SIR–GC/MS data, Figure 10, and the known structure of the series of tricyclic terpenoid hydrocarbons, present in the saturate fraction of the oil, the most abundant of which is the C_{23} member (21,22):

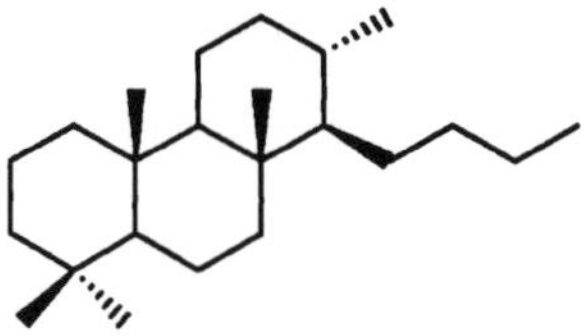

These hydrocarbons are generated from the Raney nickel reduction of the sulfides, and the most abundant member of the sulfides is also the C_{23} compound. Finally, three deuterium-exchangeable hydrogen atoms were observed for the sulfone derivative of the C_{23} sulfide, Figure 11. The structure established for the C_{40} member of the series is:

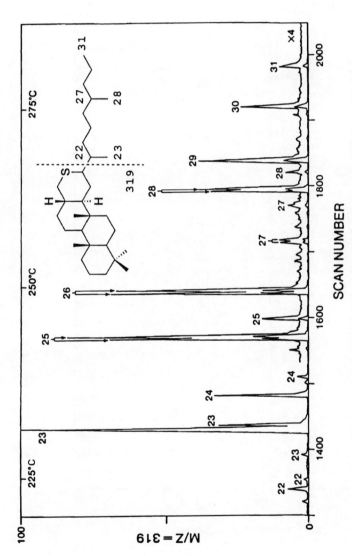

Figure 10. *m/z* = 319 SIR–GC/MS showing the distribution of the tetracyclic terpenoid sulfides in the C$_{22}$–C$_{31}$ range in Athabasca maltene. Note the presence of diastereoisomers.

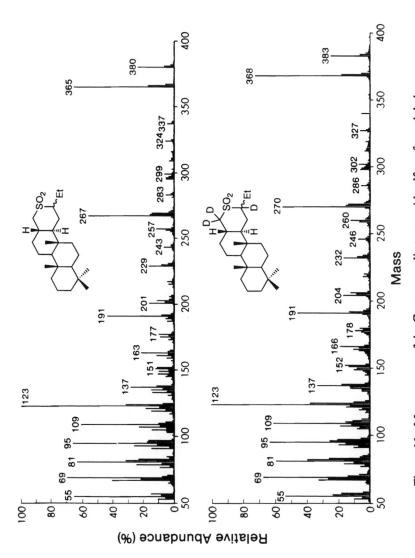

Figure 11. Mass spectra of the C_{23} tetracyclic terpenoid sulfone from Athabasca bitumen (upper panel) and the corresponding deuterated sulfone (lower panel) showing the incorporation of three atoms of deuterium. (Reproduced with permission from Ref. 9. Copyright 1988, Harwood Academic Publishers.)

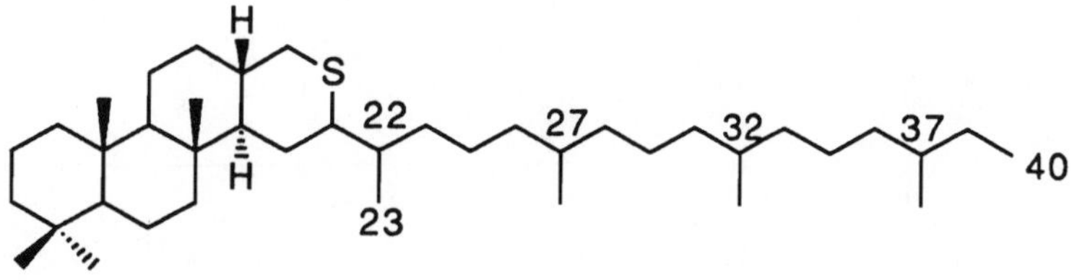

Pentacyclic Terpenoid Sulfides

This class represents a major series of the sulfides, *cf*. Figure 3, and from the appearance of steranes in the hydrocrabons generated from the reduction of the sulfides it is concluded that a portion of the pentacyclic sulfides consists of sterane sulfides in which the sulfur bridge introduced a five- or six-membered fifth ring. We have not further investigated this sulfide class although they may be related to the steroid sulfides identified by the Strasbourg group (12).

Hexacyclic Terpenoid Sulfides

This series of sulfur compounds was identified in a concentrate of sulfides obtained by chromatographic enrichment of the crude sulfide fraction (23). Reduction of the sulfides with Raney nickel gave a suite of C_{30}–C_{35} 17α(H),21β(H) hopanes. SIR–GS/MS of the sulfides showed the presence of $C_{30}H_{50}S$–$C_{35}H_{60}S$ hopane sulfides, Figure 12, with each member present in one major and several minor isomeric forms. Deuterium exchange experiments revealed three hydrogen atoms on the carbons α to the sulfur in the C_{30} members and two hydrogen atoms in the higher members, indicating that the sulfur is attached to the second carbon atom of the alkyl side chain of the hopane. Both the sulfide and sulfone spectra feature intense *m/z* 191 peaks which are unaffected by deuteration. These data taken together suggests two possible structures for the main isomers of the series, which are the following:

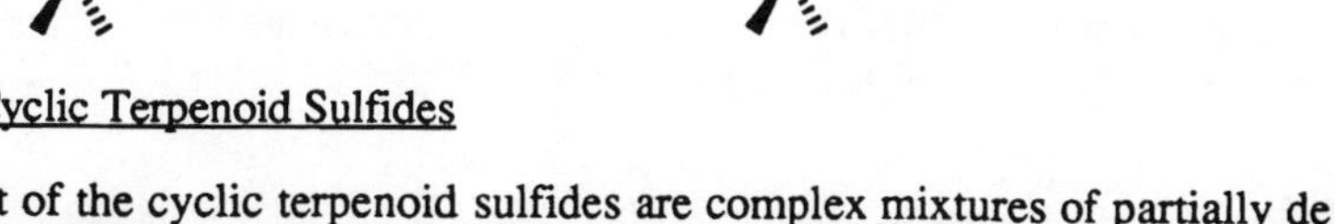

Other Cyclic Terpenoid Sulfides

The rest of the cyclic terpenoid sulfides are complex mixtures of partially degraded and isomerized derivatives of the terpenoid sulfides which elute on the capillary GC column as a broad, unresolved hump. On Raney nickel reduction this fraction yields a complex mixture of naphthenic hydrocarbons which cannot be resolved further by GC analysis.

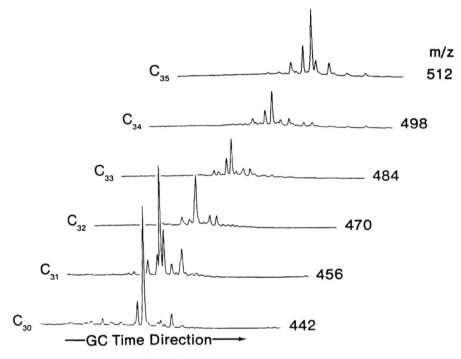

Figure 12. SIR–GC/MS traces for the molecular ions of the hopane sulfides recorded for the whole sulfide concentrate from Athabasca maltene. (Reproduced from Ref. 23 with permission. Copyright 1986, Pergamon Press Ltd.)

n-Alkyl Substituted Thiolanes and Thianes

The sulfide fraction of non-biodegraded petroleums usually contains a complement of n-alkyl substituted sulfides, thiolanes and thianes, in addition to the terpenoid class. In the biodegraded Alberta oil sand bitumens, however, n-alkyl substituted sulfides are absent. The two classes of sulfides can be separated by thiourea adduction, the n-alkyl-containing molecules being adducted. Raney nickel reduction of the Bellshill Lake sulfides (11) yielded a series of n-alkanes showing a distribution quite similar to that of the n-alkanes in the saturate fraction of this oil. In some oils the thiolane and thiane concentrations are commensurate, e.g. Houmann, Figure 13; in others, the concentration of thianes may be small compared to thiolanes.

Biogeochemistry of the Sulfides

The terpenoid class sulfides have been detected in all sulfur-containing bitumens and crude oils investigated to date, involving about three dozen samples, but their distribution has been found to be dependent on the biological and geological history of the oil. The n-alkyl-substituted sulfides, however, are not as ubiquitous owing to their susceptibility to biodegradation (24) and as a consequence they are, in general, absent from the maltene fraction of biodegraded oils. Biodegradation, however, would not attack the large asphaltene micelles and therefore the n-alkyl substituted sulfides may still be preserved as structural units in the asphaltene fraction of biodegraded oils. This is indeed the case with Alberta oil sand bitumens, where the maltene fraction is rich in terpenoid sulfides and devoid of n-alkyl substituted sulfides, while the asphaltene fraction is rich in n-alkyl sustituted sulfides (as structural elements) and nearly (but not completely) devoid of terpenoid sulfides. If the asphaltene is the partial decomposition product of the kerogen from which the oil was formed, then the low terpenoid sufide content of the asphaltene is difficult to reconcile with the high terpenoid sulfide content of the maltene. A simple concentration enhancement owing to the removal of the biodegradable portion of the oil would not explain the required order of magnitude concentration increase. Consequently, one is forced to conclude that Alberta bitumens are the residue of a mixture of two different oils, or that some sulfur incorporation into the oil occurred at a later stage, after the genesis of the oil from kerogen decomposition.

Under thermal stress the terpenoid sulfides tend to undergo side chain loss and consequently thermally more mature oils are leaner in the longer chain members of each series than are thermally less mature oils, Figure 14. Biodegradation does not appear to affect the terpenoid sulfides but has been shown to remove n-alkyl-substituted thiolanes and thianes (24), and thus the terpenoid fraction of the sulfides becomes relatively enhanced on biodegradation.

Water washing of the formation is also capable of causing alterations in the sulfide distribution by differential removal of the more water soluble, lower MW and shorter chain members, Figure 15. This effect is probably the most striking manifestation of the water washing history of the oil-bearing rock and oil sand formations.

Thianes and thiolanes react readily with molecular oxygen to form sulfoxides and aerial or water-dissolved oxygen can partially oxidize the sulfides. Consequently, the sulfoxide to sulfide class ratio may serve as an indicator of the redox conditions in the reservoir of oil sand formations.

Values of total sulfide contents determined in a series of crude oils using the analytical procedure reported in Reference 7 are given in Table II. In some cases, values of the total thiophenes (7) — this also includes benzo-, dibenzo- and other, higher thiophenes — are also listed. The broad trends that emerge from these data are that biodegraded oils are richer in sulfides than are non-biodegraded oils and that both

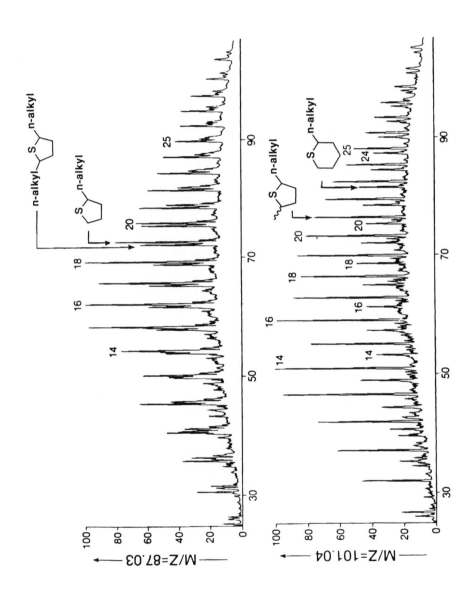

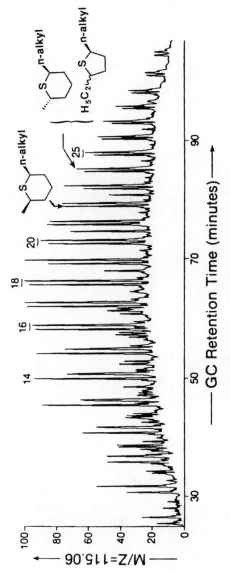

Figure 13. SIR-GC/MS data for the monocyclic sulfides from Houmann petroleum. Each fragmentogram is normalized to the most abundant peak. The relative intensities of the m/z = 87, 101 and 115 fragmentograms are 0.82:1.0:0.42, respectively. The number above each peak corresponds to the total carbon number in the molecule. (Reproduced from Ref. 11 with permission. Copyright 1989, Pergamon Press Ltd.)

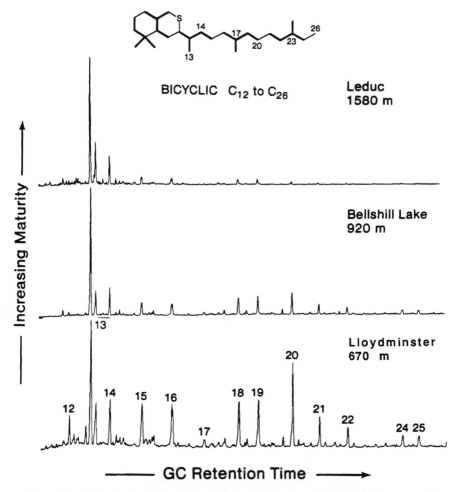

Figure 14. Variation in the carbon number distribution of the bicyclic terpenoid sulfides as a function of depth as shown by the $m/z = 183$ fragmentograms from the SIR–GC/MS experiment. All traces are normalized to the most abundant peak. The oils vary from a heavy Cretaceous oil (Lloydminster) to a light Devonian oil (Leduc). Increasing thermal maturity results in the gradual loss of the isoprenoid side chain until in the Leduc the C_{13} compound dominates the distribution. Note the intensity distribution of the peaks. Minima occur at C_{12}, C_{17}, and C_{23}. (Reproduced from Ref. 10 with permission. Copyright 1986, Pergamon Journals Ltd.)

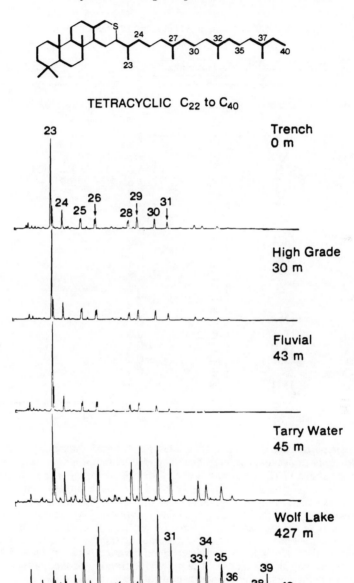

Figure 15. Variation in the carbon number distribution of the tetracyclic terpenoid sulfides as shown by the *m/z* = 319 fragmentogram from the SIR–GC/MS experiment. All traces are normalized to the most abundant peak. The carbon number distribution pattern of the tetracyclic sulfides is always dominated by the C_{23} compound except in samples of extreme water washing, *e.g.* Wolf Lake. Note the intensity distribution of the peaks. Minima occur at C_{22}, C_{27}, C_{32} and C_{37}. (Reproduced from Ref. 10 with permission. Copyright 1986, Pergamon Journals Ltd.)

the sulfide and thiophene contents in general tend to decline with increasing depth of burial.

Table II. Sulfide and Thiophene Contents of Various Petroleums

Sample	Depth (m)	Asphaltene (%)	Sulfides	Thiophenes	Biodegraded
			(% of maltene)		
Athabasca	0	17	6.9	6.4	yes
Cold Lake	550	14	3.6	2.9	yes
Peace River	558	20	16.0	7.6	yes
Lloydminster	670	13.6	2.6	5.7	yes
Wabasca	600	11	2.3		yes
Enniskillen	590	n.d.	2.2		no
Bellshill L.	920	n.d.	3.1	5.5	no
Houmann	1006	3.8	0.2		no
N. Premier	1087	n.d.	1.4		no
Boundry	1268	3.5	0.9		no
Pembina	1398	1.7	0.2		no
Nordegg	1423	17	2.2		no
Kingsford	1434	2.5	0.8		no
Leduc	1617	n.d.	0.3		no
Utikuma L.	1726	n.d.	1.2		no
Rainbow	1795	n.d.	0.3		no
Kumak	2310	n.d.	<0.1		no
Brazeau	2559	nil	<0.1		no
Judy Creek	2605	n.d.	<0.1		no
Iverson	3551	nil	<0.1		no

Thiophenes

The maltene fraction of Alberta bitumens is rich in thiophenes, dominated by long progressions of alkyl-substituted benzo- and dibenzothiophene series up to ~C_{40}, Figures 16 and 17 (5). Various higher aromatized thiophenes of serial formulae $C_nH_{2n-24}S$, $C_nH_{2n-28}S$, $C_nH_{2n-28}S_2$, etc. have also been detected. Most of these thiophenes appear to have originated from the decomposition of asphaltene and the precursor kerogen to the oil. This postulate is supported by the observation that mild thermolysis of the oil sands decreases the asphaltene content of the bitumen and increases the thiophene content of the maltene. In the few cases where the thiophene contents of the maltene have been determined it varied between 2.9 and 7.6%, which is similar to the sulfide content. The high MW fraction of the maltene, however, is probably much richer in thiophenes than in sulfides.

Sulfur in Asphaltene

Oxidation studies on Athabasca and other oil sand asphaltenes have shown the presence of aliphatic sulfides in amounts of up to 25% of the total sulfur (25). The structure of these sulfides has been established using mild thermolysis to liberate them from the polymeric framework of the asphaltene molecules. The produced pyrolysis oil contains significant concentrations of the sulfides and can be readily subjected to analysis. SIR–GC/MS traces of the homologous series of sulfides identified are

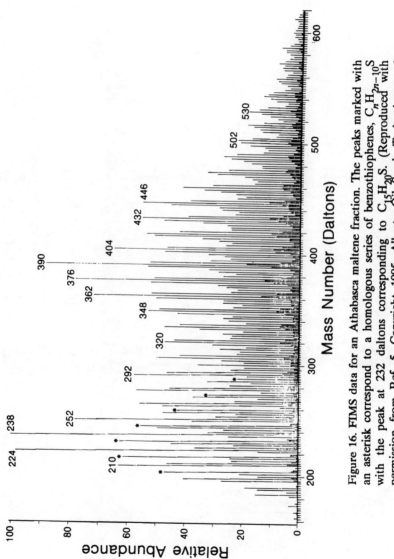

Figure 16. FIMS data for an Athabasca maltene fraction. The peaks marked with an asterisk correspond to a homologous series of benzothiophenes, $C_nH_{2n-10}S$ with the peak at 232 daltons corresponding to $C_{15}H_{20}S$. (Reproduced with permission from Ref. 5. Copyright 1985, Alberta Oil Sands Technology and Research Authority.)

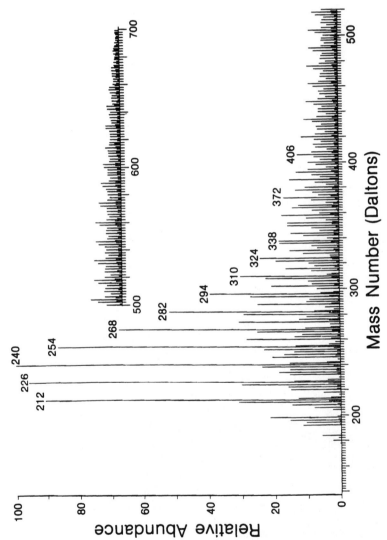

Figure 17. FIMS data for an Athabasca maltene fraction. The prominent series of peaks at 212, 226 and 240 daltons correspond to a homologous series of dibenzothiophenes, $C_nH_{2n-16}S$ with $n = 14$, 15 and 16, respectively. (Reproduced with permission from Ref. 5. Copyright 1985, Alberta Oil Sands Technology and Research Authority.)

shown in Figure 18 (26). The common structural denominator throughout these series is the α and α,α' substitution pattern, revealing an *normal* alkanoic carbon skeleton and providing evidence for an *n*-alkanoic origin for the precursor oil. This is in sharp contrast with the composition of the saturate fraction of the bitumen which is practically devoid of *normal* alkanes as a result of their microbial removal during the secondary microbiological degradation of the precursor oil, giving rise to the formation of the bitumen. On the other hand, the *n*-alkanoic moieties in the asphaltene have been protected against microbial attack by the micellar structure of the large asphaltene molecules and hence their structural integrity has been preserved. From these results, it appears likely that the precursor oil of the Alberta oil sand bitumen once had a *n*-alkanoic thiane/thiolane complement which, being biodegradable, has been removed during the secondary degradation of the oil. The Bellshill Lake oil which is rich in *n*-alkyl thianes and thiolanes has been shown to be genetically related to the oil sand bitumens of the area and could be a remnant of the precursor oil, thus lending support to the above hypothesis.

The *n*-alkyl thianes and thiolanes are thermally interconvertible, with the equilibrium shifted toward the thiolane side. Also, the thiolane ring can move along the *n*-alkyl side chain hopping by three carbon atoms until it reaches a terminal position (11). For this reason the original thiane/thiolane distribution in the asphaltene molecule before its thermal breakup could have been somewhat different from that in the pyrolysis oil.

The rest of the sulfur in oil sand asphaltenes, which is not in the form of thiane and thiolane, is present as thiophenes. Indeed, the asphaltene pyrolysis oil contains homologous series of α and α,α' *n*-alkyl substituted thiophenes, 2-, 4- and 2,4-

$$R_1, R_2 = H, CH_3 \ .... \ n\text{-alkyl}$$

n-alkyl benzo[b]thiophenes, Figures 19 and 20, as well as 1-*n*-alkyl-substituted dibenzothiophenes which have recently been identified. The sulfur substitution pattern in the thiophenes follows that established for thianes and thiolanes, further strengthening the argument in favor of an *n*-alkanoic precursor to the Alberta oil sand bitumens. The identification of the various classes of sulfur compounds in asphaltene pyrolysis oils also represents an important step in the elucidation of structural details of asphaltene molecules.

Sulfur Isotopic Studies

Presently, only preliminary results are available. Two oils have been studied (27), one was the biodegraded Athabasca bitumen and the other, the Rainbow conventional oil (*cf.* Table II). The following values of $\delta^{34}S$ were obtained:

	Sulfides	Thiophenes
Athabasca	+4.6, +4.9	+2.0, +2.0
Rainbow	+0.4, +0.4	−4.4, −4.5

Thus, the sulfides in both oils are less depleted in the heavier ^{34}S relative to the thiophenes. This difference in the sulfur isotope values between the sulfides and the thiophenes in the same oil appears to be consistent with a different origin, as

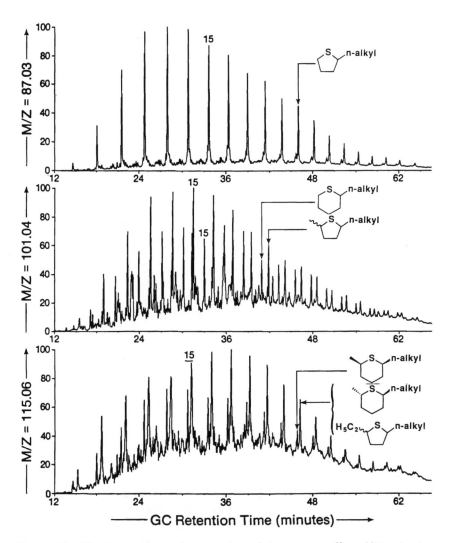

Figure 18. Distribution by carbon number of the monocyclic sulfides in the pyrolysis oil of Athabasca asphaltene as determined by SIR–GC/MS. Peaks labelled 15 correspond to compounds having 15 carbon atoms. Each fragmentogram is normalized to the most abundant peak. The relative intensities of the *m/z* = 87, 101 and 115 fragmentograms are 9.1:3.1:1.0, respectively. (Reproduced with permission from Ref. 26. Copyright 1988, Alberta Oil Sands Technology and Research Authority.)

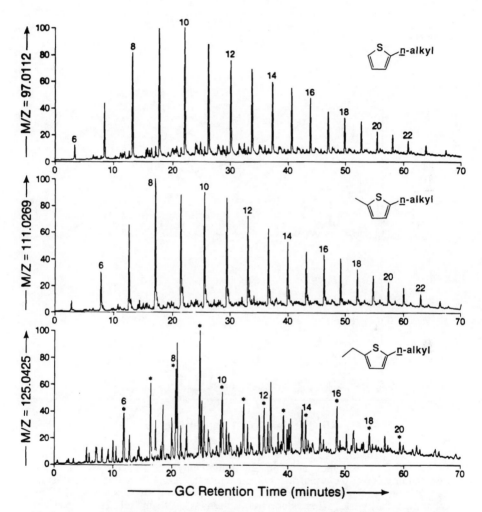

Figure 19. Distribution of the thiophenes in the pyrolysis oil of Athabasca asphaltene as determined by SIR–GC/MS. The number above each peak corresponds to the number of carbon atoms in the *n*-alkyl side chain. Each fragmentogram is normalized to the most abundant peak. (Reproduced with permission from Ref. 26. Copyright 1988, Alberta Oil Sands Technology and Research Authority.)

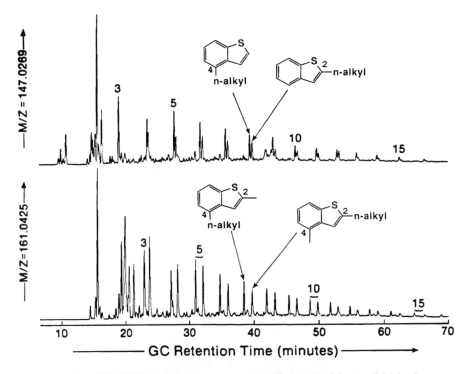

Figure 20. SIR–GC/MS fragmentograms for the *n*-alkyl benzo[b]thiophenes (m/z = 147 Daltons, upper panel) and the monomethyl benzo[b]thiophenes (m/z = 161 Daltons, lower panel). The numbers above the peaks indicate the number of carbon atoms in the *n*-alkyl side chain. Each fragmentogram is normalized to the most abundant peak. (Reproduced with permission from Ref. 26. Copyright 1988, Alberta Oil Sands Technology and Research Authority.)

suggested above, for these two classes of sulfur compounds. The fact that the mean ^{34}S values for the two oils are so different presumably relates to the original source of sulfur in the oils. Further isotopic studies are in progress.

Acknowledgments

We thank the Alberta Oil Sands Technology and Research Authority and the Natural Sciences and Engineering Research Council of Canada for financial support, and Dr. Theodore Cyr for many valuable discussions on petroleum chemistry during the course of these studies.

Literature Cited

1. Orr, W. in Oil Sand and Oil Shale Chemistry; Strausz, O.P.; Lown, E.M., Eds.; Verlag Chemie: N.Y., 1978, p.223–243.
2. Koots, J.A.; Speight, J.G. Fuel 1975, 54, 179.
3. Payzant, J.D.; Montgomery, D.S.; Strausz, O.P. Tetrahedron Lett. 1983, 24, 651–654.
4. Payzant, J.D.; Hogg, A.M.; Montgomery, D.S.; Strausz, O.P. AOSTRA J. Res. 1985, 1, 203–210.
5. Payzant, J.D.; Hogg, A.M.; Montgomery, D.S.; Strausz, O.P. AOSTRA J. Res. 1985, 1, 183–202.
6. Payzant, J.D.; Mojelsky, T.W.; Strausz, O.P. Energy and Fuels 1989, 3, 449.
7. Strausz, O.P.; Payzant, J.D.; Lown, E.M. "Isolation of Sulfur Compounds From Petroleums", this volume.
8. Payzant, J.D.; Cyr, T.D.; Montgomery, D.S.; Strausz, O.P. Tetrahedron Lett. 1985, 26, 4175–4178.
9. Payzant, J.D.; Cyr, T.D.; Montgomery, D.S.; Strausz, O.P. in Geochemical Markers; Yen, T.F.; Moldowan, J.M., Eds.; Harwood Academic Publishers: Chur, Switzerland, 1988, p. 133–147.
10. Payzant, J.D.; Montgomery, D.S.; Strausz, O.P. Org. Geochem. 1986, 9, 357–369.
11. Payzant, J.D.; McIntyre, D.D.; Mojelsky, T.W.; Torres, M.; Montgomery, D.S.; Strausz, O.P. Organic Geochem. 1989, 14, 461–473.
12. Valisolalao, J.; Perakis, N.; Chappe, B.; Albrecht, P. Tetrahedron Lett. 1984, 25, 1183–1186.
13. J.C. Schmid, PhD Thesis, University of Strasbourg, 1986.
14. Schmid, J.C.; Connan, J.; Albrecht, P. Nature 1987, 329, 54–56.
15. Brassel, S.C.; Lewis, C.A.; de Leeuw, J.W.; de Lange, F.; Sinninghe Damsté, J.S. Nature 1986, 320, 160–162.
16. Sinninghe Damsté, J.S.; ten Haven, H.L.; de Leeuw, J.W.; Schenck, P.A. Org. Geochem. 1986, 10, 791–805.
17. Sinninghe Damsté, J.S.; de Leeuw, J.W.; Kock-van Dalan, A.C.; de Zeeuw, M.A.; de Lange, F.; Rijpstra, W.I.C.; Schenck, P.A. Geochim. Cosmochim. Acta 1987, 51, 2369–2391.
18. Sinninghe Damsté, J.S.; Rijpstra, W.I.C.; de Leeuw, J.W.; Schenck, P.A. Geochim. Cosmochim. Acta 1989, 53, 1323–1341.
19. Sinninghe Damsté, J.S.; Rijpstra, W.I.C.; Kock-van Dalan, A.C.; de Leeuw, J.W.; Geochim. Cosmochim. Acta 1989, 53, 1343–1355.
20. Dimmler, A.; Cyr, T.D.; Strausz, O.P. Org. Geochem. 1984, 7, 231–238.
21. Ekweozor, C.M.; Strausz, O.P. in Advances in Organic Geochemistry 1981; Bjøroy, M. et al., Eds.; Wiley Heyden: London, England, 1983, p. 746–766.
22. Ekweozor, C.M.; Strausz, O.P. Tetrahedron Lett. 1982, 23, 2711–2714.

23. Cyr, T.D.; Payzant, J.D.; Montgomery, D.S.; Strausz, O.P. Org. Geochem.
     1986, 2, 139–143.
24. Fedorak, P.M.; Payzant, J.D.; Montgomery, D.S.; Westlake, D.W.S. Appl.
     Environ. Microbiol. 1988, 54, 1243–1248.
25   Mojelsky, T.W.; Montgomery, D.S.; Strausz, O.P. AOSTRA J. Res. 1986, 3,
     43.
26. Payzant, J.D.; Montgomery, D.S.; Strausz, O.P. AOSTRA J. Res. 1988, 4,
     117–131.
27. Krouse, R.; Montgomery, D.S.; Payzant, J.D.; Strausz, O.P. unpublished
     results.

RECEIVED March 26, 1990

Chapter 23

Identification of Alkylthiophenes Occurring in the Geosphere by Synthesis of Authentic Standards

T. M. Peakman[1] and A. C. Kock-van Dalen

Organic Geochemistry Unit, Faculty of Chemistry and Materials Science, Delft University of Technology, De Vries van Heystplantsoen 2, 2628 RZ Delft, Netherlands

Identification of naturally occurring organic sulfur compounds (OSC) is possible by co-chromatography (on capillary GC) and comparison of mass spectra (by GC-MS) with synthesized standards. Methods of preparation of various thiophenes (*e.g.* from suitable thiophene precursors and coupling with either the appropriate carboxylic acid, aldehyde or alkyl halide) will be outlined in addition to some of their spectral properties (NMR, IR, MS). Specific syntheses of biological marker OSC will be discussed. These will be exemplified by series of alkyl, branched and isoprenoidal thiophenes. These methodologies have been used for the preparation and subsequent identification of series of alkylthiophenes comprising a linearly extended phytane skeleton.

This decade has seen the publication of many articles concerning the nature of organic sulfur compounds (OSC) in sedimentary organic matter (1-20). Of particular interest has been the recognition of many thiophenes, benzo-thiophenes, thiolanes and thianes with biological marker-type structures. The precise identification of such compounds is only possible by either isolation, and subsequent structural elucidation, or by comparison of the mass spectra of, and by GC co-injections with, authentic standards synthesized in the laboratory. This approach has been particularly important in the field of OSC where many components have similar mass spectra and similar GC retention times. The distributions of, and the presence or absence of various structural types, are important in solving geochemical problems, *e.g.* source and environmental deposition of sedimentary organic matter, correlation of oils and source rocks, extents of in-reservoir biodegradation and migration processes.

To date only three research groups have been primarily concerned with the synthesis of various sulfur-containing biological marker compounds. These are

[1]Current address: Institute of Petroleum and Organic Geochemistry (ICH-5), KFA Jülich, Postfach 1913, D–5170 Jülich, Federal Republic of Germany

0097–6156/90/0429–0397$06.00/0
© 1990 American Chemical Society

the research groups of Professor Strausz (Edmonton, Canada), Dr. Albrecht (Strasbourg, France) and Dr. de Leeuw (Delft, The Netherlands). The research group of Professor Castle (South Florida, U.S.A.) has also carried out numerous syntheses of various polyaromatic sulfur-containing compounds. In this article we shall outline various methods for the preparation of alkylthiophenes. These methods will be exemplified by examples taken mainly from work carried out in Delft.

Methodologies

Alkylthiophenes have been known for over one hundred years and the traditional method of preparation was the reaction of suitable 1,4-dicarbonyl compounds with phosphorous trisulfide or pentasulfide (Figure 1).

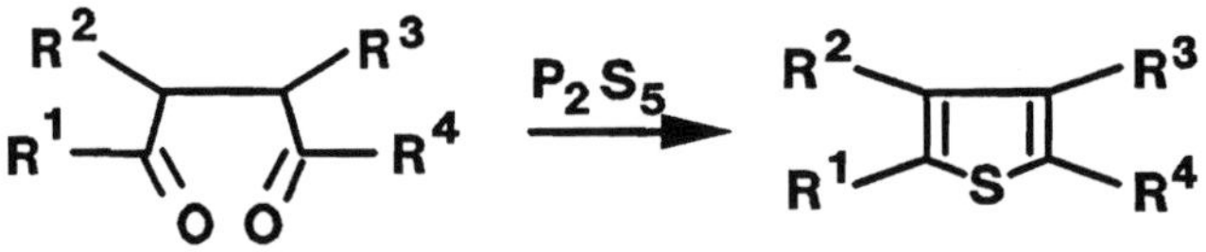

Figure 1. Traditional preparation of alkylthiophenes.

This and other older methods, which have been reviewed previously (21, 22), have been largely superceded in the laboratory by more controlled, selective reactions, starting from suitable thiophene precursors. We shall discuss two major reactions which can be used for the preparation of alkylthiophenes: electrophilic substitution of thiophenes and the formation and reactions of lithiated thiophenes. For further details and other aspects of thiophene chemistry the reader is referred to previous reviews (21, 22) and references cited therein. Accounts of yearly developments in thiophene chemistry can also be found in the relevant Royal Society of Chemistry publications.

Electrophilic Substitution of Thiophenes. Electrophilic substitution involves the reaction of electron deficient species, so-called electrophiles (E^+), with suitable substrates, in this case thiophene or alkylthiophenes, to give a 2-substituted thiophene (Figure 2).

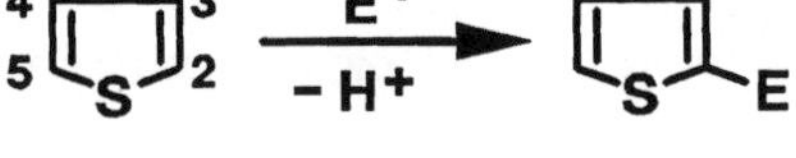

Figure 2. Reaction between thiophene and a hypothetical electrophile (E^+).

This is the typical reaction undergone by thiophenes, *i.e.* electrophilic substitution at the carbon atom adjacent to the sulfur (positions 2 and 5). This immediately raises the question "why does electrophilic substitution occur primarily at C-2 (or C-5)?" Consideration of Wheland intermediates gives an easily understandable, simplified picture of what is happening. In the case of electrophilic substitution of thiophene at C-2 three canonical forms can be drawn [Figure 3(i)] which in effect delocalize the positive charge almost around the thiophene ring. For electrophilic substitution at C-3 only two canonical forms can be drawn [Figure 3(ii)]. The more canonical forms which can be drawn implies greater stability of the hypothetical intermediate leading to substitution and hence reaction at C-2 is favored over that at C-3. This selectivity can decrease with particularly vigorous reagents [*e.g.* nitration of thiophene gives a small amount (up to 15%) of 3-nitrothiophene in addition to 2-nitrothiophene] although this problem does not affect us here.

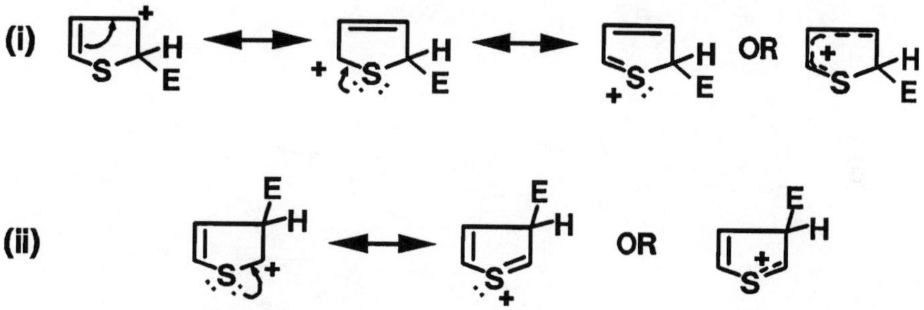

Figure 3. Consideration of Wheland intermediates for electrophilic substitution of thiophene at: (i) C-2, (ii) C-3.

In deciding the outcome of electrophilic substitution of various alkylated thiophenes a similar approach to that given above gives the correct answers. We shall illustrate this for the case of electrophilic substitution of 2-alkylthiophenes. Substitution at C-3 [Figure 4 (i)] gives rise to two canonical forms. One of these forms has the positive charge located at C-2 which is substituted with an alkyl group. The electron donating or positive inductive effect of the alkyl group will, therefore, stabilize the positive charge to some extent. Electrophilic substitution at C-4 [Figure 4(ii)] gives rise to two canonical forms neither of which has the positive charge located on the alkyl substituted carbon atom. For the case of C-5 [Figure 4(iii)] three canonical forms can be drawn and one of these has the positive charge located on the alkyl substituted carbon atom. Taking the three possibilities together it is clear that substitution at C-5 is particularly favoured, that at C-3 partly and that at C-4 not at all. In practice electrophilic substitution of C-2 alkylated thiophenes occurs at C-5 unless particularly vigorous reagents are used. Comparison of Figure 4 (iii) with Figure 3(i) indicates that 2-alkyl-thiophenes are theoretically more reactive towards electrophilic substitution than thiophene itself. This is indeed the case for small alkyl groups.

Figure 5 summarizes, in tabular form, the products of electrophilic substitution of various alkylated thiophenes. Note that in the case of 3-alkylated thiophenes a mixture of products may be obtained. The 2,3-isomer is the favoured product but steric effects may play a role here resulting in the formation of the 2,4-isomer. It should be mentioned in passing that under certain special conditions it is possible to obtain the unfavoured electrophilic substitution products (22).

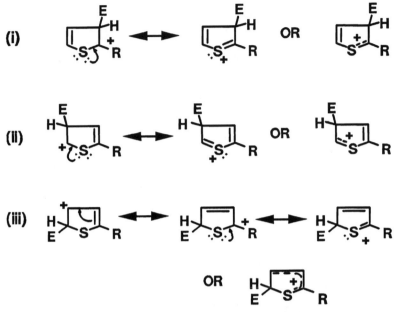

Figure 4. Consideration of Wheland intermediates for electrophilic substitution of 2-alkylthiophenes at: (i) C-3, (ii) C-4, (iii) C-5.

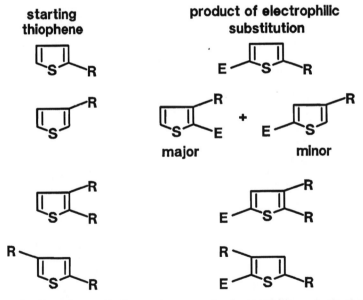

Figure 5. Summary of the outcome of electrophilic substitution of alkylthiophenes. R=alkyl group.

So far we have only been dealing with the hypothetical electrophile E^+. What, in practice, are these electrophiles? We shall consider those that are likely to prove useful in the synthesis of alkylated thiophenes.

Acylation.   One of the most useful and important methods for preparations of alkylthiophenes is acylation. This is the reaction of carboxylic acids and carboxylic acid chlorides with thiophenes in the presence of a suitable catalyst and leads to 2-acylthiophenes in good yields (typically *ca.* 60-80%; *e.g.* 23). Reduction of the ketone functionality by either a modified Wolff-Kishner procedure (8, 24) or with a mixed hydride, formed from aluminium trichloride and lithium aluminium hydride (25), yields the alkylthiophene (Figure 6). Alternatively, the ketone group can be alkylated, thereby giving access to 2-(1'-alkyl)alkylthiophenes.

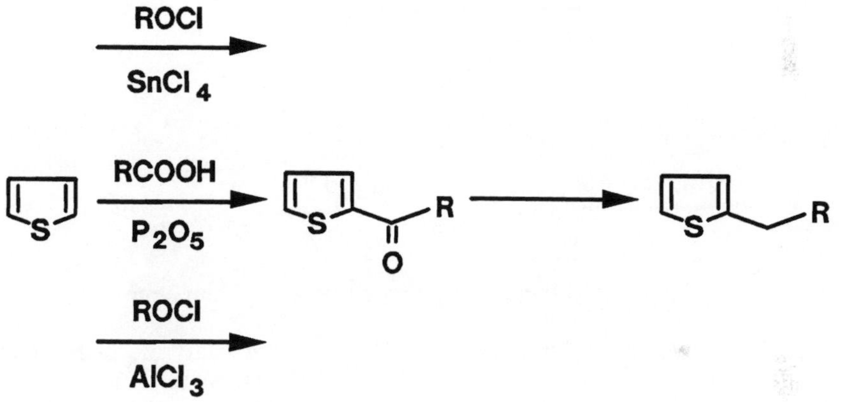

Figure 6. Preparation of 2-alkylthiophenes by acylation and reduction of the resulting ketone.

Examples of the use of acylation in the preparation of alkylthiophenes as standards for organic geochemical studies are given below:

(i) 5-nonyl-2-pentadecylthiophene (C_{28}). Investigations of the OSC of the Jurf ed Darawish oil shale (19) indicated the presence of C_{28} 2,5-dialkylthiophenes in high relative abundance. The major components were the mid-chain dialkyl-thiophenes and mass spectral interpretation suggested the presence of only two major isomers, 2-hexadecyl-5-octylthiophene and 5-nonyl-2-pentadecylthiophene. Evidence for this observation came from co-injection of the above two compounds, synthesized in the laboratory, and by comparison of their mass spectra. Reference to Figure 5 indicates that 2,5-dialkylthiophenes can be obtained from 2-alkylthiophenes which in turn can be prepared from thiophene itself (Figure 3). Thus, 5-nonyl-2-pentadecylthiophene was prepared by acylation of thiophene with nonanoyl chloride in the presence of tin tetrachloride. The resulting acylated thiophene was reduced with a mixed hydride, prepared from aluminium trichloride and lithium aluminium hydride (25), to give 2-nonylthiophene. This was further acylated with pentadecanoyl chloride, in the presence of tin tetrachloride, and the product again reduced with the mixed hydride to afford the required 5-nonyl-2-pentadecylthiophene (Figure 7).

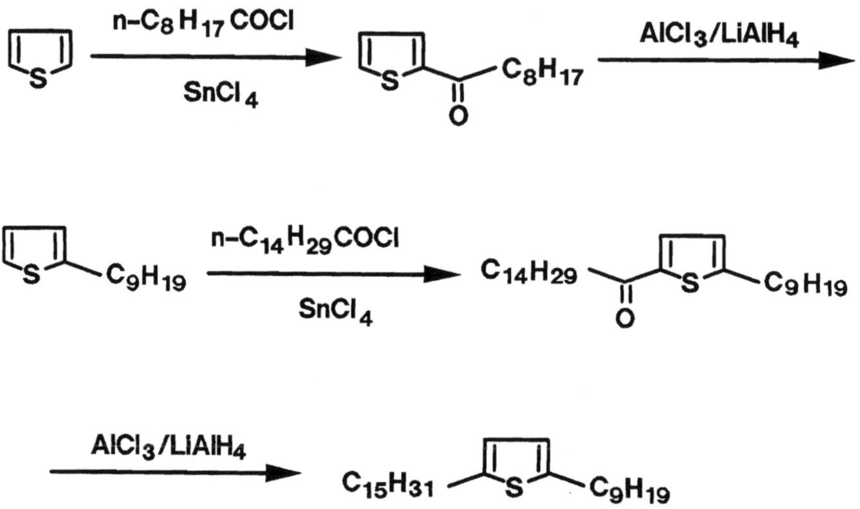

Figure 7. Synthesis of 5-nonyl-2-pentadecylthiophene.

(ii) 2,5-diheptyl-3-methylthiophene (C_{19}). Also present in high relative abundance in the solvent extract of the Jurf ed Darawish shale are three alkylthiophenes with a carbon skeleton corresponding to that of 9-methyloctadecane (19). Mass spectral studies suggested that one of these was 2,5-diheptyl-3-methylthiophene. Reference to Figure 5 indicates that such a compound can be prepared from 3-methylthiophene in two overall stages. Fortunately the symmetrical substitution of the heptyl units at C-2 and C-5 allows us to obtain a pure final product even though the intermediates will be mixtures of isomers. Thus, acylation of 3-methylthiophene gave a mixture of 3-methyl-2-(1'-oxo)heptyl- and 4-methyl-2-(1'-oxo)-heptylthiophene isomers, which were reduced with the mixed hydride, to give a mixture of 2-heptyl-3-methyl- and 2-heptyl-4-methylthiophenes. Acylation, again with heptanoyl chloride, afforded a mixture of 2-heptyl-3-methyl-5-(1'-oxo)heptyl- and 5-heptyl-3-methyl-2-(1'-oxo)-heptylthiophenes. Reduction afforded the required product (Figure 8) which was used for co-injection studies and mass spectral comparisons with the component of the bitumen.

(iii) 5-butyl-2-(2'-undecyl)thiophene (C_{19}). One of the other alkylthiophenes with the 9-methyloctadecane carbon skeleton in the bitumen of the Jurf ed Darawish oil shale appeared to be 5-butyl-2-(2'-undecyl)thiophene (19). A synthetic standard was prepared as illustrated in Figure 9. 2-Butylthiophene was acylated with decanoyl chloride and the resulting ketone methylated with methyllithium. After subsequent dehydration of the newly formed tertiary alcohol the resulting

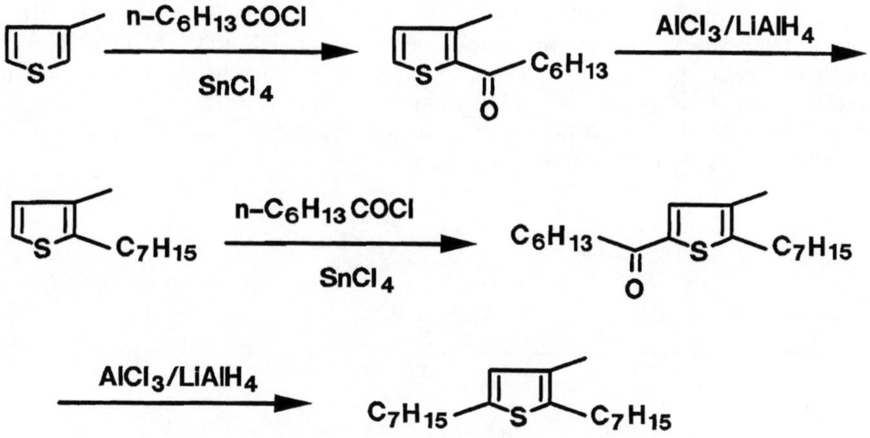

Figure 8. Synthesis of 2,5-diheptyl-3-methylthiophene.

alkene was reduced by ionic hydrogenation to give 5-butyl-2-(2'-undecyl)-thiophene which was used for co-injection and mass spectral studies. A similar synthetic approach has also been used (18) in the preparation of certain highly branched isoprenoid thiophenes related to the highly branched isoprenoid C_{20} and C_{25} hydrocarbons. This method also provides a route to complex alkanes such as the highly branced C_{20} and C_{25} hydrocarbons since they can be obtained from the alkylthiophenes by desulfurisation with Raney nickel.

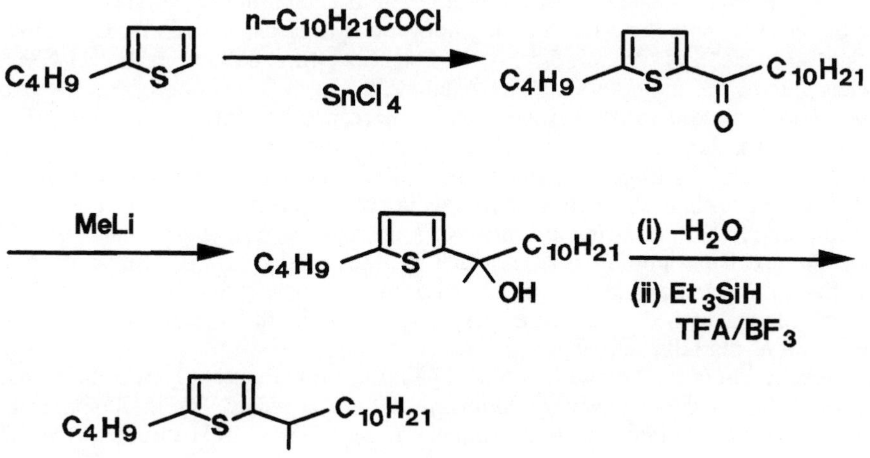

Figure 9. Synthesis of 2-butyl-5-(2'-undecyl)thiophene.

(iv) 2-(3',7'-dimethyloctyl)-5-methylthiophene (C_{15}). The final example given here is intended as an exercise in the spectroscopy (IR and NMR) and mass spectrometry of the intermediates and final products arising from the acylation method and subsequent mixed hydride reduction. Acylation of 2-methyl-thiophene with 3,7-dimethyloctanoyl chloride affords 5-methyl-2-(1'-oxo-3',7'-dimethyloctyl)thiophene (Figure 10).

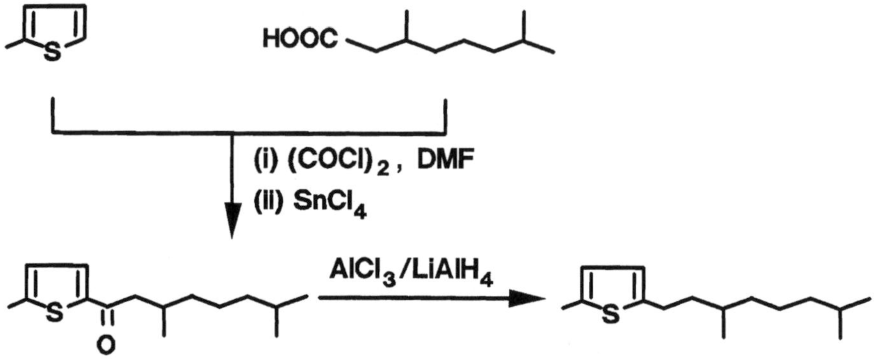

Figure 10. Synthesis of 5-methyl-2-(3',7'-dimethyloctyl)thiophene.

How do we know that we have actually obtained this compound? The IR spectrum shows a strong absorption at 1655 cm^{-1} indicative of a carbonyl group bonded to the α-carbon of a thiophene (rather than an absorption at 1680 cm^{-1} when bonded to a ß-carbon). The substitution pattern around the thiophene ring is confirmed by the 200 MHz ^{1}H NMR spectrum (Figure 11) which shows the presence of two thiophenic hydrogens coupled to each other (J=3.6 Hz) which is of the correct magnitude for coupling between hydrogens at C-3 and C-4 rather than C-2 and C-3 (J typically 5-6 Hz). Other interesting features in the ^{1}H NMR spectrum are the multiplets at δ 2.601 and 2.786 ppm which arise from the two hydrogens adjacent to the ketone group. These protons form part of an ABX system. Thus, they are strongly coupled to each other (J_{AB}=14.8 Hz) and to the adjacent tertiary hydrogen (J_{AX}=6.0 Hz, J_{BX}=8.0 Hz). The methyl groups of the alkyl chain are clearly present as doublets in the spectrum. The 6H doublet at δ 0.858 ppm (J=6.6 Hz) being indicative of a terminal isopropyl group and the 3H doublet at δ 0.947 ppm (J=6.6 Hz) for the other methyl group. The thiophenic methyl is observed as a singlet at δ 2.528 ppm. The mass spectrum (Figure 12) is in accordance with the structure. The major ions in the spectrum arise from cleavages around the ketone group.

Reduction of the ketone functionality in one step (Figure 10) with the mixed hydride affords the dialkylated thiophene. The 200 MHz ^{1}H NMR spectrum (Figure 13) clearly indicates the terminal isopropyl group (6H doublet, δ 0.862 ppm, J=6.6 Hz). The methyl group on the thiophene ring can be clearly seen as the singlet at δ 2.422 ppm. The multiplet at ca. δ 2.74 ppm arises from the two hydrogens of the alkyl chain adjacent to the thiophene ring. An interesting phenomenon is seen in the aromatic region where basically only a single resonance is apparent (δ 6.532 ppm) which integrates for two hydrogens. This

200 MHz ^{1}H nmr spectrum

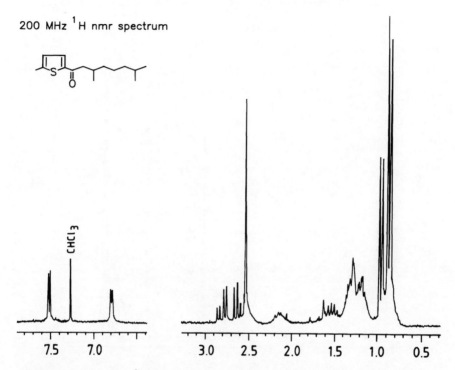

Figure 11. 200 MHz ^{1}H NMR spectrum of 5-methyl-2-(1'-oxo-3',7'-dimethyl-octyl)thiophene.

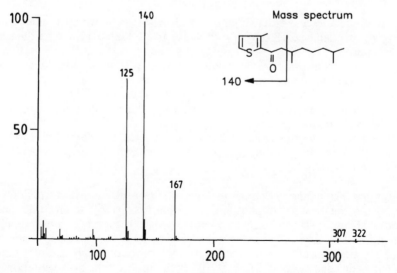

Figure 12. Mass spectrum of 5-methyl-2-(1'-oxo-3',7'-dimethyloctyl)-thiophene.

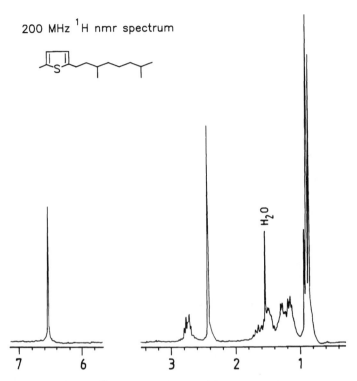

Figure 13. 200 MHz ^{1}H NMR spectrum of 5-methyl-2-(3',7'-dimethyl-octyl)thiophene.

feature is due to a second order situation which arises when two hydrogens, which are strongly coupled to each other (in this case J=3.5 Hz), have almost the same chemical shift.

The mass spectrum of the dialkylated thiophene (Figure 14) is very simple showing a molecular ion and major fragment ions at m/z 111 and 112. The base peak at m/z 111 arises from cleavage ß to the thiophene ring. The ion at m/z 112 arises from a McLafferty rearrangement whereby the C-3' tertiary hydrogen is transfered to the thiophene ring.

Other research groups have made use of the acylation method, followed by reduction of the resulting ketone group, to prepare alkylated thiophenes; *e.g.* C$_{20}$ alkylthiophenes (**8**).

<u>Formylation.</u> Thiophene undergoes electrophilic substitution with formaldehyde derivatives, in the presence of phosphoroyl trichloride (**26**, **27**), to give thiophene-2-carboxaldehyde. This can be reacted with Grignard reagents thereby extending the length of the alkyl side chain at C-2 (Figure 15). The resulting secondary alcohol can be reduced in a number of ways. One particularly useful method, which has been used to good effect (**9**, **10**), is the use of ionic hydrogenation. By varying the conditions either the thiophene and/or the corresponding thiolane can be obtained. Indeed, one of the best methods for the reduction of a thiophene to a thiolane is ionic hydrogenation (**28**). Reduction of thiophenes with

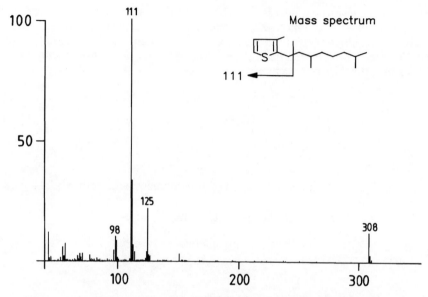

Figure 14. Mass spectrum of 5-methyl-2-(3',7'-dimethyloctyl)thiophene.

palladium has also been used (*e.g.* 14); this furnishes mainly the *cis* dialkylthiophene whilst the ionic hydrogenation method furnishes both *cis* and *trans*.

Another method for the preparation of synthetically useful thiophene-carboxaldehydes will be outlined in the section dealing with lithiated thiophenes.

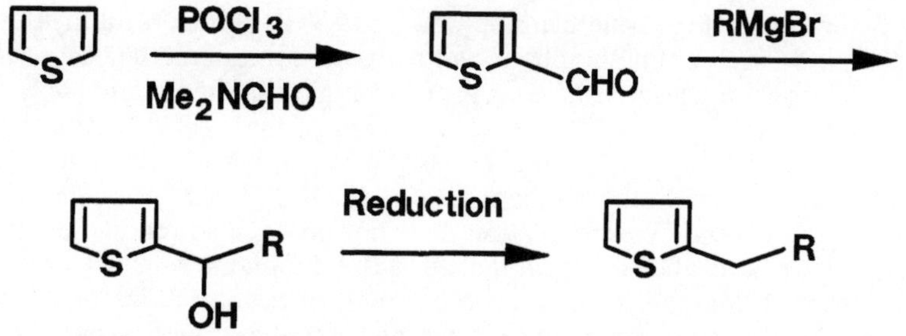

Figure 15. Preparation of 2-alkylthiophenes from thiophene-2-carboxaldehyde.

Bromination. Another useful example of electrophilic substitution of thiophene and alkylthiophenes is that of bromination. Thus, thiophene itself undergoes electrophilic substitution in the presence of bromine. The reaction proceeds systematically from 2-bromothiophene, to 2,5-dibromothiophene, to 2,3,5-tribromothiophene and finally to tetrabromothiophene (29, 30). These brominated thiophenes exhibit sufficiently different physical properties to allow their ready purification. A more controlled method, especially for monobromination, involves the use of N-bromosuccinamide (NBS) rather than bromine itself. Thus, 3-methylthiophene can be selectively converted into 2-bromo-3-methylthiophene by the use of one equivalent of NBS (e.g. 31; Figure 16). The use of brominated thiophenes in the synthesis of alkylthiophenes will be discussed in the following section.

Figure 16. Monobromination of 3-methylthiophene with NBS.

Preparation and Reactions of Lithiated Thiophenes.

Preparation. Thiophene can be metallated with *n*-butyllithium to give 2-lithiothiophene (31; Figure 17).

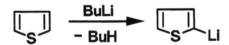

Figure 17. Reaction of thiophene with *n*-butyllithium.

The metallation proceeds by initial co-ordination of the electropositive lithium to the sulfur atom followed by abstraction of the neighbouring most acidic proton by the butyl anion. 2-Alkylthiophenes give 2-alkyl-5-lithiothiophenes whilst 3-alkylthiophenes give a mixture of 3-alkyl-2-lithio- and 4-alkyl-2-lithiothiophenes (32) with the latter predominating. This is further increased with increasing size of the 3-alkyl group [e.g. 3-methylthiophene gives 80% 2-lithio-4-methylthiophene and 20% 2-lithio-3-methylthiophene with *n*-butyl-lithium/diethylether whilst 3-isopropylthiophene gives 95% 3-isopropyl-2-lithiothiophene and only 5% 3-isopropyl-2-lithiothiophene (22)]. The formation of the 4-alkyl-2-lithiothiophene is a'so favoured by the use of bulky metallation reagents [e.g. *n*-butylithium/tetramethylethylenediamine (TMEDA) gives 93-100% (22)]. It should be pointed that the metallation does not proceed so readily, or not at all, when there are large alkyl groups attached to the thiophene ring.

Metallated thiophenes can also be readily prepared by halogen-metal exchange. Thus, 2-bromothiophene is rapidly metallated with *n*-butyllithium at low temperatures (e.g. -78°C) to give 2-lithiothiophene (Figure 18). This method has been shown to be applicable to the preparation of polylithiothiophenes. Thus, di-, tri- and tetra-brominated thiophenes afford the corresponding di-, tri- and tetra-lithiated thiophenes upon reaction with *n*-butyllithium (30; e.g. Figure 18).

The metallation of these polybrominated thiophenes occurs sequentially and, therefore, selective metallations can be performed. Thus, metallation of 2,3,5-tri-bromothiophene with one equivalent of *n*-butyllithium affords 3,5-dibromo-2-lithiothiophene (33; Figure 19). Similarly, metallation of tetra-bromothiophene with two equivalents of *n*-butyllithium gives 3,4-dibromo-2,5-

dilithiothiophene (34). These metallation reactions of brominated thiophenes are also equally applicable to iodinated thiophenes.

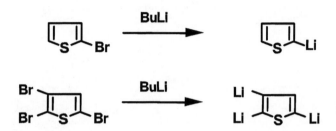

Figure 18. Formation of lithiated thiophenes from brominated thiophenes.

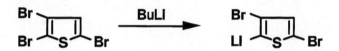

Figure 19. Selective lithiation of 2,3,5-tribromothiophene.

Reactions.

(i) With water. Lithiated thiophenes react with water to give the corresponding unsubstituted thiophene. Thus, 3,5-dibromo-2-lithiothiophene is quenched by water to give 2,4-dibromothiophene (33; Figure 20). Similarly, 3,4-dibromo-2,5-dilithiothiophene affords 3,4-dibromothiophene (34). This series of reactions of selective metallations followed by quenching with water can afford various brominated thiophenes which are not always readily obtainable by other methods, although one should be aware of the rearrangements of α-halogenated thiophenes to their ß-isomers in the presence of metal amides in liquid ammonia (*e.g.* 22). These brominated thiophenes can be useful intermediates in the synthesis of various alkylthiophenes.

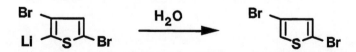

Figure 20. Reaction of 3,5-dibromo-2-lithiothiophene with water.

(ii) With aldehydes, dialkylsulfates and alkylhalides. Lithiated thiophenes undergo coupling reactions with aldehydes, dialkylsulfates and alkylhalides as illustrated in Figure 21.

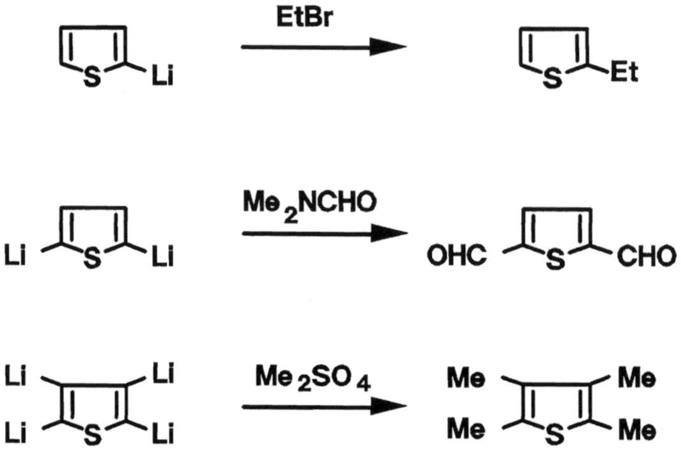

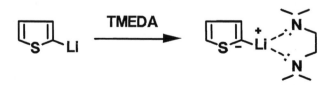

Figure 21. Examples of the reactions of lithiated thiophenes with alkylhalides, aldehydes and dialkylsulfates.

Thus, 2-lithiothiophene reacts with ethylbromide to give 2-ethylthiophene (*e.g.* 32). This method is probably the method of choice for preparing 2-alkylthiophenes although, as will be mentioned later, this is not suitable for long alkyl chains. 2,5-Dilithiothiophene reacts with dimethylformamide to give thiophene 2,5-dicarboxaldehyde (35). Such an intermediate opens up possibilities for the preparation of symmetrically disubstituted alkylthiophenes. Tetra-lithiothiophene reacts with dimethylsulfate to give tetramethylthiophene (30). This, and other related di- and tri-methylated thiophenes, prepared by the same method (30), have similarly been prepared as standards for use in pyrolysis studies of sulfur rich kerogens (15). Unfortunately, these reactions of lithiated thiophenes do not proceed satisfactorily when the thiophene ring is substituted with a long alkyl chain or when the reaction involves a long chain aldehyde or alkylhalide. The second problem can be offset to some extent by the use of a co-ordinating amine such as TMEDA (Figure 22). This helps to break up the aggregrates of the thienyllithiums which are known to exist in solution.

Figure 22. Increasing the reactivity of 2-lithiothiophene with TMEDA.

<u>The preparation of 3-methyl-2-(3',7',11'-trimethyldodecyl)thiophene.</u> This compound, which has a carbon skeleton resembling that of phytane, is often a major component of certain immature sediments (1). Previous syntheses of this compound have used the acylation method whereby 3-methylthiophene is acylated with 3,7,11-trimethyldodecanoic acid (1, 8). Reference to Figure 5 indicates that the required compound will be accompanied by the 2,4-isomer and this is indeed the case. The pure compound can, however, be prepared using a lithiated alkylthiophene and an aldehyde (Figure 23).

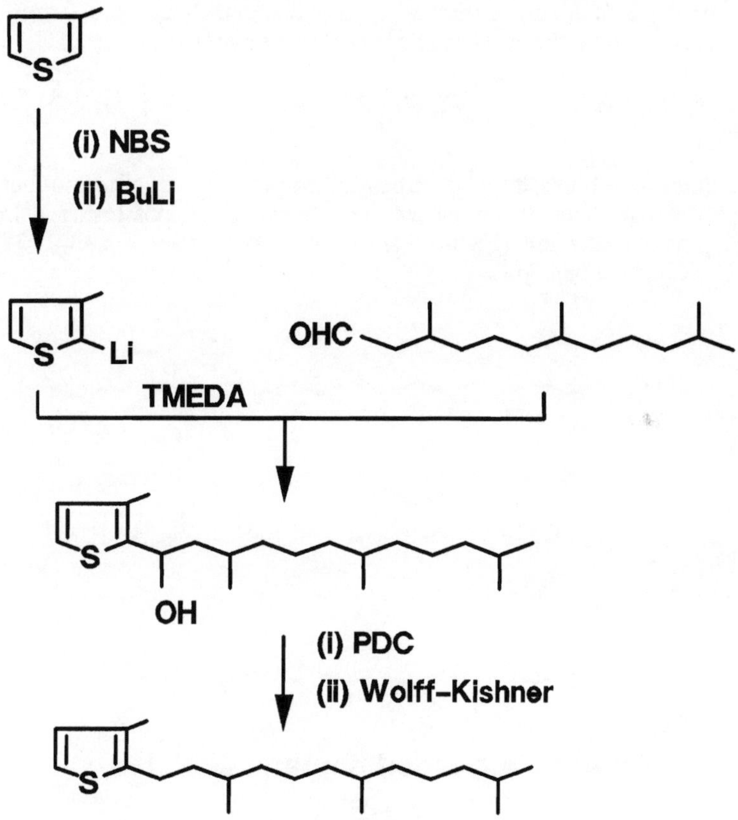

Figure 23. Synthesis of 3-methyl-2-(3',7',11'-trimethyldodecyl)thiophene.

Thus, 3-methylthiophene is brominated with NBS to give 2-bromo-3-methyl-thiophene (*cf.* Figure 16) which is then metallated with *n*-butyllithium. The lithiated thiophene is activated with TMEDA and then coupled with 3,7,11-trimethyldodecanal. The coupled product contains a secondary alcohol which is readily seen in the IR spectrum as a very broad absorption centered at 3350 cm^{-1}. The 200 MHz ^{1}H NMR spectrum and mass spectrum are in agreement with the structure.

The secondary alcohol was smoothly oxidized to a ketone in high yield with pyridinium dichromate in dimethylformamide (*cf.* 36). The IR spectrum indicated the presence of a ketone group bonded to the α-carbon of a thiophene (absorption at 1660 cm^{-1}). The 200 MHz ^{1}H NMR spectrum showed all the features expected of this structure (*cf.* Figure 12) as did the mass spectrum (*cf.* 1).

Finally, the ketone was reduced by the modified Wolff-Kishner method (*cf.* 8) to give the final product which was free of detectable amounts of the 2,4-isomer. The 200 MHz ^{1}H NMR and mass spectrum are both in accord with the structure (*cf.* 1). Full details of this synthesis and spectroscopic properties of the intermediates and final products will be reported elsewhere.

The Synthesis of Alkylthiophenes comprising a Linearly Extended Phytane Skeleton

Recent studies have realised the identification of two series of alkylthiophenes, in a bituminous marl from the northern Apeninnes, Italy, possessing a linearly extended phytane skeleton (Figure 24). Series 1 comprises C_{25}-C_{28} (37) and series 2 C_{21}-C_{26} members (38).

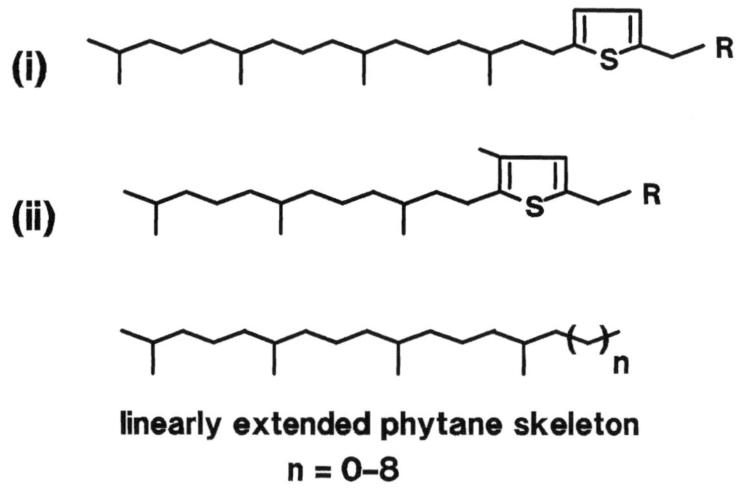

linearly extended phytane skeleton

n = 0–8

Figure 24. Alkylthiophenes in the northern Apeninnes marl possessing a linearly extended phytane skeleton: (i) series 1, R=H, Me, Et, *n*-Pr; (ii) series 2, R=H, Me, Et, *n*-Pr, *n*-Bu, *n*-Pe.

Authentic standards of both series were prepared using combinations of the techniques outlined earlier. Members of series 1 were obtained by acylation of the necessary 2-alkylthiophene with 3,7,11,15-tetramethylhexadecanoic acid and reduction of the resulting ketone (Figure 25).

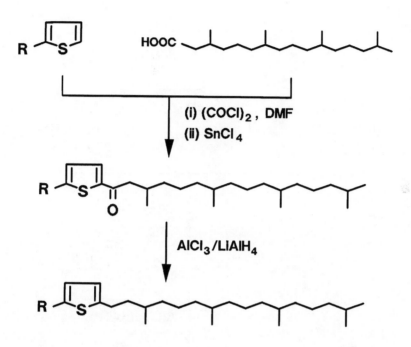

Figure 25. Synthesis of Series 1. R=Me, Et, *n*-Pr, *n*-Bu (C_{25}-C_{28}).

The C_{22}-C_{25} members of Series 2 were prepared from 3-methyl-2-(3',7',11'-trimethyldodecyl)thiophene (*cf.* Figure 23) by acylation with the appropriate carboxylic acid and reduction of the resulting ketone (Figure 26).

The C_{21} member of Series 2 was prepared by acylation of 2,4-dimethylthiophene with 3,7,11-trimethyldodecanoic acid and reduction of the resulting ketone (Figure 27). The synthesized standards were used for comparison of the mass spectra of, and for GC co-injection studies with, the alkylthiophenes occurring in the bitumen of the northern Apeninnes marl.

The identification of these alkylthiophenes may point to an unusual biosynthetic pathway since some of these molecules are comprised of isoprenoid and *n*-alkyl units.

Conclusions

In this article we have attempted not only to point out some of the most useful methods for the synthesis of alkylthiophenes but also to show·the value of synthesis of authentic standards in identifying the components of the organic extracts of sediments. It is only after the correct identification of the structures that more detailed conclusions regarding organic geochemical problems can be

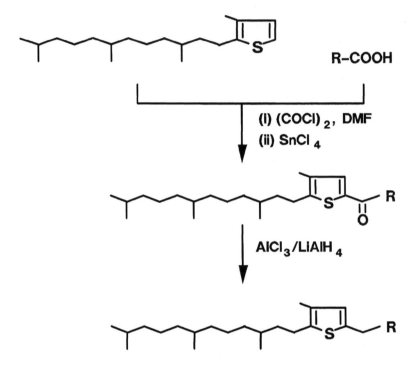

Figure 26. Synthesis of the C_{22}-C_{25} members of Series 2. R=Me, Et, *n*-Pr, *n*-Bu (C_{22}-C_{25}).

obtained. Acylation of thiophene and alkylthiophenes with carboxylic acids has been shown to be a useful method as has coupling of various functionalised groups with lithiated thiophenes. These methods have been discussed with reference to the preparation of several naturally occurring alkylthiophenes. Reduction of these thiophenes affords a ready preparation of the corresponding thiolanes and desulfurisation the corresponding alkanes. Sometimes such methods are more convenient than direct synthesis.

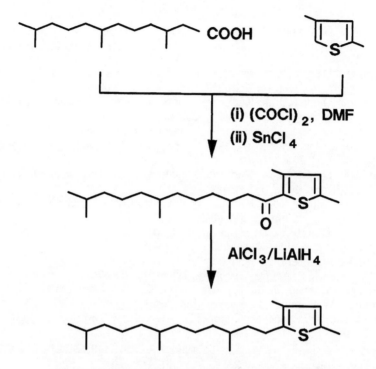

Figure 27. Synthesis of the C_{21} member of series 2.

Acknowledgments

TMP is grateful to Professor P.A. Schenck (Delft University of Technology) for a research fellowship. TMP and ACK-vD would like to thank colleagues at the Delft University of Technology for their continual help and encouragement.

Literature cited

1. Brassell, S. C.; Lewis, C. A.; de Leeuw, J. W.; de Lange, F.; Sinninghe Damsté, J. S. Nature 1986, 320, 160-62.
2. Cyr, T. D.; Payzant, J. D.; Montgomery, D. S.; Strausz, O. P. Org. Geochem. 1986, 9, 139-43.
3. Payzant, J. D.; Montgomery, D. S.; Strausz, O. P. Tetrahedron Lett. 1983, 24, 651-54.
4. Payzant, J. D.; Montgomery, D. S.; Strausz, O. P. Tetrahedron Lett. 1985, 26, 4175-78.
5. Payzant, J. D.; Montgomery, D. S.; Strausz, O. P. Org. Geochem. 1986, 9, 357-69.
6. Payzant, J. D.; Montgomery, D. S.; Strausz, O. P. AOSTRA J. Res. 1988, 4, 117-31.
7. Perakis, N. Ph.D. Thesis, University of Strasbourg, France, 1986
8. Rullkötter, J.; Landgraf, M; Disko, U. J. High Res. Chrom. and Chrom. Commun. 1988, 11, 633-38.

9. Schmid, J. C. Ph.D. Thesis, University of Strasbourg, France, 1986.
10. Schmid, J. C.; Connan, J; Albrecht, P. Nature 1987, 329, 54-56.
11. Sinninghe Damsté, J. S.; de Leeuw, J. W. Int. J. Environ. Anal. Chem. 1987, 28, 1-19.
12. Sinninghe Damsté, J. S.; ten Haven, H. L.; de Leeuw, J. W.; Schenck, P. A. Org. Geochem. 1986, 10, 791-805.
13. Sinninghe Damsté, J. S.; Kock-van Dalen, A. C.; de Leeuw, J. W.; Schenck, P. A. Tetrahedron Lett. 1987a, 28, 957-60.
14. Sinninghe Damsté, J. S.; de Leeuw, J. W.; Kock-van Dalen, A. C.; de Zeeuw, M. A.; de Lange, F.; Rijpstra, W. I. C.; Schenck, P. A. Geochim. Cosmochim. Acta 1987b, 51, 2369-91.
15. Sinninghe Damsté, J. S.; Kock-van Dalen, A. C.; de Leeuw, J. W.; Schenck, P. A. J. Chrom. 1988a, 435, 435-52.
16. Sinninghe Damsté, J. S.; Rijpstra, W. I. C.; de Leeuw, J. W.; Schenck, P. A. Org. Geochem. 1988b, 13, 593-606.
17. Sinninghe Damsté, J. S.; Eglinton, T. I.; de Leeuw, J. W.; Schenck, P. A. Geochim. Cosmochim. Acta 1989a, 53, 873-89.
18. Sinninghe Damsté, J. S.; van Koert, E. R.; Kock-van Dalen, A. C.; de Leeuw, J. W.; Schenck, P. A. Org. Geochem. 1989b 14, 555-67.
19. Sinninghe Damsté, J. S.; Rijpstra, W. I. C.; Kock-van Dalen, A. C.; de Leeuw, J. W.; Schenck, P. A. Geochim. Cosmochim. Acta 1989c, 53, 1343-55.
20. Valisolalao, J.; Perakis, N.; Chappe, B.; Albrecht, P. Tetrahedron Lett. 1984, 25, 1183-86.
21. Livingstone, R. In Rodd's Chemistry of Carbon Compounds. Volume IV, Part A. Heterocyclic Compounds; 1977; pp 219-328.
22. Meth-Cohn, O. In Comprehensive Organic Chemistry; Sammes, P.G., Ed.; Pergamon Press: Oxford, 1979; pp 789-838.
23. McGhie, J. F.; Ross, W. A.; Evans, D.; Tomlin, J. E. J. Chem. Soc. 1962, 350-55.
24. King, W. J.; Nord, F. F. J. Org. Chem. 1949, 14, 638-42.
25. Nystrom, R. F.; Berger, C. R. A. J. Amer. Chem. Soc. 1958, 80, 2896-98.
26. Campaigne, E.E.; Archer, W. L. J. Amer. Chem. Soc. 1953, 75, 989-91.
27. Weston, A.W.; Michaels, R.J. In Org. Synth. Coll. Vol. IV 1963; pp 915-18.
28. Kursanov, D. N.; Parnes, Z. N.; Loin, N. M. Synthesis 1974, 633-51.
29. Gronowitz, S.; Raznikiewicz, T. In Org. Synth. Coll. Vol. V 1973; pp 149-51
30. Janda, M.; Srogl, J.; Stibor, I.; Nemec, M.; Vopatrna, P. Synthesis 1972, 545-47.
31. Gilman, H.; Shirley, D.A. J. Amer. Chem. Soc. 1949, 71, 1870-71.
32. Ramanathan, V.; Levine, R. J. Org. Chem. 1962, 27, 1667-70.
33. Lawesson, S.-O. Arkiv. Kemi 1957a, 11, 317-24.
34. Lawesson, S.-O. Arkiv. Kemi 1957b, 11, 325-26.
35. Robba, M.; Roques, B.; Bonhomme, M. Bull. Chim. Soc. Fr. 1967, 2495-2507.
36. Corey, E. J.; Schmid, G. Tetrahedron Lett. 1979, 399-402.
37. Peakman, T. M.; Sinninghe Damsté, J. S.; de Leeuw, J. W. J. Chem. Soc., Chem.Commun. 1989, 1105-07.
38. Peakman, T. M.; Sinninghe Damsté, J. S.; de Leeuw, J. W. Geochim. Cosmochim. Acta 1989, 53, 3317-22.

RECEIVED March 5, 1990

Chapter 24

Organic Sulfur Compounds and Other Biomarkers as Indicators of Palaeosalinity

Jan W. de Leeuw and Jaap S. Sinninghe Damsté

Faculty of Chemical Engineering and Materials Science, Organic
Geochemistry Unit, Delft University of Technology,
De Vries van Heystplantsoen 2, 2628 RZ Delft, Netherlands

A number of selected molecular parameters obtained from analysis of immature crude oils and sediment extracts are evaluated as indicators of palaeosalinity. The nature of these parameters is discussed taking into account the role of intermolecular and intramolecular incorporation of sulfur into specific functionalized lipids. Specific distribution patterns of methylated chromans and C_{20} isoprenoid thiophenes and the relative abundance of gammacerane are excellent indicators for palaeosalinity, whilst other parameters such as $14\alpha(H),17\alpha(H)/14\beta(H),17\beta(H)$ -sterane ratios, the pristane/phytane ratio, the even-over-odd carbon number predominance of n-alkanes and the relative abundance of C_{35} hopanes and/or hopenes may indicate palaeohypersalinity but are affected by environmental factors other than hypersalinity and by diagenesis.

In a recent paper Evans and Kirkland (1) summarized important aspects of hypersaline (salinity > 4%) environments. Contrary to previous thoughts these environments, especially those with salinities varying between 4 and 12%, are very productive, sometimes up to 10 times more than upwelling environments. The species distribution in these environments is limited and consists in general mainly of sulfate reducing- and photosynthetic sulfur bacteria and halophilic Eubacteria, specific green algae such as *Dunaliella* species, certain diatom species and Protozoa (2-5). In extreme saline environments (brines) halophilic Archaebacteria such as *Halobacterium* may contribute significantly to the biomass.

0097–6156/90/0429–0417$07.75/0
© 1990 American Chemical Society

The surface waters are less oxygenated in these environments due to their enhanced salinities and water circulation is slowed down by its higher viscosity. Bacterial sulfate reduction can be very high ($\underline{1},\underline{6}$). These aspects favour anoxia, especially when a halocline forms. The extremely high production of biomass combined with good preservation conditions make these hypersaline environments of particular interest in organic geochemical terms. It has been suggested that a significant proportion of the petroleum reserves may derive from organic matter produced in hypersaline environments ($\underline{1}$).

The importance of these depositional environments makes it desirable that studies concerned with the reconstruction of palaeoenvironments from sediments or source rocks of oils also establish molecular parameters for palaeohypersalinity. Recently, ten Haven *et al.* ($\underline{7}$-$\underline{9}$) have summarized a number of "organic geochemical phenomena" related to hypersaline depositional environments. In addition to previously known parameters, such as an even-over-odd carbon number predominance of *n*-alkanes and a low pristane/ phytane ratio (<0.5), several new parameters were suggested. These parameters, however, are mainly based on empirical relations.

In this paper we discuss a selected set of molecular parameters which may indicate hypersaline palaeoenvironments. Attempts are made to explain the nature of most of these parameters taking into account recent results concerning the early stage diagenesis of Δ^5- and Δ^7-sterols ($\underline{10}$-$\underline{12}$) and the role and consequences of reactions of inorganic sulfur species with organic matter in Recent sediments ($\underline{13}$-$\underline{20}$). The selection of data is based on results obtained from studies of aliphatic and aromatic hydrocarbons, organic sulfur compounds and sulfur-rich high-molecular-weight substances occurring in sediment extracts and oils derived from organic matter deposited in both 'normal' marine salinity and hypersaline palaeoenvironments ($\underline{13}$-$\underline{26}$).

Chroman Distributions

In a recent paper Sinninghe Damsté *et al.* ($\underline{27}$) have described the presence of a family of methylated chromans with an isoprenoid side chain (methylated 1-methyl-1-(4,8,12-trimethyltridecyl)chromans, MTTC) in a number of sediment extracts and crude oils. A diagenetic origin from tocopherols was thought to be highly unlikely and the authors speculated about an origin from Eu- or Archaebacteria. The distribution patterns of these chromans vary considerably and a correlation was found between the chroman distributions and the palaeosalinity of the sample (Figure 1). The palaeosalinity of the samples studied was assessed on the basis of geological and biomarker data ($\underline{27}$). Chroman distributions of samples from normal marine salinity environments are dominated by the trimethyl chroman, whilst this chroman isomer is much

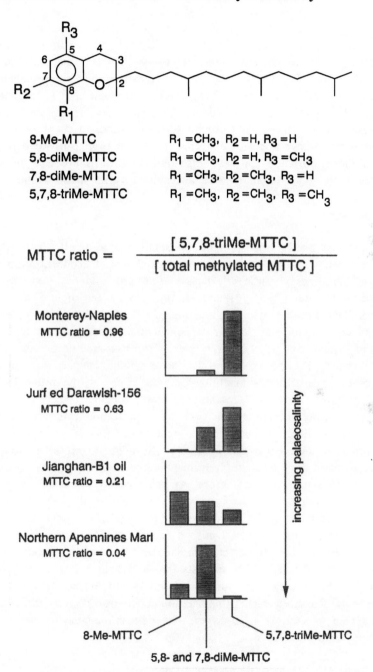

Figure 1.   Methylated 1-methyl-1-(4,8,12-trimethyltridecyl)chromans (MTTC) as indicators of palaeosalinity: structures, definition of the MTTC ratio and typical methylated MTTC distributions in sediments and petroleums. Description of the samples is given elsewhere (16).

less abundant in samples from hypersaline depositional environments. Based on this observation a parameter, the MTTC ratio, was defined as indicated in Figure 1. Values for this parameter lower than 0.5 may indicate hypersaline depositional environments. Recent (re)analysis of approximately 85 samples from normal marine salinity palaeoenvironments has strengthened the validity of this MTTC ratio: all samples contained methylated chromans dominated by the trimethyl chroman (28,29). Although these chromans or their precursors have not yet been encountered in organisms we believe that the significant variations in their distribution patterns are due to marked changes in microbial populations as a result of variations in salinity.

Gammacerane

Gammacerane occurs widespreadly in sediments and oils from different palaeoenvironments (30,31). It has been noted that its concentration relative to other triterpanes is particularly high in sediments and oils from hypersaline palaeoenvironments (7,8,30-33) and therefore may be of use in assessing palaeosalinity. For example, Mello *et al.* (31) have shown in an extensive study of source rocks and oils from Brazilian marginal basins that the ratio of gammacerane over $17\alpha(H),21\beta(H)$-hopane (as measured from m/z 191 mass chromatograms) is higher (>0.7) in marine evaporitic samples than in samples from other types of depositional environment. However, the rationale for the use of gammacerane as palaeosalinity indicator remains unclear.

Recently, ten Haven *et al.* (34) have presented evidence that tetrahymanol is the likely precursor of gammacerane. Although tetrahymanol has been found in several protozoa species of the genus *Tetrahymena*, ten Haven and co-workers suggested that tetrahymanol may be a natural product of other organisms as well. With respect to palaeosalinity assessment it is worthy of note that sediments of the hypersaline, alkaline Mono Lake contain high amounts of tetrahymanol (35). However, in two other Recent hypersaline sediments this component could not be identified (34). Similarly, Brassell *et al.* (33) were unable to detect gammacerane in a glauber salt from the Jianghan Basin (China). This suggests that the source organism(s) may be unable to tolerate very high salinity regimes, as implied by the deposition of glauber salt. Relatively high gammacerane concentrations may therefore only be observed in samples from hypersaline depositional environments with a restricted salinity range.

$14\beta(H),17\beta(H)/14\alpha(H),17\alpha(H)$-Sterane Ratio

Based on molecular mechanics studies of sterenes and on reinterpretations of steroid data in the organic geochemical and chemical literature, corrected and separate routes of early diagenesis for Δ^5- and Δ^7-sterols were postulated (11). The newly formulated diagenetic pathways are based on selective hydrogenation of double bonds in sterenes and on a limited isomerisation of double bonds *via* tertiary carbocations only. Sterane end products of the proposed diagenetic pathways for Δ^5- and Δ^7-sterols are indicated in Figure 2. The detailed analysis of steranes in extracts from relatively immature sediments and crude oils confirmed the proposed pathways in that: i) the distribution patterns of $14\beta(H),17\beta(H)$-steranes are completely different from those of the $14\alpha(H),17\alpha(H)$-steranes and ii) the $14\beta(H),17\beta(H)$-steranes are fully isomerised at C_{20} whilst the $14\alpha(H),17\alpha(H)$-steranes are not or only partly isomerised at C_{20} (10,12). A mass chromatogram of m/z 218 obtained from GC-MS analyses of the saturated hydrocarbon fractions of the Rozel Point Oil, an oil which is thought to originate from a hypersaline source rock (8,36), exemplifies this feature (Figure 3). In a study of an immature marl sample from the Northern Apennines (Italy) Peakman *et al.* (12) have recently shown that the assumed intermediate products in the isomerisation of Δ^7-sterols, the spirosterenes, indeed possess a distribution pattern similar to the $14\beta(H),17\beta(H)$-steranes and completely different from that of the $14\alpha(H),17\alpha(H)$-steranes. These considerations permit discrimination of the Δ^5- and Δ^7-sterols originally present based on measurements of ratios of $14\beta(H),$ $17\beta(H)/14\alpha(H),17\alpha(H)$ steranes in sediment extracts and crude oils. Since relatively high abundancies of Δ^7-sterols have been reported in very Recent hypersaline environments (37,38) the $14\beta(H),17\beta(H)/14\alpha(H),17\alpha(H)$ ratio can be rationalized as a palaeosalinity parameter at least in sediments or crude oils in which the $14\alpha(H),17\alpha(H)$-steranes are not fully isomerised at C_{20}.

There are, however, some problems in the direct application of such a ratio. First, relative high abundances of Δ^7-sterols have been reported in Lake Kinneret, a normal marine salinity environment (39). Hence, Δ^7-sterols are obviously not restricted to organisms living in hypersaline environments. The relative abundance of Δ^7- (and $\Delta^{8(14)}$-) sterols in hypersaline environments may be due to the absence of grazing zooplankton in these environments. In normal marine salinity environments Δ^7- and $\Delta^{8(14)}$-sterols are selectively metabolised in the guts of zooplankton resulting in a selective preservation of Δ^5-sterols in zooplankton fecal pellets, which are transported rapidly to the sediment (40,41).

Another complication is the possible interaction of inorganic sulfur species with Δ^5- and Δ^7-sterols. The occurrence of thiosteroids and of sulfur-bound steroid units in high-molecular-weight fractions of sediments and oils has been observed recently (15,16,20,26). The thiosteroids with the thiophene ring condensed to the D-ring almost exclusively possess the $14\beta(H)$ configuration (26). Thiosteroids with a thiophene unit in the side chain mainly

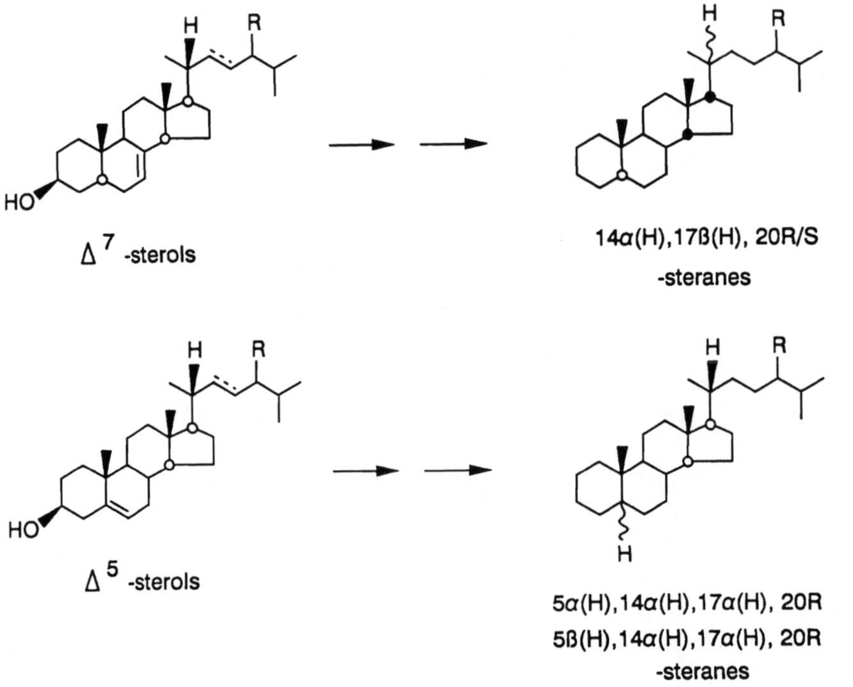

Δ^7 -sterols

14α(H),17β(H), 20R/S
-steranes

Δ^5 -sterols

5α(H),14α(H),17α(H), 20R
5β(H),14α(H),17α(H), 20R
-steranes

$$\frac{\Delta^7\text{-sterols}}{\Delta^5\text{-sterols}} \;\;\triangleq\;\; \frac{14β(H),17β(H),20R - \;+\; 14β(H),17β(H),20S - \text{steranes}}{14α(H),17α(H) - \text{steranes}}$$

(in immature sediments/oils only)

Figure 2. Sterane end products of the early diagenetic pathways of Δ^5- and Δ^7- sterols. Data from ref. 11.

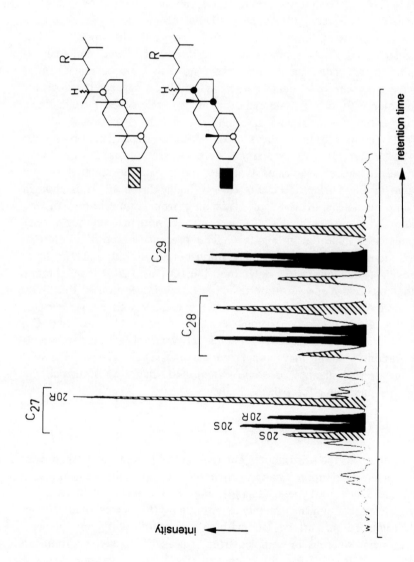

Figure 3. Mass chromatogram of *m/z* 218 of the saturated hydrocarbon fraction of the Rozel Point oil. $5\alpha(H),14\alpha(H),17\alpha(H)$- and $5\alpha(H),14\beta(H),17\beta(H)$-steranes are differentiated as shown.

have the $14\beta(H),17\beta(H),20R/S$ configuration ($\underline{26}$; Figure 4). When the sulfur atom is located in the A/B-ring part of the molecule the compounds occur exclusively in the $14\alpha(H),17\alpha(H),20R$ configuration ($\underline{26}$). After desulfurisation or flash-pyrolysis of the high-molecular-weight fraction of appropriate samples only $14\alpha(H),17\alpha(H),20R$ steranes are released and no $14\beta(H),17\beta(H),20R$ or 20S steranes are observed ($\underline{20}$; Figure 5). These observations can be speculatively interpreted as follows: $\Delta^{7,22}$-sterols can only incorporate sulfur intramolecularly in the side chain after isomerisation of the Δ^7-double bond to the D-ring or side chain; the substrate for intramolecular sulfur incorporation is obviously a diene system, as has been suggested before for other substrates with two double bonds near to each other ($\underline{18}$). Such a sequence of reactions starting from Δ^7-sterols requires the $14\beta(H)$ and $14\beta(H)$, $17\beta(H),20$ R/S configuration in these types of thiosteroids. The observation that no $14\beta(H),17\beta(H)$-steranes and/or -sterenes are released from high-molecular-weight fractions indicates that the intramolecular reaction of the diene system in the D-ring/side chain area is highly favoured. It is thought that the intermediate spirosteradienes, resulting from isomerisation of the $\Delta^{7,22}$- and/or $\Delta^{7,24(28)}$-steradienes ($\underline{12}$), incorporate inorganic sulphur species intramolecularly as indicated in Figure 6. The presence of both thiosteroids with the S-atom in the A/B ring and sulfur-bound steroid units in high-molecular-weight fractions with only the $14\alpha(H),17\alpha(H),20R$ configuration suggests that the $\Delta^{3,5}$-diene substrate resulting from dehydration of the 3-OH group of Δ^5-sterols reacts both intra- and intermolecularly with inorganic sulfur species (Figure 7). However, the natural stereochemistry at C_{14}, C_{17} and C_{20} is preserved in these reactions. This different behaviour of Δ^5- and Δ^7-sterols on incorporation of sulfur may complicate the $14\beta(H),17\beta(H)/14\alpha(H),17\alpha(H)$ ratios of the steroid alkanes to some extent and may thus influence the validity of this parameter for palaeosalinity.

C_{20} Isoprenoid Thiophene Distribution Patterns

A large variation in the distribution patterns of C_{20} isoprenoid thiophenes has been encountered upon analysis of sediments and oils ranging from normal marine saline to hypersaline ($\underline{16}$). Figure 8 shows that in samples representing a normal marine salinity environment (Monterey Shale-Naples and -El Capitan Beach, Jurf ed Darawish-45) isoprenoid thiophenes VI and VII are dominant whereas in samples from hypersaline palaeoenvironments (Sicily seep oil-E2, Rozel Point oil) isoprenoid thiophene V and the so-called midchain isoprenoid thiophenes (I-IV) are relatively abundant. C_{20} isoprenoid bithiophenes (VIII-X) only occur when the midchain isoprenoid thiophenes are relatively abundant (Figure 9).

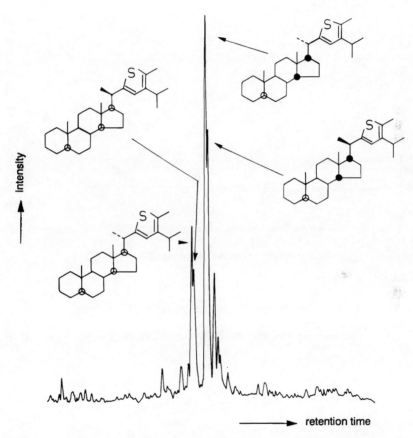

Figure 4.   Mass chromatogram of *m/z* 167 of the low molecular weight aromatic fraction (<u>15</u>) of the Rozel Point oil showing assignments of steroidal thiophenes.

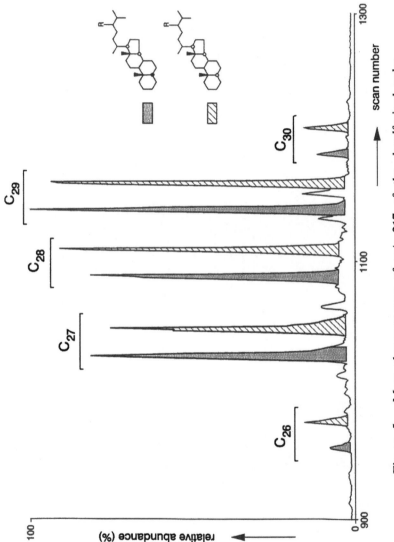

Figure 5. Mass chromatogram of *m/z* 217 of the desulfurised polar fraction of the bitumen of an outcrop section at Naples Beach (California, U.S.A.) of the Monterey Formation.

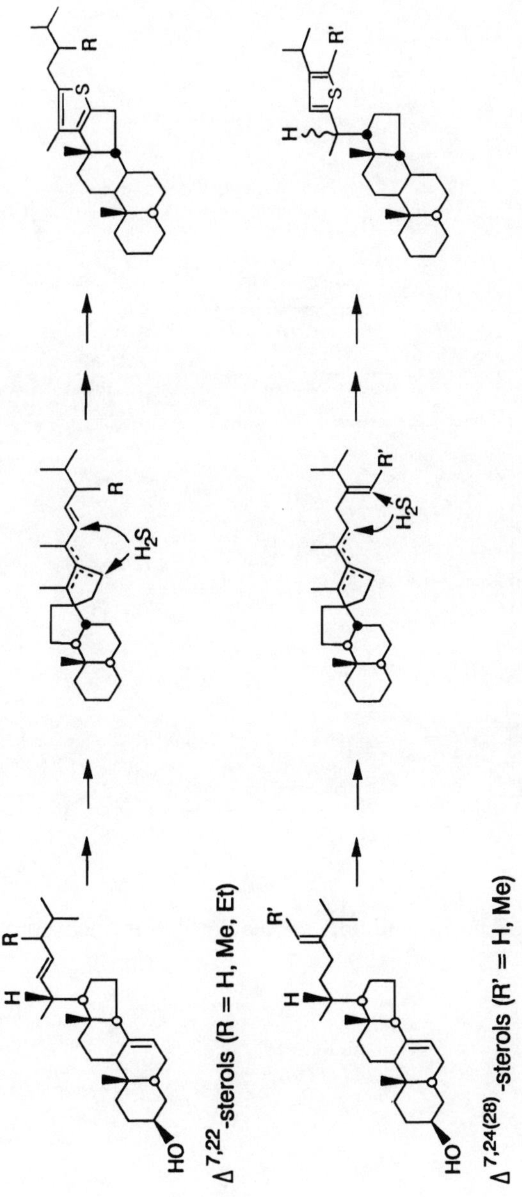

Figure 6.   Postulated intramolecular sulfur incorporation into $\Delta^{7,22}$- and $\Delta^{7,24(28)}$-sterols. Reactions in an intermolecular fashion do not seem to occur.

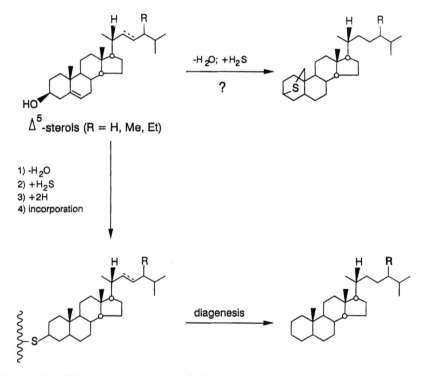

Figure 7.   Proposed intra- and intermolecular sulfur incorporation into Δ^5-sterols.

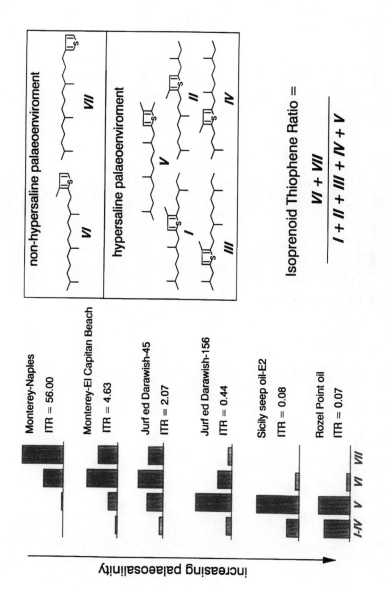

Figure 8. C_{20} isoprenoid tiophenes as indicators of palaeosalinity: structures, definition of the isoprenoid thiophene ratio (ITR) and some typical distributions (based on peak heights in m/z 308 mass chromatograms) in sediments and petroleums. Description of the samples is given elsewhere (16).

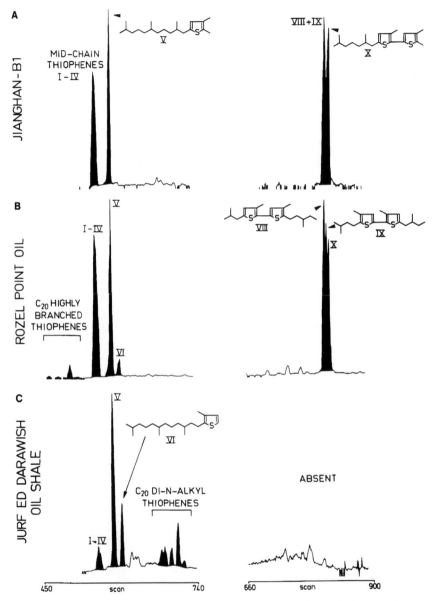

Figure 9. Mass chromatograms of m/z 308 and 334 of low molecular weight aromatic fractions (15,16) of (A) Jianghan-B1 oil (China), (B) Rozel Point oil (Utah, U.S.A.) and (C) Jurf ed Darawish Oil Shale-156 bitumen (Jordan). These examples show the co-occurrence of C_{20} mid-chain isoprenoid thiophenes I-IV with C_{20} isoprenoid bithiophenes VIII-X; if thiophenes I-IV are abundant the bithiophenes VIII-X also occur and if thiophenes I-IV are relatively minor the bithiophenes VIII-X are absent. Description of the samples is given elsewhere (16).

Isoprenoid thiophenes are formed by incorporation of sulfur into phyta-dienes and polyunsaturated phytenes (13,14). The differences in distribution patterns of C_{20} isoprenoid thiophenes may be explained as the result of differences in the abundance of the various precursors of these phytenes. Sulfur incorporation into phytol-derived phytadienes yields isoprenoid thiophenes VI and VII (Figure 10). These compounds are the common isoprenoid thiophenes found in many normal marine salinity sediments (13,24,25). Incorporation of sulfur into phytadienols and geranylgeraniol will yield *via* the same mechanistic principles isoprenoid thiophenes I-V and also isoprenoid bithiophenes in the case of geranylgeraniol (Figure 10). The nature of the observed differences in distribution patterns of the C_{20} isoprenoid thiophenes and bithiophenes may thus be explained by the restricted occurrence of geranylgeraniol and $\Delta^{2,6}$ and/or $\Delta^{2,10}$-phytenol in hypersaline environments (14) as moieties of certain bacteriochlorophylls. Consequently, an isoprenoid thiophene ratio has been defined as the ratio of the sum of thiophenes VI and VII and the sum of isoprenoid thiophenes I-V (see Figure 8), which may be applied in assessing palaeosalinity (16). Variation in this ratio from 0-40 have been observed to date (16). Values <0.5 are thought to reflect hypersaline palaeoenvironments.

Pristane/Phytane Ratios

Apart from intramolecular reactions with sulfur of intermediate phytenes resulting from (poly)unsaturated C_{20} isoprenoid alcohols strong evidence exists that phytanyl units are also 'quenched' by inorganic sulfur species in an intermolecular fashion leading to incorporation of phytanyl moieties in a high-molecular-weight matrix (20). This phenomenon is illustrated by the products of desulfurisation of polar fractions of crude oils or sediment extracts. In all cases (42) phytane is an abundant component of the desul-furisation mixture whereas pristane is only a minor product (*e.g.*, Figure 11). To explain this phenomenon it is assumed that, for example, phytolesters or phytol are quenched by H_2S or HS_x^- *via* an Sn_2 reaction or *via* the inter-mediate phytadienes, respectively (Figure 12). The resulting thiols are thought to react with the functionalities of other compounds to produce high-molecular-weight materials. Hence, relatively large amounts of the phytol precursors (*e.g.*, chlorophyll-*a*, phytylesters) probably escape from biotrans-formation and mineralisation in the upper part of sediments. Later on, on increasing diagenesis, the relatively weak C-S bonds are cleaved and phytane can be generated after hydrogenation of the intermediate phytenes or phytadienes. As a result of this selective preservation of phytol *via* sulfur

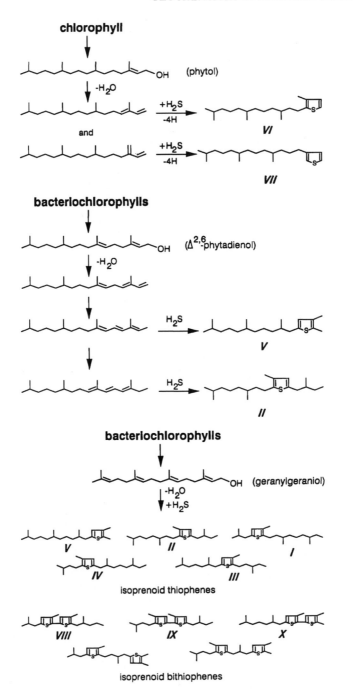

Figure 10.   Postulated formation of C_{20} isoprenoid thiophenes and bithiophenes by incorporation of sulfur into phytenes. Different substrates lead to different thiophenes which is the major control for the thiophene ratio.

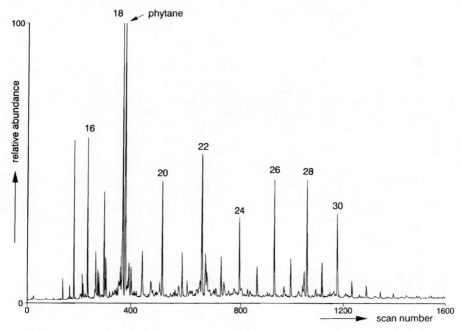

Figure 11.   Mass chromatogram of *m/z* 57 of the desulfurised polar fraction of the Sicily seep oil-E2 (Italy).

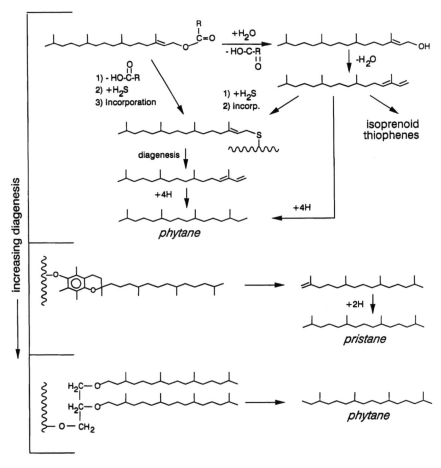

Figure 12.   Formation of phytane and pristane from high-molecular-weight substances during increasing stages of diagenesis.

quenching very low pristane/phytane ratios can be expected in sedimentary environments where H_2S (or HS_x^-) can react with functionalised lipids such as phytol. H_2S and/or HS_x^- concentrations may be high in hypersaline environments due to the presence of high amounts of relatively well-preserved organic matter. Also, the relatively low concentration, or absence, of iron is highly favourable for the genesis of organic sulfur compounds (15,16,21). In addition, the pristane/phytane ratio may be influenced by increasing diagenesis. Tocopheryl units, present in the high-molecular-weight fractions of sediments, are thought to release pristane at a latter stage of diagenesis due to their higher relative stability (43, Figure 12). In sediments originating from extremely saline environments a third effect on the pristane/phytane ratio can be expected due to the probable presence of bound dibiphytanylethers originating from membranes of halophilic Archaebacteria, which can release phytane late during diagenesis.

Relative Abundance of C_{35} Hopanes (or Hopenes)

In many samples reflecting hypersaline palaeoenvironments a series of extended hopanes and/or hop-17(21)-enes dominated by C_{35} members have been observed (7,8,31,33). It should be noted, however, that similar distribution patterns are also encountered in samples originating from normal marine salinity sediments such as those from the Brazilian marginal basins (31), the Serpiano shale (44), the Phosphoria Retort shale (45) and the Jurf ed Darawish Oil Shale (23, 46).

As mentioned above it may be expected that the original precursor, bacteriohopanetetrol (47,48), reacts with inorganic sulfur species at early stages of diagenesis preventing the biotransformation or mineralisation of this substrate (Figure 13). Intramolecular reaction products have been reported by several authors (16,23,49). Desulfurisation of polar fractions of oils or sediment extracts releases exclusively C_{35} hopanes in relatively high amounts (20,42), as illustrated by the mass chromatogram of m/z 191 for the desulfurised polar fraction of an outcrop sample of the Monterey formation at Naples Beach (Figure 14). Obviously the C_{35} bacteriohopanetetrol is also quenched intermolecularly by sulfur. Because of the relatively weak C-S bonds C_{35} hopanes and hopenes can be generated rather early during diagenesis leading to extended hopane distributions with relatively abundant C_{35} components.

Even-Over-Odd Carbon Number Predominance of *n*-Alkanes

The phenomenon of even-over-odd carbon number predominance of *n*-alkanes has long been recognised in a number of Recent and ancient sediment

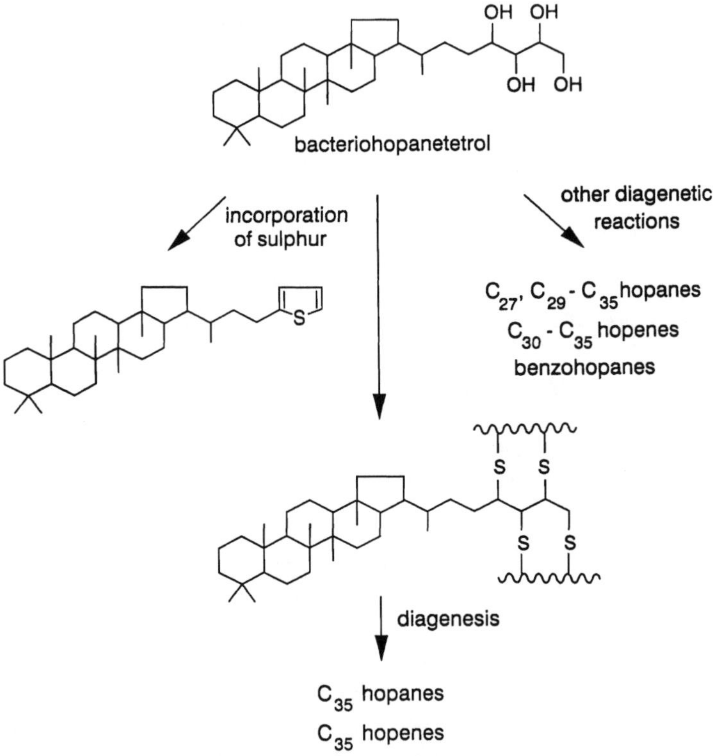

Figure 13. Proposed early diagenetic pathways of bacteriohopanetetrol.

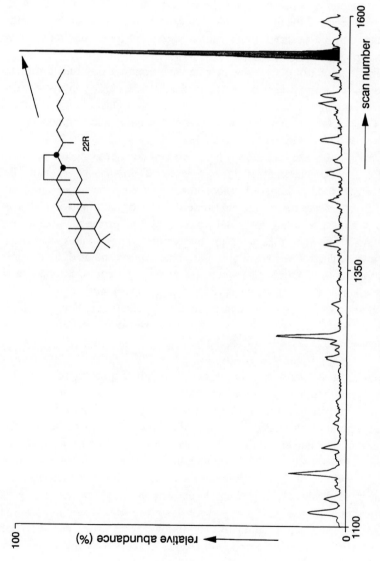

Figure 14. Mass chromatogram of *m/z* 191 of the desulfurised polar fraction of the bitumen of an outcrop section at Naples Beach of the Monterey Formation.

extracts and crude oils (50-58). Based on empirical correlations this even-over-odd carbon number predominance was associated with hypersaline palaeo-environments (50,53,54).

In most of these publications suggestions have been put forward to explain this phenomenon. Complete reduction of naturally occurring fatty acids to their corresponding alkanes have been suggested (50,51,55). Such a reduction is, however, difficult to imagine since the reduction of fatty acids to alkanes can only be realised under rather severe reaction conditons which are probably not operating in Recent sediments.

Desulfurisation of polar fractions of relative immature oils and sediment from hypersaline palaeoenvironments yields n-alkanes with strong even-over-odd carbon number predominances (20,42; e.g., Figure 11). This phenomenon is, however, not restricted to samples from hypersaline palaeoenvironments (e.g., Jurf ed Darawish Oil Shale, Monterey Shale).

In agreement with the sulfur-quenching process described in the previous paragraphs for specific functionalised lipids a selective preservation of naturally occurring even carbon numbered n-alcohols or n-alkyl moieties can be rationalized. Intermediate thiols can be formed either by addition of H_2S (or HS_x^-) to the n-alkene resulting from dehydration of the precursor alkanol or by Sn_2-reaction of H_2S or HS_x^- with an alkylester. These thiols can react with other functionalised compounds and thus become incorporated in a high-molecular-weight matrix. Upon diagenesis the relatively weak C-S bonds are cleaved and alkenes with the same (predominantly even) number of carbon atoms as their alkanol precursors are released. Hydrogenation of these alkenes leads to n-alkanes with strong even-over-odd carbon number predominances (Figure 15).

From the above it is clear that sulfur can play an important role in the selective preservation and hence enrichment of specific lipids which otherwise are prone to microbial transformation or mineralisation. In the case of inter-molecular incorporation processes via sulfur the lipids thereby preserved are released again during diagenesis. This process results in distribution patterns of compound classes which can be specific for palaeoenvironments wherein relatively high amounts of H_2S were generated due to anoxic conditions and high productivity and where the H_2S produced was available for reaction with organic matter due to the low amounts of iron present. Since hypersaline environments fall into this category it follows that the characteristic distribution patterns of specific compound classes described in this paper are observed in samples from hypersaline palaeoenvironments. It should be realised, however, that these environmental conditions are not restricted to

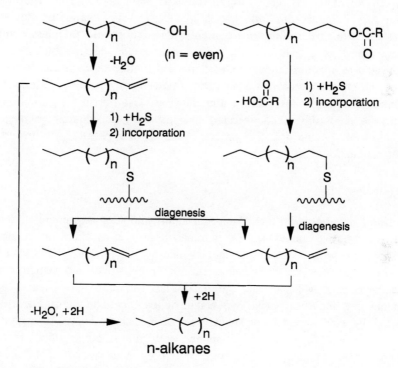

Figure 15.   Proposed early diagenetic pathways of even carbon numbered *n*-alcohols and alkylesters.

hypersaline palaeoenvironments and, therefore, these characteristic distribution patterns may also be observed in samples from non-clastic, marine environments (*e.g.*, siliceous oozes, carbonates, phosphorites).

A more critical evaluation of the above mentioned ratios and phenomena reveals the usefulness of the various palaeosalinity indicators. Distribution patterns of methylated chromans and the relative abundance of gammacerane are not influenced by sulfur incorporation reactions and may directly reflect species distributions in the palaeoenvironment. To some extent this holds for $14\alpha(H),17\alpha(H)/14\beta(H),17\beta(H)$-steraneratios as well, although incorporation of sulfur may influence this ratio and original Δ^7/Δ^5-sterol ratios do not always correlate with hypersaline environments. The isoprenoid thiophene ratio is highly useful as a palaeosalinity indicator since the distribution of the C_{20} isoprenoid thiophenes directly reflects the distribution of their substrates. The other parameters (pristane/phytane ratio, odd-over-even carbon number predominance of *n*-alkanes, relative abundance of C_{35} hopanes and/or hopenes) should be used with caution because they obviously depend on the quenching by sulfur of specific lipids, a process which is not restricted to hypersaline environments.

Conclusions

Distribution patterns of methylated chromans and C_{20} isoprenoid thiophenes and the relative abundance of gammacerane are excellent indicators for palaeosalinity whereas other parameters such as $14\alpha(H),17\alpha(H)/14\beta(H),17\beta(H)$-sterane ratios, the pristane/phytane ratio, the even-over-odd carbon number predominance of *n*-alkanes and the relative abundance of C_{35} hopanes and/or hopenes may indicate (palaeo)hypersaline depositional conditions, but are also affected by environmental conditions other than palaeosalinity. All parameters, except the MTTC, isoprenoid thiophene and gammacerane ratio, are highly dependent on maturity. It has been noted by Mello *et al.* (31) that "no single biological marker property is sufficient to characterise and assess a specific environment of deposition" and this statement also holds for assessment of palaeosalinity. We, therefore, highly recommend the use of a combination of the various palaeosalinity indicators.

Literature Cited

1. Evans, R.; Kirkland, D.W. In Evaporites and Hydrocarbons; Schreiber, B.C., Ed.; Columbia University Press: New York, 1988; p. 256.

2. Gerdes, G.; Krumbein, W.E.; Holtkamp, E. In Hypersaline Ecosystems, Ecological studies 53; Friedman, G.M.; Krumbein, W.E., Eds.; Springer-Verlag: Berlin, 1985; p. 238.

3. Ehrlich, A.; Dor, I. In Hypersaline Ecosystems, Ecological studies 53; Friedman, G.M.; Krumbein, W.E., Eds.; Springer-Verlag: Berlin, 1985; p. 296.

4. Cohen, Y.; Krumbein, W.E.; Shilo, M. Limnol. Oceanogr. 1977, 22, 621-634.

5. Krumbein, W.E.; Buchholz, H.; Franke, P.; Giani, D.; Giele, C.; Wonneberger, K. Naturwissenschaften 1979, 66, 381-389.

6. Kirkland, D.W.; Evans, R. AAPG Bull. 1981, 65, 181-190.

7. ten Haven, H.L.; de Leeuw, J.W.; Schenck, P.A. Geochim. Cosmochim. Acta 1985, 49, 2181-2191.

8. ten Haven, H.L.; de Leeuw, J.W.; Sinninghe Damsté, J.S.; Schenck, P.A.; Palmer, S.E.; Zumberge, J.E. In Lacustrine Petroleum Source Rocks; Kelts, K.; Fleet, A.J.; Talbot, M., Eds.; Geol. Soc. Spec. Publ. No. 40; Blackwell: Oxford, 1988; pp 123-130.

9. ten Haven, H.L.; de Leeuw, J.W.; Rullkötter, J.; Sinninghe Damsté, J.S. Nature (London) 1987, 330, 641-643.

10. ten Haven, H.L.; de Leeuw, J.W.; Peakman, T.M.; Maxwell, J.R. Geochim. Cosmochim. Acta 1986, 50, 853-855.

11. de Leeuw, J.W.; Cox, H.C.; van Graas, G.; van der Meer, F.W.; Peakman, T.M.; Baas J.M.A.; van de Graaf, B. Geochim. Cosmochim. Acta 1989, 53, 903-909.

12. Peakman, T.M.; ten Haven, H.L.; Rechka, J.R.; de Leeuw, J.W.; Maxwell, J.R. Geochim. Cosmochim. Acta 1989, 53, 2001-2009.

13. Brassell, S.C.; Lewis, C.A.; de Leeuw, J.W.; de Lange, F.; Sinninghe Damsté, J.S. Nature (London) 1986, 320, 160-162.

14. Sinninghe Damsté, J.S.; de Leeuw, J.W. Intl. J. Environ. Anal. Chem. 1987, 28, 1-19.

15. Sinninghe Damsté, J.S.; de Leeuw, J.W.; Kock-van Dalen, A.C.; de Zeeuw, M.A.; de Lange, F.; Rijpstra, W.I.C.; Schenck, P.A. Geochim. Cosmochim. Acta 1987, 51, 2369-2391.

16. Sinninghe Damsté, J.S.; Rijpstra, W.I.C.; de Leeuw, J.W.; Schenck, P.A. Geochim. Cosmochim. Acta 1989, 53, in press.

17. Sinninghe Damsté, J.S.; Rijpstra, W.I.C.; de Leeuw, J.W.; Schenck, P.A. Org. Geochem. 1988, 13, 593-606.

18. Sinninghe Damsté, J.S.; Rijpstra, W.I.C.; Kock-van Dalen, A.C.; de Leeuw, J.W.; Schenck, P.A. Geochim. Cosmochim. Acta 1989, 53, in press.

19. Sinninghe Damsté, J.S.; Eglinton, T.I.; de Leeuw, J.W.; Schenck, P.A. Geochim. Cosmochim. Acta 1989, 53, 873-889.

20. Sinninghe Damsté, J.S.; Eglinton, T.I.; Rijpstra, W.I.C.; de Leeuw, J.W.

In Geochemistry of Sulfur in Fossil Fuels; Orr, W.L.; White, C.M., Eds.; ACS Symposium Series; American Chemical Society: Washington, DC, 1989; this volume.

21. Sinninghe Damsté, J.S.; ten Haven, H.L.; de Leeuw, J.W.; Schenck, P.A. Org. Geochem. 1986, 10, 791-805.

22. Sinninghe Damsté, J.S.; Kock-van Dalen, A.C.; de Leeuw, J.W.; Schenck, P.A. Tetrahedron Lett. 1987, 28, 957-960.

23. Kohnen, M.E.L.; Sinninghe Damsté, J.S.; Rijpstra, W.I.C.; de Leeuw, J.W. In Geochemistry of Sulfur in Fossil Fuels; Orr, W.L.; White, C.M., Eds.; ACS Symposium Series; American Chemical Society: Washington, DC, 1989; this volume.

24. ten Haven, H.L.; Rullkötter, J.; Sinninghe Damsté, J.S.; de Leeuw, J.W. In Geochemistry of Sulfur in Fossil Fuels; Orr, W.L.; White, C.M., Eds.; ACS Symposium Series; American Chemical Society: Washington, DC, 1989; this volume.

25. Rullkötter, J.; Landgraf, M.; Disko, U. J. High Res. Chrom. & Chro. Commun. 1988, 11, 633-638.

26. Schmid, J.C. Ph.D. Thesis, University of Strasbourg, Strasbourg, 1986.

27. Sinninghe Damste, J.S.; Kock-van Dalen, A.C.; de Leeuw, J.W.; Schenck, P.A.; Sheng Guoying; Brassell, S.C. Geochim. Cosmochim. Acta 1987, 51, 2393-2400.

28. Rullkötter, J.; Sinninghe Damsté, J.S.; ten Haven, H.L.; de Leeuw, J.W.; unpublished results.

29. Kohnen, M.E.L.; Sinninghe Damsté, J.S.; Rijpstra, W.I.C.; de Leeuw, J.W.; unpublished results.

30. Moldowan, J.M.; Seifert, W.K.; Gallegos, E.J. AAPG Bull. 1985, 69, 1255-1268.

31. Mello, M.R.; Telnaes, N.; Gaglianone, P.C.; Chicarelli, M.I.; Brassell, S.C.; Maxwell, J.R. Org. Geochem. 1988, 13, 31-45.

32. Fu Jiamo; Sheng Guoying; Peng Pingan; Brassell, S.C.; Eglinton, G.; Jiang Jigang Org. Geochem. 1986, 10, 119-126.

33. Brassell, S.C.; Sheng Guoying; Fu Jiamo; Eglinton, G. In Lacustrine Petroleum Source Rocks; Kelts, K.; Fleet, A.J.; Talbot, M., Eds.; Geol. Soc. Spec. Publ. No. 40; Blackwell: Oxford, 1988; pp 299-308.

34. ten Haven, H.L.; Rohmer, M.; Rullkötter, J.; Bisseret P. Geochim. Cosmochim. Acta 1989, 53, in press.

35. Toste, A.P. Ph.D. Thesis, University of California, Berkely, 1976.

36. Meissner, F.F.; Woodward, J.; Clayton, J.L. In Hydrocarbon Source Rocks of the Greater Rocky Mountain Region; Woodward, J.; Meissner, F.F.; Clayton, J.L., Eds.; Rocky Mountain Association of Geologists: Denver; pp 1-34.

37. de Leeuw, J.W.; Sinninghe Damsté, J.S.; Klok, J.; Schenck, P.A.; Boon, J.J. In Hypersaline Ecosystems, Ecological studies 53; Friedman, G.M.; Krumbein, W.E., Eds.; Springer-Verlag: Berlin, 1985; p. 350.

38. Boon, J.J.; de Leeuw, J.W. In Cyanobacteria: Current Research; Fay, P.; van Baalen, C., Eds.; Elsevier: Amsterdam, 1987; pp 471-492.

39. Robinson, N.; Cranwell, P.A.; Eglinton, G.; Brassell, S.C.; Sharp, C.L.; Gophen, M.; Pollingher, U. Org. Geochem. 1986, 10, 733-742.

40. Harvey, H.R.; Eglinton, G.; O'Hara, S.C.M.; Corner D.S. Geochim. Cosmochim. Acta 1987, 51, 3031-3040.

41. Harvey, H.R.; O'Hara, S.C.M.; Eglinton, G.; Corner D.S. Org. Geochem. 1989, in press.

42. Sinninghe Damsté, J.S.; Rijpstra, W.I.C.; de Leeuw, J.W., in preparation.

43. Goossens, H.; de Leeuw, J.W.; Schenck, P.A.; Brassell, S.C. Nature (London) 1984, 312, 440-442.

44. McEvoy, J.; Giger, W. Org. Geochem. 1986, 10, 943-949.

45. Sinninghe Damsté, J.S., unpublished results.

46. Wehner, H.; Hufnagel, H. In Biochemistry of Black Shales; Degens, E.T. et al., Eds.; Mitt. Geol Palaeont.Inst. Univ. Hamburg 1987, 60, 381-395.

47. Ourisson, G.; Albrecht, P.; Rohmer, M. Pure Appl. Chem. 1979, 51, 709-729.

48. Ourisson, G.; Albrecht, P.; Rohmer, M. Trends Biochem. Sci. 1982, 7, 236-239.

49. Valisolalao, J.; Perakis N.; Chappe, B.; Albrecht, P. Tetrahedron Lett. 1984, 25, 1183-1186.

50. Welte, D.H.; Waples, D.W. Naturwissenschaften 1973, 60, 516-517.

51. Albaigés, J.; Torradas, J.M.; Nature (London) 1974, 250, 567-568.

52. Dembicki, H.; Meinschein, W.G.; Hattin, D.E. Geochim. Cosmochim. Acta 1976, 40, 203-208.

53. Spiro, B.; Aizenshtat, Z. Nature (London) 1977, 269, 235-237.

54. Tissot, B.; Pelet, R.; Roucahe, J.; Combaz, A. In Advances in Organic Geochemistry 1975; Campos, R.; Goni, J., Eds.; Enadimsa: Madrid, 1977; pp 117-154.

55. Sheng Guoying; Fan Shanfa; Lin Dehan; Su Nengxian; Zhou Hongming In Advances in Organic Geochemistry 1979; Douglas, A.G.; Maxwell, J.R., Eds.; Pergamon: Oxford, 1980; pp 115-121.`

56. Grimalt, J.; Albaiges, J.; Al-Saad, H.T.; Douabul, A.A.Z. Naturwissenschaften 1985, 72, 35-37.

57. Nishimura, M.; Baker, E.W. Geochim. Cosmochim. Acta 1986, 50, 299-305.

58. Grimalt, J.; Albaigés, J. Geochim. Cosmochim. Acta 1987, 51, 1379-1384.

RECEIVED March 5, 1990

Chapter 25

Alkylthiophenes as Sensitive Indicators of Palaeoenvironmental Changes

A Study of a Cretaceous Oil Shale from Jordan

Mathieu E. L. Kohnen, Jaap S. Sinninghe Damsté, W. Irene C. Rijpstra, and Jan W. de Leeuw

Faculty of Chemical Engineering and Materials Science, Organic Geochemistry Unit, Delft University of Technology, De Vries van Heystplantsoen 2, 2628 RZ Delft, Netherlands

Thirteen samples from the immature, Cretaceous Jurf ed Darawish oil shale (Jordan), spanning a depth range of 120m, were analysed quantitatively for aliphatic hydrocarbons and alkylthiophenes in the bitumens by gas chromatography-mass spectrometry after isolation of appropriate fractions. The oil shale studied consists of three different lithologically defined facies (i.e. bituminous limestone, bituminous calcareous marl and phosphorite). The quantitative and qualitative variations in the hydrocarbon biomarkers parallel the lithological variations. Considerable quantitative and qualitative variations in the various classes alkylthiophenes (i.e. 2,5-dialkylthiophenes, isoprenoid thiophenes, highly branched isoprenoid thiophenes and 3,4-dialkylthiophenes) were observed and most of the depth profiles are not related to the lithological facies, providing a fine structure in the stratigraphy. Since the precursors of most of these alkylthiophenes and the organisms biosynthesizing these precursors are known, analysis of alkylthiophenes reveals substantial information on sources of organic matter and the palaeoenvironmental conditions.

A major goal in organic geochemistry is to reconstruct depositional environments of the geological past based on qualitative and quantitative analyses of biological markers in sediments. Classes of hydrocarbons such as n-alkanes, isoprenoid alkanes, steranes and hopanes are routinely used for this purpose (1).

0097–6156/90/0429–0444$11.50/0

The application of hydrocarbons as markers for changes in palaeoenvironments is, however, somewhat limited. Williams and Douglas (2) studied Kimmeridgian sedimentary sequences with variable lithology (clay, shale, and limestone). They found no major qualitative differences between the hydrocarbon fractions of the bitumens from the three sedimentary units. Belayouni and Trichet (3) studied bitumens of a Lower Eocene sequence (Gafsa Basin, Tunisia) consisting of limestones, phosphorites, cherts, and shales. They also observed similar saturated hydrocarbon fractions from the different sedimentary units. In both studies it was suggested that the similarity of the saturated hydrocarbon fractions implies a uniform biological contribution of organic matter (phytoplankton) to the sediments. This suggestion may be biased because only a limited number of specific compounds are known and used as markers for contributions from different algal species (4).

Moldowan *et al.* (5) investigated a sediment core, spanning a depth range of 5m, of Lower Toarcian shales from W. Germany specifically at a transition zone from a rather oxidized, shallow-marine, marly sediment to an organic matter-rich, black shale. Variations in distributions of isoprenoid hydrocarbons, steranes and monoaromatic steroids were observed and were related to variations in oxidation/reduction conditions during and shortly after sedimentation.

Recently, Sinninghe Damsté *et al.* (6) suggested that organic sulfur compounds (OSC) are indicative of biosynthetic functionalised lipids present during the deposition of the sediment. OSC, in general, are formed by incorporation of hydrogen sulphide and/or polysulphides into functionalised lipids during early stages of diagenesis (6-12) and are widespread in sediments and oils (6,13). Since specific functionalised lipids can be related to specific organisms their sulphur containing counterparts may also serve as biomarkers. Preliminary studies of a limited number of samples from the Jordan oil shale deposit at Jurf ed Darawish revealed major differences in alkylthiophene distribution patterns (6,10), suggesting that alkylthiophenes are good indicators of changes in the palaeoenvironment.

In this paper we describe a detailed study of changes in depositional palaeoenvironment as revealed by relative and absolute amounts of specific alkylthiophenes in bitumens. For this purpose we analysed qualitatively and quantitatively the extractable hydrocarbons and OSC of 13 samples from the Jurf ed Darawish oil shale deposit, spanning a depth range of 120m. Effects of different degrees of thermal maturation are minimised because this small depth range.

Geological Setting

The upper Cretaceous Jurf ed Darawish oil shale (thickness *ca.* 120m; 45m-165m depth) in the Ghareb Formation is a calcareous, bituminous marl (commercially termed oil shale), which is located 130 km south of Amman, Jordan (14). The organic matter in this sequence is thermally immature ("vitrinite" reflectance = 0.28-0.31 %; L. Buxton, pers. comm.)

This deposit is part of a larger system of similar facies and age found in many localities in Jordan and Israel (15-20). Abed and Amireh (16) investigated oil shale deposits in North Jordan and the El Lajjun deposit in Central Jordan. They postulated that these bituminous limestones were deposited in an oxygenated shallow marine environment where the H_2S/O_2 interface was below the sediment/water interface ("gyttja" type of environment). Spiro and Aizenshtat (19) presented evidence that the lowermost beds of the oil shales in the Ghareb Formation from Nebi Musa, Israel were deposited in a hypersaline, stratified lagoonal environment. Bein and Amit (17) studied the upper Cretaceous sequence of the Negev in Southern Israel, which consists of cherts, phosphorites, and bituminous limestones in the Ghareb Formation. These authors proposed that after phosphorite deposition, syndepositional tectonic activity resulted in the introduction of fresh water, restriction of bottom-water circulation, and the establishment of a density-stratified water body, which consequently favoured the preservation of the organic matter. The results of these studies seem to be somewhat contradictory. However, if these oil shales are deposited in discrete restricted sub-basins such as presedimentary or synsedimentary depressions (14) these differences can be better understood.

The Jurf ed Darawish oil shale was studied in somewhat more detail with respect to geology and geochemistry (14,15) than the above mentioned deposits. The deposit occurs at the base of the Chalk-Marl Unit in the Ghareb Formation, which consists of marine marls of upper Maastrichtian age. The Chalk-Marl Unit is overlain by 50-60m of Quaternary sediments. The underlying Senonian phosphatic limestones and phosphatic cherts of the Phosphorite Unit are bituminous in the upper 10m. These phosphorites belong to the phosphorite belt, which extends along the eastern and southern coasts of the Mediterranean (18). The bituminous section of the Chalk-Marl Unit has been subdivided into a Lower Member (160-120m; calcareous bituminous marl) and an Upper Member (120-50m; bituminous limestone) (15). It is postulated by Wehner and Hufnagel (15) that the Upper and Lower Members of the sequence were deposited in a "gyttja" environment and a "sapropelic" environment with an anoxic water column, respectively. This postulation is mainly based on crossplots of trace elements (Ni, Co, Cu, V and Cr) (20), presence or lack of bioturbated sediments and microscopical observations (15).

Samples and Experimental Methods

Samples. Thirteen composite one-meter-samples of a core from a drill hole were investigated (Table I).

Table I. Samples studied and some geochemical bulk data of the Jurf ed Darawish oil shale.

depth (m)	TOC[a]	Carb.[b]	Al_2O_3[c]	P_2O_5[c]	tot-S[d]	$R_o(\%)$	Facies[e]
45-46	5.1	88	1.9	0.8	1.5	n.d.	A
50-51	4.8	83	2.0	0.7	1.5	n.d.	A
56-57	7.2	85	2.3	0.6	1.8	n.d.	A
59-60	6.9	82	2.1	1.0	1.9	n.d.	A
68-69	5.7	87	1.9	0.4	1.6	n.d.	A
83-84	6.5	83	2.1	0.5	1.6	n.d.	A
100-101	7.1	66	2.0	0.5	2.0	0.28	A
115-116	4.2	61	3.8	1.0	1.7	n.d.	B
120-121	7.0	65	6.9	1.2	2.0	n.d.	B
137-138	9.7	54	8.1	1.6	2.6	n.d.	B
144-145	14.2	53	7.2	1.8	3.5	n.d.	B
156-157	16.7	45	5.0	2.3	4.1	0.31	B
165-166	8.3	82	1.4	7.3	2.0	n.d.	C

[a] TOC in wt%, [b] HCl soluble carbonate in wt%, [c] Al_2O_3 en P_2O_5 in wt%, [d] total-S includes inorganically- and organically-bound sulfur, [e] codes refer to Figure 1.

Extraction. The powdered samples (*ca.* 60 g) were extracted sequentially with 100 ml MeOH, 100 ml MeOH/CH_2Cl_2 (1/1, v/v, x2) and with 75 ml CH_2Cl_2 (x5) using ultrasonication and centrifugation. The supernatants were combined and the bitumens were obtained after rotary evaporation of the solvent at 30°C.

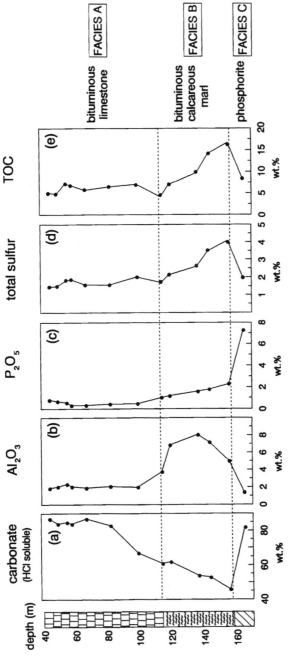

Figure 1. Lithology, variations with depth of (a) carbonate content, (b) Al_2O_3 content, (c) P_2O_5 content, (d) total S content (organically and inorganically bound sulfur) and (e) total organic carbon for the Jurf ed Darawish oil shale and the three lithologically defined Facies (bituminous limestone, bituminous calcareous marl and phophorite) are schematicaly depicted.

<u>Fractionation.</u> Aliquots of the bitumens (*ca.* 200 mg) with added standards (Table II) were separated (without prior removal of asphaltenes) into two fractions using a column (25 cm x 2 cm; column volume 35 ml) packed with alumina (activated for 2.5 h at 150°C) by elution with hexane/CH_2Cl_2 (9:1, v/v; 150 ml; "apolar fraction") and MeOH/CH_2Cl_2 (1:1, v/v; 150 ml; "polar fraction I"). Aliquots (*ca.* 10 mg) of the apolar fractions (yield typically 7-15 wt.% of bitumen) were further separated by argentatious thin layer chromatography using hexane as developer. The $AgNO_3$-impregnated silica plates (20 x 20 cm; thickness 0.25 mm) were prepared by dipping them in a solution of 1% $AgNO_3$ in MeOH/H_2O (4:1, v/v) for 45s and subsequent activation at 120°C for 1h. Four fractions (A1, R_f = 0.85-1.00; A2, R_f = 0.57-0.85; A3, R_f = 0.06-0.57; A4, R_f = 0.00-0.06) were scraped of the TLC plate and ultrasonically extracted with ethyl acetate (x3).

Other aliquots of the bitumens (*ca.* 40 mg) with an added hydrocarbon standard (I, Table II) were fractionated using a column (column volume 7 ml) packed with alumina (activated for 1.5 h at 150°C) by elution with hexane (11 ml; "saturated hydrocarbon fraction") and MeOH/CH_2Cl_2 (1:1, v/v; 20 ml; "polar fraction II"). The saturated hydrocarbon fractions and appropriate TLC fractions (A2 and A3) were analysed by GC and GC-MS.

<u>Gas chromatography.</u> Gas chromatography (GC) was performed using a Carlo Erba 5300 instrument, equipped with an on-column injector. Two column types were used: a fused silica capillary column (25 m x 0.32 mm) coated with CP Sil-5 (film thickness = 0.12 μm) with helium as carrier gas and an aluminium, high temperature, capillary column (25 m x 0.32 mm) coated with HT-5 (film thickness = 0.1 μm) with H_2 as carrier gas. Detection was performed using both a flame ionization (FID) and sulfur-selective flame photometric (FPD) detectors, using a stream splitter at the end of the capillary CP Sil-5 column. The FPD was not used with the HT-5 column. Using this column the oven was programmed from 75° to 130°C at 20°C/min, then at 4°C/min to 350°C and finally at 6°C/min to 450°C (10 min hold time). In the case of the CP Sil-5 column the oven was programmed from 75° to 130°C at 20°C/min and then at 4°C/min to 320°C (20 min hold time).

<u>Gas chromatography-mass spectrometry.</u> Gas chromatography-mass spectrometry (GC-MS) was carried out using a Hewlett-Packard 5840 gas chromatograph connected to a VG-70S mass spectrometer operated at 70 eV with a mass range *m/z* 40-800 and a cycle time of 1.8 s. The gas

chromatograph was equipped with a fused silica capillary column (25 m x 0.32 mm) coated with CP Sil-5 (film thickness = 0.1 μm). Helium was used as carrier gas. Fractions (1.0 μl) were injected "on column" at 50°C in ethyl acetate (*ca.* 6 mg/ml) and the temperature was programmed at 10°C/min to 120°C and then at 4°C/min to 320°C (30 min hold time).

Table II. General composition of the sub-fractions obtained
by argentatious TLC of the apolar fraction of the
Jurf ed Darawish oil shale bitumen

frac.	R_f	major compound classes	standards added	
A1	0.85-1.00	*n*-alkanes, isoprenoids, hopanes, steranes	I	
A2	0.57-0.85	alkylthiophenes, sterenes hopenes, alkylbenzenes C-ring monoaromatic steroids	II	
A3	0.06-0.57	alkylbenzothiophenes, thiophene-hopanoids, A- and B-ring mono-aromatic steroids, methylated 2-methyl-2-(4,8,12-trimethyl-tridecyl)-chromans, aromatic hopanoids	III	
			IV	
A4	0.06-0.00	alkylthiolanes, alkylthianes		

key: R_1 = n-C_{15} and R_2 = i-C_{15}

Quantitation. Pristane and phytane concentrations (μmol/kg bitumen) were obtained by integration of their peak areas and that of the deuterated C_{22} anteisoalkane (I, Table II) standard in the FID chromatograms. The concentration of the other compounds in the saturated hydrocarbon fraction were obtained by integration of appropriate peaks in mass chromatograms of m/z 57 (*n*-alkanes, standard), m/z 367 (extended hop-17(21)-enes), m/z 191 (hopanes), m/z 217 (steranes). Because of differences in yield of these ions for different classes of compounds the values for hop-17(21)-enes, steranes and hopanes are approximate.

The concentrations of the alkylthiophenes (μmol/kg bitumen) were obtained by integrating mass chromatograms of their molecular ions and the molecular ion of the standard, 2,3-dimethyl-5-(1',1'-d$_2$-hexadecyl)-thiophene (II, Table II). Because the alkylthiophenes and the standard show a similar ionisation potential and total ion yield, the values for the alkylthiophenes are considered as absolute concentrations.

The thiophene hopanoids in fraction A3 were quantified by integration of mass chromatograms of m/z 191 and m/z 107 for the standard, 2-methyl-2-(4,8,12-trimethyltridecyl)chroman (III, Table II).

The term "total amount of alkylthiophenes" used hereafter includes all of the identified alkylthiophenes and the major alkylthiophenes with unknown structures (<10 % of "total amount of alkylthiophenes"). It is estimated that 80 to 90 % of all the alkylthiophenes in the A2 fraction were quantified. The "relative abundance of thiophene hopanoids" is defined as the total amount of thiophene hopanoids (standardised to the chroman standard) divided by the "total amount of alkylthiophenes".

The estimation of the analytical precision is derived from three duplicate analyses. The difference between the analytical results of duplicate analyses was always less than 10% of the total variation observed throughout the whole sequence.

Synthesis of standards.   Friedel-Crafts acylation of 2,3-dimethylthiophene with hexadecanoic acid   and a subsequent reduction of the ketone with LiAlD$_4$ yielded the 2,3-dimethyl-5-(1',1'-d$_2$-hexadecyl)thiophene (II, Table II). Experimental details of these reactions have been reported elsewhere (10). Ionic hydrogenation of the 2,3-dimethyl-5-(1',1'-d$_2$-hexadecyl)thiophene in trifluoroacetic acid using triethylsilane and BF$_3$.etherate as catalyst yielded 2,3-dimethyl-5-(1',1'-d$_2$-hexadecyl)thiolane   (IV, Table II). For detailed reaction conditions see Sinninghe Damsté *et al.* (22). Raney Ni desulfurisation (9) of 2,3-dimethyl-5-(1',1'-d$_2$-hexadecyl)thiophene yielded 6,6-d$_2$-2-methyleicosane (I, Table II). The 2-methyl-2-(4,8,12-trimethyltridecyl)-chroman (III, Table II) was synthesised by a condensation reaction of phytol with phenol (23).

Whole rock analyses. The thoroughly homogenised whole rock samples were dissolved in a HF/HNO$_3$/HClO$_4$-solution and analysed for major and trace elements by emission spectroscopy (ICPES). The carbonate content was determined separately by treatment of the rock powder with HCl. Total organic carbon of the whole rock samples was determined by a Leco carbon analyzer.

Lithology and Facies Description

The bulk parameters of the samples studied (Figure 1; Table I) show notable variations throughout this sequence. The carbonate content varies over the entire depth range between 45% and 85% (Figure 1a). The Al_2O_3 content, which reflects the presence of clay minerals, shows a distinct maximum in the lower part of this oil shale deposit (Figure 1b). It is believed that the clays (kaolinite) reflect a detrital influx into the depositional basin. The P_2O_5 content (Figure 1c) is quite high (7.3 wt.%) towards the bottom part of the sequence. A deeper sample (167 m), which is presently undergoing analyses, has an even higher P_2O_5 content of 23.1 wt.%. Total sulfur, which includes organically- and inorganically-bound sulfur, also shows considerable variations and maximises in the lower part of the sequence at 4.4 wt.% (Figure 1d). The organic matter content also increases considerably in the lower part of the sequence (TOC up to 16.7%; Figure 1e). Based on these results and on literature data (14,15) this oil shale sequence was divided into three distinct facies (A, B, and C) which reflect different palaeoenvironments.

The high P_2O_5 content is characteristic for the bottom of this deposit. The lower part of the sequence (160m-170m) is therefore classified as a phosphorite (Facies C). Recent phosphatic-rich sediments are formed in nutrient-rich environments with high primary productivity (24-26). Phosphatization in Recent sediments occurs by remineralization of organic matter in anoxic pore waters during early diagenesis which releases phosphate, ultimately resulting in apatite precipitation and/or replacement of calcite by apatite (24,28). Bein and Amit (17) suggested that the upper Cretaceous phosphorites in southern Israel are due to upwelling at the boundary between the deep Tethys and the shallow shelf.

Facies B (115m-160m) is characterised by a high total sulfur content, high TOC values, absence of bioturbation (15) and a relatively high content of clastic material. All these characteristics, except the latter point, are indicative of an environment with a high preservation potential and/or high primary production. From the input of clastic material it may be expected that a part of the organic matter has a terrigenous source. Anoxic sediment and bottom water layer are favourable for preservation of organic matter (28). In environments with limited water circulation, the oxygen flux to the bottom water becomes inhibited and the H_2S concentration in the interstitial waters and the lower part of the water column increases. At the end of the period marked by the Phosphorite Unit syndepositional tectonic activity or the formation of oyster reefs (14) may have created restricted basins in which a stagnant lower water column caused enhanced preservation of the organic matter. This is supported by the suggestion that the Jordan oil shales are linked to synsedimentary depressions (14). All the characteristics mentioned above are in agreement with this palaeoenvironmental model. Anoxic bottom waters (O_2 <0.5 ml/l) inhibit bioturbation and result in conditions favourable for preservation of organic matter (28). The increased H_2S content is reflected by the increased total sulfur in the sediment.

Facies A (45m-115m) shows lower total sulfur contents, lower TOC values, negligible clastic material (relative to Facies B) and a lack of lamination probably due to bioturbation. Considering these characteristics it is inferred that the palaeoenvironment during the deposition of Facies A was probably less reducing and and/or less productive than during the deposition of Facies B.

Hydrocarbon Biomarkers

Gas chromatograms of three saturated hydrocarbon fractions, roughly representing the different facies (Figure 2), show that some relative and absolute variations exist between the three facies. *n*-Alkanes, pristane, phytane and the extended hop-17(21)-enes are indicated. The most prominent changes within different compound classes in the saturated hydrocarbon fraction are highlighted below.

The ratio of low boiling (C_{19}-C_{23}) to high boiling (C_{27}-C_{31}) *n*-alkanes and CPI values (C_{24}-C_{34}) (29) vary throughout the sequence (Figure 3a-c). The high proportion of C_{27}-C_{31} *n*-alkanes in Facies B (Figure 3a-b) compared to Facies A or C is consistent with the idea that Facies B contains relatively more terrigenous organic matter since it is thought that long chain *n*-alkanes are derived from higher plant waxes. However, the changes in the CPI values are not fully in agreement with such a relative increase of higher plants *n*-alkanes.

The pristane/phytane ratio (Figure 3d) also shows a trend which is apparently influenced by the three facies types. This ratio is thought to reflect changes in anoxicity and/or salinity (30-32).

The relative amounts of the extended hop-17(21)-enes (Figure 3e) and, especially, the relative concentrations of the C_{35} homologue (Figure 3f) show some sudden changes at the facies boundaries. It has been proposed (33-35) that this relative abundance of C_{35} extended hopenes (Facies B) reflects a hypersaline palaeoenvironment. Others consider this the typical distribution of the extended hopenes in anoxic marine sediments (32).

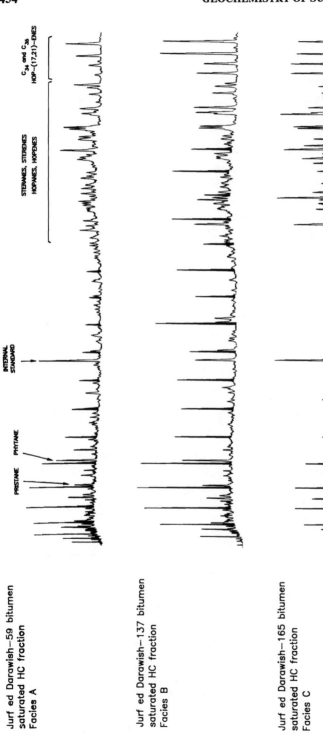

Figure 2. Gas chromatograms (CP Sil-5) of the saturated hydrocarbon fractions of the indicated samples from Facies A, B and C of the Jurf ed Darawish oil shale.

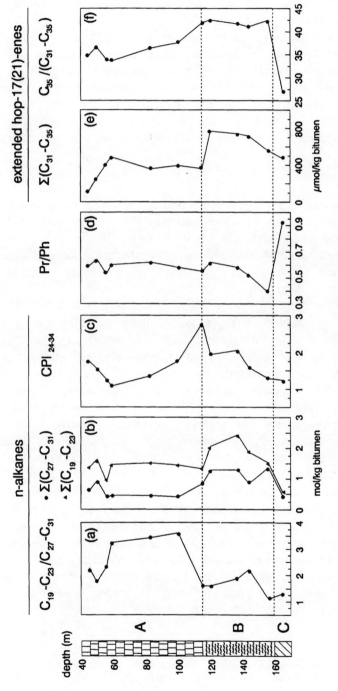

Figure 3. Depth profiles: (a) $(C_{19}-C_{23})/(C_{27}-C_{31})$ *n*-alkane ratio, (b) sum of $C_{19}-C_{23}$ and $C_{27}-C_{31}$ *n*-alkanes, (c) CPI_{24-34} (d) pristane/phytane ratio, (d) sum of $C_{31}-C_{35}$ hop-17(21)-enes and (f) $C_{35}/(C_{31}-C_{35})$ hop-17(21)-enes for the Jurf ed Darawish oil shale.

Sterane distributions (m/z 217) for three samples representing the three facies are similar (Figure 4).

In summary, variations in the distributions of hydrocarbon biological markers parallel changes in lithology. Thus the distributions of the hydrocarbon biological markers do not reveal any major variations in the composition of the organic matter related to palaeoenvironmental changes other than those revealed by the lithology itself. Therefore, the hydrocarbon biological markers do not provide any complementary information with respect to changes in the depositional palaeoenvironment.

Alkylthiophenes

Three gas chromatograms (Figure 5) of the A2 fractions representing the three different facies show qualitative and quantitative differences in OSC distributions. Peak assignments are listed in Table III.

The concentrations of the total alkylthiophenes (except thiophene hopanoids) in the bitumens show a depth profile somewhat similar to that of the total sulfur content (cf. Figures 6a and 1d). The high concentrations of alkylthiophenes in Facies B compared to A and C are consistent with the proposed incorporation of inorganic sulfur species into specific funtionalised lipids in anoxic marine environments (6-12).

The abundance of the four major classes of alkylthiophenes (i.e. thiophenes with linear, isoprenoid, highly branched isoprenoid, and mid-chain dimethylalkane carbon skeletons) varies with depth. These four different classes of alkylthiophenes are presented as a percentage of the total alkylthiophenes in the bitumens (Figure 6b-e). The reason for this is that the total amount of OSC formed during early diagenesis depends on the total amount of precursor lipids suitable for incorporation of sulfur and on the availability of inorganic sulfur species (hydrogen sulfide and/or polysulfides). By presenting specific (classes) alkylthiophenes as a percentage of the total alkylthiophenes changes in the relative concentrations of the precursors of these alkylthiophenes are recognised, because the changes in the availability of inorganic sulfur species for incorporation reactions are eliminated. As these functionalised lipids are biosynthesised by specific organisms, the amount of these lipids will reveal information concerning the relative abundance of specific organisms in the palaeoenvironment.

Alkylthiophenes with linear carbon skeletons. The alkylthiophenes with linear carbon skeletons ranging from C_{15}-C_{40} show a depth profile similar to that of the total alkylthiophenes (cf. Figures 6a and 6b). In order to

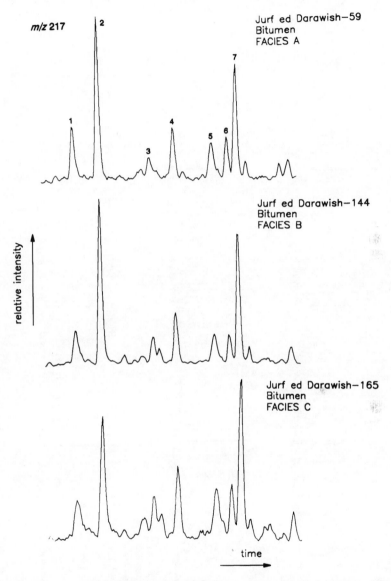

Figure 4. Mass chromatograms (*m/z* 217) for the saturated hydrocarbon fractions of the indicated samples from Facies A, B and C of the Jurf ed Darawish oil shale. Key: 1 = 20R-5β(H),14α(H),17α(H)-cholestane, 2 = 20R-5α(H),14α(H),17α(H)-cholestane, 3 = 20R-24-methyl-5β(H),14α(H),17α(H)-cholestane and 20S-24-methyl-5α(H),14β(H), 17β(H)-cholestane, 4 = 20R-24-methyl-5α(H),14α(H),17α(H)-cholestane, 5 = 20R-24-ethyl-5β(H),14α(H),17α(H)-cholestane, 6 = unknown C_{29} sterane and 20R-4α,24-dimethyl-5α(H),14α(H),17α(H)-cholestane, 7 = 20R-24-ethyl-5α(H),14α(H),17α(H)-cholestane.

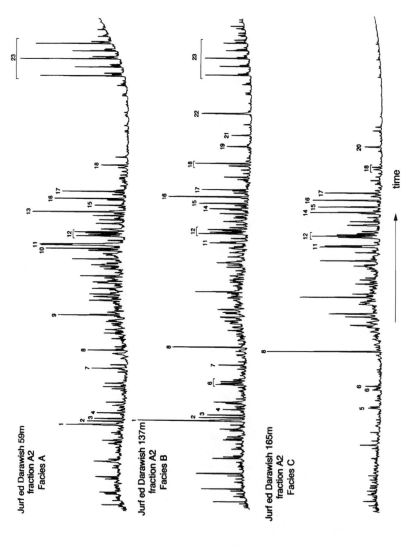

Figure 5. Gas chromatograms (HT-5) of the A2 fractions of the indicated samples from Facies A, B and C of the Jurf ed Darawish oil shale. The peak assignments are listed in Table III.

Table III. Compounds identified by GC-MS in the bitumen of
the Jurf ed Darawish oil shale

1.[a] 2-tridecyl-5-methylthiophene

2. 2-tetradecylthiophene

3. 3-methyl-2-(3,7,11-trimethyldodecyl)thiophene (**1**, Figure 14)

4. 3-(4,8,12-trimethyltridecyl)thiophene (**2**, Figure 14)

5. 2-(2'-methylbutyl)-3,5-di-(2'-(6'-methylheptyl))thiophene (**XXII**, Figure 17)

6. 2,3-dimethyl-5-(7'-(2',6',10',14'-tetramethylpentadecyl))thiophene (**XXIII**, Figure 17)

7. 2-hexadecyl-5-methylthiophene

8. 2,3-dimethyl-5-(1',1'-d_2-hexadecyl)thiophene (standard; **II**, Table II)

8. eicosylbenzene and C-ring monoaromatic steroid

10. unknown triterpenoid

11. 22S and 22R homohop-17(21)-ene

12. 22S-bishomohop-17(21)-ene, 17α(H),21β(H)-homohopane,
22R-bishomohop-17(21)-ene and unknown C_{30} hopanone
(in their elution order)

13. 22R-17β(H),21β(H)-homohopane

14. 22S-tetrakishomohop-17(21)-ene

15. 22R-tetrakishomohop-17(21)-ene

16. 22S-pentakishomohop-17(21)-ene

17. 22R-pentakishomohop-17(21)-ene

18. unknown C_{35} hopanone

19. C_{37} mid-chain 2,5-dialkylthiophenes (**XI**, Figure 9b)

20. C_{40} isoprenoid thiophene (**XX** or **XXI**, Figure 16b)

21. C_{38} mid-chain 2,5-dialkylthiophenes (**XI**, Figure 9b)

22. C_{40} mid-chain 2,5-dialkylthiophenes (**XV**, Figure 11b)

23. C_{44}-C_{48} 3,4-dialkylthiophenes (**XXIV**, Figure 18c)

---

[a] numbers refer to Figure 5.

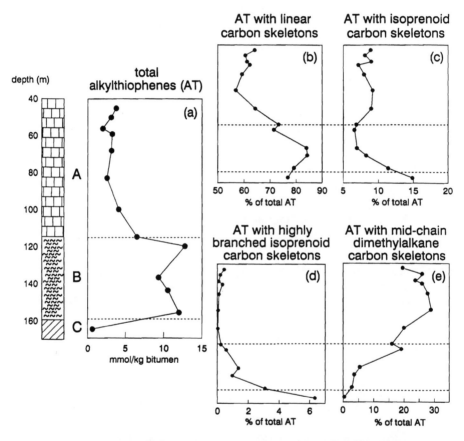

Figure 6. Depth profiles of (a) total extractable alkylthiophenes and (b) thiophenes with linear carbon skeletons, (c) thiophenes with isoprenoid carbon skeletons, (d) thiophenes with highly branched isoprenoid carbon skeletons and (e) thiophenes with mid-chain dimethylalkane carbon skeletons relative to the total alkylthiophenes for the Jurf ed Darawish oil shale.

discuss the behaviour of these thiophenes appropriately, four sub-classes consisting of different structural isomers are discriminated: 2-alkylthiophenes (V, Figure 7), 2-alkyl-5-methylthiophenes (VI, Figure 7), 2-alkyl-5-ethyl-thiophenes (VII, Figure 7) and so-called "mid-chain" 2,5-dialkylthiophenes (VIII, Figure 7). This subdivision of the linear alkylthiophenes is based on the chromatographic separation of the different structural isomers. The first three mentioned isomers are separated from each other and from the mid-chain thiophenes. The mid-chain 2,5-dialkythiophene clusters consist of several coeluting structural isomers ($\underline{10}$). The relative amounts of the different classes vary considerably (Figure 7b-e). The mid-chain 2,5-dialkylthiophenes, which are the most abundant linear thiophenes in the lower part of the sequence, show the most prominent variations in homologue (C_{15}-C_{40}) distribution patterns (Figure 8). The abundances of the C_{28}, C_{33}, C_{37}, C_{38} and the C_{40} homologues are especially notable and will be discussed in more detail.

The depth profiles for the C_{37} and C_{38} linear thiophenes are covariant and show a broad maximum in the lower part of Facies B (Figure 9a-b). This means that the precursors of these thiophenes were relatively more abundant during the deposition of that part of the sequence. It has been proposed by Sinninghe Damsté *et al.* ($\underline{6},\underline{10}$) that these thiophenes are derived by incorporation of sulfur into C_{37} and C_{38} di- and triunsaturated methyl (IX, Figure 9c) and ethyl ketones or their corresponding alkadienes and alkatrienes (X, Figure 9c), which are ubiquitous in sediments ($\underline{36}$-$\underline{42}$). These functionalised lipids are biosynthesised by algae of the class *Prymnesiophyceae* ($\underline{38},\underline{43}$). This precursor-product relation is further supported by the correlation of the depth profiles of the C_{37} and C_{38} mid-chain 2,5-dialkylthiophenes (Figure 9a-b). Figure 9c schematically shows how sulfur incorporation into these functionalised lipids results in C_{37} and C_{38} mid-chain thiophenes (XI, Figure 9c) as present in the bitumens of this oil shale. The information thus obtained about the presence and the relative abundance of this class of algae during the deposition of the sediment cannot be obtained by analyses of the saturated hydrocarbon fractions because there are no C_{37} and C_{38} alkanes and alkenes in the bitumens.

The shape of the depth profile for the C_{28} mid-chain 2,5-dialkylthiophenes (Figure 10a) differs from that of the C_{37} and C_{38} mid-chain 2,5-dialkylthiophenes, although the maxima correspond at 156m (Facies B). Averaged mass spectra of these C_{28} mid-chain 2,5-dialkylthiophenes and coinjection experiments with synthetic standards revealed ($\underline{10}$) that the C_{28} mid-chain 2,5-dialkylthiophenes in bitumen from the sample at 156m are dominated by 2-nonyl-5-pentadecylthiophene

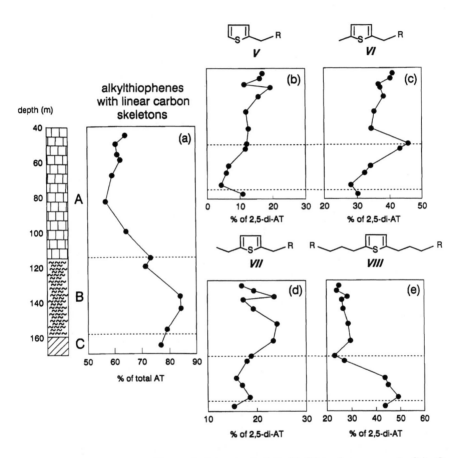

Figure 7. Depth profiles of (a) total 2,5-alkylthiophenes and (b) 2-alkylthiophenes (V), (c) 2-alkyl-5-methylthiophenes (VI), (d) 2-alkyl-5-ethylthiophenes (VII), (e) mid-chain 2,5-dialkylthiophenes (VIII) relative to the total 2,5-dialkylthiophenes for the Jurf ed Darawish oil shale.

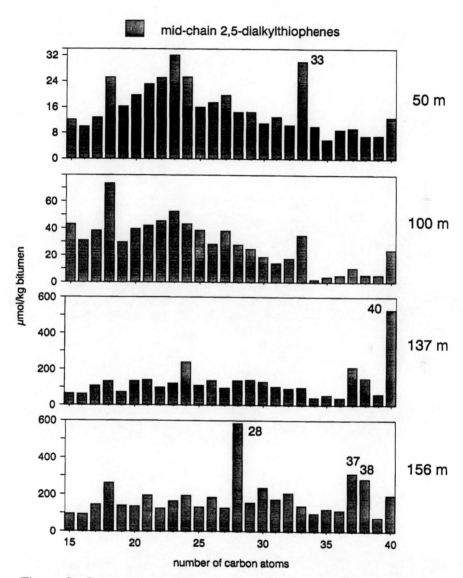

Figure 8. Carbon number distribution patterns $(C_{15}\text{-}C_{40})$ of the mid-chain 2,5-dialkylthiophenes of the Jurf ed Darawish oil shale samples indicated.

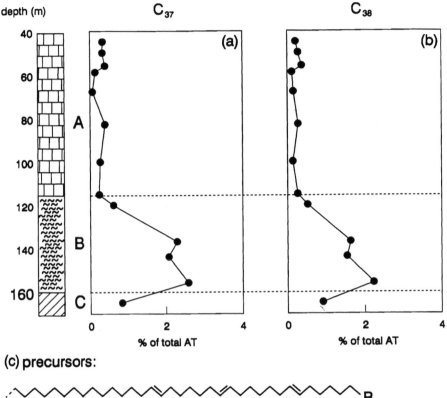

(c) precursors:

Figure 9. (a) Depth profiles of C_{37} and C_{38} mid-chain 2,5-dialkylthiophenes XI for the Jurf ed Darawish oil shale. (b) Formation of C_{37} and C_{38} mid-chain 2,5-dialkylthiophens XI by sulfur (H_2S) incorporation into triunsaturated C_{37} and C_{38} methylketones IX or hydrocarbons X.

(XII, *ca.* 25%; Figure 10b) and 2-hexadecyl-5-octylthiophene (XIII, *ca.* 25%; Figure 10b). This suggests that the precursors for these C_{28} mid-chain thiophenes were octacosa-9,12-diene (XIV, Figure 10b) and/or lipids with the same carbon skeleton (including the two double bonds) possessing an additional functional group (10). Octacosa-9,12-diene has not been reported in either organisms or sediments. Figure 10b shows how incorporation of sulfur into this hypothetical precursor yields the two major C_{28} mid-chain 2,5-dialkylthiophenes.

The plot of the relative amounts of the C_{40} mid-chain 2,5-dialkylthiophenes with depth (Figure 11a) shows a distinct maximum in the middle of facies B at 137m. The averaged mass spectrum (Figure 11b) of the C_{40} mid-chain thiophenes indicate that the major isomers have probably structures as depicted in Figure 11b (XV). Precursors of these thiophenes may be C_{40} *n*-alkadienes with the double bond positions in the center of the carbon chain, though these lipids have yet not been reported in sediments or organisms.

The depth profile of the C_{33} mid-chain 2,5-dialkylthiophenes shows a maximum in the upper part of Facies A at 56m (Figure 12a). An averaged mass spectrum (Figure 12c) reveals that these mid-chain 2,5-dialkylthiophenes are dominated by one isomer, tentatively identified as 2-tetracosyl-5-pentylthiophene (XVI, Figure 12c). Tritriaconta-6,8-diene (XVII, Figure 12b) may be a suitable precursor for these C_{33} thiophenes. This lipid and/or related compounds have not been reported in sediments or organisms. Recently, 2-tetracosyl-5-pentylthiophene (XVI, Figure 12c) has also been reported as a major thiophene in samples from offshore Morocco (ODP Site 547, Cretaceous black shales; 13) and the Nördlinger Ries deposit (44).

Extinction of the Cretaceous organisms that biosynthesised the previously mentioned hypothetical precursors for the C_{28}, C_{40} and the C_{33} mid-chain 2,5-dialkylthiophenes may explain why these functionalised lipids have not been reported in organisms.

Alkylthiophenes with isoprenoid carbon skeletons. Striking variations in the distributions of the isoprenoid thiophenes with depth are observed (Figure 13). In most of the samples from this oil shale the C_{20}, C_{26} and C_{40} isomers represent *ca.* 80 % of the total amount of isoprenoid thiophenes. Depth profiles of these major isoprenoid thiophenes will be discussed in detail.

The variation in the distribution of the various isoprenoid C_{20} thiophenes throughout the sequence is exemplified by the isoprenoid thiophene ratio (ITR; Figure 14a). This ratio is defined as depicted in

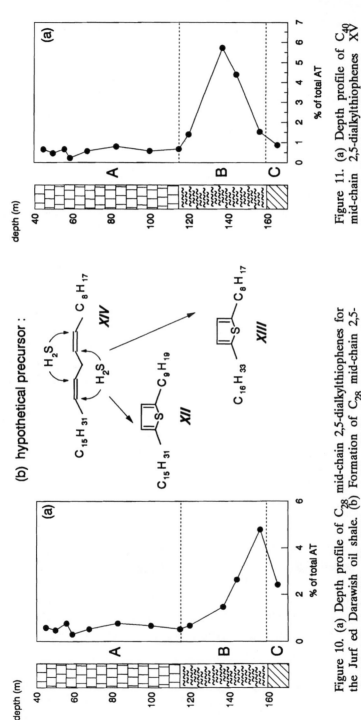

Figure 11. (a) Depth profile of C_{40} mid-chain 2,5-dialkylthiophenes XV for the Jurf ed Darawish oil shale.

Figure 10. (a) Depth profile of C_{28} mid-chain 2,5-dialkylthiophenes for the Jurf ed Darawish oil shale. (b) Formation of C_{28} mid-chain 2,5-dialkylthiophenes XII and XIII by sulfur (H_2S) incorporation into hypothetical precursor XIV.

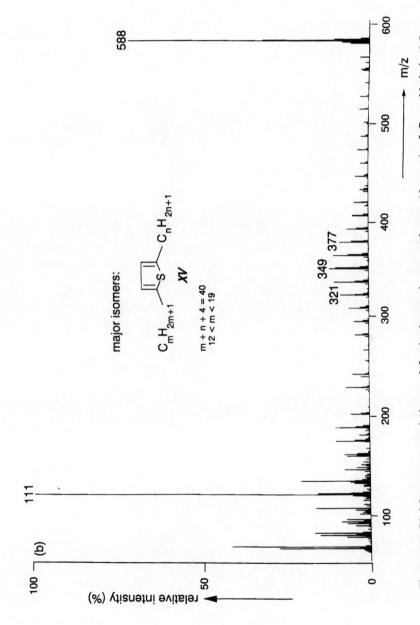

Figure 11. (b) Mass spectrum (subtracted for background; averaged over 10 scans) of C_{40} mid-chain 2,5-dialkylthiophenes XV from the sample at 137m.

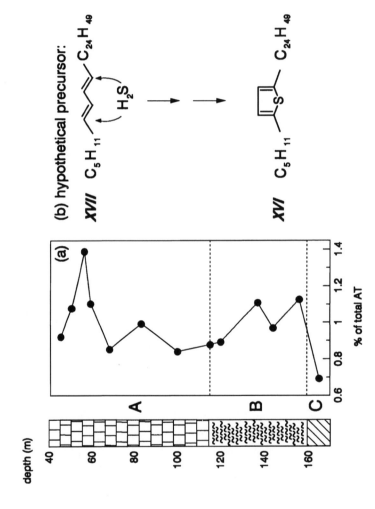

Figure 12. (a) Depth profile of C_{33} mid-chain 2,5-dialkylthiophenes for the Jurf ed Darawish oil shale. (b) Formation of 2-tetracosyl-5-pentylthiophene (XVI) by sulfur (H_2S) incorporation into hypothetical precursor XVII.

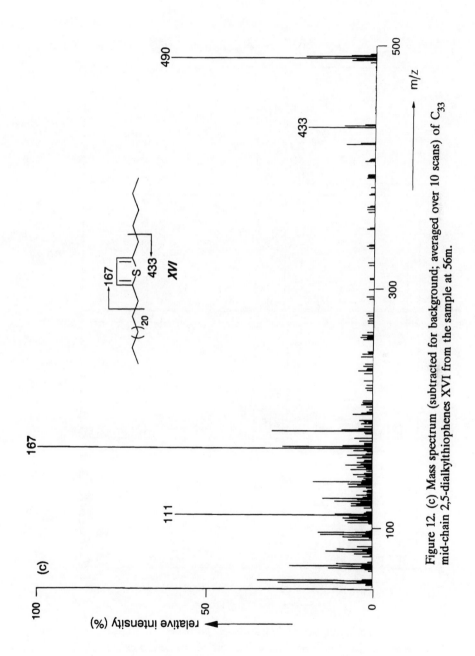

Figure 12. (c) Mass spectrum (subtracted for background; averaged over 10 scans) of C_{33} mid-chain 2,5-dialkylthiophenes XVI from the sample at 56m.

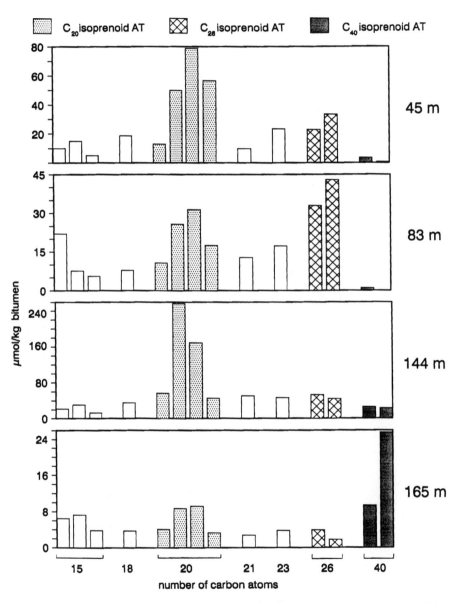

Figure 13. Carbon number distribution patterns of major isoprenoid thiophenes for the Jurf ed Darawish oil shale samples indicated.

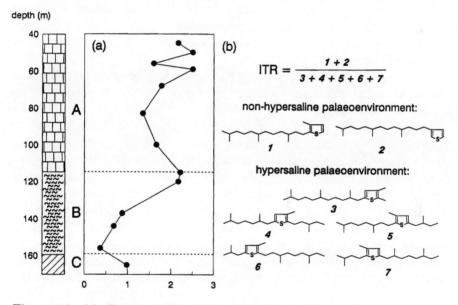

Figure 14. (a) Depth profile of isoprenoid thiophene ratio (ITR) for the Jurf ed Darawish oil shale. (b) Definition of ITR.

Figure 14b ($\underline{6}$). The changes in this ratio throughout the sequence may reflect variations in palaeosalinity ($\underline{6,32}$). ITR values smaller than 0.5 (156m) are thought to reflect hypersaline palaeoenvironments ($\underline{32}$).

The depth profiles of the two C_{26} isoprenoid thiophenes (Figure 16a-b) are similar. Both profiles show maxima (at 56m and 83m) in Facies A. Recently, the structures of both C_{26} isoprenoid thiophenes have been unambiguously elucidated by coinjection experiments and comparison of mass spectral data with synthetic standards ($\underline{45}$). The first and second eluting isomers (XVIII and XIX respectively, Figure 15) possess a regular (2,6,10,14,18-pentamethylheneicosane) and an irregular (2,6,10,14,19-pentamethylheneicosane) isoprenoid carbon skeleton, respectively. Because the depth profiles of these isomers (Figure 15a-b) are similar it is likely that one class of organisms biosynthesised both precursors. These organisms must have been relatively abundant during deposition of the middle part of Facies A.

The depth profile (Figure 16) of the C_{40} isoprenoid thiophenes shows a distinct maximum at 165m in Facies C. These thiophenes are tentatively identified by means of relative retention time and mass spectral data. Two tentative structures are proposed (XX and XXI, Figure 16) for the major C_{40} isoprenoid thiophene isomer present in the sample at 165m (Facies C). The mass spectrum of this major OSC in this sample is consistent with either a regular head-to-tail or an irregular head-to-head linked C_{40} isoprenoid thiophene. C_{40} head-to-head linked isoprenoid alkanes are building-blocks of the di-biphytanyl-diglycerol-tetraethers in Archaebacteria ($\underline{46}$). These C_{40} head-to-head linked isoprenoid alkanes are possible precursors for the C_{40} isoprenoid thiophenes, although it has not yet been reported that these putative precursors possess double bonds (or other functionalities), which make them accessible for sulfur incorporation. However, recently Volkman et al. ($\underline{47}$) showed that significant amounts of unsaturated phytenyl-glyceryl-ethers (1-6 double bonds) are present in Archaebacteria. Previously, only saturated phytanyl-glyceryl-ethers had been reported in organisms and sediments. Volkman et al. ($\underline{47}$) demonstrated that the unsaturated counterparts are decomposed by acid hydrolysis, which is the most common method used to study bacterial lipids. Bearing this in mind, the possible presence of unsaturated di-biphytenyl-diglycerol-tetraethers in Archaebacteria may be over looked merely due to the employed analytical methods. Alternatively, the regular C_{40} isoprenoid thiophenes may result from sulfur incorporation into C_{40} polyprenols or their diagenetic products, bacterial compounds proposed to be precursors of the regular C_{40} isoprenoid hydrocarbons in certain petroleums ($\underline{48,49}$).

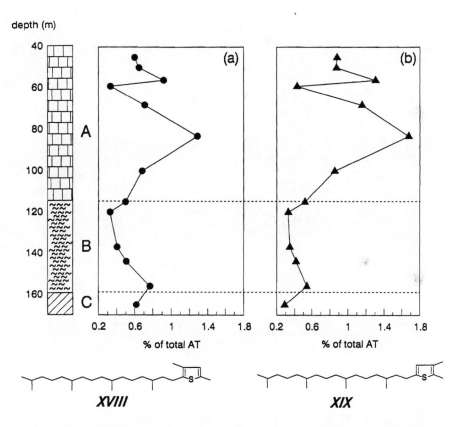

Figure 15. Depth profile of (a) C_{26} regular isoprenoid thiophene XVIII and (b) C_{26} irregular isoprenoid thiophene XIX for the Jurf ed Darawish oil shale.

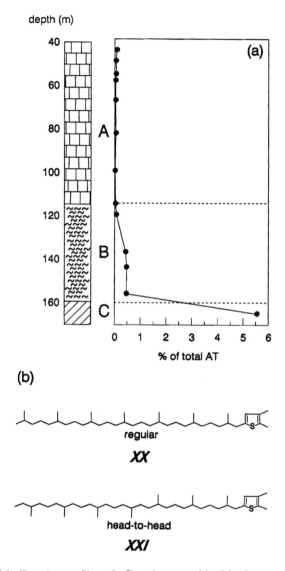

Figure 16. (a) Depth profile of C_{40} isoprenoid thiophenes for the Jurf ed Darawish oil shale. (b) Possible structures of the major C_{40} isoprenoid thiophene present in sample 165m from the Jurf ed Darawish oil shale.

Alkylthiophenes with a highly branched isoprenoid carbon skeleton. Two C_{25} thiophenes possessing a highly branched isoprenoid carbon skeleton (XXII and XXIII, Figure 17) have been previously identified (50) in the bitumen of the Jordan oil shale. Because the depth profiles of these different isomers are almost identical, only the profile of the sum of the relative concentrations of the two isomers is shown (Figure 17a). A distinct maximum is observed at 165m. Sinninghe Damsté *et al.* (50) have proposed that the C_{25} highly branched isoprenoid alkenes ubiquitous in Recent sediments (51; and references cited therein) and recently reported in field samples of sea-ice diatoms (52) are precursors for the above mentioned thiophenes.

The P_2O_5 concentration parallels the relative concentration of these highly branched C_{25} thiophenes with depth (*cf.* Figures 17a and 17b). Phosphatic sediments are often deposited in upwelling environments (24,27; Facies C) and the phytoplankton populations in such an environment tend to be dominated by diatoms (4). This is in agreement with the suggestion that C_{25} highly branched isoprenoid thiophenes are biomarkers for diatoms (10,50) as these thiophenes are relatively abundant in Facies C (Figure 17a). Microscopical examination showed trace amounts of fragmented diatom frustules in samples of Facies C (L. Buxton, pers. comm.), which supports this suggestion. Moreover, bitumen of the Monterey formation (containing also diatom frustules), which is also thought to be deposited in an upwelling environment (53), also contains relatively high amounts of C_{25} highly branched isoprenoid thiophenes (6).

Alkylthiophenes with a mid-chain dimethylalkane carbon skeleton. A novel class of thiophenes, tentatively identified as C_{44}-C_{48} 3,4-dialkylthiophenes (XXIV, Figure 18), accounts for almost 30% of the total alkylthiophenes in Facies A (Figure 6). The relative abundance of these thiophenes is much lower in Facies B and negligible in Facies C. The tentative identification is based on mass spectral and relative retention time data and Raney Ni desulfurisation experiments. Raney Ni desulfurisation of these compounds revealed that they possess a mid-chain dimethylalkane carbon skeleton. The complete structural elucidation of these new thiophenes based on comparisons with mass spectral data of synthesised model compounds will be discussed elsewhere (54). A partial gas chromatogram shows the distinct even-over-odd carbon number predominance of this homologous series (Figure 18a). The dominant isomers of the different homologues are depicted in Figure 18b. A hypothetical precursor (XXV) is depicted in Figure 18, although organisms biosynthesizing such compounds are as yet unknown.

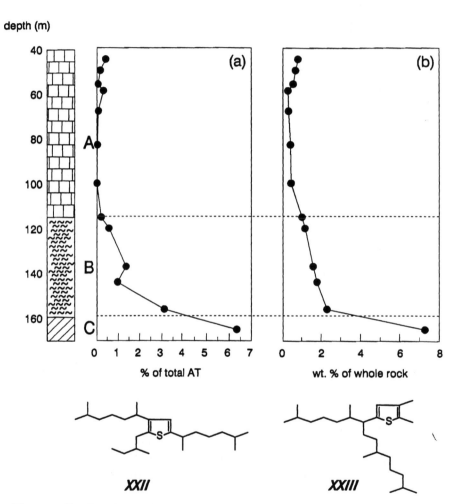

Figure 17. Depth profile of (a) highly branched isoprenoid C_{25} thiophenes XXII and XXIII and (b) P_2O_5 content for the Jurf ed Darawish oil shale.

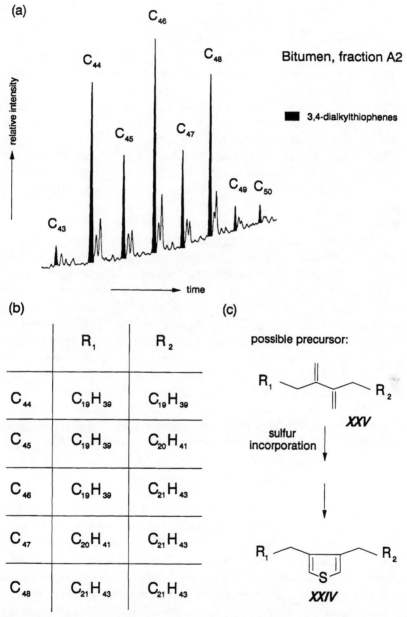

Figure 18. (a) Partial gas chromatogram (HT-5) of the Jurf ed Darawish oil shale sample from 59m showing the distribution of C_{43}-C_{50} 3,4-dialkylthiophenes XXIV. (b) Table showing the major isomers of the C_{44}-C_{48} 3,4-dialkylthiophenes present in the sample from 59m. (c) Formation of 3,4-dialkylthiophenes by sulphur incorporation into mid-chain dimethylalkadienes XXV.

Thiophene hopanoids. Two C_{35} thiophene hopanoids are present in the
A3 fraction of the bitumens. One of these isomers (XXVI, Figure 19)
has been previously identified by Valisolalao et al. (7) in black shales.
The other isomer (XXVII, Figure 19) has been tentatively identified by
Sinninghe Damsté et al. (6). Because of the lack of a suitable standard
for this fraction it was impossible to quantify these compounds as
precisely as the previously discussed alkylthiophenes. However, the relative
abundance of these hopanoid thiophenes (see experimental section) varies
considerably throughout the sequence (Figure 19a).

These C_{35} thiophene hopanoids are probably formed by sulfur
incorporation into bacteriohopanetetrol (6,7,32; Figure 19b), a widespread
membrane constituent of prokaryotes (55), and may, therefore, be
considered as bacterial markers. During the deposition of Facies C
phosphatisation of the sediments occurred which infers a high turn-over-
rate of the organic matter (24,53) probably due to bacterial reworking.
This is in agreement with the relatively high abundance of thiophene
hopanoids in Facies C. The depositional palaeoenvironment of Facies B
was highly euxinic (high total sulfur content, TOC value and hop-17(21)-
ene ratio; Figures 1 and 3; lack of bioturbation) which induces less
extensive bacterial reworking of the organic matter. This is reflected in
the low relative abundance of thiophene hopanoids in Facies B. The
reducing conditions during the deposition of Facies A were less extreme
than in Facies B (reduction of the height of the anoxic water column)
which is in agreement with the intermediate relative thiophene hopanoid
content. This striking correlation between thiophene hopanoid content and
redox conditions in the depositional environment supports the suggestion
that thiophene hopanoids are molecular markers for bacterial activity (6).

Implications for Reconstruction of Palaeoenvironments

This study of the Jurf ed Darawish oil shale sequence shows that
distributions of alkylthiophenes provide more detailed and specific
information on variations in the palaeoenvironment compared to those of
hydrocarbons. This can be explained by the following. The fate of
functionalised lipids is, in general, defunctionalisation or even
mineralisation, resulting in loss of information on the palaeoenvironment.
OSC, in general, originate from sulfur incorporation into functionalised
lipids during early diagenesis (6-12). The OSC are more stable to
microbial attack than the original functionalised lipids, allowing
preservation of the lipid substrates and even indicating the former
positions of functional groups in the lipids. Studies of OSC may thus

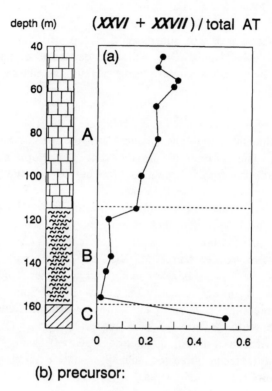

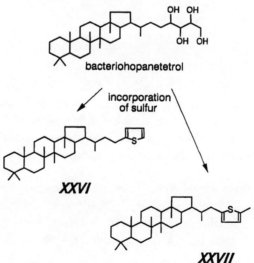

Figure 19. (a) Depth profile of C_{35} thiophene hopanoids XXVI and XXVII for the Jurf ed Darawish oil shale. (b) Formation of C_{35} thiophene hopanoids by sulfur incorporation into bacteriohopanetetrol.

provide information on specific lipids present during the deposition of the sediment. Because OSC represent specific lipids, which can be related to specific organisms, it is not surprising that the variations in absolute and relative terms of alkylthiophenes will provide more detailed and specific information than conventional hydrocarbon biomarkers on palaeo-environmental changes.

Figure 20 summarizes the most prominent alkylthiophene depth profiles. They indicate the fine structure in the changes of the palaeoenvironmental conditions in addition to the lithological variations (Figure 1a-e). The maxima in the different alkylthiophene depth profiles reflect palaeoenvironmental conditions favourable for the growth (bloom) of specific types of algae and/or bacteria.

After the deposition of the phosphorites it is suggested that restricted basins are formed in which limited bottom water circulation resulted in an enhanced preservation of the organic matter. The ITR (Figure 14a) and the relative concentration of C_{35} hop-17(21)-enes (Figure 3e) in Facies B point to a palaeoenvironment with a higher salinity than normal seawater. This is in agreement with the work of Spiro and Aizensthtat (19) which presented evidence that the lowermost beds of oil shales similar to the Jurf ed Darawish with respect to age and lithology (upper Cretaceous, Ghareb Formation; southern Israel) were deposited in a hypersaline environment. Increased salinity of the euphotic water layer can cause high productivety of biomass (56). This combined with restricted bottom water circulation and eventual density stratification resulted in high TOC values in sediments of Facies B.

Conclusions

(1) Absolute and relative amounts of alkylthiophenes vary with subtle palaeoenvironmental changes. In the case of the Jordan oil shale, alkylthiophenes provide a more detailed subdivision of lithological facies and more specific information on organisms contributing to the palaeoenvironment than conventional hydrocarbon biological markers.

(2) Further evidence confirms previous suggestions that: (i) C_{25} highly branched isoprenoid thiophenes are markers for diatoms, (ii) C_{37} and C_{38} mid-chain 2,5-dialkylthiophenes are markers for algae of the class *Prymnesiophyceae* and (iii) C_{35} hopanoid thiophenes are markers for bacterial reworking.

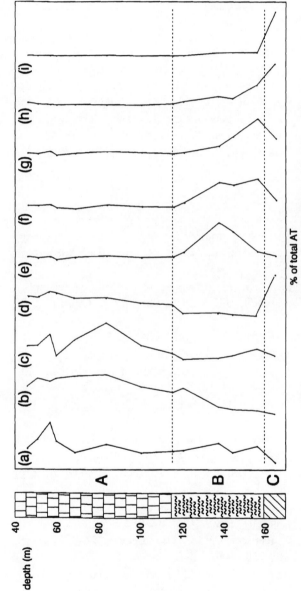

Figure 20. Summary of the most prominent alkylthiophene depth profiles showing the fine structure in the changes of the palaeoenvironmental conditions in addition to the lithological variations for the Jurf ed Darawish oil shale. Key: (a) C_{33} mid-chain 2,5-dialkylthiophenes, (b) C_{44}-C_{48} 3,4-dialkylthiophenes, (c) C_{26} isoprenoid thiophenes, (d) C_{35} thiophene hopanoids, (e) C_{40} mid-chain 2,5-dialkylthiophenes, (f) C_{37} mid-chain 2,5-dialkylthiophenes, (g) C_{28} mid-chain 2,5-dialkylthiophenes, (h) C_{25} highly branched isoprenoid thiophenes and (i) C_{40} isoprenoid thiophenes.

(3) Identification of OSC may lead to the recognition of unknown, biosynthesised functionalised lipids (e.g. C_{44}- C_{48} mid-chain dimethylalkadienes).

Acknowledgments

We thank Drs. H. Wehner and H. Hufnagel for providing the samples of the Jurf ed Darawish oil shale. We gratefully acknowledge Drs. K.E. Peters and Z. Aizenshtat and Mrs. L. Buxton for carefully reading and reviewing the manuscript. We thank Mrs. A.H. Knol-Kalkman, Mrs. A.C. Kock-van Dalen and Mr. W. Pool for analytical assistance. This work was partly supported by the Netherlands Foundation for Earth Science Research (AWON) with financial aid from the Netherlands Organisation for Scientific Research (NWO).

Literature Cited

1. Johns, R.B. (Ed.) Biological markers in the sedimentary record; Methods in Geochemistry and Geophysics, 24, Elsevier: Amsterdam, 1986.
2. Williams, P.F.V.; Douglas, A.G. In Advances in Organic Geochemistry 1979; Douglas, A.G.; Maxwell, J.R., Eds.; Pergamon: Oxford, 1980; pp 211-217.
3. Belayouni, H.; Trichet, J. Org. Geochem. 1984, 6, 741-754.
4. Volkman, J.K. In Lacustrine Petroleum Source Rocks; Kelts, K.; Fleet,, A.J.; Talbot, M., Eds.; Geol. Soc. Spec. Publ. No. 40, 103-122; Blackwell.
5. Moldowan, J.M.; Sundararaman, P.; Schoell, M. Org. Geochem. 1986, 10, 915-926.
6. Sinninghe Damsté, J.S.; Rijpstra, W.I.C.; de Leeuw, J.W.; Schenck, P.A. Geochim. Cosmochim. Acta 1989, 53, 1323-1341.
7. Valisolalao, J.; Perakis, N.; Chappe, B.; Albrecht, P. Tetrahedron Lett. 1984, 25, 1183-1186.
8. Brassell, S.C.; Lewis, C.A.; de Leeuw, J.W.; de Lange, F.; Sinninghe Damsté, J.S. Nature (London) 1986, 320, 160-162.
9. Sinninghe Damsté, J.S.; Rijpstra, W.I.C.; de Leeuw, J.W.; Schenck, P.A. Org. Geochem. 1988, 593-606.
10. Sinninghe Damsté, J.S.; Rijpstra, W.I.C; Kock-van Dalen, A.C.; de Leeuw, J.W.; Schenck, P.A. Geochim. Cosmochim. Acta 1989, 53, 1343-1355.

11. Kohnen, M.E.L.; Sinninghe Damsté, J.S.; Kock-van Dalen, A.C.; ten Haven, H.L.; de Leeuw, J.W. Geochim. Cosmochim. Acta 1990, submitted.

12. Kohnen, M.E.L.; Sinninghe Damsté, J.S.; ten Haven, H.L.; de Leeuw, J.W. Nature (London) 1989, 341, 640-641.

13. ten Haven, H.L.; Rullkötter, J.; Sinninghe Damsté, J.S.; de Leeuw, J.W. In Geochemistry of sulfur in fossil fuels; Orr, W.L.; White, C.M., Eds.; ACS Symposium Series; American Chemical Society: Washington, DC, 1990; this volume.

14. Hufnagel, H. Geol. Jb., Rh A 1984, 75 295-311.

15. Wehner, H.; Hufnagel, H. In Biogeochemistry of Black Shales; Degens, E.T. *et al.*, Eds.; Mitt. Geol. Palaeont. Int. Univ. Hamburg 1987, 60, 381-395.

16. Abed, A.M.; Amireh, B. J. Petr. Geol. 1983, 5, 261-274.

17. Bein, A.; Amit, O. Sedimentology 1982, 29, 81-90.

18. Amit, O.; Bein, A. Chem. Geol. 1982, 37, 277-287.

19. Spiro, B.; Aizenshtat, Z. Nature (London) 1977, 269, 235-237.

20. Spiru, B.; Dinur, D., Aizenshtat, Z. Chem. Geol. 1983, 39, 189-214.

21. Rösler, H.J.; Beuge, P.; Adamski, B. Z Angew. Geol. 1977, 23, 53-56.

22. Sinninghe Damsté, J.S.; de Leeuw, J.W.; Kock-van Dalen, A.C.; de Zeeuw, M.A.; de Lange, F.; Rijpstra, W.I.C.; Schenck, P.A. Geochim. Cosmochim. Acta 1987, 51, 2369-2391.

23. Sinninghe Damsté, J.S.; Kock-van Dalen, A.C.; de Leeuw, J.W.; Schenck, P.A.; Sheng Guoying; Brassell, S.C. Geochim. Cosmochim. Acta 1987, 51, 2393-2400.

24. Burnett, W.C.; Veeh, H.H.; Soutar, A. In Marine Phosphorites-Geochemistry, Occurrence, Genesis; Bentor, Y.K., Ed.; S.E.P.M Spec. Publ. 1980, 29, pp 61-72.

25. Burnett, W.C. Geol. Soc. Am. Bull. 1977, 88, 813-823.

26. Baturin, G.N. Oceanology 1972, 12, 849-855.

27. Kolodny, Y. In Marine Phosphorites-Geochemistry, Occurrence, Genesis; Bentor, Y.K., Ed.; S.E.P.M Spec. Publ. 1980, 29, p 249.

28. Demaison, G.J.; Moore, G.T. AAPG Bull. 1980, 64, 1179-1209.

29. Bray, E.E.; Evans, E.D. Geochim. Cosmochim. Acta 1961, 22, 2-15.

30. Didyk, B.M.; Simoneit, B.R.T.; Brassell, S.C.; Eglinton, G. Nature (London) 1978, 272, 216-222.

31. ten Haven, H.L., de Leeuw, J.W.; Rullkötter, J.; Sinninghe Damsté, J.S. Nature (London) 1987, 330, 641-643.

32. de Leeuw, J.W.; Sinninghe Damsté, J.S. In Geochemistry of Sulfur in Fossil Fuels; Orr, W.L.; White, C.M., Eds.; ACS Symposium Series; American Chemical Society: Washington, DC, 1990; this volume.

33. Mckirdy, D.M.; Aldridge, A.K.; Ypma, P.M.J. In Advances in Organic Geochemistry 1981; Bjorøy, M. et al., Eds.; Wiley, 1983; pp 99-107.

34. ten Haven, H.L., de Leeuw, J.W.; Schenck, P.A. Geochim. Cosmochim. Acta 1985, 49, 2181-2191.

35. ten Haven, H.L., de Leeuw, J.W.; Sinninghe Damsté, J.S.; Schenck, P.A.; Palmer, S.E.; Zumberge, J.E. In Lacustrine Petroleum Source Rocks; Kelts, K.; Fleet, A.J.; Talbot, M., Eds.; Geol. Soc. Spec. Publ. No. 40; Blackwell, 1989; pp 123-130.

36. Boon, J.J.; van der Meer, F.W.; Schuyl, P.J.; de Leeuw, J.W.; Schenck, P.A.; Burlingame, A.L. Init. Rep. Deep Sea Drilling Project 1978, 40, 627-637.

37. de Leeuw, J.W.; de Meer, F.W.; Rijpstra, W.I.C.; Schenck, P.A. In Advances in Organic Geochemistry 1979; Douglas, A.G.; Maxwell, J.R., Eds.; Pergamon: Oxford, 1980; pp 211-217.

38. Marlowe, I.T.; Brassell, S.C.; Eglinton, G.; Green, J.C. Org. Geochem. 1984, 6, 135-141.

39. Farrimond, P.; Eglinton, G.; Brassell, S.C. Org. Geochem. 1986, 10, 897-903.

40. Nichols, P.D.; Johns, R.B. Org. Geochem. 1986, 9, 25-30.

41. Brassell, S.C.; Brereton R.G.; Eglinton, G.; Grimalt, J.; Liebezeit, G.; Marlowe, I.T.; Pflaumann, U.; Sarnthein, U. Org. Geochem. 1986, 10, 649-660.

42. Brassell, S.C.; Eglinton, G.; Marlowe, I.T.; Sarnthein, U.; Pflaumann, U. Nature (London) 1986, 320, 129-133.

43. Volkman, J.K.; Eglinton, G.; Corner, E.D.S.; Sargent, J.R. In Advances in Organic Geochemistry 1979; Douglas, E.G.; Maxwell, J.R., Eds.; Pergamon: Oxford, 1980; pp 219-227.

44. Rullkötter, J.; Littke, R.; Schaefer, R.G. In Geochemistry of Sulfur in Fossil Fuels; Orr, W.L.; White, C.M., Eds.; ACS Symposium Series; American Chemical Society: Washington, DC, 1990; this volume.

45. Peakman, T.M.; Sinninghe Damsté, J.S.; de Leeuw, J.W. J. Chem. Soc. Commun. 1989, 16, 1105-1107.

46. De Rosa, M.; Gambacorta, A.; Gliozzi, A. Microbiol. Rev. 1986, 50, 70-80.

47. Volkman, J.K.; Neill, G.P.; Blackman, A.J.; Franzmann, P.D. In Advances in Organic Geochemistry 1989; Durand, B., Ed.; Pergamon: Oxford, 1990; in press.
48. Albaigés, J. In Advances in Organic Geochemistry 1979; Douglas, E.G.; Maxwell, J.R., Eds.; Pergamon: Oxford, 1980; pp 19-28.
49. Albaigés, J.; Borbon, J,; Walker II, W. Org. Geochem. 1985, 8, 293-297.
50. Sinninghe Damsté, J.S.; van Koert, E.R.; Kock-van Dalen, A.C.; de Leeuw, J.W.; Schenck, P.A. Org. Geochem. 1989, 14, 555-567.
51. Robson, J.N.; Rowland, S.J. Nature (London) 1986, 324, 561-563.
52. Nichols, P.D., Volkman, J.K.; Palmisano, A.C.; Smith, G.A.; White, D.C. J. Phycol. 1988, 24, 90-96.
53. Isaacs, C.M.; Pisciotto, K.A.; Garrison, R.E. Dev. Sedimentol. 1983, 36, 247-282.
54. Kohnen, M.E.L.; Peakman, T.M.; Sinninghe Damsté, J.S.; de Leeuw, J.W. In Advances in Organic Geochemistry 1989; Durand, B., Ed.; Pergamon: Oxford, 1990; in press.
55. Ourisson, G.; Albrecht, P.; Rohmer, M. Trends Biochem. Sci. 1982, 7, 236-239.
56. Evans, R,; Kirkland, D.W. In Evaporites and Hydrocarbons; Schreiber, B.C., Ed.; Columbia University Press: New York, 1988; p.256.

RECEIVED March 25, 1990

Chapter 26

Characterization of Organically Bound Sulfur in High-Molecular-Weight, Sedimentary Organic Matter Using Flash Pyrolysis and Raney Ni Desulfurization

Jaap S. Sinninghe Damsté, Timothy I. Eglinton[1], W. Irene C. Rijpstra, and Jan W. de Leeuw

Faculty of Chemical Engineering and Materials Science, Organic Geochemistry Unit, Delft University of Technology, De Vries van Heystplantsoen 2, 2628 RZ Delft, Netherlands

Organically-bound sulfur in three types of high-molecular-weight organic matter (kerogen, asphaltenes and resins) obtained from three organic S-rich sedimentary rock samples has been studied. Kerogen, asphaltene and resin fractions were isolated and analysed by flash pyrolysis-gas chromatography-mass spectrometry. The resin fractions were desulfurised with Raney Ni and the products obtained were characterised by gas chromatography-mass spectrometry. These experiments revealed that $S_{org.}$ in high-molecular-weight sedimentary organic matter at least partially occurs in moieties with linear, isoprenoid, branched, steroid, hopanoid and carotenoid carbon skeletons. In kerogen and asphaltenes evidence was obtained for the presence of these types of moieties as thiophenes, which are thought to be mainly bound to the macromolecular matrix by multiple C-S bonds. In resins S-containing moieties are thought to be primarily bound by only one C-S bond though they may contain additional intramolecular S-linkages. The different properties of S-rich kerogens and asphaltenes on the one hand and S-rich resins on the other hand may be explained by differences in the degree of (sulfur) cross-linking and thus by differences in molecular size and in degree of condensation. These S-rich geopolymers are formed by incorporation of inorganic sulfur species into functionalised lipids in an intermolecular fashion during early diagenesis. In depositional environments with high amounts of available S-donors these reactions may be regarded as an important pathway for the generation of high-molecular-weight material ("natural vulcanisation").

[1]Current address: Fye Laboratory, Department of Chemistry, Woods Hole Oceanographic Institution, Woods Hole, MA 02543

0097–6156/90/0429–0486$11.75/0

Organically-bound sulfur in fossil fuels and related sedimentary organic matter is present in both low-molecular-weight (MW < 800) organic sulfur compounds (OSC) and in complex macromolecules (*i.e.*, kerogen, coal, asphaltenes and resins). Major advances in the characterisation of low-molecular-weight OSC in bitumens and crude oils have been reported during the last decade with more than a thousand novel OSC identified during this period (1-17). Many of these represent new classes of sulfur-containing biological markers because their carbon skeletons are identical to the well-known hydrocarbons (*e.g.*, *n*-alkanes, isoprenoid alkanes, steranes, hopanes) occurring in the geosphere and their inferred biochemical precursors. Evidence has been presented that these OSC are formed by incorporation of inorganic sulfur species (hydrogen sulfide, polysulfides) into functionalised lipids during the early stages of diagenesis (2,8,9,11-18).

So far, less attention has been given to the characterisation at a molecular level of organically-bound sulfur in high-molecular-weight substances in sediments and crude oils, although the major part of the organic sulfur is bound in these fractions. Kerogen, which normally represents at least 90% of the organic matter in immature sediments, may contain more than 14% (by weight) organic sulfur (19). Organic sulfur in kerogen and other sedimentary macromolecules is thought to be present in sulfide, di- and/or polysulfide moieties and as a constituent of aromatic heterocycles (20), but reliable data on this subject is scant.

It is generally accepted (19,21-23) that the early generation of petroleum from sulfur-rich kerogens is due to the abundant presence of weak sulfur-sulfur (average bond energy *ca.* 250 kJ/mol) and carbon-sulfur bonds (*ca.* 275 kJ/mol) relative to carbon-carbon bonds (*ca.* 350 kJ/mol). The presence of these weak bonds lowers the maximum of the activation energy distribution for the thermal degradation of the kerogen. Indeed, sulfur-rich kerogens exhibit a much lower Rock-Eval T_{max} than sulfur-lean kerogens of the same maturity (24,25). However, the form in which organic sulfur is present in kerogen is of major importance in this respect since mono-, di- and polysulfide moieties contribute significantly to the relative thermal instability whilst the presence of thiophene and condensed thiophene units has little or no effect. Therefore, determination of the kinds of sulfur-bonding in sedimentary high-molecular-weight substances, *i.e.* relative abundances of different sulfur functional groups/bond types, is of great interest for petroleum geochemistry. This kind of information is, however, difficult to obtain. Recently, George and Gorbaty (26) reported an X-ray absorption spectroscopic method for the determination of the forms and the amounts of organically-bound sulfur in asphaltenes. Although this method will be of great use for the assessment of the abundance of thermally labile *versus* thermally stable forms of organic sulfur in macromolecules, it does not provide information on moieties bound by sulfur in the macromolecules. In this paper two techniques are described which can give such information, although these methods both suffer from other disadvantages.

The first method is flash pyrolysis-gas chromatography-mass spectrometry, which has been shown to be a powerful technique for the characterisation of bio- and geopolymers, such as lignins (27,28), humic substances (29,30), coal and its maceral fractions (31-33), asphaltenes (34-36) and kerogen (31,35-43). This method induces thermal dissociation of macromolecules with a minimum of secondary reactions. At lower Curie point temperatures ($\leq$ 610°C) the products formed are mainly generated by cleavage of one chemical bond (per product). Some subsequent internal rearrangement reactions associated with these bond cleavages (*e.g.*, elimination of H_2O, H_2S) do occur. Only recently, it has been realised that the sulfur compounds produced on flash pyrolysis of macromolecular substances may provide insight into the form and molecular arrangement in which organic sulfur is bound into these macromolecules. This has led to a number of studies (24,44-49) in which flash pyrolysates were analysed by gas chromatography in combination with sulfur-selective flame photometric detection. Sinninghe Damste *et al.* (48,49) and Eglinton *et al.* (24) have shown that flash pyrolysis of kerogens, coals and asphaltenes may yield abundant OSC, many of which provide useful information with respect to the characterisation of sulfur-containing moieties present in these substances. Payzant *et al.* (50) have similarly shown that the oil obtained from off-line pyrolysis of the Athabasca asphaltene contains abundant series of OSC.

The second method is sulfur-selective chemical degradation (*i.e.*, Raney Ni desulfurisation) of high-molecular-weight substances. It achieves a reductive cleavage of carbon-sulfur bonds, whereby units, bound in macromolecules by sulfur-carbon bond(s) only, are released and can be characterised by gas chromatography-mass spectrometry (7,18). This method can only be applied to high-molecular-weight substances which are soluble in organic solvents (*i.e.*, resins). Asphaltenes tend to form aggregates in certain organic solvents, which causes problems with desulfurisation.

In this paper organically-bound sulfur in three types of high-molecular-weight organic matter (kerogen, asphaltenes and resins) obtained from three organic sulfur-rich sedimentary rock samples has been studied. Kerogen, asphaltene and resin fractions were isolated and characterised by the two described techniques.

Materials and Methods

Samples.   Three immature sulfur-rich sedimentary rock samples were selected for study of both the kerogens and extractable bitumens. Selected properties of the specific samples together with references to previous descriptions and investigations of these deposits are as follows:

Jurf ed Darawish-156: This is a bituminous calcareous marlstone from the Jurf ed Darawish Oil Shale (Upper Cretaceous) located 130 km south of Amman (Jordan) (51), which was deposited in an anoxic, marine environment. This sample is a composite sample from a core at 156-157 m with 17.3% TOC and R_o = *ca.* 0.30%. The organic sulfur content of the isolated kerogen was not measured but the atomic S/C = 0.084 and the pyrite-S/organic-S ratio is presumably low since microscopical examination indicated a low pyrite content. Detailed information on the hydrocarbon biological markers (51), the OSC (15,16,52) and kerogen pyrolysate (48,49) have been published elsewhere.

Monterey-25: This sample is from the Upper Miocene Monterey Formation (California, U.S.A.), a phosphatic-siliceous-carbonate rock sequence, and was collected from an outcrop near Vandenberg Air Force Base (Santa Barbara County). This Formation and similar samples have been described by Isaacs (53) and Orr (19). This sample (TOC = 17.3%) also contains immature (R_o = *ca.* 0.30%), sulfur-rich (atomic $S_{org.}/C$ = 0.054) kerogen.

Northern Apennines Marl: This sample is from Miocene strata in the Perticara basin (Italy) which consists of gypsum deposits interbedded with bituminous marl layers deposited during the Mediterranean salinity crisis (54). Detailed information on the hydrocarbons and OSC in the bitumen of this sample have been reported elsewhere (9,16,54).

Separation methods.   Extraction and fractionation (Figure 1) methods have been described elsewhere (16). The kerogen from Jurf ed Darawish-156 was concentrated by decarbonating the extracted whole rock followed by re-extraction. The kerogen from the Monterey-25 was fully isolated using standard HCl/HF techniques. Asphaltene fractions were obtained and purified from bitumens by repeated precipitations with *n*-heptane.

Desulfurisation.   Aliquots of polar and High-Molecular-Weight Aromatic (HMWA; Figure 1) fractions were desulfurised using Raney Ni as described elsewhere (18). The mixture after desulfurisation was separated over a small (5 x 0.5 cm) alumina column using hexane/CH_2Cl_2 (9:1) to elute the saturated and aromatic hydrocarbons and CH_2Cl_2 to elute more polar

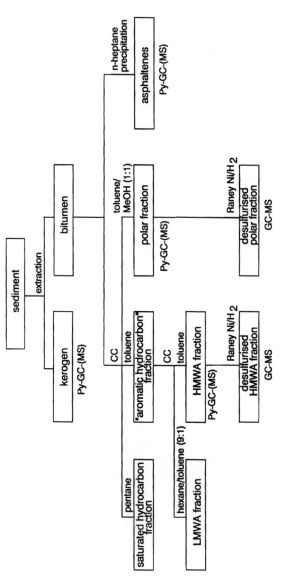

Figure 1. Analytical flow diagram.

compounds. The first fraction was analysed by GC-MS under conditions described previously (16).

Flash pyrolysis-gas chromatography-mass spectrometry. Flash pyrolysis-GC-MS analyses were performed as described in detail previously (48), although another mass spectrometer (VG-70S) was used. Electron impact mass spectra were obtained at 70 eV with a cycle time of 1.7 s and a mass range m/z 50-800 at a resolution of 1000. Data acquisition was started 1 min. after pyrolysis.

Organically-Bound Sulfur in Kerogen

Structure of Sulfur-Containing Moieties. In one of our recent studies it was shown that H_2S, thiophene and C_1-C_6 alkylated thiophenes are the major sulfur-containing products in flash pyrolysates of immature kerogens (48,49). H_2S is formed by thermal degradation of cyclic and acyclic sulfide and polysulfide moieties in the kerogen (49). Thermal degradation of these moieties already starts when using ferromagnetic wires with a Curie temperature of 358°C ("358°C pyrolysate"). This thermal instability is probably due to the lower amount of energy required to cleave S-S and S-C bonds in comparison with C-C bonds (49). Unless these cleavages result in the release of GC-amenable molecules from the macromolecular substances (as in the case of polar fractions; see below) little information on the moieties bound by sulfur is obtained. This is the case for kerogens. Cleavage of the C-S bond in aromatic sulfur-containing moieties (thiophene and benzo-thiophene units) in macromolecules requires much more energy and thermal degradation results mainly in saturated alkylated thiophenes and benzo-thiophenes predominantly by cleavage of the relatively weak C-C bond at the β position to the aromatic ring (49). This thermal degradation reaction starts only using ferromagnetic wires with higher Curie temperatures (from 510°C). H_2S in flash pyrolysates thus reflects the amount of thermally less stable forms of organic sulfur, whereas the alkylthiophenes represent the thermally more stable form of organic sulfur. However, the amounts of H_2S generated upon pyrolysis cannot be used as a measure of the total amount of thermally less stable forms of organic sulfur since H_2S can also be formed from pyrite present in the kerogen. Currently we are undertaking studies to determine whether the H_2S evolved during flash pyrolysis of asphaltenes eorrelates to their (poly)sulfide sulfur content.

Figure 2a shows the total ion chromatogram (TIC) of the flash pyrolysate of a sulfur-rich (Type II-S) kerogen from the Jurf ed Darawish Oil Shale. The major peaks are alkylated thiophenes, shown more clearly in

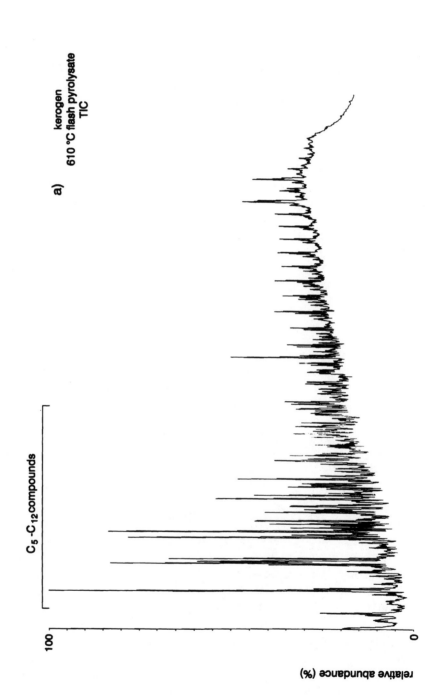

a)
kerogen
610 °C flash pyrolysate
TIC

$C_5 - C_{12}$ compounds

relative abundance (%)

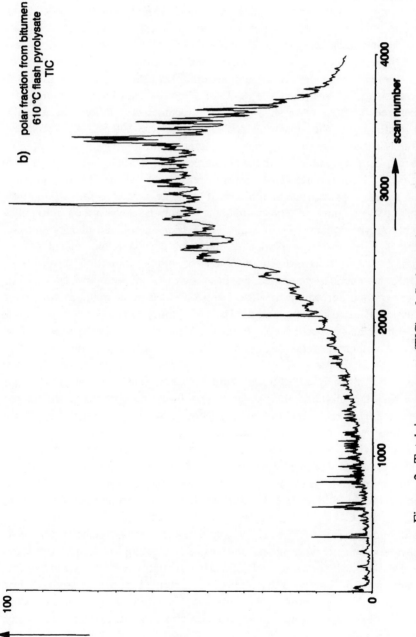

Figure 2. Total ion currents (TIC) of flash pyrolysates of several high-molecular-weight fractions of the Jurf ed Darawish-156.

Figure 3a and b. Other immature (R_0 < 0.5%), sulfur-rich kerogens also show an abundance of thiophene and C_1-C_6 alkylated thiophenes in their flash pyrolysates (e.g., Figure 4a and b). Pyrolysates of kerogens with lower organic sulfur contents (e.g., Type II kerogens) contain smaller amounts of these thiophenes in both relative and absolute terms. However, even in sulfur-poor, hydrogen-rich Type I kerogens the C_1-C_6 alkylated thiophenes are present (49). Eglinton et al. (24) have recently shown that the amount of one of these thiophenes (i.e. 2,3-dimethylthiophene) relative to an aromatic hydrocarbon (1,2-dimethylbenzene) and an aliphatic hydrocarbon (non-1-ene) as measured by flash pyrolysis-GC shows a distinct correlation with organic sulfur content of the macromolecular substances. The C_1-C_6 alkylated thiophenes in the pyrolysates typically represent, however, only a small portion of the total organic sulfur of the macromolecules (< 10%; 24). George and Gorbaty (26) have reported that in asphaltenes more than 50% of the total organic sulfur is present as thiophenes. It appears therefore that upon flash pyrolysis only a part of the thiophenic sulfur is transformed into GC-amenable alkylthiophenes. The distinct correlation mentioned above indicates that this is a representative portion of the total organic sulfur content of the kerogen. Low yields have been observed for other compound classes (e.g., hydrocarbons, phenols) generated on flash pyrolysis of kerogens as well. In these cases good agreement has also been obtained in comparison with bulk geochemical data (e.g., elemental analysis, ^{13}C nmr) (55).

Detailed mass spectroscopic studies combined with the synthesis of a number of authentic standards has led to the identification of the major C_1-C_6 alkylated thiophenes present in flash pyrolysates of kerogens (48). A study (49) of approximately 50 different immature kerogens (including 13 coals) showed that the alkylated thiophenes in the flash pyrolysates are dominated by a limited number of the theoretically possible isomers, although the isomer distributions vary. This is exemplified by the Py-GC-MS analyses of the Jurf ed Darawish-156 kerogen and an immature, sulfur-rich kerogen of the Monterey Shale as shown in Figures 3 and 4. Peak identifications (numbers) refer to compounds listed in Table I. The distributions of the alkylated thiophenes are in both cases dominated by only a limited number of isomers; e.g., (i) in the C_1-cluster 2-methylthiophene is much more abundant than 3-methylthiophene, (ii) of the six C_2-thiophenes 3-ethylthiophene and 3,4-dimethylthiophene are very minor and (iii) the C_3-thiophenes are dominated by 2-ethyl-5-methylthiophene and 2,3,5-trimethylthiophene. However, significant variations in isomer distributions are observed between samples. For example, despite their similar inferred palaeoenvironment of deposition the Jurf ed Darawish-156 kerogen pyrolysate contains relatively larger amounts of 2,3,4,5-tetramethylthiophene (compound 19, Figure 3b) and 2-ethyl-3,4,5-methylthiophene (compound 23) and lesser amounts of 2-methyl-3-isopropylthiophene (compound 13) compared with the Monterey kerogen pyrolysate.

Closer examination of the alkyl substitution patterns of the dominant alkylated thiophene isomers (48) showed that they bear a strong similarity to those of OSC recently identified in immature bitumens and crude oils (7-17). A considerable portion of the alkylated thiophenes has a 2-alkyl or 2,5-dialkyl substitution pattern indicating that these thiophenes possess a linear carbon skeleton. OSC with this type of carbon skeleton (*i.e.*, 2,5-dialkylthiophenes, 2,5-dialkylthiolanes, 2,6-dialkylthianes) are often major compounds in sulfur-rich, immature bitumens and crude oils (7,9,10,12,15-17). Three other groups of OSC have been assigned in pyrolysates of kerogens, *i.e.* those with isoprenoid, branched (*iso* and *anteiso*) and steroidal side-chain carbon skeletons (49). The four groups of alkylated thiophenes account for approximately 90% (by weight) of the total amount of alkylated thiophenes formed upon pyrolysis in all samples studied. Since C_1-C_6 alkylated thiophenes are primarily formed by β-cleavage, they can be ascribed to specific thiophene moieties within the kerogen. Based on the structures of the alkylated thiophenes it is considered that these moieties have mainly linear, branched, isoprenoid and steroidal side-chain carbon skeletons (Figure 5). The observed differences in abundances and isomer distributions of the C_1-C_6 alkylated thiophenes in different kerogen pyrolysates presumably reflect differences in both relative and absolute abundances of these specific thiophene moieties within the kerogens.

Origin of Sulfur-Containing Moieties. The only difference between thiophene moieties in kerogen and the alkylthiophenes present in immature crude oils and bitumens is that in kerogen they are attached *via* one or more bonds between the alkyl side chain(s) of the thiophene moieties and the kerogen matrix. It is therefore likely that the thiophene moieties present in immature kerogens are formed in an identical way to alkylated thiophenes and other OSC in immature bitumens and crude oils; namely by incorporation of inorganic sulfur species into functionalised lipids (2,8,9,11-18). By reaction with sulfur these lipids may become part of the kerogen structure or may remain as extractable bitumen. The former process may (partially) lead to formation of kerogen.

Less explicit statements can be made concerning the origin of other sulfur-containing moieties (*e.g.*, cyclic and acyclic (poly)sulfide moieties), since the Py-GC-MS method fails to properly characterise these functional groups. However, it is noteworthy that small amounts of alkylthiolanes and -thianes with linear carbon skeletons have been identified in kerogen pyrolysates (48). This suggests the presence of thiolane and thiane moieties in the kerogen

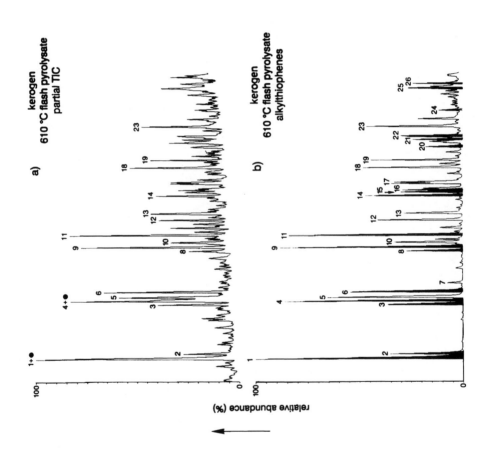

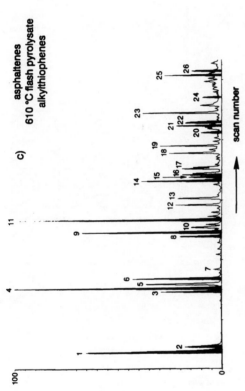

Figure 3. Partial TIC (black dots indicate alkylbenzenes) and accurate (window 0.04 dalton), summed mass chromatograms of *m/z* 97.03 + 98.03 + 111.04 + 112.04 + 125.06 + 126.06 + 139.08 + 140.08 + 153.10 + 154.10 + 167.11 + 168.11 of flash pyrolysates of several high-molecular-weight fractions of Jurf ed Darawish-156. Peak numbers refer to Table I. Black and stippled peaks indicate alkylthiophenes with linear and isoprenoid carbon skeletons, respectively.

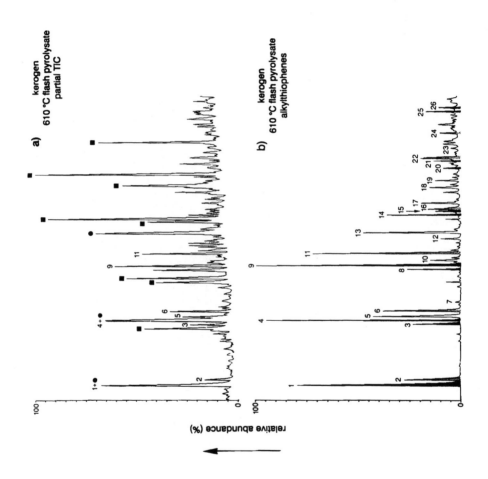

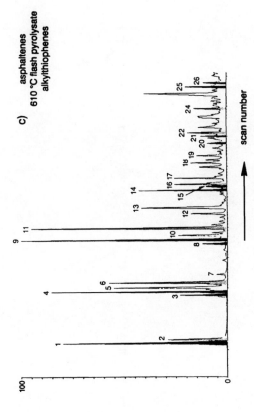

Figure 4. Partial TIC (black dots indicate alkylbenzenes, black squares indicate alkylpyrroles) and accurate (window 0.04 dalton) summed mass chromatograms of m/z 97.03 + 98.03 + 111.04 + 112.04 + 125.06 + 126.06 + 139.08 + 140.08 + 153.10 + 154.10 + 167.11 + 168.11 of flash pyrolysates of several high-molecular-weight fractions of Monterey-25. Peak numbers refer to Table I. Black and stippled peaks indicate alkylthiophenes with linear and isoprenoid carbon skeletons, respectively.

Table I. Major alkylthiophenes identified in pyrolysates of the Jurf
ed Darawish Oil Shale and Monterey kerogens and asphaltenes [a]

	compound	possible origin (precursor moieties)[b]
1	2-methylthiophene	L
2	3-methylthiophene	I
3	2-ethylthiophene	L
4	2,5-dimethylthiophene	L
5	2,4-dimethylthiophene	B,S
6	2,3-dimethylthiophene	I,B
7	3,4-dimethylthiophene	S
8	2-propylthiophene	L
9	2-ethyl-5-methylthiophene	L
10	2-ethyl-4-methylthiophene	B
11	2,3,5-trimethylthiophene	I,B
12	2,3,4-trimethylthiophene	S
13	3-isopropyl-2-methylthiophene	S
14	2-methyl-5-propylthiophene and	L
	2,5-diethylthiophene	L
15	2-butylthiophene	L
16	2-ethyl-3,5-dimethylthiophene and	I
	3-ethyl-2,5-dimethylthiophene	
17	5-ethyl-2,3-dimethylthiophene and	B
	an ethyldimethylthiophene	
18	an ethyldimethylthiophene	
19	2,3,4,5-tetramethylthiophene	
20	2-ethyl-5-propylthiophene	L
21	2-butyl-5-methylthiophene	L
22	2-pentylthiophene and	L
	2,3-dimethyl-5-propylthiophene	I,B
23	2-ethyl-3,4,5-trimethylthiophene	
24	2,3-dimethyl-5-isobutylthiophene	I
25	2-methyl-5-pentylthiophene	L
26	2-hexylthiophene	L

[a]   numbers refer to Figures 3 and 4.
[b]   L = derived from alkylthiophene moieties with linear alkyl carbon
skeletons; B = derived from alkylthiophene moieties with *iso* and *anteiso*
alkyl carbon skeletons; I = derived from alkylthiophene moieties with
isoprenoid alkyl carbon skeletons; S = derived from alkylthiophene
moieties with steroidal side chain carbon skeletons.

matrix. These compounds, however, are believed to mainly yield H_2S upon thermal degradation through cleavage of the carbon-sulfur bond in the ring. A similar origin as proposed for the thiophene moieties in kerogen for these putative thiolane and thiane moieties seems probable. The, not unlikely, formation of acyclic (poly)sulfide moieties in a similar fashion would be an important pathway for the generation of high-molecular-weight material (*e.g.*, kerogen) from low-molecular-weight precursors.

Bonding of Alkylthiophene Units. The presence of acyclic (poly)sulfide moieties in kerogen is further supported by the preferential elimination of alkylthiophene and alkane units from sulfur-rich kerogen (Monterey) upon maturation reported elsewhere in this volume (25). From the structures of the alkylthiophene units in kerogen (Figure 5) it is not easy to understand why these units are preferentially eliminated (as compared with alkylbenzene units) upon maturation since the energy required for cleavage of the C-C bond β to the thiophene and benzene system is approximately the same. Therefore, the attachment of these alkylthiophene units to the kerogen has to be different as compared to that of the alkylbenzene units. Three model structures may be envisaged in an attempt to explain the above observations (Figure 6). Alkylthiophene units linked by a single C-C bond to the macromolecular matrix (model 1, Figure 6a) would generate low-molecular-weight alkylthiophenes upon both flash pyrolysis and maturation, but does not explain the preferential elimination of alkylthiophene units. Alkylthiophene units linked by a single C-S bond to the macromolecular matrix (model 2, Figure 6b) may explain the preferential elimination of alkylthiophene units over alkylbenzene units from sulfur-rich kerogens upon maturation by cleavage of the relatively weak C-S bond. However, this bond is also the most susceptible to cleavage upon flash pyrolysis and low-molecular-weight alkylthiophenes would, therefore, not be major pyrolysis products. This model may explain, however, the occurrence of series of longer-chain alkylthiophenes in pyrolysates of sulfur-rich kerogens (48,49). Alkylthiophene units linked by multiple acyclic C-S bonds to the macromolecular matrix (model 3, Figure 6c) explain both observations; upon flash pyrolysis (high temperature, short reaction time) it is unlikely that multiple, relatively weak, C-S bonds are broken, resulting in the release of higher-molecular-weight alkylthiophenes. Instead, cleavage of the somewhat stronger, but single, C-C bond β to the thiophene unit is more likely and would therefore give rise to low-molecular-weight alkylthiophenes. Upon maturation under natural conditions (relatively low temperature, very long reaction time) the cleavage of the multiple C-S bonds is more likely than cleavage of the C-C bond β to the thiophene unit and this may explain the preferential elimination of thiophene units from

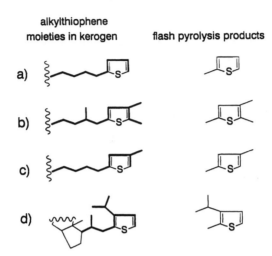

Figure 5. Proposed structures of alkylthiophene moieties in kerogens and asphaltenes and their presumed flash pyrolysis products. Examples are give for alkylthiophene moieties with (a) linear, (b) isoprenoid, (c) branched and (d) steroidal side-chain carbon skeletons. Carbon skeletons are indicated with bold lines.

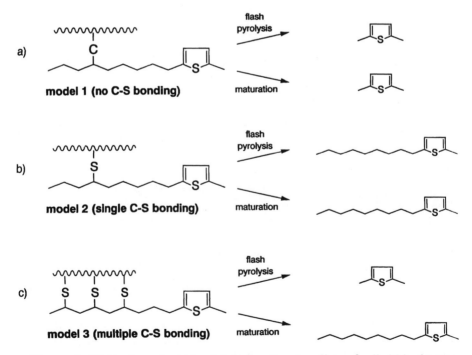

Figure 6. Three hypothetical models for the bonding of alkylthiophene units in kerogen and their products on flash pyrolysis and natural maturation.

sulfur-rich kerogen upon maturation. It is therefore likely that in sulfur-rich kerogens, such as the Monterey kerogen, alkylthiophene units are mainly bonded by several acyclic sulfur linkages to the kerogen matrix. This model is also consistent with the observation that the abundance of C_1-C_6 alkylated thiophenes in pyrolysates shows a distinct correlation with the total organic sulfur content of kerogens and other geopolymers despite the fact that they represent only a small portion of the total organic sulfur content of the kerogen (24). It is also the most satisfactory way of explaining the extremely high organic sulfur contents of Type II-S kerogens (up to 14%, 19).

Organically-Bound Sulfur in Asphaltenes

Asphaltenes isolated from immature sediments and crude oils yield the same types of sulfur compounds as immature kerogens on flash pyrolysis do; *i.e.* H_2S, thiophene and C_1-C_6 alkylated thiophenes and small amounts of benzothiophene and C_1-C_4 alkylated benzothiophenes. Furthermore, the distributions of the C_1-C_6 alkylated thiophenes are dominated by the same isomers as in the kerogen pyrolysates as illustrated by these distributions in flash pyrolysates of the asphaltene fractions isolated from the bitumens of the Jurf ed Darawish Oil Shale and Monterey Shale described above (Figures 3c and 4c). Moreover, a comparison of the mass chromatograms of the asphaltene pyrolysates with those of the corresponding kerogen pyrolysates (*cf.* Figures 3b and 3c and 4b and 4c, respectively) indicates that similar fingerprints are obtained. For example, compounds 18, 19 and 23 are relatively abundant in both the kerogen as well as the asphaltene pyrolysate of the Jurf ed Darawish Oil Shale. These compounds are much less abundant in the pyrolysates of the corresponding fractions from the Monterey Shale, where compound 13 is more abundant. Minor differences in the distributions of the C_1-C_6 alkylated thiophenes of the corresponding kerogen and asphaltene pyrolysates also occur. In both cases compounds 6 and 11, both derived from thiophene moieties with isoprenoid carbon skeletons, are relatively more abundant in the asphaltene pyrolysates. It is noteworthy within this context that Solli *et al.* (36) noted a general feature of asphaltenes; their pyrolysates contain higher abundances of isoprenoids than their corresponding kerogen pyrolysates.

The structure and origin of the thiophene moieties in asphaltenes is believed to be identical to those in kerogens. Furthermore, the similar C_1-C_6 alkylated thiophene fingerprints of pyrolysates of asphaltenes and kerogen from the same sediments testifies to the concept of asphaltenes as being small "soluble" fragments of kerogen having comparable structures (19,20,34,35,56). Asphaltenes isolated from crude oils do exhibit a small difference compared to those isolated from bitumens; their pyrolysates contain somewhat more abundant alkylthiophenes with extended alkyl side-chains (*i.e.* $>C_6$). This probably indicates that asphaltenes from crude oils contain relatively higher amounts of "Model 2" (Figure 6) alkylthiophene units.

Organically-Bound Sulfur in Resins

The resin fraction of a crude oil or bitumen is defined as the high-molecular-weight (MW > *ca.* 800) fraction soluble in light hydrocarbons (20). With conventional separation methods for bitumens and crude oils the resins mainly elute in the so-called polar (or NSO or hetero compound) fraction (Figure 1). However, the "aromatic hydrocarbon" fraction of immature bitumens and crude oils may also contain substantial amounts of high-molecular-weight, low volatile (non-GC amenable) substances, especially in sulfur-rich samples. These substances can be separated from lower-molecular-weight aromatic (LMWA) hydrocarbons and OSC by a simple column chromatographic step (Figure 1) since they are somewhat more polar. This yields the so-called "high-molecular-weight aromatic" (HMWA) fraction (12,16). From an analytical point of view resins are easier to characterise than kerogens and asphaltenes since they are completely soluble in the appropriate organic solvents required for chemical degradation.

Jurf ed Darawish-156.   Flash pyrolysis of polar fractions of immature bitumens and crude oils generates pyrolysates which differ markedly from those of kerogens and asphaltenes. Whilst pyrolysates of asphaltenes and kerogens are dominated by volatile compounds (< C_{12}), pyrolysates of polar fractions comprise a complex, partially unresolved mixture of mainly $C_{15}+$ components. This is illustrated in Figure 2b which shows the TIC from flash pyrolysis (610°C) of the polar fraction from the extracted bitumen of the Jurf ed Darawish-156 sample. This pyrolysate is not dominated by C_1-C_6 alkylated thiophenes as are the kerogen (Figure 2a) and asphaltene (not shown) pyrolysates. Another major difference from flash pyrolysates of kerogens and asphaltenes is that similar results are obtained for the polar fraction when wires with a Curie temperature of 358°C are used (Figure 2c). This indicates that either (i) the polar fraction is comprised of mainly low-molecular-weight compounds (MW < 800) which simply evaporate or (ii) that the polar fraction contains mainly high-molecular-weight substances which are thermally much more labile than the kerogens and asphaltenes or (iii) that both phenomena occur.

Some of the major compounds in the 358°C pyrolysate were identified as series of 2-alkylthiolanes, -thianes, 2,5-dialkylthiolanes and 2,6-dialkylthianes, all OSC with linear carbon skeletons. Figure 7 shows partial mass chromatograms of m/z 87 and 101 visualizing the distributions of some of these series. All these series are characterised by a strong even-over-odd carbon number predominance with (sub)maxima at C_{18}, C_{28} and/or C_{30}. The m/z 101 mass chromatogram also reveals the presence of saturated and monounsaturated C_{20} isoprenoid thiolanes. Since these compounds have a polarity which results in their presence exclusively in the LMWA fraction upon fractionation of the bitumen with column chromatography (Figure 1), these compounds have to be formed by thermal degradation of resin-like material. Further support for this theory came from the identification of monounsaturated thiolanes and thianes; the presence of a double bond strongly suggesting a pyrolytic origin (*cf.* n-alk-1-enes). On the basis of mass spectra of these compounds it is thought that their double bond positions are not located at the end of the long alkyl side-chain (*cf.* n-alk-1-enes) but in the alkyl side-chain near the ring system. These results indicate that alkylthiolane and -thiane units with linear carbon skeletons are bonded by one (poly)sulfide moiety to the macromolecular matrix of resins. It is unlikely that these units are bound by more than one acyclic (poly)sulfide linkage to the macromolecular matrix since their release would then require the cleavage of more than one bond.

Raney Ni desulfurisation of this polar fraction yields 28 wt.% hydrocarbons dominated by n-alkanes, also with a strong even-over-odd carbon number predominance and with (sub)maxima at C_{18}, C_{28} and C_{30} (Figure 8a, 18). Since Raney Ni reductively cleaves carbon-sulfur bonds selectively, this experiment provides further support for the proposed structure of the resin fraction as mainly consisting of a network of predominantly straight-chain hydrocarbons with relatively high abundances of octadecane, octacosane and triacontane. These units are connected to each other by one or more (poly)-sulfide linkages. It is, as yet unknown, whether these linkages in this geopolymer are mainly mono-, di- and/or polysulfide moieties. Kohnen *et al.* (17) have recently shown that OSC with di- and trisulfide groups occur in immature sediments. It is therefore not unlikely that these groups also occur in macromolecules. In addition to the (poly)sulfide linkages connecting the several units to eachother, the flash pyrolysis results show that these units themselves consist mainly of linear thiolanes and thianes, which may be considered as intramolecular sulfide linkages. These units are also desulfurised which explains the high amounts of n-alkanes obtained. Desulfurisation of the polar fraction also yields phytane (and very small amounts of pristane) and 22R pentakishomohopane, indicating the presence of these units in these sulfur-rich macromolecules. Raney Ni desulfurisation of the HMWA fraction of the same sample also yields hydrocarbons (43 wt.%) dominated by n-

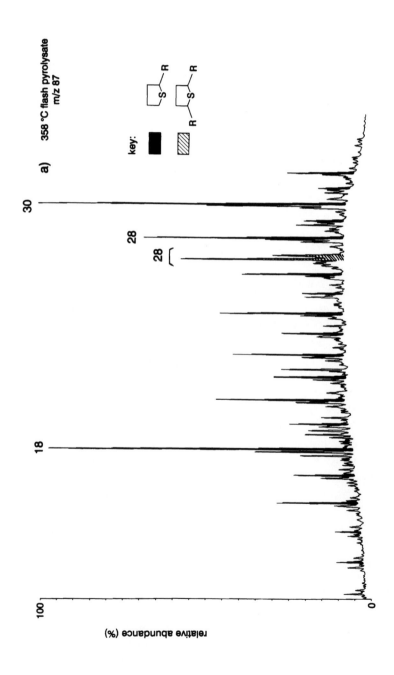

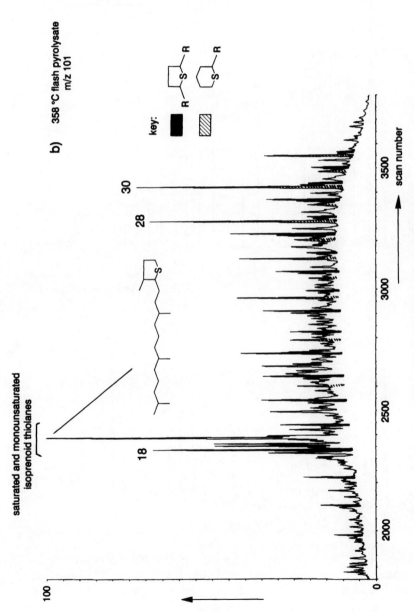

Figure 7. Partial mass chromatograms of the flash pyrolysate of the polar fraction of the Jurf ed Darawish-156 bitumen. Numbers indicate total number of carbon atoms.

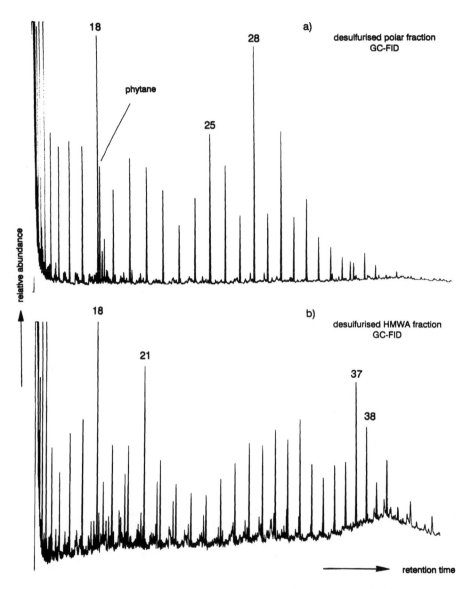

Figure 8. Gas chromatograms of the hydrocarbons obtained by desulfurisation of the polar and HMWA fractions of the Jurf ed Darawish-156 bitumen. Numbers indicate total number of carbon atoms of *n*-alkanes.

alkanes with (sub)maxima at C_{18}, C_{21} and C_{37} (Figure 8b, 18). However, the distribution pattern of the n-alkanes obtained by desulfurisation of the polar fraction is completely different, indicating that the resins in the HMWA fraction are structurally different from those present in the polar fraction.

<u>Northern Apennines Marl.</u> Figure 9 shows the partial TIC of the pyrolysates of the polar fraction of the Northern Apennines Marl bitumen by using ferromagnetic wires of both 358 and 610°C. The composition of both pyrolysates is similar but differs markedly from those of the polar fraction of the Jurf ed Darawish-156 bitumen, although the pyrolysates of both samples are dominated by $C_{15}+$ components. The major products identified in both pyrolysates are phytenes and phytane, saturated and monounsaturated C_{20} isoprenoid thiolanes and thiophenes, saturated and monounsaturated, linear C_{22} thiolanes and thianes, Δ^2- and Δ^3-cholestenes and cholestane and an, as yet unknown, C_{27} sulfur-containing steroid, tentatively identified as cholestane-3,19-episulfide (7). The generation of these relatively apolar compounds at 358°C again indicates that structurally related units are bonded to the macromolecular matrix of the resin fraction by probably one (poly)sulfide linkage. The pyrolytic origin of these compounds is also further supported by the generation of their unsaturated counterparts.

The selective generation of (poly)sulfide-linked molecules from the resin macromolecules at lower pyrolysis temperatures is exemplified by Figure 10, which shows mass chromatograms of m/z 57 of the 358 and 610°C pyrolysates. At · 358°C, phytenes and phytane dominate this mass chromatogram, whilst at 610°C in addition to the phytenes and phytane a series of n-alkanes and isoprenoid alkanes (C_{12}-C_{18}) and prist-1-ene are observed. These latter compounds are probably formed mainly by other cleavages than those of carbon-sulfur bonds. The higher amounts of phytane relative to the phytenes in the 610°C pyrolysate is also notable. The ratio cholestane/cholestenes also increased in the 610°C pyrolysate; this probably indicates that a change in the composition of the products from carbon-sulfur bond cleavage occurs with increasing pyrolysis temperature.

Apart from phytanyl moieties bound by a (poly)sulfide linkage to the macromolecular matrix such phytanyl moieties also occur with an additional intramolecular sulfide linkage. A mass chromatogram of m/z 101 of the 358°C pyrolysate (Figure 11b) shows the presence of a C_{20} isoprenoid thiolane and its monounsaturated counterparts which are major pyrolysis products. C_{20} isoprenoid thiophenes also occur in the pyrolysates but are less abundant.

Units consisting of linear carbon skeletons occur less abundantly in the resins of the Northern Apennines Marl bitumen than in those of the Jurf ed Darawish-156 bitumen. Mass chromatograms of m/z 87 and 101 (Figure 11)

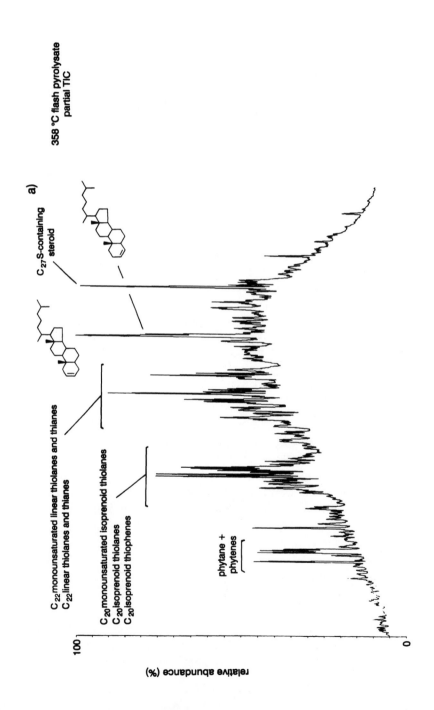

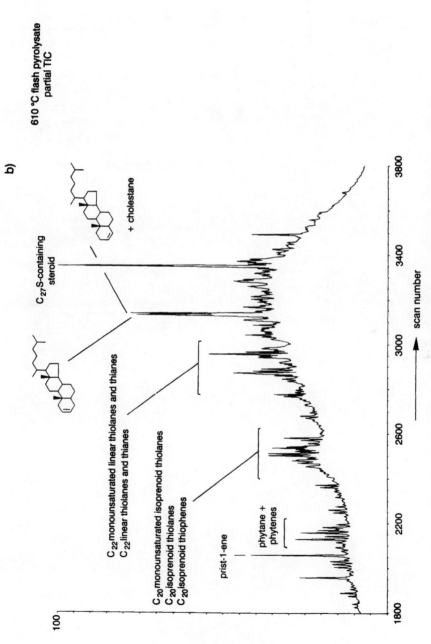

Figure 9. Partial total ion chromatograms of the flash pyrolysates (358 and 610°C) of the polar fraction of the Northern Apennines Marl bitumen.

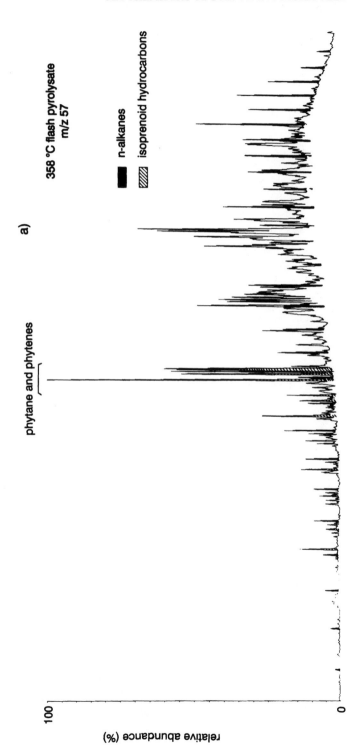

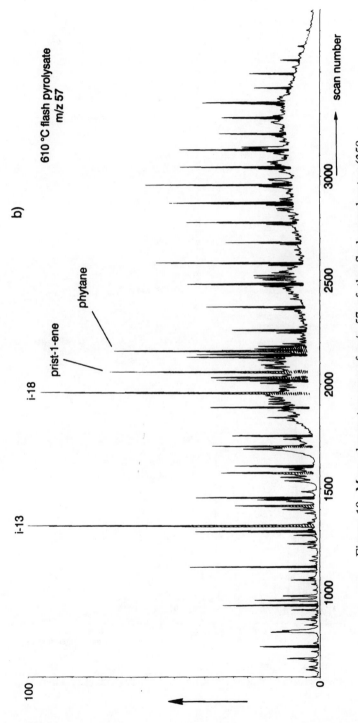

Figure 10. Mass chromatograms of *m/z* 57 of the flash pyrolysates (358 and 610°C) of the polar fraction of the Northern Apennines Marl bitumen.

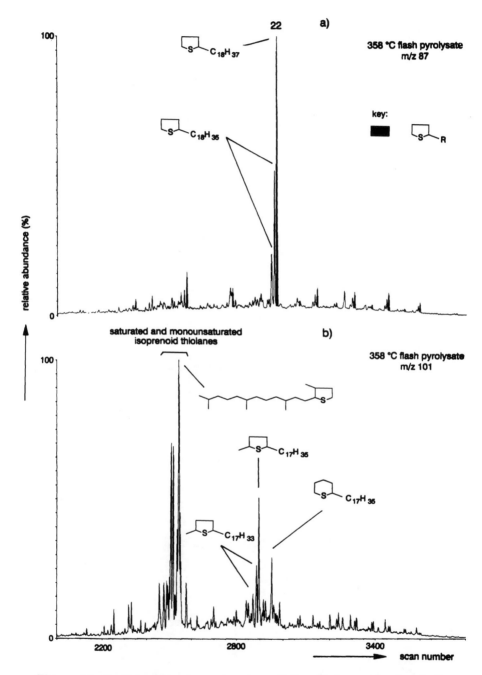

Figure 11. Partial mass chromatograms of the flash pyrolysate of the polar fraction of the Northern Apennines Marl bitumen. Numbers indicate total number of carbon atoms.

reveal the presence of a series of 2-alkylthiolanes and -thianes and 2-alkyl-5-methylthiolanes and unsaturated counterparts highly dominated by C_{22} members.

The dominance of Δ^2- and, tentatively identified, Δ^3-cholestenes and $5\alpha(H),14\alpha(H),17\alpha(H),20R$-cholestane in the 358°C pyrolysate (Figure 9a) indicates that relatively high amounts of cholestanyl units bound by a (poly)sulfide linkage to the macromolecular matrix occur in the Northern Apennines Marl resins. These (poly)sulfide linkages are probably connected to position 3 of the cholestane units since high amounts of Δ^2- and, less stable (57), Δ^3-cholestenes are formed upon pyrolysis. This is consistent with the work of Schmid (7) who reported that steroid units in a high-molecular-weight fraction of the Rozel Point seep oil were bonded *via* sulfur bridges at position 3 of the steroid units because of the formation of 3-thiosteroids upon heating (200°C, 48h) of this fraction and by treating it with LiAlH$_4$. Furthermore, desulfurisation of this fraction with deuterated Raney Ni yielded steranes which had incorporated one deuterium atom in the A-ring (58).

The flash pyrolysis results are in full agreement with those obtained by sulfur-selective chemical degradation of this resin fraction. Raney Ni desulfurisation of this fraction yielded saturated and aromatic hydrocarbons (11 wt.%) dominated by phytane, docosane, $5\beta(H),14\alpha(H),17\alpha(H),20R$-cholestane, $17\alpha(H),21\beta(H),22R$-pentakishomohopane and β-carotane (Figure 12a). A mass chromatogram of m/z 57 (Figure 11b) shows the distribution of the n-alkanes with a strong even-over-odd carbon number predominance and with (sub)-maxima at C_{18}, C_{22}, C_{28} and C_{37} and the strong dominance of phytane over pristane. A partial mass chromatogram of m/z 113 (inset in Figure 12b) reveals the presence of a number of C_{25}-C_{27} isoprenoid hydrocarbons with the indicated carbon skeletons. Alkylthiophenes with these carbon skeletons occur in the LMWA fraction of the Northern Apennines Marl bitumen (59), but the ratio of the various isomers is different. A mass chromatogram of m/z 217 (Figure 13a) shows the distribution of the steranes formed by desulfurisation, which are dominated by $5\beta(H),14\alpha(H),17\alpha(H),20R$-steranes. The reason for this is not completely understood since flash pyrolysis yielded predominantly $5\alpha(H),14\alpha(H),17\alpha(H)20R$-cholestane. Desulfurisation of polar fractions of other samples yielded steranes which were not dominated by $5\beta(H),14\alpha(H),17\alpha(H),20R$-stereoisomers (60,61). The m/z 191 mass chromatogram (Figure 13b) reveals the exclusive presence of C_{35} extended hopanes. Besides β-carotane several other, tentatively identified, mono- and diaromatic carotenoids are also present in the desulfurised polar fraction as revealed by the partial mass chromatogram of m/z 134 (Figure 13c). The diaromatic carotenoids have been reported to occur as such in bitumens of Toarcian shales (62) and are thought to be derived from the highly unsaturated

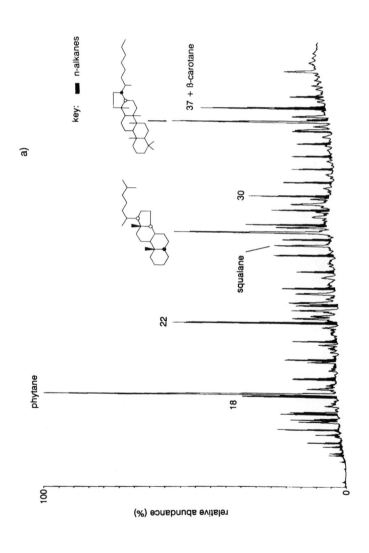

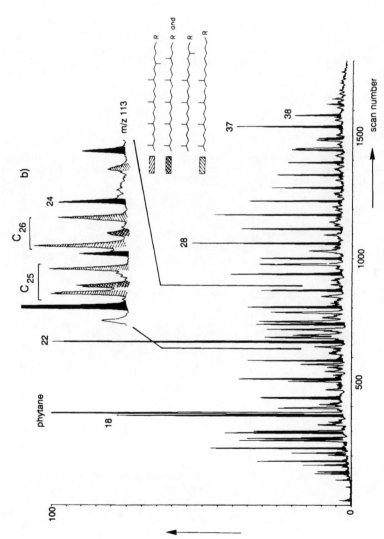

Figure 12. Total ion chromatogram (a) and *m/z* 57 mass chromatogram (b) of the hydrocarbons obtained upon desulfurisation of the polar fraction of the Northern Apennines Marl bitumen. Numbers indicate total number of carbon atoms of *n*-alkanes. The inset in Figure 12b shows a partial mass chromatogram of *m/z* 113.

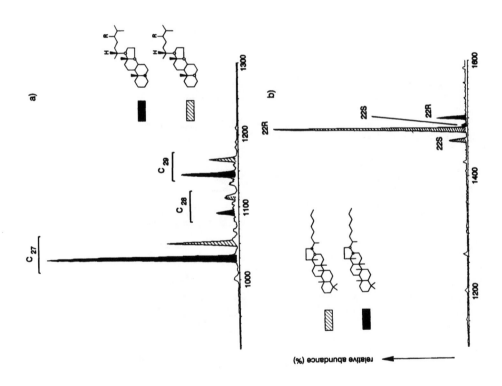

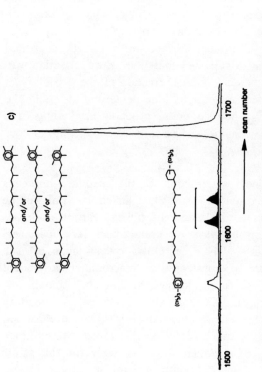

Figure 13. Partial mass chromatograms of (a) *m/z* 217, (b) *m/z* 191 and (c) *m/z* 134 of the desulfurised polar fraction of the Northern Apennines Marl bitumen.

carotenoids renieratene, isorenieratene and renierapurpurin, which occur in photosynthetic bacteria and marine sponges (63,64) or their symbiotic bacteria (65,66).

In summary, Raney Ni desulfurisation of the polar fraction of the Northern Apennines Marl further supports the presence of (poly)sulfide-linked phytanyl, docosanyl and cholestanyl moieties (some of them with additional intramolecular sulfur linkages) in the resin fraction as proposed from the pyrolysis experiments. In addition, a number of other structural units are revealed; e.g. pentakishomohopane, carotenoids, n-alkanes and isoprenoid alkanes. The reason why these structural units are not revealed by the pyrolysis experiments may be: (i) their much lower relative abundance (e.g., other n-alkanes and isoprenoid alkanes), (ii) their attachment in the macromolecules by more than one (poly)sulfide linkage, which make their release from the macromolecule by flash pyrolysis unlikely (e.g., pentakishomohopane, carotenoids). The latter explanation is strongly supported by Mycke et al. (58). These authors reported that upon desulfurisation with deuterated Raney Ni of a high-molecular-weight fraction of the Rozel Point seep oil C_{35} hopanes were formed with up to four deuterium atoms in the side-chain. This indicates that C_{35} hopane units (probably originating from bacteriohopanetetrol) are bonded in sulfur-rich macromolecules via up to four sulfur bridges.

Origin of Sulfur-Rich Resins.   The resin fractions of the examined bitumens comprise macromolecules which contain specific units linked by (poly)sulfide bridges to the macromolecular matrix. These units possess carbon skeletons identical to the well-known geologically occurring hydrocarbons (i.e., n-alkanes, isoprenoid alkanes, steranes, hopanes). These units also contain intramolecular sulfur linkages and may therefore be considered as macromolecularly-bound OSC. The structures of these OSC (alkylthiolanes, thianes and -thiophenes) are identical to those occurring in GC-amenable fractions of immature bitumens and crude oils (2-16). Since low-molecular-weight OSC are formed by incorporation of inorganic sulfur species into specific functionalised lipids during early diagenesis (2,8,9,11-18) a similar origin is likely for the sulfur cross-linked geopolymer. Instead of incorporation of inorganic sulfur species into functionalised lipids in an intramolecular fashion, leading to the formation of low-molecular-weight OSC (15), these reactions also occur on an intermolecular basis giving rise to the sulfur cross-linked geopolymer. The formation of these sulfur-rich high-molecular-weight substances during early diagenesis by inter- and intramolecular sulfur cross-linking of functionalised lipids is supported by several observations from the present study.

Firstly, the composition of both the flash pyrolysates and the desulfurisation mixtures obtained from the resin fractions described here and elsewhere (60,61) indicate that specific functionalised lipids react with H_2S and/or polysulfides, yielding sulfur-rich high-molecular-weight substances. All three mixtures of hydrocarbons obtained after desulfurisation of resin fractions show a strong even-over-odd carbon number predominance for the *n*-alkanes, a very low pristane/phytane ratio and a hopane distribution dominated by the C_{35} homologue. This is interpreted as reflecting the reaction of *n*-alcohols, chlorophyll-derived phytenes and bacteriohopanetetrol, respectively, with inorganic sulfur species during the early stages of diagenesis. These compounds would otherwise be mainly mineralised or diagenetically transformed to other compounds (*e.g.*, hydrocarbons). The relative abundance of bound C_{22} linear alkylthiolanes and -thianes in the resin fraction of the Northern Apennines Marl bitumen may be ascribed to a high abundance of docosahexaenoic acid and/or docosahexaenol during deposition of this sediment. The presence of sulfur-bound carotenoids in the resin fraction also has to be explained by incoporation of labile carotenoids such as β-carotene and the aromatic carotenoids renieratene, isorenieratene and/or renierapurpurin into high-molecular-weight material by reactions with inorganic sulfur species during early diagenesis. It is not known at present whether these bound aromatic carotenoids also contain intramolecular sulfur linkages. Thermal maturation of such thiophene-containing units in sulfur-rich macromolecules may explain the distribution patterns of the C_{13}-C_{31} aryl isoprenoids identified in bitumen and oils (67,68). The relative abundance of the C_{37} and, to a lesser extent, C_{38} *n*-alkanes in the desulfurisation products of the Northern Apennines Marl and Jurf ed Darawish-156 polar fractions and, especially, of the HMWA fraction of the Jurf ed Darawish-156 is probably due to reaction with inorganic sulfur species of C_{37} and C_{38} unsaturated ketones or their corresponding alkadienes and alkatrienes present in sediments (69-77).

Secondly, in the case of the bound steroids it was possible to locate the position of the (poly)sulfide linkages in these structural units; *i.e.* mainly at position 3 of the steroid units. This strongly supports the argument that functionalised lipids are incorporated into these macromolecules by reaction of H_2S and/or polysulfides with functionalities, since sterols or corresponding sterenes are the only likely precursors for the steroid units in these macromolecules.

Other sulfur-selective chemical degradation techniques (work in progress) will probably yield additional information to further resolve remaining questions with respect to structure, origin and thermal behaviour of these sulfur-rich geopolymers.

Relationship between Molecular Weight and Organic Sulfur in Sulfur-Rich Sedimentary Organic Matter

Based on both our analyses of the different sulfur-rich high-molecular-weight fractions and on previous observations from studies on low-molecular-weight OSC (1-17) we propose that there is a direct relationship between the type (and degree) of sulfur cross-linking and molecular weight for sulfur-rich sedimentary organic matter (Figure 14). The zonations in this simplified diagram are somewhat arbitrary, however, the chemical and physical properties of the various organic matter fractions and their interrelationships may, at least partially, be explained in this general framework.

Sulfur is incorporated into sedimentary organic matter during early diagenesis through reaction of functionalised lipids with inorganic sulfur species. The extent to which intra- *versus* intermolecular sulfur cross-linking occurs determines the molecular weight of the formed products and, thus, the relative proportions of low-molecular-weight OSC (MW < 800), sulfur-rich resins, asphaltenes and kerogen. To give an indication of the type of cross-linking involved the positions of the three models proposed for the bonding of alkylthiophene units (Figure 6) are plotted along the X-axis. In kerogen relatively abundant intermolecular sulfur cross-links are thought to be present (*cf.* Model 3) and kerogen, therefore, plots in the upper-right corner of the diagram. Low-molecular-weight OSC (*cf.* Model 1) contain only intramolecular sulfur linkages and plot, consequently, in the lower-left corner of the diagram. Asphaltenes and resins may be seen as intermediates in this respect. In this context the differences in solubility between the different fractions may be explained by the extent of sulfur cross-linking.

Maturation will give rise to cleavage of relatively weak C-S and S-S bonds and to aromatisation of thiolane moieties (which contain cyclic, intramolecular sulfide linkages). The latter reaction achieves conversion of a thermally labile to a more stable form of organic sulfur. This conversion can only happen with intramolecular sulfur linkages. The result of both aforementioned reactions is a shift of the composition of the sulfur-containing substances from the upper-right corner to the lower-left corner of the diagram. A concomitant loss of organic sulfur will occur also during maturation.

Conclusions

(1) Immature kerogens and asphaltenes contain thiophene units in their macromolecular structure, which have mainly linear, isoprenoid, branched and steroidal side-chain skeletons. These units and possibly other sulfur-

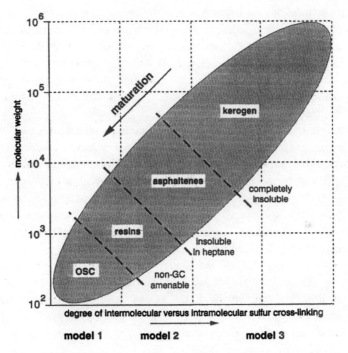

Figure 14. Relationship between the degree of intermolecular *versus* intramolecular sulfur cross-linking and molecular weight for sulfur-rich sedimentary organic matter.

containing moieties, such as cyclic and acyclic (poly)sulfide moieties, are formed by incorporation of inorganic sulfur species into functionalised lipids during early diagenesis. In depositional environments with high amounts of available sulfur donors these reactions may be regarded as an important pathway for the generation of high-molecular-weight material ("natural vulcanisation"). The fact that sulfur-containing moieties in asphaltenes are similar to those in kerogen supports the concept of asphaltenes as being small "soluble" fragments of kerogen having comparable structures.

(2) The resin fractions of organic sulfur-rich bitumens are for a substantial part composed of monomers with linear, isoprenoid, steroid, hopanoid and carotenoid carbon skeletons connected to each other by (poly)sulfide linkages. These structural units may contain additional intramolecular sulfur linkages. This sulfur-rich geopolymer is also formed by sulfur incorporation into functionalised lipids in an intermolecular fashion during early diagenesis.

(3) The formation of sulfur-rich macromolecules is, therefore, analogous to the formation of low-molecular-weight OSC in immature sediments, which are formed by similar reactions but in an intramolecular fashion.

(4) The different properties of sulfur-rich kerogen and asphaltenes, on the one hand, and sulfur-rich resins on the other hand (*e.g.*, differences in solubility and flash pyrolysis behaviour) may be explained only by differences in degree of (sulfur) cross-linking and thus by differences in molecular size and in degree of condensation.

(5) Raney Ni desulfurisation of resins may be used in fingerprinting oils and bitumens since certain so-called biomarkers may be selectively incorporated (and preserved) in the resins (*e.g.*, carotenoids, unsaturated C_{37} and C_{38} ketones or alkenes). This may provide additional information on the palaeoenvironment of deposition as well as on the structure of these complex materials.

Literature cited

1. Payzant, J.D.; Montgomery, D.S.; Strausz, O.P. Tetrahedron Lett. 1983, 24, 651-654.
2. Valisolalao, J.; Perakis N.; Chappe, B.; Albrecht, P. Tetrahedron Lett. 1984, 25, 1183-1186.
3. Payzant, J.D.; Cyr, T.D.; Montgomery, D.S.; Strausz, O.P. Tetrahedron Lett. 1985, 26, 4175-4178.
4. Payzant, J.D.; Montgomery, D.S.; Strausz, O.P. Org. Geochem. 1986, 9, 357-369.

5. Cyr, T.D.; Payzant, J.D.; Montgomery, D.S.; Strausz, O.P. Org. Geochem. 1986, 9, 139-143.
6. Perakis, N. Ph.D. Thesis, University of Strasbourg, Strasbourg, 1986.
7. Schmid, J.C. Ph.D. Thesis, University of Strasbourg, Strasbourg, 1986.
8. Brassell, S.C.; Lewis, C.A.; de Leeuw, J.W.; de Lange, F.; Sinninghe Damsté, J.S. Nature (London) 1986, 320, 160-162.
9. Sinninghe Damsté, J.S.; ten Haven, H.L.; de Leeuw, J.W.; Schenck, P.A. Org. Geochem. 1986, 10, 791-805.
10. Schmid, J.C., Connan, J.; Albrecht, P. Nature (London) 1987, 329, 54-56.
11. Sinninghe Damsté, J.S.; de Leeuw, J.W. Intl. J. Environ. Anal. Chem. 1987, 28, 1-19.
12. Sinninghe Damsté, J.S.; de Leeuw, J.W.; Kock-van Dalen, A.C.; de Zeeuw, M.A.; de Lange, F.; Rijpstra, W.I.C.; Schenck, P.A. Geochim. Cosmochim. Acta 1987, 51, 2369-2391.
13. Sinninghe Damsté, J.S.; Kock-van Dalen, A.C.; de Leeuw, J.W.; Schenck, P.A. Tetrahedron Lett. 1987, 28, 957-960.
14. Rullkötter, J.; Landgraf, M.; Disko, U. J. High Res. Chrom. & Chro. Commun. 1988, 11, 633-638.
15. Sinninghe Damsté, J.S.; Rijpstra, W.I.C.; Kock-van Dalen, A.C.; de Leeuw, J.W.; Schenck, P.A. Geochim. Cosmochim. Acta 1989, 53, 1343-1355.
16. Sinninghe Damsté, J.S.; Rijpstra, W.I.C.; de Leeuw, J.W.; Schenck, P.A. Geochim. Cosmochim. Acta 1989, 53, 1323-1341.
17. Kohnen, M.E.L.; Sinninghe Damsté, J.S.; ten Haven, H.L.; de Leeuw, J.W. Nature (London) 1989, 341, 640-641.
18. Sinninghe Damsté, J.S.; Rijpstra, W.I.C.; de Leeuw, J.W.; Schenck, P.A. Org. Geochem. 1988, 13, 593-606.
19. Orr, W.L. Org. Geochem. 1986, 10, 499-516.
20. Tissot, B.P.; Welte, D.H. Petroleum Formation and Occurrence; Springer: Heidelberg, 1984.
21. Jones, R.W. In Petroleum Geochemistry and Source Rock Potential of Carbonate Rocks; Palacas, J.G., Ed.; AAPG Studies in Geology #18, AAPG: Tulsa, 1984; p. 163-180.
22. Lewan, M.D. Phil. Trans. Roy. Soc. London 1985, A315, 123-134.
23. Tannenbaum, E.; Aizenshtat, Z. Org. Geochem. 1985, 8, 181-192.
24. Eglinton, T.I.; Sinninghe Damsté, J.S.; Kohnen, M.E.L.; de Leeuw, J.W. Fuel 1989 (submitted).
25. Eglinton, T.I.; Sinninghe Damsté, J.S.; Kohnen, M.E.L.; de Leeuw, J.W.; Larter, S.R.; Patience, R.L. In Geochemistry of Sulfur in Fossil Fuels;

Orr, W.L.; White, C.M., Eds.; ACS Symposium Series; American Chemical Society: Washington, DC, 1989; this volume.

26. George, G.N.; Gorbaty, M.L. J. Am. Chem. Soc. 1989, 111, 3182-3186.

27. Saiz-Jimenez, C.; de Leeuw, J.W. Org. Geochem. 1986, 10, 869-876.

28. Saiz-Jimenez, C.; de Leeuw, J.W. Org. Geochem. 1984, 6, 417-422.

29. Saiz-Jimenez, C.; de Leeuw, J.W. Org. Geochem. 1984, 6, 287-293.

30. Saiz-Jimenez, C.; de Leeuw, J.W. J. Anal. Appl. Pyrol. 1986, 9, 99-119.

31. van Graas, G.; de Leeuw, J.W.; Schenck, P.A. J. Anal. Appl. Pyrol. 1980, 2, 265-276.

32. Nip, M.; de Leeuw, J.W.; Schenck, P.A.; Meuzelaar, H.L.C.; Stout, S.A.; Given P.H.; Boon, J.J. J. Anal. Appl. Pyrol. 1985, 8, 221-239.

33. Nip, M.; de Leeuw, J.W.; Schenck, P.A. Geochim. Cosmochim. Acta 1989 submitted.

34. Behar, F.; Pelet, R. J. Anal. Appl. Pyrol. 1985, 7, 121-135.

35. Philp, R.P.; Gilbert, T.D. Geochim. Cosmochim. Acta 1985, 49, 1421-1432.

36. Solli, H.; Leplat, P. Org. Geochem. 1986, 10, 313-329.

37. Larter, S.R.; Solli, H.; Douglas, A.G. J. Chromatogr. 1978, 167, 421-431.

38. van de Meent, D.; Brown, S.C.; Philp, R.P.; Simoneit, B.R.T. Geochim. Cosmochim. Acta 1980, 44, 999-1013.

39. van Graas, G.; de Leeuw, J.W.; Schenck, P.A.; Haverkamp, J. Geochim. Cosmochim. Acta 1981, 45, 2456-2474.

40. Solli, H.; Bjorøy, M.; Leplat, P.; Hall, K. J. Anal. Appl. Pyrol. 1984, 7, 101-119.

42. Larter, S.R. In Analytical Pyrolysis; Voorhees, K.J., Ed.; Butterworths: London, 1984; p. 212-275.

42. Larter, S.R.; Senfle, J.T. Nature (London) 1985, 318, 27-280.

43. Horsfield, B. Geochim. Cosmochim. Acta 1989, 53, 891-901.

44. Boudou, J.P.; Boulege, J.; Malechaux L.; Nip, M.; de Leeuw, J.W.; Boon, J.J. Fuel 1987, 66, 1558-1569.

45. Eglinton, T.I.; Philp, R.P.; Rowland, S.J. Org. Geochem. 1988, 12, 33-41.

46. Philp, R.P.; Bakel, A. Energy Fuels 1988, 2, 59-64.

47. Philp, R.P.; Bakel, A.; Galvez-Sinibaldi, A.; Lin, L.H. Org. Geochem. 1988, 13, 915-926.

48. Sinninghe Damsté, J.S.; Kock-van Dalen, A.C.; de Leeuw, J.W.; Schenck, P.A. J. Chromatogr. 1988, 435, 435-452.

49. Sinninghe Damsté, J.S.; Eglinton, T.I.; de Leeuw, J.W.; Schenck, P.A. Geochim. Cosmochim. Acta 1989, 53, 873-889.

50. Payzant, J.D.; Montgomery, D.S.; Strausz, O.P. AOSTRA J. Res. 1988, 4, 117-131.

51. Wehner, H.; Hufnagel, H. In Biochemistry of Black Shales; Degens, E.T. et al., Eds.; Mitt. Geol Palaeont.Inst. Univ. Hamburg 1987, 60, 381-395.

52. Kohnen, M.E.L.; Sinninghe Damsté, J.S.; Rijpstra, W.I.C.; de Leeuw, J.W. In Geochemistry of Sulfur in Fossil Fuels; Orr, W.L.; White, C.M., Eds.; ACS Symposium Series; American Chemical Society: Washington, DC, 1989; this volume.

53. Isaacs, C.M. OGJ 1984, 82, 75-81.

54. ten Haven, H.L.; de Leeuw, J.W.; Schenck, P.A. Geochim. Cosmochim. Acta 1985, 49, 2181-2191.

55. Larter, S.R.; Horsfield, B. In Organic Geochemistry; Engel, M.; Macko, S., Eds., in press.

56. Bandurski, E. Energy Sources 1982, 6, 47-66.

57. de Leeuw, J.W.; Cox, H.C.; van Graas, G.; van der Meer, F.W.; Peakman, T.M.; Baas, J.M.A.; van de Graaf, B. Geochim. Cosmochim. Acta 1989, 53, 903-909.

58. Mycke, B.; Schmid, J.C.; Albrecht, P. In Geochemistry of Sulfur in Fossil Fuels; Orr, W.L.; White, C.M., Eds.; ACS Symposium Series; American Chemical Society: Washington, DC, 1989; this volume.

59. Peakman, T.M.; Sinninghe Damsté, J.S.; de Leeuw, J.W. Chem. Commun. 1989, 1105-1107.

60. de Leeuw, J.W.; Sinninghe Damsté, J.S. In Geochemistry of Sulfur in Fossil Fuels; Orr, W.L.; White, C.M., Eds.; ACS Symposium Series; American Chemical Society: Washington, DC, 1989; this volume.

61. Sinninghe Damsté, J.S.; Rijpstra, W.I.C.; de Leeuw, J.W., in preparation.

62. Schaefle, J.; Albrecht, P.; Ourisson, G. Tetrahedron Lett. 1977, 3673-3676.

63. Liaaen-Jensen, S. In Marine Natural Products; Faulkner, D.J.; Fenical, W.H., Eds.; Academic: New York, 1978; pp 1-73.

64. Liaaen-Jensen, S. In Photosynthetic Bacteria; Clayton, R.K.; Sistrom, W.R., Eds.; Plenum, 1978; pp 233-248.

65. Imhoff, J.F.; Trüper, H.G. Microbiol. Ecol. 1976, 3, 1-9.

66. Imhoff, J.F.; Trüper, H.G. Zbl. Bakt. I. Abt Orig. 1980, C1, 61-69.

67. Ostroukhov, S.B.; Arefev, O.A.; Makushina, V.M.; Zabrodina, M.N.; Petrov, Al.A. Neftekhimiya 1982, 22, 723-788.

68. Summons, R.E.; Powell, T.G. Geochim. Cosmochim. Acta 1987, 51, 557-566.

69. Boon, J.J.; van der Meer, F.W.; Schuyl, P.J.W.; de Leeuw, J.W.; Schenck, P.A.; Burlingame, A.L. Init. Repts. DSDP 1978, 40, 627-637.

70. de Leeuw, J.W.; van der Meer, F.W.; Rijpstra, W.I.C.; Schenck, P.A. In Advances in Organic Geochemistry 1981; Douglas, A.G.; Maxwell, J.R., Eds.; Pergamon: Oxford, 1982; pp 211-217.

71. Volkman, J.K.; Eglinton, G.; Corner, E.D.S.; Sargent, J.R. In <u>Advances in Organic Geochemistry 1981</u>; Douglas, A.G.; Maxwell, J.R., Eds.; Pergamon: Oxford, 1982; pp 219-227.

72. Marlowe, I.T.; Brassell, S.C.; Eglinton, G.; Green, J.C. In <u>Advances in Organic Geochemistry 1983</u>; Schenck, P.A.; de Leeuw, J.W.; Lijmbach, G.W.M., Eds.; Pergamon: Oxford, 1984; pp 135-141.

73. Cranwell, P.A. <u>Geochim. Comochim. Acta</u> 1985, <u>49</u>, 1545-1551.

74. Brassell, S.C.; Brereton, R.G.; Eglinton, G.; Grimalt, J.; Liebezeit, G.; Marlowe, I.T.; Pflaumann, U.; Sartnthein, U. <u>Org. Geochem</u>. 1986, <u>10</u>, 649-660.

75. Brassell, S.C.; Eglinton, G.; Marlowe, I.T.; Sartnthein, U.; Pflaumann, U. <u>Nature (London)</u> 1986, <u>320</u>, 129-133.

76. Farrimond, P.; Brassell, S.C.; Eglinton G. <u>Org. Geochem</u>. 1986, <u>10</u>, 897-903.

77. ten Haven, H.L.; Baas, M.; Kroot, M.; de Leeuw, J.W.; Schenck, P.A.; Ebbing, J. <u>Geochim. Cosmochim. Acta</u> 1987, <u>51</u>, 803-810.

RECEIVED March 5, 1990

Chapter 27

Analysis of Maturity-Related Changes in the Organic Sulfur Composition of Kerogens by Flash Pyrolysis—Gas Chromatography

Timothy I. Eglinton[1,2,4], Jaap S. Sinninghe Damsté[1], Mathieu E. L. Kohnen[1], Jan W. de Leeuw[1], Steve R. Larter[2], and Richard L. Patience[3]

[1]Faculty of Chemical Engineering and Materials Science, Organic Geochemistry Unit, Delft University of Technology, De Vries van Heystplantsoen 2, 2628 RZ Delft, Netherlands
[2]Department of Geology, University of Oslo, P.O. Box 1047, Blindern, N–0316, Oslo 3, Norway
[3]BP Research Centre, Sunbury-on-Thames, TW16 7LN, United Kingdom

Analysis of several sedimentary rock sequences by Pyrolysis-Gas Chromatography (Py-GC) has revealed that thiophene precursors are preferentially removed from kerogen during maturation. This was illustrated by a decrease with increasing maturity in the TR ratio: [2,3-dimethylthiophene]/[n-non-1-ene + 1,2-dimethyl-benzene]. Quantitative Py-GC data for a series of Monterey Fm. kerogens were used to determine *pseudo*-activation energies for the loss of precursors for these species during maturation. Preliminary kinetic calculations gave mean activation energies for the loss of precursors for 2,3-dimethylthiophene, n-non-1-ene and 1,2-dimethylbenzene to be 181, 193 and 203 kJ/mol respectively. The low value for the loss of the thiophene precursors supports the argument that $S_{org.}$ is, at least partly, responsible for the early petroleum generation from S-rich kerogens. Specific changes observed in the distributions of organic sulfur pyrolysis products with increasing maturity include: an increase of alkylbenzothiophenes relative to alkylthiophenes; a shift in the carbon number distributions within these compound classes; and preferential removal of alkylthiophenes with linear carbon skeletons. It is postulated that cyclisation and subsequent aromatisation of kerogen-bound alkylthiophene precursors possessing linear carbon skeletons could account for these maturity-related changes in pyrolysis product distributions. Both the quantitative and qualitative changes in $S_{org.}$ composition were supported by results from the analysis of a suite of artificially matured kerogens.

NOTE: Organic Sulfur in Macromolecular Sedimentary Organic Matter, Part IV
[4]Current address: Fye Laboratory, Department of Chemistry, Woods Hole Oceanographic Institution, Woods Hole, MA 02543

One of the main reasons for the recent interest in organic sulfur in sedimentary organic matter is the growing evidence that it plays a role in the petroleum generation process.   Organically-bound sulfur has been implicated in the formation of low API gravity, asphaltene-rich petroleums at a low level of thermal exposure (1).   Recent studies indicate that organically-bound sulfur in kerogen is formed very early during diagenesis mainly *via* incorporation of inorganic sulfur into functionalised lipids which become part of the kerogen (2).   Indeed, sulfur may facilitate cross-linking of low molecular weight structural units to form insoluble, high molecular weight substances including sulfur-rich kerogen (*i.e.* "natural vulcanisation") (2,3).

It has been recently shown that the most abundant sulfur-containing pyrolysis products of kerogens isolated from immature organic-rich sedimentary rocks are low molecular-weight alkylthiophenes (*i.e.* typically containing 1 to 6 carbon atoms attached to the thiophene ring) (3,4). The abundance of thiophenes relative to aromatic and aliphatic hydrocarbons in flash pyrolysates of kerogens and asphaltenes has been shown to be strongly correlated with organic sulfur content (5).   This suggests that thiophene moieties are a major form of kerogen-bound organic sulfur.   We have therefore proposed a ratio (termed "TR") based on the relative abundance of alkylthiophenes in pyrolysates as a rapid means of estimating organic sulfur content in kerogens (5).   Figure 1 shows a ternary plot of the relative abundance of a thiophene (2,3-dimethylthiophene) in relation to two hydrocarbon products (1,2-dimethylbenzene and *n*-non-1-ene) from pyrolysis-gas chromatography of a wide variety of immature (vitrinite reflectance, R_o < 0.5%) kerogens and coals.   The figure shows that those samples designated "Type II-S" (*i.e.* $S_{org.}/C$ > 0.04; (1)) are easily distinguished from other kerogens. Moreover, the other types of kerogens are also discriminated in such a diagram as expected from previous studies (6,7).   However, maturity also effects sample position in the diagram (5).

The aim of the present investigation was to study the variation in organic sulfur composition of kerogens with maturity.   Samples from four well-known sedimentary sequences (Upper Jurassic shales of the North Sea, Toarcian shales of the Paris Basin, Tertiary coals of the Mahakam Delta and Miocene Monterey shales of offshore California) were analysed by flash pyrolysis-gas chromatography (Py-GC) using both flame ionisation (FID) and sulfur-specific flame photometric detection (FPD) to yield both qualitative and quantitative information on the amounts and types of sulfur-containing pyrolysis products.   In addition, a suite of residues from artificial maturation (hydrous pyrolysis) of Kimmeridge kerogen (8) was analysed for comparison.   For one sequence (Monterey Fm.) quantitative pyrolysis data were used together with information on the geothermal history of the basin to determine values for *pseudo*-kinetic parameters relating to the loss of thiophene precursors from the kerogen.   These

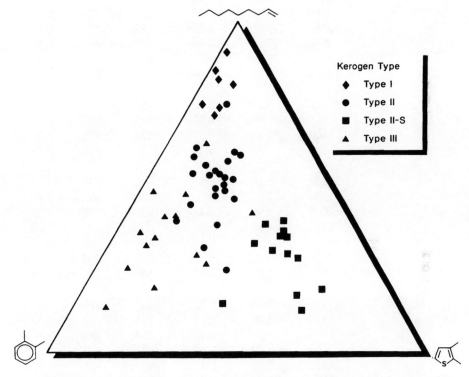

Figure 1. Ternary plot showing the relative abundances of 2,3-dimethylthiophene, *n*-non-1-ene and 1,2-dimethylbenzene (determined from FID chromatograms after flash pyrolysis, 610°C, 10s) of immature kerogens and coals (modified after Eglinton *et al.* (5)).

Table I. Sample details

Name and Location	Depth (m)	TOC (%)	R_o (%)	T_{max} (°C)	HI	TR^a
Mahakam Delta,	993	53.20	0.35	408	116	0.33
Indonesia	1390	66.20	0.38	420	275	0.14
	1920	69.00	0.43	424	278	0.15
	2020	71.10	0.47	420	291	0.13
	2480	74.30	0.52	435	274	0.07
	2600	73.60	0.58	435	277	0.11
	2950	77.50	0.59	442	264	0.12
	3100	77.60	0.65	445	261	0.09
	3140	75.20	0.89	456	116	0.01
Paris Basin,	500	6.50	0.32	427	672	0.23
France[*]	600	8.00	0.33	424	523	0.29
	600	8.10	0.37	424	736	0.26
	700	5.80	0.41	426	601	0.19
	1016	7.10	0.36	427	692	0.25
	1280	6.50	0.38	430	787	0.24
	1320	7.20	0.36	425	662	0.28
	2065	2.90	0.43	436	536	0.27
	2207	n.d.	n.d.	n.d.	n.d.	0.05
	2130	1.30	0.40	435	128	0.04
	2430	7.80	0.42	440	598	0.24
	2633	1.80	0.52	448	80	0.16
Draupne Fm.	1326	2.70	0.51	424	210	0.61
North Sea	1550	4.90	0.45	425	319	0.33
	2065	5.10	0.40	424	422	0.44
	2132	6.70	0.50	426	427	0.37
	2322	7.00	0.47	421	303	0.31
	2911	7.20	0.50	427	344	0.20
	3902	3.20	0.77	429	53	0.11
	4130	4.30	0.76	452	301	0.09
Monterey Fm.	1326	3.80	n.d.	393	458	1.58
CA, U.S.A.	2362	6.40	n.d.	389	553	1.37
	3651	3.60	n.d.	418	503	0.64
	4063	3.50	n.d.	421	411	0.54
	5002	4.10	n.d.	426	335	0.16
	5242	1.70	n.d.	424	238	0.07

[a]TR = [2,3-dimethylthiophene]/[1,2-dimethylbenzene + n-non-1-ene].
n.d. = not determined.   [*]Maximum depths are indicated.

values are discussed in terms of the role of organic sulfur in the decomposition of kerogens and the generation of petroleum.

The results are described in several sections. The first section deals with gross variations in organic sulfur with maturity as determined by Py-GC. The relationship between organic sulfur and Rock-Eval T_{max} is discussed in the second section. In the third section, yields from flash pyrolysis are used to determine kinetic parameters for the loss of precursors for organic sulfur and other compounds from the kerogen during maturation. The last section describes the maturity-related variations in organic sulfur pyrolysis products.

Materials and Methods

<u>Sample description and preparation.</u> The samples analysed in this study are listed in Table I together with available geochemical data. Table II provides general burial history information for the four sedimentary sequences.

Table II. Burial history information for the four sedimentary sequences studied

Sequence	Age (Ma)	Temp. Gradient (°C/Km)	Max. Temp. (°C)
Mahakam	< 25	34	132
Kimmeridge	150	33-35	155
Paris Basin	180-190	34	100
Monterey	< 25	28-35	150

Nine coals from the Mahakam Delta, Kalimantan, Indonesia which were supplied by Institute Francais du Petrole (IFP) formed one sample set. The samples were taken from two wells penetrating this Miocene deltaic sequence, seven samples came from one borehole (Handil 627). The Mahakam Delta has been extensively studied (9-12). It is believed that the coals represent a relatively uniform sequence containing organic matter varying in properties due only to maturity. The kerogen in these samples is considered to be Type III, derived primarily from terrestrial (equiotorial forest) material deposited in the delta. The organic sulfur content of the coals is low (S_{org}/C < 0.01). The vitrinite reflectance values for the sequence range from immature to mature (R_o = 0.35% - 0.89%). The temperature history of the basin is relatively simple with the threshold of oil generation placed at *ca.* 3000 meters. Maturity-related geochemical transformations have been described for related samples from the

Mahakam Delta (12-17). The coal samples were solvent-extracted prior to analysis.

The second suite also supplied by IFP comprised twelve Toarcian shale samples from the Paris Basin, France. These consisted of both outcrop and core samples from the same stratigraphic zone across the basin. These sediments have also been the subject of numerous geochemical investigations (18-20) and are believed to have been deposited under shallow, normal salinity marine conditions, giving rise to a predominantly carbonate lithology and oil-prone Type II/I-II kerogen. Although not measured for the samples used in the present study, the organic sulfur content of the Paris basin kerogens is generally quite low, with atomic S/C ratios ranging from 0.025 (immature) to less than 0.01 (mature). The thermal history of the Paris Basin is more complex than that of the Mahakam Delta due to substantial uplift in the eastern part of the basin. Oil generation is believed to occur below *ca.* 1500 meters. The samples used in the present study were analysed as extracted rocks without prior kerogen isolation. Maturity-related geochemical transformations have been described for related samples from the Paris Basin (21-25).

A set of eight core samples of Upper Jurassic shales from the Norwegian continental shelf of the North Sea were analysed. The samples covered a depth range of *ca.* 1000 - 4000 meters. The North Sea basins have been the subject of numerous geochemical studies (26-32) and sophisticated models describing their thermal history have been constructed. These shales typically contain Type II kerogen of marine origin, however, significant facies variations are observed across the basin. In particular, samples from the basin flanks (*i.e.* the shallowest samples) recieved a significant input of land-derived organic matter as evidenced by relatively low Rock-Eval hydrogen indices (Table I). Although variable, the upper limit of oil generation is typically estimated to be at *ca.* 3000 meters. The core samples were analysed as extracted rocks.

Six samples (provided by BP) from four separate wells penetrating the Miocene Monterey Fm., Santa Barbara Channel Basin were studied. The drill cuttings were obtained from depths varying between *ca.* 1000 and 5300 meters. Previous geochemical investigations of the Monterey Fm. sediments (33-35) indicate that they have been deposited in an upwelling environment and are comprised mainly of biogenic silica, carbonate and phosphate with very little clastic material. The total organic carbon (TOC) contents of these particular sediments varies from *ca.* 2-5%, the most immature samples containing Type II-S kerogen (*e.g.* sample from 1326 m, $S_{org}/C = 0.07$). Because of the very low terrestrial input, reliable determination of vitrinite reflectance was not possible. Independent measurments of maturity (R. Patience, unpublished results) show that these samples range from immature through peak oil generation, the present-day temperature of the deepest sample being approximately 150°C and the upper limit of petroleum generation inferred at *ca.* 3500 meters.   The samples were analysed as isolated kerogens.

The starting material for the artificial maturation experiments was kerogen isolated from an outcrop sample of Upper Jurassic Kimmeridgian shales from Dorset, U.K.   The particular horizon sampled (Blackstone band) is a highly bituminous (TOC = *ca.* 45%) calcareous shale containing immature (R_o = 0.35%, T_{max} = 421°C), sulfur-rich kerogen (S_{org}/C = 0.047).   Several organic geochemical studies on this and related samples from this area have been conducted previously (36-41).   The Dorset Kimmeridge shales are geochemically different from their stratigraphic equivalents in the North Sea, so direct comparisons between the two are not valid.   This is exemplified by the much higher abundance of organic sulfur in the Dorset compared to North Sea Kimmeridge equivalents. Detailed information concerning the conditions employed for the artificial maturation experiments are described elsewhere (8,41).   Briefly, aliquots (*ca.* 250mg) of isolated kerogen were heated in sealed stainless steel vessels in the presence of excess water (*i.e.* "hydrous pyrolysis", 42,43).   In each case the duration of heating at the alloted temperature was 72 hours.   It is suggested that these conditions closely mimic natural petroleum generation (42,43).   Selected geochemical information for the solvent-extracted residues are shown in Table III.

Table III Artificial Maturation (Hydrous Pyrolysis) of Kimmeridge Kerogen (Dorset, U.K.)

Name/Location	Temp. (°C)	%R_o	H/C	TR
Kimmeridge	Unheated	0.35	1.24	0.86
Blackstone,	200	0.18	1.20	0.88
Dorset, U.K.	250	0.20	1.12	0.70
	280	0.28	1.14	0.61
	310	0.58	1.02	0.40
	330	0.95	1.00	0.28
	340	1.13	0.67	0.29
	350	1.24	0.67	0.27
	360	1.86	0.63	0.34

Kerogens were isolated from the Monterey and Dorset Kimmeridge rock samples according to standard procedures involving successive HCl/HF treatments (41).   Elemental analysis of the Monterey Fm. kerogens was performed by the Analytical Division at BP Research Centre (Sunbury, U.K.).   Elemental compositions of the Mahakam coals were provided by IFP.   No differentiation was made in these analyses between total S and organic S, however correction for inorganic S was made for the Mahakam coal samples on the assumption that all Fe was in the form of FeS_2 (pyrite). Rock-Eval pyrolysis was performed either at the University of Oslo, at BP Reseach Centre or at IFP. Vitrinite reflectance determinations (*ca.* 20 measurements/sample) were made at the Organic Geochemistry Unit, University of Newcastle upon Tyne.

Pyrolysis-Gas Chromatography. Flash pyrolysis analyses were performed according to previously described methods (3). In brief, samples were pyrolysed (10 s) on a wire with a Curie temperature of 610°C. The products were separated on a fused silica column (25 m x 0.32 mm i.d.) coated with a methyl silicone phase (CP Sil-5, Chrompack, film thickness 0.45 μm) using a temperature programme from 0°C (5 min.) to 300°C (15 min.) at 3°C/min. The separated products were detected simultaneously by both a flame ionisation detector (FID) and a sulfur-selective flame photometric detector (FPD) using a stream splitter at the end of the chromatographic column. Peak data (sampling rate 6 points/sec.) were acquired and processed using Maxima Chromatography software (Waters Assoc.). Peak assignments were made on the basis of GC retention data and were confirmed using pyrolysis-gas chromatography-mass spectrometry (4). Any partial co-elution of components was compensated for during peak integration.

Absolute quantitation of individual components in the FID chromatograms was achieved using the method described by Larter and Senftle (44) in which a known amount of polymer standard (poly-*p-t*-butylstyrene, Polysciences Inc.) dissolved in dichloromethane was mixed with the sample prior to pyrolysis. This polymer degrades quantitatively and reproducibly to its monomer, *p-t*-butylstyrene (85% conversion under the pyrolysis conditions employed). The calibration of the polymer standard on the Curie-point pyrolysis system will be described separately (45). Yields were determined from triplicate Py-GC analyses. The analyses were performed under only one set of pyrolysis conditions since the relative and absolute amounts of individual pyrolysis products are known to vary with heating conditions. These relationships are currently being investigated.

Variation in Abundance of Pyrolysis Products with Organic Sulfur Content and Maturity

Partial pyrolysis-gas chromatograms of representative immature kerogens or coals from the four sequences studied are shown in Figure 2. The abundance of thiophenes relative to aliphatic and aromatic hydrocarbons in the partial FID chromatograms differs markedly for the four samples shown. This is reflected by the ratio of the peak area of 2,3-dimethylthiophene relative to those due to 1,2-dimethylbenzene and *n*-non-1-ene (*i.e.* TR = [2,3-dimethylthiophene]/[1,2-dimethylbenzene+*n*-non-

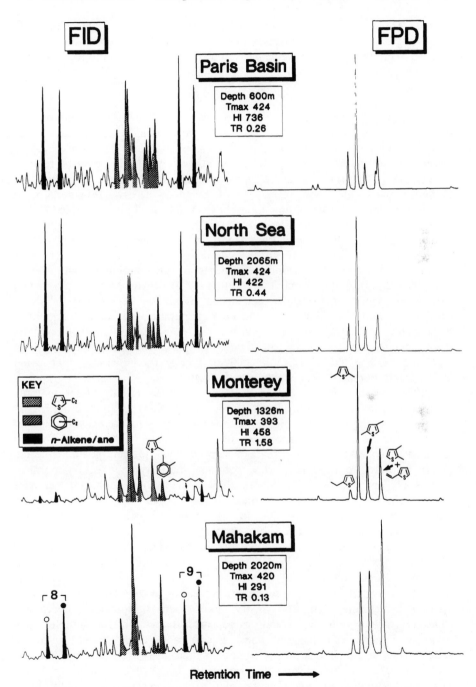

Figure 2. Partial chromatograms (n-C_8 to n-C_9 region) from flash pyrolysis of immature kerogen/coal samples. Open and closed circles in FID chromatograms (left) indicate n-alk-1-enes and n-alkanes respectively. See Materials and Methods for pyrolysis and chromatographic conditions.

1-ene], Table I). This ratio, which is measured from the FID chromatograms after flash pyrolysis, broadly reflects organic sulfur content (5) and therefore indicates large variations in $\%S_{org.}$ for these four sample sets. A detailed description of the relationship between $S_{org.}$ and TR is presented elsewhere (5).

Natural Maturity Sequences. Figure 3 shows partial FID chromatograms from flash pyrolysis of a suite of Monterey Formation kerogens taken from different depths in the Santa Barbara Basin, offshore California. With increasing depth, there is a preferential loss of the C_2-thiophenes relative to the n-C_8 and n-C_9 alkene/alkane doublets and, most noticeably, the C_2-benzenes in the pyrolysates. This trend is clearly reflected by a decrease in the TR ratio from almost 1.6 in the shallowest sample to less than 0.1 in the deepest (Figure 4). Figure 5 shows the variation in absolute yields of 2,3-dimethylthiophene and the two hydrocarbons from flash pyrolysis of kerogens from the Monterey depth sequence. There is a sharp decrease in the absolute amounts of the thiophene with depth. The decrease is more dramatic than that of the hydrocarbons (as would have been predicted from the decreasing thiophene ratio) and therefore provides additional evidence for the early generation of sulfur compounds from sulfur-rich kerogens (1,46). While the yield of 2,3-dimethylthiophene drops to very low values in the deeper samples, hydrogen indices remain high (*e.g.* 335mg HC/g TOC at 5002m, Table I), indicating that some hydrocarbon potential remains in the kerogen.

The variation in the TR with depth for three other natural maturity sequences (Mahakam Delta, Paris Basin, North Sea) is plotted with that for the Monterey Fm. in Figure 4. While the geothermal histories of these sequences are different, there is in each case a strong relationship between maximum temperature experienced by the sample and depth. Depth is therefore used here as a common maturity axis for comparative purposes. Each of these sequences show a slight but significant decrease in this ratio with increasing depth. The extent of the decrease is, however, less dramatic than that for the Monterey sequence due mainly to their lower initial organic sulfur contents.

Figure 6 shows the evolution of samples from the four sequences within a ternary diagram. With the exception of the North Sea samples, a general trend towards the lower left of the diagrams with increasing maturity is evident for each sequence, indicating a relative enrichment in 1,2-dimethylbenzene compared to the alkene and thiophene. There are some reversals within the overall trend, most noticeably for samples lean in organic sulfur (Mahakam, Paris Basin) and these probably reflect combinations of small-scale facies variations and analytical errors. For the Mahakam coals the main process appears to be depletion of n-non-1-ene. Loss of thiophene (in addition to loss of the alkene) is slightly more predominant in the Paris Basin suite. Thiophene loss appears to precede

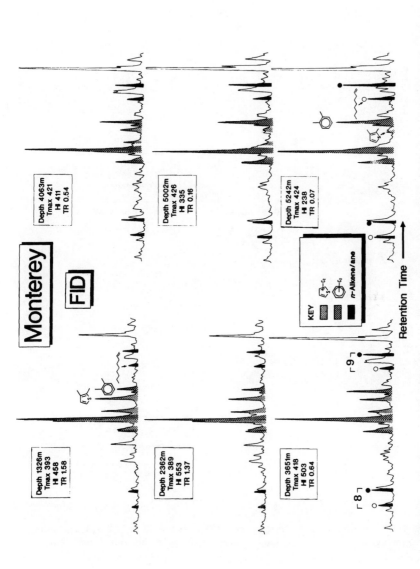

Figure 3. Partial FID chromatograms (*n*-C$_8$ to *n*-C$_9$ region) from flash pyrolysis of Monterey Fm. kerogens, Santa Barbara basin. Open and closed circles indicate *n*-alk-1-enes and *n*-alkanes respectively. Depth of burial, Rock-Eval pyrolysis T$_{max}$, Hydrogen Index (HI), and the thiophene ration (TR) are shown.

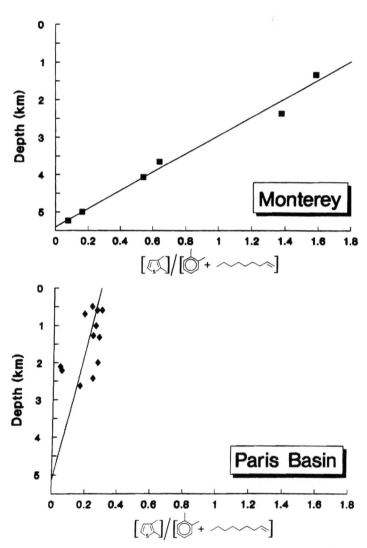

Figure 4. Variation in the thiophene ratio (TR) with depth for samples from four sedimentary sequences. For samples from the Paris Basin, maximum depths of burial from Mackenzie *et al.* (23) were used.

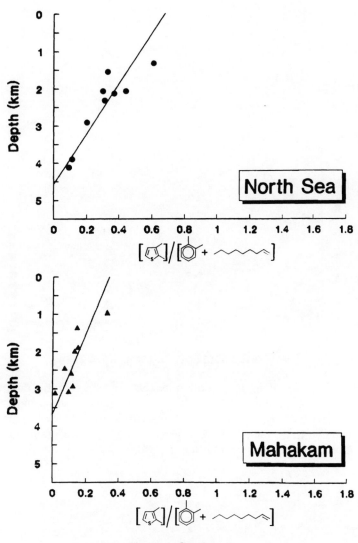

Figure 4. Continued.

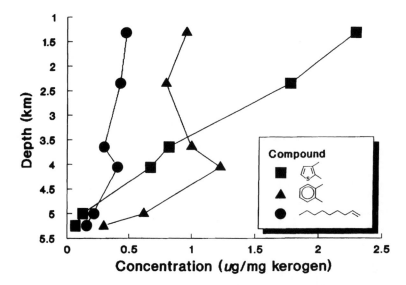

Figure 5.   Variation in absolute yields of 2,3-dimethylthiophene, *n*-non-1-ene and 1,2-dimethylbenzene (μg/mg kerogen) with depth for the Monterey Fm. kerogens.   Yields determined from FID chromatograms after co-pyrolysis of the kerogens with known amounts of poly-*p-t*-butylstyrene (Larter and Senftle, 44).

the loss of the alkene in the North Sea sequence, initially resulting in a relative enrichment of the latter. This trend, *i.e.* an early increase in "aliphaticity" with maturity supports previous observations from Py-GC of kerogens from natural maturity sequences (24,47) and is consistent with the evolution tracks of the principal kerogen types (48).

The trend with depth for the Monterey samples is dominantly due to the loss of thiophene but the maturity trend within the ternary plot (Figure 6) suggests that a significant depletion of the alkene occurs in association with thiophene loss.

Artificial Maturation. Laboratory maturation studies provide a means to determine the influence of temperature on kerogen composition, since other variables (*e.g.* source input) can be eliminated. In order to study the behaviour of organically bound sulfur under these controlled conditions, Py-GC-FID/FPD was performed on a suite of solvent-extracted residues from sealed vessel (hydrous pyrolysis) experiments aimed at simulating maturation over the range involved in petroleum generation.

The unheated Kimmeridge kerogen sample used for the artificial maturation experiments (chromatogram not shown) gave rise, on flash pyrolysis, to abundant thiophenes both in relative and in absolute terms (5). A clear trend has already been observed in flash pyrolysates of these samples in which the ratio, 2-methylthiophene/toluene decreases with increasing artificial maturation temperature or time (8). Figure 7 shows the variation in the TR ratio for extracted residues from artificial maturation at each of a variety of temperatures ranging from 200°C to 360°C (duration of heating, 72 hours). The ratio steadily decreases with increasing maturation temperature up to 330°C. The ratio for residues heated above this temperature remains approximately constant. These results support the findings of Eglinton *et al.* (8). From a ternary plot (Figure 7), it is evident that up to temperatures of 330°C, the main process operative is loss of the thiophene. In the residues from artificial maturation at temperatures above 330°C substantial loss of the alkene also appears to take place, thus giving the appearance of no further depletion of the thiophene and a relatively constant value for TR. The residues from artificial maturation at temperatures between 310°C and 340°C are therefore somewhat analogous to the changes observed in the natural North Sea maturity sequence. The temperatures involved in these experiments are, however, by no means equivalent (being necessarily much higher in the laboratory experiments in order to promote kerogen degradation in a reasonable period of time).

Relationship between Maturity, Rock-Eval T$_{max}$ and Organic Sulfur content

The relationship between Rock-Eval T$_{max}$ and depth for the different sedimentary sequences is depicted in Figure 8. For a given sequence

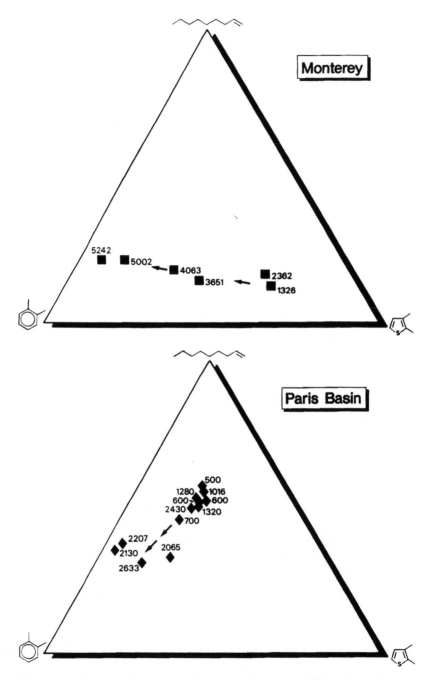

Figure 6. Ternary plots showing the variation in relative abundances of
2,3-dimethylthiophene, *n*-non-1-ene and 1,2-dimethylbenzene with depth
(meters) for samples from the four sedimentary sequences. Maximum
depths of burial are given for samples from the Paris Basin.

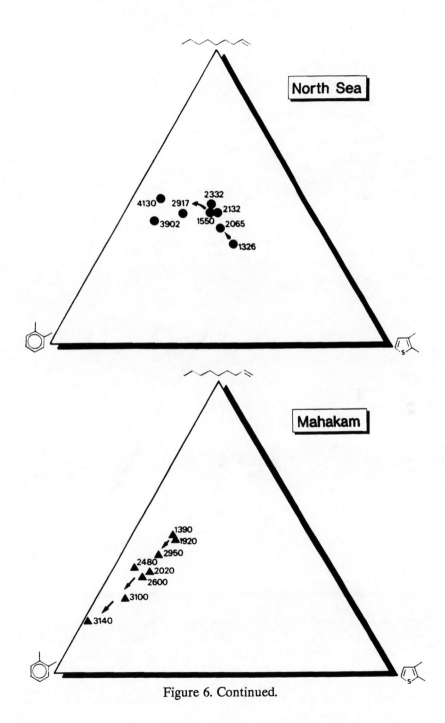

Figure 6. Continued.

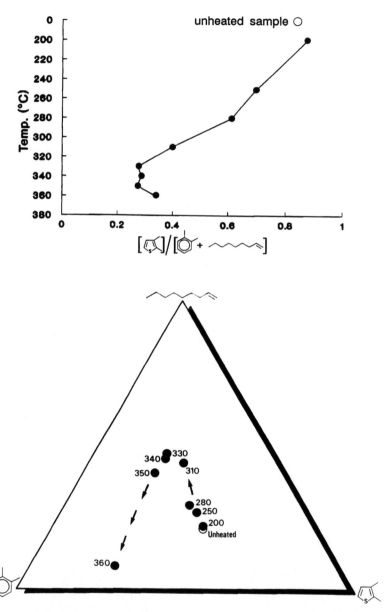

Figure 7. Variation in the thiophene ratio (TR; upper diagram) and Ternary plot showing the variation in the relative abundances of 2,3-dimethylthiophene, *n*-non-1-ene and 1,2-dimethylbenzene (lower diagram) with artificial maturation temperature for a suite of kerogen residues obtained from hydrous pyrolysis (72 hrs) of Kimmeridge kerogen (Dorset, U.K.). NB. Integrated peak areas for 2,3-dimethylthiophene also include 2-vinylthiophene.

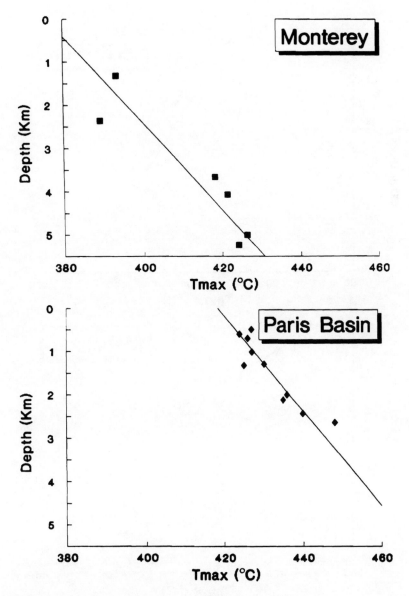

Figure 8. Relationship between Rock-Eval pyrolysis T_{max} and depth for samples from the four sedimentary sequences (modified after Eglinton et al., 5). *Continued on next page.*

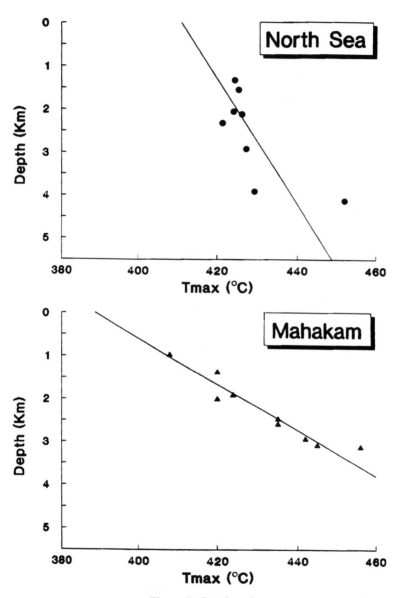

Figure 8. Continued.

T_{max} increases with increasing depth. However, examination of a large number of samples differing widely in organic sulfur content (as measured by the thiophene ratio), but all of low maturity gives the relationship in Figure 9. Samples with the highest thiophene ratio (organic sulfur content) give rise to the lowest T_{max} values. Rock-Eval analyses of solvent-extracted, isolated kerogens as well as whole rock samples demonstrate that the low T_{max} values from thiophene-rich samples are not due to bitumen adsorption effects (*i.e.* tightly held bitumen contributing to the "S_2" peak). The samples shown in Figure 9 are all at the pre-oil generation stage and fall within a relatively narrow reflectance range (R_o = 0.3-0.5%). The figure shows that organic sulfur content influences the thermal lability of the kerogen (as measured by T_{max}). These results indicate that T_{max} should be used cautiously as a source of maturity information for rock sequences with marked facies variations.

Kinetic Analysis of Quantitative Pyrolysis Data

In order to better understand the factors controlling the loss of thiophenic and other moieties from the kerogens during maturation *pseudo*-kinetic parameters for loss of the different precursor species from the kerogens were determined. It is now generally agreed that kinetic analysis of geochemical processes involved in kerogen maturation requires a distributed activation energy solution (49-51). Attempts to simulate complex reaction systems using single reactions invariably lead to absurd, heating rate dependent parameters of no chemical significance (49). It is preferable to monitor rates of reactant (kerogen-bound species) loss, rather than product (bitumen species) generation. This is because primary migration from mature, organic-rich rocks is pervasive and efficient (52,53). Thus, bitumen species concentrations are not a reliable indicator of production rates, and it follows that kinetic parameters derived from bitumen analyses will reflect these errors.

In the present study the approach of Larter (51) was used. The initial concentrations of kerogen-bound species (precursors for 2,3-dimethylthiophene, 1,2-dimethylbenzene and *n*-non-1-ene) were determined based on analysis of immature Monterey Fm. kerogens using quantitative Py-GC methods. Concentrations of these species in the kerogen were calculated assuming that Py-GC yields are directly proportional to the actual precursor species concentrations in the kerogens (54) and that this proportionality is invariant with maturity. Concentrations of precursor species for mature kerogens were determined in the same manner but were normalised to initial kerogen concentration by allowing for loss of material during maturation. This correction was performed on the assumption that the amount of inert material in the immature and mature kerogens was constant (52,53). Estimates of inert organic matter were based on "Rock-Eval" hydrogen index determinations.

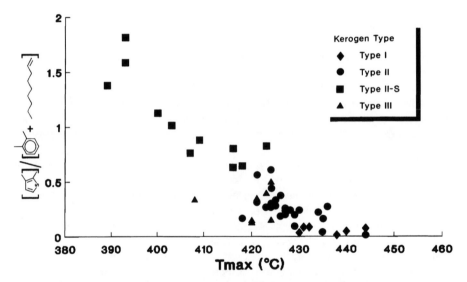

Figure 9. Relationship between the thiophene ratio, TR, and Rock-Eval T_{max} for samples of varying kerogen type. NB. All samples are immature with respect to oil generation (i.e. vitrinite reflectance, R_0 < 0.5%).

It was assumed that kerogen decomposition was governed kinetically by first order reactions proceeding with, initially, a Gaussian distribution of activation energies with a mean activation energy E_m and a standard deviation σ. Using the method of Larter (51) the kinetic parameters are derived using regression methods by comparing the quadratic errors between the measured concentrations of components (corrected to initial kerogen concentration) with a computed concentration based on geologically determined subsurface heating rates and maximum burial temperatures. When this was performed it was found that the six data points available from the Monterey Fm. were insufficient to adequately constrain all three parameters (*i.e.* too few data points, too much data scatter). Therefore a single, mean pre-exponential frequency factor of 1×10^{13}/sec. was assumed for all three reactions and the best fit E_m and σ values were re-computed. These values are shown in Table IV.

Table IV. Kinetic Parameters for the Loss of Thiophene and Hydrocarbon Precursors from Monterey Fm. Kerogens

Compound	E_m (kJ/mol)	σ ($\%E_m$)
2,3-Dimethylthiophene	181	8
1,2-Dimethylbenzene	203	8
n-Non-1-ene	193	12

Pre-exponential factor, $A = 1 \times 10^{13}$/sec. (assumed)

These preliminary calculations confirm the conclusion obtained by inspection, *i.e.* that the loss of thiophenic species from kerogens requires a lower average activation than does hydrocarbon generation. However, all reactions show significant distributions about the mean with σ ranging from 8 to 12% E_m (Table IV). With these limited data it is probable that this standard deviation represents an error assessment rather than a true estimate of the actual activation energy distribution. With the data available it is unwise to pursue these results too far and more rigorous analysis will await additional data to better confine the kinetic solutions. Nevertheless, several independent observations appear to support the kinetic results from the present study. Gransch and Posthuma (46) noted that the early generation products from artificial maturation of a sulfur-rich kerogen (La Luna, Venezuela) contained much more sulfur than did the later products. This trend was also observed by Eglinton *et al.* (41) for the organic-soluble products from artificial maturation of Kimmeridge kerogen.

As the thiophene ring is rather stable, the early release of these species suggests that the weak linkages are in the alkyl chains bonding the thiophenes to the kerogen.   As there is evidence for the presence of aliphatic carbon-sulfur bonds in condensed organic matter (55) it is quite possible that these weak linkages may be related to such.

Comparison of these kinetic parameters with those for bulk "hydrocarbon" generation from source rocks containing Type II kerogen, using similar distributed activation energy kinetic models (49,50) is risky because our data are insufficient to adequately confine the frequency factors.   Assuming, however, that the frequency factor estimate is approximately correct, the value for loss of-the thiophene is significantly lower than that of the average activation energy of petroleum generation, even from Type II-S kerogens (56).   This might be expected, since thiophene loss is probably partly responsible for a shift of the activation energy distribution to lower values.   However, the activation energy for loss of n-non-1-ene (Table IV) is also substantially lower than that compared to E_a values for generation from highly aliphatic (Type I) kerogens, which are typically in the order of 250 kJ/mol (56,57).   This may reflect the limited data set used for the present study.   Nevertheless, a concomitant loss of the alkene in association with the thiophene was also apparent in Figure 6.   This suggests that in the Monterey kerogens the nature of the alkene/alkane precursors (as exemplified by n-non-1-ene) is different from those in the other kerogens studied.   It has been postulated that one precursor of alkenes/alkanes in sulfur-rich kerogens may result from natural vulcanisation of n-alkyl-containing substrates (58,59). Due to the relative weakness of the carbon-sulfur bond it is likely that alkenes/alkanes would be generated at a lower thermal stress from these precursors   than   from   kerogens   containing   alkylchains   linked   by predominantly carbon-carbon bonds.   While Type II-S kerogens differ structurally from Type II kerogens, the general similarity between overall yields of n-hydrocarbon moieties on pyrolysis (60) suggests that we do not yet fully understand the influence of organic sulfur on alkyl group incorporation into kerogens.   We are currently extending these kinetic studies to a larger sample set and to different sedimentary sequences to determine the importance of these different influences.

Variation in distributions of organic sulfur pyrolysis products with maturity

Natural Maturity sequences.   Figure 10 shows FPD chromatograms of the pyrolysates of seven Mahakam coal samples studied by flash pyrolysis. Peak identifications are listed in Table V.

Even in these relatively sulfur-lean samples, organic sulfur pyrolysis products are detected throughout the maturity range studied.   The FPD chromatogram of the most immature sample is dominated by C_1-C_4 alkylthiophenes.   Compared to a wide variety of kerogens, pyrolysates of the Mahakam coals tend to contain relatively high amounts of thiophenes substituted at positions 3 and 4 (61).

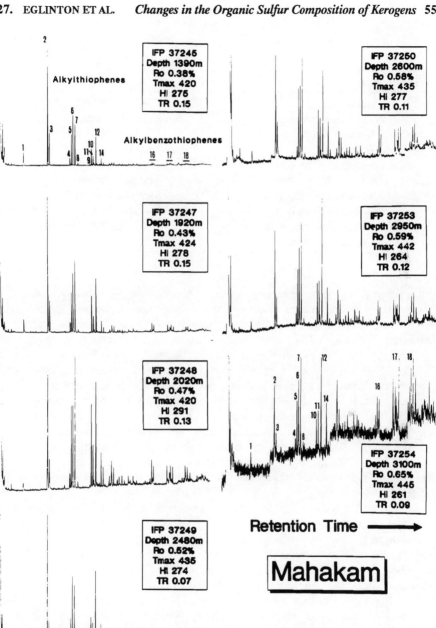

Figure 10. FPD chromatograms from flash pyrolysis of Mahakam coals. Sample depth, vitrinite reflectance, Rock-Eval T_{max}, hydrogen index (HI) and the thiophene ratio (TR) are shown. Peak identifications are given in Table V.

Table V. Peak identifications for FPD chromatograms.

Peak	Compound	Peak	Compound
1	Thiophene	10	2-Ethyl-5-methylthiophene
2	2-Methylthiophene	11	2-Ethyl-4-methylthiophene
3	3-Methylthiophene	12	2,3,5-Trimethylthiophene
4	2-Ethylthiophene	13	2-Methyl-5-vinylthiophene
5	2,5-Dimethylthiophene	14	2,3,4-Trimethylthiophene
6	2,4-Dimethylthiophene	15	Benzo[b]thiophene
7	2,3-Dimethylthiophene	16	Methylbenzo[b]thiophenes
	(+ 2-Vinylthiophene)	17	C_2-Benzo[b]thiophenes
8	3,4-Dimethylthiophene	18	C_3-Benzo[b]thiophenes
9	2-Propylthiophene		

Several distinct maturity-related trends emerge through detailed examination of the traces (Figure 10). There are variations in both the relative abundance of different compound classes and within individual compound classes. The most marked trend is an increase in abundance of alkylbenzothiophenes relative to alkylthiophenes. This is clearly seen in Figure 11 where the ratio of these two compound classes is plotted with respect to depth. There is also an apparent shift in the internal carbon number distributions of both the alkylbenzothiophenes and alkylthiophenes. Specifically, there appears to be a preferential loss of short-chain alkylthiophenes and alkylbenzothiophenes (*e.g.* methylthiophenes decrease relative to dimethylthiophenes). The exception to this depth trend is the sample from 2480m which appears less mature (on the basis of sulfur compound distribution) than the adjacent sample of shallower depth (2020m).

A number of trends in isomer distributions for a given compound class are also evident with maturity. On the basis of previous observations (3) it appears that alkylthiophenes derived from precursors with a linear carbon skeleton are preferentially depleted relative to those derived from branched thiophenic precursors in the coal. For example, there is a decrease in the amount of 2-ethyl-5-methylthiophene relative to 2-ethyl-4-methylthiophene (Peaks 10 and 11 respectively, Figure 10) with increasing depth. The trend with maturity is illustrated in Figure 12 where the ratio of short-chain thiophenes of "branched" origin (*i.e.* at least one substitution at position 3 or 4 on the thiophene ring) relative to those of "linear" origin (*i.e.* substituted at positions 2 and/or 5 only) is plotted *versus* depth. Isomer trends within the alkylated benzothiophenes were generally more difficult to see because of their lower relative abundance.

With regard to the Monterey kerogens there are some similar trends to those observed for the Mahakam coals although the transformations are

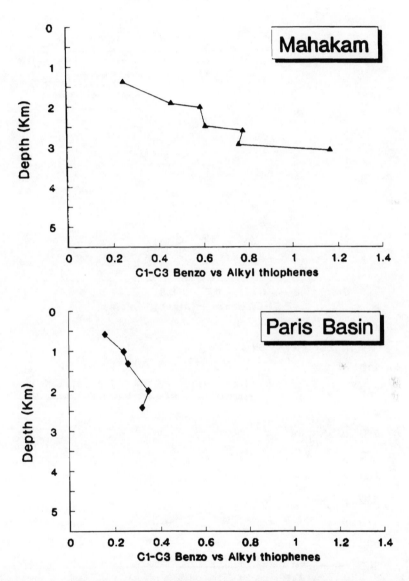

Figure 11. Variation in the ratio of ΣC_1-C_3 alkylbenzo[b]thiophenes (Peaks 16-18, Figs. 10 and 13) relative to ΣC_1-C_3 alkylthiophenes (Peaks 2-14, Figs. 10 and 13) with respect to depth or temperature of artificial maturation. Maximum depths of burial are used for samples from the Paris Basin. Values determined from peak height data, corrected for quadratic response of FPD. *Continued on next page.*

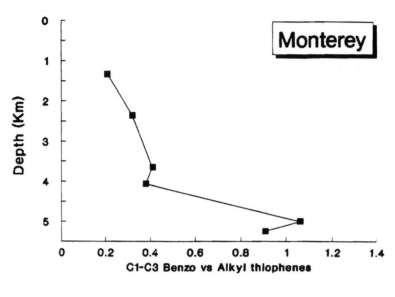

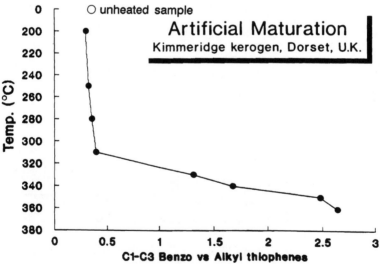

Figure 11. Continued.

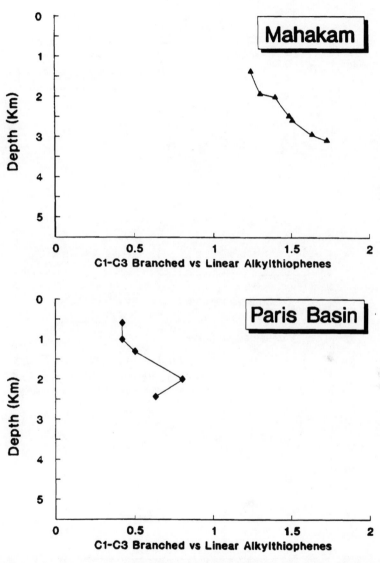

Figure 12. Variation in the ratio of C_1-C_3 alkylthiophenes derived from precursors with branched carbon skeletons (Peaks 3,6,7,8,11,12,14, Figs. 10 and 13) relative to those with linear carbon skeletons (Peaks 2,4,5,9,10,13, Figure 10 and 13) with respect to depth or temperature of artificial maturation. Note: For samples from the Paris Basin maximum depths of burial were used. Values determined from peak height data, corrected for quadratic response of FPD. *Continued on next page.*

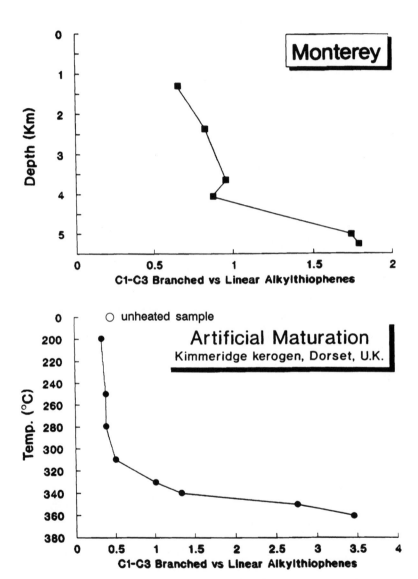

Figure 12. Continued.

generally smaller and less consistent. Examples include a carbon number shift in the alkylthiophenes, an increase in alkylbenzothiophenes relative to alkylthiophenes (Figure 11) and an increase in the ratio of "branched" *versus* "linear" thiophenes (Figure 12).

The trends described above can, to a certain extent, also be discerned in the Paris Basin maturity sequence (see Figures 11 and 12) but facies variations, which have been previously noted for these samples (62), also appear to significantly influence the distributions.

The samples from the North Sea were analysed as extracted whole rocks. The low TOC content of these samples prevented analysis by dual FID/FPD. Descriptions of maturity-related changes in kerogens isolated from these shales will be given elsewhere.

Artificial Maturation. Figure 13 shows FPD chromatograms from flash pyrolysis of the series of artificially matured kerogens. Analyses of these samples revealed several trends consistent with those observed in the natural maturity sequences. The dominance of 2-methyl-5-alkylthiophenes in the pyrolysate of the unheated sample appears to be characteristic of immature kerogens of marine origin (3,61). With increasing artificial maturation temperature there is an increase in the amounts of alkylbenzothiophenes relative to alkylthiophenes (Figure 11) and a depletion of the C_1-substituted homologues of both compound classes. Similar observations were reported previously for a number of these samples (8). There is also a change in the isomer distribution in favour of those derived from moieties with "branched" relative to "linear" carbon skeletons as noted for the natural maturity sequences (Figure 12).

The fact that the above tranformations are evident both in the artificial as well as the natural maturation series strongly suggests that they are temperature dependent and are not simply facies-related. One process which could explain the three maturity-related trends (*i.e.* increase in alkylbenzothiophene/thiophene ratio; preferential loss of C_1-substituted homologues; depletion of thiophenes with a "linear" as opposed to a "branched" origin) is cyclisation and subsequent aromatisation of alkylthiophene precursors in the kerogen with linear carbon skeletons. This would explain the increased importance of alkylbenzothiophenes in the pyrolysates of mature kerogens and, since alkylthiophenes with linear carbon skeletons are commonly the dominant isomers, their conversion to alkylbenzothiophenes would cause the observed carbon number shift. The preferential aromatisation of linear thiophene precursors could arise partly because 2,3-disubstituted thiophene units cannot aromatise. Other branched precursors, such as 2,4-disubstituted thiophene units, may aromatise but would give rise, on pyrolysis, to alkylbenzothiophenes with substitution patterns which differ from those commonly observed in kerogen pyrolysates (3,4). In contrast, alkylbenzothiophenes with substitution patterns consistent with aromatisation of linear alkylthiophenes are usually the dominant

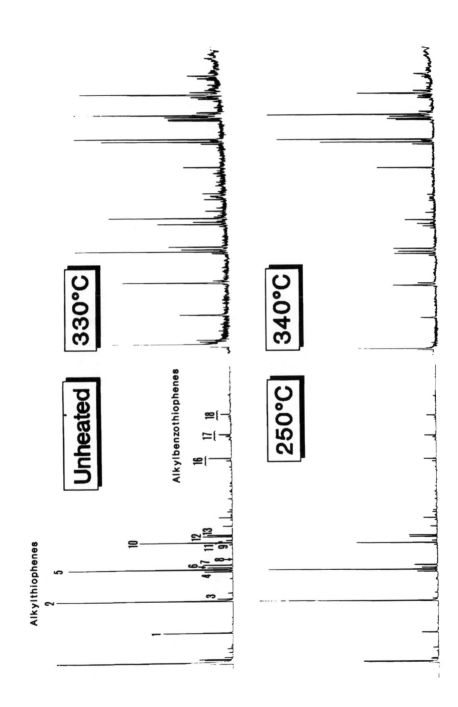

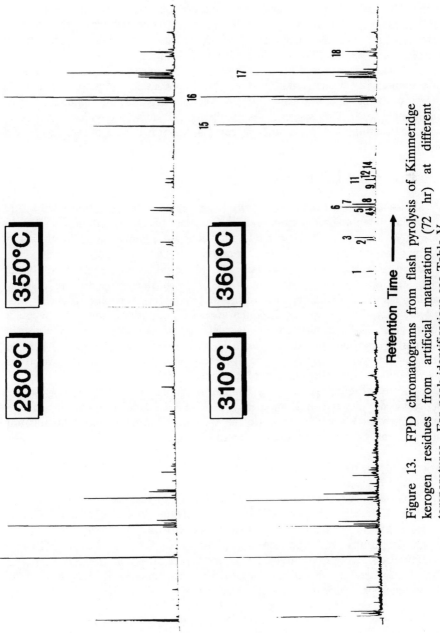

Figure 13. FPD chromatograms from flash pyrolysis of Kimmeridge kerogen residues from artificial maturation (72 hr) at different temperatures. For peak identifications see Table V.

isomers present in kerogen pyrolysates and immature oils (4,63-65), thus lending support to this hypothesis.   The reasons why 2,4-disubstituted thiophene units do not appear to aromatise is not completely understood. It may be that either the substitution pattern on the thiophene ring affects the aromatisation reaction (*cf.* differences in mass spectral behaviour of various thiophene isomers) or that these units originate partly from steroidal side-chains (3) and thus cannot aromatise.

While the absolute amounts of alkylthiophenes decreased drammatically with increasing maturity in the Monterey sequence there was only a slight increase in the ratio of alkylbenzothiophenes *versus* alkylthiophenes.   This suggests that while cyclisation/aromatisation may occur with maturation, the dominant process is loss of labile, organically-bound sulfur.

Conclusions and Geochemical Implications

Pyrolysis-gas chromatography shows that precursors for thiophenes are preferentially eliminated from kerogen during burial maturation.   Results from the preliminary kinetic study implicate thiophenic sulfur in the early thermal degradation of sulfur-rich kerogen.   Whilst this analysis is far from conclusive, further kinetic studies of this type, with larger and better calibrated data sets may enable us to make a more objective assessment of the chemical characteristics of bonds being cleaved during maturation processes.   Monitoring the disappearance of thiophenes and other specific compounds (*e.g.* phenols) from pyrolysates of samples over a maturity range may provide important information regarding the kinetics of kerogen degradation, particularly for Type II-S kerogens.   The systematic variations with depth in the distributions of thiophenes released on pyrolysis do not hold much promise as maturity indicators because of their strong source dependence.   Nevertheless, they may provide useful information on changes in kerogen structure with maturity as well as allowing further classification of kerogens according to organic sulfur composition.

Acknowledgments

The following organisations are thanked for their kind donation of samples: Institute Francais du Petrole, British Petroleum, Norsk Hydro.   We are grateful to the analytical departments at Bristol University and at BP Research Centre for the elemental analyses. Dr. A. Pepper (BP) is acknowledged for provision of burial history information for the Monterey Fm. kerogens.   L. Buxton (Geolab Nor) is thanked for selected TOC determinations.   Dr. J.M. Jones (University of Newcastle) is thanked for vitrinite reflectance analyses.   We thank Dr. A.K. Burnham for supplementary kinetic analyses.   W. Pool is thanked for technical assistance. Drs. K. Peters, W.L. Orr and C.M. White are thanked for critical reviews.

Literature cited

1. Orr, W.L. Org. Geochem. 1986, 10, 499-516.
2. Sinninghe Damsté, J.S.; Rijpstra, W.I.C.; de Leeuw, J.W.; Schenck, P.A. Geochim. Cosmochim. Acta 1989, 53, 1343-1355.
3. Sinninghe Damsté, J.S.; Eglinton, T.I.; de Leeuw, J.W.; Schenck, P.A. Geochim. Cosmochim. Acta 1989, 53, 873-889.
4. Sinninghe Damsté, J.S.; Kock-van Dalen, A.C.; de Leeuw, J.W.; Schenck P.A. J. Chromatogr. 1988, 435, 435-452.
5. Eglinton, T.I.; Sinninghe Damsté, J.S.; Kohnen M.E.L.; de Leeuw J.W. Fuel 1989 (submitted).
6. Larter, S.R.; Douglas, A.G. In Advances in Organic Geochemistry, 1979; Douglas, A.G.; Maxwell J.R., Eds.; Pergamon Press: Oxford, 1980, pp. 579-584.
7. Larter, S.R. In Analytical Pyrolysis - Methods and Applications; Voorhees, K. Ed.; Butterworths: London, 1984, pp. 212-275.
8. Eglinton, T.I.; Philp, R.P.; Rowland, S.J. Org. Geochem. 1988, 12, 33-41.
9. Magnier, P.L.; Oki, T.; Kartaaipoetra, L. Proc. 9th World Pet. Congr. 1975, 2, 239-250.
10. Vandenbroucke, M.; Durand, B. In Advances in Organic Geochemistry, 1981. Bjorøy, M. et al. Eds. 1983, pp. 147-155. Wiley and Sons Ltd.
11. Schoell, M.; Teschner, M.; Wehner, H.; Durand, B.; Oudin, J.C.; In Advances in Organic Geochemistry, 1981. Bjorøy, M. et al. Eds.; Wiley and Sons, 1983, pp. 156-163.
12. Mackenzie, A.S.; Mackenzie, D. Geol. Mag. 1983, 120, 417-528.
13. Combaz, A.; Matharel, M. Bull. Amer. Assoc. Petrol. Geol. 1978, 62, 1684-1695.
14. Boudou, J.P. Org. Geochem. 1984, 6, 431-437.
15. Boudou, J.P.; Durand, B.; Oudin, J.L. Geochim. Cosmochim. Acta 1984, 48, 2005-2010.
16. Hoffman, C.F.; Mackenzie, A.S.; Lewis, C.A.; Maxwell, J.R.; Oudin, J.L.; Durand, B.; Vandenbroucke M. Chem. Geol. 1984, 42, 1-23.
17. Monthioux, M.; Landais, P.; Monin, J-C. Org. Geochem. 1985, 8, 275-292.
18. Durand, B.; Espitalie, J.; Nicaise, G.; Combaz, A. Rev. Inst. Francais du Petr. 1972, 865-884.
19. Alpern, B.; Cheymol, D. Rev. Inst. Francais du Petr. 1978, 515-535.
20. Mackenzie, A.S.; Hoffman, C.F.; Maxwell, J.R. Geochim. Cosmochim. Acta 1981, 45, 1345-1355.
21. Louis, M.C. In Advances in Organic Geochemistry, 1964 Hobson, G.D.; Louis, M.C. Eds.; Pergamon Press: Oxford, 1966, pp. 85-94.
22. Tissot, B.; Califert-Debyser, Y.; Deroo, G.; Oudin, J.L. Bull. Amer. Assoc. Petrol. Geol. 1971, 58, 499-506.

23. Mackenzie, A.S.; Patience, R.L.; Maxwell, J.R.; Vandenbroucke, M.; Durand, B. Geochim. Cosmochim. Acta 1980, 44, 1709-1721.
24. van Graas, G.; de Leeuw, J.W.; Schenck, P.A.; Haverkamp, J. Geochim. Cosmochim. Acta 1981, 45, 2465-2474.
25. Espitalie, J. In Petroleum Geology of North West Europe; Brooks, J.; Glennie, K.W. Eds.; Graham and Trotman: London, 1987, pp. 71-86.
26. Thomas, B.M.; Moller-Pedersen, P.; Whiticar, M.F.; Shaw, N.D.; In Petroleum Geochemistry in Exploration of the Norwegian Shelf. Thomas, B.M. et al. Eds.; Graham and Trotman: London, 1985, 3-26.
27. Pearson, M.J.; Watkins, D.; Small, J.S.; In Developments in Sedimentology 35 van Olphen, H.; Veriale, F. Eds.; Elsevier: Amsterdam, 1982, pp. 665-675.
28. Dypvik, H.; Rueslatten, H.G.; Throndsen, T. Bull. Amer. Assoc. Petrol. Geol. 1979, 63, 2222-2226.
29. Barnard, P.C.; Cooper, B.S. In Petroleum Geology of the Continental Shelf of N.W. Europe; Illing, L.V.; Hobson, G.D. Eds.; Institute of Petroleum: London, 1981, pp. 169-175.
30. Illing, L.V.; Hobson, G.D. Eds. Petroleum Geology of the Continental shelf of N.W. Europe; Institute of Petroleum, London 1981,
31. Goff, J.C. J. Geol. Soc. Lond. 1983, 140, 445-474.
32. Mackenzie, A.S.; Maxwell, J.R.; Coleman, M.L.; Deegan C.E. Proc. 11th World Petrol. Cong. 1983, 2, 45-56.
33. Isaacs, C.M. Oil and Gas J. 1984, 82, 75-81.
34. Curiale, J.A.; Cameron, D.; Davis, D.V. Geochim. Cosmochim. Acta 1985, 49, 271-288.
35. Kruge, M.A. Org. Geochem. 1986, 10, 517-530.
36. Farrimond, P.; Comet, P.; Eglinton, G.; Evershed, R.P.; Hall, M.A.; Park, D.W.; Wardroper, A.M.K.; Mar. Petrol. Geol. 1984, 1, 340-354.
37. Williams, P.F.V.; Douglas, A.G.; Fuel 1986, 65, 1728-1734.
38. Williams, P.F.V.; Mar. Petrol. Geol. 1986, 3, 258-281.
39. Gallois, R.W. Rep. Inst. Geol. Sci. 1978, 78, 13.
40. Douglas, A.G.; Williams, P.F.V. In Organic maturation studies and fossil fuel exploration. Brooks, J. Ed.; Academic Press, 1981, 255-269.
41. Eglinton, T.I.; Douglas, A.G.; Rowland, S.J. Org. Geochem. 1988, 13, 655-663.
42. Lewan, M.D.; Winters, J.C.; McDonald J.H. Science 1979, 203, 897-899.
43. Lewan, M. Phil. Trans. R. Soc., Lond. 1985, A315, 123-134.
44. Larter, S.R.; Senftle, J.T. Nature (London) 1985, 318, 277-280.
45. Kohnen, M.E.L.; Eglinton, T.I.; Sinninghe Damsté, J.S.; de Leeuw, J.W. 1989 In Prep.

46. Gransch, J.A.; Posthuma, J. In Advances in Organic Geochemistry, 1973; Tissot, B.; Bienner, F. Eds.; Editions Technip: Paris, 1974, pp. 727-739.
47. Solli, H.; van Graas, G.; Leplat, P.; Krane, J. Org. Geochem. 1984, 6, 351-358.
48. Tissot, B.P.; Welte, D.H. Petroleum Formation and Occurrence (Second Edition). Springer-Verlag: Berlin, 1984, 699pp.
49. Braun, R.L.; Burnham, A.K. Energy and Fuels, 1987, 1, 153-161.
50. Quigley, T.M.; Mackenzie, A.S.; Gray, J.R. In Migration of hydrocarbons in sedimentary basins Doligez, B. Ed.; Editions Technip: Paris, 1987, pp. 649-655.
51. Larter, S.R. Geol. Rund. 1989, 78, 1-11.
52. Cooles, G.P.; Mackenzie, A.S.; Quigley, T.M. Org. Geochem. 1986, 10, 235-246.
53. Larter, S.R. Marine and Petrol. Geol. 1988, 5, 193-204.
54. Larter, S.R.; Horsfield, B. In Organic Geochemistry; Engel, M.; Macko, S. Eds. 1990, In Press
55. Boudou, J.P.; Boulegue, J.; Malechaux, L.; Nip, M.; de Leeuw, J.W.; Boon, J.J. Fuel 1987, 66, 1558-1569.
56. Tissot, B.P.; Pelet, R.; Ungerer, P. Bull. Amer. Assoc. Petrol. Geol. 1987, 71, 1445-1466.
57. Burnham, A.K.; Braun, R.L.; Gregg, H.R.; Samoun, A.M. Energy and Fuels 1987, 1, 452-458.
58. de Leeuw, J.W.; Sinninghe Damsté, J.S. In Geochemistry of Sulfur in Fossil Fuels; Orr, W.L.; White, C.M., Eds. ACS Symposium Series; American Chemical Society: Washington, DC, 1989; this volume.
59. Sinninghe Damsté, J.S.; Eglinton, T.I.; Rijpstra, W.I.C.; de Leeuw, J.W. In Geochemistry of Sulfur in Fossil Fuels; Orr, W.L.; White, C.M., Eds. ACS Symposium Series; American Chemical Society: Washington, DC, 1989; this volume.
60. Larter, S.R. In Petroleum Geochemistry in Exploration of the Norwegian Shelf; Thomas, B.M. et al. Eds. Graham and Trotman: London, 1985, pp. 269-286.
61. Eglinton, T.I.; Sinninghe Damsté, J.S.; de Leeuw, J.W.; Boon, J.J. 1989 Abstract, 14th International Meeting on Organic Geochemistry, Paris Sept. 1989.
62. Goossens, H.; Due, A.; de Leeuw, J.W.; van de Graaf, B.; Schenck P.A. Geochim. Cosmochim. Acta 1988, 52, 1189-1193.
63. Sinninghe Damsté, J.S.; de Leeuw, J.W.; Kock-van Dalen, A.C.; de Zeeuw, A.; de Lange, F.; Rijpstra, W.I.C.; Schenck P.A. Geochim. Cosmochim. Acta 1987, 51, 2369-2391
64. Sinninghe Damsté, J.S.; Rijpstra, W.I.C.; de Leeuw, J.W.; Schenck, P.A. Geochim. Cosmochim. Acta 1989, 53, 1323-1341.
65. Perakis, N. Ph.D. Thesis, University of Strasbourg, Strasbourg, 1986.

RECEIVED March 5, 1990

SPECIAL STUDIES

Chapter 28

Isotopic Study of Coal-Associated Hydrogen Sulfide

J. W. Smith[1] and R. Phillips[2]

[1]Division of Exploration Geoscience, Commonwealth Scientific and Industrial Research Organisation, P.O. Box 136, North Ryde, N.S.W. Australia 2113
[2]Capricorn Coal Management Pty., Ltd., Private Bag, Middlemount, Queensland, Australia 4745

As much as 0.1% by weight of an Australian bituminous coal occurs as hydrogen sulfide. The gas is tightly held within the coal structure and is only detectable by odour on breakage of the coal. $^{34}S/^{32}S$ ratios 1) discount sulfate reduction as the mechanism for sulfide production and 2) suggest the gas is generated at an early stage of diagenesis by the cracking of thermally unstable S-S and C-S linkages in organic compounds.

Occurrences of hydrogen sulfide in coal seam gases are rare. Within Australia, only hydrogen sulfide at the Collinsville Colliery in the Bowen Basin has previously been investigated isotopically (1). The widespread invasion of that mine by carbon dioxide, derived externally to the coal, precluded definition of precise hydrogen sulfide sources. The opportunity to gain a better understanding of the factors leading to hydrogen sulfide generation within coal seams was provided by its release in mining operations at the Southern Colliery, Qld. This mine, unlike many in the Bowen Basin, is unaffected by externally generated carbon dioxide and thus transport into the mine workings of hydrogen sulfide in association with that gas could be disregarded.

At the Southern Colliery, located in the Bowen Basin, Central Queensland, Figure 1, coal is won from the Permian, German Creek Measures at a depth of some 200 to 350 feet. The immediate roof of the seam is comprised of highly quartzose sandstone, strongly cemented by secondary silica overgrowths and most probably of marine origin. The coal is of bituminous rank with a vitrinite reflectance of 1.4%. The quantity of H_2S is insignificant in comparison with those encountered in sour natural gas wells, but, as the gas is released in confined workings, it poses a real discomfort and health hazard in mining operations.

Interesting facets of this gas occurrence are that 1) it is limited to particular sections of the workings, and 2) in hydrogen sulfide-rich zones, this gas is scarcely detectable by odour until breakage of the coal.

0097–6156/90/0429–0568$06.00/0
© 1990 American Chemical Society

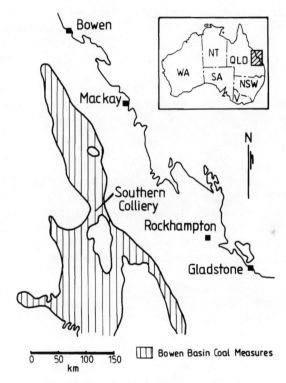

Figure 1. Location of Southern Colliery

In efforts to locate and identify the source of this gas, and more importantly, to be able to predict when and where it might be encountered, geological and scientific staff at the mine have examined coal from the two zones both chemically and petrographically. Specific factors which might be related to the occurrence of the gas have not been identified. Certainly the total sulfur content of the coal is variable to 3.0%, due to the random occurrence of massive pyrite, but, in general, total sulfur contents are less than 1.1%, with organic sulfur contents in the range of 0.4% to 0.8%. In addition, geological factors, including faults, intrusions etc., within the mine, cannot be related to the presence of H_2S. Measurements of the hydrogen sulfide content of the coal were made by either tumbling and crushing lump coal in a closed drum containing steel cubes and drawing the gas through a Draeger tube, or by crushing the coal directly in silver nitrate solution. This latter method showed the coal to contain as much as 700 litres of hydrogen sulfide/1000 kg, that is, approximately 0.1% of the coal by weight may be present as hydrogen sulfide. In an attempt to gain some understanding of the source of this gas, the sulfur isotopic composition of the H_2S and other forms of sulfur (including acid-volatile sulfide) in the coal were measured in samples from the H_2S - rich and H_2S -free zones.

Experimental

The H_2S was sampled and collected as silver sulfide (Ag_2S) by drawing gas released in the gas desorption tests through silver nitrate solution. Sulfate and pyrite sulfur were determined by the sequential treatment of 4-5 g of coal (-72 B.S. Sieve) with boiling 10% hydrochloric acid and boiling 10% nitric acid, each for 30 minutes, respectively. Sulfate in solution after each acid treatment was recovered as barium sulfate ($BaSO_4$). The organic sulfur in the residual acid-treated coal was determined as $BaSO_4$ after ignition of the coal with Eschka mixture. In determining acid-volatile sulfide as Ag_2S, 100 g of coal were reacted with boiling 10% hydrochloric acid for 2 hours and the gaseous products carried through a silver nitrate solution in a stream of nitrogen.

Ag_2S was mixed with cuprous oxide and converted to SO_2 by heating at $900°C$ (2). $BaSO_4$ was directly converted to SO_2 by thermal decomposition in quartz at $1600°C$ (3). Product SO_2 was freed from water and CO_2 before isotopic measurement on a Micromass 602D mass spectrometer. $^{34}S/^{32}S$ ratios are reported to a precision of $\pm0.2°/oo$ relative to troilite from the Canyon Diablo meteorite using the normal $\delta^{34}S°/oo$ notation.

Results and Discussion

The concentration and isotopic composition of sulfur in all measured forms are given in Table I.

Previous studies (4-6) have illustrated the difficulties in relating the isotopic composition of the organic and pyritic forms of sulfur in coals. This is particularly so where large additions of secondary pyrite further distort the frail relationship, if

Table I. Concentrations and Isotopic Compositions
of the Forms of Sulfur in Coal

Mine Zone		Organic S		Pyritic S		Sulfate S		Acid Volatile S		
		%	$\delta^{34}S$ ‰	%	^{34}S ‰	%	$\delta^{34}S$ ‰	ppm	$\delta^{34}S$	H_2S
	1	0.62	+14.2	2.89	+13.7	0.10	+2.1	2	n.d.	
H_2S -	2	0.61	+14.0	0.21	+0.4	0.03	-6.8	11	+10.4	+9.3
Rich	3	0.56	+ 7.8	0.34	+9.7	0.05	+1.5	n.d.	n.d	+7.9
H_2S -	4	0.44	+6.0	0.08	+11.7	<0.01	n.d.	<1	n.d.	
Free	5	0.46	+6.7	1.42	+5.8	<0.01	n.d.	n.d.	n.d.	

any, which exists between primary organic and primary pyritic sulfur even in low concentrations (7). In these respects, samples 1 and 2 and 4 and 5, are of interest as these demonstrate the isotopic stability of the organic sulfur in the face of large variations in the concentration and isotopic composition of associated pyrite (Table I).

Sulfate sulfur was only observed in coals from the H_2S -rich zones. It is generally present in low concentration and is consistently depleted in ^{34}S relative to associated pyrite and organic sulfur. On this evidence, if derived from the pyrite, sulfate cannot be the only sulfur containing product of the weathering of pyrite, another ^{34}S-enriched product must exist. Perhaps this product is elemental sulfur, or a precursor thereof, since elemental sulfur occurring in weathered pyritic zones in coal is frequently more enriched in ^{34}S than co-existing pyrite (4).

In Figure 2, the isotopic data for the organic sulfur in Permian Australian coals containing less than 1% of total sulfur are shown (7). The three coal samples from the Southern Colliery which have sulfur contents within this range are included in Figure 2. Sample No. 4, from the H_2S -free zone, has a $\delta^{34}S$ value in accord with these earlier data. However coals from the H_2S -rich zone are generally more enriched in ^{34}S. Sample No. 3 has a $\delta^{34}S$ value at the isotopic limit previously seen for coals of this type and sample No. 2 possesses, in our experience, an exceptional enrichment in ^{34}S. The higher ^{34}S contents of these two samples could be explained in terms of a very limited invasion by sea water of an extremely reactive organic deposit in a low iron environment. In such circumstances where competition by pyrite formation is small, a large part of the generated marine, ^{34}S-enriched, H_2S , might be incorporated into the organic material. In the few samples of coal examined, those from the H_2S -rich zones tend to have both higher organic sulfur and $\delta^{34}S$ values than those from the H_2S -free areas (Figure 3). The analysis of more samples is clearly required to confirm these apparent relationships. They may not be real, but, certainly the range of isotopic compositions is suggestive of variations in depositional environments and/or contributing sources.

The reduction of sulfate is most commonly the mechanism for sulfide generation. This reaction has a pronounced kinetic isotope effect which results in the initial production of ^{34}S-reduced H_2S . In this case, the ^{34}S-enrichment of the H_2S relative to the possible parent sulfate, (Table I), virtually excludes this reaction as a possible mechanism for H_2S generation. An unlikely circumstance is that the early formed H_2S was lost from the system and that the H_2S collected represents that generated from very strongly ^{34}S-enriched sulfate at the final stages of reduction. Moreover, if it is accepted that organic sulfur is largely derived by the incorporation of H_2S into parent organic materials then, on the same basis, existing sulfate is prohibited as the source of the relatively ^{34}S enriched organic sulfur.

As shown in Table I, the quantities of hydrogen sulfide released on acid treatment never exceeded 11 ppm in the samples tested. Swaine reports the range of zinc and lead contents in Permian Australian coals to be 12-73 and 2-60 ppm respectively (8). In this work analysis of the hydrochloric acid extracts of samples 1 and 2 gave comparable maximum contents of zinc and lead as 8 and 2 ppm, respectively. The low yield of sulfide, 2 ppm, from the pyrite-rich coal sample, No. 2, also demonstrates that no acid- volatile, iron sulfide is associated with the pyrite. From these data it is evident that significant hydrogen sulfide is not generated from the possible reaction of acid mine waters with acid- volatile sulfides. Indeed if such reactions were common, the occurrence of hydrogen sulfide in mine gases would not be a rarity since, in mines where sphalerite - and pyrite-rich coals are won, such as in parts of the Illinois Basin, this problem would also be encountered. The similarity in isotopic composition of the acid released sulfide ($\delta^{34}S$ + 10.4°/oo and that in mine gas ($\delta^{34}S$ + 9.3 and +7.9°/oo) suggests the former to represent residual free hydrogen sulfide.

During the maturation of coals, thermal molecular stabilization re-arrangements result in the loss, in particular, of H_2O, CO_2, CH_4 and higher alkanes. There is no reason, or evidence to believe, that H_2S might not be similarly lost. Indeed, the strength of single bonds between sulfur atoms (255 kJ/mol) and between sulfur and carbon atoms (272 kJ/mol) is much less than that between carbon atoms (348 kJ/mol), therefore, the loss of sulfur-containing compounds is favoured when -S-S- linkages occur in the parent coal. The isotopic compositions of the organic sulfur ($\delta^{34}S$ + 7.8 to +14.2°/oo) and of the free and the residual acid volatile H_2S ($\delta^{34}S$ + 7.9 to + 10.4°/oo) do not deny such a relationship.

This coal appears unique in respect of the remarkable quantities of hydrogen sulfide tightly held within the coal structure. If the size and polarity of the hydrogen sulfide molecule are held to be largely responsible for the retention of this gas, then similarly, the factor of molecular size argues against diffusion of gas into the coal. Therefore, the view advanced is that the entrapped hydrogen sulfide

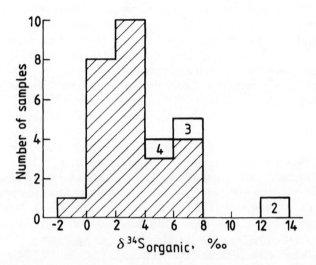

Figure 2. Isotopic composition of organic sulfur in Permian Australian coals containing <1% total sulfur.

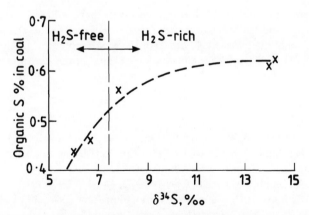

Figure 3. Variation of isotopic composition with content of organic sulfur.

is a product of coal maturation. A series of events postulated to explain both this phenomenon and the isotopic data is that hydrogen sulfide generated by the reduction of a limited quantity of marine sulfate reacted with, and was organically combined into, a deposit of immature reactive land plant material. Whatever the mode of combination of sulfur in the original plant material and whatever the mode of reaction of the sulfide with this organic material, certainly the organic sulfur compounds present in the coal are unusual. This is manifested by the regeneration of hydrogen sulfide by the 'cracking' of organic structures as the coal matures. An explanation of this is the presence in the coal of a large proportion of the organic sulfur in the form of relatively thermally unstable S-S linkages. No direct evidence of such structures is presented and confirmation must await detailed chemical investigation.

Other information which may have relevance in gaining an under standing of the genesis of this coal follow.

Four of the five colleries in which hydrogen sulfide in seam gas has been studied (Southern and Collinsville) or reported (Oaky Creek, Hebburn II and Gregory) are located in the Bowen Basin. Is this a coincidence or a reflection on either the composition of available plant material and/or the depositional environment? With regard to the former, (Moelle, K.H.R., University of Newcastle, personal communication, 1989) has reported the detection of plant fossils of the families Neuropteris , Lonchopteris and Sphenopteris in European sapropelic mudstone/torbanites of Carboniferous age where hydrogen sulfide concentrations occur.

Acknowledgments

The interest and input of Prof. A.J. Hargraves in the early stages of this study are gratefully acknowledged. Thanks are particularly due to the scientific staff (Capricorn Coal Management Pty. Ltd.) for the development of methods for the determination of H_2S and for the provision of data on the distribution of H_2S within Southern Colliery.

Literature Cited

1. Smith, J.W.; Gould, K.W.; Rigby, D. Org. Geochem. 1982, 3 , 111-131.
2. Kaplan, I.R.; Smith, J.W.; Ruth, E. Proc. Apollo II Lunar Sci. Conf ., 1970, 2, 1317-1329.
3. Bailey, S.A.; Smith, J.W. Anal. Chem . 1972, 44 , 1542-1543.
4. Smith, J.W.; Batts, B.D. Geochim. Cosmochim. Acta 1974, 38 , 121-133.
5. Price, F.T.; Shieh, Y.N. Econ. Geol . 1979, 74 , 1445-1461.
6. Westgate, L.M.; Anderson, F. Int. J. Coal Geol . 1984, 4, 1-20.
7. Hunt, J.W.; Smith, J.W. Chem. Geol. (Isotope Geoscience) 1985, 58 , 137-144.
8. Swaine, D.J. Aust. Bur. Miner. Resourc. Geol. Geophys. Bull. 1989, 231 , 297-300.

RECEIVED February 28, 1990

Chapter 29

Pyrolysis of High-Sulfur Monterey Kerogens

Stable Isotopes of Sulfur, Carbon, and Hydrogen

Erdem F. Idiz[1], Eli Tannenbaum[2], and Isaac R. Kaplan

Department of Earth and Space Sciences and Institute of Geophysics
and Planetary Physics, University of California—Los Angeles,
Los Angeles, CA 90024-1567

Sulfur, carbon and hydrogen stable isotope ratios of
pyrite, kerogens, and bitumens of two high-sulfur
Monterey formation samples from the onshore Santa
Maria Basin in California were determined. Kerogens
from these were pyrolyzed at 300°C for periods of 2, 10
and 100 hours in closed systems and the yields and
isotopic compositions of S-containing fractions (residual
kerogens, bitumens and hydrogen sulfide) were
determined.
    With increasing thermal stress, H/C and S/C
elemental ratios of kerogens and S, C and H isotope
ratios of pyrolysis products show systematic maturation
trends. S isotope fractionation between solid, liquid and
gaseous pyrolysates falls within a narrow range of about
2 per mil and mimics the variation observed in natural
samples. These results confirm the utility of S isotopes
towards source rock-oil correlations in sedimentary
basins.

Laboratory pyrolysis of sedimentary organic matter has been extensively
used to simulate thermal maturation in attempts to understand the
process of kerogen cracking and hydrocarbon generation. However,
with the exception of heating experiments on oils ([1]) and asphaltenes
([2]) no such work has been published to date on the sulfur isotopic
composition of parent kerogens and their S-containing pyrolysis
products (bitumen and hydrogen sulfide) during simulated maturation.

NOTE: This chapter is the Institute of Geophysics and Planetary Physics Contribution No. 3355
[1]Current address: Koninklijke–Shell Exploration and Production Laboratories, Volmerlaan 6,
  Rijswijk 2288GD, Netherlands
[2]Current address: 21 Yoni Havani Street, Tel Aviv 63302, Israel

Two Type II-S kerogens (as defined by Orr (1)) from the onshore Santa Maria Basin Monterey formation were pyrolyzed in this study to determine (a) the distribution of sulfur and its isotopic composition among the various products formed during artificial maturation, and (b) maturation trends reflected in the sulfur isotopic and elemental S/C ratios of kerogens, and in the variation of C and H isotopes. In addition, S isotopes in pyrites, kerogens and bitumens from the two Monterey shale samples were examined to speculate on the mode of S incorporation into Santa Maria Basin sediments.

Experimental

Two Monterey Formation samples (M4110 and M4140) were obtained as drill cuttings from a well in the Santa Maria Valley West field. The pertinent characteristics of these samples are listed in Table I.

Table I. Characteristic of Two Monterey Formation Shale Samples Used in the Pyrolysis Experiments

	M4110	M4140
LITHOLOGY	Silty shale	Silty foram shale
PYRITE	2%	2%
TOC	6.4%	6.1%
DEPTH	4110 ft	4140 ft
KEROGEN Tmax °C	398	400
HI	590	610
Ro%	0.48	0.5
Total S	10.4%	10.2%
$\delta^{34}S$	+15.9⁰/oo	+17.1⁰/oo
BITUMEN Wt. %	2	1.8
SAT%	3	3
ARO%	9	11
NSO%	78	76
PR/PH	0.5	0.5
PR/n-C_{17}	1.1	1
PH/n-C_{18}	1.4	1.6

The samples were ground and extracted with the solvent mixture 9:1 dichloromethane/methanol for 48 hours in a soxhlet system. The extracted samples were then demineralized to isolate kerogens using a detailed procedure described in (3). The kerogen was next solvent extracted a second time and treated with 0.2 N nitric acid at room temperature to separate pyrite from the organic matter and the solution was kept for determination of pyrite S content and isotopic composition. X-ray diffraction was used to verify the removal of pyrite from the kerogen.

Kerogen samples from M4110 (0.3-1.5 g) were placed into pyrex tubes (3 cm diameter, 25 cm length), evacuated and sealed on a vacuum line. These were heated in an oven at 300°C for periods of 2, 10 and 100 hours. After each run, hydrogen sulphide, bitumen and residual kerogens were separated for analysis. The reaction vessel was placed on a vacuum line and the hydrogen sulfide cryogenically transferred to another vessel containing a frozen solution of $AgNO_3$. This tube was then sealed and the $AgNO_3$ allowed to thaw, reacting with H_2S to form solid Ag_2S. The bitumen formed by pyrolysis was separated by solvent extraction from the residual kerogen. Elemental S was removed from the bitumens using an activated copper column. C_1-C_4 range hydrocarbons and carbon monoxide were cryogenically separated as one fraction and were combusted to carbon dioxide in a copper furnace at 800°C and isolated for isotope analysis.

Sample M4140 was pyrolyzed using the same technique, but after each step of pyrolysis, the residual kerogens were sealed and pyrolyzed again. This was done in order to minimize the effects of secondary cracking of products during longer heating times.

C, H and N isotopic and elemental analyses of bitumens and kerogens and pyrolysates were obtained using the technique described by ($\underline{4}$). Moisture and ash percentages were determined by noting weight loss after heating at 100°C for two hours and then combusting at 550°C respectively.

Pyrite sulfur, isolated by nitric acid oxidation of kerogens, was precipitated as $BaSO_4$ with a 10% $BaCl_2$ solution, filtered and dried. This was then used to quantify the yield of pyrite sulfur, and was converted to SO_2 for isotope analysis on a vacuum line. The S contents and isotope compositions of the Ag_2S fractions (representing the hydrogen sulfide formed during pyrolysis) were determined by digesting in aqua regia and bromine and precipitating out as $BaSO_4$ in a similar way to the pyrite procedure. Organic S from pyrolysates, kerogens and bitumens were determined by Parr bomb oxidation and conversion to $BaSO_4$.

All collected gas samples were analyzed for isotope ratios. Carbon and hydrogen measurements were made on a Varian MAT 250 triple collecting MS and sulfur and nitrogen on a Nuclide RMS 6-60 dual collecting MS. Atmospheric N, Chicago Pee Dee belemnite, SMOW and Canyon Diablo troilite were used for isotope standards. Analytical errors for isotope ratio measurements were as follows: $\delta^{13}C\pm0.05$ per mil, $\delta D\pm3$ per mil, $\delta^{34}S\pm0.4$ per mil, $\delta^{13}N\pm0.4$ per mil, and for elemental analyses, $C\pm0.5\%$, $H\pm1\%$, $S\pm0.5\%$, $N\pm0.5\%$.

Results

Starting Materials. Pyrite isolated from the two kerogens constitutes approximately 2 wt % and has $\delta^{34}S$ values of +8.7 per mil. These values are enriched by 20-30 per mil compared to pyrite formed from near-surface biogenically reduced sulfate in porewaters of "normal" near-surface marine sediments ($\underline{5}$). Bitumens extracted from the two samples contain 8.0 and 8.7 wt % organic sulfur with $\delta^{34}S$ values of +17.3 and +18.5 per mil respectively. Both pyrite free kerogens contained 10.3 wt % organic sulfur with isotope values of +15.9 per mil and +17.0 per mil respectively.

Pyrolysates. The quantities of products from the pyrolysis experiments on the two kerogen samples, along with elements and isotope analysis results are listed in Tables 2 and 3. Of the four S-containing fractions recovered from the pyrolysis reactions, elemental S was found to be in yields too low for accurate isotopic analysis, although its presence was apparent from blackening of activated copper. In the "normal" pyrolysis series (M4110) the cumulative amount of bitumen and hydrogen sulfide increased at the expense of kerogen with longer heating times. The interval between 0-2 hours (Fig. 1) showed the greatest rate of bitumen generation (76 mg/hr/g kerogen). After this step, the rate of generation dropped to 3.9 mg/hr/g kerogen and 2.1 mg/hr/g kerogen respectively for the intervals of 2-10 and 10-100 hours heating. The bitumen generated in the pyrolysates contains between 12-14 wt % S and after 100 hours of heating. The S-containing bitumen is the fraction accounting for over 40% of the original organically bound S of the parent kerogen (Fig. 2). Hydrogen sulfide accounts for about 28% of the original S-content. The most pronounced rate of hydrogen sulfide generation also occurs during the 0-2 hour period (4.8 mg/hr/g kerogen) after which the production rate drops below 1 mg/hr/g kerogen for the remaining intervals.

For the stepwise pyrolysis (M4140), the rate of bitumen generation is greatest in the 0-2 hour interval (58.5 mg/hr/g kerogen) and drops significantly in the next two intervals to 16.1 mg/hr/g kerogen and 1.3 mg/hr/g kerogen. Hydrogen sulfide generation shows similar behavior, with the highest rate of generation (3 mg/hr/g kerogen) at the shortest interval.

In both types of pyrolyses, the most striking feature is that the pattern of bitumen and hydrogen sulfide generation parallel each other (Fig. 1), as does the distribution of S content in these two fractions (Fig. 2). In both cases, the rate of generation of hydrogen sulfide relative to bitumen reaches a maximum after 10 hours of heating, then decreases. The differences in pyrolysate yields after two hours heating time suggests different cracking behavior of the two kerogens used.

C, H, and S isotopic analyses of kerogens, bitumens and gaseous products are listed in Table 3. Kerogen H/C ratios show a continuous decrease with longer heating times and are accompanied by progressively heavier C, H and S isotope values (Figs. 3-5). The variation of $\delta^{13}C$ is small (0.2 per mil), whereas that of δD changes by 15 per mil for M4110 and 25 per mil for M4140. The modest increase in ^{13}C in kerogens with maturation has been previously noted ([6-9]).

Bitumen associated with the original shale and in the pyrolysates shows a similar trend for $\delta^{13}C$ values in both samples (Fig. 3). These values are 0.3 to 0.8 per mil lighter in the bitumens than the corresponding kerogens, as commonly seen in other source rocks ([10-11]). In addition, the variation of $\delta^{13}C$ of bitumens generated during pyrolysis is much larger than that seen in the residual kerogens. The isotopic compositions of C_1-C_4 gases generated during both types of pyrolysis are significantly lighter than the other pyrolysates (Table 3), and show trends similar to those observed by ([9]).

S isotope values of kerogens, bitumens and hydrogen sulfide are seen in Figure 6. Kerogens show progressively heavier $\delta^{34}S$ values with increasing heating time, with varying degrees in both types of pyrolysis. The $\delta^{34}S$ of the bitumens generated show striking similarities in both sets of experiments, with little isotopic fractionation until the longer heating times. In both types of pyrolyses, hydrogen

Table II. Yields and Sulfur Contents of Starting Materials and Pyrolysates and Their Production Rates

Sample	Heating Hours	Bitumen	Bitumen %S	H_2S	Residual Kerogen	Kerogen %S	Total	Production Rates Bitumen	Production Rates H_2S	Bitumen/H_2S
M4110	2	152	12	10	790	10	952	76	5	15.2
	10	183	14	15	734	9	932	3.9	0.6	6.5
	100	374	14	37	501	7	912	2.1	0.2	10.5
M4140	0-2	117	13	6	847	10	970	59	3	20
	2-10	129	13	15	753	8	897	16.1	1.9	8.5
	10-100	120	10	8	789	7	917	1.3	0.1	13

All quantities in mg product/g kerogen.
Production rates as mg product/g kerogen/hour.

Table III. Elemental and Isotopic Compositions of Starting Materials
and Pyrolysates Generated During Artificial Maturation

Sample	Heating Hours	H/C	S/C	N/C	$\delta^{13}C$ $^{o}/oo$	δD $^{o}/oo$	$\delta^{34}S$ $^{o}/oo$
M4110-K	0	1.08	0.054	0.051	-21.95	-141	+15.9
	2	0.95	0.050	0.051	-21.95	-135	+16.3
	10	0.86	0.046	0.046	-21.84	-133	+16.5
	100	0.76	0.037	0.044	-21.80	-124	+17.7
M4110-B	0	1.36	-	-	-22.88	-	+17.3
	2	1.31	-	-	-22.58	-	+15.2
	10	1.23	-	-	-22.40	-	+15.2
	100	1.17	-	-	-22.15	-	+16.2
M4110-H	2	-	-	-	-	-	+15.2
	10	-	-	-	-	-	+16.3
	100	-	-	-	-	-	+15.3
M4110-G	2	-	-	-	-30.91	-	-
	10	-	-	-	-34.72	-	-
	100	-	-	-	-36.40	-	-
M4140-K	0	1.11	0.056	0.050	-22.20	-147	+17.1
	0-2	0.94	0.055	0.046	-22.16	-138	+17.3
	2-10	0.87	0.044	0.046	-22.13	-125	+18.0
	10-100	0.78	0.039	0.047	-22.02	-120	+18.2
M4140-B	0	1.24	-	-	-23.01	-	+18.5
	0-2	1.18	-	-	-22.78	-	+16.7
	2-10	1.09	-	-	-22.52	-	+16.6
	10-100	1.01	-	-	-22.51	-	+17.2
M4140-H	0-2	-	-	-	-	-	+15.4
	2-10	-	-	-	-	-	+17.2
	10-100	-	-	-	-	-	+17.1
M4140-G	0-2	-	-	-	-33.83	-	-
	2-10	-	-	-	-35.67	-	-
	10-100	-	-	-	-36.61	-	-

K: Kerogen; B: Bitumen; H: Hydrogen Sulfide; G: C_1-C_4 Hydrocarbon Gases.
Isotope values in per mil.
0 Hours heating connotates original unheated kerogen and extracted bitumen.

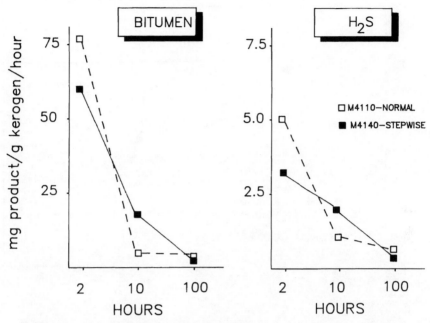

Figure 1: Production rates of bitumen and hydrogen sulfide in pyrolysates of the two types of pyrolysis.

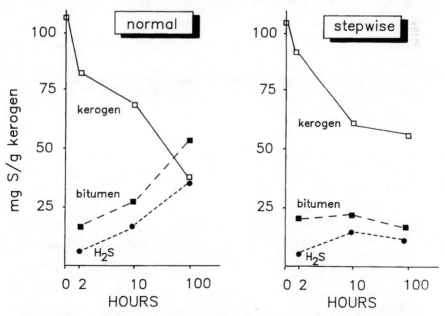

Figure 2: Sulfur contents of residual kerogens, bitumens and hydrogen sulfide generation during pyrolysis. 0 hours heating denotes starting kerogens. All values are in mg S and are normalized to 5 g starting kerogen.

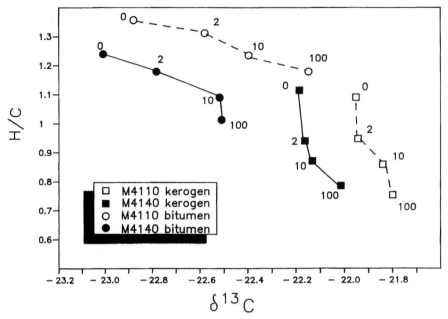

Figure 3: Variation of carbon isotope ratios with H/C ratios in residual kerogens and bitumens generated during pyrolysis. 0 hours denotes starting kerogens and original sedimentary bitumen. 2, 10 and 100 denote heating times.

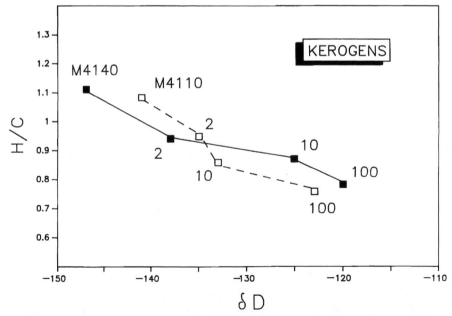

Figure 4: Variation of hydrogen isotope ratios with H/C ratios in original and residual kerogens during pyrolysis.

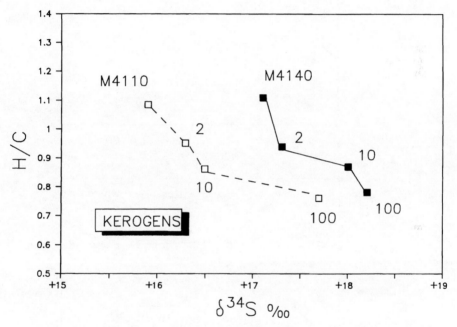

Figure 5: Variation of sulfur isotope ratios with H/C ratios in original and residual kerogens during pyrolysis.

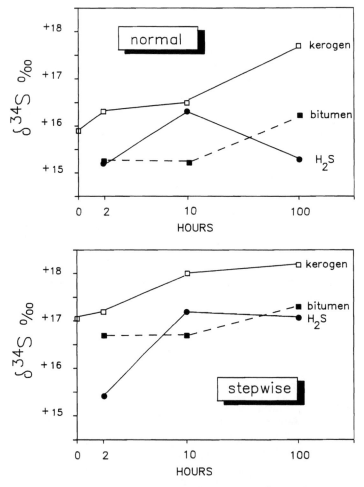

Figure 6:   Variation of sulfur isotope ratios of parent and residual kerogens, bitumens and hydrogen sulfide during pyrolysis.

sulfide, which is isotopically heavier than the corresponding bitumens, is generated at the 10 hour heating interval. It becomes the isotopically lightest component at 100 hours during normal pyrolysis, but is essentially the same as the bitumen for the stepwise pyrolysis (Table 4). An isotopic and elemental mass balance of these three components accounts for most of the S in the original kerogen and gives calculated $\delta^{34}S$ values very close to the parent materials (Table 4). The low total sulfur value and heavier $\delta^{34}S$ of the calculated value for M4140 at 10-100 hours heating may be due to some loss of H_2S during the capture process.

Pyrite-Organic Sulfur Relationships. The isotopic value of S in pyrite and sedimentary organic matter is controlled by several mechanisms of which the most important are the rate of sulfate reduction and the amount of sulfate reservoir reduced. Goldhaber and Kaplan (5) have shown that there is an inverse correlation between rate of sulfate reduction and the magnitude of isotope fractionation between seawater sulfate and hydrogen sulfide. Similarly, increasing amounts of reduction, especially in a closed system, will result in hydrogen sulfide with isotope values progressively closer to seawater sulfate due to simple Rayleigh distillation (12-15). By this principle, the last formed hydrogen sulfide in a depleting reservoir will be isotopically heavier than the starting value of the seawater sulfate.

The Santa Maria basin is thought to have been a tectonically controlled, large partially anoxic hemi-pelagic basin, separated from the offshore and current circulation by a sill (16). The enriched organic matter content of the basin shales is a result of high productivity and preservation of this organic matter by oxygen-free bottom waters due to density stratification and lack of mixing with overlying waters (16, 17). Isotopically heavy pyrite and organic matter, along with high organic S content are common in the petroleum source rocks of the Miocene Santa Maria Basin (1). This suggests that near-complete reduction of marine sulfate resulted in the generation of isotopically heavy hydrogen sulfide. Early formed hydrogen sulfide was probably incorporated into pyrite until the availability of reactive iron species was depleted, after which remaining hydrogen sulfide was incorporated into sedimentary organic matter. This scheme would explain the overall +8 to +10 per mil enrichment of organic sulfur in kerogens relative to pyrite. To explain the high contents of organically bound S adds an additional constraint. There is probably insufficient sulfate in sediment pore waters to account for the high S contents of the Monterey kerogens by simple sulfate reservoir depletion. Hence, a mechanism invoking diffusion of sulfate into the sediment column from the overlying sea water, in excess of that in the interstitial water is required followed by sulfate reduction and incorporation of reduced sulfur species into the organic matter. A steady state condition where the rate of sulfate reduction is equal or greater than sulfate replenishment is one mechanism to account for both the high organic S contents and the heavy isotopic signatures.

The bitumen extracted from the Monterey shale samples have $\delta^{34}S$ values heavier than their corresponding kerogens. It is difficult to explain this observation within the framework of normal kerogen cracking, as suggested by our experiments. This bitumen, consisting of up to 78 wt % asphaltenes, is indigenous based on other criteria (18)

Table IV. Mass Balance Calculations for Sulfur Isotope and Total Sulfur Data. Yields are normalized to 1 g starting kerogen for both types of pyrolysis. For the stepwise pyrolysis, the "parent" kerogen for each step is the residual kerogen of the preceding stage of heating.

Sample	Heating Hours	Kerogen mg S	Kerogen $\delta^{34}S$ o/oo	Bitumen mg S	Bitumen $\delta^{34}S$ o/oo	H_2S mg S	H_2S $\delta^{34}S$ o/oo	Total mg S	Parent* $\delta^{34}S$ o/oo
M4110-K	0	104	+15.9	-	-	-	-	104	+15.9
	2	79	+16.3	18	+15.2	9	+15.2	106	+16.0
	10	66	+16.5	26	+15.2	14	+16.3	106	+16.2
	100	35	+17.7	52	+16.2	35	+15.3	122	+16.3
M4140-K	0	102	+17.1	-	-	-	-	102	+17.1
	0-2	85	+17.3	15	+16.7	6	+15.4	106	+17.0
	2-10	60	+18.0	17	+16.6	14	+17.2	91	+17.6
	10-100	55	+18.2	12	+17.2	8	+17.1	75	+17.9

* Calculated isotope value of parent kerogens.
All isotope values in per mil.
All quantities in mg S/g kerogen.

and may have never been part of the original kerogen. If this deduction is correct, we speculate that the last, residual unreacted hydrogen sulfide, containing the heaviest $\delta^{34}S$ signatures, may have been the component incorporated into the asphaltenic bitumen.

Maturation Trends

C and H isotopes. Peters et al. (19) calculated the theoretical changes in C and H isotopic compositions of kerogens during thermal maturation and proposed a model whereby the H/C ratios and isotopic compositions of kerogens are controlled by Rayleigh fractional distillation of methane generated during cracking. According to this model, with progressive maturation, kerogens become more enriched in the heavier isotopes of C and H due to loss of lighter isotopes (^{12}C and ^1H) in methane. Experimental work by Chung (20) and Rohrback (21) seem to support this hypothesis. Such a scheme would explain the C and H isotope behavior of the kerogens we pyrolyzed. The fact that bitumens released during cracking of these kerogens are isotopically lighter is probably due to their greater content of aliphatic (lipid) components (10). The greater variability of $\delta^{13}C$ values in the bitumens suggests significant secondary cracking of these bitumens and loss of light isotopes through methane generation. This observation appears to support the suggestions of Peters et al. (19) that significant methane generation from kerogens primarily occurs through a liquid hydrocarbon intermediate, and that this reduces the effective isotopic fractionation in the organic matter of the residual kerogen composition.

S isotopes. In order to understand the behavior of S isotopes during kerogen cracking, it is necessary to have some knowledge on the form and incorporation of S into kerogen. As noted previously, it is now generally accepted that the ultimate source of this sulfur is microbially reduced seawater sulfate. Because early formed hydrogen sulfide preferentially reacts with metallic species, pyrite is most probably the first form of reduced sulfur incorporated into the sediment (22). The incorporation of S into sedimentary organic matter proceeds either through reaction of hydrogen sulfide with low molecular weight components (23), and metabolites (24-27), or by reaction of polysulfides directly with higher molecular weight condensed structures such as humic substances (28). It is possible that both these processes occur during diagenesis, complicated further by the possibility of microbial reworking of the substrate. In either case, even if there is a constant input of progressively isotopically heavy reduced S species into the organic matter, the very nature of kerogen (polycondensed, randomly structured organic compounds) may preclude the systematic incorporation of S into the kerogen network. In this case, the overall isotopic signature of the kerogen would be an average value, possibly reflecting a wide range of $\delta^{34}S$ values actually within its framework, and depending on the timing of incorporation of the reduced sulfur into individual constituents.

The H/C vs S/C data from our study indicate that there is progressive loss of S relative to C in kerogens with increasing maturation (Fig. 7). The relative ease of breaking C-S bonds and its implications for early generation of oil from high-S kerogens has been previously noted ($\underline{1}$, $\underline{29\text{-}31}$). This early cleaving of C-S bonds results in heavy crude oils rich in resins and asphaltenes, with abundant organic S moieties, which themselves undergo further decomposition to yield hydrocarbons. Our study has shown that the residual kerogens, after prolonged heating, still contain up to 30 wt % of their original S content, and that early formed asphaltene-rich bitumen is relatively enriched in S content.

The narrow range of isotope values exhibited in the pyrolysis products in this study, and in kerogens, oils and gases from natural basins ($\underline{32}$, $\underline{33}$) can be interpreted as a direct result of the disordered random incorporation of S into organic matter as described above. In both types of pyrolysis experiments, residual kerogens and bitumens show depletion of lighter isotopes of sulfur (Fig. 8), as is expected from kinetic isotope effects associated with the relative reactivity of the C-^{32}S and C-^{34}S bonds ($\underline{2}$). The variable isotopic composition of H$_2$S evolved during pyrolysis suggests multiple sources for labile S, as noted by Krouse ($\underline{2}$). Since it is difficult to envisage generating H$_2$S which is isotopically heavier than its organic source, the H$_2$S in both pyrolysis experiments at 10 hours heating must have been generated predominantly from the kerogen. For the other two heating intervals, it is difficult to resolve the relative contributions of kerogen and/or bitumen.

The results of our experiments show that the range of variation of δ^{34}S values between the pyrolysates and the source kerogen is about 2 per mil. This value is in agreement with measured δ^{34}S values in oils and source rocks from different sedimentary basins ($\underline{1}$, $\underline{32\text{-}39}$). These results thus confirm the utility of S isotope measurements in oils and kerogens as a useful correlation tool. Furthermore, in our experiments, the relatively small enrichment of ^{32}S in hydrogen sulfide relative to the ^{34}S/^{32}S content in the kerogens and bitumens from which it is sourced, is in agreement with values measured for H$_2$S associated with the maturing of natural oils ($\underline{34}$).

Conclusions

In a series of experiments performed on two shale samples of Monterey Shale, the effects of dry pyrolysis was studied at 300°C for periods of 2-100 hours on the δ^{34}S and δ^{13}C of the starting and residual kerogen and bitumen and on the hydrogen sulfide and hydrocarbon gases. The results show that during initial heating, the dominant δ^{34}S change is in the preferential loss of ^{34}S and ^{12}C by the bitumen. Light carbon is lost primarily as C$_1$-C$_5$ gases. The reason for ^{34}S loss during initial pyrolysis followed later by ^{32}S loss is explained as being due to incorporation of C-S compounds by the asphaltenic bitumen during late stage diagenesis from a pool of ^{34}S-enriched reduced sulfur species. The organic compounds formed (e.g. thiolanes) may be unstable to pyrolysis and will release the isotopically heavy H$_2$S which will mix with H$_2$S released from the kerogen. The total sulfur isotope effects measure were very small, resulting in a total measured $\Delta \delta^{34}$S $\approx$ ±1.5°/oo. These results further support the use of ^{34}S/^{32}S in correlating crude oil with kerogen and in oil-oil correlations.

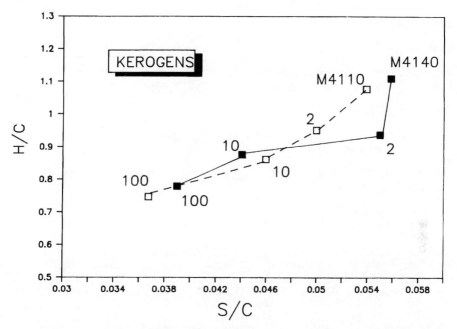

Figure 7: H/C vs S/C elemental ratios of parent and residual kerogens during pyrolysis.

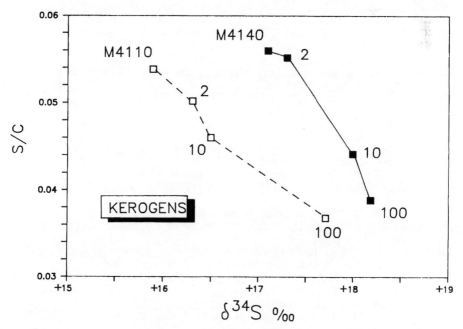

Figure 8: Variation of S isotope ratios vs S/C elemental ratios of parent and residual kerogens during pyrolysis.

Acknowledgments

Samples were gratefully supplied by Global Geochemistry Corporation, Canoga Park, California. D. Winter conducted C and H isotope determinations. Fruitful discussions with B. Durand, E. Lafargue, R. Pelet, G. Lijmbach and A. Kornacki were much appreciated.

Literature Cited

1. Orr, W.L. In Advances in Organic Geochemistry 1985; D. Leythauser and J. Rullkotter, Eds.; Technip: Paris, 1986; pp. 499-516.
2. Krouse, H.R.; Ritchie, R.G.S.; Roche, R.S. J. Anal. Appl. Pyrolysis 1987, 12, 19.
3. Tannenbaum, E.; Huizinga, B.J.; Kaplan, I.R. Am Assoc. Geol. Bull. 1986, 65, 1527.
4. Minagawa, M.; Winter, D.A.; Kaplan, I.R. Anal. Chem. 1984, 56, 1859.
5. Goldhaber, M.B.; Kaplan, I.R. Mar. Chem. 1980, 9, 97.
6. Ishiwatari, R.; Rohrback, B.G., Kaplan, I.R. Am. Assoc. Petrol. Geol. Bull. 1978, 62, 687.
7. Chung, H.M.; Sackett, W.M. In Advances in Organic Geochemistry 1979; A.G. Douglas and J.R. Maxwell, Eds.; Technip: Paris, 1980; pp. 705-710.
8. Lewan, M.D. Geochim. Cosmochim. Acta 1983, 47, 1471.
9. Arneth, J.D.; Matzigkeit, U. In Advances in Organic Geochemistry 1985; D. Leythauser and J. Rullkotter, Eds.; Technip: Paris, 1986; pp. 1067-1071.
10. Galimov, E.M. In Advances in Organic Geochemistry 1973; B. Tissot and F. Bienner, Eds.; Technip: Paris, 1973; pp. 439-452.
11. Welte, D.H.; Hagemann, H.W.; Hollerbach, A.; Leythauser, D.; Stahl, W. 9th World Pet. Cong. Proc. 2, 1975, pp. 179-191.
12. Berner, R.A. Phil. Trans. R. Soc. Lond. 1985, A315, 25.
13. Orr, W.L. In Handbook of Geochemistry; K.H. Wedpohl, Ed.; Springer, 1974a; volume 11, Sec. 16-L.
14. Krouse, H.R. J. Geochem. Expl. 1977, 7, 189.
15. Hayes, J.M. Spectra 1982, 8, 3.
16. Bertucci, P.F. 192nd Am. Chem. Soc. National Mtg., 1986, Anaheim, California (abstract).
17. Williams, L.A. 192nd Am. Chem. Soc. National Mtg., 1986, Anaheim, California (abstract).
18. Idiz, E.F.; Monin, J.C.; Behar, F. (in preparation).
19. Peters, K.E.; Rohrback, B.G.; Kaplan, I.R. Am. Assoc. Pet. Geol. Bull. 1981, 65, 501.
20. Chung, H.M. Ph.D. Thesis, Texas A&M University, 1976.
21. Rohrback, B.G. Ph.D. Thesis, University of California, Los Angeles, 1979.
22. Goldhaber, M.B.; Kaplan, I.R. In The Sea, Marine Chemistry; E.D. Goldberg, Ed.; Wiley: New York, NY, 1974; pp. 569-655.
23. Mango, F. Geochim. Cosmochim. Acta 1983, 47, 1433.
24. Sinninghe-Damste, J.S.; Ten Haven, H.L.; DeLeeuw, J.W.; Schenck, P.A. In Advances in Organic Geochemistry 1985; D. Leythauser and J. Rullkotter, Eds.; Technip: Paris, 1985; pp. 791-805.

25. Sinninghe-Damste, J.S.; DeLeeuw, J.W.; van Dalen, A.C.K.; DeZeeuw, M.A.; Lange, F.; Rijpstra, W.I.C.; Schenck, P.A. Geochim. Cosmochim. Acta 1987, 51, 2369.
26. Sinninghe-Damste, J.S.; van Koert, E.R.; Kock-van Dalen, A.C.; DeLeeuw, J.W.; Schenck, P.A. Org. Geochem. 1989, 14, 555.
27. Brassell, S.C.; Lewis, C.A.; DeLeeuw, J.W.; deLange, F.; Sininhe-Damste, J.S. Nature 1986, 320, 160.
28. Dinur, D.; Spiro, B.; Aizenshtat, Z. Chem. Geol. 1980, 31, 37.
29. Gransch, J.A.; Posthuma, J. In Advances in Organic Geochemistry 1973; B. Tissot and F. Bienner, Eds.; Technip: Paris, 1973; pp. 729-739.
30. Tannenbaum, E.; Aizenshtat, Z. Org. Geochem. 1985, 8, 181.
31. Lewan, M. Phil. Trans. Royal Soc. Lond. 1985, A315, 123.
32. Orr, W.L. Am. Assoc. Pet. Geol. Bull. 1974b, 58, 2295.
33. Thode, H.G. Am. Assoc. Pet. Geol. Bull. 1981, 65, 1527.
34. Thode, H.G.; Monster, J.; Dunford, H.B. Am. Assoc. Pet. Geol. Bull. 1958, 42, 2619.
35. Thode, H.G.; Monster, J. Am. Assoc. Pet. Geol. Bull. 1970, 54, 627.
36. Thode, H.G.; Rees, C.E. Endeavour 1970, 29, 24.
37. Vredenburgh, L.D.; Cheney, E.S. Am. Assoc. Pet. Geol. Bull. 1971, 55, 1954.
38. Monster, J. Am. Assoc. Pet. Geol. Bull. 1972, 56, 941.
39. Hirner, A.V.; Graf, W.; Treibs, R.; Melzer, A.N.; Hahn-Weinheimer, P. Geochim. Cosmochim. Acta 1984, 48, 2179.

RECEIVED March 5, 1990

Chapter 30

Sulfur Isotope Data Analysis of Crude Oils from the Bolivar Coastal Fields (Venezuela)

B. Manowitz[1], H. R. Krouse[2], C. Barker[3], and E. T. Premuzic[1]

[1]Department of Applied Science, Brookhaven National Laboratory,
Upton, NY 11973
[2]Department of Physics, University of Calgary,
Calgary, Alberta T2N 1N4, Canada
[3]Geosciences Department, University of Tulsa, Tulsa, OK 74104

Oils in the Bolivar Coastal Fields of
Venezuela have been divided into five
major oil classes believed to reflect
largely variations caused by biodegradation in the reservoirs. Classes are
based on variations in composition of
hydrocarbons, NSO components (including asphaltenes) and API gravity, as
well as other considerations including
the geological framework and reservoir
depths and temperatures. Based on
sulfur isotope data reported here, the
oils fall into two groups, a non- or
little-biodegraded group with $\delta^{34}S$
values averaging +7.56 $^\circ/_{oo}$ and a
heavily biodegraded group in which $\delta^{34}S$
values average +5.1 $^\circ/_{oo}$. Thermal
alteration effects also are probable
with reservoir temperatures ranging
from 20°C to 209°C. Various possibilities for explaining the isotopic
data are considered. The relatively
narrow range in $\delta^{34}S$ values suggests
reasonably uniform source-rock character and rather minor isotopic changes
from all alteration processes.

The Bolivar Coastal Fields (BCF) of eastern Lake
Maracaibo, Venezuela, contain five classes of oil as
reflected by their API gravities, $C_{15}+$ saturates-and-
aromatics contents as well as their total nitrogen,
sulfur, and oxygen (NSO) compositions. Biodegradation
appears to have had a major role in controlling the

0097–6156/90/0429–0592$06.25/0
© 1990 American Chemical Society

compositions of the various classes of oil. The geology
and geochemistry of crude oils and source rocks in the
BCF have been described (1-3).

In the Cretaceous, the area was part of the platform
of a large geosyncline, and by the Eocene it was near a
coast where a series of large sandy deltas was deposited, with terrestrial sediments on the south and thick
marine shales to the north. At this time, conditions
for oil generation in the shales and migration to the
sands may have been established, and the subsequent
Oligocene faulting, uplift, and erosion could have
allowed meteoric water to penetrate into reservoirs.
During the Miocene and Pliocene, the basin was tilted
first west and then south, and filled with continental
sediments from the rising Andes. Tilting is still continuing and oil is moving up along the Oligocene unconformity, forming surface seeps. Most of the oil fields
are located in sands above this unconformity or in fault
blocks immediately below it.

Thirty crude oils from the BCF were collected (1)
along two parallel and generally southwest-northeast
trends. The areal extent of the BCF showing locations
of wells sampled is shown in Figure 1. These oils were
characterized by their API gravity, percent saturates,
aromatics, NSO and asphaltene compounds, gas chromatograms for whole oils, C_4-C_7 fractions, and aromatics.
Concurrently, 24 associated waters were also sampled and
analyzed for Ca^{++}, Mg^{++}, Na^+, HCO_3, CO_3^-, SO_4^-, pH, and
total dissolved solids (TDS) (1).

In the present work, twenty-seven of these oils were
separately analyzed for sulfur content and sulfur
isotope ratio ($\delta^{34}S$). The samples were oxidized in a
Parr Instrument Company bomb. Sulfate in washings from
the bomb were precipitated with Ba^{2+}. The $BaSO_4$ precipitate served for gravimetric determination of the
S-content conversion to SO_2 for mass spectrometry (4).
The $^{34}S/^{32}S$ abundance ratios are presented in the usual
$\delta^{34}S$ notation.

Oil Classes

As described (1), the oils were divided into five
classes on the basis of their chemical compositions
(Table I). These classes are consistent with reservoir
assignment (Table II).

The four oils in Class 1, from Wells CL-20, CL-99,
VLC-531, and VLC-642, are dark olive green and are very
light, with API gravities ranging from 36.6° to 42.2°.

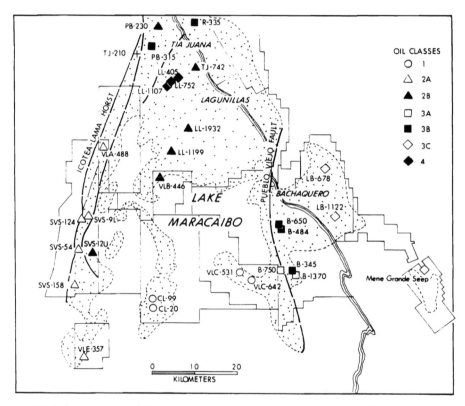

Figure 1. Areal extent of Bolivar Coastal Fields
showing locations of wells sampled (Reprinted with
permission from Ref. 1. Copyright 1983 Bockmeulen et
al.)

Table I. Oil Classes and Their Compositional Ranges

Oil Class	$C_{15}+$ (%)	Saturates (%)	Aromatics (%)	NSOs (%)	Benzene + Toluene (%)	C_7/MCH
1	55-65	69-74	14-19	7-13	12-17	1.18-1.29
2A	67-76	47-55	26-32	15-21	4.8-8.7	1.05-1.55
2B	75-83	39-46	30-33	16-21	3.7-7.6	0.68-0.93
2C	84.6	34.5	34.6	18.4	1.5	0.40
3A	86-88	27-34	33-35	22-25	2.5-2.7	0.30-0.35
3B	81-8	24-33	35-38	21-28	1.1-3.5	0.16-0.34
3C	91-95	21-27	38-41	23-26	0.8-3.7	0.05-0.07
4	79-82	35-44	31-33	16-23	3.2-4.3	0.67-0.78
5	84.0	51.1	30.1	15.5	5.8	0.25

Table II. Oil Classes and Their Reservoirs

Class	Oils	Reservoirs
1	CL-20,CL-99	Cretaceous Cogollo
	VLC-531,VLC-642	Eocene C
2A	VLA-488S,VLE-357,SVS-12U	Eocene C
	SVS-158	Miocene Santa Barbara
2B	SVS-54,SVS-9L,SVS-124,	Eocene B-6
	VLB-446,LL-1199,LL-1932	
2C	PB-230,TJ-742	Eocene B-6
3A	B-345,B-750,B-1307	Miocene Bachaquero
3B	B-484,B-650	Miocene Bachaquero
	PB-315,R-335	Miocene Lower Lagunillas
3C	LB-678	Miocene Bachaquero
	LB-1122	Miocene Lagunillas and LaRosa
4	LL-405,LL-752,LL-1107	Miocene Lower Lagunillas and LaRosa
5	TJ-210	Miocene Lower Lagunillas and LaRosa

The oils in this Class have a lower content of aromatics in the $C_{15}+$ fraction, and a higher percentage of benzene and toluene in the C_4-C_7 range compared with the rest of the oils. The n-heptane (C_7)-to-methylcyclohexane (MCH) ratio is greater than one.

The Class 2 oils have fewer saturates than those in Class 1, more aromatics and NSOs, and about half as much benzene and toluene in the C_4-C_7 light fraction (Table I). They can be subdivided into those oils with C_7/MCH ratios greater than one (Class 2A), and those for which

the ratio is less than unity (Class 2B). API gravities
for Class 2A range from 27° to 33°.

The Class 2B oils, in which methylcyclohexane is the
predominant light hydrocarbon, extend the compositional
trends that distinguish Classes 1 and 2A, and the ranges
of values for the distinguishing characteristics show
very little overlap (Table I).

Oils TJ-742 and PB-230 fall within the range of the
Class 2B oils in gross characteristics, but show lower
pristane/phytane ratios. In TJ-742, the normal distribution of n-alkanes is absent and none are prominent
above C_{24}, suggesting a moderately to severely bacterially degraded oil. The relatively high isoprenoid-to-
n-alkane abundance is also consistent with incipient or
continuing bacterial activity, which is further supported by the lower C_{10}-C_{14} n-alkane content. The oil,
however, is currently from a reservoir at a depth of
about 6,000 ft (1,829 m), and the temperature is above
176°F (80°C), which would seem to exclude or limit continuing bacterial activity. It is probably also
affected by "water-washing," as shown by the very low
content of benzene and toluene.

The Class 3 oils are all degraded, with API
gravities ranging from 14° to 18°. This Class has been
subdivided into three sub-classes for convenience in
discussing various degrees of degradation. The heavy
C_{15}+ fraction increases in content down the Classes (3A
to 3C), as does the aromatic content of the oil. With
increasing degradation, the C_7/MCH ratio decreases
steadily. The amount of benzene and toluene decreases,
but the NSO and asphaltene contents are variable. The
Class 3A oils show the presence of most of the n-
alkanes, although they are depleted in the C_9-C_{13} range.
In the Class 3B oils, most of the n-alkanes beyond C_{12}
have been degraded, pristane and phytane are absent, and
the content of the other light isoprenoids is reduced.
The Class 3C oils were the most degraded, with the C_{15}+
fraction accounting for up to 95% of the oils.

The Class 3 oils (B-345, B-650, B-760, B-1307, LB-
678, and LB-335) are all found in Miocene reservoirs in
the South Bachaquero area. In this area, Class 3A oils
are deeper and further south than 3B oils while the most
degraded Class 3C oils are further north and shallower.

The Class 4 oils are distinguished by their unusual
n-alkane distribution, which is unlike that for any of
the other classes; these oils contain no n-alkanes
heavier than about C_{15}, are not condensates and have API
gravities that range from 22.9° to 24.7° (Table III).
The compounds in the light fractions, including the

Table III. Correlation of Sulfur Data with Other Parameters-BCF Oils

	$\delta^{34}S$ ($^o/_{oo}$)	%S	Reserv. T°C	Grav. API°	Asphal. %	(TOL +BZ) %	Age of Reservoir Rock
CLASS 1							
CL20	+8.95	0.38	205	36.6	1.7	14.5	Cretaceous Cogollo
CL99	+9.49	0.31	209	40.2	2.1		Cretaceous Cogollo
VLC531	+6.97	0.16	177	40.0	3.4		Eocene C
VLC642	+6.89	0.05	173	40.0	1.3		
	8.07Av	0.22Av	191Av	39.2Av	2.1 Av		
CLASS 2A							
VLA488	+8.80	1.38	116	27.5	4.7	6.75	Eocene C
SUS9L	+8.57	0.80	117	31.6	3.5		Eocene B-6
SUS54	+6.24	1.09	130	30.8	2.6		Eocene B-6
SUS158	+7.82	0.89	119	32.6	1.4		Miocene Santa Barb.
VLE357	+7.74	1.91	137	28.9	6.5		
	7.83Av	1.21Av	124Av	30.3Av	3.7Av		
CLASS 2B							
TJ742	+8.01	2.57	97	16.9	12.5	5.65	Eocene B-6
LL1932	+6.03	1.42	96	26.9	7.9		Eocene B-6
LL1199	+5.31	1.59	104	23.2	11.9		Eocene B-6
VLB446	+5.79	0.88	113	22.1	9.6		Eocene B-6
SUS12U	+9.30	1.32	113	30.5	5.9		Eocene C
	6.88Av	1.56Av	105Av	23.9Av	9.6Av		
CLASS 3A							
B750	+4.30	2.39	84	17.0	13.3	2.6	Miocene B
B1307	+4.35	2.46	87	16.9	13.3		Miocene B
	4.32Av	2.42Av	86Av	17.0Av	13.3Av		
CLASS 3B							
B650	+4.81	2.50	66	15.7	12.9	2.3	Miocene B
B484	+4.29	2.47	71	14.1	12.3		Miocene B
B345	+4.61	2.27	87	16.8	11.5		Miocene B
PB315	+6.71	2.08	63	17.9	9.4		Miocene LL
	5.10Av	2.33Av	72Av	16.1Av	11.5Av		
CLASS 3C							
LB678	+4.25	2.90	42	10.6	12.6	2.25	Miocene B
LB112	+6.66	1.69	68	11.6	11.0		Miocene LL
SEEP	+4.08	2.71	20	11.7	16.8		
	5.0Av	2.40Av	43Av	11.3Av	13.5Av		
CLASS 4							
LL405	+3.99	2.00	68	24.7	8.5	3.75	Miocene LL
LL752	+6.66	2.17	70	23.2	8.0		Miocene LL
LL1107	+6.44	1.74	71	22.9	8.8		Miocene LL
	5.7Av	1.97Av	70Av	23.6Av	8.4Av		
CLASS 5							
TJ210	+7.61	1.47	63	24.4	3.3	5.8	Miocene LL

light aromatics, appear to be similar to those in the
other oils. The properties of the bulk fractions might
otherwise be considered very similar to those of Class
2B. The Class 4 oils are all found in the Miocene L-5
reservoir in the Lagunillas field, nearshore in the
northern part of the lake.

Oil from Well TJ-210 is put in a class by itself and
is different from the other classes. It is produced from
the same sands of the L-5 reservoir as Class 4 oils,
although a short distance away. It is depleted in
n-alkanes in the C_9-C_{13} range, but shows an abundant distribution from C_{14}-C_{35}. Pristane and phytane are both
prominent. It does not appear to be unduly affected by
water washing as it contains an appreciable amount of
benzene and toluene. The distribution of components in
the C_4-C_7 range shows a larger proportion of naphthenic
hydrocarbons compared to the rest of the oils. Its
gross chemical composition would otherwise suggest a
Class 2A-type oil. The oil contains many more saturates
and n-alkanes than any other of the Miocene-reservoir
crudes analyzed.

Relationships Among Oil Classes

Grouping of the analyzed oils into five categories was
done for convenience, and although it was based on gross
chemical composition it does not necessarily have any
genetic implications. Typical whole oil gas chromatograms representative of each class of oils are given
in Figure 2. Differences in oil composition can can be
due not only to different source rocks, but also the
effects of maturation and alteration that occur in the
reservoirs (or during migration to them). The most
obvious differences among the oils in the BCF are in
their normal alkane distributions. Since these
compounds are the most susceptible to biodegradation, it
suggests that bacteria may have been one factor in
developing the observed oil chemical compositions.

Support for a common source for the Class 1 and
Class 2 oils is provided by the similarity in the
relative abundance of the isoprenoids, similarities in
the minor peaks in the total oil gas chromatograms, the
nature of the aromatic compounds, and GCMS data for biomarkers ($\underline{5}$). The nC_9 to nC_{11} sections of the gas
chromatographic traces for typical oils from Classes 1,
2A, and 2B are given in Figure 3 which shows that the
patterns of the minor peaks are almost identical. The
aromatic traces ("fingerprints") for these three oils
(Figure 4) show close similarities, although the

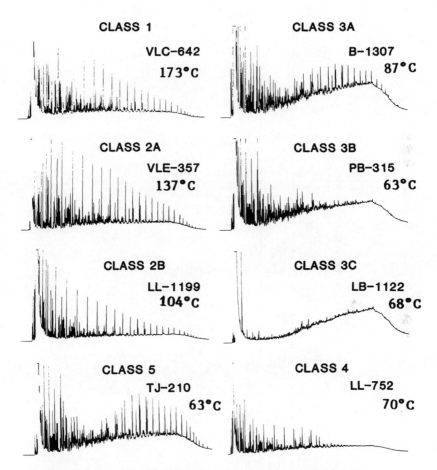

Figure 2. Whole oil gas chromatograms for representative oils from each of different classes (Reprinted with permission from Ref. 1. Copyright 1983 Bockmeulen et al.)

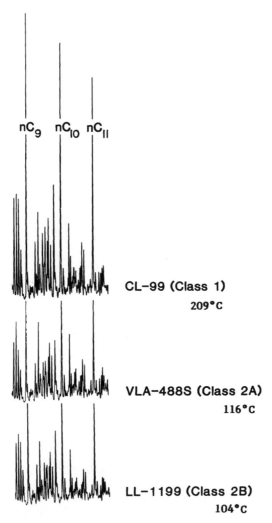

Figure 3. Details of gas chromatograms in range from nC₉ to nC₁₁ for representative oils from Classes 1, 4, and 5 (Reprinted with permission from Ref. 1. Copyright 1983 Bockmeulen et al.)

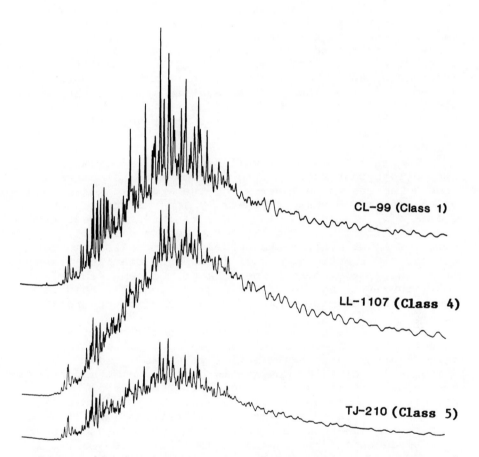

Figure 4. Gas chromatograms for aromatic fractions of representative oils from Classes 1, 2A, and 2B (Reprinted with permission from Ref. 1. Copyright 1983 Bockmeulen et al.)

aromatic response for the deep Cretaceous oils is somewhat higher, probably reflecting a slightly greater
maturity. Because aromatic compounds resist biodegradation more than n-alkanes, their distribution in the C_{15}
fraction will be either unaffected or considerably less
affected as degradation proceeds. Further, in the light
fractions, the small aromatic molecules can be removed
by the water washing that accompanies biodegradation,
and this leads to a reduced concentration of benzene and
toluene. Multiringed cyclic compounds and many highly
branched saturates are also resistant to bacterial
degradation and remain unchanged in relative amounts
during the early stages of degradation, so that the
similarity in the minor peaks (such as those shown in
Figure 3) suggests a common ancestor for the oils now
distinguished as Classes 1 and 2 on the basis of normal
alkane distribution.

The Class 3 oils show various stages of normal
alkane loss that suggest progressive bacterial degradation. The gas chromatograms for the aromatic fractions
of the 3A and 3B classes generally resemble those for
the 2A and 2B oils, but show some minor differences from
Class 1 (Figure 6). The lighter ends are difficult to
correlate, but minor peaks between the normals show very
similar distributions. Thus, it appears that the Class
1, 2, and 3 oils could be members of the same family,
but with differing degrees of degradation. If this is
true, then presumably they share a common source rock
(1,3).

Class 4 oils present a somewhat different pattern.
In terms of C_9-C_{11} and corresponding range of aromatic
compounds, there is a similarity to Class 1 oils
(Figures 4 and 5), an observation which would be consistent with a common source. However, there is a
difference in the normal alkane distribution when Class
4 oils are compared to Class 1 or Class 2 oils. Class 4
oils have a lower benzene and toluene content and appear
more biodegraded relative to Class 1 oils (see Tables I
and III). The Class 4 oils are all in a Miocene reservoir, the L-5, which is in permeable contact with the
Eocene B sands below. Other oils in the area do not
show the same normal alkane distribution, but these
differences are insufficient to invoke a separate source
rock. The TJ-210 oil (Class 5) is also reservoired in
an L-5 sand and is unusual in that it is depleted in
normal alkanes between C_9 and C_{13} (Figure 2). It is possible that the Class 5 and the Class 4 oils are related
by some process of natural distillation as reported (6)
for Trinidad oils. In this process the light ends (C_{15}+)

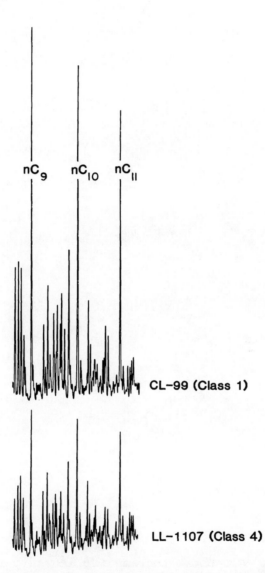

Figure 5. Details of gas chromatograms in range from nC_9 to nC_{11} for representative oils from Classes 1 and 4 (Reprinted with permission from Ref. 1. Copyright 1983 Bockmeulen et al.)

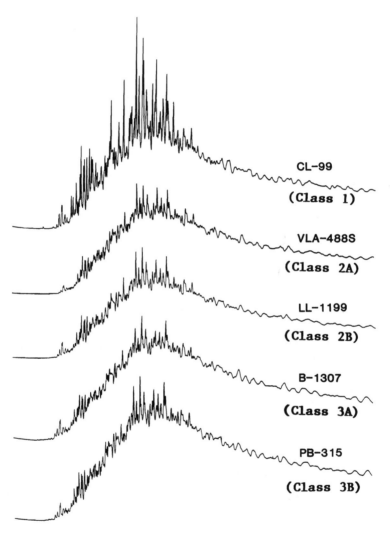

Figure 6. Gas chromatograms for aromatic fractions of
representative oils from Classes 1, 2A, 2B, 3A and 3B
(Reprinted with permission from Ref. 1. Copyright
1983 Bockmeulen et al.)

that characterize the Class 4 oils would be separated
from a Class 1 parent to leave a residue resembling TJ-
210. This type of process has been demonstrated
experimentally by heating a sample of the CL-99 or VLC-
642 Class 1 oil at 100°C in a flask fitted with a vertical sandpacked column. After several days, the material distilled up into the sand column resembled the
Class 4 oils while the residue in the flask had a composition very close to that of oil TJ-210.

In summary, all the BCF oils analyzed showed very
similar patterns for minor components and aromatics, and
appeared to be related. Major compositional trends
reflect variations in the amounts of normal alkanes.
The sequence of their removal is the same as that
observed in many other areas and attributed to bacterial
degradation. A few oils show unusual normal alkane distributions that are not well explained by currently
understood degradation or migration processes.

In order to clarify some of the above discussed
trends we have analyzed oils from the BCF for δ^{34}S and S
contents and compared the results with other known
parameters.

Analysis of δ^{34}S and S Data. Table III lists the sulfur
data along with other parameters. The δ^{34}S values are
self-consistent with the classes of oil identified by
Bockmeulen et al. ([1]). δ^{34}S values correlate well with
other class parameters in that they are consistent with
1) chemical composition as indicated by API° and percent
asphaltenes, 2) age of producing formation, and 3) evidence of biodegradation. Separating the data into two
groups, non- or slightly-biodegraded (Class 1, 2A, and
2B) and heavily biodegraded (Class 3A, 3B, 3C, and 4),
the δ^{34}S values for the non- or slightly-biodegraded
group average +7.56. On the assumption that the variability in sampling can be about 2 del units, these data
are remarkably consistent and are a strong indication
that all of the oils in Classes 1, 2A, and 2B are from
the same source. The δ^{34}S values for oils in Classes 3A,
3B, 3C, and 4 average +5.1. The average differences in
δ^{34}S are still not large, only 2.5 °/$_{oo}$, providing evidence that all of the oils are from the same source.
Class 3 oils, however, do seem to have lower ^{34}S values
in that only two of the nine samples cause the large
range difference. Class 5 seems to be an intermediary
oil, falling between Classes 2 and 4 (see also Figure
10).

There are several ways of interpreting the differences between the del values of oils in Classes 1 and 2

and Class 3. The lower values in Class 3 could be
interpreted as representing biodegradation of the oils.
Biodegradation generally does not involve major sulfur
bond changes. The sulfur compounds preferentially
survive, and the degraded oils should retain the isotopic value of the undegraded precursor oil. If organic
sulfur bonds are ruptured, normal kinetic isotope
effects favoring the lighter ^{32}S would increase the $\delta^{34}S$
value of the unreacted material rather than the decrease
observed. An alternative explanation is that the
observed change in del value is due to microbial action
on the sulfate in associated waters resulting in lighter
sulfur being added to the oil. Sulfate tends to have
significantly positive $\delta^{34}S$ values, and the natural isotope selectivity during its reduction ranges from negligible to very large. Hence, oil that has incorporated
products of reduction may acquire ^{34}S enrichments or
depletions. Orr (7) presented data where high temperature maturation and sulfate interactions enriched the
oil in ^{34}S. Thus, there is another explanation for the
differences in the data. The formation of $S°$ and H_2S by
thermochemical sulfate reduction and sulfur introduction
at higher temperatures is possible for Class 1 and 2
oils if adequate sulfate is available. This could make
Class 1 and 2 oils isotopically heavier. In addition,
differences in thermal maturity can result in variations
of one or two $\delta^{34}S$ units.
    The sulfur content of the oils is self-consistent
within the classes of oils and consistent with
anticipated thermal alteration effects. One would
expect biodegradation to slow down or end at oil temperatures of about 70°-80°C. In general, the data follow
these trends. With rising temperature, there is a
tendency for crude oils in reservoirs to become specifically lighter and contain an increasing amount of low
molecular weight hydrocarbons at the expense of high
molecular weight constituents. This could be due simply
to thermal alteration or due to the precipitation of
asphaltene out of the oils by gas and/or low boiling
hydrocarbon solubility changes.
    From the data, the average sulfur content (%S) of
each class of oil increases with increasing asphaltene
content (Figure 8), whereas both sulfur and asphaltene
content decrease linearly with increasing temperature
(Figure 7). This implies that both asphaltenes and
sulfur were removed from the oil during thermal alteration. Similarly, the average sulfur content of each
class of oil decreases linearly with increasing API
gravity (Figure 8), suggesting that sulfur compounds are

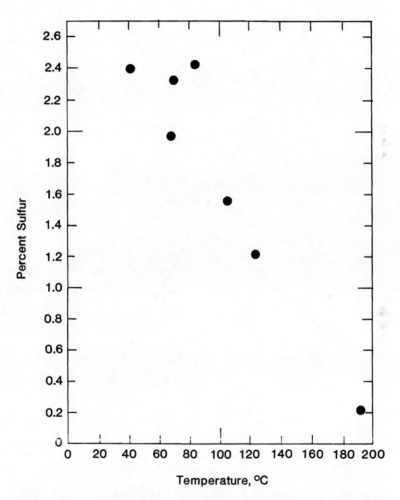

Figure 7. Variation of Sulfur Content of BCF Oils with Temperature.

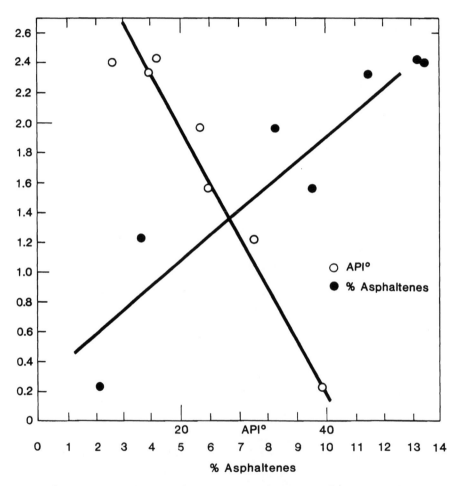

Figure 8.   Change in API Gravity and in Percent
Asphaltenes of BCF Oils as a Function of Percent
Sulfur.

not metabolized by microbial action and tend to accumulate so that degraded crudes have an increased sulfur content.

Further, the initial attack of aerobic bacteria is on the saturate paraffins and straight chain substituents of polycyclics (C_{15}+). This is consistent with the increase in C_{15}+ and a decrease in saturates components in all of the oils analyzed at lower temperatures due to higher bacterial activity. The fact that long chain substituents on the polycyclics, e.g., steranes, are affected precludes the possibility that changes in the oil, including sulfur compounds composition, are due solely to water washing (8). Thus, the correlations between all the parameters considered indicates a possibility of a combined effect of biodegradation and water washing. Increased water washing decreases the toluene and benzene contents (along with other light ends), thereby increasing the sulfur content of the undissolved oil as shown in Figure 9.

Another explanation for the lower δ^{34}S values in the degraded oil is that "water washing" selectively removed thiophene and dibenzothiophene type compounds soluble in the alcohol, acid, or phenol solutions present in the biodegraded oil-water interface. This mechanism requires that these aromatics have higher δ^{34}S values so that their removal will deplete the average δ^{34}S values of the oil. This should be observed as an increased ratio of saturated sulfur compounds over aromatic in the biodegraded oil.

Alternatively, if reduction of sulfate in associated waters occurred, the active sulfur (H_2S) would have preferentially reacted with the saturates producing thiols and sulfides. This mechanism, too, should result in an increased ratio of aliphatic sulfur compounds over aromatic in the biodegraded oil.

The fact that all of the isotopic ratios are relatively close suggests reasonably uniform source rocks and relatively minor isotopic changes by all of the interrelated alteration processes involved (9). An additional useful oil-source rock correlation is the pristane/phytane ratio. Except in the most severely degraded oils, the isoprenoids pristane (pr) and phytane (ph) are present and pr/ph ratios range from 0.75 to 1.48. For example, a plot of pr/ph versus δ^{34}S (Figure 10) shows that the Class 1 oils overlap the range of the Class 2 (A and B) oils, while the reservoir ages show a clear separation, with the Miocene and Cretaceous oils being quite distinct from those in Eocene reservoirs.

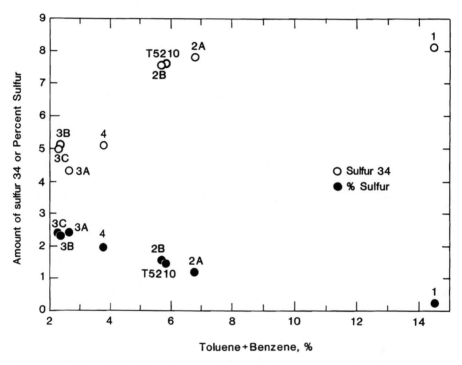

Figure 9. Total Toluene and Benzene vs. Sulfur 34 and vs. Percent Sulfur.

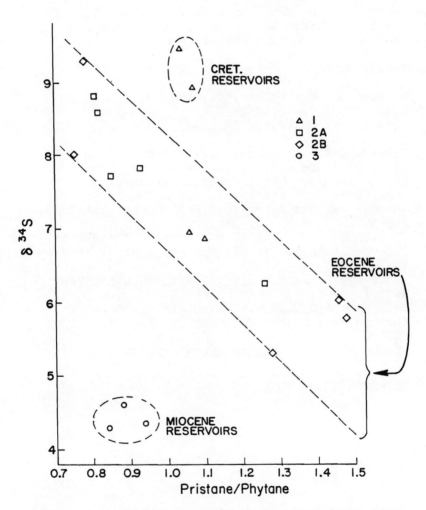

Figure 10.   Variation of $\delta^{34}S$ with Pristane/Phytane. There are only data for three Miocene reservoirs, because the oils in other Miocene reservoirs are highly degraded, therefore pristane and phytane are absent.

Acknowledgments

This research was supported in part by the U.S. Department of Energy under Contract No. DE-AC02-76CH00016. The Stable Isotope Laboratory at the University of Calgary is supported by the National Science and Engineering Research Council of Canada.

Literature Cited

1. Bockmeulen, H.; Barker, C.; Dickey, P. AAPG Bull. 1983, 67(2), 242-70.
2. Blaser, R.; White, C. AAPG Mem. 1984, 35, 229-52.
3. Talukdar, S.; Gallango, O.; Chin-A-Lien, M. In Advances in Geochemistry; Leythaeuser, D.; Rullkotter, J., Eds.; Pergamon Press, Inc.: New York, 1985; pp. 261-79.
4. Yanagisawa, F.; Sakai, H. Anal. Chem. 1983, 55, 985-87.
5. Flory, D.; Lichenstein, H.; Bieman, K.; Biller, J.E. and Barker, C. Oil and Gas Journal 1983, 91-98.
6. Ross, L.; Ames, R. Oil and Gas Journal 1988, 86(39), 72-6.
7. Orr, W. AAPG Bull. 1974, 50(11), 2295-2318.
8. Lafarge, E.; Barker, C. AAPG Bull. 1988, 72(3), 263-78.
9. Krouse, R.H. J. Chemical Exploration 1987, 7, 189-211.

RECEIVED March 5, 1990

Chapter 31

Distribution of Organic Sulfur Compounds in Mesozoic and Cenozoic Sediments from the Atlantic and Pacific Oceans and the Gulf of California

H. Lo ten Haven[1,3], Jürgen Rullkötter[1], Jaap S. Sinninghe Damsté[2], and Jan W. de Leeuw[2]

[1]Faculty of Chemical Engineering and Materials Science, Organic Geochemistry Unit, Delft University of Technology, De Vries van Heystplantsoen 2, 2628 RZ Delft, Netherlands
[2]Organic Geochemistry Unit, Delft University of Technology, De Vries van Heystplantsoen 2, NL–2628 RZ Delft, Netherlands

Gas chromatography - mass spectrometry data of the "aromatic hydrocarbon" fractions of nearly 100 Deep Sea Drilling Project (DSDP) and Ocean Drilling Program (ODP) sediment samples have been re-examined for the occurrence of organic sulfur compounds (OSC). Approximately 70% of the samples contain OSC with varying distribution patterns, although C_{20} isoprenoid thiophenes are invariably present.

Cenozoic samples, mostly from high productivity areas, often contain a large variety of OSC with the C_{25} and/or C_{27} 2-alkyl-thiophenes sometimes being particularly abundant. In others, two stereoisomeric C_{25} highly-branched isoprenoid thiophenes are the dominant compounds, although the saturated counterparts were tentatively identified only in a few cases. In samples from the Peruvian upwelling area a C_{20} isoprenoidal cyclic trisulfide is often the most significant organic sulfur component.

In Mesozoic samples, mostly Cretaceous and Jurassic black shales, the abundance and variance of OSC is generally low. A few samples of Jurassic age show a distribution pattern of C_{20} isoprenoid thiophenes which may indicate increased salinity during deposition. A C_{20} isoprenoid thiolane is the most abundant compound in black shales from the Falkland Plateau, whereas at other locations the C_{35} thienylhopane dominates. In one case two C_{33} mid-chain 2,5-dialkylthiophenes were identified.

[3]Current address: Institut Français du Pétrole, BP 311, F–92506, Rueil Malmaison, Cedex, France

Treib's discovery of porphyrins in geological samples (1) is considered to be the beginning of molecular organic geochemistry, relating organic compounds occurring in the geosphere to natural precursors in the biosphere; this is known as the biological marker concept. The identification of a thiophene-containing hopanoid (2) marks the start of a new epoch, featuring studies related to the organic geochemistry of sulfur compounds. Several papers, dealing with the occurrence, distribution, and structural identification of organic sulfur compounds (OSC) have been published recently (e.g., 3-8). Some of the most common OSC have previously been tentatively identified as alkylated thiophenes in sediments recovered by the Deep Sea Drilling Project (DSDP) (9-12). Here we extend the report on the occurrence and distribution of OSC in deep-sea sediments by a re-examination of gas chromatography - mass spectrometry (GC-MS) data of the "aromatic hydrocarbon" fractions of almost 100 samples from various parts of the world's oceans (Figure 1), collected and analyzed during 12 years participation of the Institute of Petroleum and Organic Geochemistry (KFA Jülich, F.R.G.) in the Deep Sea Drilling Project and its successor project, the Ocean Drilling Program (ODP).

Samples and Experimental Methods

The samples re-investigated are listed in Table I, together with information on their sub-bottom depth, age, lithology, and organic-carbon content. The locations of the drilling sites are shown in Figure 1. Offshore California, in the Gulf of California and on the Falkland Plateau the present geothermal gradients are particularly high (ca. 70-150°C/km), whereas in the other holes moderate geothermal conditions were encountered (ca. 20-40°C/km) (13).

The procedures for sample preparation and the operating conditions of the GC-MS system have been described elsewhere (9-12, 14-16; ten Haven et al., Proc. ODP, Sci. Results, 112, in press). All GC-MS data had been stored on magnetic tapes, but in some cases could not be retrieved. In these cases, indicated by an asterisk in Table I, old hardcopy files were checked for the occurrence of OSC. Mass chromatograms of characteristic ions (m/z 87, 97, 98, 101, 111, 115, 125, 129, 139, 195, 265, 331, 341), in combination with published information on relative retention times (3-5, 8), were used to identify the various compounds.

To confirm the identification of 2-tricosylthiophene, this compound was synthesized following standard procedures (17) with thiophene and tricosanoic acid as starting products.

The polar fraction of one sample from the Peru continental margin (679, E24359) was subjected to Raney Nickel desulfurization (5). The hydrocarbon fraction isolated from the desulfurized mixture was analyzed by GC-MS.

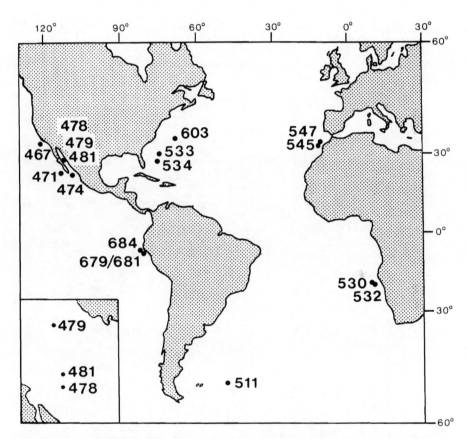

Figure 1. Map showing locations of DSDP/ODP drilling sites in the Atlantic and Pacific Oceans and in the Gulf of California (inset).

Table I. Background information on DSDP/ODP samples investigated, abundance of extractable OSC and most abundant OSC

Lab. code	Hole, core, section, interval (cm)	Depth (mbsf)	Age	Lithology	C_{org} (%)	Ref.	A**	B***
E8444*	467-13-4, 132-148	116.4	late Pliocene	calcareous claystone	3.70	9	++	7
E8445*	467-36-2, 133-150	331.9	early Pliocene	silty nannofossil clay	5.40	9	+	7
E8446	467-54-2, 100-115	502.6	late Miocene	silty claystone	4.37	9	+++	8
E8447	467-63-2, 130-133	588.3	late Miocene	calcareous claystone	1.46	9	++	2
E8448	467-97-2, 105-109	911.1	middle Miocene	calcareous claystone	1.63	9	+++	8
E8449	467-104-1, 145-150	976.5	middle Miocene	silty claystone	2.30	9	++	8
E8450	471-3-2, 120-137	21.8	Pleistocene	silty nannofossil clay	0.86	9	o	
E8451	471-13-7, 112-130	124.2	late Miocene	diatomaceous clay	0.72	9	+	7
E8452*	471-34-2, 103-107	316.0	late(?) Miocene	silty claystone	0.81	9	++	8
E8453	471-44-1, 100-118	409.6	middle Miocene	silty claystone	1.12	9	++	8
E8454*	471-57-3, 100-126	536.1	middle Miocene	silty claystone	0.78	9	+	8
E8455*	471-69-3, 135-150	650.4	m. Miocene (?)	silty claystone	0.70	9	+	8
E8456*	474-6-5, 0-15	46.0	early Pleistocene	silty nannofossil clay	1.24	10	+++	7
E8457	474A-7-2, 110-125	223.1	early Pleistocene	silty clay	1.16	10	+++	7
E8458	474A-21-6, 48-63	352.0	early Pleistocene	silty claystone	1.49	10	++	7
E8459	474A-32-2, 135-150	450.7	l. Plio./e. Pleist.	silty claystone	1.78	10	++	7
E8460	474A-40-3, 120-140	528.8	early Pliocene	silty claystone	0.52	10	++	8
E8461	478-6-3, 120-137	45.8	late Pleistocene	diatomaceous ooze	1.37	10	o	
E8462	478-11-4, 120-136	94.8	late Pleistocene	diatomaceous mud	1.59	10	+	2
E8463*	478-17-4, 120-139	151.8	late Pleistocene	diatomaceous mud	1.12	10	o	
E8464*	478-28-4, 120-140	248.3	late Pleistocene	diatomaceous claystone	2.19	10	++	8
E8465	478-35-5, 110-125	305.2	late Pleistocene	sand	0.50	10	+	8

E8466	479-9-2, 110-125	72.2	late Pleistocene	diatomaceous mud	2.36	10	+	2
E8467	479-17-5, 120-135	152.8	late Pleistocene	diatomaceous mud	2.19	10	o	
E8468	479-27-4, 135-150	246.4	early Pleistocene	diatomaceous mud	2.78	10	+	8
E8469	479-34-5, 110-125	314.2	early Pleistocene	diatomaceous mud	3.13	10	+	8
E8470	479-39-4, 110-125	359.2	early Pleistocene	diatomaceous silty clay	3.80	10	++	8
E8471	479-47-4, 110-122	436.2	e. Pleistocene (?)	claystone	1.15	10	+	8
E8472	481-8-2, 130-140	36.1	late Pleistocene	diatomaceous mud	1.54	10	o	
E8473	481A-10-2, 110-125	130.2	late Pleistocene	diatomaceous mud	5.76	10	++++	2
E8474	481A-22-4, 126-150	246.4	late Pleistocene	diatomaceous mud	1.92	10	+++	8
E8475	481A-26-5, 120-135	286.8	late Pleistocene	claystone	1.31	10	+++	8
E11369	511-16-1, 135-150	139	early Oligocene	diatomaceous ooze	0.60	14	o	
E11373	511-57-2, 37-41 & 135-140	504	Albian	black shale	0.69	14	o	
E11374	511-58-1, 9-14 & 120-125	513	Aptian	black shale	1.88	14	+	4
E11376	511-60-2, 78-82	533	Barremian/Aptian	black shale	3.84	14	+++	4
E11377	511-61-1, 139-143	542	Barremian/Aptian	black shale	5.17	14	+++	4
E11378	511-62-5, 135-150	554	late Jurassic	black shale	5.07	14	+++	4
E11379	511-64-4, 135-150	571	late Jurassic	black shale	4.66	14	+++	4
E11380	511-65-1, 41-46	577	late Jurassic	black shale	4.46	14	+	4
E11381	511-66-1, 23-28	586	late Jurassic	black shale	5.02	14	++	4
E11382*	511-67-1, 114-118	597	late Jurassic	black shale	5.43	14	++	4
E11384*	511-69-2, 145-150	617	late Jurassic	black shale	4.11	14	++	4
E11385	511-70-3, 135-150	627	late Jurassic	black shale	4.31	14	++	4
E13096*	530A-9-2, 120-130	203.8	late Miocene	mud-clast conglomerate	1.04	11	+++	7
E13097	530A-13-5, 120-137	246.3	late Miocene	nannofossil ooze	0.44	11	++++	7
E13101*	530A-43-1, 130-140	525.4	late Paleocene	mudstone	0.10	11	+	10
E13104	530A-89-5, 120-130	965.3	Coniacian	claystone	0.05	11	o	

Continued on next page.

Table I, continued

Lab. code	Hole, core, section, interval (cm)	Depth (mbsf)	Age	Lithology	C_{org} (%)	Ref.	A**	B***
E13105	530A-93-5, 135-145	997.4	Coniacian	claystone/black shale	2.30	11	o	
E13106	530A-97-3, 105-110	1030.1	Cenomanian	black shale	10.60	11	o	
E13107	530A-99-5, 0-10	1050.1	Cenomanian	mudstone/black shale	0.73	11	o	
E13109	530A-103-3, 110-120	1084.3	Alb./Cenoman.	mudstone	2.79	11	o	
E13110	530A-105-1, 90-100	1095.0	Alb./Cenoman.	mudstone	0.82	11	o	
E13111	532-1-2, 110-120	2.7	late Pleistocene	nannofossil/foram. ooze	2.76	11	o	
E13115	532-29-2, 90-100	123.3	late Pliocene	nannofossil marl	3.92	11	+++	6
E13116	532-37-2, 110-120	153.3	early Pliocene	nannofossil/foram. ooze	1.92	11	+++	6
E13435	533A-11-5, 117-136	225.9	late Pliocene	silty nannofossil mud	2.24	15	++	7
E13440	534A-34-1, 141-150	832.5	Albian	claystone	1.70	15	o	
E13442	534A-44-2, 130-140	925.9	Aptian	claystone	1.82	15	o	
E13443	534A-50-2, 132-142	975.4	Barremian	claystone	1.71	15	o	
E13445	534A-77-2, 120-130	1213.8	Valangian	claystone/limestone	0.71	15	o	
E13450	534A-125-5, 127-131	1619.8	Callovian	nannofossil claystone	2.81	15	o	
E12730	545-31-1, 49-65	284.6	Cenomanian	nannofossil clay	1.01	12	+	1
E12731	545-34-1, 70-81	313.3	Cenomanian	claystone clast	1.90	12	++	1
E12733	545-38-1, 65-80	351.2	Albian	nannofossil claystone	1.30	12	+	1
E12735	545-43-2, 130-140	400.9	Albian	nannofossil claystone	2.29	12	+	1
E12740	547A-16-3, 90-105	216.5	Cretac. slump (?)	clayey nannofos. chalk	0.30	12	o	
E12743	547A-20-3, 60-75	254.2	Cretac. slump (?)	clayey nannofos. chalk	3.96	12	++	11
E12747	547A-35-2, 15-30	394.7	Campanian	clayey nannofos. chalk	0.06	12	o	
E12748	547A-41-1, 30-48	440.9	Cenomanian	claystone	0.67	12	+	1
E12750*	547A-56-1, 23-28	583.3	Cenomanian	nannofossil claystone	0.74	12	+	1
E12751	547A-66-2, 115-130	680.7	Albian	nannofossil mudstone	0.69	12	+	1

E12753	547B-5-6, 83-101	770.9	Albian	nannofossil claystone	1.29	12	+++	5
E12755	547B-6-1, 77-84	772.8	Albian	nannofossil claystone	1.87	12	++	10
E12756	547B-15-2, 8-12	847.6	Jurassic	black shale	4.75	12	+	10
E12757	547B-15-2, 14-17	847.7	Jurassic	black shale	1.13	12	o	
E12759	547B-18-1, 100-117	874.1	Jurassic	black shale	0.70	12	+	2
E12760	547B-20-2, 92-104	893.5	Jurassic	black shale	1.63	12	++	3
E12761	547B-22-1, 13-18	905.2	Jurassic	black shale	1.03	12	+	3
E12764	547B-30-3, 59-75	981.2	Triassic (?)	sandy mudstone	0.28	12	o	
E18099	603B-32-1, 30-32	1109.8	Coniac./Turon.	clayey sandstone	0.14	16	o	
E18100	603B-34-1, 77-99	1128.2	Cenomanian	black claystone	9.32	16	+	1
E18101	603B-34-3, 68-70	1132.6	Cenomanian	black claystone	14.50	16	++	10
E24353	679C-2-4, 30-37	13.9	Quaternary	diatom./foram. mud	3.72	17	+	2
E24355	679C-4-4, 30-37	32.9	Quaternary	diatom./foram. mud	3.21	17	+	2
E24357	679C-6-4, 30-37	51.9	Quaternary	diatomaceous mud	2.57	17	+	2
E24359	679C-8-2, 30-37	67.9	late Pliocene	diatomaceous mud	7.52	17	+	12
E24361	681C-1-1, 30-37	0.4	Quaternary	diatomaceous mud	4.65	17	+	2
E24363	681C-2-6, 5-12	13.5	Quaternary	diatomaceous mud	1.32	17	+	12
E24365	681C-4-4, 30-36	29.8	Quaternary	diatomaceous mud	6.51	17	+	2
E24369	681C-9-4, 30-37	77.3	Quaternary	mud, silt, silty sand	0.53	17	+	2
E24371	684C-1-1, 30-37	0.4	Quaternary	diatomaceous mud	3.56	17	o	
E24375	684C-3-6, 30-37	25.2	Pliocene	diatomaceous mud	5.89	17	o	
E24376	684C-4-3, 30-37	30.2	Pliocene	silty diatomaceous mud	7.46	17	+++	12
E24377	684C-4-7, 30-37	36.2	Pliocene	diatomaceous mud	4.09	17	+++	12

* Only old output data available. ** Occurrence: o = absent, + = present, ++ = minor, +++ = abundant, ++++ = very abundant.
*** Most important organic sulfur compound (numbers refer to Figure 3).

Results and Discussion

Approximately 70% of the samples re-investigated contain OSC (Table I). In Cenozoic samples they are more common and more diverse than in Mesozoic sediments (Table II). Most of the Cenozoic samples were collected from areas of present-day high surface-water bioproductivity, which presumably prevailed already when these sediments were deposited ([18]). The samples of Mesozoic age are predominantly so-called black shales, for which the exact paleodepositional environment (oxygen depletion due to restricted water circulation or due to high productivity) is still a matter of debate ([19]).

The maturity of the organic matter in all samples is low, although differences were noted based on the extent of isomerization at chiral centers of certain unsaturated steroids: diasterenes and spirosterenes at C-20, and monoaromatic anthrasteroids at C-14 ([13], [20]). In areas with a high geothermal gradient these isomerization reactions have gone to completion at a shallower sub-bottom depth whereas in areas with a low geothermal gradient they are sometimes not even completed at much greater depth ([13]).

There is no correlation between the presence or abundance of extractable OSC in the "aromatic hydrocarbon" fraction and the organic-carbon content of a sediment, *e.g.*, in a Miocene nannofossil ooze from the Angola Basin (530, E13097) with a C_{org} content of 0.44% the "aromatic hydrocarbon" fraction consists almost entirely of OSC, whereas no OSC occur in the "aromatic hydrocarbon" fraction of a Cretaceous black shale (530, E13106) from the same hole with a C_{org} content of 10.60%. Clearly, the competition between hydrogen sulfide and/or polysulfides on one hand and available reactive iron on the other determines whether sulfur is incorporated in sedimentary organic matter. The absence of extractable OSC in the "aromatic hydrocarbon" fraction, however, does not imply that sulfur has not been incorporated at all, because sulfur-containing moieties may, of course, occur in the kerogens, asphaltenes and polar fractions of the samples. The investigation of these fractions is beyond the scope of this paper.

Partial reconstructed ion chromatograms (RIC) of three samples show how variable the distributions of OSC are (Figure 2; due to different capillary colums and different GC-MS operating conditions the elution times of identical compounds differ). In almost all samples the aromatic hydrocarbon, perylene, is dominantly present, but in the upper trace 2-tricosylthiophene (**7**; confirmed by coinjection of an authentic standard) is by far the most abundant OSC, while in the other two examples this compound does not occur at all. Two stereoisomeric C_{25} highly-branched isoprenoid thiophenes (**8**) dominate in a late Pleistocene claystone from the Gulf of California (481, E8474), while 3-methyl-2-(3,7,11-trimethyldodecyl)thiolane (**4**) is abundant is in a late Jurassic black shale from the Falkland Plateau (511, E11381). The most abundant OSC of each sample is given in Table I and their structures are shown in Figure 3. Reference to these structures in the text is made by bold numbers. The abundances of these OSC

Table II. Relative abundance of the major OSC in the "aromatic hydrocarbon" fraction of DSDP/ODP sediment extracts *

Site	Sample	Compounds **								others ***
		1	2	3	4	5	6	7	8	
467	E8444	+	++	-	-	-	+	++	-	
	E8445	+	+	-	-	-	+	++	+	
	E8446	+++	+++	+	-	-	+	++++	++++	9
	E8447	++	+++	+	++	+++	o	+	++	
	E8448	+	++++	o	+	+	o	+	++++	9
	E8449	+	+	+	-	+	+	+	+++	
471	E8451	o	o	o	o	o	o	+	o	
	E8452	+	+	o	-	+	o	o	++	9
	E8453	+	+	+	+	+	+	+	+++	9
	E8454	+	+	-	-	-	-	-	++	
	E8455	-	-	-	-	-	-	-	+	
474	E8456	+	+	-	-	-	++++	++++	-	
	E8457	+	+	+	o	+	+	++++	+	
	E8458	o	+	o	o	o	o	++	++	
	E8459	+	+	o	+	+	+++	++	+++	
	E8460	+	+	o	+	o	o	+	++	
478	E8462	+	+	o	o	o	o	o	o	
	E8464	+	++	-	-	-	-	-	++	
	E8465	+	+	o	o	o	o	o	+	

Continued on next page.

Table II, continued

Site	Sample	Compounds[**]								others[***]
		1	2	3	4	5	6	7	8	
479	E8466	o	+	o	o	o	o	o	o	
	E8468	o	o	o	o	o	o	o	+	
	E8469	+	+	o	+	o	o	o	++	
	E8470	o	o	o	o	o	o	o	+	
	E8471	+	+	o	o	o	o	o	+	
481	E8473	++	+++	+	+++	+	+	++	+	9, 13
	E8474	+++	++++	+	o	++	o	o	++++	
	E8475	++	++	+	+	+	o	o	+++	
511	E11374	o	o	o	++	o	o	o	o	
	E11376	+	o	o	+++	o	o	o	o	
	E11377	+	+	o	++++	++	o	o	o	
	E11378	o	o	o	++++	o	o	o	o	
	E11379	+	+	o	++++	+	o	o	o	
	E11380	o	o	o	+	o	o	o	o	
	E11381	o	o	o	++++	+	o	o	o	
	E11382	o	o	o	++++	o	o	o	o	
	E11384	-	-	-	++++	-	-	-	-	
	E11385	-	-	-	+++	-	-	-	-	

					10	10	11											10	10
·	o	·	o	+	o	o	o	o	o	o	o	o	·	o	o	o	o	o	o
+	+	·	+	+	+	o	o	o	o	+	o	o	·	o	o	o	o	+	o
·	+	·	+	+	+	o	o	o	o	+	o	o	·	o	o	o	o	+	o
·	o	·	+	+	+	o	+	o	o	+	o	+	·	+	+	o	+	+	o
·	o	·	+	o	o	o	o	o	o	o	o	o	·	o	o	o	o	o	o
+	o	·	+	+	o	o	+	o	o	+	o	o	·	+	o	o	+	+	o
+	+	·	+	+	+	+	+	+	+	+	+	+	+	+	+	o	+	+	+
+	+	·	+	+	+	+	+	+	+	+	+	+	+	+	o	+	+	+	+
E13096	E13097	E13101	E13115	E13116	E13435	E12730	E12731	E12733	E12735	E12743	E12748	E12750	E12751	E12753	E12755	E12756	E12759	E12760	E12761
530			532		533	545				547									

Continued on next page.

Table II, continued

Site	Sample	Compounds**								others***
		1	2	3	4	5	6	7	8	
603	E18100	++	++	+	o	+	o	o	o	
	E18101	++	++	+	+	+	o	o	o	10
679	E24353	+	++	o	o	o	o	o	o	12
	E24355	+	++	o	o	o	o	o	o	12
	E24357	+	+	o	o	+	o	o	o	12
	E24359	+	+	o	o	+	o	o	o	12
681	E24361	+	++	o	o	o	o	o	o	
	E24363	o	+	o	o	o	o	o	o	12
	E24365	+	+	o	o	o	o	o	o	12
	E24369	+	+	o	o	o	o	o	o	12
684	E24376	+	++	o	o	o	o	o	o	12
	E24377	+	++	o	o	+	o	o	o	12

* expressed as percentages of the relative intensity in the total ion current: - = not determined (only old output data available, o = < 1%, + = 1 - 10%, ++ = 10 - 30%, +++ = 30 - 60%, ++++ = 60 - 100%.
** Numbers refer to Figure 3
*** abundance higher then 10%.

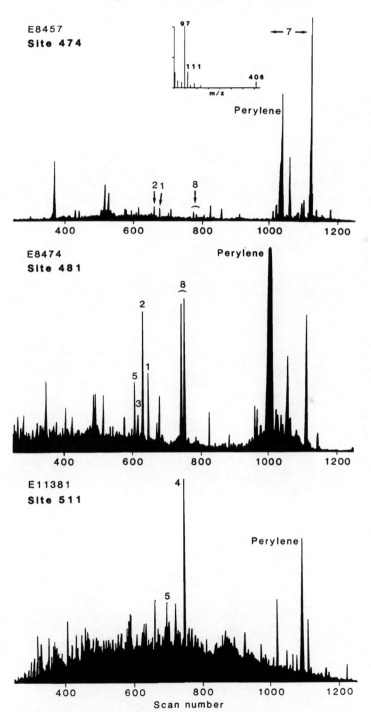

Figure 2. Partial reconstructed ion chromatograms of the "aromatic hydrocarbon" fraction of three samples. Numbers refer to Figure 3.

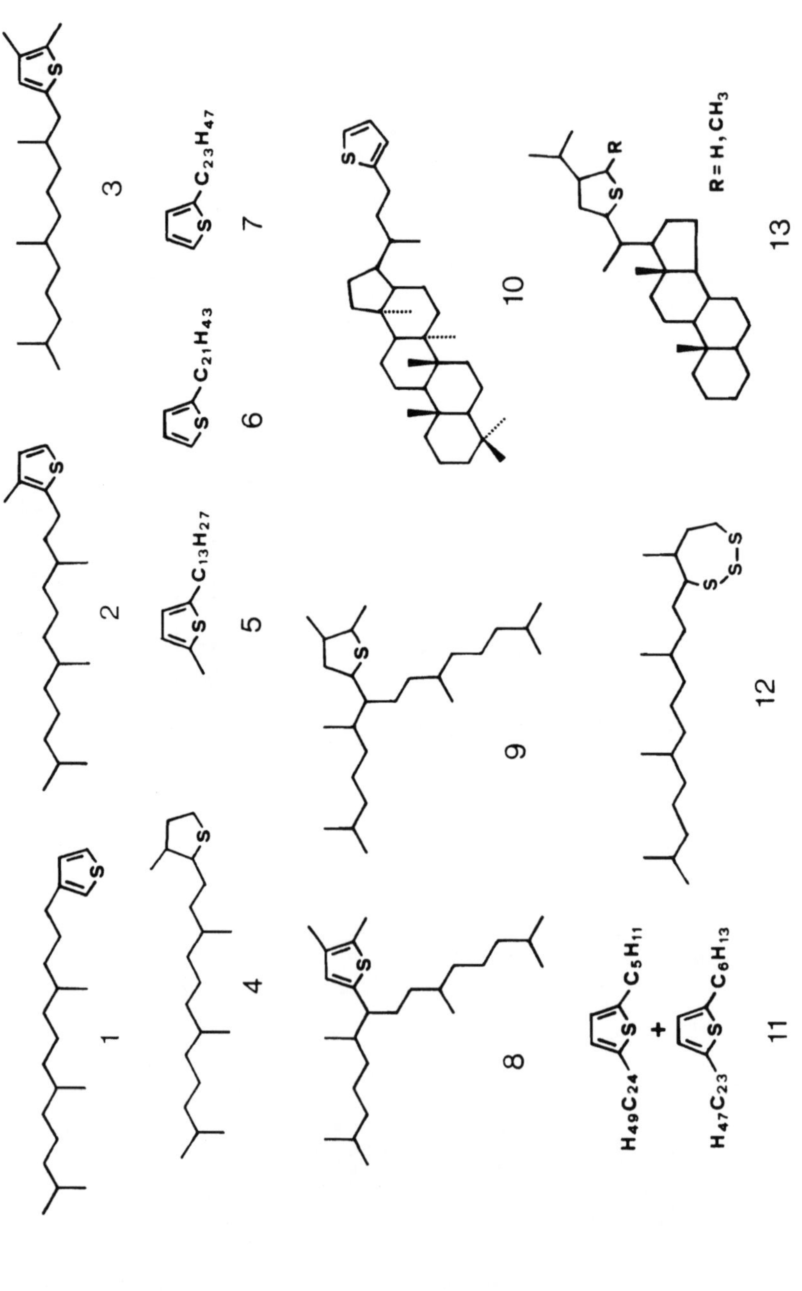

Figure 3. Selected structures of OSC discussed in text.

relative to the major component of the reconstructed total ion current is given in Table II.

Highly-Branched Isoprenoid Thiophenes. Several samples of Cenozoic age contain two stereoisomeric C_{25} highly-branched isoprenoid thiophenes (8; 2,3-dimethyl-5-(7'-(2',6',10',14'-tetramethylpentadecyl))thiophenes), and in a few cases the saturated thiolane analogs (9) were tentatively identified (Table II). These compounds are thought to be the result of sulfur incorporation into corresponding diolefinic hydrocarbons (21), which are constituents of certain diatom species (22), and, hence, were suggested to be potential biological markers for diatoms (21). Indeed, we have found the C_{25} highly-branched isoprenoid thiophenes and thiolanes in sediments with abundant diatom tests from the upwelling areas off southern California and Baja California (Sites 467, 471, 474, 478, 479, 481), but in samples from the upwelling regimes off Peru (Sites 679, 681, 684) and Namibia (Sites 530, 532) these compounds are absent or present in trace amounts only, with the exception of the deepest sample from Site 532.

In surface sediments off Peru the diolefinic precursor hydrocarbons are present in relatively high abundance, but a strong decrease in concentration with depth has been observed (23). If the reaction of sulfur species with the highly-branched diolefins is responsible for this decrease and no corresponding thiophenes and/or thiolanes can be found in the upper sedimentary sequence of any of the holes drilled offshore Peru (the depth range studied by us was from 0 to 80 meters below sea floor, mbsf), then intermolecular cross-linking with the formation of macro molecular substances may have occurred. The C_{25} highly-branched isoprenoid thiophenes may then be liberated only at a later stage of diagenesis. There are indications for this in the holes drilled off California and in the Gulf of California, where in the upper 200 m of the sedimentary column the highly-branched isoprenoid thiophenes were also not detected, with the exception of sample E8473 (130.2 mbsf) from Hole 481A. In these holes there seems to be an increase in relative abundance of these sulfur compounds at greater depths (Table II). In order to study this further, the polar fraction of one sample from the Peru upwelling area was treated with Raney Ni. The hydrocarbon fraction isolated from the desulfurized mixture was dominated by phytane, but also contained considerable amounts of the C_{25} highly-branched isoprenoid alkane, indicating that the mechanism proposed above may be valid. The analysis of other samples form deeper levels and of more desulfurization experiments of polar fractions are pending to confirm this.

2-Alkylthiophenes. We have shown in Figure 2 that 2-tricosylthiophene (7) is very abundant in several samples studied. In a few cases (Tables I and II) this is accompanied by 2-henicosylthiophene (6) in high relative concentration. A similarly high abundance of these compounds has been observed previously in a sample of Cenozoic age (173 mbsf) from Hole 532 (3). The distribution pattern

of 2-alkylthiophenes in nine samples is given in Figure 4, which is based on peak
height measurements in mass chromatograms of m/z 97+98. It includes the
relative abundance of 3-(4,8,12-trimethyltridecyl)thiophene (1), which has a base
peak at m/z 98 (8). Prominent C_{25} and C_{27} thiophenes occur in many of the
Cenozoic but not in Mesozoic sediments studied (e.g., samples E12731, E12743
of Cretaceous age from the Mazagan Escarpment, Sites 545 and 547; Figure 4).
Instead, the older samples often contain a homologous series of 2-
alkylthiophenes over a wider carbon number range (Figure 4). It is noteworthy
that the distribution of 2-methyl-5-alkylthiophenes, according to mass
chromatograms of m/z 111, does not reveal any predominance of C_{25} and/or
C_{27} compounds. Hence, the precursors of the C_{25} and C_{27} 2-alkylthiophenes and
the diagenetic reaction leading to their formation must be highly specific.
Suitable precursors of these two 2-alkylthiophenes can be proposed based on the
model of quenching of labile functionalised lipids by inorganic sulfur species
(24), but such precursors have not yet been reported to occur in the biosphere.

Isoprenoid Thiophenes and Thiolanes. The two C_{20} isoprenoid thiophenes, 3-
(4,8,12-trimethyltridecyl)thiophene (1) and 3-methyl-2-(3,7,11-trimethyldodecyl)-
thiophene (2), are the only OSC which occur in almost all the sediments
included in this study. In several samples another C_{20} isoprenoid thiophene
isomer, 2,3-dimethyl-5-(2,6,10-trimethylundecyl)thiophene (3), is also present,
but usually only in trace amounts. Only in the Jurassic black shales from Hole
547B at the Mazagan Escarpment on the Moroccan margin (Figure 1) the
concentration of compound 3 exceeds those of the other two C_{20} isoprenoid
thiophenes. This is shown in Figure 5 by a mass chromatogram of m/z 308, the
parent ion of C_{20} thiophenes. A distribution pattern like the one shown for the
Jurassic sample has been interpreted to indicate increased salinity in the water
column during deposition (25). For the Moroccan continental margin this
interpretation is supported to some extent by local geological features (26).
After rifting started in the Triassic, a series of small basins were created,
separated by faulted and rotated basement blocks. In some of these basins halite
with minor potash salts was deposited in shallow marine salt ponds, although not
in the basin drilled at Site 547. Possibly in this basin the salinity never reached
the level of oversaturation with respect to gypsum and/or halite, but
nevertheless the salinity may have been enhanced.
    The occurrence of a C_{20} isoprenoid thiolane, 3-methyl-2-(3,7,11-
trimethyldodecyl)thiolane (4), is more restricted (Table II), but it is the most
important compound of the "aromatic hydrocarbon" fraction of several Mesozoic
black shales from the Falkland Plateau (e.g., late Jurassic sample E11381 from
Site 511; Figure 2). In these samples, additional isoprenoid thiolanes with a base
peak at m/z 115 occur in great abundance. Thiolanes have tentatively been
suggested as intermediates in the formation of the corresponding thiophenes
(24). This is in contradiction to the observation of abundant isoprenoid thiolanes
in samples which have experienced more thermal stress and thus are

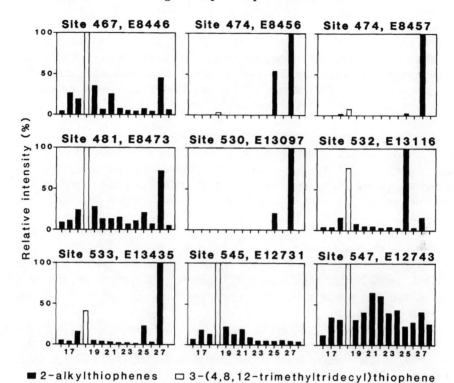

Figure 4. Histograms, based on mass chromatograms of m/z 97+98, showing distribution of 2-alkylthiophenes and 3-(4,8,12-trimethyltridecyl)thiophene.

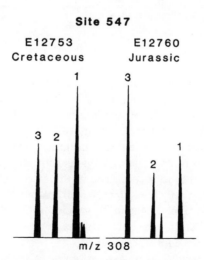

Figure 5. Partial mass chromatograms of m/z 308, showing the distribution of C_{20} isoprenoid thiophenes. Numbers refer to Figure 3.

diagenetically more advanced than others from the Atlantic Ocean, which contain thiophenes and almost no thiolanes. Thus, it may also be possible that reduction of thiophenes becomes effective at a later stage of diagenesis. Alternatively, thiolanes may have been liberated from cross-linked high-molecular-weight material and have overwhelmed the thiophenes at the higher temperature, *e.g.*, in the Falkland Plateau.

Other Characteristic Organic Sulfur Compounds. The thienylhopane (10), which has previously been found in deep-sea sediments (2), in some cases occurs as the most important OSC. However, we cannot exclude the possibility that its occurrence is more widespread than reported in Table II, because sometimes the conditions of GS-MS data aquisition were not appropriate to detect this compound.

In a Cretaceous slump clast from Site 547 off Morocco (E12743), recovered in the Eocene section, two coeluting C_{33} mid-chain thiophenes (11; m/z 490(55%), 433(15%), 419(18%), 181(92%), 167(73%), 111(100%)) were detected. Similarly high abundances of C_{33} mid-chain thiophenes and thiolanes were observed in a sample from the Nördlinger Ries (27) and in the Jurf ed Darawish oil shale (28).

Two thiolane steroids, for which the structures are also presented in Figure 3 (13), were only found in a late Pleistocene mud from the Gulf of California (481, E8473). This sample contains the largest variety of OSC of all samples investigated.

Most sediments from the Peruvian upwelling area contain a C_{20} isoprenoid trisulfide (12), as well as the corresponding isoprenoid disulfides. These compounds have been found to date only in one sediment from the northern Apennines and an incorporation of polysulfides has been invoked to explain their occurrence (29).

Conclusions

The brief discussion above shows the difficulty of a straightforward paleoenvironmental assessment based on OSC. Their occurrence seems not to be related to any special type of environment; on the contrary, they are widespread. However, several of the encountered compounds cannot be related chemotaxonomically to specific biological contributions yet, because information on the lipid composition of appropriate organisms is still scant. Nevertheless, it is clear that much information is contained in the extractable OSC, information which is not evident when only the aliphatic hydrocarbon fraction is investigated. Furthermore, high-molecular-weight sulfur species not directly amenable to GC-MS analysis is another source of information, but chemical treatment (desulfurization) is required for its disclosure.

For the paleoenvironmental assessment of deep-sea sediments the study of the concentrations and distribution patterns of various OSC is just another fragment needed to reconstruct a larger mosaique centering around the question of oxygen depletion in the water column and the subsequently favored preservation of organic matter, either as a consequence of reduced water circulation ("Black Sea model", stagnant conditions), or by high biological productivity in the surface water ("oxygen minimum model", dynamic situation). Organic sulfur compounds may play a key role in unraveling this controversy because the occurrence of reactive inorganic sulfur species (*e.g.*, polysulfides) are sensitive to variations in the degree of oxygen depletion as long as these variations also have an effect in the upper part of the sediment column. It may be necessary, however, to develop a more specific sampling strategy for the investigation of OSC, which may differ from that for the collection of the sample set presented in this study.

Acknowledgments

Samples were made available by the support of the Deep Sea Drilling Project and the Ocean Drilling Program. This overview work was supported by the Deutsche Forschungsgemeinschaft (DFG), grant No. We 346/27, while the original data collection have been financially supported by the DFG with grant Nos. We 346/21, We 346/23, We 346/24 and We 346/25.

Literature Cited

1. Treibs, A. Ann. Chem. 1936, 49, 682-686.
2. Valisolalao, J.; Perakis, N.; Chappe, B.; Albrecht, P. Tetrahedron Lett. 1984, 1183-86.
3. Brassell, S. C.; Lewis, C. A.; de Leeuw, J. W.; de Lange, F.; Sinninghe Damsté, J. S. Nature 1986, 329, 160-62.
4. Sinninghe Damsté, J. S.; ten Haven, H. L.; de Leeuw, J. W.; Schenck, P. A. In Advances in Organic Geochemistry 1985; Leythaeuser, D.; Rullkötter, J., Eds.; Pergamon: Oxford, 1986; pp 791-805.
5. Sinninghe Damsté, J. S.; de Leeuw, J. W.; Kock-van Dalen, A. C.; de Zeeuw, M. A.; de Lange, F.; Rijpstra, W. I. C.; Schenck, P. A. Geochim. Cosmochim. Acta 1987, 51, 2369-91.
6. Sinninghe Damsté, J. S.; de Leeuw, J. W. Int. J. Environ. Anal. Chem. 1987, 28, 1-19.
7. Schmid, J. C.; Connan, J.; Albrecht, P. Nature 1987, 329, 54-56.
8. Rullkötter, J.; Landgraf, M.; Disko, U. J. High Res. Chrom. & Chrom. Comm. 1988, 11, 633-38.
9. Rullkötter, J.; von der Dick, H.; Welte, D. H. In Init. Repts. DSDP 63; Haq, B.; Yeats, R. S. et al., Eds.; U.S. Government Printing Office: Washington, DC, 1981; pp 819-36.
10. Rullkötter, J.; von der Dick, H.; Welte, D. H. In Init. Repts. DSDP 64; Curray, J. R.; Moore, D. G. et al., Eds.; U.S. Government Printing Office: Washington, DC, 1982; pp 837-53.
11. Rullkötter, J.; Mukhopadhyay, P. K.; Welte, D. H. In Init. Repts. DSDP 75; Hay, W. W.; Sibuet, J. C. et al., Eds.; U.S. Government Printing Office: Washington, DC, 1984; pp 1069-87.

12. Rullkötter, J.; Mukhopadhyay, P. K.; Schaefer, R. G.; Welte, D. H. In Init. Repts. DSDP 79; Hinz, K.; Winterer, E. L. et al., Eds.; U.S. Government Printing Office: Washington, DC, 1984; pp 775-806.
13. ten Haven, H. L.; Rullkötter, J.; Welte, D. H. Geol. Rundsch. 1989, 78, 841-50.
14. von der Dick, H.; Rullkötter, J.; Welte, D. H. In Init. Repts. DSDP 71; Ludwig, W. J.; Kraskeninnikov, V. A. et al., Eds.; U.S. Government Printing Office: Washington, DC, 1984; pp 1015-32.
15. Rullkötter, J.; Mukhopadhyay, P. K.; Disko, U.; Schaefer, R. G.; Welte, D. H. Mitt. Geol. Paläont. Inst. Univ. Hamburg 1986, 60, 179-203.
16. Rullkötter, J.; Mukhopadhyay, P. K.; Welte, D. H. In Init. Repts. DSDP 93; van Hinte, J. E.; Wise, S. W. et al., Eds.; U.S. Government Printing Office: Washington, DC, 1987; pp 1163-76
17. Peakman, T. M.; Kock-van Dalen, A. C. In Geochemistry of Sulfur in Fossil Fuels; Orr, W. L.; White, C. M., Eds.; ACS Symposium Series; American Chemical Society: Washington, DC; this volume.
18. Parrish, J. T.; Curtis, R. L. Palaeogeog. Palaeoclimat. Palaeoecol. 1982, 40, 31-66.
19. Stein, R.; Rullkötter, J; Welte, D. H. Chem. Geol. 1986, 56, 1-32.
20. Brassell, S. C.; McEvoy, J.; Hoffmann, C. F.; Lamb, N. A.; Peakman, T. M.; Maxwell, J. R. In Advances in Organic Geochemistry 1983; Schenck, P. A.; de Leeuw, J. W.; Lijmbach, G. W. M., Eds.; Pergamon: Oxford, 1984; pp 11-23.
21. Sinninghe Damsté, J. S.; van Koert, E. R.; Kock-van Dalen, A. C.; de Leeuw, J. W.; Schenck, P. A. Org. Geochem. 1989, 14, 555-67.
22. Nichols, P. D.; Volkman, J. K.; Palmisano, A. C.; Smith, G. A.; White, D. C. J. Phycol. 1988, 24, 90-96.
23. Volkman, J. K.; Farrington, J. W.; Gagosian, R. B.; Wakeham, S. G. In Advances in Organic Geochemistry 1981; Bjorøy, M. et al., Eds.; Wiley: Chichester, 1983; pp 228-40.
24. Sinninghe Damsté, J. S.; Rijpstra, W. I. C.; Kock-van Dalen, A. C.; de Leeuw, J. W.; Schenck, P. A. Geochim. Cosmochim Acta 1989, 53, 1343-55.
25. Sinninghe Damsté, J. S.; Rijpstra, W. I. C.; de Leeuw, J. W.; Schenck, P. A. Geochim. Cosmochim. Acta 1989, 53, 1323-41.
26. Winterer, E. L.; Hinz,K. In Init. Repts. DSDP 79; Hinz, K.; Winterer, E. L. et al., Eds.; U.S. Government Printing Office: Washington, DC, 1984; pp 893-919.
27. Rullkötter, J.; Littke, R.; Schaefer, R. G. In Geochemistry of Sulfur in Fossil Fuels; Orr, W. L.; White, C. M., Eds.; ACS Symposium Series; American Chemical Society: Washington, DC; this volume.
28. Kohnen, M. E. L.; Sinninghe Damsté, J. S.; Rijpstra, W. I. C.; de Leeuw, J. W. In Geochemistry of Sulfur in Fossil Fuels; Orr, W. L.; White, C. M., Eds.; ACS Symposium Series; American Chemical Society: Washington, DC; this volume.
29. Kohnen, M. E. L.; Sinninghe Damsté, J. S.; ten Haven, H. L.; de Leeuw, J. W. Nature 1989, 341, 640-41.

RECEIVED March 5, 1990

Chapter 32

Carbon Isotope Fractionation during Oxidation of Light Hydrocarbon Gases

Relevance to Thermochemical Sulfate Reduction in Gas Reservoirs

Y. Kiyosu[1], H. R. Krouse[2], and C. A. Viau[3]

[1]Department of Earth Sciences, Nagoya University, Nagoya, Japan
[2]Department of Physics and Astronomy, University of Calgary, Calgary, Alberta T2N 1N4, Canada
[3]Shell Canada Limited, Calgary, Alberta T2P 3S6, Canada

Relative enrichments of ^{13}C in light hydrocarbon gases imply that they have served as thermochemical sulfate reducing agents during the production of H_2S in some deep Paleozoic carbonate reservoirs of western Canada. This hypothesis was tested with laboratory measurements of carbon isotope selectivity during oxidation of light hydrocarbons. Although some experiments were carried out with sulfate as the oxidant, the accompanying slow reaction rates prompted the use of metal oxides. In flow and sealed tube experiments, $^{12}CH_4$ was oxidized 1.021 to 1.011 times faster than $^{13}CH_4$ over the temperature range 300 to 900°C. In sealed tube experiments, product CO_2 was depleted in ^{13}C by 8 and 4°/oo with respect to reactants C_2H_6 and C_3H_8 respectively in the temperature range 300 to 400°C. CH_4, relatively enriched in ^{13}C, was produced as an intermediate during C_2H_6 oxidation.

It has long been recognized that the high concentrations of H_2S in "sour gas" in deep carbonate reservoir rocks arise from thermochemical sulfate reduction (TSR) (1-4). The remaining problem is to define the reducing agents and the specific chemical reactions. Organic matter has been implicated in general terms (4) and a reaction sequence involving H_2S and sulfate proposed (1,2). Light hydrocarbon gases were identified as sulfate reducing agents in the Burnt Timber and Limestone Fields (6) in the Upper Devonian Crossfield Member (Wabamum Group) and the Leduc Formation, as well as in the Mississippian Turner Valley Formation in the deep (3,500-4,200 m) Foothills region of southwestern Alberta, Canada (7). This identification is based on the fact that these gases are more enriched in ^{13}C than expected from data obtained from other Devonian gas deposits of western Canada where H_2S was not as abundant. In particular, the $\delta^{13}C$ values for ethane and propane in deeper sour gas (>3500 m depth) are about 10°/oo higher than those in shallower sweeter gas (Figure 1). The shifts to more negative

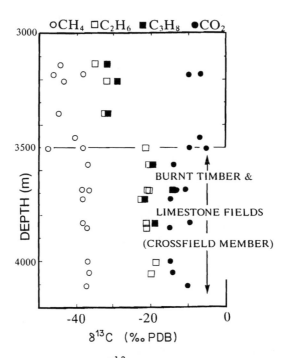

Figure 1.   Comparison of $\delta^{13}C$ values of CO_2 and light hydrocarbon gases in the Burnt Timber and Limestone Fields (Crossfield Member) to other Devonian gas occurrences at immediately shallower depths in western Canada.

and positive $\delta^{13}C$ values for CO_2 and CH_4 respectively in the deeper sour gas are smaller. This phenomenon has been recently observed in other sour gas formations in western Canada with even greater ^{13}C depletions in the CO_2 and enrichments in the hydrocarbon gases. The shifts are attributed to kinetic isotope effects whereby ^{12}C-containing gases were oxidized preferentially, producing ^{12}C-enriched CO_2 thereby enriching the unreacted gases in ^{13}C.

It is difficult to assess the extent of redox reactions in gas reservoirs because of several input and removal processes. For example, CO_2 may be generated by oxidation of organic matter or dissolution of carbonate and removed by carbonate mineral formation. Mass and isotope balances on individual gases would be possible if the kinetic isotope effects and their $\delta^{13}C$ values prior to oxidation were known. The "original" $\delta^{13}C$ values might be inferred from data in Figure 1. However, it should be possible to make use of the fact that isotope data are available for three light hydrocarbon gases. If kinetic isotope effects were known for the oxidation of each gas, it might be possible to use the field data to better characterize the in situ redox reactions.

Whereas isotope selectivity during bacterial oxidation of CH_4 has been measured (8,9), there are few data relevant to TSR. Consequently, laboratory experiments were conducted to measure carbon isotope fractionation accompanying the oxidation of methane, ethane, and propane.

Experimental

Experiments were conducted using both flow and sealed reactor techniques. In the former, methane was passed over a solid oxidant. For each run, a quartz tube with an O.D. of 7 mm, wall thickness 1 mm, length 3.5 cm and containing about 2 mmoles of the oxidant was placed in another quartz tube of I.D. 10 mm and 50 cm length. This outer tube was placed in a furnace and methane flowing at 13.4 ± 0.8 ml/min was passed through it. Initial experiments were conducted with anhydrite, but because of low reaction rates, hematite-anhydrite mixtures were tried. Flow experiments were later conducted with other oxidants such as hematite, cupric oxide, and manganese dioxide. The produced carbon dioxide was precipitated as barium carbonate in a solution of BaOH. The precipitate was weighed as well as the residual oxide to determine the extent of reaction. The barium carbonate was then reacted with phosphoric acid to generate CO_2 for isotopic analysis.

In closed tube experiments, vertical 7 mm OD, 12 cm long pyrex tubes containing 0.2 gm CuO, were evacuated. Then methane, ethane, or propane at slightly less than atmospheric pressure was transferred into them. The gases were condensed by applying liquid N_2 to the bottoms of the pyrex tubes. A gas-oxygen torch was used to collapse and seal the tubes which were then placed in ovens at constant temperatures for times that were selected to give different extents of reaction. After reaction, the contents of the tubes were separated by gas chromatography and the relative concentrations of the gases determined. CO_2 was analysed directly whereas unreacted or generated hydrocarbon gases were combusted to CO_2 prior to carbon isotope analyses by a mass spectrometer constructed from basic Micromass 903 components. The reproducibility of $\delta^{13}C$ values was $\pm 0.2^{\circ}/oo$. In all experiments, the $\delta^{13}C$ values of initial gases were obtained by quantitative combustion to CO_2 with CuO at 800°C.

Results

In the flow experiments, the oxidation rates with anhydrite were extremely slow and data could only be obtained within a reasonable time at temperatures above 600°C (Table I).

Table I. Rate Constants, k, for Anhydrite Reduction
         by Methane in Flow Experiments

Temp. °C	$k(s^{-1})$	
	$CaSO_4$	$CaSO_4 + Fe_2O_3$
930	6.1×10^{-4}	
915	4.4×10^{-4}	
900	2.5×10^{-4}	
864	8.5×10^{-5}	
850	2.4×10^{-5}	
813		7.7×10^{-4}
800	9.8×10^{-6}	
728		3.1×10^{-4}
700		1.8×10^{-4}
650		2.1×10^{-5}
600		2.5×10^{-6}

Arrhenius plots ($\ln k$ versus T^{-1}) for the data of Table I give
$CaSO_4$; $\ln k = -(4.4 \pm 0.4)10^4/T + (28.8 \pm 0.3)$, $r^2 = 0.96$
$CaSO_4:Fe_2O_3$; $\ln k = -(2.6 \pm 0.4)10^4/T + (17.7 \pm 0.7)$, $r^2 = 0.93$
Extrapolation to 200°C predicts that with $CaSO_4$ alone, the order of 10^{17} years would be required for 10 percent oxidation. Interestingly, added hematite increased the reduction rate of anhydrite (not just methane oxidation) by about a factor of 60 as shown in Table I.

In some experiments, carbon dioxide was generated without coproduction of hydrogen sulfide. However, treatment of the solid remaining after the experiment with HCℓ, evolved hydrogen sulfide stoichiometrically consistent with carbon dioxide production. This implies that under the experimental conditions, CaS was probably formed. The importance of this observation is that lack of hydrogen sulfide generation does not necessarily mean no sulfate reduction. This should be kept in mind for future experiments, or indeed for interpreting published results.

In view of the slow reaction rates with anhydrite, flow experiments were conducted with manganese dioxide, hematite, and cupric oxides. Experiments with the latter two oxidants are described elsewhere (10).

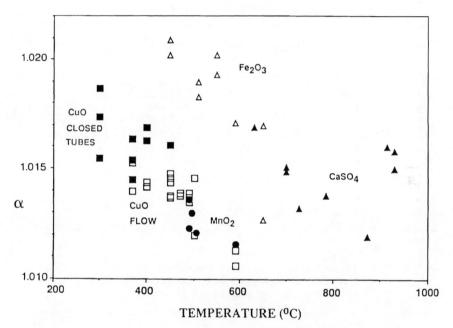

Figure 2.   Temperature dependence of the ratio of isotopic
rate constants, $\alpha = k_{12}/k_{13}$ for methane oxidation by various
oxidants.

Values of the ratio of the rate constants, $\alpha = k_{12}/k_{13}$, are
plotted as a function of temperature in Figure 2. It is seen that
(1) the α-values decrease with temperature as expected
theoretically and (2) the kinetic isotope effects are not the same
for the different oxidants. There appears to be one trend for
hematite and anhydrite and another for cupric oxide and manganese
dioxide.

Lower temperature closed tube experiments with CH_4 and CuO are
summarized in Table II. The α-values have the same temperature
dependence as the flow experiments with CuO (Figure 2). This
suggests minimal exchange of carbon isotopes between methane and
product carbon dioxide.

In closed tube experiments of ethane oxidation by CuO, methane
was a reaction intermediate (Table III).

Table II. Percentages and Carbon Isotope Composition of Reaction
           Components During Methane Oxidation by CuO. The initial
           $\delta^{13}C$ value of methane is $-34.5^{o}/oo$

Temp (°C)	Time (hour)	VOLUME PERCENT		$\delta^{13}C$ (°/oo, PDB)		k_{12}/k_{13}
		CO_2	CH_4	CO_2	CH_4	
300	48	12.2	87.8	−52.0	−32.6	1.0187
300	92	12.3	87.7	−48.8	−32.5	1.0155
300	120	14.1	85.9	−50.4	−31.9	1.0174
370	24	28.2	71.8	−48.0	−29.0	1.0164
370	24	29.8	70.2	−47.2	−28.1	1.0154
370	48	32.0	68.0	−47.7	−28.3	1.0164
370	72	36.1	63.9	−45.8	−28.1	1.0145
400	17	27.1	72.9	−47.7	−28.4	1.0163
400	48	27.1	62.9	−47.6	−26.8	1.0169
450	17	39.4	60.6	−45.1	−22.5	1.0161

Table III. Percentages and Carbon Isotope Composition of Reaction
            Components During Ethane Oxidation by CuO. The initial
            $\delta^{13}C$ value of ethane is $-27.9^{o}/oo$

Temp. (°C)	Time (hour)	VOLUME PERCENT			$\delta^{13}C$ (°/oo, PDB)			k_{12}/k_{13}
		CO_2	C_2H_6	CH_4	CO_2	C_2H_6	CH_4	
300	72	10.6	89.2	tr	−34.6	−27.2	−	1.0071
300	72	15.6	85.0	tr	−35.3	−26.0	−	1.0084
300	120	15.9	84.2	tr	−34.6	−27.0	−	1.0072
300	15	9.6	90.4	−	−34.6	−27.5	−	1.0069
330	21	11.0	89.0	−	−34.1	−27.3	−	1.0055
370	3	21.0	79.0	tr	−34.4	−27.1	−	1.0069
370	6	19.0	80.5	0.5	−34.4	−26.7	−	1.0071
370	11	24.1	75.1	0.8	−34.2	−26.2	−	1.0071
370	15	23.5	74.9	1.6	−32.6	−25.5	−	1.0058
370	24	27.1	69.1	3.8	−34.9	−23.5	−17.0	1.0079
370	48	37.9	53.4	8.7	−32.9	−21.3	−17.3	1.0060
370	48	24.1	68.8	7.1	−34.8	−22.5	−21.3	1.0077
400	2	28.2	71.2	0.6	−33.9	−26.6	−	1.0068

The $\delta^{13}C$ values of this methane are not as negative as other
components in the system. Isotope data for the different
components are plotted as a function of extent of ethane reaction,
at 370°C, in Figure 3.

Data for closed tube experiments of propane oxidation by CuO
are given in Table IV.

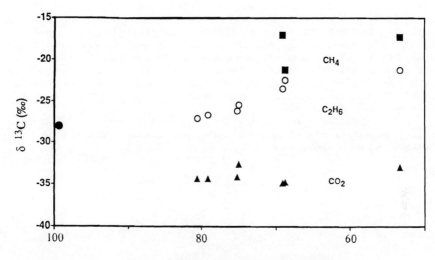

Figure 3. Carbon isotope composition of remaining C_2H_6, product CO_2, and CH_4 intermediate during oxidation of C_2H_6 by CuO at 370°C.

Table IV. Percentages and Carbon Isotope Composition of Reaction Components During Propane Oxidation by CuO. The initial $\delta^{13}C$ value of propane was $-33.2°/oo$.

Temp. (°C)	Time (hour)	VOLUME PERCENT			$\delta^{13}C$ (°/oo, PDB)		k_{12}/k_{13}
		CO_2	C_3H_8	CH_4	CO_2	C_3H_8	
280	48	35.0	65.0	–	-37.4	-31.7	1.0050
280	72	42.3	57.7	–	-37.9	-32.1	1.0052
280	72	37.2	62.8	–	-37.6	-33.0	1.0045
300	17	32.2	67.8	–	-37.3	-33.0	1.0042
300	25	39.3	60.7	–	-35.8	-31.8	1.0032
300	45	45.1	54.9	–	-37.1	-32.1	1.0044
330	15	47.0	52.1	0.9	-37.4	-31.7	1.0049
370	6	52.1	47.9	1.7	-37.1	-31.7	1.0046
370	11	59.0	39.8	2.1	-35.1	-30.5	1.0034

Although some methane is produced, it is insufficient for isotopic analysis. Neither ethane or ethene was detected.

Discussion

The oxidation of methane was very slow under the experimental conditions employed. The slowest rates are those with anhydrite as oxidant. Because the ratio of the rate constants, α, is dependent upon the oxidant, it is difficult to estimate the carbon isotope selectivity during sulfate reduction at temperatures relevant to TSR in sour gas occurrences. However, the effects are substantial with the cupric oxide-manganese dioxide and hematite-anhydrite trends in Figure 2 giving extrapolated α-values of about 1.02 and 1.04 respectively at 200°C.

The appearance of methane relatively enriched in ^{13}C during ethane oxidation may seem surprising but can be explained by the isotopic reaction sequence

$$^{12}CH_3{}^{12}CH_3 \xrightarrow{k_1} {}^{12}CO_2 + {}^{12}CH_4$$

$$^{13}C^{12}CH_6 \quad \overset{k_2}{\underset{k_3}{\rightrightarrows}} \quad \begin{array}{l} {}^{12}CO_2 + {}^{13}CH_4 \\ {}^{13}CO_2 + {}^{12}CH_4 \end{array}$$

where $k_1 > k_2 > k_3$, and

$$^{12}CH_4 \xrightarrow{k_5} {}^{12}CO_2$$

$$^{13}CH_4 \xrightarrow{k_6} {}^{13}CO_2$$

where $k_5 > k_6$

In these competitive reactions, $k_5 > k_6$ on the basis of experiments with methane alone. Whereas k_1 must be larger than either k_2 or k_3, there is a question as to which of k_2 or k_3 is larger. If $k_3 > k_2$, then methane would be produced with more negative $\delta^{13}C$ values than the CO_2. This would mean that the observed $\delta^{13}C$ values for CH_4 would be due to an upward shift of more than 20°/oo being caused solely by $k_5 > k_6$. Therefore, it seems more likely that $k_2 > k_3$ in which case both steps, production and oxidation of CH_4, promoted its enrichment in ^{13}C.

The presence of the intermediate methane in the ethane and propane oxidation experiments, coupled with failure to detect ethane or ethene intermediates in the propane experiments constitutes indirect evidence that the reaction rate constants are in the order $k_{propane} > k_{ethane} > k_{methane}$. This order is confirmed by comparing Tables II to IV. It is difficult to make a quantitative comparison because the rate constants calculated on the assumption of first order kinetics, decrease with percent reaction. This is presumably the consequence of the closed reactor conditions and for this reason, rate constants are not given in the tables. However, on average, the assumption of first order kinetics with respect to disappearance of reactant hydrocarbon yields $k_{methane}$ slower than k_{ethane} by less than an order of magnitude and k_{ethane} in turn slower than $k_{propane}$ by about an order of magnitude.

The α-values for ethane oxidation were much smaller than those of methane ranging from 1.006 to 1.008 in the temperature range 300 to 400°C. The α-values for propane were slightly lower at about 1.004.

Although the laboratory data are insufficient for quantitative evaluation of TSR by light hydrocarbon gases, they provide an understanding of the in situ $\delta^{13}C$ distributions. The large shifts in $\delta^{13}C$ values of the ethane and propane are consistent with their relatively faster oxidation rates and therefore larger percentages of reaction. The $\delta^{13}C$ value of the methane has not increased as much because a much lower percentage has been oxidized. The $\delta^{13}C$ values of the carbon dioxide have not been shifted drastically to those of the isotopically lightest methane because of partial generation from ethane, propane, and presumably higher molecular weight hydrocarbons and subsequent moderation by exchange with carbonate minerals.

It must be realized that the in situ reaction may not be simply sulfate reduction by organic matter, since reduction by H_2S may play a dominant role (1,2). While it is gratifying that the laboratory measurements of kinetic isotope effects during oxidation of light hydrocarbons qualitatively account for the in situ carbon isotope distribution, shifts in $\delta^{13}C$ values of the light hydrocarbon gases might arise from other reactions, particularly with sulfur compounds.

Acknowledgments

The authors acknowledge support for this project from Shell Canada Ltd., Shell Research International Company (SIRM) and the Natural Sciences and Engineering Research Council of Canada.

Literature Cited

1. Orr, W.L. Am. Assoc. Petrol. Geol. Bull. 1974, 50, 2295-2318.
2. Orr, W.L. in Advances in Organic Geochemistry 1975, Madrid; Enadisma, E.; Campos, R.; Goni, J., Eds.; 1977, 571-597.
3. Krouse, H.R. J. Geochem. Explor. 1977, 7, 189-211.
4. Mekhtiyeva, V.L.; Brizanova, L.Ya., Geologiya Nefti Gaza 1980, N3, 32-39.
5. Bely, V.M.; Vinogradov, V.I. Geologiya Nefti Gaza 1972, N2, 37-41.
6. Eliuk, L.S.; Hunter, D.F. In 13th Can. Soc. Petrol. Geol. Core Conf. Ed.; Krause, F.; G. Burrowes, 1987, 39-62.
7. Krouse, H.R.; Viau, C.A.; Eliuk, L.S.; Ueda, A.; Halas, S. Nature 1988, 333, 415-419.
8. Barker, J.F.; Fritz, P. Nature 1981, 293, 289-291.
9. Coleman, D.D.; Risatti, J.B., Schoell, M. Geochim. Cosmochim. Acta 1981, 45, 1033-1037.
10. Kiyosu, Y.; Krouse, H.R. EPSL 1989, 95, 302-306.

RECEIVED March 8, 1990

BIBLIOGRAPHY AND INDEXES

BIBLIOGRAPHY

The following bibliography contains most of the literature citations
for the 32 chapters in this volume.

Abed A.M., and Amireh B. (1983) Petrography and geochemistry of some Jordanian oil shales from North Jordan. *Journal of Petroleum Geology* 5, 261-274.

Aczel T., Williams R.B., Chamberlain N.F., and Lumpkin H.E. (1981) Composition of asphaltenes from coal liquids. In *Chemistry of Asphaltenes* (eds. J.W. Bunger, and N.C. Li), Advances in Chemistry Series **195**, pp. 237-251. American Chemical Society.

Aczel T. (1972) Applications of high resolution mass spectrometry to the analysis of complex aromatic mixtures derived from petroleum and coal liquids. *Rev. Anal. Chem.* **1**, 226-261.

Aitken J., Heeps T., and Steedman W. (1968) Organic sulfur in coal: Model compound studies. II. The pyrolysis of thianthrene and dibenzothiophene. *Fuel* **47**, 353-357.

Aizenshtat Z., Stoler A., Cohen Y., and Nielsen H. (1983) The geochemical sulphur enrichment of recent organic matter by polysulfides in the Solar-lake. In *Advances in Organic Geochemistry 1981* (ed. M. Bjorøy), pp. 279-288 Wiley

Aksenov V.S., and Kamyanov V. F. (1981) Regularities in composition of native sulfur compounds. In *Organic Sulfur Chemistry* (eds R Kh Freidlina, and A.E. Skarova), pp. 1-13. Pergamon Press.

Albaigés J., Borbon J., and Walker W., II (1985) Petroleum isoprenoid hydrocarbons derived from catagenetic degradation of Archaebacterial lipids *Org. Geochem.* **8**, 293-297.

Albaigés J. (1980) Identification and geochemical significance of long chain acyclic isoprenoid hydrocarbons in crude oil. In *Advances in Organic Geochemistry 1979* (eds E.G. Douglas, and J.R. Maxwell), pp 19-28. Pergamon Press.

Albaigés J., and Torradas J.M. (1974) Significance of the even carbon *n*-paraffin preference of a Spanish crude oil. *Nature* **250**, 567-568.

Allen R.M., and vanderSande J. (1984) Analysis of sub-micron mineral matter in coal via scanning transmission electron microscopy. *Fuel* **63**, 24-29.

Alpern B., and Cheymol D. (1978) Reflectance et fluorescence des organoclasts du Toarcian du bassin de Paris en function de la profondeur et de la temperature. *Revue de L'institut Français du Pétrole* **33(4)**, 515-535.

Altschuler Z.S., Schnopfe M.M., Silber C.C., and Simon F.O. (1983) Sulfur diagenesis in Everglades peat and the origin of pyrite in coal. *Science* **221**, 221-227.

Amit O., and Bein A. (1982) Organic matter in Senonian phosphorites from Israel - origin and diagenesis. *Chem. Geol.* **37**, 277-287.

Amphlett M.J., and Callely A.G. (1969) The degradation of 2-thiophenecarboxylic acid by a *Flavobacterium* species. *Biochem. J.* **112**, 12-13

Anderson G.C. (1958) Some limnological features of a shallow saline meromictic lake *Limnol. Oceanogr.* **3**, 259-270.

Annunziata R., and Barbarella G. (1984) [33]S chemical shifts and line widths of selected sulphones, sulphoximides, sulphimides, sulphides and sulphonium ions. *Org. Mag. Res.* **22**, 250-254

Arneth J.D., and Matzigkeit U. (1986) Variations in the carbon isotopic composition and production yield of various pyrolysis products under open and closed system conditions. In *Advances in Organic Geochemistry, 1985* (eds. D. Leythauser, and J. Rullkötter), pp. 1067-1071. Pergamon Press.

Arpino P.J., Ignatiadis I., and DeRycke G. (1987) Sulphur containing polynuclear aromatic hydrocarbons from petroleum. *J. Chromatogr.* **390**, 329-348.

Asinger F., and Leuchtenberger W. (1974) 2H-Imidazole from ketones and ammonia. *Justus Liebigs Ann. Chem.*, 1183-1189.

Asinger F., and Offermanns H. (1967) Synthesis with ketones, sulfur and ammonia or amines at room temperature. *Angew. Chem. internat. Edit.* **6**, 907-919

Asinger F., Schaefer W., Halcour K., Saus A., and Triem, H. (1964) The course of the Wellgerodt-Kindler reaction of alkyl aryl ketones. *Angew. Chem. internat. Edit.* **3**, 19-28.

Asinger F., Schroeder L., and Hoffman, S. (1961) Concomitant action of elementary S and gaseous NH_3 on ketones (XXXIII). Interaction of α-mercapto ketones and β -aminovinyl ketones. *Justus Liebigs Ann. Chem.* **648**, 83-95

Asinger F., Thiel M., and Lipfert G. (1959) Concomitant reaction of elementary S and gaseous NH_3 on ketones (XXXIII). 1,2,4-trithiolanes and 1,2,4,5-tetrathiones. *Justus Liebigs Ann. Chem.* **627**, 195-212

Asinger F., Thiel M., Lipfert G , Plessmann R.E., and Menning J (1958) 1,2,4-Trithiolanes (trithioözonides) and duploditheoketones. *Angew Chem.* **70**, 372.

Asinger F., Thiel M., and Pallas E. (1957) Simultaneous interaction of elementary S and gaseous NH_3 with diethylketone. *Justus Liebigs Ann. Chem.* **602**, 37-49.

Asinger F., Thiel M., and Schroeder L. (1957) Concommitant action of elementary S and gaseous NH_3 on ketones (VII). Dehydrogenation of Δ^3-thiazolines with the formation of thiazoles. *Justus Liebigs Ann. Chem.* **610**, 49-56.

Atlas R.M., Boehm P.D., and Calder J.A. (1981) Chemical and biological weathering of oil, from the Amoco Cadiz spillage, within the lottoral zone. *Estuarine Coastal Shelf Sci.* **12**, 589-608.

Attar A., Villoria R.V., Verona D.F., and Parisi S. (1984) Sulfur functional groups in heavy oils and their transformations in steam injected enhanced oil recovery. *ACS Div. Petr. Chem. Preprints* 29(4), 1212-1222.

Attar A., Sulfur groups in coal and their determination. (1984) DOE Report DOE, PC/30145-T1 (DE84007770).

Attar A., and Hendrickson G.G. (1982) Functional groups and heteroatoms in coal. In *Coal Structure* (ed. R.A. Meyers), pp. 131-198. Academic Press.

Attar A., and Dupuis R. (1981) Data on the distribution of organic sulfur functional groups in coals. In *Coal Structure* (eds. M.L. Gorbaty, and K. Ouchi), ACS Advances in Chemistry Series **192**, pp. 239-256. Amercian Chemical Society

Attar A. (1979) Sulfur groups in coal and their determinations. In *Analytical Methods for Coal and Coal Products* (ed..C. Karr) Vol III, Chap. 56, pp 585-624 Academic Press.

Attar A. (1978) Chemistry, thermodynamics and kinetics of reactions of sulphur in coal-gas reactions. A review. *Fuel* **57**, 201-211.

Babenzien H.D., Genz I., and Köhler M. (1979) Oxydativer abbau von dibenzylsulfid. *Z. Allg. Mikrobiol.* **19**, 527-533.

Badin E.J. (1984) Coal combustion chemistry-correlation aspects. In *Coal Science and Technology* (ed. L.L. Anderson), Vol. **6**, pp. 40-78. Elsevier.

Bailey A., and Blackson J. (1984) Examination of organic-rich sediments structurally maintained using low-viscosity resin impregnation. *Scanning Electron Microsc.*, 1475-1481.

Bailey S.A., and Smith J.W. (1972) Improved method for the preparation of sulfur dioxide from barium sulfate for isotope ratio studies. *Anal. Chem.* **44**, 1542-1543.

Baker E.W., and Lauda J.W. (1977) Porphyrins in the geologic record. In *Biological Markers in the Sedimentary Environment* (ed. R.B. Johns), pp. 125-224. Elsevier.

Baker I., Huang B., and Schulson E.M. (1988) The effect of born on the lattice properties of Ni_3Al. *Acta Met.* **36**, 493-499.

Barnard P.C., and Cooper B.S. (1981) Oils and source rocks of the North Sea area. In *Petroleum Geology of the Continental Shelf of North West Europe* (eds. Illing L.V., and Hobson G.D.), pp. 169-175. Heydon and Son.

Bandurski E. (1982) Structural similarities between oil-generating kerogens and petroleum asphaltenes. *Energy Sources* **6**, 47-66.

Barker J.F., and Fritz P. (1981) Carbon isotope fractionation during microbial methane oxidation. *Nature* **293**, 289-291.

Baturin G N. (1972) Phosphorous in interstitial waters of sediments of the southeastern Atlantic. *Oceanology* **12**, 849-855.

Bearse A.E., Cox J.L., and Hillman M. (1975) *Production of Chemicals by Oxidation of Coal.* Battelle Energy Program Rep., Columbus, Ohio.

Begheijn L.Th., van Breeman N., and Velthorste E.J. (1978) Analysis of sulfur compounds in acid sulfate solids and other recent marine soils. *Commun. in Soil Science and Plant Analysis* **9**, 873-882.

Behar F., and Pelet R. (1988) Hydrogen transfer reactions in the thermal cracking of asphaltenes. *Energy & Fuels* **2**, 259-264.

Behar F., and Vandenbroucke M. (1987) Chemical modeling of kerogen. *Org. Geochem.* **11**, 15-24.

Behar F., and Pelet R. (1985) Asphaltene characterization by pyrolysis and chromatography. *J. Anal. Appl. Pyrolysis* **7**, 121-135.

Bein A., and Amit O. (1982) Depositional environments of the Senonian chert, phosphorite and oil shale sequence in Israel as deduced from their organic matter composition. *Sedimentology* **29**, 81-90.

Belayouni H., and Trichet J. (1984) Hydrocarbons in phosphatized and non- phosphatized sediments from the phosphate basin of Gafsa. In *Advances in Geochemistry 1983* (eds. P.A. Schenck, J.W. de Leeuw, and G.W.M. Lijmbach), *Org. Geochem.* **6**, 741-754.

Belyi V.M., and Vinogradov V.I. (1972) Isotopic composition of sulfur and genesis of highly concentrated hydrogen sulfide gases of oil-gas-bearing areas. *Geologiya Nefti Gaza*, 37-41.

Benson L.V., and Mifflin M.D. (1986) Reconnaissance bathymetry of basins occupied by pleistocene Lake Lahontan, Nevada and California, U.S. Geological Survey, Water Resources Investigation Report 85-4262, pp. 14.

Benson L.V., and Spencer R.J. (1983) A hydrochemical reconnaissance study of the Walker river basin, California and Nevada, U.S. Geological Survey Open-File Report 83-740, pp. 53.

Benson L.V. (1981) Paleoclimatic significance of lake-level fluctuations in the Lahontan basin. *Quaternary Research* **16**, 390-403.

Bergholm A. (1955) Oxidation of pyrite. *Jernkonotorets Annaler Arg.* **139**, 531-549.

Berner R.A. (1985) Sulphate reduction, organic matter decomposition and pyrite formation. *Phil. Trans. R. Soc. Lond.* **A 315**, 25-38.

Berner R.A. (1984) Sedimentary pyrite formation: An update. *Geochim. Cosmochim. Acta* **48**, 605-615.

Berner R.A., and Raiswell R. (1983) Burial of organic carbon and pyrite sulfur in sediments over Phanaerozoic time: a new theory. *Geochim. Cosmochim. Acta* **47**, 855-862.

Berner R.A. (1974) Iron sulfides in pleistocene deep Black Sea sediments and their paleo-oceanographic significance, In *The Black Sea--Geology, Chemistry, and Biology.* (eds. E.T. Degens, and D.A. Ross), pp. 524-531. The American Association of Petroleum Geologists.

Berner R.A. (1970) Sedimentary pyrite formation. *Amer. J. of Sci.* **268**, 1-23.

Berner R.A. (1969) The synthesis of framboidal pyrite. Econ. Geol. 64, 383-384.

Berner R.A. (1964) Iron sulfides formed from aqueous solution at low temperatures and atmopsheric pressure. Jour. Geol. 72, 293-306.

Berner R.A. (1963) Electrode studies of hydrogen sulfide in marine sediments. Geochim. Cosmochim. Acta 27, 563.

Berteloot J. (1947) Presence of native sulfur in coal. Variations in the sulfur content from top to bottom of a coal seam. *Ann. Soc. Geol. Nord.* **67**, 195-206; *Chem. Abstr.*, 1950, **44**, 818a.

Bertucci P.F. (1986) Geologic setting and petroleum potential of the Monterey formation, southern California. (Abstr.) 192nd Amer. Chem. Soc. National Meeting, Anaheim,California,.

Beyer M., Ebner H.G., Assenmacher H., and Frigge J. (1987) Elemental sulphur in microbiologically desulphurized coals. *Fuel* **66**, 551-555.

Bianconi A. (1988) XANES spectroscopy. In *X-Ray Absorption* (eds. D.C. Konigsberger, and R. Prins), pp. 573-662. John Wiley & Sons.

Blackson J.H., and Bailey A. (1985) Low-viscosity resin impregnation of hydrated unconsolidated sediments. *EMSA Bulletin* **15**, 115-117.

Blaser R., and White C. (1984) Source rock and carbonization study, Maracaibo Basin, Venezuela. *Amer. Assoc. Petr. Geol. Memoirs* **35**, 229-252.

Block E. (1978) Reactions of organosulfur compounds. Academic Press.

Blumer M., Guillard R.R.L., and Chare D. (1971) Hydrocarbons of marine phytoplankton. *Mar. Biol.* **8**, 183-189.

Bluth V.S. (1989) Early diagenesis of iron in coastal marine sediments: Chincoteague Bay and Wallops Island, Virginia. M.S. Thesis, The Pennsylvania State University.

Bockmeulen H., Barker C., and Dickey P., 1983, Geology and geochemistry of crude oils, Bolivar coastal fields, Venezuela. *Amer. Assoc. Petr. Geol. Bull.* **67(2)**, 242-270.

Boduszynski M.M. (1988) Composition of heavy petroleum. 2. Molecular characterization. *Energy & Fuels* **2**, 597-613.

Boduszynski M.M. (1987) Composition of heavy petroleum. I. Molecular weight, hydrogen deficiency, and heteroatom concentrations as a function of atmospheric equivalent boiling point up to 1400°F (760°C). *Energy & Fuels* **1**, 2-11.

Bodzek D., and Marzec A. (1981) Molecular components of coal and coal structure. *Fuel* **60**, 47-51.

Bohonos N., Chou T.W., and Spanggord R.J. (1977) Some observations on biodegradation of pollutants in aquatic systems. *Jpn. J. Antibiot.* **30** (suppl), 275-285.

Bomberger D.R., and Deul M. (1964) Study of fine coal cleaning process by automatic microscopy. *Trans. Soc. Mining Eng., AIME* **229**, 65-70.

Boon J.J., and de Leeuw J.W. (1987) Organic geochemical aspects of cyano- bacterial mats. In *Cyanobacteria: Current Research* (eds. P. Fay, and C. van Baalen), pp. 471-492. Elsevier.

Boon J.J., van der Meer F.W., Schuyl P.J.W., de Leeuw J.W., Schenck P.A., and Burlingame A.L. (1978) Organic geochemical analyses of core samples from Site 362, Walvis Ridge, DSDP Leg 40. *Init. Rep. DSDP* **40**, pp. 627-637.

Bordwell F.G., and McKellin W.H. (1951) The reduction of sulfones to sulfoxides. *J. Amer. Chem. Soc.* **73**, 2251-2253.

Borgward R.H., Bruce K.R., and Blake J. (1987) An investigation of product-layer diffusivity for CaO sulfation. *Ind. Eng. Chem. Res.* **26**, 1993-1998.

Boudou J.P., Espitalié J., and Marquis F. (1990) Continuous detection during heating of coal and kerogen. In *New Methodologies for Coal Characterization* (ed. H. Charcosset). Elsevier.

Boudou J.P., Boulegue J., Malechaux T., Nip M., de Leeuw J.W., and Boon J.J. (1987) Identification of some sulfur species in a high sulfur coal. *Fuel* **66**, 1558-1569.

Boudou J.P. (1984) Chloroform extracts of a series of coals from Mahakam delta. In *Advances in Geochemistry 1983* (eds. P.A. Schenck, J.W. de Leeuw, and G.W.M. Lijmbach), *Org. Geochem.* **6**, 431-437.

Boudou J.P., Durand B., and Oudin J.L. (1984). Diagenetic trends of a tertiary low-rank coal series. *Geochim. Cosmochim. Acta* **48**, 2005-2010.

Boulègue J., Lord C.J. III, and Church T.M. (1982) Sulfur speciation and associated trace metals (Fe, Cu) in the pore waters of Great Marsh, Delaware. *Geochim. Cosmochim. Acta* **46**, 453-464.

Braekman-Danheux C. (1985) Pyrolysis-gas chromatography-mass spectrometry of model compounds from coal hydrogenates. *J. Anal. Appl. Pyrolysis* **7**, 315-322.

Brassell S.C., Sheng Guoying, Fu Jiamo, and Eglinton G. (1988) Biological markers in lacustrine Chinese oil shales. In *Lacustrine Petroleum Source Rocks* (eds. K. Kelts, A.J. Fleet, and M. Talbot), pp. 299-308. Blackwell.

Brassell S.C., Lewis C.A., de Leeuw J.W., de Lange F., and Sinninghe Damsté J.S. (1986) Isoprenoid thiophenes: Novel products of sediment diagenesis? *Nature* **320**, 160-162.

Brassell S.C., Eglinton G., Marlowe I.T., Sarnthein U., and Pflaumann U. (1986) Molecular stratigraphy - a new tool for climatic assessment. *Nature* **320**, 129-133.

Brassell S.C., Brereton R.G., Eglinton G., Grimalt J., Liebezeit G., Marlowe I.T., Pflaumann U., and Sarnthein U. (1986) Palaeoclimatic signals recognized by chemometric treatment of molecular stratigraphic data. In *Advances in Organic Geochemistry 1985* (eds. D. Leythaeuser, and J. Rullkötter), *Org. Geochem.* **10**, 649-660.

Brassell S.C., McEvoy J., Hoffman C.F., Lamb N.A., Peakman T.M., and Maxwell J.R. (1984) Isomerisation, rearrangements and aromatisation of steroids in distinguishing early stages of diagenesis. In *Advances in Organic Geochemistry 1983* (eds. P.A. Schenck, J.W. de Leeuw, and G.W.M. Lijmbach), pp. 11-23. Pergamon.

Breger I.A., ed. (1963) *Organic Geochemistry*. Macmillan.

Braun R.L., and Burnham A.K. (1987) Analysis of chemical reaction kinetics using a distribution of activation energies and simpler models. *Energy & Fuels* **1**, 153-161.

Broecker W.S., and Peng T. (1982) *Tracers in the Sea*. Lamont-Doherty Geological Observatory Publication.

Britton L.N. (1984) Microbial degradation of aliphatic hydrocarbons. In *Microbial Degradation of Organic Compounds* (ed. D.T. Gibson), pp. 89-129. Marcel Dekker, Inc.

Brown E.V. (1975) The Willgerodt reaction. *Synthesis*, 358-375.

Bruemmer S.M., Fluhr C.B., Beggs D.V., Wert C.A., and Fraser H.L. (1980) An analytical microscopy study of the high temperature carbide formed in a V-5-Ti-C alloy. *Met. Trans.* **11A**, 693-699.

Brupbacher R.H., Sedberry J.E., and Willis W.M. (1973) The coastal marshlands of Louisiana, chemical properties of the soil materials. *Louisiana Agricultural Experiment Station Bulletin No.* 672.

Buchanan D., and Chavan D. (1989) Coal desulfurization using perchloroethylene. *Proceedings: Fourteenth Annual EPRI Fuel Science Conference;* Electric Power Research Institute.

Burnett W.C., Veeh H.H., and Soutar A. (1980) Uranium series, oceanographic and sedimentary evidence in support of recent formation of phosphate nodules off Peru. In *Marine Phosphorites -Geochemistry, Occurrence, Genesis* (ed. Y.K. Bentor), pp. 61-72, S.E.P.M Spec. Publ. 29. Society Econ. Paleontol. Mineral.

Burnett W.C. (1977) Geochemistry and origin of phosphorite deposits from off Peru and Chile. *Geol. Soc. Am. Bull.* **88**, 813-823.

Burnham A.K., Braun R.L., Gregg H.R., and Samon A.M. (1987) Comparison of methods for measuring kerogen pyrolysis rates and fitting kinetic parameters. *Energy & Fuels* **1**, 452-458.

Burns B.J., Hogarth J.T.C., and Milner C.W.D. (1975) Properties of Beaufort basin liquid hydrocarbons. *Bull. Can. Petr. Geol.* **23**, 295-303.

Bustin R.M., and Lowe L.E. (1987) Sulphur, low temperature ash and minor elements in humid-temperature peat of the Fraser river delta, British Columbia. *J.Geol. Soc. (London)* **144**, 435-450.

Calkins W.H. (1987) Investigation of organic sulfur-containing structures in coal by flash pyrolysis experiments. *Energy & Fuels* **1**, 59-64.

Calvert J. et al., (1963) Appendix C, *Atmospheric Deposition Processes; Acid Deposition, Atmospheric Processes in Eastern North America*. National Academy Press.

Campaigne E.E., and Archer W.L. (1953) The use of dimethylformamide as a formylation reagent. *J. Amer. Chem. Soc.* **75**, 989-991.

Chantret F. (1969) L'analyse thermique différentielle associée à l'analyse en continue de l'anhydride sulfureux et de l'oxyde de carbone. Applications qualitatives et quantitatives à l'étude des roches sédimentaries. *Bull. Soc. Fr. Mineral. Crist.* **92**, 462-467.

Chatterjee N.N. (1942) Free sulphur in some weathered tertiary coal specimens of India. *Quart. Jour. Geol. Mining Metall. Soc. India* **14**, 1-7.

Chivers T. (1977) Polychalcogenide anoins. In *Homoatomic Rings, Chains and Macromolecules or Main Group Elements* (ed. A.L. Rheingold), pp. 499-537. Elsevier.

Chou C.-L. (1984) Relationship between geochemistry of coal and the nature of strata overlying the Herrin Coal in the Illinois Basin, U.S.A. *Memoir Geol. Soc. China* **6**, 269-280.

Chou M.-I., Lake M.A., and Griffin R.A. (1988) Flash pyrolysis of coal, coal maceral, and coal-derived pyrite with on-line characterization of volatile sulfur components. *J. Anal. Appl. Pyrolysis* **13**, 199-207.

Chukhrov F.V., Ermilova L.P., Churikov V.S., and Nosik L.P. (1980) The isotopic composition of plant sulfur. *Org. Geochem.* **2**, 69-75.

Chukhrov F.V., and Ermilova L.P. (1970) Zur schwefel-isotopenzusammensetzung in Konkretionen. *Ber. Deut. Ges. Geol. Wiss.* **15**, 255-267.

Chung H.M., and Sackett W.M. (1980) Carbon isotope effects during pyrolitic formation of early methane from carbonaceous materials. In *Advances in Organic Geochemistry, 1979* (eds. A.G. Douglas, and Maxwell, J.R.), pp. 705-710. Pergamon Press.

Chung H.M. (1976) Isotope fractionation during maturation of organic matter. Ph.D. dissertation, Texas A&M University.

Clark C.P., Freeman G.B., and Hower J.C. (1984) Non-matrix corrrected organic sulfur determination by energy dispersive x-ray spectroscopy for western Kentucky coals and residues. *Scanning Electron Microsc.* **2**, 537-545.

Claypool G.E., Holser W.T., Kaplan I.R., Sakai H., and Zak T. (1980) The age curves of sulfur and oxygen isotopes in marine sulfate and their mutual interpretation. *Chem. Geol.* **28**, 199-260.

Cline J.D., and Richards F.A. (1972) Oxygen deficient conditions and nitrate reduction in the eastern tropical north Pacific Ocean. *Limnol. Oceanogr.* **17**, 885-900.

Cobb J.C. (1981) Geology and geochemistry of sphalerite in coal. Ph.D. dissertation, Univ. of Illinois.

Cohen A.D., Spackman W., and Dolsen P. (1984) Occurrence and distribution of sulfur in peat-forming environments of southern Florida. *International Journal of Coal Geology* **4**, 73-96.

Cohen Y., Krumbein W.E., and Shilo M. (1977) Solar Lake (Sinai). 3. Bacterial distribution and production. *Limnol. Oceanogr.* **22**, 621-634.

Coleman D.D., Risatti J.B., and Schoell M. (1981) Fractionation of carbons and hydrogen isotopes by methane-oxidizing bacteria. *Geochim. Cosmochim. Acta* **45**, 1033-1037.

Coleman H.J., Hopkins R.L,. and Thompson C.J. (1971) Highlights of some 50 man-years of petroleum sulfur studies by the Bureau of Mines. *Int. J. Sulfur Chem.* **6**, 41-62.

Combaz A., and de Matharel M. (1978) Organic sedimentation and genesis of petroleum in the Mahakam delta, Borneo. *Amer. Assoc. Petr. Geol. Bull.* **62**, 1684-1695.

Connan J. (1984) Biodegradation of crude oils in reservoirs. In *Adv. Pet. Geochem.* (eds. J. Brooks, and D.H. Welte), Vol. **1**, pp. 299-335. Academic Press.

Canfield D.E., Raiswell R., Westrich J.T., Reaves C., and Berner R.A. (1986) The use of chromium reduction in the analysis of reduced inorganic sulfur in sediments and shales. *Chem. Geol.* **54**, 149-155.

Carlton R.W. (1985) Image analysis of pyrite in Ohio coal: Relation between pyrite grain-size distribution and pyritic sulfur reduction. In *Processing and Utilization of High Sulfur Coals* (ed. Y.A. Attia), pp. 3-17. Elsevier.

Carmack M., and Spielman M.A. (1946) the Willgerodt reaction. In *Organic Reactions* (ed. R. Adams), pp. 83-107. Wiley.

Carroll D. (1958) Role of clay minerals in the transportation of iron. *Geochim. Cosmochim. Acta* **14**, 1-27.

Casagrande. D.J. (1987) Sulphur in peat and coal. In *Coal and Coal-Bearing Strata: Recent Advances* (ed. A.C. Scott), pp. 87-105. Geol. Soc. Special Publication No. 32.

Casagrande D.J. (1985) Distribution of sulfur in progenitors of low-sulfur coal: Origins of organic sulfur. In *Proceedings Ninth International Congress on Carboniferous Stratigraphy and Geology, 1979, Compte Rendu* (ed. A.T. Cross), **4**, 299-307. Southern Illinois University Press.

Casagrande D.J., and Price F.T. (1981) Sulfur isotope distribution in coal precursors (abstr.). *Geol. Soc. Amer. Abs. Prog.* **13**, 423-424.

Casagrande D.J., and Ng L. (1979) Incorporation of elemental sulphur in coal as organic sulphur. *Nature* **282**, 598-599.

Casagrande D J., Idowu G., Friedman A , Rickert P., Siefert K., and Schlenz D. (1979) H_2S incorporation in coal precursors: origins of organic sulphur in coal. *Nature* **282**, 599-600.

Casagrande D.J., and Siefert K. (1977) Origin of sulfur in coal: Importance of ester sulfate content. *Science* **195**, 675-676.

Casagrande D.J., Siefert K., Berschinski C., and Sutton N. (1977) Sulfur in peat forming systems of the Okefenokee swamp and Florida Everglades: Origin of sulfur in coal. *Geochim. Cosmochim. Acta* **41**, 161-167.

Casey W.H., Guber A., Bursey C., and Olsen C.R. (1986) Chemical controls on ecology in a coastal wetland. *EOS Trans. Amer. Geoph. Union* **67**, 1305-1311.

Castex H., Roucache J., and Boulet R. (1974) Le soufre thiophenique dans les petroles et les extraits de roches. *Revue de l'Institute Francais du Petrole* **31**, 3-38.

Cernic-Simic S. (1962) A study of factors that influence the behaviour of coal during carbonization. *Fuel* **41**, 141.

Cerniglia C.E. (1984) Microbial transformation of aromatic hydrocarbons. In *Petroleum Microbiology* (ed. R.M. Atlas), pp. 99-128. MacMillan Publishing, Co.

Chakrabarty S., and du Plessis M. (1985) *Evaluation of Alberta plains coals for pyrolysis and liquefaction processes. Coal Report 85-1.* Albert Research Council, Energy Resources Division, Coal Research Department.

Chambers L.A., and Trudinger P.A. (1979) Microbiological fractionation of stable sulfur isotopes: A review and critique. *Geomicrobiology Journal* **1**, 249-293.

Chang H.M., Allan G.G., Sarkanen K.V., and Ludwig C.H., eds., (1971) Oxidation. In *Lignins*. Wiley.

Cooles G.P., Mackenzie A.S., and Quigley T.M. (1986) Calculation of petroleum masses generated and expelled from source rocks. In *Advances in Organic Geochemistry 1985* (eds. D. Leythaeuser, and J. Rullkotter), *Org. Geochem.* 10, 235-246.

Corbett L.W. (1969) Composition of asphalt based on generic fractionation using solvent deasphaltening, elution-adsorption chromatography and densimetric characterization. *Anal. Chem.* 41, 576-579.

Corey E.J., and Schmidt G. (1979) Useful procedures for the oxidation of alcohols involving pyridinium dichromate in aprotic media. *Tetrahedron Lett.*, 399-402.

Coshell L. (1983) Cyclic depositional sequences in the Rundle oil shale deposit. *Proceedings of the 1st Australian Workshop on Oil Shale*, Lucas Heights, N.S.W.

Couch G.R. (1987) *Biotechnology and Coal*. Chap. 4, pp. 23-38. International Energy Agency.

Cranwell P.A. (1985) Long-chain unsaturated ketones in recent lacustrine sediments. *Geochim. Cosmomochim. Acta* 49, 1545-1551.

Cripps R.E. (1973) The microbial metabolism of thiophen-2-carboxylate. *Biochem. J.* 134, 353-366.

Crisp P.T., Ellis J., Hutton A.C., Lorth J., Martin F.A., and Saxby J.D. (1987) *Australian oil shales: a compendium of geological and chemical data*, North Ride NSW, Australia, CSIRO Institute of Energy and Earth Resources.

Curiale J.A., Cameron D., and Davi, D.V. (1985) Biological marker distribution and significance in oils and rocks of the Monterey formation, California. *Geochim. Cosmochim. Acta* 49, 271-288.

Cyr T.D., Payzant J.D., Montgomery D.S., and Strausz O.P. (1986) A homologous series of novel hopane sulfides in petroleum. *Org. Geochem.* 9, 139-143.

Dade W.B. (1983) Bryozoans of the modern Wallops-Chincoteague coast, Virginia. M.S. thesis, The Pennsylvania State University.

Damberger H.H., Harvey R.D., Ruch R.R., and Thomas J., Jr. (1984) Coal characterization. In *The Science and Technology of Coal and Coal Utilization* (eds. B.R. Cooper and W.A. Ellington), Chap. 2, pp. 7-45. Plenum.

Damberger H.H., Nelson W.J., and Krausse H.F. (1980) Effect of geology on roof stability in room-and-pillar mines in the Herrin (No. 6) coal of Illinois. *Proceedings, First Conference on Ground Control Problems in the Illinois Coal Basin* (eds. Y.P. Chugh, and A. van Besien), pp. 14-32. Southern Illinois University.

Davison W., Lishman J.P., and Hilton J. (1985) Formation of pyrite in freshwater sediments; Implications for C/S ratios. *Geochim. Cosmochim. Acta* 49, 1615-1620.

Dean R.A., and Whitehead E.V (1967) Status of work in separation of sulfur compounds in petroleum and shale oil. In *Proc. 7th World Petroleum Congress: Panel Discussion 23*, 165-175.

Dean-Raymond D., and Batha R. (1975) Biodegradation of some polynuclear aromatic petroleum components by marine bacteria. *Dev. Ind. Microbiol.* 16, 97-110.

Del Mazza D., and Reinecke M.G. (1981) Reactions of a $\alpha.\beta$-unsaturated ketones with hydrogen sulfide. γ-Keto sulfide or tetrahydrothiopyrananols? *J. Org. Chem.* 46, 128-134.

Demaison G.J., and Moore G.T. (1980) Anoxic environments and oil source bed genesis. *Amer. Assoc. Petr. Geol. Bull.* 64, 1179-1209.

Demaison G.J. (1977) Tar sands and supergiant oil fields. *Amer. Assoc. Petr. Geol. Bull.* **61**, 1950-1961.

Dembicki H., Jr., Meinschein W.G., and Hatten D.E. (1976) Possible ecological and environmental significance of the predominance of even-carbon number C_{20}- C_{30} n-alkanes. *Geochim. Cosmochim. Acta* **40**, 203-208.

Demi I., and Harvey R.D. (1989) Distribution of organic sulfur in macerals of selected Illinois Basin coals (abstr.) *Geol. Soc. Amer. Abs. Prog.* **21**, A162.

Deno N.C., Curry K.W., Greigger B.A., Jones A.D., Rakitsky W.G., Smith K.A., Wagner K., and Minard R.D. (1980) Dihydroaromatic structure of Illinois No. 6 Monterey coal. *Fuel* **59**, 694-700.

Deno N.C., Greigger B.A., Stroud S.G. (1978) New method for elucidating structures of coal. *Fuel* **57**, 455-459.

Derbyshire D.H. (1950) An oxidation involving the hydroxyl cation $(OH)^{+}$. *Nature* **165**, 401-402.

Deroo G., Tissot B., McCrossan R.G., and Der F. (1974) Geochemistry of the heavy oils of Alberta. In *Oil Sands, Fuel of the Future* (ed. L.V. Hills), *Can. Soc. Petr. Geol. Mem.* **3**, 148-167.

De Roo J., and Hodgson G.W. (1978) Geochemical origin of organic sulfur compounds. Thiophene derivatives from ethylbenzene and sulfur. *Chem. Geol.* **22**, 71-78.

De Rosa M., Gambacorta A., and Gliozzi A. (1986) Structure, biosynthesis, and physicochemical properties of archaebacterial lipids. *Microbiol. Rev.* **50**, 70-80.

Didyk B.M., Simoneit B.R.T., Brassell S.C., and Eglinton G. (1978) Organic geochemical indicators of palaeoenvironmental conditions of sedimentation. *Nature* **272**, 216-222.

Dimmler A., Cyr T.D., and Strausz O.P. (1984) Identification of bicyclic terpenoid hydrocarbons in the saturate fraction of Athabasca oil sand bitumen. *Org. Geochem.* **7**, 231-238.

Dinur D., Sprio B., and Aizenshtat Z. (1980) The distribution and isotopic composition of sulfur in organic-rich sedimentary rocks. *Chem. Geol.* **31**, 37-51.

Dollar M., Bernstein I.M., Daeubler M., and Thompson A.W. (1989) The effect of cyclic loading on the dislocation structure of fully pearlitic steel. *Met. Trans.* **20A**, 447-451.

Donnell J.R. (1980) Potential contribution of oil shale to U.S., world energy needs. *Oil Gas J.* **78** (41), 218-224.

Douglas A.G., and Williams P.F.V. (1981). Kimmeridge oil shale. A study of organic maturation. In *Organic Maturation Studies and Fossil Fuel Exploration* (ed. J. Brooks), pp. 255-269. Academic Press.

Dressler M. (1986) *Selective gas chromatographic detectors.* Elsevier.

Drushel H.V. (1970) Sulfur compounds in petroleum - known and unknown. *ACS Division of Petroleum Chemistry Preprints* **15** (2), C12-42.

Drushel H.V. (1969) Sulfur compound type distribution in petroleum using in-line reactor or pyrolysis combined with gas chromatography and a microcoulometric sulfur detector. *Anal. Chem.* **41**, 569-576.

Duran J.E., Mahasay S.R., and Stock L.M. (1986) The occurrence of elemental sulphur in coals. *Fuel* **65**, 1167-1168.

Durand B. (1980) Sedimentary organic matter and kerogen: Definition and quantitative importance of kerogen. In *Kerogen - Insoluble Organic Matter from Sedimentary Rocks* (ed. B. Durand), pp. 13-34. Editions Technip.

Durand B., and Monin J.C. (1980) Elemental analysis of kerogen. In *Kerogen - Insoluble Organic Matter from Sedimentary Rocks* (ed. B. Durand), pp. 113-142. Editions Technip.

Durand B., Espitalie J., Nicaise G., and Combaz A. (1972) Etude de la matière organique insoluble (kerogene) des argiles du Toarcian du Bassin de Paris. *Revue de L'institute Français du Pétrole. Ann. Combust. Liquides* **27**, 865-884.

Dutta S.N., Dowerah D., and Frost D.C. (1983) Study of sulfur in Assam coals by x-ray photoelectron - spectroscopy. *Fuel* **62**, 840-841.

Dypvik H., Rueslatten W.G., and Throndse T. (1979) Liptinite organic matter from north Atlantic Kimmeridge shales. *Amer. Assoc. Pet. Geol. Bull.* **63**, 2222-2226.

Dyrkacz G.R., Bloomquist A.A., Ruscic L., and Horwitz E.P. (1984) Variations in properties of coal macerals elucidated by densitry gradient separation. In *Chemistry and Characterization of Coal Macerals* (eds. R.E. Winans and J.C. Crelling), pp. 65-77, ACS Symposium Series **252**. American Chemical Society.

Dyrkacz G.R., and Horowitz E.P. (1982) Separation of coal macerals. *Fuel* **61**, 3-12.

Dyrkacz G.R., Bloomquist A.A., and Horowitz E.P. (1981) Laboratory scale separation of coal macerals. *Sep. Sci. Technol.* **16**, 1571-1588.

Eckart V., Köhler M., and Hieke W. (1986) Microbial desulfurization of petroleum and heavy petroleum fractions. 5. Anaerobic desulfurization of Romashkino-crude oil. *Zentralbl. Bakeriol. II* **141**, 291-300.

Eckart V., Hieke W., Bauch J., and Gentzsch H. (1982) Microbial desulfurization of petroleum and heavy petroleum fractions. 3. The change of chemical composition of fuel-D-oil by microbial aerobic desulfurization. *Zentralbl. Bakeriol. II* **137**, 152-160.

Eckart V., Hieke W., Bauch J., and Bohlmann D. (1981) Microbial desulfurization of petroleum and heavy petroleum fractions. 2. Studies on microbial aerobic desulfurization of Romashkino-crude oil. *Zentralbl. Bakeriol. II* **136**, 152-160.

Eckart V., Hieke W., Bauch J., and Gentzsch H. (1980) Microbial desulfurization of petroleum and heavy petroleum fractions. 1. Studies on microbial aerobic desulfurization of Romashkino-crude oil. *Zentralbl. Bakeriol. II* **135**, 674-681.

Edmonson W.T. (1963) Pacific coast and Great Basin. In *Limnology in North America* (ed. D.G. Frey), pp. 371-392. University of Wisconsin Press.

Eglinton G., and Murphy M.T.J., eds. (1969) *Organic Geochemistry Methods and Results.* Springer-Verlag.

Eglinton G., and Hamilton R.J. (1967) Leaf epicuticular waxes. *Science* **156**, 322-1335.

Eglinton T.I, Sinninghe Damsté J.S., Kohnen M.E.L., de Leeuw J.W., Larter S.R., and Patience R.L. (1990) Analysis of maturity-related changes in the organic sulphur composition of kerogens by flash pyrolysis-gas chromatography. In *Geochemistry of Sulfur in Fossil Fuels* (eds. W.L. Orr, and C.M. White), ACS Symposium Series (this volume). American Chemical Society.

Eglinton T.I, Sinninghe Damsté J.S., de Leeuw J.W., and Boon J.J. (1989) Organic sulphur in macromolecular sedimentary organic matter. II. Multivariate analysis of distributions of organic sulphur pyrolysis products (Abstr.). *14th International Meeting on Organic Geochemistry.*

Eglinton T.I., Douglas A.G., and Rowland S.J. (1988) Release of aliphatic, aromatic and sulphur compounds from Kimmeridge kerogen by hydrous pyrolysis: A quantitative study. In *Advances in Organic Geochemistry 1987* (eds. L. Novelli, and L. Mattavelli), *Org. Geochem.* **13**, 655-663.

Eglinton T.I., Philp R.P., and Rowland S.J. (1988) Flash pyrolysis of artificially matured kerogens from the Kimmeridge clay, in U. K. *Org. Geochem.* **12**, 33- 41.

Ehlers E.G., and Stiles D.V. (1975) Melanterite-Rozenite equilibrium. *Amer. Mineral* **50**, 1457-1461.

Ehrlich A., and Dor I. (1985) Photosynthetic microorganisms of the Gavish Sabkha. In *Hypersaline Ecosystems*, Ecological Studies Vol. **53** (eds. G.M. Friedman, and W.E. Krumbein), pp. 296-321. Springer-Verlag.

Ekweozor C.M., and Strausz O.P. (1983) Tricyclic terpanes in the Athabasca oil sands: Their geochemistry. *Advances in Geochemistry 1981*. (eds. M. Bjorøy, C. Albrecht, C. Cornford, K. de Groot, G. Eglinton, E. Galimov, D. Leythaeuser, R. Pelet, J. Rullkötter, and G. Speers), pp. 746-766 Wiley Heyden.

Ekweozor C.M., and Strausz O.P. (1982) 18,19-Bisnor 13βH, 14αH-cheilanthane: A novel degraded sesterterpenoid-type hydrocarbon from the Athabasca oil sands. *Tetrahedron Lett.* **23**, 2711-2714.

Eliuk L.S., and Hunter D.F. (1987) Wabamum group structural thrust fault fields: The limestone-burnt timber example. In Devonian Lithofacies and Reservoir Styles in Alberta, 13th CSPG Core Conference (eds. F.F. Krouse, and O.G. Burowes).

Energy Information Administration (1989) Estimation of U.S. Coal Reserves by Coal Type, Heat and Sulfur Content. pp. 57 DOE/EIA-0529.

Ensley B.D., Jr. (1984) Microbial metabolism of condensed thiophenes, In *Microbial Degradation of Organic Compounds* (ed. D.T. Gibson), pp. 309-317. Marcel Dekker.

Erdman J.G. (1965) Petroleum- its origin in the earth. In *Fluids in Subsurface Environments, AAPG Memmor 4* (eds. A. Young, and J.E. Galley), pp. 20-52. AAPG Press.

Erten M.H. (1985) Comparison of sulfur absorption capabilities of pure lime and limestone in coal combustion zone and flue gas desulfurization systems. In *Coal Science and Technology. Processing and Utilization of High Sulfur Coals* (ed. A. Attia), Vol. **9**, pp. 451-466. Elsevier.

Espitalie J. (1987) Organic geochemistry of the Paris basin. In *Petroleum Geology of North West Europe* (eds. J. Brooks, and K.W. Glennie), pp. 71- 86. Graham and Trotman.

Espitalie J., Marquis F., and Barsony I.(1984) Geochemical logging. In *Analytical Pyrolysis* (ed. K.J. Voorhees), pp. 276-304. Butterworth and Co.

Eugster H.P., and Hardie L.A. (1978) Saline lakes. In *Lakes--Chemistry, Geology, Physics* (ed. A. Lerman), pp. 237-294. Springer-Verlag.

Eugster H.P., and Surdam R.C. (1973) Depositional environment of the Green River formation of Wyoming. *Geological Society Amer. Bulletin* **84**, 1115-1120.

Evans G., Schmidt V., Bush P., and Nelson H. (1969) Stratigraphy and geologic history of the sabkha, Abu Dhabi, Persian Gulf. *Sedimentology* **12**, 145-159.

Evans R., and Kirkland D.W. (1988) Evaporitic environments as a source of petroleum. In *Evaporites and Hydrocarbons* (ed. B C. Schreiber), pp. 256-299. Columbia University Press.

Evans T.J. (1974) *Bituminous Coal in Texas*. Handbook No. 4, Texas Bureau of Economic Geology.

Farrimond P., Brassell S.C., and Eglinton G. (1986) Alkenones in Cretaceous black shales, Blake-Bahama basin, western north Atlantic. In *Advances in Organic Geochemistry 1985* (eds. D. Leythaeuser, and J. Rullkötter). *Org. Geochem.* **10**, 897-903.

Faure R., Vincent E.J., Ruiz J.M., and Lena L. (1981) Sulfur-33 nuclear magnetic resonance sudies of sulphur compounds with sharp resonance lines. ^{33}S NMR spectra of some sulphones and sulphonic acids. *Org. Mag. Res.* **15**, 401-403.

Fedorak P.M., Payzant J. D., Montgomery D.S., and Westlake D.W.S. (1988) Microbial degradation of n-alkyl tetrahydrothiophenes found in petroleum. *Appl. Environ. Microbiol.* **54**, 1243-1248.

Fedorak P.M., and Westlake D.W.S. (1986) Fungal metabolism of n-alkylbenzenes. *Appl. Environ. Microbiol.* **51**, 435-437.

Fedorak P.M., and Westlake D.W.S. (1983) Selective degradation of biphenyl and methylbiphenyls in crude oil by two strains of marine bacteria. *Can. J. Microbiol.* **29**, 497-503.

Fedorak P.M., and Westlake D.W.S. (1981) Microbial degradation of aromatics and saturates in Prudhoe Bay crude oil as determined by glass capillary gas chromatography. *Can. J. Microbiol.* **27**, 432-443.

Fedorak P.M., and Westlake D.W.S. (1981) Degradation of aromatics and saturates in crude oil by soil enrichments. *Water, Air, Soil Pollut.* **21**, 255-230.

Feijtel T.C. (1986) *Biochemical cycling of metals in Barataria basin.* Ph.D. dissertation, Louisiana State University.

Filby R.H., and van Berkyl G.J. (1987) Geochemistry of metal complexes in petroleum, source rocks, and coals: An overview. In *Metal Complexes in Fossil Fuels* (eds. R.H. Filby, and J.F. Branthaver), ACS Symposium Series **344**, pp. 2-39. American Chemical Society.

Finnerty W.R, Shockley K., and Attaway H. (1983) Microbial desufurization and denitrification of hydrocarbons. In *Microbially Enhanced Oil Recovery* (eds. J.E. Zajic, D.C. Cooper, T.R. Jack, and N. Kosaric), pp. 83-91. PennWell Publishing Company.

Fischer U. (1989) Enzymatic steps and dissimilatory sulfur metabolism by whole cells of Anoxyphotobacteria. In *Biogenic Sulfur in the Environment* (eds. E.S. Saltzman, and W.J. Cooper) ACS Symposium Series **393**, pp. 262-279. American Chemical Society.

Flory D., Lichtenstein H., Bieman K., Biller J.E., and Barker C. (1983) Computer process uses entire GC-MS data. *Oil Gas J.* **81**, 91-98.

Foght J.M., and Westlake D.W.S. (1988) Degradation of polycyclic aromatic hydrocarbons and aromatic heterocycles by a *Pseudomonas* species. *Can. J. Microbiol.* **34**, 1135-1141..

Förstner U., and Rothe P. (1977) Bildung und diagenese der karbonatsedimente im riessee (nach dem profil der forschungsbohrung Nördlingen 1973). *Geol. Bavarica* **75**, 49-58.

Forward G.C., and Whiting D.A. (1969) Reaction of benzylideneacetone with hydrogen sulfide and ammonia; Heterocyclic analogues of bicyclo [3,3,1] nonone. *J. Chem. Soc. (C)* , 1647-1652.

Francois R. (1987) A study of the extraction conditions of sedimentary humic acids to estimate their true in situ sulfur content. *Limnol. Ocenogr.* **32**, 964-972.

Francois R. (1987) A study of sulfur enrichment in the humic fraction of marine sediments during early diagenesis. *Geochim. Cosmochim. Acta* **51**, 17-27.

Frank P., Hedman B., Carlson R.M.K., Tyson T., Roe A.L., and Hodgson K.O.(1987) A large reservoir of sulfate and sulfonate residues within plasma cell from *Ascidia ceratodes*, revealed by x-ray absorption near-edge structure spectroscopy. *Biochemistry* **26**, 4975-4979.

Frankie K.A., and Hower J.C. (1987) Variations in pyrite size, form, and microlithotype association in the Springfield (No. 9) and Herrin (No. 11) coals, western Kentucky. *Intern. Jour. Coal Geol.* **7**, 349-364.

Frankie K.A., and Hower J.C. (1985) Pyrite/marcasite size, form, and microlithotype association in western Kentucky prepared coals. *Fuel Process. Tech.* **10**, 269-283.

Frazier D.E., and Osanik A. (1969) Recent peat deposits - Louisiana coastal plain. In *Environments of Coal Deposition* (eds. E.C. Dapples, and M.E. Hopkins). Spec. Paper Geol. Soc. American **114**, pp. 63-85.

Friedman I., and Redfield A.C. (1971) A model of the hydrology of the lakes of the lower Grand Coulee, Washington. *Water Resources Research* **7**, 874-898.

Friedman S.A. (1974) *Investigation of the coal reserves in the Ozarks section of Oklahoma and their potential uses. Final report to the Ozark regional commission.* Oklahoma Geological Survey.

Friocourt M.P., Berthou F., and Picart D. (1982) Dibenzothiophene deriviatives as organic markers of oil pollution. *Toxicol. Environ. Chem.* **5**, 205-215.

Fromm E., and Haas F. (1912) Concerning the effect of alkali sulfides on benzalacetone. *Justus Liebigs Ann. Chem.* **394**, 291-300.

Fromm E., and Hubert E. (1912) Concerning the effect of alkali sulfides on benzal-acetophenone. *Justus Liebigs Ann. Chem.* **394**, 301-309.

Fromm E., and Achert O. (1903) Uber schwefelhaltige benzylderivate and deren zerzelzung durch trockne destillation. *Ber.* **36**, 534-546.

Fuentes F.A. (1984) Diauxic growth of *Pseudomonas aeruginosa* PRG-1 on glucose and benzothiophene. Abst. #N26. *Proc. 84th Ann. Meeting American Society for Microbiology.*

Fuex A.N. (1977) The use of stable carbon isotopes in petroleum exploration. *J. Geochem. Exploration* **7**, 155-188.

Fu Jiamo, Sheng Guoiying, Peng Pingan, Brassell S. C., Eglinton G., and Jiang Jigang (1986) Pecularities of salt lake sediments as potential source rocks in China. In *Advances in Organic Geochemistry 1985* (eds. D. Leythaeuser, and J. Rullkötter), *Org. Geochem.* **10**, 119-126.

Furfari S., and Cyprès R. (1982) Hydropyrolysis of a high-sulphur high-calcite Italian Sulcis coal. 2. Importance of the mineral matter on the sulphur behaviour. *Fuel* **61**, 453-459.

Galimov E.M. (1973) Organic geochemistry of carbon isotopes. In *Advances in Organic Geochemistry 1973* (eds. B. Tissot, and F. Bienner), pp. 439-452. Paris: Technip.

Gallois R.W. (1978) A pilot study of oil shale occurrence in the Kimmeridge clay. *Rep. Inst. Geol. Sci.* (U.K.) **78**, 1-26.

Galloway, J.N. (1987) The western Atlantic ocean experiment. In *The Chemistry of Acid Rain, Sources and Atmospheric Processes,* (eds. Johnson R.W., and Gordon, G.E.) ACS Symposium Series **349**, pp. 39-57, American Chemical Society.

Gal'pern G.C. (1985) Thiophenes occurring in petroleum, shale oil and coals (ed S. Gronowitz), In *The Chemistry of Heterocyclic Compounds. Thiophene and its Derivatives*. Vol. 1, pp. 325-351. Wiley.

Gal'pern G.C. (1976) Heteroatomic components of petroleum. *Russian Chemical Reviews* **45**, 701-720.

Gal'pern G.C. (1971) Organosulfides of petroleum. *International Journal of Sulfur Chemistry* **B6**, 115-130.

Garcia-Suarez A.B., and Schobert H.H. (1988) Hydrodesulfurization of Spanish lignite. *ACS Div. Fuel Chem. Preprints* **33**, 241-246.

Ge P., and Wert C.A. (1990) Spacial variation of organic sulfur in coal. In *Geochemistry of Sulfur in Fossil Fuels* (eds. W.L. Orr, and C.M. White), ACS Symposium Series (this volume). American Chemical Society.

Gehman H.M., Jr. (1962) Organic matter in limestones. *Geochim. Cosmochim. Acta* **26**, 885-897.

George G.N., Gorbaty M.L., and Kelemen S.R. (1990) Sulfur K edge x-ray adsorption spectroscopy of petroleum asphaltenes and model compounds. In *Geochemistry of Sulfur in Fossil Fuels* (eds. W.L. Orr, and C.M. White), ACS Symposium Series (this volume). American Chemical Society.

George G.N., and Gorbaty M.L. (1989) Sulfur K-edge x-ray absorption spectroscopy of petroleum asphaltenes and model compounds. *J. Amer. Chem. Soc.* **111**, 3182-3186.

George G.N., Byrd J., and Winge D.R. (1988) X-ray absorption studies of yeast copper metallothionein. *J. Biol. Chem.* **263**, 8199-8203.

Gerdes G., Krumbein W.E., and Holtkamp E. (1985) Salinity and water activity related zonation of microbial communities and potential stromatolites of the Gavish Sabkha. In *Hypersaline Ecosystems, Ecological Studies* **53** (eds. G.M. Friedman, and W.E. Krumbein), pp. 238-266. Springer-Verlag.

Giblin A.E., and Howarth R.W. (1984) Porewater evidence for a dynamic sedimentary iron cycle in salt marshes. *Limnol. Oceanogr.* **29**, 46-63.

Gibling M.R., Zentilli M., and McCready R.G.L. (1989) Sulphur in Pennsylvanian coals of Atlantic Canada: Geologic and isotopic evidence for a bedrock evaporite source. *Intern. Jour. Coal Geol.* **11**, 81-104.

Gibson D.T., and Subramanian V. (1984) Microbial degradation of aromatic hydrocarbons. In *Microbial Degradation of Organic Compounds* (eds. D.T. Gibson), pp. 181-253. Marcel Dekker.

Gilman H., and Shirley D.A. (1949) Metalation of thiophene by butyllithium. *J. Amer. Chem. Soc.* **71**, 1870-1871.

Gilman H., and Jacoby A.L. (1938) Dibenzothiophene: Orientation and deriviatives. *J. Org. Chem.* **3**, 108-119.

Given P.H. (1984) An essay on the organic geochemistry of coal. In *Coal Science* (eds. M.L. Gorbaty, J.W. Larsen, and I. Wender), pp. 63-252. Academic Press.

Given P.H. (1984) Concepts of coal structure in relation to combustion behavior. *Prog. Energy Combust. Sci.* **10**, 149-154.

Gluskoter H.J., Shimp N.F., and Ruch R.R. (1981) Coal analyses, trace elements, and mineral matter. In *Chemistry of Coal Utilization, Second Supplementary Volume* (ed. M.A. Elliott.), Chap. 7, pp. 369-424. Wiley.

Gluskoter H.J., Ruch R.R., Miller W.G., Cahill R.A., Dreher G.B., and Kuhn J.K. (1977) Trace elements in coal: Occurrence and distribution. *Illinois State Geological Survey Circular* **499**, 154.

Gluskoter H.J. (1977) Inorganic sulfur in coal. In *Energy Sources, Vol. 3*, pp. 125-131. Crane Russak & Co.

Gluskotter H. (J965) Electronic Low-temperature ashing of bituminous coal. *Fuel* **44**, 285-291.

Gluskoter H.J., and Simon J.A. (1964) Sulfur in Illinois coals. *Illinois State Geological Survey Circular* **432**, 28.

Goff J.C. (1983) Hydrocarbon generation and migration from Jurassic source rocks in the East Shetland basin and Viking Graben of the northern North Sea. *J. Geol. Sci. Lond.* **140**, 445-474.

Goldhaber M.B., Tuttle M.L., and Baedecker M.J. (1984) Response of carbon-sulfur sedimentary record of Great Salt Lake, Utah to changing depositional environments. *Geological Society of America Annual Meeting*, Reno, Abstracts, No. 33625.

Goldhaber M.B., and Kaplan I.R. (1980) Mechanisms of sulfur incorporation and isotope fractionation during early diagenesis in sediments of the Gulf of California. *Marine Chem.* **9**, 97-106.

Goldhaber M.B., and Kaplan I.R. (1974) The sulfur cycle. In *The Sea* (ed. E.D. Goldberg), Vol. 5, pp. 569-655. Wiley.

Goldstein J.I. (1979) Principles of thin film x-ray microanalysis. In *Introduction to Analytical Microscopy* (eds. J.J. Hren, J.I. Goldstein, and D.C. Joy), pp. 83-120. Plenum Press.

Goossens H., de Leeuw J.W., Schenck P.A., and Brassell S.C. (1984) Tocopherols as likely precursors of pristane in ancient sediments and crude oils. *Nature* **312**, 440-442.

Gorbaty M.L., and George G.N. (1990) Sulfur K edge x-ray absorption spectroscopy of petroleum asphaltenes and model compounds. In *Geochemistry of Sulfur in Fossil Fuels* (eds. W.L. Orr and C.M. White), ACS Symposium Series (this volume). American Chemical Society.

van Graas G., de Leeuw J.W., Schenck P.A., and Haverkamp J. (1981) Kerogen of Toarcian shales of the Paris Basin. A study of its maturation by flash pyrolysis techniques. *Geochim. Cosmochim. Acta* **45**, 2456-2474.

van Graas G., de Leeuw J.W., and Schenck P.A. (1980) Characterization of coals and sedimentary organic matter by Curie-point pyrolysis-mass spectrometry. Part I. *J. Anal. Appl. Pyrolysis* **2**, 265-276.

van Graas G., de Leeuw J.W., and Schenck P.A. (1980) Analysis of coals of different rank by Curie-Point pyrolysis-mass spectrometry and Curie-point pyrolysis-gas chromatography-mass spectrometry. *Adv. Org. Geochem., Phys. Chem. Earth* **12**, 485-494.

Gransch J.A., and Posthuma J. (1974) On the origin of sulfur in crudes. In *Advances in Organic Geochemistry 1973* (eds. B. Tissot, and F. Bienner), pp. 729-739. Editions Technip.

Grbic'-Galic' D. (1990) Anaerobic microbial transformations of nonoxygenated aromatics and alicyclic compounds in soil, subsurface, and freshwater sediments. In *Soil Biochemistry* (eds. J.M. Bollag, and G. Stotzky), Vol. **8**, pp. 117-189. Marcel Dekker.

Grbic'-Galic' D. (1989) Microbial degradation of homocyclic and heterocyclic hydrocarbons under anaerobic conditions. *Dev. Ind. Microbiol.* **30**, 237-253.

Greer R.T. (1979) Organic and inorganic sulfur in coal. *Scanning Electron Microsc.*, 477-486.

Greer R.T. (1978) Evaluation of pyrite particle size, shape, and distribution factors. *Energy Sources* **4**, 23-51.

Greer R.T. (1977) Coal microstructure and pyrite distribution. In *Coal Desulfurization*. ACS Symposium Series **64**, pp. 3-15. American Chemical Society.

Greer R.T. (1976) Colloidal pyrite growth in coal. *Colloidal and Interface Science* **4**, 411-423.

Grimalt J., and Albaiges J. (1987) Sources and occurrence of C_{12}-C_{22} *n*-alkane distributions with even carbon-number preference in sedimentary environments. *Geochim. Cosmochim. Acta* **51**, 1379-1384.

Grimalt J., Albaiges J., Al-Saad H.T., and Douabul A.A.Z. (1985) n-Alkane distributions in surface sediments from the Arabian Gulf. *Naturwissenschaften* **72**, 35-37.

Gronowitz S., and Raznikiewicz T. (1973) 3-Bromothiophene. In *Org. Synt. Coll.* Vol. V, pp. 149-151. John Wiley and Sons.

Gruner D.S., and Hood W.C. (1971) Three iron sulfate minerals from coal mine refuse dumps in Perry County, Illinois. *Trans. IL State Acad. Sci.* **64**, 156-158.

Guenther A., Lamb B., and Westberg H. (1989) U.S. national biogenic sulfur emission inventory. In *Biogenic Sulfur in the Environment* (eds. E.S. Saltzman, and W.J. Cooper), ACS Symposium Series **393**, pp. 14-30. American Chemical Society.

Guilianelli J.L., and Williamson D.L. (1982) Comparison of microwave and radiofrequency low-temperature ashing of coal using ^{57}Fe Mössbauer spectroscopy of sulphur forms. *Fuel* **61**, 1267-1272.

Gundlach E.R., Boehm P.D., Marchand M., Atlas R.M., Ward D.M., and Wolfe D.A. (1983) The fate of Amoco Cadiz oil. *Science* **221**, 122-129.

Gwynn J.W. (1980) *Great Salt Lake: A scientific, historical and economic overview.* Utah Department of Natural Resources Bulletin 116.

Hackley K.C., Buchanan D.H., Coombs K., Chaven C., and Kruse C.W. (1990) Solvent extraction of elemental sulfur from coal and a determination of it source using stable sulfur isotopes. *Fuel Process. Tech.* **24**, 431-436.

Hackley K.C., and Anderson T. (1986) Sulfur isotope variation in low sulfur coal from the Rocky Mountain region. *Geochim. Cosmochim. Acta* **50**, 1703-1713.

Hall T.A. (1979) Biological x-ray microanalysis. *J. Microscopy* **117**, 146-163.

Hall T.A., Anderson H.C., and Appleton T. (1973) The use of thin specimens for x-ray microanalysis in biology. *J. Microscopy* **99**, 177-182.

Hamrla M. (1960) K ragvoju in stratigrafiji produktivnik liburnijskih plasti Primorskega krasa. *Rudarsko-Metalurski Zbornik* **3**, 203-216.

Hamrla M. (1959) On the conditions of origin of the coal beds in the Karst region. *Geologija* **5**, 180-264.

Hamrla M. (1955) Petrographic composition of some specimens of Rasa coal, regarding their varying coking ability. *Geologija* **3**, 181-197.

Harris D.L., and Evans S.A. (1982) Sulfur-33 nuclear magnetic resonance spectroscopy of simple sulfones. Alkyl substituent induced chemical shift effect. *J. Org. Chem.* **47**, 3355-3358.

Harris L.A., Yust C.S., and Crouse R.S. (1977) Direct determination of pyritic and organic sulphur by combined coal petrography and microprobe analysis (CPMA) - a feasibility study. *Fuel* **56**, 456-457.

Harris L.A., and Yust C.S. (1976) Transmission electron microscopy observation of porosity in coal. *Fuel* **55**, 233-236.

Harvey H.R., Eglinton G., O'Hara S.C.M., and Corner E.D.S. (1987) Biotransformation and assimilation of dietary lipids by *Calanus* feeding on a dinoflagelate. *Geochim. Cosmochim. Acta* **51**, 3031-3040.

Harvey R.D., and Demaris P.J. (1987) Size and maceral association of pyrite in Illinois coals and their float-sink fractions. *Org. Geochem.* **2**, 343-349.

Harvey R.D., and Dillon J.W. (1985) Maceral distributions in Illinois coals and their Paleoenvironmental implications. *Intern. Jour. Coal Geol.* **5**, 141-165.

Hatch J.R., Gluskoter H.J., and Lindahl P.C. (1976) Sphalerite in coals from the Illinois Basin. *Econ. Geol.* **71(3)**, 613-624.

ten Haven H.L., Rullkötter J., Sinninghe Damsté J.S., and de Leeuw J.W. (1990) Distribution of organic sulphur compounds in mesozoic and cenozoic deep-sea sediments from the Atlantic and Pacific Oceans and the Gulf of California. In *Geochemistry of Sulfur in Fossil Fuels* (eds. W.L. Orr and C.M. White), ACS Symposium Series (this volume). American Chemical Society.

ten Haven H.L., Rullkötter J., and Welte D.H. (1989) Steroid biological marker hydrocarbons as indicators of organic matter diagenesis in deep sea sediments: Geochemical reactions and influence of different heat flow regimes. *Geol. Rundsch.* **78**, 841-850.

ten Haven H.L., Rohmer M., Rullkötter J., and Bisseret P. (1989) Tetrahymanol, the most likely precursor of gammacerane, occurs ubiquitously in marine sediments. *Geochim. Cosmochim. Acta* **53**, 3073-3079.

ten Haven H.L., de Leeuw J.W., Sinninghe Damsté J.S., Schenck P.A., Palmer S.E., and Zumberge J.E. (1988) Application of biological markers in the recognition of palaeo hypersaline environments. In *Lacustrine Petroleum Source Rocks* (eds. K. Kelts, A. J. Fleet, and M. Talbot), *Geol. Soc. Spec. Publ. No. 40*, pp. 123-130. Blackwell.

ten Haven H.L., Baas M., Kroot M., de Leeuw J.W., Schenck P.A., and Ebbing, J. (1987) Late quaternary Meditterranean sapropels.III. Assessment of source of input and palaeotemperature as derived from biological markers. *Geochim. Cosmochim. Acta* **51**, 803-810.

ten Haven H.L., de Leeuw J.W., Rullkötter J., and Sinninghe Damsté J.S. (1987) Restricted utility of the pristane/phytane ratio as a palaeoenvironmental indicator. *Nature* **330**, 641-643.

ten Haven H.L., de Leeuw J.W., Peakman T.M., and Maxwell J.R. (1986) Anomalies in steroid and hopanoid maturity indices. *Geochim. Cosmochim. Acta* **50**, 853-855.

ten Haven H.L., de Leeuw J.W., and Schenck P.A. (1985) Organic geochemical studies of a messinian evaporoilic basin, northern Apennines (Italy) I: Hydrocarbon biological markers for a hypersaline environment. *Geochim. Cosmochim. Acta* **49**, 2181-2191.

Hayatsu R., Scott R.G., and Winans R.E. (1982) Oxidation of coal. In *Oxidation in Organic Chemistry, Part D* (ed. W.S. Trahanovsky), pp. 279-354. Academic Press.

Hayatsu R., Winans R.E., Scott R.G., Moore L.P., and Studier M.H. (1978) Oxidative degradation studies of coal and solvent-refined coal. In *Organic Chemistry of Coal* (ed. J.W. Larsen), ACS Symposium Series **71**, American Chemical Society.

Hayatsu R., Winans R.E., Scott R.G., Moore L.P., and Studier M.H. (1978) Trapped organic components and aromatic units in coals. *Fuel* **57**, 541-548.

Hayatsu R., Scott R.G., Moore L.P., and Studier M.H. (1975) Aromatic units in coal. *Nature* **257**, 378-380.

Hayes J.M. (1982) Fractionation: An introduction to isotope meausrements and terminology. *Spectra* **8**, 3-8.

Haynie F.H. (1980) In *Durability of Building Materials and Components*, ASTM STP-691, pp. 157-175, American Society for Testing Materials.

Hedman B., Frank P., Gheller S.F., Roe A.L., Newton W.E., and Hodgson K.O. (1988) New structural insights into the iron-molybdenum cofactor from azotobacter vinelandii nitrogenase through sulfur K and molybdenum L x-ray absorption edge studies. *J. Amer. Chem. Soc.* **110**, 3798-3805.

Hedman B., Frank P., Penner-Hahn J.E., Roe A.L., Hodgson K.O., Carlson R.M.K., Brown G., Cerino J., Hettel R., Troxel T., Winick H., and Yang J. (1986) Sulfur K-edge x-ray absorption studies using the 54-pole wiggler at SSRL in undulator mode. *Nucl. Instrum. Methods Phys. Res. Sect.* **A246**, 797-803.

Hendrey G.R., Hoogendyk C.G., and Gmur N.F. (1984) Analysis of trends in the chemistry of surface waters in the United States, Annual Report for NAPAP Projects EI-8 and E2-11. BNL 34956. Brookhaven National Lab.

Hengel T.D., and Walker P.L. (1984) Catalysis of lignite char gasification by exchangeable calcium and magnesium. *Fuel* **63**, 1214-1220.

Herod A.A., and Smith C.A. (1985) Release of oxygen and sulphur compounds from coal. *Fuel* **64**, 281-283.

Hidy G.M. (1987) Subcontinental air pollution phenomena. In *The Chemistry of Acid Rain, Sources and Atmospheric Processes*, (eds. R.W. Johnson, and G.E. Gordon), ACS Symposium Series **349**, pp. 10-27. American Chemical Society.

Hippo E.J., Crelling J.C., Sarvela D.P., and Mukerjee J. (1987) Organic sulfur distribution and desulfurization of coal maceral fractions. In *Processing and Utilization of High Sulfur Coals II* (eds. Y.P. Chugh, and R.D. Caudle), pp. 13-22. Elsevier.

Hirner A.V., Graf W., Treibs R., Melzer A.N., and Hahn-Weinheimer P. (1984) Stable sulfur and nitrogen isotopic compositions of crude oil fractions from southern Germany. *Geochim. Cosmochim. Acta* **48**, 2179-2186.

Ho T.Y., Rogers M.A., Drushel H.V., and Koons C.B (1974) Evolution of sulfur compounds in petroleum. *Amer. Assoc. Petr. Geol. Bull.* **58**, 2338-2348.

Hoffman C.F., Mackenzie A.S., Lewis C.A., Maxwell J.R., Oudin J.L., Durand B., and Vandenbrouke M. (1984) A biological marker study of coals, shales and oils from the Mahakam delta, Kalimantan, Indonesia. *Chem. Geol.* **42**, 1-23.

Hoffmann M.R. (1977) Kinetics and mechanism of oxidation of hydrogen sulfide by hydrogen peroxide in acidic solution. *Environ. Sci. Technol.* **11**, 61-66.

Hollerbach A., Hufnagel H., and Wehner H. (1977) Organisch-geochemische und petrologische Untersuchungen und den see-sedimenten aus der forschungsbohrung Nördlingen 1973. *Geol. Bavarica* **75**, 139-153.

Holser W.T., and Kaplan I.R. (1966) Isotope geochemistry of sedimentary sulfates. *Chem. Geol.* **1**, 93-135.

Horsfield B. (1989) Practical criteria for classifying kerogens: Some observations from pyrolysis-gas chromatography. *Geochim. Cosmochim. Acta* **53**, 891-901.

Hou C.T., and Laskin A.I. (1976) Microbial conversion of dibenzothiophene. *Dev. Ind. Microbiol.* **17**, 351-362.

Howarth R.W. (1978) Pyrite: Its rapid formation in a salt marsh and its importance in ecosystem metabolism. *Science* **203**, 49-51.

Hsieh K.C., and Wert C.A. (1985) Direct measurement of organic sulfur in coal. *Fuel* **64**, 255-262.

Hsieh K.C., and Wert C.A. (1985) Measurement of organic sulfur in coal. *Proc. Intern. Conf. Coal Sci.*, 826-829.

Hsieh K.C., and Wert C.A. (1981) Sulfide crystals in coal. *Materials Sci. Eng.* **50**, 117-125.

Huang E.Y.K., and Pulsifer A.H. (1977) Coal desulfurization during gaseous treatment. In *Coal Desulfurization. Chemical and Physical Methods* (ed. T.D. Wheelock), ACS Symposium Series **64**, pp. 290-303. American Chemical Society.

Huffman G.P., Huggins F.E., Mitra S., Shah N., Pugmire R.J., Davis B., Lytle F.W., and Greegor R.B. (1989) Investigation of the molecular structure of organic sulfur in coal by XAFS spectroscopy. *Energy & Fuels* **3**, 200-205.

Huffman G.P., Huggins F.E., Shah N., Bhattacharyya D., Pugmire R.J., Davis B., Lytle F.W., and Greegor R.B. (1988) EXAFS investigation of organic sulfur in coal. *ACS Div. Fuel Chem. Preprints* **33**, 200-208.

Huffman G.P., Huggins F.E., Shah N., Bhattacharyya D., Pugmire R.J., Davis B., Lytle F.W., and Greegor R.B. (1987) Investigation of the atomic and physical structure of organic and inorganic sulfur in coal. In *Processing and Utilization of High Sulfur Coals II*, (eds. Y.P. Chugh, and R.D. Caudle), pp. 3-12. Elsevier.

Huffman G.P., and Huggins F.E. (1978) Mössbauer studies of coal and coke: Quantitative phase identification and direct determination of pyritic and iron sulphide sulfur content. *Fuel* **57**, 592-604.

Hufnagel H. (1984) Oil shale in Jordan. *Geol. Jb., Rh A* **75**, 295-311.

Hughes W.B. (1984) Use of thiophenic organosulfur compounds in characterizing crude oils derived from carbonate versus siliciclastic sources. In *Petroleum Geochemistry and Source Rock Potential of Carbonate Rocks, AAPG Studies in Geology #18* (ed. J.G. Palacas), pp. 181-196. AAPG Press.

Hunt J.W., and Smith J.W. (1985) $^{34}S/^{32}S$ ratios of low-sulfur Permian Australian coals in relation to depositional environments. *Chem. Geol. (Isotope Geoscience)* **58**, 137-144.

Hunt J.M. (1979) *Petroleum Geochemistry and Geology.* W. H. Freeman.

Huntington J.G., Mayo F.R., and Norman A.K. (1979) Mild oxidation of coal models. *Fuel* **58**, 31-36.

Hurley R.W., and White E.W. (1974) New soft x-ray method for determining the chemical forms of sulfur in coal. *Anal. Chem.* **46**, 2234-2237.

Hussain Z., Umbach E., Shirley D.A., Stohr J., and Feldhaus J. (1982) Performance and application of a double crystal monochromator in the energy region $800 \leq h\nu \leq 4500$ eV. *Nuclear Instrum. Methods Phys. Res.* **195**, 115-131.

Ignasiak B.S., Fryer J.F., and Jodernik P. (1978) Polymeric structure of coal. 2. Structure and thermoplasticity of sulphur-rich Rasa lignite. *Fuel* **57**, 578-584.

Illing L.V., and Hobson G.D. (eds.) *Petroleum Geology of the Continental Shelf of North-West Europe*. Heydon and Son.

Imhoff J.F., and Trüper H.G. (1980) *Chromatium purpuratum* sp. nov., a new species of *Chromatiacae*. *Zbl. Bakt. I. Abt Orig.* **C1**, 61-69.

Imhoff J.F., and Trüper, H.G. (1976) Marine sponges as habitats of anaerobic photo-synthetic bacteria. *Microbiol. Ecol.* **3**, 1-9.

Ingamells C.O. (1970) Lithium metaborate flux in silicate analysis. *Analytica Chimica Acta* **52**, 323-334.

Ingram L.L., Ellis J., Crisp P.T., and Cook A.C. (1983) Comparative study of oil shale and shale oil from the Mahogany zone, Green River formation USA, and Kerosene creek seam, Rundle formation, Australia. *Chem. Geol.* **38**, 185-212.

International electric research exchange, effects of SO_2 and its derivatives on health and ecology (1981) Vol. 1 - Human Health. Vol. 2 - Natural Ecosystems, Agriculture, Forestry, and Fisheries. Available from Electric Power Research Institute, Research Reports Center, P.O. Box 50490, Palo Alto, CA 94303.

Isaacs C.M. (1984) The Monterey - key to offshore California boom. *Oil Gas J.* **82**, 75-81.

Isaacs C.M., and Pisciotto K.A., and Garrison R.E. (1983) Facies and diagenesis of the Miocene Monterey formation, California: A summary. *Dev. Sedimentol.* **36**, 247-282.

Isbister J.D., Wyza R., Lippold J., DeSouza A., and Anspach G. (1988) Bioprocessing of coal. In *Reducing Risks from Environmental Chemicals through Biotechnology* (ed. G.S. Omenn), pp. 281-293. Plenum Press.

Ishiwatari R., Rohrback B.G., and Kaplan I.R. (1978) Hydrocarbon generation by thermal alteration of kerogen from different sediments. *Amer. Petr. Geol. Bull.* **62**, 687-692.

Janda M., Srogl J., Stibor I., Nemec M., and Vopatrna P. (1972) Polylithiothiophenes and a convenient sysnthesis of polymethylthiophenes. *Synthesis*, 545-547.

Jankowski B., and Littke R. (1986) Das organische material der ölschiefer von messel. *Geowiss. in unserer Zeit* **4**, 73-80.

Jankowski B. (1981) *Die geschichte der sedimentation im nördlinger ries und randecker maar*. Bochumer Geol. und Geotechn. Arb. Vol. **6**, Ruhr-University, Bochum.

Johns R.B.(ed.) (1986) *Biological markers in the sedimentary record*. Methods in Geo-chemistry and Geophysics Series, Vol. **24**. Elsevier.

Johnson R.C. (1985) Early cenozoic history of the Uinta and Piceance creek basins, Utah and Colorado, with special reference to the development of Eocene Lake Uinta. In *Cenozoic Paleogeography of Western United States* (eds. R.M. Flores, and S.S. Kaplan), Rocky Mountain Paleontology Symposium 3, pp. 247-276. Society of Economic Paleontologists and Minerologists.

Johnson R.C. (1981) Stratigraphic evidence for a deep Eocene Lake Uinta, Piceance Creek basin, Colorado. *Geology* **9**, 55-62.

Johnson R.W., and Gordon G.E., eds. (1987), *The Chemistry of Acid Rain, Sources and Atmospheric Processes*, ACS Symposium Series **349**, American Chemical Society.

Johnson S.Y., Otton J.K., and Macke D.L. (1987) Geology of holocene surficial uranium deposit of the north fork of Flodelle Creek, northeastern Washington. *Geological Society of America Bulletin* **98**, 77-85.

Jones R.B., McCourt C.B., and Swift P. (1981) XPS studies of nitrogen and sulphur in coal. *Proceedings International Conference on Coal Science, Dusseldorf* pp. 657-662.

Jones R.W. (1984) Comparison of carbonate and shale source rocks. In *Petroleum Geochemistry and Source Rock Potential of Carbonate Rocks* (ed. J. G. Palacas) *AAPG Studies in Geology* **18**, pp. 163-180. AAPG.

Joyce W.F., and Uden P.C. (1983) Isolation of thiophenic compounds by argentation chromatography. *Anal. Chem.* **55**, 540-543.

Jüntgen H., and van Heek K.H. (1970) Reaktionsabläufe unter nicht-isothermen Bedingungen. *Fortschritte der chemischen Forschung* **13**, 601-699.

Kanagawa T., and Kelly D.P. (1987) Degradation of substituted thiophenes by bacteria isolated from activated sludge. *Microb. Ecol.* **13**, 47-57.

Kaplan I.R. (1983) Stable isotope of sulfur, nitrogen and deuterium in recent marine environments. In *Stable Isotopes in Sedimentary Geology* (eds. M.A. Arthur, T.F. Anderson, I.R. Kaplan, J. Veizer, and L.S. Land), Chap. 2, pp. 2.1-2.108. SEPM.

Kaplan I.R., Smith J.W., and Ruth E. (1970) Carbon and sulfur concentration and isotopic composition in Apollo II lunar samples. *Proc. Apollo II Lunar Sci. Conf.* **2**, 1317-1329.

Kaplan I.R., and Rittenberg S.C. (1964) Microbial fractionation of sulfur isotopes. *Jour. Gen. Microbiol.* **34**, 195-212.

Kaplan I.R., Emery K.O., and Rittenberg S.C. (1963) The distribution and isotopic abundance of sulphur in recent sediments off southern California. *Geochim. Cosmochim. Acta* **27**, 297-331.

Kargi F. (1987) Biological oxidation of thianthrene, thioxanthene and dibenzothiophene by the thermophilic organism *Sulfobus acidocaldarius*. *Biotechnol. Lett.* **9**, 478-482.

Kargi F., and Robinson J.M. (1984) Microbial oxidation of dibenzothiophene by thermophilic organism *Sulfobus acidocaldarius*. *Biotechnol. Bioeng.* **26**, 687-690.

Kavcic R. (1958) Sulfur isotope abundance study in high sulfur coals. *Slovenskega, Kemijskeg a Drustva* **5**, 7-11.

Kavcic R. (1954) The determination of the sulfur linkage in the Rasa coal. *Bulletin Scientific, Yugoslavie* **49**, 5809.

Kendall C.G. St. C., and d'E. Skipworth Bt P.A. (1968) Recent algal mat of a Persian Gulf lagoon, *Journal of Sedimentary Petrology* **38(4)** 1040-1058.

Kent B.H. (1986) Evolution of thick coal deposits in the Powder River Basin, northern Wyoming. *Geol. Soc. Amer. Special Paper* **210**, 105-122.

Kiene R.P., and Capone D.G. (1988) Microbial transformations of methylated sulfur compounds in anoxic salt marsh sediments. *Microb. Ecol.* **15**, 275-291.

Kiene R.P., Ormland R.S., Catena A., Miller L.G., and Capone D.G. (1986) Metabolism of reduced methylated sulfur compounds in anaerobic sediments and by a pure culture of an estuarin methanogen. *Appl. Environ. Microb.* **52**, 1037-1045.

King G.M., Klug M.J., Wiegert R.G., and Chalmers A.G. (1982) The relationship between soil water movement, sulfide concentration, and Spartina alterniflora production in a Georgia salt marsh. *Science* **218**, 61-63.

King W.J., and Nord F.F. (1949) Studies in the thiophene series. V. Wolff-Kishner reductions. *J. Org. Chem.* **14**, 638-642.

Kinoshita K., Honda T., and Tanaka N. (1966) Mirabilite and thenardite produced from refuse heap. *Jour. Mining Inst. Kyushu (Kyushu Kozan Gakkaishi)* **34**, 377-380.

Kinsman D.J.J., and Park R.K. (1976) Algal belt and coastal sabkha evolution, Trucial Coast, Persian Gulf. *Stromatolites* (ed. M.H. Walter), Developments in Sedimentology Series **20**, pp. 421-433. Elsevier.

Kirkland D.W., and Evans R. (1981) Source-rock potential of evaporitic environment. *Amer. Assoc. Petr. Geol. Bull.* **65**, 181-190.

Kiyosu Y., and Krouse H.R. (1989) Carbon isotope effect during abiogenic oxidation of methane. *Earth and Planetary Science Letters* **95**, 302-306.

Kleinman L.I. (1983) A regional scale modeling study of the sulfur oxides with a comparison to ambient and wet deposition monitoring data. *Atmos. Environ.* **17(6)**, 1107-1121.

Kneller W.A., and Maxwell G.P. (1985) Size, shape, and distribution of microscopic pyrite in selected Ohio coals. In *Processing and Utilization of High Sulfur Coals* (ed. Y.A. Attia), pp. 41-65. Elsevier.

Kodama K. (1977) Co-metabolism of dibenzothiophene by *Pseudomonas jianii. Agric. Biol. Chem.* **41**, 1305-1306.

Kodama K. (1977) Induction of dibenzothiophene oxidation by *Pseudomonas jianii. Agric. Biol. Chem.* **41**, 1193-1196.

Kodama K., Nakatani S., Umehara K., Shimizu K., Minoda Y., and Yamada K. (1973) Identification of microbial products from dibenzothiophene and its proposed oxidation pathway. *Agric. Biol. Chem.* **37**, 45-50.

Kodama K., Nakatani S., Umehara K., Shimizu K., Minoda Y., and Yamada K. (1970) Microbial conversion of petro-sulfur compounds. Part III. Isolation and identification of products from dibenzothiophene. *Agric. Biol. Chem.* **34**, 1320-1324.

Köhler M., Genz I.L., Schicht B., and Eckart V. (1984) Microbial desulfurization of petroleum and heavy petroleum fractions. 4. Anaerobic degradation of organic sulfur compounds of petroleum. *Zentralbl. Mikrobiol.* **139**, 239-247.

Kohnen M.E.L., Sinninghe Damsté J.S., Rijpstra W.I.C., and de Leeuw J.W. (1990) Alkylthiophenes as sensitive indicators of palaeoenvironmental changes: A study of a Cretaceous oil shale from Jordan. In *Geochemistry of Sulfur in Fossil Fuels* (ed. W.L. Orr, and C.M. White), ACS Sympsoium Series (this volume). American Chemical Society.

Kohnen M.E.L., Sinninghe Damsté J.S., ten Haven H.L., and de Leeuw J.W. (1989) Early incorporation of polysulphides in sedimentary organic matter. *Nature* **341**, 640-641.

Kolodny Y. (1978) The origin of phosphorite deposits in the light of occurrences of recent sea-floor phosphorites. In *Marine Phosphorites- Geochemistry, Occurrence, Genesis* (ed. Y.K. Bentor), pp. 249, SEPM Spec. Publ.29.

Kong R.C., Lee M.L., Iwao M., Tominaga Y., Pratap R., Thompson R.D., and Castle R.N. (1984) Determination of sulfur heterocycles in selected synfuels. *Fuel* **63**, 702-708.

Koots J.A., and Speight J.G. (1975) Relation of petroleum resins to asphaltene. *Fuel* **54**, 179-184.

Kossenberg M., and Cook A.C. (1961) Weathering of sulphide minerals in coal: Production of ferrous sulphate heptahydrate. *Mineral Mag.* **32**, 829-830.

Kosters E.C., Chumra G.L., and Bailey A. (1987) Sedimentary and botanical factors influencing peat accumulation in the Mississippi Delta. *Journal of the Geological Society, London* **144**, 423-434.

Kosters E.C. (1983) *Louisiana Peat Resources*. Dept. of Natural Resources, Louisiana Geological Survey, Baton Rouge, LA.

Kosters E.C., and Bailey A. (1983) Characteristics of peat deposits in the Mississippi delta plain. *Transactions Gulf Coast Association of Geological Societies* **33**, 311-325.

Kreulen D.J.W. (1952) Sulphur coal of Istria. *Fuel* **31**, 462-467.

van Krevelen D.W. (1961) *Coal*. Elsevier.

Krouse H.R., and Yonge C.J. (1990) Sulfur isotope studies of H_2S evolved during pyrolysis of bitumens, oil, coal, shales, and extracts. In *Geochemistry of Sulfur in Fossil Fuels* (eds. W.L. Orr, and C.M. White), ACS Symposium Series (this volume). American Chemical Society.

Krouse H.R., Viau C.A., Eliuk L.S., Ueda A., and Halas S. (1988) Chemical and isotopic evidence of thermochemical sulphate reduction by light hydrocarbon gases in deep carbonate reservoirs. *Nature, 333*, 415-419.

Krouse H.R., Ritchie R.G.S., and Roche R.S. (1987) Sulfur isotope composition of H_2S evolved during the non-isothermal pyrolysis of sulfur-containing materials. *J. Anal. Appl. Pyrolysis* **12**, 19-29.

Krouse H.R. (1977) Sulfur isotope studies and their role in petroleum exploration. *J. Geochem. Exploration* **7**, 189-211.

Kruge M.A. (1986) Biomarker geochemistry of the Miocene Monterey formation, west San Joaquin Basin California: Implications for petroleum generation. In *Advances in Organic Geochemistry 1985* (eds. D. Leythaeuser, and J. Rullkötter), *Org. Geochem.* **10**, 517-530.

Krumbein W.E., Buchholz H., Franke P., Giani D., and Wonneberger K. (1979) O_2 and H_2S coexistence in stromatolites. A model for the origin of mineralogical lamination in stromatolites and banded iron formations. *Naturwissenschaften* **66**, 381-389.

Kuczynski W., and Andrzejak A. (1961) A note on organic matter extracted from brown coals after treatment with acids. *Fuel* **40**, 203-206.

Kuhn J.K., Kohlenberger L.B., and Shimp N.F. (1973) Comparison of oxidation and reduction methods in the determination of forms of sulfur in coal. *Environmental Geology Note 66*, Illinois State Geological Survey.

Kurita S., Endo T., Nakamura H., Yagi T., and Tamiya N. (1971) Decomposition of some organic sulfur compounds in petroleum by anaerobic bacteria. *J. Gen. Appl. Microbiol.* **17**, 185-198.

Kursanov D.N., Parnes Z.N., and Loin N.M. (1974) Applications of ionic hydrogenation to organic synthesis. *Synthesis,* 633-651.

Laborde A.L., and Gibson D.T. (1977) Metabolism of dibenzothiophene by a *Beijerinckia* species. *Appl. Environ. Microbiol.* **34**, 783-790.

LaCount R.B., Anderson R.R., Friedman S., and Blaustein B.D. (1987) Sulfur in coal by programmed temperature oxidation. *Fuel* **66**, 909-913.

LaCount R.B., Anderson R.R., Friedman S., and Blaustein B.D. (1986) Sulfur in coal by program-temperature oxidation. *ACS Div. Fuel Chem. Preprints* **31**, 70-78.

Lacount R.B., Gapen D.K., King W.P., Dell D.A., Simpson F.W., and Helms C.A. (1981) Thermal oxidative degradation of coal as a route to sulfur functionality: An initial study. In *New Approaches in Coal Chemistry*, ACS Symposium Series **169**, American Chemical Society.

Lafarge E., and Barker C.(1988) Effect of water washing on crude oil composition. *Amer. Assoc. Petr. Geol. Bull.* **72(3)**, 263-278.

LaLonde R.T. (1990) Overview of polysulfide reactions relative to the formation of organosulfur and other organic compounds in the geosphere. In *Geochemistry of Sulfur in Fossil Fuels* (eds. W.L. Orr and C.M. White), ACS Symposium Series (this volume). American Chemical Society.

LaLonde R.T., Ferrara L.M., and Hayes M.P. (1987) Low-temperature polysulfide reactions of ene carbonyls: A reaction model for the geologic origin of S-heterocycles. *Org. Geochem.* **11**, 563-571.

LaLonde R.T., Codacovi L.M., Cun-heng H., Cang-fu X., Clardy J., and Krishnan B.S. (1986) Polysulfide-promoted conversions of ene carbonyls. Arrangements of o-methoxy-substituted chalcone units incorporated into product thiolanes. *J. Org. Chem.* **51**, 4899-4905.

LaLonde R.T., Florence R.A., Horenstein B.A., Fritz R.C., Silveira L., Clardy L., and Krishnan B.S. (1985) Variable reaction pathways for the action of polysulfide on Michael acceptors. *J. Org. Chem.* **50**, 85-91.

LaLonde R.T., Horenstein B.A., Schwendler K., Fritz R.C., Florence R.A., Ekiel I., and Smith I.C.P. (1983) Formation of 2,4-diaryl-3,5-diarylthiolanes from chalcones and polysulfide Scope, conditions and the thiolane NMR. *J. Org. Chem.* **48**, 4049-4052.

LaLonde R.T. (1982) Thiolan and monothio-β-diketone formation through the use of a nucleo-electrophilic thiating agent. *J. Chem. Soc. Chem. Commun.*, 401-402.

Larsen J.W., Pan C.S., and Shawver S. (1989) Effects of demineralization on the macromolecular structure of coals. *Energy & Fuels* **3**, 557-561.

Larter S.R. (1989) Chemical models of vitrinite reflectance evolution. *Geol. Rund.* **78**, 1-11.

Larter S.R. (1988) Some pragmatic perspectives in source rock geochemistry. *Marine and Petrol. Geol.* **5**, 193-204.

Larter S.R. (1985) Integrated kerogen typing in the recognition and quantitative assessment of petroleum source rocks. In *Petroleum Geochemistry in Exploration of the Norwegian Shelf.* pp. 269-286. Graham and Trotman.

Larter S.R., and Senftle J.T. (1985) Improved kerogen typing for petroleum source rock analysis. *Nature* **318**, 277-280.

Larter S.R. (1984) Application of analytical pyrolysis techniques to kerogen characterization and fossil fuel exploration/exploitation. In *Analytical Pyrolysis* (ed. K.J. Voorhees), pp. 212-275. Butterworths.

Larter S.R., and Douglas A.G. (1980) A pyrolysis-gas chromatographic method for kerogen typing. In *Advances in Organic Geochemistry, 1979* (eds. A.G. Douglas, and J.R. Maxwell), pp. 579-584. Pergamon Press.

Larter S.R., Solli H., and Douglas A.G. (1978) Analysis of kerogens by pyrolysis-gas chromatography-mass spectrometry using selective ion monitoring. *J. Chromatogr.* **167**, 421-431.

Later D.W., Lee M.L. Bartle K.D., Kong R.C., and Vassilaros D.L. (1981) Chemical class separation and characterization of organic compounds in synthetic fuels. *Anal. Chem.* **53**, 1612-1620.

Lauf R.J. (1982) Microstructure of coal fly ash particles. *Amer. Cer. Soc. Bull.* **61**, 487-490.

Lawesson S.O. (1957) Thiophene chemistry. II. A new method of preparing 2,4-dibromothiophene and its Grignard and lithium reagents. Two stage mechanism in the metalation of 2,4-dibromothiophene with butyllithium. *Arkiv. Kemi* **11**, 317-324.

Lawesson S.O. (1957) Thiophene chemistry. III. A new method of preparing 3,4-dibromothiophene and its Grignard and lithium reagents. Two stage metalation of 3,4-dibromothiophene. Substitution reactions of 4-bromo-3-thiophenecarboxylic acid. *Arkiv. Kemi* **11**, 325-326.

Leatherman S.P. (1982) *Barrier Island Handbook.* University of Maryland.

de Leeuw J.W., and Sinninghe Damsté J.S. (1990) Organic sulphur compounds and other biomarkers as indicators of palaeosalinity: A critical evaluation. In *Geochemistry of Sulfur in Fossil Fuels* (eds. W.L. Orr, and C.M. White),ACS Symposium Series (this volume). American Chemical Society.

de Leeuw J.W., Cox H.C., van Graas G., van de Meer F.W., Peakman T.M., Baas J.M.A. and van de Graaf B. (1989) Selective hydrogenation of sterenes during early diagenesis. *Geochim. Cosmochim. Acta* **53**, 903-909.

de Leeuw J.W. (1986) Sedimentary lipids and polysaccharides as indicators for sources of input, microbial activity and short-term diagenesis (ed. M.L. Sohn), *Organic Marine Geochemistry*, pp. 33-61. American Chemical Society.

de Leeuw J.W., Sinninghe Damsté J.S., Klok J., Schenck P.A. and Boon, J.J. (1985) Bio-geochemistry of Gavish Sabkha sediments I. Studies on neutral reducing sugars and lipid moieties by gas chromatography-mass spectrometry. In *Ecological studies 53, Hypersaline Ecosystems* (eds. G.M. Friedman, and W.E. Krumbein), pp. 350-367. Springer Verlag.

de Leeuw J.W., van der Meer F. W., Rijpstra W.I.C., and Schenck P.A. (1980) On the occurrence and structural identification of long chain unsaturated ketones and hydrocarbons in sediments. In *Advances in Organic Geochemistry 1979* (eds. A.G. Douglas, and J.R. Maxwell), pp. 211-217. Pergamon Press.

Lee M.L., Wiley C., Castle R.N., and White C.M. (1980) Separation and identification of sulfur heterocycles in coal-derived products. In *Polynuclear Aromatic Hydrocarbons: Chemistry and Biological Effects* (eds. A. Bjorseth, and A.J. Dennis), pp. 59-73. Battelle Press.

Le Perchec P., Thomas M., Fixari B., and Bigois M. (1989) Quantitative and simultaneous detection of hydrogen, carbon, sulfur and nitrogen elements in volatile and nonvolatile fractions of heavy oils by oxidative pyroanalysis. *ACS Div. Petr. Chem. Preprints* **34**, 261-267.

Leventhal J.S. (1983) An interpretation of carbon and sulfur relationships in Black Sea sediments as indicators of environments of deposition. *Geochim. Cosmochim. Acta* **47**, 133-137.

Lewan M.D. (1985) Evaluation of petroleum generation by hydrous pyrolysis experi-mentation. *Phil. Trans. R. Soc. Lond. A* **315**, 123-134

Lewan M.D. (1983) Effects of thermal maturation of stable isotope compositions as determined by hydrous pyrolysis of Woodford shale. *Geochim. Cosmochim. Acta* **47**, 1471-1479.

Lewan M.D., Winters J.C., and McDonald J.H. (1979) Generation of oil-like pyrolysates from organic rich shales. *Science* 203, 879-899.

Leythaeuser D., Radke M., and Willsch H. (1988) Geochemical effects of primary migration of petroleum in Kimmeridge source rocks from Brae field area, North Sea. II: Molecular composition of alkylated naphthalenes, phenanthrenes and dibenzo-thiophenes. *Geochim. Cosmochim. Acta* 52, 2879-2891.

Liaan-Jensen S. (1978) Marine carotenoids. In *Marine Natural Products* (eds. D.J Faulkner and W.H. Fenical), pp. 1-73. Academic Press.

Liaan-Jensen S. (1978) Chemistry of carotenoid pigments. In *Photosynthetic Bacteria* (eds. R.K. Clayton, and W.R. Sistrom), pp. 233-248. Plenum Press.

Lipfert F.W. (1989) Air Pollution and Materials Damage. In. *Handbook of Environmental Chemistry*, Vol. 4, Part B, (ed. O. Hutzinger). Springer-Verlag.

Lipfert F.W., Dupuis L.R., and Schaedler J.S. (1985) Methods for mesoscale modeling for materials damage assessment. Brookhaven National Laboratory Report to US Environmental Protection Agency.

Littke R., Baker D.R., and Leythauser D. (1988) Microscopic and sedimentologic evidence for the generation and migration of hydrocarbons in Toarcian source rocks of different maturities. In *Advances in Organic Geochemistry 1987* (eds. L. Mattavelli, and L. Novelli), *Org. Geochem.* 13, 549-559. Pergamon Press.

Liu, C.L., Hackley K.C., and Coleman D.D. (1987) Use of stable sulphur isotopes to monitor directly the behaviour in coal during thermal desulphurization. *Fuel* 66, 683-687.

Livingstone R. (1977) Compounds with five-membered rings having one heteroatom from Group VI; Sulphur and its analogues. In *Rodd's Chemistry of Carbon Compounds. Volume IV, Part A. Heterocyclic Compounds*, pp. 219-328. Elsevier.

Lloyd W.G., Riley J.T., Kuehn K.W., and Kuehn D.W. (1988) Chemistry and reactivity of micronized coal. Final Report, US DOE Contract No. DE-FG22-85PC80514.

Lord C.J. (1982) A selective and precise method for pyrite determination in sedimentary materials. *J. Sed. Petrology* 52, 664-666.

Louis M. (1966) Etudes geochimiques sur les "Schistes Cartons" du Toarcien du Bassin de Paris. In *Advances in Organic Geochemistry 1964* (eds. G.D. Hobson, and M.C. Louis), pp. 85-94. Pergamon Press.

Love L.G., Coleman M.L., and Curtis C.D. (1983) Diagenetic pyrite formation and sulphur isotope fractionation associated with a Westphalian marine incursion, northern England. *Trans. Royal Soc. Edinburgh: Earth Sciences* 74, 165-182.

Love L.G., and Murray J.W. (1963) Biogenic pyrite in recent sediments of Christchurch Harbour, England. *Amer. J. Sci.* 261, 433-448.

Lowe L.E., and Bustin R.M. (1985) Distribution of sulfur forms in six facies of peat of the Fraser River Delta. *Canadian Journal of Soil Science* 65, 531-541.

Lowry H.H., ed. (1963) *Chemistry of Coal Utilization*. Supplementary Vol. Wiley.

Luther G.W., and Church T.M. (1988) Seasonal cycling of sulfur and iron in pore waters of a Delaware salt marsh. *Marine Chemistry* 23, 295-309.

Luther G.W. III, Giblin A.E., and Varsolona R. (1985) Polarographic analysis of sulfur species in marine porewaters. *Limnol. Oceanogr.* 30, 727-736.

Luther G.W., Giblin A., Howarth R.W., and Ryans R.A. (1982) Pyrite and oxidized mineral phases formed from pyrite oxidation in salt marsh and estuarine sediments. *Geochim. Cosmochim. Acta* **46**, 2665-2669.

Lyons W.B., and Gaudette H.E. (1979) Sulfate reduction and the nature of organic matter in estuarine sediments. *Org. Geochem.* **1**, 151-155.

Lytle F.W., Greegor R.B., Sandstrom D.R., Marques E.C., Wong J., Spiro C.L., Huffman G.P., and Huggins F.E. (1984) Measurement of a soft x-ray absorption spectra with a fluorescent ion chamber detector. *Nucl. Instrum. Meth.* **226**, 542-548.

McCartney J.T., O'Donnell H.J., and Ergun S. (1969) Pyrite size distribution and coal-pyrite particle association in steam coals. *U.S. Bur. Mines Rept. Investigation* **7231**.

McGhie J.F., Ross W.A., Evans D., and Tomlin J.E. (1962) The use of thiophene as a chain extender. Part II. Synthetic branched-chain alkanoic acids. *J. Chem. Soc.*, 350-355.

McGowan C.W., and Markuszewski R. (1988) Direct determination of sulfate, sulphide, pyritic and organic sulphur in a single sample of coal by selective, step-wise oxidation with perchloric acid. *Fuel* **67**, 1091-1095.

McGowan C.W., and Markuszewski R. (1987) Fate of sulfur compounds in coal during oxidative dissolution in perchloric-acid. *Fuel Process. Technol.* **17**, 29-40.

McIntyre D.D., and Strausz O.P. (1987) Sulphur-33 nuclear magnetic resonance of oxidized petroleum asphaltene. *Magnetic Resonance in Chemistry* **25**, 36-38.

Mckirdy, D.M., Aldridge A.K., and Ypma P.M.J. (1983) A geochemical comparison of some crude oils from pre-Ordovician carbonate rocks. In *Advances in Organic Geochemistry 1981* (ed. M. Bjorøy), pp. 99-107. Wiley.

McKirdy D.M., and Kantsler A.J. (1980) Oil geochemistry and potential source rocks of the Officer Basin, South Australia. *Aust. Petr. Explor. Assoc. (APEA) J.* **20**, 68-86.

Maciel G., Bartuska V.J., and Miknis F.P. (1979) Characterization of organic material in coal by proton-decoupled ^{13}C nuclear magnetic resonance with magic-angle spinning. *Fuel* **58**, 391- 394.

Mycke B., Schmid J.C., and Albrecht P. (1989) A novel cross-linked polymer in crude oil: Origin and geochemical significance. *Abstracts 197th American Chemical Society Meeting*; GEOC Abstract 24.

Mackenzie A.S., and McKenzie D. (1983) Isomerization and aromatisation of hydro-carbons in sedimentary basins formed by extension. *Geol. Mag.* **120**, 417-528.

Mackenzie A.S., Maxwell J.R., Coleman M.L., and Deegan C.E. (1983) Biological marker and isotope studies of North Sea crude oils and sediments. *Proc. 11th World Petrol. Cong.* **2**, 45-56.

Mackenzie A.S., Hoffman C.F., and Maxwell J.R. (1981) Molecular parameters of maturation in the Toarcian shales, Paris Basin, France-III. Change in aromatic steroid hydrocarbons. *Geochim. Cosmochim. Acta.* **45**, 1345-1355.

Mackenzie A.S., Patience R.L., Maxwell J.R., Vandenbroucke M., and Durand B. (1980) Molecular parameters of maturation in the Toarcian shales, Paris Basin, France - I. Changes in the configurations of acyclic isoprenoid alkanes, steranes and triterpanes. *Geochim. Cosmochim. Acta* **44**, 1709-1721.

Mackowsky M.-Th. (1968) Mineral matter in coal. In *Coal and Coal-Bearing Strata* (eds. D. Murchinson, and T.A. Westoll), pp. 309-321. Oliver and Boyd.

MacPhee J.A., and Nandi B.N. (1981) ^{13}C N.M.R. as a probe for the characterization of the low-temperature oxidation of coal. *Fuel* **60**, 169-170.

Madec M., and Espitalié,J. (1985) Determination of organic sulfur in sedimentary rocks by pyrolysis. *J. Anal. Appl. Pyrolysis* **8**, 201-219.

Madesclaire M. (1986) Synthesis of sulfoxides by oxidation of thioethers. *Tetrahedron* **42**, 5459-5495.

Magnier P.L., Oki T., and Kartaaipetra L. (1975) The Mahakam delta, Kalimantan, Indonesia. *Proc. 9th World Pet. Congr.* **2**, 239-250.

Maijgren B., Hubner W., Norrgard K., and Sunduall S.B. (1983) Determination of organic sulfur in raw and chemically cleaned coals by scanning electron microscopy and energy dispersive x-ray spectrometry. *Fuel* **61**, 1076-1078.

Major D.W., Mayfield C.I., and Barke, J.F. (1988) Biotransformation of benzene by dentrification in aquifer sand. *Ground Water* **26**, 8-14.

Maka A., McKinley V.L., Conrad J.R., and Fannin K.F., (1987) Degradation of benzothiophene and dibenzothiophene under anaerobic conditions by mixed cultures. *Proc. 87th Ann. Meeting American Society for Microbiology.* Abst. #O-54.

Malik K.A., and Claus D. (1976) Microbiol degradation of dibenzothiophene. *Proceedings 5th Intl. Fermentation Symposium.* Berlin. Abst. #23.03.

Mango F.D. (1983) The diagenesis of carbohydrates by hydrogen sulfide. *Geochim. Cosmochim. Acta* **47**, 1433-1441.

Manowitz B., and Lipfert F.W. (1990) Environmental aspects of the combustion of sulfurbearing fuels. In *Geochemistry of Sulfur in Fossil Fuels* (eds. W.L. Orr, and C.M. White), ACS Symposium Series (this volume). American Chemical Society.

Manowitz B., Krouse H.R., Barker C., and Premuzic E.T. A sulfur isotope data analysis of crude oils from the Bolivar coastal area, Venezuela. In *Geochemistry of Sulfur in Fossil Fuels* (eds. W.L. Orr and C.M. White), ACS Symposium Series (this volume). American Chemical Society.

Marcilio N.R., Charcosset H., Tournayan L., Nickel B., and Jeunet A. (1989) The catalytic effect of calcium on air gasification of a subbituminous coal char. *ACS Div. Fuel Chem. Preprints* **34**, 29E-29E.

Markuszewski R. (1988) Some thoughts on the difficulties in the analysis of sulfur forms in coal. *Jour. Coal Quality* **7**, 1-4.

Marlowe I.T., Brassell S.C., Eglinton G., and Green J.C. (1984) Long chain unsaturated ketones and esters in living algae and marine sediments. In *Advances in Geochemistry 1983* (eds. P.A. Schenck, J.W. de Leeuw and G.W.M. Lijmbach), *Org. Geochem.* **6**, 135-141.

Martin R.L., and Grant J.A. (1965) Determination of sulfur compound distributions in petroleum by gas chromatography with a microcoulometric detector. *Anal. Chem.* **37**, 644-649.

Martin R.L., and Grant J. A. (1965) Determination of thiophenic compound types in petroleum samples. *Anal. Chem.* **37**, 649-657.

Martin T.H., and Hodgson G.W. (1977) Geochemical origin of organic sulfur compounds: Precursor products in the reactions of phenylalanine and benzylamine with elemental sulfur. *Chem. Geol.* **20**, 9-25.

Martin, T.H. and Hodgson, G.W. (1973) Geochemical origin of organic sulfur compounds: Reaction of phenylalanine with elemental sulfur. *Chem. Geol.* **12**, 189-208.

Mauger R.L. (1972) A sulfur isotope study of bituminous sands from the Uinta basin, Utah. International Geology Congress, 24th, Montreal 1972, *Comptes Rendus*, Section **5**, 19-27.

Mayer R., Hiller G., Nitzschke M., and Jentzsch J. (1963) Base-catalyzed reactions of ketones with hydrogen sulfide. *Angew. Chem. internat. Edit.* **2**, 370-373.

van de Meent D., Brown S.C., Philp R.P., and Simoneit B.R.T. (1980) Pyrolysis-high resolution gas chromatography and pyrolysis gas chromatography-mass spectrometry of kerogens and kerogen precursors. *Geochim. Cosmochim. Acta* **44**, 999-1013.

Mehmet Y. (1971) The occurrence and origin of sulfur in crude oils - a review. *Alberta Sulfur Research Quart. Bull.* **6**, 1-17.

Meissner F.F., Woodward J., and Clayton J.L. (1984) Stratigraphic relationships and distribution of source rocks in the greater Rocky Mountain region. In *Hydrocarbon Source Rocks of the Greater Rocky Mountain Region* (ed. J. Woodward, F.F. Meissner, and J.L. Clayton), pp. 1-34. Rocky Mountain Association of Geologists.

Mekhtiyeva V.L., and Brizanova L. Ya. (1980) Abiogenic reduction of sulfates in the Earth's crust. *Geologiya Nefti Gaza N3*, 32-39 (in Russian). Translation (1982) *Internat. Geol. Rev.* **24**, 439-444.

Mello M.R., Telnaes N., Gaglianone P.C., Chicarelli M.I., Brassell S.C., and Maxwell J.R. (1988) Organic geochemical characterization of depositional palaeoenvironments of source rocks and oils in Brazilian marginal basins. In *Advances in Organic Geochemistry 1987* (eds. L. Mattavelli and L. Novelli), *Org. Geochem.* **13**, 31-45.

Merritt R.D. (1986) Paleoenvironmental and tectonic controls in major coal basins of Alaska. *Geol. Soc. Amer. Special Paper* **210**, 173-200.

Messenger L., and Attar A. (1979) Thermodynamics of the transformation of oxygen-containing and sulfur-containing functional groups during coal liquefaction in hydrogen and hydrogen donor. *Fuel* **58**, 655-660.

Meth-Cohn O. (1979) Thiophenes. In *Comprehensive Organic Chemistry* (ed. P.G. Sammes), pp. 789-838 Pergamon Press.

Meyer B., Peter L., and Spitzer K. (1977) Charge distribution in neutral and ionic elementary sulfur chains and rings. In *Homoatomic Rings, Chains and Macromolecules of Main-Group Elements* (ed. A.L. Rheingold), pp. 477-497. Elsevier.

Meyers R.A., Hart W.D., and McClanathan L.C. (1981) Gravimelt process for near complete chemical removal of sulfur. *Coal Process. Technol.* **7**, 89-93.

Mihelcic J.R., and Luthy R.G. (1988) Degradation of polycycilic aromatic hydrocarbons under various redox conditions in soil-water systems. *Appl. Environ. Microbiol.* **54**, 1182-1187.

Miller W.G. (1974) Relationships between minerals and selected trace elements in some Pennsylvanian age coals of northwestern Illinois. M.S. Thesis, University of Illinois.

Minagwa M., Winter D.A., and Kaplan I.R. (1984) Comparison of Kjeldahl and combustion methods for measurements of nitrogen isotope ratios in organic matter. *Anal. Chem.* **56**, 1859-1861.

Moers M.E.C., de Leeuw J.W., Cox H.C., and Schenck P.A. (1988) Interaction of glucose and cellulose with hydrogen sulfide and polysulfides. *Org. Geochem* **13**, 1087-1091.

Mojelsky T.W., Montgomery D.S., and Strausz O.P. (1986) Pyrolytic probes into the aliphatic core of oxidized Athabasca asphaltene non-distillable ester residue. *AOSTRA Journal of Research* **3**, 43-51.

Moldowan J.M., Seifert W.K., and Gallegos E.J. (1986) Relationship between petroleum composition and depositional environment of petroleum source rocks. *Amer. Assoc. Petr. Geol. Bull.* **69**, 1255-1268.

Moldowan J.M., Sundararaman P., and Schoell M. (1986) Sensitivity of biomarker properties to depositional environment and/or source input in the lower Toarcian of SW-Germany. In *Advances in Organic Geochemistry 1985* (eds. D. Leythaeuser, and J. Rullkötter), *Org. Geochem.* **10**, 915-926.

Monster J. (1972) Homogeneity of sulfur and carbon isotope $^{34}S/^{32}S$ and $^{13}C/^{12}C$ in petroleum. *Amer. Assoc. Petr. Geol. Bull.* **56**, 941-949.

Monthioux M., Landais P., and Monin J-C. (1985) Comparison between natural and artificial maturation series of humic coals from the Mahakam Delta, Indonesia. *Org. Geochem.* **8**, 275-292.

Monticello D.J., Bakker D., and Finnerty W.R. (1985) Plasmid-mediated degradation of dibenzothiophene by *Pseudomonas* species. *Appl. Environ. Microbiol.* **49**, 756-760.

Monticello D.J., and Finnerty W.R. (1985) Microbiological desulfurization of fossil fuels. *Ann. Rev. Microbiol.* **39**, 371-389.

Mormile M.R., and Atlas R.M. (1988) Mineralization of the dibenzothiophene products 3-hydroxy-2-formyl benzothiophene sulfone. *Appl. Environ. Microbiol.* **54**, 3183-3184.

Morse J.W., and Cornwell J.C. (1987) Analysis and distribution of iron sulfide minerals in recent anoxic marine sediments. *Marine Chemistry* **22**, 55-69.

Morse J.W., Millero F.J., Cornwell J.C., and Rickard D. (1987) The chemistry of hydrogen sulfide and iron sulfide systems in natural waters. *Earth-Science Reviews* **24**, 1-42.

Moza A.K., and Neavel R.C. (1986) Estimation of coal washability using x-ray radiography. *Fuel* **65**, 547-551.

Mukherjee D., Dunn L.C., and Houk K.N. (1979) Efficient guaiazulene and chamazulene synthesis involving [6 + 4] cycloaddition. *J. Amer. Chem. Soc.* **101**, 251-252.

Nakai N., and Jensen M.L. (1964) The kinetic isotope effect in the bacterial reduction and oxidation of sulfur. *Geochim. Cosochim. Acta* **28**, 1893-1912.

National Acid Precipitation Assessment Program (1987), Interim Assessment, The causes and effects of acidic deposition, Vols. I-IV, U.S. Gov't Printing Office.

National Research Council Canada (1977). Sulphur and its inorganic derivatives in the Canadian environment, NRCC No. 15015. Available from publications, NRCC/ CNR, Ottawa, Canada, K1A OR6.

Narayan R., Kullerud G., and Woods K.V. (1988) A new perspective on the nature of "organic" sulfur in coal. *ACS Div. Fuel Chem. Preprints* **33(1)**, 193-197.

Neill P.H., Xia Y.J., and Winans R.E. (1989) Identification of the heteroatom containing compounds in the benzene/methanol extracts of the Argonne Premium Coal Samples. *ACS Div. Fuel Chem. Preprints* **34(3)**, 745-751.

Nelson W.J. (1987) Coal deposits of the United States. *Intern. Jour. Coal Geol.* **8**, 355-365.

Ngassoum M.B., Faure R., Ruiz J.M., Lena L., Vincent E.J., and Neff B. (1986) ^{33}S NMR. A tool for analysis of petroleum oils. *Fuel* **65**, 142-143.

Nichols P.D., Volkman J.K, Palmisano A.C., Smith G.A., and White D.C. (1988) Occurrence of an isoprenoid C_{25} diunsaturated alkene and high neutral lipid content in Antarctic sea-ice diatom communities. *J. Phycol.* **24**, 90-96.

Nichols P.D. and Johns R.B. (1986) The lipid chemistry of sediments from the St. Lawrence estuary. Acyclic unsaturated long chain ketones, diols and ketone alcohols. *Org. Geochem.* **9**, 25-30.

Nielsen, H. (1979) Sulfur isotopes. In *Lectures in Isotope Geology* (eds. E. Jäger, and J.C. Hunziker), pp. 283-312. Springer-Verlag.

Nielsen H. (1974) Isotopes in nature (sulfur). In *Handbook of Geochemistry* (ed. K.H. Wedpohl) Vol. II-I, Sec. 16-B. Springer-Verlag.

Nip M., de Leeuw J.W., and Schenck P.A. (1988) The characterization of eight maceral concentrates by means of Curie point pyrolysis-gas chromatography and Curie point pyrolysis-gas chromatography-mass spectrometry. *Geochim. Cosmochim. Acta* **52**, 637-648.

Nip M., Tegelaar E.W., Brinkhuis H., de Leeuw J.W., Schenck P.A., and Holloway P.J. (1986) Analysis of modern and fossil plant cuticles by Curie point PyGC and Curie point Py-GC-MS: Recognition of a new highly aliphatic and resistant bipolymer. *Advances in Organic Geochemistry 1985*, (eds. D. Leythaeuser, and J. Rullkotter), pp. 769-778. Pergamon Press.

Nip M., de Leeuw J.W., Schenck P.A., Meuzelaar H.L.C., Stout S.A., Given P.H., and Boon J.J. (1985) Curie-point pyrolysis-mass spectrometry, Curie-point pyrolysis-gas chromatography-mass spectrometry and fluorescence microscopy as analytical tools for the characterization of two uncommon lignites. *J. Anal. Appl. Pyrolysis* **8**, 221-239.

Nishimura M., and Baker E.W. (1986) Possible origin of n-alkanes with a remarkable even-to-odd predominance in recent sediments. *Geochim. Cosmochim. Acta* **50**, 299-305.

Nishioka M. (1988) Aromatic sulfur compounds other than condensed thiophenes in fossil fuels: Enrichment and identification. *Energy & Fuels* **2**, 214-219.

Nishioka M., Lee M.L., and Castle R.N. (1986) Sulfur heterocycles in coal-derived products. Relation between structure and abundance. *Fuel* **65**, 390-396.

Nishioka M., Campbell R.M., Lee M.L., and Castle R.N. (1986) Isolation of sulfur heterocycles from petroleum- and coal-derived materials by ligand exchange chromatography. *Fuel* **65**, 270-273.

Nishioka M., Lee M.L., Kudo H., Muchiri D.E., Baldwin L.J., Pakray S., Stuart J.G., and Castle R.N. (1985) Determination of hydroxylated thiophenic compounds in a coal liquid. *Anal. Chem.* **57**, 1327-1330.

Nishioka M., Campbell R.M., West W.R., Smith P.A., Booth G.M., Lee M.L., Kudo H., and Castle R.N. (1985) Determination of aminodibenzothiophenes in a coal liquid. *Anal. Chem.* **57**, 1868-1871.

Norton G.A., Mroch D.R., Chriswell C.D., and Markuszewski R. (1987) Chemical cleaning of coal with high organic sulfur using fused caustic. *Proceedings of the 2nd International Conference on Processing and Utilization of High Sulfur Coals*, pp. 213-223.

Nriagu J.O., ed. (1978) *Sulfur in the Environment*, Wiley-Interscience.

Nriagu J.O., and Coker R.D. (1976) Emission of sulfur and Lake Ontario sediments. *Limnol. Oceanogr.* **21**, 485-489.

Nuttle, W.K. and Hemond, H.F. (1988) Salt marsh hydrology: Implications for bio-geochemical fluxes to the atmosphere and estuaries. *Global Biogeochem. Cycles.* **2**, 91-114.

Nystrom,R.F., and Berger C.R.A. (1958) Reduction of organic compounds by mixed hydrides. II. Hydrogenolysis of ketones and alcohols. *J. Amer. Chem. Soc.* **80**, 2896-2898.

Oae S. (1985) Historical development of sulfur bonding: A view of an experimental organosulfur chemist. In *Organic Sulfur Chemistry* (eds. F. Bernardi, I.G. Csizmadia, and A. Mangini), pp. 1-67. Elsevier.

Oremland R.S., Whiticar M.J., Strohmaier F.E., and Kiene R.P. (1988) Bacterial ethane formation from reduced ethylated compounds in anoxic sediments. *Geochim. Cosmochim. Acta* **52**, 1895-1904.

Orr W.L. (1986) Kerogen/asphaltene/sulfur relationships in sulfur-rich Monterey oils. In *Advances in Organic Geochemistry 1985* (eds. D. Leythaeuser, and J. Rullkötter). *Org. Geochem.* **10**, pp. 499-516.

Orr W.L. (1978) Sulfur in heavy oils, oil sands and oil shale. In *Oil Sand and Oil Shale Chemistry* (eds. O.P. Strausz, and E.M. Lown), pp. 223-243. Verlag Chemie.

Orr W.L. (1977) Geologic and geochemical controls on the distribution of hydrogen sulfide in natural gas. In *Advances in Organic Geochemistry 1975* (eds. E.R. Campos, and J. Goni), pp. 571-597. Enadisma, Madrid.

Orr W.L. (1975) Sulfur in petroleum and related fossil organic materials. *ACS Div. Petr. Chem. Preprints* **21**, 417-421.

Orr W.L. (1974) Changes in sulfur content and isotopic ratios of sulfur during petroleum maturation: Study of Big Horn Basin Paleozoic oils. *Amer. Assoc. Petr. Geol. Bull.* **50**, 2295-2318.

Orr W.L. (1974) Biogeochemistry of sulfur. In *Handbook of Geochemistry* (ed. K.H. Wedpohl) Vol. II-I, Sec. 16-L. Springer-Verlag.

Ourisson G., Albrecht P., and Rohmer M. (1982) Predictive microbial biogeochemistry from molecular fossils to procaryotic membranes. *Trends Biochem. Sci.* **7**, 236-239.

Ourisson G, Albrecht P., and Rohmer M. (1979) The hopanoids. Palaeochemistry and biochemistry of a group of natural products. *Pure Appl. Chem.* **51**, 709-729.

Ostroukhov S.B., Arefev O.A., Makushina V.M., Zabrodina M.N., and Petrov Al. A. (1982) Monocyclic aromatic hydrocarbons with isoprenoid chains. *Neftekhimiya* **22**, 723-788.

Palacas J.G., ed. (1984) *Petroleum Geochemistry and Source Rock Potential of Carbonate Rocks*, AAPG Studies in Geology No. 18. American Association of Petroleum Geologists.

Panagiotidis T., Richter E., and Jüntgen H. (1988) Structural changes of an anthracite char during the reaction with sulphur dioxide. *Carbon* **26**, 89-95.

Pan Suixian, Cheng Baozhou, et al. (1987) *Sedimentary Environments of Taiyuan Xishan Coal Basin*. Ministry of Coal Industry Press (Beijing), pp. 631 (In Chinese, with an English abstr.).

Paris B. (1977) The direct determination of organic sulphur in raw coals. In *Coal Desulfurization. Chemical and Physical Methods* (ed. T.D. Wheelock), Vol. **66**, pp. 22-34, American Chemical Society.

Parrish J.T., and Curtis R.L. (1982) Atmospheric circulation, upwelling and organic-rich rocks in the Mesozoic and Cenozoic eras. *Palaeogeog. Palaeoclimat. Palaeoecol.* **40**, 31-66.

Patterson J.H. (1988) Minerology and chemistry over a cycle of oil shale deposition in the Brick Kiln Member, Rundle deposit, Queensland, Australia. *Chem. Geol.* **68**, 207-219.

Patton J.S., Rigler M.W., Boehm P.D., and Fiest D.L. (1981) Ixtoc 1 oil spill: Flaking of the surface mousse in the Gulf of Mexico. *Nature* **290**, 235-238.

Paulik F., Pauli, J., and Arnold M. (1982) Simultaneous TG, DTG, DTA, and EGA technique for the determination of carbonate, sulphate, pyrite and organic material in minerals, soils and rocks. *J. Thermal Anal.* **25**, 327-340.

Payzant J.D., Mojelsky T.M., and Strausz O.P. (1989) Improved methods for the selective isolation of the sulfide and thiophenic classes of compounds form petroleum. *Energy & Fuels* **3**, 449-454.

Payzant J.D., McIntyre D.D., Mojelsky T.W., Torres M., Montgomery D.S., and Strausz O.P. (1989) The identification of homologous series of thiolanes and thianes possessing a linear carbon framework from petroleum and their interconversion under simulated geological conditions. *Org. Geochem.* **14**, 461-473.

Payzant J.D., Cyr D.S., Montgomery D.S., and Strausz O.P. (1988) Studies on the structure of the terpenoid sulfide type biological markers in petroleum. *Geochemical Biomarkers* (eds. T.F. Yen, and J.M. Moldowan), pp. 133-147. Harwood Academic Publishers.

Payzant J.D., Montgomery D.S., and Strausz O.P. (1988) The identification of homologous series of benzo[b]thiophenes, thiophenes, thiolanes and thianes possessing a linear carbon framework in the pyrolysis oil of Athabasca asphaltene. *AOSTR J. Res.* **4**, 117-131.

Payzant J.D., Cyr T.D., Montgomery D.S., and Strausz O.P. (1988) Studies on the structure of the terpenoid sulfide type biological markers in petroleum. In *Geochemical Biomarkers* (eds. T.F. Yen, and J.M. Moldowan), pp. 133-147. Harwood Academic Press.

Payzant J.D., Montgomery D.S., and Strausz,O.P. (1986) Sulfides in petroleum. *Org. Geochem.* **9**, 357-369.

Payzant J.D., Hogg A.M., Montgomery D.S., and Strausz O.P. (1985) Field ionization mass spectrometric study of the maltene fraction of Athabasca bitumen. Part II-the aromatics. *AOSTRA Journal of Research* **1**, 183-202.

Payzant J.D., Montgomery D.S., and Strausz O.P. (1985) The synthesis of 1-β(H), 6β(Me)-2,2,6-trimethyl-8-ethylbicyclo-4,3,0,7-thianonane and the occurence of the isomeric bicyclic terpenoid sulfides occurring in petroleum. *Tetrahedron Lett.* **26**, 4175-4178.

Payzant J.D., Hogg A.M., Montgomery D.S., and Strausz O.P. (1985) Field ionization mass spectrometric study of the maltene fraction of Athabasca bitumen. Part III-The polars. *AOSTRA Journal of Research* **1**, 203-210.

Payzant J.D., Montgomery D.S., and Strausz O.P. (1983) Novel terpenoid sulfoxides and sulfides in petroleum. *Tetrahedron Lett.* **24**, 651-654.

Peakman T.M., and Kock-van Dalen A.C. (1990) Identification of alkylthiophenes occurring in the geosphere by synthesis of authentic standards. In *Geochemistry of Sulfur in Fossil Fuels* (eds. W.L. Orr, and C.M. White), ACS Syposium Series (this volume). American Chemical Society.

Peakman T.M., Sinninghe Damsté J.S., and de Leeuw J.W. (1989) The identification and geochemical significance of a series of alkylthiophenes comprising a linearly extended phytane skeleton in sediments and crude oils. *Geochim. Cosmochim. Acta* **53**, 3317-3322.

Peakman T.M., ten Haven H.L., Rechka J.R., de Leeuw J.W., and Maxwell J.R. (1989) Occurrence of (20R)- and (20S)-Δ8(14) and Δ14 5α(H)-sterenes and the origin of 5α(H),14ß(H),17ß(H)-steranes in an immature sediment. *Geochim. Cosmochim. Acta* **53**, 2001-2009.

Peakman T.M., Sinninghe Damsté J.S., and de Leeuw J.W. (1989) Organic geochemical evidence for series of C25-C28 sulphur containing lipids comprising regular and irregular extended phytane skeletons. *J. Chem. Soc. Chem. Commun.*, 1105-1107.

Pearson M.J., Watkins D., and Small J.S. (1982) Clay diagenesis and organic maturation in northern North Sea sediments. In *Developments in Sedimentology* (eds. H. van Olphen, and F. Veriale), Vol. 35, pp. 665-675. Elsevier.

Peer E.L. (1979) United States refineries and their adaptability to process heavy oils. In *Proceedings of the First International Conference on the Future of Heavy Crude Oils and Tar Sands, Alberta 1979*, pp. 651-654. Unitar McGraw-Hill.

Pelet R., Behar F., and Monin J.C. (1986) Resins and asphaltenes in the generation and migration of petroleum. In *Advances in Organic Geochemistry 1985* (eds. D. Leythaeuser, and J. Rullkötter.), *Org. Geochem.* **10**, 481-498.

Perakis N. (1986) Separation et detection selective de composes soufres dans les fractions lourdes des petroles. Geochimie des benzo[b]thiophenes. Ph.D. dissertation. Univ. of Strasbourg.

Perry J.J. (1984) Microbiol metabolism of cyclic alkanes. In *Petroleum Microbiology* (ed. R.M. Atlas), pp.61-98. MacMillan.

Peters K.E., Rohrback B.G., and Kaplan I.R. (1981) Carbon and hydrogen stable isotope variations in kerogen during laboratory simulated thermal maturation. *Amer. Assoc. Petr. Geol. Bull.* **65**, 501-508.

Pfennig N., and Widdel F. (1982) The bacteria of the sulphur cycle. *Phil. Trans. R. Soc. London, series B* **298**, 433-441.

Philp R.P., and Bakel A. (1988) Production of heteroatomic compounds by pyrolysis of asphaltenes, coals and source rocks. *Energy & Fuels* **2**, 59-64.

Philp R.P., Bakel A., Galvez-Sinibaldi A., and Lin L.H. (1988) A comparison of organo-sulphur compounds produced by pyrolysis of asphaltenes and those present in related crude oils and tar sands. In *Advances in Organic Geochemistry 1987* (eds. L. Mattavelli and L. Novelli), *Org. Geochem.* **13**, 915-926.

Philp R.P., and Gilbert T.D. (1985) Source rock and asphaltene biomarker characterization by pyrolysis-gas chromatography-mass spectrometry-multiple ion detection. *Geochim. Cosmochim. Acta* **49**, 1421-1432.

Philp R.P., Gilbert T., and Russell N. (1982) Characterization of Victorian brown coals by pyrolysis techniques combined with gas chromatography and gas chromatography-mass spectrometry. *Austr. Coal Geol.* **4**, 228-243.

Plesnicar B., and Trahanovsky W.S., ed. (1978) Oxidations with peroxy acid and other peroxides. In *Oxidation in Organic Chemistry Part C*, Chap. III, Academic Press.

Pohl J., and Gall H. (1977) Bau and Entstehung des Ries-Kraters. *Geol. Bavarica* **76**, 159-175.

Postgate J.R. (1984) *The Sulphate-Reducing Bacteria.* Cambridge University Press.

Price, F.T., and Shieh Y.N. (1979) The distribution and isotopic distribution of sulfur in coals from the Illinois Basin. *Econ. Geol.* **74**, 1445-1461.

Pryor W.A., (1962) Oxidation by polysulfide to form aldehydes, carboxylic acids, or carboxyamides. *Mechanisms of Sulfur Reactions*, Chap. 7, pp. 127-138. McGraw-Hill.

Przewocki K., Malinski E., and Szafranek J. (1984) Elemental sulfur reactions with toluene, dibenzyl, z-stilbene and their geochemical significance. *Chem. Geol.* 47, 347-360.

Purser B.H. (1980) Sedimentation et diagenese des carbonates neritiques recents: Paris. *Edition Technip* 2, 389.

Pye K. (1981) Marshrock formed by iron sulphide and siderite cementation in saltmarsh sediments. *Nature*, 294, 660-662.

Quigley T.M., Mackenzie A.S., and Gray J.R. (1987) Kinetic theory of petroleum generation. In *Migration of Hydrocarbons in Sedimentary Basins* (ed. Doligez) pp. 649-655. Editions Technip.

Raasch M.S. (1985) Thiophene 1,1-dioxides, sesquioxides and 1-oxides. In *The Chemistry of Heterocyclic Compounds. Thiophene and its Derivatives*, (ed. S. Gronowitz), Vol. 1, pp. 571-627. J. Wiley and Sons.

Radke M. (1988) Applications of aromatic compounds as maturity indicators in source rocks and crude oils. *Marine Petrol. Geol.* 5, 224-236.

Radke M. (1987) Organic geochemistry of aromatic hydrocarbons. In *Advances in Organic Geochemistry* (eds. J. Brooks and D. Welte), Vol. 2, pp. 141-207. Academic Press.

Radke M., Willsch H., and Welte D.H. (1986) Maturity parameter based on aromatic hydrocarbons: Influence of organic matter type. In *Advances in Organic Geochemistry 1985* (eds. D. Leythaeuser, and J. Rullkötter), *Org. Geochem.* 10, 51-63.

Radke M., Welte D.H., and Willsch H. (1982) Geochemical study of a well in the Western Canada Basin: Relation of the aromatic distribution pattern to maturity of organic matter. *Geochim. Cosmochim. Acta* 46, 1-10.

Radke M., Willsch H., and Welte D.H. (1980) Preparative hydrocarbon group type determination by automated medium pressure liquid chromatography. *Anal. Chem.* 52, 406-411.

Radke M., Sittardt H.G., and Welte D.H. (1978) Removal of soluble organic matter from source rock samples with a flow-through extraction cell. *Anal. Chem* 50, 663-665.

Radunz H.E. (1977) Newer preparative aspects of the Michael addition. *Kontakte*, 3-10.

Rafter T.A. (1962) Sulfur isotope measurements on New Zealand, Australian, and Pacific Islands specimens. In *Biogeochemistry of Sulphur Isotopes* (ed. M.L. Jensen), Proc. National Science Foundation Symposium, pp. 42-60, Yale University.

Raiswell R. (1982) Pyrite texture, isotopic composition and the availability of iron. *Am. Journ. Sci.* 282, 1244-1263.

Rall H.T., Thompson C.J., Colemam H.J., and Hopkins R.L. (1972) *Sulfur Compounds in Crude Oil; U. S. Bureau of Mines. Bulletin 659*, 187 pp. US Government Printing Office.

Ramanathan V., and Levine R. (1962) The alkylation and arylation of 2-thienyllithium and the reactions of 3-methylthiophene with organometallic compounds. *J. Org. Chem.* 27, 1667-1670.

Ramirez-Rojas A.J. (1988) Geochemistry of nutrient elements in water and sediment of the Tuy River Basin, Venezuela. Ph.D. dissertation, The Pennsylvania State University.

Raymond R., Jr. (1982) Electron probe microanalysis, a means of direct determination of organic sulfur in coal. In *Coal and Coal Products: Analytical Characterization Techniques* (ed. E.L. Fuller, Jr.), pp. 191-203, ACS Symposium Series **205**. American Chemical Society.

Raymond R., and Hagan R.C. (1982) Relationship between pyrite formation and organic sulfur content of coal as revealed by electron microscopy. *Scanning Electron Microsc.* 619-627.

Raymond R., Jr., and Gooley R. (1979) Electron probe microanalyzer in coal research. In *Analytical Methods for Coal and Coal Products* (ed. C. Karr), Vol. III, Chap. 48, pp. 337-356. Academic Press.

Raymond R., Jr., and Gooley R. (1978) A review of organic sulfur analysis in coal and a new procedure. *Scanning Electron Microsc.*, 93-108.

Raymond R., Relationship between pyrite formation and organic sulfur content of coal by electron microscopy. *Proc. Int. Coal Sci.* **198**, 857-862.

Record F.A., Bebenick D.V., and Kindya R.J. (1982) *Acid Rain Information Book*, Noyes Data Corp.

Reggel L., Wender I., and Raymond R. (1970) Catalytic dehydrogenation of coal: Part 4. A comparison of exinite, micrinite, and fusinite with vitrinite. *Fuel* **49**, 281-286.

Reinecke M.G., Morton D.W., and Del Mazza D. (1983) The scope of the synthesis of tetrahydrothiophenes from α.8-unsaturated carbonyl compounds and sodium polysulfide. *Synthesis*, 160-161.

Reinson G.E. (1984) Barrier-island and associated stand-plain systems. In *Facies Models* (ed. R.G. Walker), pp. 119-140, Geoscience Canada.

Retcofsky H.L., and Friedel R.A. (1972) Sulfur-33 magnetic resonance spectra of selected compounds. *J. Amer. Chem. Soc.* **94**, 6579-6584.

Retcofsky H.L., and Friedel R.A. (1970) Sulfur-33 magnetic resonance spectrum of thiophene. *Applied Spectroscopy* **24**, 379-380.

Richard J.J., Vick R.D., and Junk G.A. (1977) Determination of elemental sulfur by gas chromatography. *Environ. Sci. Tech.* **11**, 1084-1086.

Richards F.A, Anderson J.J., and Cline J.D. (1971) Chemical and physical observations in Golfo Dulce. An anoxic basin on the Pacific coast of Costa Rica. *Limnol. Oceanogr.* **16**, 43-50.

Richardson C.V. (1988) *Keystone Coal Industry Manual*. Maclean Hunter Publishing Co.

Rickard D.T. (1975) Kinetics and mechanism of pyrite formation at low temperatures. *Amer. J. Sci.* **275**, 636-652.

Rickard D.T. (1974) Kinetics and mechanism of the sulfidation of goethite. *Amer. J. Sci.* **274**, 941-952.

Rickard D.T. (1969) The chemistry of iron sulphide formation at low temperatures. *Stockholm Contrib. Geol.* **20**, 67-95.

Roadifer R.E. (1987) Size distribution of the world's largest known oil and tar deposits. In *Exploration for Heavy Crude Oil and Natural Bitumens. AAPG Studies in Geology 25* (ed. R. F. Meyer), pp. 3-23. AAPG Press.

Robba M., Roques B., and Bonhomme M. (1967) Synthèse de thiéno-[2,3-d]pyridazines et de thiéno-[3,4-d]pyridazines à partir de dérivés thiophéniques. *Bull. Chim. Soc. Fr.*, pp. 2495-2507.

Roberts D.L. (1988) The relationship between macerals and sulphur content of some African Permian coals. *Intern. Jour. Coal Geol.* **10**, 399-410.

Roberts W.M.B., Walker A.L., and Buchanan A.S. (1969) The chemistry of pyrite formation in aqueous solution and its relation to the depositional environment. *Mineral Deposita* **4**, 18-29.

Robinson N., Cranwell P.A., Eglinton G., Brassell S.C., Sharp C.L., Gophen M., and Pollingher U. (1986) Lipid geochemistry of Lake Kinneret. In *Advances in Geochemistry 1985* (eds. D. Leythaeuser, and J. Rullkötter), *Org. Geochem.* **10**, 733-742.

Robson J.N., and Rowland S.J. (1986) Identification of novel widely distributed sedimentary acyclic sesterterpenoids. *Nature* **324**, 561-563.

Rohrback B.G. (1979) Analysis of low molecular weight products generated by thermal decomposition of organic matter in recent sedimentary environments. Ph.D. dissertation, Univ. California.

Rose K.D., and Francisco M.A. (1988) A two step chemistry for highlighting heteroatom species in petroleum materials using ^{13}C NMR spectroscopy. *J. Amer. Chem. Soc.* **110**, 637-638.

Rösler H.J., Beuge P., and Adamski B. (1977) Dar verhalten chemischer elemente bei der diagenese und metamorphose. *Z. Angew. Geol.* **23**, 53-56.

Ross L., and Ames R. (1988) Stratification of oils in Columbus Basin off Trinidad. *Oil Gas J.* **86(39)**, 72-76.

Rowley P.G., and Brown T. (1982) Australian oil shale development. In *Oil Shale--The Environmental Challenges II* (ed. K.K. Petersen), Proc. International Symposium 1981, pp. 1-28. Colorado School of Mines Press.

Roy M.M. (1965) Chlorination of coal. *Brennst. Chemie* **46**, 407-411.

Rucj R.R., Gluskoter H.J., and Shimp N.F. (1974) Occurrence and distribution of potentially volatile trace elements in coal. Environmental Geology Note 72, Illinois State Geological Survey.

Rullkötter J., Littke R., and Schaefer R.G. (1990) Characterization of organic matter in sulfur-rich lacustrine sediments of Miocene age (Nördlinger Ries, Southern Germany. In *Geochemistry of Sulfur in Fossil Fuels* (eds. W.L. Orr, and C.M. White), ACS Symposium Series (this volume). American Chemical Society.

Rullkötter J., and Orr W.L. (1989) The effects of thermal maturation on the distribution of sulfur compounds in sulfur-rich Monterey crude oils. Abstracts 197th ACS Meeting, GEOC.

Rullkötter J., Landgraff M., and Disko U. (1988) Gas chromatographic and mass spectrometric characterization of isomeric alkylthiophenes (C_{20}) and their occurrence in deep sea sediments. *J. High Res. Chrom. & Chrom. Commun.* **11**, 633-638.

Rullkötter R.J., Mukhopadhyay P.K., and Welte D.H. (1987) Geochemistry and petrography of organic matter from Deep Sea Drilling Project site 603, lower continental rise of Cape Hatteras. In *Init. Reprts. DSDP* (eds. J.G. Hinte, and S.W. Wise, Jr.), Vol. **93**, pp. 1163-1176. U.S.Government Printing Office.

Rullkötter J., Mukhopadhyay P.K., Disko U., Schaefer R.G., and Welte D.H. (1986) Facies and diagenesis of organic matter in deep sea sediments from the Blake Outer Ridge and the Blake Bahama Basin, western North Atlantic. *Mitt. Geol. Paläont. Inst. Univ. Hamburg* **60**, 179-203.

Rullkötter J., Mukhopadhyay P.K., Schaefer R.G, and Welte D.H. (1984) Geochemistry and petrography of organic matter in sediments from Deep Sea Drilling Project Sites 545 and 547, Mazagan Escarpment. In *Init. Reprts. DSDP* (eds. W. Hinz, and E.L. Winterer), Vol. **79**, pp. 775-806. U.S.Government Printing Office.

Rullkötter J., Mukhopadhyay, and Welte D.H. (1984) Geochemistry and petrography of organic matter in sediments from Holes 530A, Angola Basin, and hole 532, Walvis Ridge, Deep Sea Drilling Project. In *Init. Reprts. DSDP* (eds. W.W. Hay, and J.C. Sibuet), Vol. **75**, pp. 1069-1087. US Government Printing Office.

Rullkötter J., von der Dick H., and Welte D.H. (1982) Organic petrography and extractable hydrocarbons of sediments from the Gulf of California, Deep Sea Drilling Project Leg 64. In *Init. Repts. DSDP* (eds. J.R. Curray, and D.G. Moore), Vol. **64**, pp. 837-853. US Government Printing Office.

Rullkötter J., von der Dick H., and Welte D.H. (1981) Organic petrography and extractable hydrocarbons of sediments from the eastern north Pacific ocean, Deep Sea Drilling Project Leg 63. In *Init. Repts. DSDP* (eds. B. Haq, and R.S. Yeates), Vol. **63**, pp. 819-836. US Government Printing Office.

Ryder R.T., Fouch T.D., and Elison J.H. (1976) Early tertiary sedimentation in western Uinta basin, Utah. *Geological Society of America Bulletin* **87**, 496-512.

Rye R.O., Luhr J.F., and Wasserman M.D. (1984) Sulfur and oxygen isotopic systematics of the 1982 eruptions of El Chichon Volcano, Chiapas, Mexico. *Journal of Volcanology and Geothermal Research* **23**, 109-123.

Sagardía F., Rigau J.J., Martínez-Lahoz A., Fuentes F., López C., and Flores W. (1975) Degradation of benzothiphene and related compounds by a soil *Pseudomonas* in an oil-aqueous environment. *Appl. Microbiol.* **29**, 722-725.

Saiz-Jiminez C., and de Leeuw J.W. (1986) Lignin pyrolysis: Their structures and their significance as biomarkers. *Org. Geochem.* **13**, 869-876.

Saiz-Jimenez C., and de Leeuw J.W. (1986) Chemical characterization of soil organic matter fractions by analytical pyrolysis-gas chromatography- mass spectrometry. *J. Anal. Appl. Pyrolysis* **9**, 99-119.

Saiz-Jimenez C., and de Leeuw J.W. (1984) Pyrolysis-gas chromatography-mass spectrometry of isolated, synthetic and degraded lignins. In *Advances in Organic Geochemistry 1983* (eds. P.A. Schenck, J.W. de Leeuw, and G.W.M. Lijmbach), *Org. Geochem.* **6**, 417-422.

Saiz-Jimenez C., and de Leeuw J.W. (1984) Pyrolysis-gas chromatography-mass spectrometry of soil polysaccharides, soil fulvic acid and polymaleic acid. In *Advances in Organic Geochemistry 1983* (P.A. Schenck, J.W. de Leeuw and G.W.M. Lijmbach, eds.), *Org. Geochem.* **6**, 287-293.

Salisbury F.B., and Ross C.W., *Plant Physiology, 3rd Ed.* Wadsworth Publishing.

Saltzman E.S., and Cooper W.J., eds. (1989) *Biogenic Sulfur in the Environment.* ACS Symposium Series **393**, pp. 672. American Chemical Society.

Sanders J.E., and Imbrie J. (1963) Continuous cores of calcareous Bahamian sands made by Vibrodrilling. *Geol. Soc. Amer. Bull.* **74**, 1287-1292.

Sariaslani F.S., Harper D.B., and Higgins I.J. (1974) Microbial degradation of hydro-carbons: Catabolism of 1-phenyl-alkanes by *Nocardia salmonicolor. Biochem. J.* **140**, 31-45.

Schaefle J., Ludwig B., Albrecht P., and Ourisson G. (1977) Hydrocarbures aromatiques d'orgine geologique. II Nouveaux carotenoids aromatiques fossles. *Tetrahedron Lett.*, 3673-3676.

Schafer H.N.S. (1980) Pyrolysis of brown coal. *Fuel* **59**, 295-304.

Schafer H.N.S. (1979) Pyrolysis of brown coals. *Fuel* **58**, 6667-6679.

Schank K. (1988) Synthesis of open-chain sulfones. *The Chemistry of Sulfones and Sulfoxides* (eds. S. Pataim, Z. Rappoport, and C. Stirling), pp. 165-232. J. Wiley and Sons.

Schicho R.N., Brown S.H., Kelly R.M., and Olson G.J. (1988) Microbial oxidation and reduction of sulfur in coal for speciation and desulfurization. *ACS Div. Fuel Chem. Preprints* **34(4)**, 554-560.

Schmid J.C., Connan J., and Albrecht P. (1987) Occurrence and geochemical significance of long-chain dialkythiacyclopentanes. *Nature* **329**, 54-56.

Schmid J.C. (1986) *Marquers biologiques soufres dans les petroles.* Ph.D. dissertation, Univ. of Strasbourg.

Schmidt C.E., and Sprecher R.F. (1989) Analysis of coal-dervied materials with low volatility by low-voltage, high-resolution mass spectrometry in conjunction with direct-insertion probe techniques. In *Novel Techniques in Fossil Fuel Mass Spectrometry* (eds. T.R. Ashe, and K.V. Wood), pp. 116-132. ASTM STP 1019, ASTM.

Schmidt C.E., Sprecher R.F., and Batts B.D. (1987) Low-voltage, high-resolution mass spectrometric methods for fuel analysis: Application to coal distillates. *Anal. Chem.* **59**, 2027-2033.

Schnitzer M., and Skinner S.I.M. (1974) Low temperature oxidation of humic substances. *Can. J. Chem.* **52**, 1072.

Schoell M., Tescher M., Wehner H., Durand B., and Oudin J.C. (1983) Maturity related biomarker and stable isotope variations and their application to oil/source rock correlations in the Mahakam Delta, Kalimantan. In *Advances in Organic Geochemistry 1981* (ed. Bjorøy M.), pp 156-163. Wiley and Sons Ltd.

Schou L., and Myhr M.B. (1988) Sulphur aromatic compounds as maturity parameters. In *Advances in Organic Geochemistry 1987* (eds. L. Novelli, and L. Mattavelli), *Org. Geochem.* **13**, 61-66.

Schouten J.C., Hakvoort G., Valkenburg P.J.M., and van den Bleek C.M. (1987) An approach with use of EGA to the mechanism of sulfur release during coal combustion. *Thermochim. Acta* **114**, 171-178.

Schouten J.C., Blomaert F.Y., Hakvoort G., and van den Bleek C.M. (1987) A thermal analysis study on the release of sulfur during coal combustion, In *Intern. Conf. Coal Sci.* (eds. J.A. Moulijn, K.A. Nater, and H.A.G. Chermin), pp. 837-840. Elsevier.

Schultz J.L., Kessler T., Friedel R.A., and Sharkey A.G. (1972) High-resolution mass spectrometric invetigation of heteroatom species in coal-carbonization products. *Fuel* **51**, 242-246.

Schwartz S.E. (1989) Acid deposition: Unraveling a regional phenomenon. *Science* **243**, 753-763.

Schwartz S.E., and Freiberg J.E. (1981) Mass-transport limitation to the rate of reaction of gases in liquid droplets: Application to the oxidiation of SO_2 in aqueous solutions. *Atmos. Environ.* **15**, 1129-1144.

Schwartzenbach G., and Fischer A. (1960) The acidity of the sulfanes and the composition of aqueous polysulfide solutions. *Helv. Chim. Acta* **43**, 1365-1390.

Seifert W.K., and Moldowan J.M. (1986) Use of biological markers in petroleum maturation. In *Biological Markers in the Sedimentary Record* (ed. R. B. Johns), pp. 261-286. Elsevier.

Sekyama H., Kosugi N., Kuroda H., and Ohta T. (1986) Sulfur K-edge x-ray absorption spectroscopy of Na_2SO_4, Na_2SO_3, $Na_2S_2O_3$ and $Na_2S_2O_x$ (x = 5-8). *Bull. Chem. Soc. Jpn.* **59**, 575-579.

Sentfle J., Larter S. Bromley B., and Brown J.H. (1986) Quantitative chemical characterization of vitrinite concentrates using pyrolysis-gas chromatography. Rank variation of pyrolysis products. *Org. Geochem.* **9**, 345-350.

Shahab Y.A., and Siddiq A.A. (1988) The synthesis of carbon-sulfur bulk compounds. *Carbon* **26**, 801-802.

Sheng Guoying, Fu Jimo, Brassell S.C., Gowar A.P., Eglinton G., Sinninghe Damsté J.S., de Leeuw J.W., and Schenck P.A. (1987) Sulphur-containing compounds in sulfur-rich crude oils from hypersaline lake sediments and their geochemical implications. *Geochem.* **6**, 115-156.

Sheng Guoying, Fan Shanfa, Lin Dehan, Su Nengxian, and Zhou Hongming (1980) The geochemistry of n-alkanes with an even-odd predominance in the Tertiary Shahejie formation of northern China. In *Advances in Organic Geochemistry 1979* (eds. A.G. Douglas, and J.R. Maxwell), pp. 115-121. Pergamon Press.

Sheppard D., Wong W.S., Uehara C.F., Nadel J.A., and Boushey H.A. (1988) *Am. Rev. Resp. Dis.* **122**, 873-878.

Silverman S.R. (1965) Migration and segregation of oil and gas. In *Fluids in Subsurface Environments, AAPG Memmor 4* (eds. A. Young, and J. E. Galley), pp. 53-65. AAPG Press.

Singer M.E., and Finnerty W.R. (1984) Microbial metabolism of straight-shain and branched alkanes. In *Petroleum Microbiology* (ed. R.M. Altas), pp. 1-60. MacMillan Publishing Co.

Sinninghe Damsté J.S., Eglinton T.I, Rijpstra W.I.C., and de Leeuw J.W. (1990) Molecular characterisation of organically-bound sulfur in sedimentary high-molecular-weight organic matter using flash pyrolysis and Raney Ni desulfurization. In *Geochemistry of Sulfur in Fossil Fuels* (eds. W.L. Orr, and C.M. White), ACS Symposium Series (this volume). American Chemical Society.

Sinninghe Damsté J.S., Eglinton T.I., de Leeuw J.W., and Schenk P.A. (1989) Organic sulphur in macromolecular sedimentary organic matter: I. Structure and origin of sulfur containing moieties in kerogen, asphaltenes and coal as revealed by flash pyrolysis. *Geochim. Cosmochim. Acta* **53**, 873-889.

Sinninghe Damsté J.S., van Koert E.R., Kock-van Dalen A.C., de Leeuw J.W., and Schenck P.A. (1989) Characterization of highly branched isoprenoid thiophenes occurring in sediments and immature crude oils. *Org. Geochem.* **14**, 555-567.

Sinninghe Damsté J.S., and de Leeuw J.W. (1989) Analysis and geochemical significance of organically-bound sulfur in the geosphere: state of the art and future research. *Abstracts Fourteenth International Meeting on Organic Geochemistry*, Paris, Keynote Paper **10**.

Sinninghe Damsté J.S., Rijpstra W.I.C., Kock-van Dalen A.C., de Leeuw J. W., and Schenck P. A. (1989) Quenching of labile functionalised lipids by inorganic sulphur species: Evidence for the formation of sedimentary organic sulphur compounds in the early stages of diagenesis. *Geochim. Cosmochim. Acta* **53**, 1343-1355.

Sinninghe Damsté J. S., Rijpstra W.I.C., de Leeuw J.W., and Schenck P.A. (1989) The occurrence and identification of a series of organic sulphur compounds in oils and sediment extracts: II. Their presence in samples from hypersaline and non-hypersaline palaeoenvironments and possible applications as source, palaeoenvironmental and maturity indicators. *Geochim. Cosmochim. Acta* **53**, 1323-1341.

Sinninghe Damsté J.S. (1988) Organically-bound sulphur in the geosphere: A Molecular approach. Ph.D. dissertation. Delft University of Technology, The Netherlands.

Sinninghe Damsté J.S., Kock-van Dalen A.C., de Leeuw J.W., and Schenck P.A. (1988) Identification of homologous series of alkylated thiophenes, thiolanes, thianes and benzothiophenes present in pyrolysates of sulphur-rich kerogens. *J. Chromatogr.* **435**, 435-452.

Sinninghe Damsté J.S., Koch-van Dalen A.C., de Leeuw J.W., and Schenk P.A. (1988) Identification of a homologous series of alkylated thiophenes, thiolanes, thianes, and benzothiophenes present in pyrolysates of sulfur-rich kerogens. *J. Chromatogr.* **435**, 435-452.

Sinninghe Damsté J.S., Rijpstra W.I.C., de Leeuw J.W., and Schenck P.A. (1988) Origin of organic sulphur compounds and sulphur-containing high molecular weight substances in sediments and immature crude oils. In *Advances in Organic Geochemistry 1987* (eds. L. Mattavelli, and L. Novelli), *Org. Geochem.* **13**, 593-606.

Sinninghe Damsté, J.S., Kock-van Dalen A.C., de Leeuw J.W., Schenck P.A., Sheng Guoying, and Brassell S.C. (1987) The identification of mono-, di- and trimethyl 2-methyl-2-(4,8,12-trimethyltridecyl)chromans and their occurrence in the geosphere. *Geochim. Cosmochim. Acta* **51**, 2393-2400.

Sinninghe Damsté J.S., and de Leeuw J.W. (1987) The origin and fate of isoprenoid C_{20} and C_{15} sulfur compounds in sediments and oils. *Int. J. Environ. Anal. Chem.* **28**, 1-19.

Sinninghe Damsté J.S., Kock-van Dalen A.C., de Leeuw J.W., and Schenck P.A. (1987) The identification of 2,3-dimethyl-5-(2,6,10-trimethyl-undecyl)thiophene, a novel sulfur containing biological marker. *Tetrahedron Lett.* **28**, 957-960.

Sinninghe Damsté J.S., de Leeuw J.W., Kock-van Dalen A.C., de Zeeuw M.A., de Lange F., Rijpstra W.I.C., and Schenck P.A. (1987) The occurrence and identification of series of sulphur compounds in oils and sediment extracts: I. A study of Rozel Point oil (USA). *Geochim. Cosmochim. Acta.* **51**, 2369-2391.

Sinninghe Damsté J.S., ten Haven H.L., de Leeuw J.W., and Schenck P.A. (1986) Organic geochemical studies of a Messinian evaporitic basin, Northern Apennines (Italy) II. Isoprenoid and *n*-alkylthiophenes and thiolanes. *Org. Geochem.* **10**, 791-805.

Siskin M., and Aczel T. (1981) Pyrolysis studies on the structure of ethers and phenols in coals. *Proceedings, 1981 International Conference on Coal Science*, pp. 651-656.

Skelton R.J., Chan H.-C.K., Farnsworth P.B., Markides K.E., and Lee M.L. (1989) Radio frequency plasma detector for sulfur selective capillary gas chromatographic analysis of fossil fuels. *Anal. Chem.* **61**, 2292-2298.

Smith H.M. (1968) Quantitative and qualitative aspects of crude oil composition. United States Bureau of Mines Bulletin **642**, pp. 1-136.

Smith J.W., Gould K.W., and Rigby D. (1982) The stable isotope geochemistry of Australian coals. *Org. Geochem.* **3**, 111-131.

Smith J.W,. and Batts B.D. (1974) The distribution and isotopic composition of sulfur in coal. *Geochim. Cosmochim. Acta* **38**, 121-133.

Smith R.L., and Oremland R.S. (1987) Big Soda Lake (Nevada). 2. Pelagic sulfate reduction. *Limnol. Oceanogr.* **32**, 794-803.

Solli H., and Leplat P. (1986) Pyrolysis-gas chromatography of asphaltenes and kerogens from source rocks and coals. A comparative structural study. In *Advances in Organic Geochemistry 1985* (eds. D. Leythaeuser, and J. Rullkötter), *Org. Geochem.* **10**, 313-329.

Solli H., Bjorøy M., Leplat P., and Hall K. (1984) Analysis of organic matter in small rock samples using combined thermal extraction and pyrolysis-gas chromatography. *J. Anal. Appl. Pyrolysis* **7**, 101-119.

Solli H., van Graas G., Leplat P., and Krane J. (1984) Analysis of kerogens of Miocene shales in a homogenous sedimentary column. A study of maturation using flash pyrolysis techniques and carbon-13 CP-MAS NMR. In *Advances in Geochemistry 1983* (eds. P.A. Schenck, J.W. de Leeuw, and G.W.M. Lijmbach), *Org. Geochem.* **6**, 351-358.

Solomon P.R., and Manzione A.V. (1977) New method for sulphur concentration measurements in coal and char. *Fuel* **56**, 393-396.

Speight J.G. (1988) Evidence for the types of polynuclear aromatic systems in nonvolatile fractions of petroleum. In *Polynuclear Aromatic Compounds* (ed. L.B. Ebert), ACS Advances in Chemistry Series **217**, pp. 201-215. American Chemical Society.

Speight J.G. (1987) Initial reactions in the coking of residua. *ACS Div. Petr. Chem. Preprints* **32(2)**, 413-418.

Speight J.G., and Moschopedis S.E. (1981) On the molecular nature of petroleum asphaltenes. In *Chemistry of Asphaltenes* (eds. J.W. Bunger, and N.C. Li), ACS Advances in Chemistry Series **195**, pp. 1-15. American Chemical Society.

Spielholtz G.I., and Diehl H. (1966) Wet ashing of coal with perchloric acid mixed with periodic acid for the determination of sulphur and certain other constituents. *Talanta* **13**, 991-1002.

Spiro C.L., Wong J., Lytle F.W., Greegor R.B., Maylotte D.H., and Lamson S.H. (1984) X-ray absorption spectroscopic investigation of sulfur sites in coal: Organic sulfur identification. *Science* **226**, 48-50.

Spiro B., Dinur,D., and Aizenshtat Z. (1983) Bacterial sulphate reduction and calcite precipitation in hypersaline deposition of bituminous shales. *Chem. Geol.* **39**, 189-214.

Spiro B., and Aizenshtat Z. (1977) Bacterial sulphate reduction and calcite precipitation in hypersaline deposition of bitumenous shales. *Nature* **269**, 235-237.

Stach E., Mackowsky M-Th, Teichmuller M., Taylor G.H., Chandra D., and Teichmuller R. (1975) *Stach's Textbook of Coal Petrology.* Gebruder Borntraeger.

Stanek J., Buryan P., and Macak J. (1986) Distribution of organic functional groups of sulfur in brown coal. *Acta Mont.* **73**, 225-235. *Chem. Abstr.* (1988) **106**, 70070d.

Stein R., Rullkötter J., Littke R., Schaefer R.G., and Welte D.H. (1988) Organofacies reconstruction and lipid geochemistry of sediments from the Galicia Margin, Northeast Atlantic (ODP Leg 103). In *Proc. ODP, Sci. Res.* Vol. 103 (eds. G. Boillot , E.L. Winterer), pp.567-585. Ocean Drilling Program.

Stein R., Rullkötter J., and Welte D.H. (1986) Accumulation of organic-carbon-rich sediments in the late Jurassic and Cretaceous Atlantic ocean: A synthesis. *Chem. Geol.* **56**, 1-32.

Steudel R., Holdt G., Gobel T. and Hazen W. (1987) Chromatographic separation of higher polythionates $S_nO^{2-}_6$ (n = 3-22) and their detection in cultures of *thiobacillus ferroöxidans*; molecular composition of bacterial sulfur secretions. *Ang. Chem. intern. Edit.* **26**, 151-153.

Stock L.M., Wolny R., and Bal B. (1989) Sulfur distribution in American bituminous coals. *Energy & Fuels* **3**, 651-661.

Straszheim W.E., Greer R.T., and Markuszewski R. (1983) Direct determination of organic sulphur in raw and chemically desulphurized coals. *Fuel* **62**, 1070-1075.

Strausz O.P. (1989) Recent advances in the chemistry of Alberta bitumens and heavy oils. *Proc. 4th UNITAR/UNDP Conf. on Heavy Crude and Tar Sands.* Vol. 2, pp. 607-628. Edmonton, Alberta. Alberta Oil Sands Technology and Research Authority.

Strawinski R.J. (1951) Purification of substances by microbial action. US Patent 2,574,070.

Strawinski R.J. (1950) Method of desulfurizing crude oil. US Patent 2,521,761.

Strickland J.D.H., and Parsons T.R. (1977) *A Practical Handbook of Seawater Analysis* (eds. J.D.H. Strickland, and T.R. Parsons), pp. 11-34, Alger Press Ltd.

Suggate R.P. (1959) New Zealand coals. Their geological setting and its influence on their properties. pp. 1-113, *New Zealand Department of Scientific and Industrial Research Bulletin* **134**.

Summons R.E., and Powell T.G. (1987) Identification of aryl isoprenoids in source rocks and crude oils: Biological markers for green sulphur bacteria. *Geochim. Cosmochim. Acta* **51**, 557-566.

Sutherland, J.K. (1975) Determination of organic sulphur in coal by microprobe. *Fuel* **54**, 132.

Swaine D.J. (1989) Trace elements in the Permian coals. Aust. But. Minet. Resource. *Geol. Geophys. Bull.* **231**, 297-300.

Sweeney R.E., and Kaplan I.R. (1973) Pyrite framboid formation: Laboratory synthesis and marine sediments. *Econ. Geol.* **68**, 618-634.

Sweeney R.E. (1972) Pyritization during diagenesis of marine sediments. Ph.D. dissertation, University of California, Los Angeles.

Takata T., Ishibashi K., and Ando W. (1985) Photosensitized oxygenation of cyclic sulfides. Selective C-S bond cleavage. *Tetrahedron Lett.* **25**, 4609-4612.

Talukdar S., Gallango O., and Chin-A-Lien M. (1986) Generation and migration of hydrocarbons in the Maracaibo Basin, Venezuela: An integrated basin study. *Org. Geochem.* **10**, 261-279.

Tanaka H., and Yokoyama M. (1960) Sulfur-containing chelating agents. I. Synthesis of ß-mercaptoketones and their copper chelates. *Chem. Pharm. Bull. (Tokyo)* **8**, 275-279.

Tannenbaum E., Huzinga B.J., and Kaplan I.R. (1986) The role of minerals in the thermal alteration of organic matter. II. A material balance. *Amer. Assoc. Petr. Geol. Bull.* **65**, 1527-1535.

Tannenbaum E., and Aizenshtat Z. (1985) The formation of immature asphalt from organic-rich carbonate. I. Correlation of maturation indicators. *Org. Geochem.* **8**, 181-192.

Tegelaar E.W., de Leeuw J.W., Largeau C., Derenne S., Schulten H.R., Muller R., Boon J.J., Nip J.C., and Sprenkels J.C.M. (1989) Scope and limitations of several pyrolysis methods in the structural elucidation of a macromolecular plant constituent in the leaf of Agave Americana L. *J. Anal. Appl. Pyrolysis* **15**, 29-54.

Thiel M., Asinger F., and Fedtke M. (1958) Concomitant action of elementary S and gaseous NH₃ on ketones (XIII) simulaneous action of S and NH₃ on methylbenzylketone. *Justus Liebigs Ann. Chem.* **615**, 77-84.

Thode H.G. (1981) Sulfur isotope ratios in petroleum research and exploration: Williston Basin. *Amer. Assoc. Petr. Geol. Bull.* **65** 1527-1535.

Thode H.G., and Monster J. (1970) Sulfur isotope abundances and genetic relations of oil accumulations in Middle East basins. *Amer. Assoc. Petr. Geol. Bull.* **54**, 627-637.

Thode H.G., and Rees C.E. (1970) Sulfur isotope geochemistry and Middle East oil studies. *Endeavour* **29**, p. 24-28.

Thode H,G. (1965) Sulfur isotope geochemistry of petroleum, evaporites, and ancient seas. In *Fluids in Subsurface Environments*, AAPG Memmor 4 (eds. A. Young, and J.E. Galley), pp. 367- 377. AAPG Press.

Thode H.G., Monster J., and Dunford H.B. (1958) Sulfur isotope abundances in petroleum and associated materials. *Amer. Assoc. Petr. Geol. Bull.* **42**, 2619-2641.

Thomas B.M., Moller-Pedersen P., Whiticar M.F., and Shaw N.D. (1985) In: *Petroleum Geochemistry and Exploration of the Norwegian Shelf.* (ed. B.M. Thamas) pp. 3-26. Graham and Trotman.

Thompson C.J. (1981) Identification of sulfur compounds in petroleum and alternative fossil fuels. In *Organic Sulfur Chemistry* (eds. R.Kh. Freidlina, and A.E. Skarova), pp. 189-208. Pergamon Press.

Timmer J.M., and van der Burgh N. (1984) Automated determination of organic sulphur in coal: Use of scanning electron microscopy and energy dispersive x-ray microanalysis. *Fuel* **63**, 1645-1648.

Tissot B.P., Pelet R. and Ungera P. (1987) Thermal history of sedimentary basins, maturation indices and kinetics of oil and gas generation. *Amer. Assoc. Petr. Geol. Bull.* **71**, 1445-1466.

Tissot B.P. (1984) Recent advances in petroleum geochemistry applied to hydrocarbon exploration. *Amer. Assoc. Petr. Geol. Bull.* **68**, 545-563.

Tissot B.P., and Welte D.H. (1984) *Petroleum Formation and Occurence*, 2nd edition. Springer-Verlag.

Tissot B.P., Pelet R., Roucache J., and Combaz A. (1977) Alkanes as geochemical fossil indicators of geological environments. In *Advances in Organic Geochemistry 1975* (eds. R. Campos, and J. Goni), pp. 117-154. Enadisma, Madrid.

Tissot B.P., Durand B., Espitalie J., and Combaz A. (1974) Influence of nature and diagenesis of organic matter in formation of petroleum. *Amer. Assoc. Petr. Geol. Bull.* **58**, 499-506.

Tissot B.P., Califert-Debyser Y., Deroo G., and Oudin J.L (1971) Origin and evolution of hydrocarbons in early Toarcian shales, Paris Basin, France. *Amer. Assoc. Petr. Geol. Bull.* 55, 2177-2193.

Toste A.P. (1976) The steroid molecule: its analysis and utility as a chemotaxonomic marker and a fine geochemical probe into the earth's past. Ph.D. dissertation. University of California, Berkely.

Treibs A. (1936) Chlorophyll- und Häminderivate in Organischen Mineralstoffen. *Ann. Chem.* 49, 682-686.

Tromp P.J.J., Moulijn J.A., and Boon J.J. (1986) Probing the influence of K_2CO_3 and Na_2CO_3 addition on the flash pyrolysis of a lignite and bituminous coal with Curie-point pyrolysis techniques. *Fuel* 65, 960.

Trudgill P.W. (1984) Microbial degradation of the alicyclic ring: structural relationships and metabolic pathways. In *Microbial Degradation of Organic Compounds* (ed. D.T. Gibson), pp. 131-180. Marcel Dekker.

Tseng B.H., Ge Y-P, Hsieh K.C., and Wert C.A. (1987) Loss of organic sulfur in coal during heating. In *Processing and Utilization of High Sulfur Coals II* (eds. Y.P. Chugh, and R.D. Caudle), pp. 33-40. Elsevier.

Tseng B.H., Buckentine M., Hsieh K.C., Wert C.A., and Dyrkacz G.R. (1986) Organic sulfur in coal. *Fuel* 65, 385-389.

Tuttle M.L. (1988) Geochemical evolution and depositional history of sediment in modern and ancient saline lakes: Evidence from sulfur geochemistry. Ph.D. dissertation, Colorado School of Mines.

Tuttle M.L., Goldhaber M.B., and Rice C.A. (1987) Sulfur diagenesis in alkaline, sulfur-rich Soap Lake, Washington: A modern geochemical analogue for Green river oil shales? *Geol. Soc. Amer. Abstracts with Program* 19, pp. 873.

Tuttle M.L., Goldhaber M.B., and Williamson D.L. (1986) An analytical scheme for determining forms of sulphur in oil shales and associated rocks. *Talanta* 33, 953-961.

Tuttle M.L., and Goldhaber M.B. (1986) Sulfur geochemistry and its implications for the depositional and early diagenetic history of the Green River Formation. *Geol. Soc. Amer. National Meeting*, San Antonio, Abstracts 18, no. 108714.

Vairavamurthy A., and Mopper K. (1989) Mechanistic studies of organosulfur (thiol) formation in coastal marine sediments. In *Biogenic Sulfur in the Environment* (eds. E.S. Saltzman, and W.J. Cooper), pp. 231- 242, ACS Symposium Series 393. American Chemical Society.

Vairavamurthy A., and Mopper K. (1987) Geochemical formation of organosulfur compounds (thiols) by addition of H_2S to sedimentary organic matter. *Nature* 329, 623-625.

Valisolalao J., Perakis N., Chappe B., and Albrecht P. (1984) A novel sulfur containing C_{35} hopanoid in sediments. *Tetrahedron Lett.* 25, 1183-1186.

Vandenbroucke M., and Behar F. (1988) Geochemical characterization of the organic matter from some recent sediments by a pyrolysis technique. In *Lacustrine Petroleum Source Rocks* (eds. A.J. Fleet, K. Kelts, and M.R. Talbot), Blackwell. *Geological Society Special Publication* 40, 91-101.

Vandenbrouke M., and Durand B. (1983) Detecting migration phenomena in a geologic series by means of C_1-C_{35} hydrocarbon amounts and distributions. In *Advances in Organic Geochemistry 1981* (ed. M. Bjorøy), pp. 147-155. Wiley and Sons Ltd.

Vetter R.D., Matrai P.A., Javor B., and O'Brien J. (1989) Reduced sulfur compounds in the marine environment. Analysis by high-performance liquid chromatography. In *Biogenic Sulfur in the Environment* (eds. E.S. Saltzman, and W.J. Cooper), ACS Symposium Series 393, pp. 243-261. American Chemical Society.

Vinogradov V.I., and Kizilsthein L.J. (1969) Ob isotopnom sostave sulfidoy sery v uglh'akh Donezkogo Basseina. *Litol. Polez. Iskopaiem* 5, 149-151.

Viswanadham R.K., and Wert C.A. (1976) Electron microscopic study of precipitation in the system niobium-carbon. *J. Less Common Metals* 48, 135-150.

Vogel T.M., and Grbic'-Galic' D. (1986) Incorportion of oxygen from water into toluene and benzene during anaerobic fermentative transformation. *Appl. Environ. Microbiol.* 52, 200-202.

Volkman J.K. (1988) Biological marker compounds as indicators of the depositional environments of petroleum source rocks. In *Lacustrine Petroleum Source Rocks* (eds. K. Kelts, A.J. Fleet, and M. Talbot), pp. 103-122, Geol. Soc. Spec. Publ. No. 40. Blackwell.

Volkman J.K., Farrington J.W., Gagosian R.B., and Wakeham S.G. (1983) Lipid composition of coastal marine sediments from the Peru upwelling region. In *Advances in Organic Geochemistry 1981* (ed. M. Bjorøy), pp. 228-240. Wiley.

Volkman J.K., Eglinton G., Corner E.D.S., and Sargent J.R. (1980) Novel unsaturated straight chain C_{37} - C_{39} methyl and ethyl ketones in marine sediments and a coccolithophore *Emiliania huxleyi*. In *Advances in Organic Geochemistry 1979* (eds. A.G. Douglas, and J.R. Maxwell), pp. 219-227. Pergamon Press.

von der Dick H., Rullkötter J., and Welte D.H. (1984) Content, type, and thermal evolution of organic matter in sediments from the eastern Flakland plateau, Deep Sea Drilling Project, Leg 71. In *Init. Repts. DSDP 71* (eds. W.J. Ludwig, V.A. Kraskeninnikov), pp. 1015-1032. US Government Printing Office.

Vorres K.S., and Janikowski S.K. (1987) The eight coals in the Argonne Premium Coal Sample Program. *ACS Div. Fuel Chem. Preprints* 32(1), 492-499.

Vredenburgh L.D., and Cheney E.S. (1971) Sulfur and carbon isotopic investigation of petroleum, Wind River Basin, Wyoming. *Amer. Assoc. Petr. Geol. Bull.* 55, 1954-1975.

Wagner G.A. (1977) Spaltspurendatierung an apatit und titanit aus dem ries: Ein beitrag zum alter und zur wärmegeschichte. *Geol. Bavarica* 75, 349-354.

Walker J.D., Colwell R.R., and Petrakis L. (1975) Microbial petroleum degradation: Application of computerized mass spectrometry. *Can. J. Microbiol.* 21, 1760-1767.

Weber W.P., Stromquist P., and Ito T.I. (1974) Observation on the mechanism of reduction of sulfones to sulfides. *Tetrahedron Lett.* 30, 2595-2598.

Welte D.H., Hagemann H.W., Hollerbach A., Leythauser D., and Stahl W. (1975) Correlation between petroleum and source rock. *9th World Pet. Cong. Proc.* 2, 179-191.

Westlake D.W.S. (1983) Microbial activities and changes in the chemical and physical properties of oil. *Proc. 1982 International Conference on Microbial Enhancement of Oil Recovery* (eds. E.C. Donaldson, and J.B. Clark), pp. 102-111, Bartlesville Energy Technology Center.

William J.A., Bjorøy M., Dolcater D.L., and Winters J.C. (1986) Biodegradation of south Texas Eocene oils-effects on aromatics and biomarkers. *Org. Geochem.* 10, 451-461.

Williams P.A., Catterall F.A., and Murray K. (1975) Metabolism of naphthalene, 2-methylnaphthalene, salicylate, and benzoate by *Pseudomonas* P_G: Regulation of tangential pathways. *J. Bacteriol.* **124**, 679-685.

Williams L.A. (1986) Depositional and diagenetic setting of fine-grained Miocene Monterey formation source and reservoir rocks in California basins, 192nd Amer. Chem. Soc. National Meeting, Anaheim, California, (Abstr.)

Wehner H., and Hufnagel H. (1987) Some characteristics of the inorganic and organic composition of oil shales from Jordan. In *Biogeochemistry of Black Shales* (ed. E.T. Degens), *Mitt. Geol. Palaeont. Inst Univ. Hamburg* **60**, 381-395.

Weitkamp A.W. (1959) I. The action of sulfur on terpenes. The limonene sulfides. *J. Amer. Chem. Soc.* **81**, 3430-3434.

Welte D.H., and Waples D.W. (1973) Uber die bevorzugung geradzahliger n-alkane in sedimentgesteinen. *Naturwissenschaften* **60**, 516-517.

Wert C., Ge Y., Tseng B.H., and Hsieh K.C. (1988) Analysis of organic sulfur in coal by use of transmission electron microscopy. *J. Coal Quality* **7**, 118-121.

Wert C.A., Hsieh K.C., Buckentin M., and Tseng B.H. (1988) Application of transmission electron microscopy to coal. *Scanning Microscopy* **88**, 83-96.

Wert C.A., Hsieh K.C., Tseng B.-H., and Ge Y.-P. (1987) Applications of transmission electron microscopy to coal chemistry. *Fuel* **66**, 914-920.

Wert C.A., and Hsieh K.C. (1983) Minerals in coal - a transmission electron microscopy study. *Scanning Electron Microsc.*, 1123-1136.

Westgate L.M., and Anderson T.F. (1984) Isotopic evidence for the origin of sulfur in the Herrin (No. 6) coal member of Illinois. *Intern. Jour. Coal Geol.* **4**, 1-20.

Weston A.W., and Michaels R.J. (1963) 2-Thiophenaldehyde. In *Org. Synth. Coll. Vol. IV*, pp. 915-918. John Wiley and Sons.

Westrich J.T., and Berner R.A. (1984) The role of sedimentary organic matter in bacterial sulfate reduction: The G model tested. *Limnol. Oceanogr.* **29**, 236-249.

Wetzel R.G. (1983) *Limnology*. Saunders College Publishing.

Whelan J.F., Cobb J.C., and Rye R.O. (1988) Stable isotope geochemistry of sphalerite and other mineral matter in coal beds of Illinois and Forest City basins. *Econ. Geol.* **83**, 990-1007.

Whelan J.K., Solomon P.R., Deshpande G.V., and Carangelo R.M. (1988) Thermogravimetric Fourier infrared spectroscopy (TG-FTIR) of petroleum source rocks, initial results. *Energy & Fuels* **2**, 65-73.

Whelan J.K., Hunt J.M., and Berman J. (1980) Volitile C_1-C_7 organic compounds in surface sediments from Walvis Bay. *Geochim. Cosmochim. Acta* **44**, 1767-1785.

White C.M., Douglas L.J., Anderson R.R., Schmidt C.E., and Gray R.J. (1990) On the nature of organosulfur constituents in Rasa coal. In *Geochemistry of Sulfur in Fossil Fuels* (eds. W.L. Orr, and C.M. White), ACS Symposium Series (this volume). American Chemical Society.

White C.M., Douglas L.J., and Hackett M. (1988) Formation of polycyclic thiophenes from reaction of selected polycyclic aromatic hydrocarbons with elemental sulfur and/or pyrite under mild conditions. *Energy & Fuels* **2**, 220-223.

White C.M., Douglas L.J., Perry M.B., and Schmidt C.E. (1987) Characterization of extractable organosulfur constituents from Bevier seam coal. *Energy & Fuels* 1, 222-226.

White C.M. (1983) Determination of polycyclic aromatic hydrocarbons in coal-derived materials. In *Handbook of Polycyclic Aromatic Hydrocarbons* (ed. A. Bjorseth), pp. 525-616. Marcel Dekker.

White C.M., and Lee M.L. (1980) Identification and geochemical significance of some aromatic components of coal. *Geochim. Cosmochim. Acta* 44, 1825-1832.

White D. (1913) Physiographic conditions attending the formation of coal. *U.S. Bureau of Mines Bulletin*, 38, pp. 52-84.

Willey C., Iwao M., Castle R.N., and Lee M.L. (1981) Determination of sulfur heterocycles in coal liquids and shale oils. *Anal. Chem.* 53, 400-407.

Williams E.G., and Keith M.L. (1963) Relationship between sulfur in coals and the occurrence of marine roof beds. *Econ. Geol.* 58, 720-729.

Williams P.F.V. (1986) Petroleum geochemistry of the Kimmeridge clay of onshore southern and eastern England. *Mar. Petr. Geol.* 3, 258-281.

Williams P.F.V., and Douglas A.G. (1986) Organic geochemistry of the British Kimmeridge clay 2. Acyclic isoprenoid alkanes in Kimmeridge shale oils. *Fuel* 65, 1728-1734.

Williams P.F.V., and Douglas A.G. (1980) The effects of lithologic variation on organic geochemistry in the Kimmeridge Clay of Britain. In *Advances in Organic Geochemistry 1979* (eds. A.G. Douglas, and J.R. Maxwell), pp. 211-217, Pergamon Press.

Winans R.E., Scott R.G., Neill P.H., Dyrkacz G.R., and Hayatsu R. (1986) Characterization and pyrolysis of separated coal macerals. *Fuel Process. Tech.* 12, 77-88.

Winans R.E., Scott R.G., Neill P.H., Dyrkacz G.R., McBeth R.L., and Hayatsu R. (1985) Characterization and reactivity of hydroxyls and ethers in coal macerals. *Proceedings, 1985 International Conference on Coal Science*, pp. 687-690.

Winans R.E., Hayatsu R., Scott R.G., and McBeth R.L. (1984) Reactivity and characterization of coal macerals. In *Chemistry and Characterization of Coal Macerals* (eds. R.E. Winans, and J.C. Crelling) ACS Symposium Series 252, pp. 137-155. American Chemical Society.

Winterer E.L., and Hinz K. (1984) The evolution of the Mazagan continental margin: A synthesis of geophysical and geological data with results of drilling during Deep Sea Drilling Project, Leg 79. In *Init. Repts. DSDP 79* (eds. E.L. Winterer, and K. Hinz), pp. 893-919. US Government Printing Office.

Wolf M. (1977) Kohlenpetrographische Untersuchung der See-Sedimente det Forschungsbohrung Nördlingen 1973 und Vergleich mit anderen Untersuchungsergebnissen aus dem Ries. *Geol. Bavarica* 75, 127-138.

Wong J., Sprio C.L., Maylotte D.H., Lytle F.W., and Greegor R.B. (1984) EXAFS and Xanes studies of trace elements in coal. In *EXAFS and Near-Edge Structure III* (eds. K.O. Hodgson, B. Hedman, and J.E. Penner-Hahn), pp. 362-367. Springer.

Wood G.H. R., Kehn T.M., Carter M.D., and Culbertson W.C. (1983) Coal resource classification system of the U.S. Geological Survey, *U.S. Geological Survey Circular 891*.

Yanagisawa F., and Sakai H. (1983) Thermal decomposition of barium sulfate-vanadium pentaoxide-silica glass mixtures for preparation of sulfur dioxide in sulfur isotope ratio measurements. *Anal. Chem.* 55, 985-987.

Yang Xilu, Pan Suixian, and Cheng Baozhou (1987) Sedimentary environments of Taiyuan Xishan coal basin. *Coal Science Technology (Beijing)* **6**, 16-26. (In Chinese with an English abstr.).

Yen T.F., and Moldowan J.M. eds. (1988) *Geochemical Biomarkers*. Harwood Academic Publishers.

Yergey A.L., Lampe F.W., Vestal M.L., Day A.G., Fergusson G.J., Johnston W.H., Synderman I.S., Essenhigh R.H., and Hudson J.E. (1974) Nonisothermal kinetics studies of the hydrodesulfurization of coal. *Ind. Eng. Chem. Process Des. Develop.* **13**, 233-240.

Yurovski A.Z. (1960) *Sulfur in Coals*, English translation available from the US Department of Commerce, National Technical Information Service, Springfield, VA 22161, USA (Ref. No. TT 70-57216).

Zeyer J., Kuhn E.P., and Schwarzenbach R.P. (1986) Rapid microbial mineralization of toluene and 1,3-dimethylbenzene in the absence of molecular oxygen. *Appl. Environ. Microbiol.* **52**, 944-947.

Zhang Z.G., Kyotani T., and Tomita A. (1989) Dynamic behaviour of surface oxygen complexes during O_2 chemisorption and subsequent temperature-programmed desorption of calcium-loaded coal chars. *Energy & Fuels* **3**, 566-571.

Zielinski R.A., Otton J.K., Wanty R.B., and Pierson C.T. (1987) The geochemistry of water near a surficial organic-rich uranium deposit, Northeastern Washington State, U.S.A. *Chem. Geol.* **62**, 263-289.

Zinder S.H., and Brock T.D. (1978) Production of methane and carbon dioxide from methane thiol and dimethylsulphide by anaerobic lake sediments. *Nature* **273**, 226-228.

Zobell C.E. (1963) Organic geochemistry of sulfur. In *Organic Geochemistry* (ed. I.A. Breger), pp. 579-595. McMillan.

Author Index

Anderson, R. R., 261
Bailey, A. M., 186
Bakel, A. J., 326
Barker, C., 592
Blackson, J. H., 186
Boudou, J., 345
Calkins, W. H., 287
Chou, C.-L., 30
Crelling, J. C., 296
de Leeuw, J. W., 417,444,486,529,613
Douglas, L. J., 261
Eglinton, T. I., 486,529
Fedorak, P. M., 93
Galvez-Sinibaldi, A., 326
Ge, E., 316
George, G. N., 220
Goldhaber, M. B., 114
Gorbaty, M. L., 220
Gray, R. J., 261
Hippo, E. J., 296
Huc, A. Y., 170
Idiz, E. F., 575
Kaplan, I. R., 575
Kelemen, S. R., 220
Kenig, F., 170
Kiyosu, Y., 633
Klein, M. T., 287
Kock-van Dalen, A. C., 397
Kohnen, M. E. L., 444,529
Kosters, E. C., 186
Krouse, H. R., 592,633
Kruge, M. A., 296
Kump, L. R., 204
LaLonde, R. T., 68
Larter, S. R., 529
Lee, C. C., 231
Lipfert, F. W., 53

Littke, R., 149
Lown, E. M., 83,366
Manowitz, B., 53,592
Morrison, J. L., 204
Neill, P. H., 249
Orr, W. L., 2
Palmer, S. R., 296
Patience, R. L., 529
Payzant, J. D., 83,366
Peakman, T. M., 397
Phillips, R., 568
Philp, R. P., 326
Premuzic, E. T., 592
Rice, C. A., 114
Rijpstra, W. I. C., 444,486
Riley, J. T., 231
Ruba, G. M., 231
Rullkötter, J., 149,613
Schaefer, R. G., 149
Schmidt, C. E., 261
Sherrill, J. F., 186
Sinninghe Damsté, J. S.,
    2,417,444,486,529,613
Smith, J. W., 568
Stock, L. M., 241
Strausz, O. P., 83,366
Tannenbaum, E., 575
ten Haven, H. L., 613
Torres-Ordoñez, R. J., 287
Tuttle, M. L., 114
Viau, C. A., 633
Wert, C., 316
White, C. M., 261
White, T. S., 204
Winans, R. E., 249
Wolny, R., 241

Affiliation Index

Argonne National Laboratory, 249
BP Research Centre, 529
Brookhaven National Laboratory, 53,592
Capricorn Coal Management Pty., 568
Centre National de la Recherche
    Scientifique, 345
Commonwealth Scientific and Industrial
    Research Organisation, 568
Delft University of Technology,
    2,397,417,444,486,529,613
Dow Chemical Company, 186

Ebasco Services, Inc., 186
Exxon Research and Engineering
    Company, 220
Illinois State Geological Survey, 30
Institut Français du Pétrole, 170
Institute of Petroleum and Organic
    Geochemistry, 149,613
Mobil Research & Development
    Corporation, 2
Nagoya University, 633
Pennsylvania State University, 204

Pittsburgh Energy Technology Center, 261
Ralph Gray Services, 261
Shell Canada Limited, 633
Southern Illinois University, 296
State University of New York, 68
University of Alberta, 83,93,366
University of Calgary, 592,633
University of California—Los Angeles, 575
University of Chicago, 241

University of Delaware, 287
University of Illinois, 316
University of Oklahoma, 326
University of Oslo, 529
University of Southwestern Louisiana, 186
University of Tulsa, 592
University of Utrecht, 186
U.S. Geological Survey, 114
Western Kentucky University, 231

Subject Index

A

Acid aerosol, description, 58
Acidification effects of airborne sulfur
    compounds, 62–63,64f
Acid washing, effect on coal
    desulfurization, 351,355f
Acylation, in preparation of alkylthiophenes,
    401–407
Airborne sulfur compounds
    acidification effects, 62–63,64f
    atmospheric visibility effects, 63,65
    human health effects, 61–62
    materials degradation effects, 62
n-Alkanes, predominance of even-number
    compounds, 435,438,439f,440
n-Alkyl monocyclic sulfides, structures,
    97,98f
n-Alkyl-substituted sulfides,
    biogeochemistry, 383,386–387f,388t
n-Alkyl-substituted thianes, GC–MS,
    383,384–385f
n-Alkyl-substituted thiolanes, GC–MS
    383,384–385f
Alkylthiophene(s), traditional method of
    preparation, 398f
2-Alkylthiophene(s), distribution in deep-sea
    sediments, 627–628,629f
Alkylthiophene(s) as sensitive indicators of
    palaeoenvironmental changes
    carbon number distribution patterns of
        isoprenoid carbon skeleton
        alkylthiophenes, 465,470f
    carbon number distribution patterns of
        linear carbon skeleton
        alkylthiophenes, 461,463f
    composition of subfractions, 449,450t
    depth profiles of alkylthiophenes,
        456,460f
    depth profiles of highly branched
        isoprenoid carbon skeleton
        alkylthiophenes, 475,476f
    depth profiles of hydrocarbon biomarkers,
        453,455f

Alkylthiophene(s) as sensitive indicators of
    palaeoenvironmental changes—Continued
    depth profiles of isoprenoid carbon
        skeleton alkylthiophenes, 465,470–474
    depth profiles of linear carbon skeleton
        alkylthiophenes, 456,461–462,464–469
    depth profiles of thiophene hopanoids,
        478,479f
    extraction procedure, 447
    facies description, 452–453
    fractionation procedure, 449
    GC chromatograms of alkylthiophenes,
        456,458f,459t
    GC chromatograms of saturated hydrocarbon
        fractions, 453,454f
    GC of midchain dimethylalkane carbon
        skeleton alkylthiophenes, 475,477f
    GC–MS procedure, 449–450
    geological setting, 446–447
    hydrocarbon biomarkers, 453–457
    implications for reconstruction of
        palaeoenvironments, 478,480
    lithology, 448f,452
    procedure for synthesis of standards, 451
    procedure for whole rock analyses, 451
    quantitative procedure, 450–451
    sample properties, 447t,448f
    sterane distributions, 456,457f
    summary of depth profiles, 480,481f
Alkylthiophene(s) possessing linearly extended
    phytane skeleton, synthesis,
    412,413–415f
Alkylthiophene preparation
    acylation, 401–407
    bromination, 407,408f
    electrophilic substitution of thiophenes,
        398–400f,401
    formylation, 406,407f
    preparation and reaction of lithiated
        thiophenes, 408–412
    structures, 412f
    synthesis of linearly extended
        phytane skeleton, 412–415f
    traditional method, 398f

Ambient concentration levels of sulfur
species
clustering of tall stacks, 61
low-level sources, 60
tall stacks, 60
Analytical pyrolysis, use in studying
coals, 250
Aromatic compounds, identification in
coals, 39
Asphaltenes
characterization of organic sulfur,
503–504
definitions, 14,16
sulfur content, 388,391,392–394f
transformation sequence to
hydrocarbons, 16
ASTM Method D 2492, description, 231–232
Atmospheric removal of sulfur species,
wet and dry processes, 58,59t
Atmospheric transport of sulfur
species by wind, 59–60
Atmospheric visibility effects of airborne
sulfur compounds, 63,65
Average organic sulfur concentration,
determination via transmission electron
microscopy, 319,320–321f

B

Bacteriohopanetetrol
MS, 435,437f
proposed early diagenetic pathways,
435,436f
Barrier-island system, types of depositional
environments, 204,206
Basinal brines, source of sulfur in
high-sulfur coals, 44
Benzothiophene
aromatic intermediates of metabolism,
104,105f
identification in coals, 39
microbial metabolism in petroleum,
101,103f,104,105f
oxidation products, 101,103f,104
Bicyclic terpenoid sulfides
distribution of carbon number vs. depth,
383,386f
structure, 371,373–376
Biodesulfurization, removal of sulfur
compounds from petroleum, 96
Biodesulfurization of petroleum and
petroleum fractions, laboratory
studies, 100
Bitumens, sulfur content, 7
Bolivar coastal fields
oil content, 592–593
sulfur isotopic data analysis of crude
oils, 593–611

Brominated thiophenes, formation of
lithiated thiophenes, 408,409f
Bromination, preparation of
2-alkylthiophenes, 407,408f
Bulk sediment properties of sulfur-rich
lacustrine sediments, 152–157
5-Butyl-2-(2-undecyl)thiophene, synthesis,
402,403f

C

C_{13} terpenoid sulfone, MS, 374,376f
C_{20} isoprenoid thiophene(s)
MS, 424,430f
postulated formation, 431,432f
structures, 424,429f
typical distributions, 424,429f
C_{20} isoprenoid thiophene distribution
patterns as indicators of palaeosalinity,
424,429–432
Carbon isotopes, studies, 5
Carbon isotopic fractionation during
oxidation of light hydrocarbon gases
appearance of ^{13}C-enriched methane during
ethane oxidation, 640
carbon isotopic composition vs. extent of
ethane reaction, 638,639f
estimation of carbon isotopic
selectivity, 640
experimental procedure, 635–636
order of reaction rate constants, 640–641
percentages and carbon isotopic
composition during ethane oxidation,
637,638t
percentages and carbon isotopic
composition during methane oxidation,
637,638t
percentages and carbon isotopic
composition during propane oxidation,
638,639t
rate constants for anhydrite reduction by
methane, 636t
temperature dependence of ratio of
isotopic rate constants, 637f
Chalcone
formation of 2,4-dibenzoyl-3,5-diphenyl-
thiolane, 72–73
formation of S heterocycles, 73–75
Chloride salinities, content in sediment
sites, 196t,199
Chroman distributions as indicators of
palaeosalinity, 418,419f,420
Coal
comparison of pyrolysis products with
·those of sulfur model compounds, 288
definition, 16
geochemistry of sulfur, 30–47
sources of sulfur, 55–56,57f
sulfur characterization, 16

Coal-associated hydrogen sulfide, isotopic
    study, 588–574
Coal desulfurization, importance, 345
Coal desulfurization by programmed
    temperature heating
  effect of acid washing, 351,355f
  effect of adding pure mineral pyrite,
    351,356
  effect of coal rank on sulfur evolution,
    359,360f
  effect of low-temperature ashing,
    356,358f,359
  effect of maturation on distribution
    of sulfur compounds, 359,361f
  effect of organic solvent extraction,
    356,357t
  effect of particle size, 348,349f,351,353f
  effect of treatments on sulfur gas
    evolution, 351,354f
  experimental procedures, 345–346,347–348t
  peak temperature of sulfur-containing
    model compounds, 348t
  properties of coal samples, 346,347t
Coal desulfurization by programmed-
    temperature oxidation
  effect of acid washing, 351,355f
  effect of adding pure mineral pyrite,
    351,356
  effect of atmosphere composition,
    348,351,352f
  effect of coal rank on sulfur evolution,
    359,360f
  effect of low-temperature ashing,
    356,358f,359
  effect of maturation on distribution
    of sulfur compounds, 359,361f
  effect of organic solvent extraction,
    356,357t
  effect of particle size, 348,350f
  effect of treatments on sulfur gas
    evolution, 351,354f
  experimental procedures, 345–346,347–348t
  properties of coal samples, 346,347t
Coal lithotypes
  descriptions, 35
  relation between organic sulfur and
    pyritic plus sulfate sulfur, 35,37f
  vertical variation, 35,36f
Coal macerals, pyrolysis studies, 329
Coal sulfur forms
  classification, 231,287
  determination by ASTM Method D 2492,
    231–232
Croweburg coals, pyrolysis—GC
  characterization of organic sulfur
    compounds, 340,341f
Curie point pyrolysis, in characterization
    of organic sulfur compounds in coal, 297

Cyclic terpenoid sulfides
  capillary GC trace, 371,372f
  detection, 367–372
  ^{1}H-NMR spectrum, 371,373f
  IR spectrum, 367,368f
  relative abundances vs. carbon number,
    367,370f
  single ion recording GC–MS spectrum,
    374,375f
  structure of bicyclics, 371,373–376
  structure of hexacyclics, 381,382f
  structure of pentacyclics, 381
  structure of tetracyclics,
    378,379–380f,381
  structure of tricyclics, 374,377f,378
Cyclic terpenoid sulfoxides
  detection, 367–372
  field ionization MS, 367,369f
  IR spectrum, 367,368f
  mass measurements, 367t
  relative abundances vs. carbon number for
    rings 1–6, 367,369f

D

Desulfurization, problems, 3
Diagenic synthesis, environmental conditions
    for laboratory simulation, 68–80
Dialkyl sulfides, identification in
    coals, 39
Dibenzothiophenes
  distributions, 12–13
  identification in coals, 39
  microbial metabolism in petroleum,
    104,105f,106,107f
  numbering system, 89
  products from microbial metabolism,
    106,107f
  proposed microbial metabolic pathways,
    104,105f
  specific excision of sulfur atom by
    aerobic bacterium, 104,106,107f
2,4-Dibenzoyl-3,5-diphenylthiolane,
    formation from chalcone, 72–73
Dibenzyl sulfide, microbial metabolism in
    petroleum, 106,108f
3,5-Dibromo-2-lithiothiophene, reaction
    with water, 409f
2,5-Diheptyl-3-methylthiophene, synthesis,
    402,403f
Dimethyl disulfide, microbial metabolism in
    petroleum, 109
Dimethyl sulfide, microbial metabolism in
    petroleum, 109
Direct determination of total organic sulfur
    in coal
  extraction method, 232

Direct determination of total organic sulfur
  in coal—*Continued*
microprobe analysis, 232
nitric acid extraction, 233–238
use of perchloric acid as selective
  oxidizing agent, 232
Distribution of organic sulfur-containing
  structures in high-organic-sulfur coals
analyses, 289t
analysis of gases and tars, 291
calculation of residence time, 291
estimated distribution of organic
  sulfur, 293,294t
experimental procedures,
  288,289t,290f,291
schematic representation of fluidized-bed
  pyrolyzer, 289,290f,291
sulfur yield from New Zealand coal,
  292,293f
sulfur yield from Spanish lignite,
  291,292f
Distribution of sulfur in fossil fuels and
  shales, controlling factors, 186
Dry atmospheric removal processes for sulfur
  species
application, 58–59
description, 58
influencing factors, 59

E

Electrophilic substitution of thiophenes
consideration of Wheland intermediates,
  398,399–400f
reaction, 398f
summary of products, 399,400f
Elemental analysis of Rasa coal, procedure,
  264–265
Elemental sulfur
importance in formation of pyritic and
  organic sulfur compounds, 34
occurrence, 34–35
Elemental sulfur in bituminous coals
analysis, 241–242,243t
composition of coals, 241,242t
content after intentional exposure to
  environment, 246t
experimental data, 242,244t
nonpristine coals, 244,246
occurrence, 241
origins, 246–247
pristine coals, 244,245f
sulfur content of coals, 241,243t
Environmental aspects of combustion
  of sulfur-bearing fuels
effects of airborne sulfur compounds,
  61–65
fate of sulfur emissions, 56,58–61
sources of sulfur emissions, 53–57

Even-over-odd carbon number predominance of
  n-alkanes, explanation,
  435,438,439f,440
Excess-oxidant oxidation, description,
  302–303

F

Fate of sulfur emissions
ambient concentration levels, 60–61
atmospheric removal mechanisms, 58,59t
atmospheric transformations, 58
atmospheric transport processes, 59–60
Five-membered S-heterocycle formation
conditions of laboratory simulation, 68–69
low-temperature nonoxidative–reductive
  processes, 72–75
low-temperature oxidation–reduction,
  69,70f,71–72
structure, 75–79
Flash pyrolysis–GC, in analysis of
  maturity-related changes in organic
  sulfur composition of kerogens, 530–562
Formylation, in preparation of
  2-alkylthiophenes, 406,407f
Fossil fuels, removal of sulfur, 56
Freshwater environment, sulfur geochemistry,
  126–127
Fuels
SO$_2$ emission 54t
sulfur content, 54t
types, 54t

G

Gammacerane, indicator of
  palaeosalinity, 420
Geochemical model of pyrite and organic
  sulfur compound formation in high-sulfur
  coals
model, 44,45f
organic sulfur compound formation, 46
pyrite formation, 44,46
source of sulfur, 44
Geochemistry, comparison to petroleum
  processing, 3
Geochemistry of sulfur in coal
abundance of sulfur in major U.S. coal
  basins, 31,32–33t
distribution of sulfur in coal lithotypes
  and macerals, 35,36–37f,38
forms of sulfur, 31,34–35
importance of studies, 30
model for formation of pyrite and organic
  sulfur compounds, 44,45f,46
origin of sulfur, 40–44
sulfur-containing organic compounds in
  coal, 38–40

Geochemistry of sulfur in petroleum systems
advances in molecular structure of organic
sulfur compounds, 17–21
areas of interest, 4
early reviews, 4–5
expectations for future, 24
focus, 4–6
limitations imposed by analytical
methods, 4–5
origin of sulfur in fossil fuels, 21–24
scope, 5
sulfur in kerogens and asphaltenes,
13–14,15t,16
sulfur in natural gas, 16–17
sulfur in petroleum and related bitumens,
7–13
Geology and geochemistry of oil and gas
occurrences, monographs, 6
Great Salt Lake, sulfur geochemistry,
132,133f,134–135
Green River Formation oil shale, sulfur
geochemistry, 138–140,143f

H

Harshorne coals, pyrolysis–GC
characterization of organic sulfur
compounds, 333,338f
Heteroaromatic compounds, identification in
coals, 39
Hexacyclic terpenoid sulfides
GC–MS, 381,382f
structure, 381
Highly branched isoprenoid thiophenes,
distribution in deep-sea sediments, 627
High molecular weight organic sulfur
compounds
importance of determination of types of
sulfur bonding, 487–488
molecular characterization methods, 488
High-sulfur crude oils, refinement, 4
High-sulfur Monterey kerogens, pyrolysis,
575–589
Human health effects of airborne sulfur
compounds, 61–62
Hydrocarbon(s)
application as indicators of
palaeoenvironmental changes, 445
microbial metabolism, 94,95f
Hydrocarbon distribution in sulfur-rich
lacustrine sediments,
157,158–161f,162
Hydrogen sulfide
generation within coal seams, 568–574
occurrence in natural sediment systems, 22
reaction with organic matter, 5–6
role in sulfur enrichment of peat, 328

Hypersaline environment
influencing parameters, 418
species distribution, 417

I

Illinois Basin Coal Sample Program
composition of coals, 241,242t
sulfur content of coals, 241,243t
Indanethiol, identification in coals, 38
Inorganic sulfur, forms, 287
Iron Post coals, pyrolysis–GC
characterization of organic sulfur
compounds, 333,339f
Iron sulfide formation in modern salt marsh
sediments
abundance of sulfate reducers,
209,211f,212
alkalinity of pore fluids vs. salinity of
water, 209,210f
analytical methods, 209
concentrations of dissolved sulfate,
209,210f
diagenetic overprinting, 215
organic carbon and pyrite relationships,
212,215
organic carbon profiles of
subenvironments, 211f,212
pH of pore fluids, 209,210f
photomicrographs, 212,214f
pore fluid–sediment chemistries of
samples, 209,210–211f,212
pyrite morphology, 215–216
salinity distribution of samples,
209,210f
sample collection, 206
sample processing, 206
stratigraphic columns of cored
subenvironments, 206,208f
subenvironments, 206,207f
sulfide characterization data, 212,213t
Isolation of sulfur compounds from
petroleums
methods, 83
reasons for interest, 83
separation of sulfides from
maltene fraction, 84–88
separation of thiophenes from
maltene fraction, 85f,88–91
Isoprenoid thiolanes, distribution in
deep-sea sediments, 628,630
Isoprenoid thiophene(s), distribution in
deep-sea sediments, 628,629f,630
Isoprenoid thiophene ratio, definition,
424,428f
Isotopic study of coal-associated hydrogen
sulfide
concentrations of sulfur forms, 570,571t

Isotopic study of coal-associated hydrogen
  sulfide—*Continued*
  experimental procedure, 570
  isotopic composition of sulfur forms,
    570,571*t*
  isotopic composition vs. organic sulfur
    content, 571–572,573*f*
  mechanism for H$_2$S generation, 572,574
  occurrence of sulfate sulfur, 571
  organic sulfur composition, 571–572,573*f*

J

Jurf ed Darawish—156
  GC of hydrocarbons from desulfurization,
    505,508*f*,509
  molecular characterization of organic
    sulfur, 504–509
  partial mass chromatograms of flash
    pyrolyzate, 504–505,506–507*f*

K

Kerogens
  characterization, 487
  characterization of organic sulfur,
    491–502
  definition, 13
  elemental analysis, 177*t*,178*f*,179
  examples of organic sulfur content,
    14,15*t*
  factors influencing composition, 13
  forms of organic sulfur, 14
  maturity-related changes in organic sulfur
    composition, 530–562
  preparation, 176
  pyrolysis, 575–589
  sulfur content, 13–14

L

Laboratory geochemical simulation,
  approach and goal, 68
Lacustrine environments, sulfur models,
  141–146
Light hydrocarbon gases, identification as
  sulfate-reducing agents, 633,634*f*,635
Lithiated thiophenes
  preparation, 408–409*f*
  reactions with aldehydes, dialkyl sulfates,
    and alkyl halides, 409,410*f*
  reactions with water, 409*f*
2-Lithiothiophene, reactivity with tetra-
  methylethylenediamine, 410*f*
Low molecular weight organic sulfur
  compounds, identification, 487

Low-temperature ashing, effect on coal
  desulfurization, 356,358*f*,359
Low-temperature nonoxidative–reductive
  processes of sulfur and carbon, in
  formation of S-heterocycles, 72–75
Low-temperature oxidation–reduction
  processes of sulfur and carbon
  3-thiazoline formation
  effect of temperature on product
    formation, 71
  factors influencing product formation, 71
  formation of heterocyclic compound from
    α-hydrogen ketones, 69,70*f*,71
  requirement of α-hydrogen for product
    formation, 71–72

M

Maceral(s)
  definition, 35
  determination of sulfur, 35,38
Maceral content of Rasa coal, determination
  using Leitz Orthodux microscope, 264
Materials degradation effects of airborne
  sulfur compounds, 62
Maturity-related changes in organic sulfur
  composition of kerogens
  absolute yields of thiophene and
    hydrocarbons vs. depth, 538,542*f*
  artificial maturation, 543,546*f*
  artificial maturation of Kimmeridge
    kerogen, 535*t*
  burial history as source of information
    for sedimentary sequences, 533*t*,534–535
  flame photometric detection chromatograms
    for artificial maturation,
    559,560–561*f*
  flame photometric detection chromatograms
    for natural maturity sequences,
    552,553*f*,554*t*
  geochemical implications, 562
  kinetic analysis of quantitative pyrolysis
    data, 549,551*t*,552
  natural maturity sequences, 538–545
  partial chromatograms from flash pyrolysis
    of immature samples, 536,537*f*,538
  partial flame ionization detection
    chromatograms, 538,539*f*
  pyrolysis–GC procedure, 536
  pyrolysis product distribution for
    artificial maturation,
    559,560–561*f*,562
  pyrolysis product distribution in natural
    maturity sequences, 552–559
  ratio of alkylbenzothiophenes to
    alkylthiophenes, 554,555–556*f*,559
  ratio of branched alkylthiophenes to linear
    alkylthiophenes, 554,557–558*f*,559

Maturity-related changes in organic sulfur
    composition of kerogens—*Continued*
  relative abundances of thiophene and
      hydrocarbons vs. artificial maturation
      temperature, 543,546f
  relative abundances of thiophene and
      hydrocarbons vs. depth,
      538,543,544–545f
  Rock–Eval T_{max} vs. depth,
      543,547–548f,549
  sample characteristics, 532t,533
  sample description and preparation,
      532–536
  thiophene ratio vs. artificial
      temperature, 543,546f
  thiophene ratio vs. depth, 538,540–541f
  thiophene ratio vs. Rock–Eval T_{max},
      549,550f
McAlester coals, pyrolysis–GC
    characterization of organic sulfur
    compounds, 333,338f
Mean maximum reflectance of vitrinite,
    determination, 265
3-Mercaptopropionic acid, mechanism
    of formation, 23–24
Methanethiol, microbial metabolism in
    petroleum, 109
Methyldibenzothiophenes, distributions,
    12–13
5-Methyl-2-(3,7-dimethyloctyl)thiophene
  ^{1}H-NMR spectrum, 404,406f
  MS, 406,407f
  synthesis, 404f
5-Methyl-2-(1-oxo-3,7-dimethyloctyl)thiophene
  ^{1}H-NMR spectrum, 404,405f
  MS, 404,405f
3-Methyl-2-(3,7,11-trimethyldodecyl)-
    thiophene, synthesis, 411f,412
Microbial mats
  description, 171
  pyrolyzates and extracts, 179t,180f,181
Microbial metabolism of organic sulfur
    compounds in petroleum
  benzothiophene, 101,103f,104,105f
  biodesulfurization, 96
  dibenzothiophenes, 104,105f,106,107f
  dibenzyl sulfide, 106,108f
  environmental and reservoir observations,
      96–97,98f
  experimental procedure, 93–94
  future research, 109
  GC analysis of sulfide fraction of crude
      oil, 100,102f
  laboratory studies, 97,99f,100,102f
  other compounds, 106,109
  thiophenes, 101,102–103f
Microorganisms
  metabolism of hydrocarbons, 94,95f
  metabolism of organic sulfur compounds in
      petroleum, 93–109

Mineral sulfur, presence in coal, 232
Mississippi River Delta Plain, composition
    of sediment, 186
Molecular characterization of organic sulfur
    in high molecular weight sedimentary
    organic matter
  desulfurization procedure, 489,491
  flash pyrolysis–GC–MS procedure, 491
  organic sulfur in asphaltenes, 503–504
  organic sulfur in kerogens, 491–503
  organic sulfur in resins, 504–521
  relationship between molecular weight and
      organic sulfur, 522,523f
  samples, 489
  separation methods, 489,490f
Multiple-heteroatom-containing sulfur
    compounds in high-sulfur coal
  composition of coal sample, 250,251t
  contributions from pyrite, 257–258,259f
  distribution of ions containing S and O,
      256f
  experimental procedures, 250–251
  formula for sulfur-additional heteroatom,
      257,258t
  ion peaks found at m/z 268, 251,252t
  list of species, 252,253t
  occurrence, 249–250
  peaks containing S–O, 255f,256
  selected ions containing SO_2, 257f
  structures of species, 252–253
  sulfur-additional heteroatoms,
      256,257f,258t
  sulfur-containing ion distribution from
      high-resolution MS, 251,254f
  sulfur–nitrogen, 254–255
  sulfur–oxygen, 255–256f
  sulfur–sulfur, 254

N

Naphthalenethiol, identification in
    coals, 38
Natural gas, H_2S content, 16–17
Nitric acid extraction method for direct
    determination of total organic sulfur in
    coal
  analytical values for coals after
      extraction, 235t
  characterization of coals, 233,234t
  elemental ratios of cleaned and extracted
      coals, 237,238t
  experimental procedures, 233,235
  forms of sulfur data for coals, 233,235t
  organic sulfur data, 236,237t
  precision, 239
  sulfur values for coals before and after
      cleaning, 236t
  sulfur values from float–sink separations,
      237,238t

Nitrogen, importance in petroleum
processes, 6
Nonpristine coals, elemental sulfur, 244
Nonvolatile petroleum materials, methods for
determination of organic sulfur, 220–221
5-Nonyl-2-pentadecylthiophene, synthesis,
401,402f
Northern Apennines Marl
mass chromatograms of flash pyrolyzates,
509,512–513f
molecular characterization of organic
sulfur, 509–520
partial mass chromatogram,
515,518–519f,520
partial total ion chromatograms of flash
pyrolyzates, 509,510–511f,514f,515
total ion chromatogram and mass
chromatogram of hydrocarbons,
515,516–517f

O

Oil, sources of sulfur, 56
Oklahoma coals
categories, 329,330f
geologic column showing coal beds,
329,331f,332
Organic geochemistry
establishment, 5
goal, 444
Organic matter characterization in
sulfur-rich lacustrine sediments of
Miocene age
bulk sediment properties, 152–157
experimental procedures, 152
geological background, 150,151f
hydrocarbon distributions,
157,158–161f,162,167
hydrogen indexes vs. total organic carbon,
157,167
kerogen composition, 156–157
mineralogy, 154
organic sulfur compounds,
162–163,164–166f,167
sample selection, 150
sulfur content, 154,155f,156t
total organic carbon contents,
152,153t,154
Organic solvent extraction, effect on coal
desulfurization, 356,357f
Organic sulfur
determination by ASTM Method D 2492, 232
determination methods, 31,34
electron optical values vs. ASTM values,
319,321,324t
Organic sulfur composition of kerogens,
maturity-related changes, 530–562

Organic sulfur compound(s)
distribution of types in crude oils, 4
examples of compounds and precursors,
18,20f,21
examples with structures related to
well-known biological markers, 18,19f
formation, 21, 445
geochemical model of formation in
high-sulfur coal, 44,45f,46
in sulfur-rich lacustrine sediments,
162–163,164–166f
indicator of biosynthetic functionalized
lipids, 445
microbial metabolism in petroleum, 93–109
structure determination by GC–MS, 17
types found in coal, 38–40
Organic sulfur compound distribution in
deep-sea sediments
2-alkylthiophenes, 627–628,629f
background information on samples,
614,616–619t
experimental methods, 614
highly branched isoprenoid thiophenes, 627
isoprenoid thiophenes and thiolanes,
628,629f,630
locations of drilling sites, 614,620f
partial reconstructed ion chromatograms of
aromatic hydrocarbon fraction,
615,625f
relative abundances in aromatic
hydrocarbon fractions, 615,621–624t
sample preparation, 614
structures of compounds, 615,626f
thienylhopane, 630
thiolane steroids, 630
Organic sulfur compounds in coal
characterication by peroxyacetic
acid oxidation, 297–313
characterization methods, 296–297
Organic sulfur compounds in geochemical
samples
characterization and identification, 327
factors influencing distribution, 326
structures, 327,344
Organic sulfur compounds in Oklahoma coals,
characterization by pyrolysis–GC,
326–341
Organic sulfur constituents in Rasa coal
advantages of characterization, 261–262
ash analysis of coal, 267–268,269t
carbon aromaticity, 268
compound types determined in
pyridine–toluene extract,
280,281t,282–283
effect of formation environment on sulfur
content, 268–270
elemental analysis, 271t
experimental procedure, 263–266

Organic sulfur constituents in Rasa coal—
 Continued
 isotopic abundances, 270–271
 low-voltage, high-resolution MS, 272–282
 multiplet at *m/e* 288, 272,273*t*
 partial low-voltage, high-resolution MS,
  274,275–279*t*,280
 petrographic studies, 266
 proximate and ultimate analysis of coal,
  266,267*t*
 resolution required to achieve base-line
  separation at *m/e* 288, 272,273*t*
 vitrinite reflectance values, 268
Organic sulfur containing structures,
 distribution in high organic sulfur
 coals, 288–294
Organic sulfur in asphaltenes,
 characterization, 503–504
Organic sulfur in kerogen
 bonding of alkylthiophene units,
  501,502*f*,503
 identification of alkylthiophenes in
  pyrolyzates, 494–495,500*t*
 origin of sulfur-containing moieties,
  495,501
 partial total ion currents and summed MS,
  491,494,496–499*f*
 proposed structures of alkylthiophene
  moieties, 495,502*f*
 structure of sulfur-containing moieties,
  491–502
 total ion currents of flash pyrolyzates,
  491,492–493*f*,494
Organic sulfur in peat, sources, 328
Organic sulfur in resins
 Jurf ed Darawish—156, 504–509
 molecular characterization, 504–521
 Northern Apennines Marl, 509
 origin of sulfur-rich resins, 520–521
Organic sulfur in sedimentary organic
 matter, role in petroleum generation, 530
Organic sulfur in vicinity of pyrite
 particle, measurement via transmission
 electron microscopy, 321,322–323*f*
Organosulfur, definition, 115
Origin of sulfur in coal
 stratigraphic evidence, 40–41
 use of trace elements in studying, 42,44
Oxidant-starved oxidation, description,
 303*t*,304
Oxygen, importance in petroleum processes, 6

P

Palaeoenvironmental changes, alkylthiophenes
 as sensitive indicators, 445–481
Palaeosalinity indicators
 C$_{20}$ isoprenoid thiophene distribution
  patterns, 424,429–432

Palaeosalinity indicators—*Continued*
 chroman distributions, 418,419*f*,420
 predominance of even-carbon-number alkanes,
  435,438,439*f*,440
 gammacerane, 420
 pristane/phytane ratios,
  431,433–434*f*,435
 relative abundance of C$_{35}$ hopanes,
  435,436–437*f*
 14β(H),17β(H),14α(H),17α(H)-sterane
  ratio, 420
Palaeosoil of mangrove, pyrolyzates and
 extracts, 181,182*f*
Peat, sources of sulfur, 328
Peat deposits, formation requirements,
 327–328
Pentacyclic terpenoid sulfides,
 structure, 381
Peroxyacetic acid oxidation,
 description, 297
Peroxyacetic acid oxidation in characterization
 of organic sulfur compounds in coal
 distribution of oxidation products, 311
 effects of sulfur forms on separation
  of model compounds, 299
 excess-oxidant oxidation, 302–303
 Fourier-transform IR analysis, 301
 Fourier-transform IR and NMR analysis, 304
 GC analysis, 301–302
 GLC–flame ionization detection/flame
  photometric detection and GC–MS
  analysis of oxidation products,
  305,307–312
 identification of sulfur compounds,
  311,312*t*,313
 instrumentation, 299–300
 mild oxidation of coal and macerals, 299
 NMR analysis, 301
 oxidant-starved oxidation, 303*t*,304
 sample preparation, 298*t*
 solvent extraction, 300*t*,301
 sulfur-33 NMR, 304–305,306*f*
 treatment of model compounds, 300
Petrographic studies of Rasa coal,
 procedure, 264
Petroleum
 classifications based on distribution of
  sulfur types, 12
 desulfurization using thermal and
  thermocatalytic treatments, 3
 examples of nonthiophenic sulfur compounds
  8,10*f*
 examples of thiophenic sulfur compounds,
  8,11*f*
 identification of sulfur compounds,
  7–8,9*t*
 isolation of sulfur compounds, 83–91
 microbial metabolism of organic sulfur
  compounds, 93–109

Petroleum—*Continued*
sulfide and thiophene contents, 383,388*t*
sulfur content, 7
Petroleum alteration, examples of processes, 6
Petroleum processing, comparison to
geochemistry, 3
Phenyl sulfide, microbial metabolism in
petroleum, 106
Phytane
formation, 431,434*f,*435
MS, 431,433*f*
Polysulfides
importance in organic sulfur compounds, 23
presence in geosphere, 80
Porphyrins in geological samples,
discovery, 614
Pristane
formation, 431,434*f,*435
MS, 431,433*f*
Pristane/phytane ratios, indicator of
palaeosalinity, 431,433–434*f,*435
Pristine coals, elemental sulfur, 244,245*f*
Programmed-temperature heating
function, 345
use in coal desulfurization, 345–361
Programmed-temperature oxidation
function, 345
use in characterization of organic
sulfur compounds in coal, 297
use in coal desulfurization, 345–361
Programmed-temperature reduction,
use in characterization of organic
sulfur compounds in coal, 297
Pure mineral pyrite, effect of addition on
coal desulfurization, 351,356
Pyrite
contribution to organic polycyclic
aromatic sulfur compounds,
257–258,259*f*
formation, 40, 215
geochemical model of formation in
high-sulfur coal, 44,45*f,*46
morphology, 215–216
photomicrographs, 212,214*f*
relationship with iron sulfides, 212,213*t*
relationship with organic carbon, 212,215
sulfur source, 22–23
Pyrite in recent sediments
chemical analysis, 192
chloride salinities, 196*t,*199
cross section of sites, 188,190–191*f*
formation, 201
framework for main environments,
187–188,189*f*
location map for sites, 188,189*f*
micrographs, 196,197*f*
organic matter and sulfur forms, 192,195*f*
organic matter, total sulfur, and total
iron, 192,194*f*

Pyrite in recent sediments—*Continued*
petrographic work, 192
pH, sulfate, and chloride concentrations,
192,193*f*
sampling procedure, 188
site selection, 188,189–191*f*
total sulfur vs. chloride salinity and
organic matter, 199,200*f,*201
X-ray microanalysis, 196,198*f*
Pyritic sulfur
determination, 31
forms, 31
pyrolysis products, 288
role in formation of organic sulfur
structures in coal, 288
size distribution, 31
Pyrolysis–GC characterization of organic
sulfur compounds in Oklahoma coals
calorific value vs. organic sulfur
compound concentration, 340
Croweburg coals, 340,341*f*
experimental materials and procedures,
332,334–335*t*
flame photometric detector pyrogram of
coal, 332,336*f*
Hartshorne coal, 333,338*f*
identification of compounds via retention
time comparison, 332,337*t*
Iron Post coal, 333,339*f*
McAlester coal, 333,338*f*
Stigler coal, 333,339*f*
Pyrolysis of high-sulfur Monterey kerogens
C and H isotopic maturation trends, 587
carbon isotopic ratios vs. H/C ratios,
578,582*f*
characteristics of samples, 576*t*
elemental and isotopic compositions of
starting materials and pyrolyzates,
577–578,580*t*
experimental procedures, 576–577
H/C vs. S/C elemental ratios,
587–588,589*f*
hydrogen isotopic ratios vs. H/C ratios,
578,582*f*
mass balance calculations for sulfur
isotopic and total sulfur data,
585,586*t*
production rates of bitumen and H_2S in
pyrolyzates, 578,581*f*
pyrite–organic sulfur relationships,
585,587
sulfur contents of residual kerogens,
bitumens, and H_2S generation,
578,581*f*
sulfur isotopic maturation trends,
587–588,589*f*
sulfur isotopic ratios of kerogens,
bitumens, and H_2S, 578,584*f,*585
sulfur isotopic ratios vs. H/C ratios, 578,583*f*

Pyrolysis of high-sulfur Monterey kerogens—
Continued
sulfur isotopic ratios vs. S/C elemental
ratios, 588,589*f*
yields and sulfur contents of starting
materials and pyrolyzates, 577–578,579*t*

R

Rasa coal
characteristics, 262
forms of organic sulfur, 262–263
nature of organosulfur constituents,
261–283
occurrence, 262
Relative abundance of C_{35} hopanes, indicator
of palaeosalinity, 435,436–437*f*
Resins, characterization of organic sulfur,
504–521
Resin fraction of crude oil or bitumen,
definition, 504
Rundle Formation oil shale, sulfur
geochemistry, 140–141

S

Sabkha, definition, 171
Scrubbers, use in removal of sulfur from
fossil fuels, 56
Sedimentary organic matter, sulfur
source, 22
Sedimentary sulfur chemistry, controls,
115,117–118
Sedimentary sulfur isotopy, controls,
118–119,120*f*
S heterocycle(s)
element sources in geosphere, 80
presence in geosphere, 79–80
S heterocycle structure
effect of polysulfides on product, 79
effect of solvent on product, 77,79
thiolanes, 75,76*f*,77
thiophenes, 77,78*f*,79
SO_2
mechanisms of removal from the atmosphere,
58,59*t*
transformations in the atmosphere, 58
Soap Lake
depth plots, 136,137*f*,138
sulfur geochemistry, 135–136,137*f*,138
weight percent sulfide sulfur vs. weight
percent reactive iron, 135–136,137*f*
Sources of sulfur emissions
residential heating fuels, 54,55*t*
sulfur in fossil fuels, 55–56,57*f*
Southern Colliery, location, 568,569*f*

Spatial variation of organic sulfur in coal,
measurement via transmission electron
microscopy, 316–325
Steranes
end products of early diagenetic pathways
of sterols, 420–421,422*f*
indicator of palaeosalinity, 421,424–428
MS of saturated hydrocarbon fraction,
421,423*f*
Sterols
postulated intramolecular sulfur
incorporation, 424,427*f*
proposed intra- and intermolecular sulfur
incorporation, 424,428*f*
Stigler coals, pyrolysis–GC characterization
of organic sulfur compounds, 333,339*f*
Sulfate, source of sulfur in fossil
fuels, 21
Sulfate particles
mechanisms of removal from the atmosphere,
58,59*t*
transformations in the atmosphere, 58
Sulfate-reduction reaction, 117
Sulfate sulfur, forms, 34
Sulfide(s)
GC–flame ionization detection
chromatograms of fractions from
petroleums, 84,86,87*f*
generalized structures of compounds
from petroleums, 86,88
identification in coals, 39
in petroleums, 84,86*t*
isolation methods, 83
separation from maltene fraction of
petroleum, 84,85*f*
Sulfide–mineral formation, variables,
115,117–118
Sulfur
areas of environmental interest, 2–3
basinal brines as source in high-sulfur
coals, 44
deposition flux, 59*t*
distribution in coal lithotypes and
macerals, 35,36–37*f*,38
factors controlling distribution in fossil
fuels and shales, 186
forms, 31,34–35
formation of pyrite during early
diagenesis, 40
geochemistry in coal, 30–47
geochemistry in petroleum systems, 2–17
importance in petroleum processes, 6
isotope composition in coal, 41–42
kerogens and asphaltenes, 13–14,15*t*,16
natural gas, 16–17
nature in peat, 40
origin in coal, 40–44
origin in fossil fuels, 21–24
petroleum and related bitumens, 7–13

Sulfur—*Continued*
  removal from fossil fuels, 56
  sources, 53
  sources in coal, 41–42
  sources in coal and oil, 55–56,57f
  stratigraphic evidence of origin in coal,
    40–41
  use of trace elements in studying origin,
    42,44
Sulfur–carbon chemistry of diagenesis
  limitations of geochemical processes,
    68–69
  low-temperature nonoxidative–reductive
    processes, 68
  low-temperature oxidation–reduction, 60,
    70f,71–72
  S-heterocycle structure, 75–79
Sulfur compounds in petroleums, isolation,
    83–91
Sulfur-containing bituminous coal, pyrolysis
    products, 287
Sulfur-containing compounds in Alberta
    petroleums
  n-alkyl-substituted thiolanes and thianes,
    383,384–385f
  biogeochemistry, 383,386–387f,388t
  detection of cyclic terpenoid sulfides and
    sulfoxides, 367–372
  discovery, 366
  structures of cyclic terpenoid sulfides,
    372–382
  sulfur in asphaltene, 388,391,392–394f
  sulfur isotopic studies, 391,395
  thiophenes, 388,389–390f
Sulfur-containing organic compounds in coal
  identification methods, 38
  quantification methods, 39
  types, 38
Sulfur content of sedimentary organic
    matter, environmental factors,
    170–171
Sulfur emissions
  environmental effects, 61–65
  fate, 56,58–61
  sources, 53–57
  trends, 54–55
Sulfur geochemistry in lakes
  analysis of freshwater environment,
    126–127
  analysis of Great Salt Lake,
    132,133f,134–135
  analysis of Green River Formation oil
    shale, 138–140,143f
  analysis of Rundle Formation oil shale,
    140–141
  analysis of Soap Lake, 135–136,137f,138
  analysis of Walker Lake,
    127,130,131f,132
  analytical methods, 124,125f

Sulfur geochemistry in lakes—*Continued*
  case studies, 119,121f,122–123t
  characteristics of oil shale deposits,
    119,122t
  controls on sedimentary sulfur chemistry,
    115,117–118
  controls on sedimentary sulfur isotopy,
    118–119,120f
  influencing factors, 115
  morphometric and geochemical
    characteristics of modern lakes,
    122,123t
  oil shale locations, 119,121f,122
  sampling methods, 122,124
  schematic representation, 115,116f
  sulfur analysis by induction furnace–IR
    detection system, 124,125f
  sulfur models for lacustrine environments,
    141–146
Sulfur heterocycles, microbial oxidation,
    97,99f,100,102f
Sulfur in asphaltene, GC–MS,
    388,391,392–394f
Sulfur incorporation into recent organic
    matter in carbonate environment
  analysis of kerogen, 177t,178f,179
  distribution of hydrocarbons, 183
  effect of sulfate reduction on organic
    matter incorporation, 175
  extraction and extract separation, 176
  factors influencing distribution, 183–184
  GC and GC–MS procedures, 176
  geological setting, 171,173–174f,175
  incorporation during early diagenesis,
    181,183
  kerogen preparation, 176
  location map of carbonate sedimentation,
    171,172f
  pyrolysis procedure, 176–177
  pyrolyzates and extracts of microbial mat,
    179t,180f,181
  pyrolyzates and extracts of paleosoil of
    mangrove, 181,182f
  sampling of lagoonal muds, 176
  sampling of mangrove soils, 175–176
  sampling of microbial mats, 175
  S/C atomic ratios, 181
  schematic map of recent sedimentary
    system, 171,173f
  schematic section through intertidal zone
    and sabkha, 171,174f
Sulfur in fossil fuels, applications, 4
Sulfur in petroleum
  distribution, 366
  effect on physical properties, 366
Sulfur in recent sediments
  chemical analysis, 192
  chloride salinities, 196t,199
  cross section of sites, 188,190–191f

Sulfur in recent sediments—*Continued*
  framework for main environments,
    187–188,189f
  location map for sites, 188,189f
  micrographs, 196,197f
  organic matter and sulfur forms, 192,195f
  órganic matter, total sulfur, and total
    iron, 192,194f
  petrographic work, 192
  pH, sulfate, and chloride concentrations,
    192,193f
  sampling procedure, 188
  site selection, 188,189–191f
  total sulfur vs. chloride salinity and
    organic matter, 199,200f,201
  X-ray microanalysis, 196,198f
Sulfur isotopes, studies, 5
Sulfur isotopic composition
    determination, 118
  Rayleigh fractionation curves vs. percent
    of initial sulfate reduced, 119,120f
  values for coals from western United
    States, 328–329
Sulfur isotopic data analysis of crude oils
    from Bolivar coastal fields
  areal extent of fields showing well
    locations, 593,594f
  correlation of sulfur data with
    parameters, 596,597t,598
  ^{34}S values, 605–606
  GC for aromatic fractions,
    598,601f,602,604f
  oil classes and compositional ranges,
    593,595t,596,598
  oil classes and reservoirs,
    593,595t,596,598
  relationships among oil classes, 598–611
  sulfur content values, 606
  sulfur content vs. asphaltene content,
    606,608f,609
  sulfur content vs. temperature, 606,607f
  toluene and benzene vs. sulfur, 609,610f
  variation of $\delta^{34}S$ with pristane/phytane
    ratio, 609,611f
  whole oil GC, 598,599f
Sulfur isotopic studies of petroleums,
    391,395
Sulfur K-edge X-ray absorption near-edge
    structure spectroscopy of petroleum
    asphaltenes and model compounds
  calibration plot for quantification of
    dibenzothiophene–dibenzyl sulfide
    mixtures, 225,227f
  data collection, 222
  experimental materials, 221
  first inflection energies, 222,224t
  quantification of organic sulfur forms,
    225,229t
  sample preparation, 221–222

Sulfur K-edge X-ray absorption near-edge
    structure spectroscopy of petroleum
    asphaltenes and model compounds—*Continued*
  spectra of asphaltenes and Rasa coal,
    225,228f
  spectra of dibenzothiophene–dibenzyl
    sulfide mixtures, 225,226f
  spectra of model compounds,
    222,223f,225
Sulfur models for lacustrine environments
  depositional and diagenetic behavior of
    sulfur, 141,142t,143f,144
  quantitative model of organic sulfur,
    144,145f,146
Sulfur-rich lacustrine sediments of miocene
    age, characterization of organic matter,
    149–167
Sulfur-rich resins, origin, 520–521

T

Terminal electron acceptor, description, 94
Tetracyclic terpenoid sulfides
  GC–MS, 378,379f
  MS, 378,380f,381
  structures, 378
  variation of carbon number distribution,
    383,387f
Thermochemical sulfate reduction in gas
    reservoirs, carbon isotopic
    fractionation during oxidation of light
    hydrocarbon gases, 633–641
Thianthrene, microbial metabolism in
    petroleum, 106
3-Thiazolines, synthesis from ketones, 71
Thienylhopane, distribution in deep-sea
    sediments, 630
Thiolane(s), structure. 75,76f,77
Thiolane steroids, distribution in deep-sea
    sediments, 630
Thiophenes
  correlation of abundance with organic
    sulfur content, 530,531f
  flame ionization–MS spectrum,
    388,389–390f
  GC–flame ionization detection
    chromatograms of fractions from
    petroleums, 89,90f,91
  identification in coals, 38–39
  in petroleums, 84,86t
  isolation methods, 83–84
  microbial metabolism in petroleum,
    101,102–103f
  presence in geosphere, 79–80
  rate of oxidation by *m*-chloroperbenzoic acid,
    88t,89
  reaction with *n*-butyllithium, 408f

Thiophenes—*Continued*
  separation from maltene fraction of
    petroleum, 85*f*,88–89
  structure, 77,78*f*,79
Thiosteriods
  MS of desulfurized polar fraction,
    424,426*f*
  MS of low-molecular-weight fraction,
    421,424,425*f*
Thioxanthene, microbial metabolism in
  petroleum, 106
Total organic sulfur, direct determination
  in coal, 231–239
Transmission electron microscopy
  applications, 317,319–323
  calculation of sulfur concentration, 317
  determination of average organic sulfur
    concentration, 319,320–321*f*
  determination of organic sulfur in
    vicinity of pyrite particle,
    321,322–323*f*
  determination of spatial variation of
    organic sulfur, 319,320*f*
  method, 316–317,318*f*
  schematic representation of apparatus,
    317,318*f*
  typical energy-dispersive X-ray spectrum,
    317,318*f*
2,3,5-Tribromothiophene, selective
  lithiation, 408,409*f*

Tricyclic terpenoid sulfides, structure,
  374,377*f*,378

V

Vitrinite, organic sulfur content, 35

W

Walker Lake, sulfur geochemistry,
  127,130,131*f*,132
Wallops Island, Virginia
  iron sulfide formation, 206–216
  location map, 204,205*f*
Western U.S. coals, sulfur isotopic
  composition, 328–329
Wet processes for removal of sulfur
  species from the atmosphere
  description and application, 58
  influencing factors, 59

X

X-ray absorption fine structure
  spectroscopy, in characterization of
  organic sulfur compounds in coal, 297
X-ray absorption spectroscopy, for direct
  measurement of organic sulfur, 221

Printed and bound in the UK by
CPI Antony Rowe, Eastbourne